Offset Lithographic Technology

by

Kenneth F. Hird

South Holland, Illinois

THE GOODHEART-WILLCOX COMPANY, INC.

Publishers

Library of Congress Catalog Card Number 89-29923
International Standard Book Number 0-87006-785-0

1234567890-91-0987654321

```
Library of Congress Cataloging in Publication Data

Hird, Kenneth F.
    Offset lithographic technology / by Kenneth F.
Hird.
        p.  cm.
    Includes index.
    ISBN 0-87006-785-0
    1. Photolithography. 2. Offset lithography
I. Title.
TR940.H572    1991
686.2'325--dc20                        89-29923
                                           CIP
```

Introduction

The growth of the printing industry has been immense. Electronics and the computer have completely changed the complexion of the industry. Not only are presses electronically monitored and controlled, but most hand and machine composition has been replaced by photo and computer typesetting. Photography is now controlled by electronic densitometers, exposure computers, and automatic processors. Most color separations and corrections are done electronically on scanners, and halftone output films are generated by lasers. A large amount of stripping is now being done by computer-aided design (CAD) equipment. Plates are being processed automatically and some are being exposed directly from paste-ups by lasers without any intermediate films. Press key-less ink fountains are being preset automatically from data derived from scanning the printing plates before mounting on the press. Computers are used to analyze production information on presses. Bindery equipment is being operated almost automatically by computers for some publication printing.

These advances and improvements, along with the conventional methods, have made offset lithography a mature combination of processes and techniques. There are many career opportunities for those who have the necessary skills and educational background.

OFFSET LITHOGRAPHY TECHNOLOGY has been designed and organized to teach you about offset lithography methods and processes as they are used by professionals in the industry. Keep in mind that some of the content of this textbook may be made obsolete by new materials, changes in technology, educational emphasis, or by new consumer demands. However, it is hoped that the principles and thought processes will be transferred and applied to these new methods.

SCOPE OF THE TEXTBOOK

The textbook is divided into nine major sections. Each section contains one or more chapters that describe a sequential part of offset lithography.

In Section I, Introduction to the Printing and Publishing Industry, the chapters describe the offset lithography sector of the printing and publishing industry. Each of the various career opportunities within the industry are examined. Information related to job application and job interviewing is covered in detail. Emphasis is placed on safety and health practices that everyone must observe when engaged in offset lithography production activities. Information is also given describing the purpose and function of the Occupational Safety and Health Act regulations.

Section II, Estimating and Trade Customs, describes the specific procedures followed when estimating the cost of a printed product. The estimator determines the cost based on a number of criteria. A quotation is prepared for the customer. The section concludes with a thorough examination of the printing industry trade customs, and legal restrictions.

Section III, Typography, Design and Layout, describes the developmental stage of the actual production printing sequence. The principles of good typography are covered in detail. The nomenclature and characteristics of type are given special attention. The design, layout, and actual creation of printed products are described and fully illustrated.

Section IV, Image Generation and Assembly, deals with the actual production sequence. Image generation includes the creation of visual elements needed to manufacture the product. Image generation includes the various hand-mechanical and computer-assisted typesetting processes, including desktop publishing. Image assembly is also described and illustrated in detail. The section concludes with information on proofing and proofreading. This important phase of image generation and assembly is emphasized because of its important role in the prepress production process.

Section V, Image Conversion, covers procedures used to convert assembled images to photographic materials. This section concentrates on the photographic conversion practices including process cameras and film processing, line photography, halftone photography, special effects photography, and full-color process photography.

In Section VI, Stripping and Platemaking, details the last step prior to reproducing the image on paper. In the offset lithography process, negatives and/or positives are stripped up in preparation for platemaking. The stripping process is gradually moving from a hand-mechanical process to one of automation. The image carrier or offset lithography plate is prepared from stripped negatives and/or positives. Automated equipment is used for step-and-repeat work. Plates are made from many different kinds of materials including paper, plastic, and aluminum.

Section VII, Paper and Ink, describes two very important industries that supply materials essential to offset lithography. Paper is the main ingredient in almost every printed product. It is available in many weights, sizes, colors, textures, and thicknesses. The paper manufacturing process is detailed, as is the extensive classification system used by paper merchants. This section explains how the ink must be compatible with the paper or other substrate. The chemistry of lithographic ink is also described in detail.

Section VIII, Image Transfer and Finishing Operations, is primarily concerned with the process of physically placing ink on paper. In the offset press process, this includes the control of paper and other similar substrates as the image is transferred to that material from the offset plate or image carrier. The chapters in this section describe the principles and concepts of paper feed, registration, transfer, and delivery using various sheet-feed and web-fed press designs. Typical press operating problems that can be encountered are also summarized.

Once the image has been transferred to the paper, other operations are generally required to finish the printed product. Finishing operations covered in the final chapter of this section include folding, trimming, collating, binding, perforating, scoring, and packaging.

Section IX, In-Plant Printers, Quick Printers, and Brokers, describes three very distinct and unique areas of the printing industry. These areas are often overlooked in the study of offset lithography. The in-plant printer is generally responsible for the printing needs of a large firm. Unlike the commercial offset lithography printer, in-plant printers exist solely to meet the needs of their own organizations.

Quick printers are categorized as either franchised or independent. They are located in almost every busy commercial area of cities and towns. As their name implies, they are specialists in fast turnarounds of printed and copied materials.

The section concludes with a description of printing brokers. These are business people who strike out on their own in printing sales. Most printing brokers work out of an office. They generally do not own their own equipment or facilities. Instead, the broker deals with several commercial printers to get the printing completed. Their profits are derived from the markup applied to each job.

ABOUT THE TEXTBOOK

The textbook is divided into nine major sections. Each section contains one or more chapters. Section overviews provide an introduction to the content of the chapters. Each chapter begins with a list of instructional objectives. These objectives tell the student and instructor what will be learned by studying the chapter.

The subject matter follows a logical "building block" approach, with all key terms emphasized in *italics* and defined the first time they are used. The textbook uses design features that facilitate learning: large type, white space between paragraphs, and short line lengths. Numerous photographs, drawings, and other illustrations are included throughout the textbook to assist the student in understanding the material that has been presented. A second color is used to clarify illustrations and stress safety. Four-color is included in the chapter covering color process photography and in other sections.

Each chapter includes a summary of the material covered. Important technical terms are listed for review. A good mix of review questions are provided with each chapter. The questions are designed to test your comprehension of the chapter material.

An appendix is provided at the end of the textbook and includes related information and tables of metric and customary units of measurement. A glossary summarizes the terminology presented in the textbook and is provided for student and instructor reference.

ABOUT THE AUTHOR

Kenneth F. Hird is a professor in the Department of Technology at California State University, Los Angeles. He and his wife, Nancy, have owned a printing firm in San Dimas, California. His career spans fifteen years in the printing industry and 30 years in graphic arts education. He learned the printing trade as an apprentice for Campbell Press in San Jose, California. He began his teaching career at William Howard Taft High School in Woodland Hills, California. After nine years at the high school, the author moved on to Kirkwood Community College in Cedar Rapids, Iowa, where he established a new printing trades program. During his teaching career, he has been involved in industrial arts, trade and industry programs, printing management, and graphic communications technology. He has had three other textbooks published in the graphic arts/printing technology field.

CONTENTS

IMPORTANT SAFETY NOTICE

The theory, procedures, and safety rules provided in this book are typical to the industry. However, in most cases they are general and do *not* apply to all situations. For this reason, it is very important that you refer to the manufacturer's instructions when using any product or machine. These factory directions will provide the details needed to work safely while producing high-quality printed products.

Section I

Introduction to the Printing and Publishing Industry

The printing and publishing industry is made up of many related occupational groups. These include printers of books, newspapers, magazines, packaging, and all types of commercial printing. Offset lithography is the most widely used of the four major printing processes within the industry. The other three processes include letterpress, gravure, and screen printing. Offset lithography accounts for over 70 percent of all printing and publishing in the United States.

Career opportunities in the printing and publishing industry, with emphasis on offset lithography, are excellent. People are needed for positions in management, production planning, production control, sales, creative areas such as design and photography, and all technical areas, such as typesetting, process camera, electronic publishing, scanner, and press areas.

The printing and publishing industry adheres to rigid safety and health regulations. The Occupational Safety and Health Act (OSHA) governs the safety and health of all employees within the vast printing and publishing industry. These regulations cover such important areas as guards on equipment, safety clothing, eye protection, toxic materials and waste removal, personal cleanliness, and general comfort of employees.

Section I contains three chapters. Chapter 1 describes the offset lithography sector of the printing and publishing industry including its size and function in the United States. Chapter 2 summarizes the career opportunities available in the offset lithography sector. Chapter 3 describes the important safety and health components, issues, and regulations that affect all personnel in offset lithography.

At the completion of Section I, you should have a better understanding of the offset lithography sector of the printing and publishing industry and the contributions it makes to our technological society.

Consumers come into contact with a variety of printed products. Many of the products are printed using the offset lighography process.

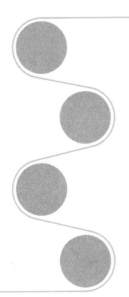

Chapter 1

Introduction to Offset Lithography

When you have completed the reading and assigned activities related to this chapter, you will be able to:
O Explain the theory of offset lithography.
O List the basic steps in the offset lithography printing process.
O Describe how photography influenced the development of offset lithography.
O Identify several types of business enterprises that make up the printing industry.
O Describe why the printing industry is considered to be a growth industry.
O Describe the primary difference between direct and indirect lithography.
O Name typical products that are manufactured by the offset lithography process.
O Identify specific reasons why the printing industry is considered to be one of the ten largest manufacturing industries in the United States.

Printing is the process of manufacturing visual products that are intended to communicate a message through permanent graphic images. Printing places an inked image, like the words on this page, onto a *substrate* such as paper, plastic, metal, glass, or cloth. Everyone is a consumer of printed products. You are a consumer of printing when reading a newspaper or magazine. You are a consumer of printing when you buy food wrapped in printed containers. You are a consumer of printed goods when you read the instruments in your automobile, use electronic devices, or fill out an application form for employment. You are a con-

sumer of a printed, published product even as you read this textbook.

Printing processes allow a high-quality image to be duplicated in large quantities at a reasonable cost. The production of these graphic products is generally customized and client initiated. Many professional and technical people are required to produce these challenging and creative products. Offset printing is the most widely used printing process in the printing industry.

PURPOSE OF THE PRINTING INDUSTRY

The primary purpose of the printing industry is to create and manufacture products that communicate visually. As an example, most businesses would be unable to function without some form of graphic communications. The need for graphic communications is basic in a technological society.

The printing industry is constantly developing more efficient ways to meet the needs of our growing population. Industry has made numerous and significant break throughs in technology. The printing industry is now one of the largest service organizations in the world. Its sales income and number of employees rank it as the sixth largest industry in the United States.

Printing is a *growth industry*. This means that each year it manufactures more products than it did the previous year. As a result, present and future employment opportunities are excellent. Chapter 2 discusses the various employment opportunities in the printing industry. The United States Bureau of Labor estimates that the industry will need 60,000 new workers each year from now until the year 2000 because of expanding technology. More than two-thirds of all printing plants employ fewer than 20 people, making the industry one in which rapid advancement is common.

According to the United States Department of Commerce, more than 54,000 individual firms that employ one and one-half million people produce printed products. This figure represents the largest number of American firms engaged in any one type of manufacturing process. The ten largest printing centers in the United States are shown in Fig. 1-1.

PRINTING INDUSTRY STRUCTURE

The printing industry is frequently referred to as the printing, publishing, and packaging industry. As a modern, high-technology enterprise, the printing industry is made up of many related occupational groups including printers of books, newspapers, and magazines. The industry also includes equipment manufacturers, sales and service groups, suppliers of materials, and specialized trade shops, Fig. 1-2.

Commercial printing

Commercial printing involves a large variety of printing products. Commercial printing does not specialize in a specific product. It serves many customers by manufacturing many different products, such as catalogs, advertising, business forms, and process color. If a commercial printer does not have all of the equipment or expertise to complete the entire job, parts of the job may be subcontracted to a trade shop.

Publishing

Publishing is composed of many thousands of companies that manufacture daily and weekly newspapers, Fig. 1-3. The larger group of companies that produces periodicals, such as *Life* and *Business Week,* are also included. There are many companies that produce and market books; the book you are now reading was published by a private company that produces textbooks. Keep in mind that printers do not generally make decisions to publish books or magazines. The publisher is responsible for that decision. The commercial printer, however, is hired by the publisher to manufacture textbooks and other kinds of books and magazines.

Packaging

Packaging refers to labels, tags, cartons, wrappers, plastic bags, and other containers. The hundreds of different containers we use every day are produced by packaging printers. The package items are used to identify products, decorate the product, or give instructions.

IN-PLANT PRINTING

One of the seldom seen areas of printing is in-plant printing. This area has shown tremendous growth over the last decade. *In-plant printing* is defined as any printing operation that is owned by and serves the needs of a single company or corporation. These operations

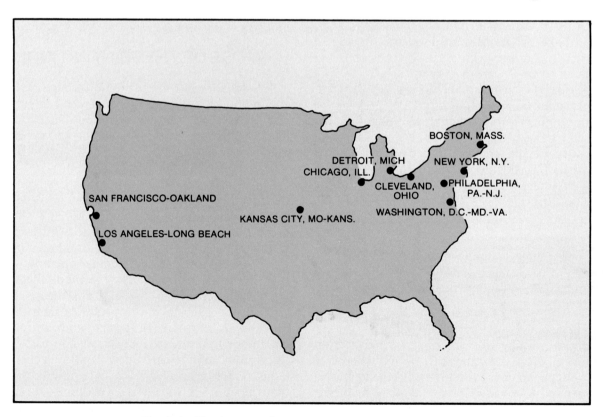

Fig. 1-1. The largest printing centers in the United States.

STRUCTURE OF THE PRINTING INDUSTRY

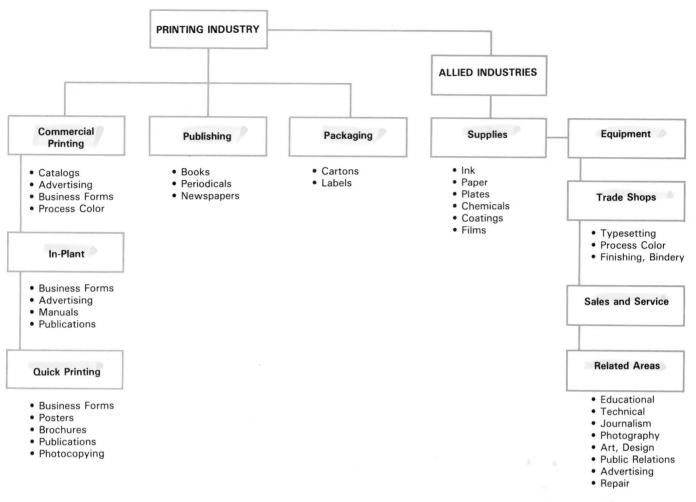

Fig. 1-2. The printing industry is composed of commercial, publishing, packaging, in-plant, quick printing, and allied industries. (Printing Industries Association, Southern California)

Fig. 1-3. Publishing includes daily and weekly newspapers, periodicals, and books.

typically produce in-house business forms, time cards, brochures, instruction manuals for products they manufacture, and other business-related materials. In-plant printing relies on other areas of the industry for printed work that cannot be produced internally by sub-contracting to commercial or trade shops.

QUICK PRINTING

One of the fastest growing areas in the printing industry is *quick printing,* Fig. 1-4. These shops are equipped with offset duplicators that are small versions of the larger offset presses. Quick printers also make use of the xerographic (commonly called photocopying) process. A quick printer can turn out large quantities of printed materials with a small camera-plate-maker unit, one or more duplicators, a paper cutter, and simple bindery equipment. Press runs can be as small as one hundred or as large as several thousand. Quick printers often use trade shops and commercial printers to produce work they are not equipped to handle.

Fig. 1-4. Quick printing is one of the fastest growing sectors of the printing industry. (National Association of Quick Printers)

Fig. 1-5. Equipment is one of the most important ingredients in a printing firm. (Heidelberg Eastern, Inc.)

SUPPLIES

Every type of printing facility requires *supplies* in order to print materials. Supplies account for over one-half of the production cost for most printed products. Supplies such as paper, ink, offset plates, chemicals, photographic films, and solvents are a few of the items required on a daily basis. Printers generally order supplies from printing supply firms specializing in these products. Printers are especially careful when selecting and using printing supplies since they are expensive.

EQUIPMENT

Equipment makes up a large and important investment for every type of printing facility. Offset duplicators and presses, cameras, folders, and paper cutters are necessary pieces of equipment, Fig. 1-5. Careful consideration must be given to the purchase of new or used equipment. The money, or capital, is invested to purchase the equipment, most of which is automated.

TRADE SHOPS

A printing firm that is not equipped to do a particular phase of the printing process relies on an outside source called a *trade shop*. These shops specialize in work such as typesetting, artwork, process color separations, stripping, presswork, and finishing operations. The initiating printing firm subcontracts services from a trade shop and adds a percentage of markup to the job when billing the customer.

SALES AND SERVICE

The printing industry requires many people in *sales and service*. Salespeople sell printed products to a variety of business firms. Other salespeople sell printing supplies and equipment to printers. Consulting firms and advertising agencies that prepare designs for reproduction also provide services for printers. It should be noted that the printing industry is experiencing a shortage of qualified sales and service people.

RELATED AREAS

Several *related areas* of the printing industry contribute various services that are essential to the printing manufacturing process. Artists, designers, and photographers provide the graphic ideas for printing. Journalists prepare copy and other written communications. Educators prepare people for work in the industry. Technicians and scientists plan and produce new processes, equipment, and graphic arts materials.

MAJOR PRINTING PROCESSES

The printing industry has expanded dramatically for economic and efficiency purposes. Computers, robotics, and new photographic processes are now in common use. Even the printing processes have a new emphasis. Workers are learning these new processes on the job and in classrooms.

There are four major printing processes commonly used today. These include offset lithography, letterpress, gravure, and screen printing, Fig. 1-6. Each process incorporates a different principle.

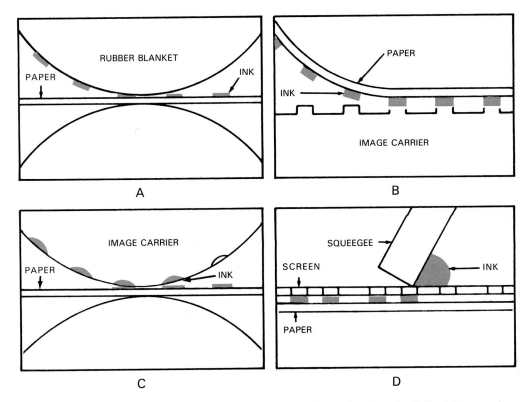

Fig. 1-6. Four major printing processes are commonly used today. A—Offset lithography. B—Letterpress. C—Gravure. D—Screen printing. (A.B. Dick Co.)

1. *Offset lithography* uses a flat (planographic) printing surface. As you will learn, the ink only adheres to the image or printing area and not to the non-image area of the *image carrier* (printing plate), Fig. 1-7.
2. *Letterpress* uses a raised (relief) printing surface. The non-image or non-printing area of the image carrier or plate is below the printing surface, Fig. 1-8.
3. *Gravure* printing uses a sunken (intaglio) printing surface or image carrier to hold and deposit ink on the substrate, Fig. 1-9.
4. *Screen printing* uses a stencil attached to a porous mesh to form the image carrier. Ink is forced through this mesh and onto the substrate surface to print an image, Fig. 1-10.

Offset lithography is the fastest-growing of the four

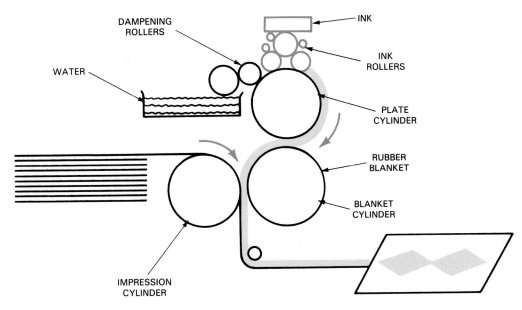

Fig. 1-7. Offset lithography is based on the principle that water and grease or oil do not mix. The printing plate is flat or planographic. (National Association of Printing Ink Manufacturers)

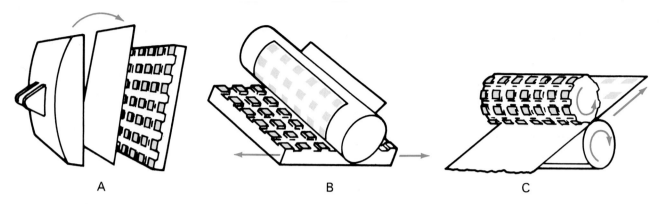

Fig. 1-8. Letterpress prints from a raised or relief printing surface. Three types of letterpress printing presses include A—platen, B—cylinder, and C—web. (Eastman Kodak Co.)

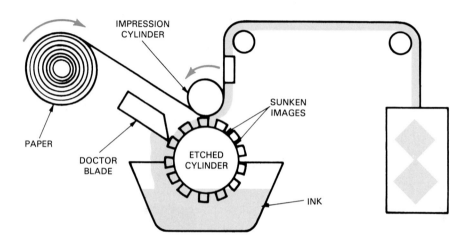

Fig. 1-9. Gravure prints from sunken images on the printing surface of a copper cylinder. The process is also referred to as intaglio. (National Association of Printing Ink Manufacturers)

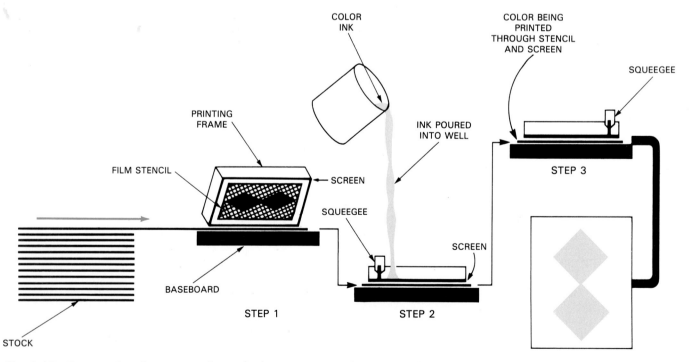

Fig. 1-10. Screen prints from a stencil attached to a porous mesh screen to form the image carrier. Ink is forced through the stencil image area and mesh screen onto the substrate surface. (National Association of Printing Ink Manufacturers)

Fig. 1-11. Offset lithography printing methods produce over 70 percent of all publications and other commercially printed products. (Sun Chemical Corp.)

major printing processes. It also is the most commonly used method of printing, Fig. 1-11. United States government statistics indicate that offset lithography is used to produce over 70 percent of all commercially printed products. This figure indicates a significant increase in the use of offset lithography in recent years. Greater use of offset lithography is forecast in the future.

The offset lithography printing process is also known by other names, such as *photo-offset lithography, photo-lithography, offset, photo-offset,* and *lithography.* For simplicity, the term *OFFSET LITHOGRAPHY* is used throughout this book.

Keep in mind that this book is concerned primarily with offset lithography. The other major printing processes are discussed, however. Such information is needed to understand the development of offset lithography. The information also will help you understand the ways in which related industries serve offset lithography.

Lithography principles

Offset lithography also is called *planography* because the printing areas are flush (flat) or on the same plane as the surface of the printing plate. Offset lithography is based on the principle that water and grease (or oil) do NOT readily mix. The image area accepts ink (oil) only and the non-image area accepts only water, not ink.

The printing, or image area on the flat offset printing plate is made by photographic and chemical processes. This photochemical process makes the image area receptive to the grease-based ink used on the printing press. The remainder of the plate, the non-image area, is receptive only to water. The non-image area of the plate is referred to as a *hydrophilic surface.* The non-image area is wet down with water. Ink is then applied to the plate. The ink sticks only to the image area. The image area of the plate is referred to as a *oleophilic surface.*

When the plate is on the offset printing press, it receives water first, then ink. This process is repeated over and over again. This is achieved through the use of two separate roller systems on the press—one for water and one for ink.

The image on the plate is NOT printed directly onto paper. Instead, the image is transferred from the plate on the *plate cylinder* to a rubber blanket on another cylinder, called the *blanket cylinder.*

The *rubber blanket* then transfers or offsets the image to the paper by gentle pressure from another cylinder, called the *impression cylinder.* This process is shown in Fig. 1-12.

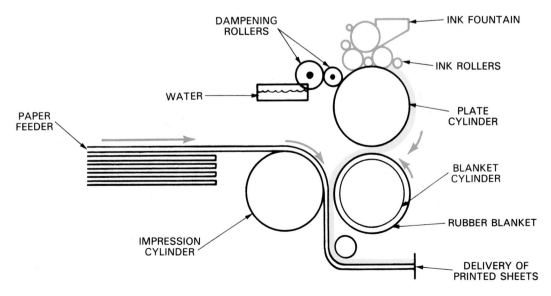

Fig. 1-12. In offset printing, the image on the plate is transferred to the rubber blanket on the blanket cylinder. The rubber blanket then transfers or offsets the image to the paper.

HISTORY OF OFFSET LITHOGRAPHY

The printing process known as offset lithography dates back almost 200 years. Alois Senefelder of Munich, Germany, discovered lithography in 1798, Fig. 1-13. Senefelder's invention was based on the principle that oil or grease and water do NOT readily mix.

Working on a slab of porous limestone, Senefelder sketched a design with a greasy crayon-like material that was absorbed by the stone, Fig. 1-14. Then, the entire surface of the stone was wet with a mixture of water and gum arabic. Only the blank areas absorbed this solution. The design area repelled it. A grease-based ink was then applied to the design. Since grease and water do not mix, the ink did not adhere to the moist blank areas. When a sheet of paper was pressed against the surface of the stone, a print of the design was made.

The lithographic stone used by Senefelder was made of Bavarian limestone. Such stone had qualities that were especially good for lithographic printing. It absorbed grease readily, was porous enough to hold moisture, and provided a smooth printing surface.

At first, words had to be lettered on these stones in reverse. They would then be readable on the printed sheets, Fig. 1-15. Later, a new method was tried. The artist first drew the image on a thin sheet of paper. The image was then transferred to the stone. It could then be printed from the stone. Both methods had advantages and disadvantages. For example, the transfer method did not always give a clear image. Thus, the most popular method was drawing the image directly onto the stone.

As might be expected, the lithographic stone was heavy and bulky. Before being used, the stones had to be carefully cut and processed. The top of the stone had to be very smooth and level. The bottom of the stone also had to be smooth and parallel with the top. The stone had to be grained with sand and then with pumice and emery stone.

The first lithographic presses

The first lithographic press was known as a *flatbed press* because the lithographic stone rested on a flatbed in the press. Fig. 1-16 shows such a press. The press could produce about 1000 impressions a day. Later, automatic dampening and inking rollers were made for the presses. A cylinder for carrying the paper also was added. The lithographic stone was placed on the flatbed of the press. This bed moved back and forth under the dampening and inking rollers and then under the cylinder that carried the paper. The cylinder turned at the same speed as the bed moved. It pressed the paper

Fig. 1-13. Alois Senefelder is credited with discovering lithography in 1798.

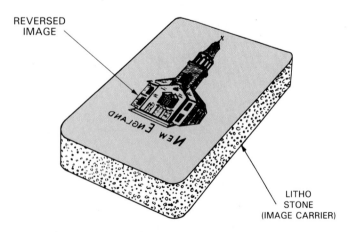

Fig. 1-14. Alois Senefelder sketched designs in reverse with a greasy crayon-like material on Bavarian limestone.

Fig. 1-15. Words lettered in reverse on lithographic stones are readable on the printed sheets.

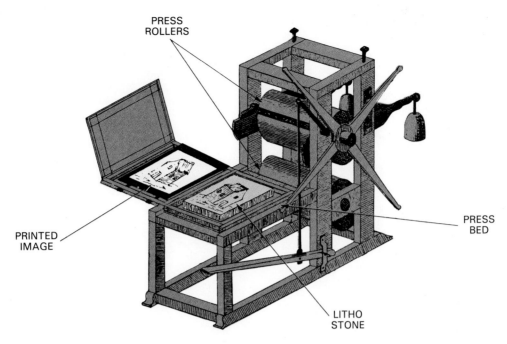

PRESS
ROLLERS

PRESS
BED

PRINTED
IMAGE

LITHO
STONE

Fig. 1-16. The first lithographic press was known as a flatbed press because the lithographic stone rested on a flatbed in the press.

firmly against the stone to make the impression. These presses were the forerunners of the high-speed offset presses used today.

Black-and-white lithographic printing processes were too slow and expensive to compete with the well-established letterpress printing process. As mentioned in letterpress printing, the raised-type characters are inked, then pressed against the paper. However, lithography became more popular with the invention of aluminum and zinc printing plates. These plates could absorb grease and water just like the lithographic stone. More importantly, they were light and flexible. Thus, they could easily be wrapped around a printing press cylinder.

The rotary press resulted from experiments with lightweight metal plates, Fig. 1-17. The *rotary press* was made of two cylinders that rotated in the same direction. The plate was attached to one cylinder and the paper to the other. They were set to meet at the same rate of speed. Thus, the plate would transfer an even impression to the paper. On a rotary press, a printer could produce about 1000 impressions per hour.

It is important to remember that all the lithographic printing up to this time had been done by *direct lithography*. The printed impressions were made on the paper "directly" from the stone or metal stone or plate. However, an *indirect* or *offset* lithographic printing process had been in use for years. It was used by tinmakers to print on tin for commercial purposes.

Ira Rubel, a New Jersey lithographer, developed an offset press designed for indirect lithography, Fig. 1-18. On Rubel's press, the inked plate rotated against a rubber-covered cylinder, called the blanket cylinder. The rubber blanket cylinder then pressed the image

against the paper. The image was printed as it appeared. It no longer needed to be reversed.

Invention of photography

In 1839, a process for producing permanent photographic images was developed. This was done by the French painter, Louis Jacques Mende Daguerre. It was, however, George Eastman (founder of Eastman Kodak Company) who first made photography practical, Fig. 1-19. In the late 1800s, he standardized camera sizes and produced roll film to fit the cameras, Fig. 1-20.

Fig. 1-17. A rotary press similar to this could print between 4000 and 5000 impressions per hour.
(Gutenberg Museum, Mainz)

17

Fig. 1-18. Ira Rubel is shown standing beside the offset press he designed for indirect lithography. (Smithsonian Institute)

Many technological advances have occurred in the design of cameras and lenses. New materials and processes also have been developed, Fig. 1-21. In addition, photography has expanded into the area of color film. Today, photographs are made from flexible film, either from rolls or sheets.

With the development of photography came greater use of offset lithography. This was made possible by the invention of the *halftone screen* in 1852. The first halftone screen was made of glass. It was made from two exposed glass negatives. Each negative contained

Fig. 1-19. George Eastman, founder of Eastman Kodak Company, is credited with making photography a practical technology. (Eastman Kodak Co.)

Fig. 1-20. In the late 1800s, George Eastman standardized camera sizes and produced roll film for home cameras. (Eastman Kodak Co.)

Fig. 1-21. Highly sophisticated cameras, lenses, and photographic materials are commonly used today. (Minolta Corp.)

Fig. 1-22. The development of photography promoted greater use of offset lithography. This was made possible by the invention of the glass halftone screen, a segment of which is shown magnified. (Eastman Kodak Co.)

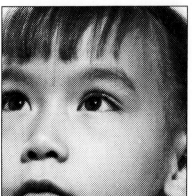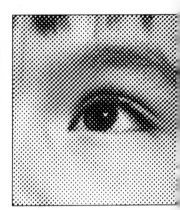

Fig. 1-23. A halftone is made up of many small dots of varying sizes, which print in a solid color, giving the illusion of light, medium, and dark areas in the picture. (Eastman Kodak Co.)

equally spaced scribed lines. The negatives were cemented together with their lines at right angles, Fig. 1-22.

The original photograph was photographed through the glass screen. This produced a negative made up of many small dots, called a *halftone*. On the printing press, each dot prints in a solid color, such as black. There are no gray dots, Fig. 1-23.

Note! Chapter 16 contains detailed information on halftone photography.

Photo-offset lithography

Used with the halftone screen, photography added to the versatility of offset lithography. It was now possible to reproduce both type and photographs (halftones). Thus, a new term developed: *photo-offset lithography*. Today, photo-offset lithography depends on highly technical photographic processes.

Many small printing shops began offset operations by adding *offset duplicators*, Fig. 1-24. These small offset presses have been improved over the years. For some printing, the work compares in quality to that of larger offset presses.

With improvements in photographic processes and offset printing plate materials, more sophisticated offset press designs followed. Multicolor *sheet-fed* (individual pieces of paper) offset presses and *web-fed* (large roll of paper) offset presses were designed, Figs. 1-25 and 1-26. Most printers, both small and large, turned to offset production methods.

Fig. 1-24. Duplicators like this are small offset presses used by many quick printers and in-plant operations. (A.B. Dick Co.)

PRINTING UNITS

DELIVERY END
OF PRESS

SHEETS OF PAPER
IN FEEDER

Fig. 1-25. Modern, high-speed, sheet-fed offset presses are commonly used by commercial offset printers.
(Heidelberg U.S.A.)

Fig. 1-26. Web offset press used for newpaper printing.

Fig. 1-27. Offset lithography technology continues to rely heavily on computers, lasers, cathode ray tubes, and robotics.

New and emerging technologies

Today, offset lithography relies on quality chemicals, films, inks, papers, plates, and related materials. In addition, skilled personnel perform high-quality color printing on fast precision presses. Together, these technologies form the basis for the manufacture of books, magazines, newspapers, commercial advertising, packaging, and thousands of other printed products.

Offset lithography technology is expanding rapidly. This is due largely to advances in science, such as lasers, computers, cathode ray tubes, and robotics, Figs. 1-27 and 1-28. Examples of how computers have made so many advances into the offset lighography industry include the following:

1. *Color electronic prepress systems* — Computers are being used to speed up layout and design, leaving more time for the art director's creative tasks, Fig. 1-29. The new design and imaging systems increase output and decrease the time and expense of

Fig. 1-28. An operator retouches a photograph on the 3M Color Workstation, a prepress tool for color correction, retouching, and image manipulation. (3M)

designing. Designers and printers can reduce production costs and expedite the time-consuming process of preparing multiple design samples for clients. These systems allow the operator to visualize printed ideas.

2. *Digital typesetters.* Computers are used to process and output copy. This translates into greater speed, more flexibility, and lower costs than previous typesetting methods. New desktop publishing software and hardware systems are drastically changing the way in which type is composed, Fig. 1-30. These computers are generally linked to output devices such as laser printers or phototypesetters.

3. *Color scanners.* Color scanners are electronic dot-generating machines that use computer-directed lasers for enlarging, recording, screening, and color correcting images. The scanner system stores and retrieves the scanned sets of separations in digital form. This allows for quick production of duplicate sets of color separations. Many scanners now operate as filmless systems. This makes it possible to go directly from original color copy to the printing of images directly on offset press plates. A digital-controlled color separation toning machine has been developed for the purpose of accurately positioning separation negatives in signature format. The process solves many problems associated with color registration and signature fit and accuracy, Fig. 1-31.

4. *Process cameras.* New microprocessor-equipped camera systems have revolutionized the photographic processes, Fig. 1-32. These camera systems enable skilled operators to produce large volumes of film negatives and positives with consistent quality.

5. *Digital color proofing.* This method of checking image quality before going to press will play a crucial role in tomorrow's electronic prepress systems. Images are checked on a display terminal screen, while still in electronic form. This procedure can save time and materials since irregularities can be detected before going to press, Fig. 1-33.

6. *Image assembly (stripping).* Much of the work associated with stripping, or film assembly, is now done on computerized equipment. Film masking materials are automatically cut to form a flat according to a layout for the offset plate with the aid of the computer, Fig. 1-34.

7. *Computerized press scanners.* Computers measure the job's ink area coverage and adjust all press inking keys (ink feed controls) automatically in a few seconds. This technology is also being applied to keyless inking systems.

A

B

Fig. 1-29. Design and prepress systems are being integrated so designers can make more efficient use of their time. (HELL, Harris Graphics Corp.)

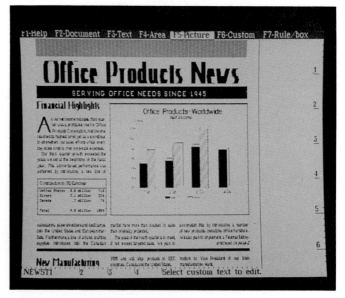

Fig. 1-30. New desktop publishing software allows you to preview your design on screen.

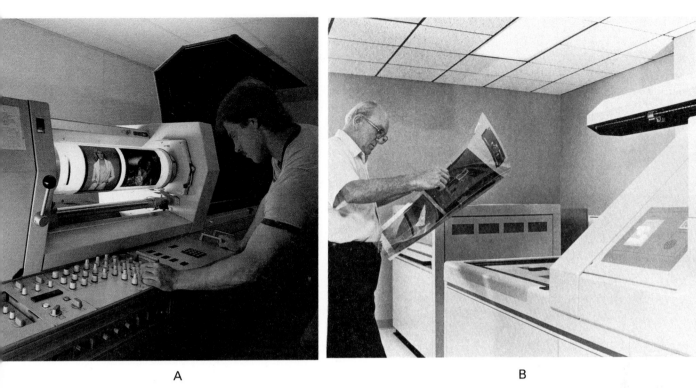

A   B

Fig. 1-31. This electronic color separation scanner uses a laser and fiber optics to form halftone dots in negatives. B—Operator checks separation films before exposing all four films onto a single sheet of film.   (Eastman Kodak Co.)

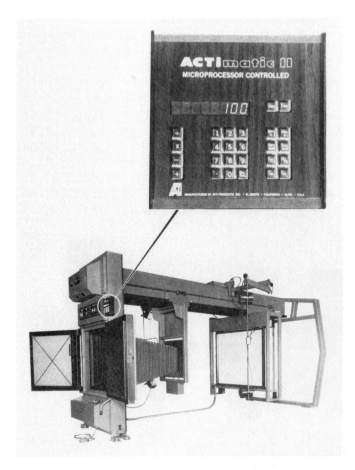

Fig. 1-32. Computerized process cameras enable skilled operators to produce large volumes of film negatives and positives with consistent quality.   (Acti Manufacturing)

Fig. 1-33. An electronic prepress system checks images on a display terminal screen while still in electronic form, thus saving time and materials.

Fig. 1-34. Image assembly (stripping) is no longer a labor-intensive operation. Computerized film assembly equipment automatically cuts masking materials to form a complete flat.

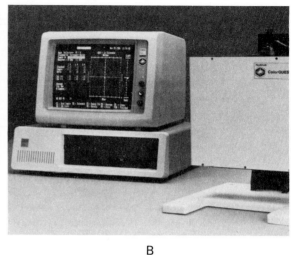

A   B

Fig. 1-35. A—A press information system is used to monitor ink and paper consumption and waste during a press run. B—Quality software and hardware provides fast, accurate color measurement data to operators. (Rockwell International Corp., Graphic Systems Div.; Hunter Lab Corp.)

8. *Press information systems.* Computerized systems used to monitor ink and paper consumption and waste during the press run are improving efficiency. These systems are fully computerized and are especially useful in monitoring complex functions on high-speed web-fed presses.

9. *Quality control.* Computers are used to monitor and control color and appearance of printed sheets. These systems consist of an optical sensor, quality control software, and a computer for data processing. The system provides fast, accurate color measurement data for purposes of quality control, Fig. 1-35.

10. *Scanning densitometers.* These devices assist in obtaining print quality control by monitoring or measuring ink film thickness, ink trapping, and ink/water balance.

11. *In-line finishing systems.* These components place printed sheets on platforms called *pallets* and transport them through the various finishing operations, such as collating, stitching, and trimming.

12. *Computerized cutters and folders.* Operator-programmed instructions determine how cutting and folding are done. In addition, robotic technology is now being used to cut and trim paper. Paper is automatically moved by a robot from a pallet to the cutter. The operator then makes the necessary cuts and/or trims. Another robot *palletizes* (stacks) the finished paper, Fig. 1-36.

Most of the new technologies are helping to lower production costs in the printing industry. This expansion will bring about rapid change and affect personnel at all levels and in every department. It is predicted that by the end of this century, many revolutionary new devices and processes will rival or replace the so-called

Fig. 1-36. Traditional labor-intensive finishing operations are now being handled by fully automated, computerized, high-speed equipment. (Heidelberg U.S.A.)

sophisticated offset lithography processes being used today.

To learn more about the industry

Magazines are more than just good sources of information. They provide an excellent method of getting acquainted with the whole industry. You can find many of these magazines at your school or community library. If you want to write to any magazine, look for the address in the masthead. Here is a list of some magazines that will help you keep up with the technology in the printing industry.

1. GRAPHIC ARTS MONTHLY covers a broad area of printing and has a business/management focus. It is published 13 times a year (twice in October).
2. AMERICAN PRINTER is another monthly with a little of everything including some articles which cover production procedures.
3. INSTANT AND SMALL COMMERCIAL PRINTER covers the one end of the industry its title describes.
4. IN-PLANT PRINTER and ELECTRONIC PUBLISHER is a bi-monthly, and is filled with practical advice.
5. QUICK PRINTING carries ideas to owners and operators of quick printing shops. It is a monthly publication.

POINTS TO REMEMBER

1. The primary purpose of the printing industry in the United States is to create and produce products that communicate visually.
2. The printing industry is made up of many related occupational groups. These include printers of books, newspapers, magazines, in-plant operations, quick printers, and commercial operations.
3. Everyone is a consumer of printed communications.
4. The printing industry is the sixth largest industrial groups in the United States. It employs over one and one-half million people.
5. Workers in the printing industry are distributed among approximately 54,000 printing establishments.
6. The United States Bureau of Labor estimates that the printing industry will need 60,000 new workers each year in the future because of expanding technology.
7. There are four major printing processes in general use; these include offset lithography, letterpress, gravure, and screen printing.
8. Offset lithography is called planography because the printing areas are flush (flat) or on the same plane as the surface of the printing plate.
9. Offset lithography is based on the principle that water and grease (or oil) do NOT readily mix.
10. Alois Senefelder, of Munich, Germany, discovered lithography in 1798.
11. Alois Senefelder used a Bavarian limestone to print from because such stones had qualities that were especially good for lithographic printing.
12. The first lithographic press was known as a flatbed press. It produced about 1000 impressions a day.
13. Presses were later developed that would be the forerunners of the high-speed presses used today.
14. All of the early lithographic printing was done by direct lithography.
15. Ira Rubel developed an offset press that was designed for indirect lithography.
16. George Eastman was instrumental in making photography practical. In the late 1800s he standardized camera sizes and produced roll film to fit the cameras.
17. With the development of photography came greater use of offset lithography.
18. The invention of the halftone screen in 1852 was also responsible for the increased use of offset lithography.
19. Offset lithography technology is expanding rapidly. This is due largely to advances in science, such as lasers, computers, cathode ray tubes, and robotics.
20. By the end of this century, it is predicted that revolutionary new devices and processes will rival or even replace the traditional offset lithography processes used today.

KNOW THESE TERMS

Graphic communication, Offset lithography, Letterpress, Gravure, Screen printing, Planography, Image area, Non-image area, Hydrophilic surface, Oleophilic surface, Plate cylinder, Blanket cylinder, Impression cylinder, Flatbed press, Rotary press, Direct lithography, Indirect lithography, Halftone screen, Offset duplicator.

REVIEW QUESTIONS

1. The printing, publishing, and packaging industry is the _____ largest technology group in the United States.
   a. Second.
   b. Third.
   c. Fifth.
   d. Sixth.

2. There are four major printing processes in the printing and publishing industry. List and briefly describe each.
3. List the ten related occupational groups that make up the printing, publishing, and packaging industry.
4. Offset lithography is called a _____ process.
   a. Intaglio.
   b. Planographic.
   c. Relief.
   d. Screen.
5. The printing, or _IMAGE_ _AREA_, on the flat offset printing plate is made by _PHOTOGRAPHIC_ and _CHEMICAL_ processes.
6. Offset lithography is based on the principle that _OIL_ and _WATER_ do NOT readily mix.
7. Another term for offset lithography is:
   a. Photography.
   b. Flexography.
   c. Gravure.
   d. Planography.
8. Limestone absorbs _grease_ easily.
9. The process of stone lithography was discovered in 1798 by:
   a. Ottmar Mergenthaler.
   b. Ira Rubel.
   c. Alois Senefelder.
   d. Johann Gutenberg.
10. The original lithographic stones were made of _Bavarian Limestone_.
11. The first lithographic presses were known as _____ presses.
12. Ira Rubel developed a press designed for:
    a. Direct printing.
    b. Indirect printing.
    c. Gravure printing.
    d. Flexographic printing.
13. Photography, used together with the _____ screen, added to the versatility of offset lithography.
14. The rapid expansion of offset lithography is due primarily to advances in science. True or false?
15. Most of the new technologies in the printing industry are helping to lower _prices_.

SUGGESTED ACTIVITIES

1. Using your local or school library, write a page describing the contributions to offset lithography made by Senefelder and Rubel.
2. Visit a local offset lithography firm and observe the several kinds and sizes of offset presses. Ask printing technicians about the functions of the presses, such as: kind of work done, sheet sizes, etc. Prepare a short report on your findings.
3. In your school, home, or community, find samples of offset printing. Prepare a bulletin board display with the samples as the theme.
4. George Eastman (founder of Eastman Kodak Company) first made photography practical. Prepare an outline on his experiences and adventures in the field of photography.
5. Research and prepare a visual example of the halftone screen process.

Chapter 2

Careers in Offset Lithography

When you have completed the reading and assigned activities related to this chapter, you will be able to:
O Describe the various categories of workers and their job responsibilities in the offset lithography industry.
O List and give a brief description of the career opportunities in the offset lithography industry.
O Prepare a personal resumé using the format described in this chapter.
O List factors which you like and dislike about each of the various job classifications described in this chapter.
O Describe the primary differences between management and creative jobs in the offset lithography industry.

Have you given much thought to career goals and the kind of work you would like to do? Most people spend about one-third of their lives working for a living. This means they work approximately eight hours a day or about 2,000 hours a year. Since a large portion of your life is spent working, it is important to select a career that is not only interesting, but also challenging and profitable. Offset lithography is a career field that has all of these features.

CAREERS

The offset lithography industry offers many challenging *careers* (job areas) for men and women. The industry offers opportunities to work and achieve personal goals in several areas. Managerial, creative, and technical jobs are available.

The scope of modern offset lithography has widened. The industry now includes many new occupational areas. For example, a person can work with the chemistry of inks and papers or design lenses and optical systems. A person might design a color brochure or operate a printing press, Fig. 2-1.

Fig. 2-1. The offset lithography industry includes many new and emerging occupational fields within managerial, creative, and technical job classifications. (Sir Speedy, Inc.)

The applied and pure sciences are basic to offset lithography. These include chemistry and mathematics. Anyone planning a career in offset lithography needs a basic understanding of these sciences.

Industry needs

The population of the United States continues to increase at a modest rate. Each person in our society is a potential consumer of printed products. As a result, the demand for printed communications will continue to increase every year. This means that people with managerial, creative, and technical skills will be needed by the offset lithography industry. Personnel will be hired in management, sales, copy writing, design, typesetting, photography, platemaking, presswork, and finishing and binding operations.

For a career in offset lithography, a technical background is essential. The industry is changing rapidly due to technological progress. With this change comes the need for broader educational backgrounds for workers. Many exciting job opportunities are available in the offset lithography industry, Fig. 2-2.

While traditional printing industry centers have been located in the East and Midwest, U.S. Government statistics indicate that the industry is growing fastest in the South and West, Fig. 2-3. Those areas now have a 44 percent share of the total industry, and more than 50 percent of the commercial printing sector.

Fig. 2-4 shows the organization of a medium-sized printing plant. The occupational titles described reflect the more visible areas of the industry.

CAREER PROFILE

Bruce Jones is the owner of IPF Printing in Burbank, California. Bruce started out in the printing industry when he was 15 years old, learning the business in his high school print shop at Burbank High School. He took all the graphic arts classes offered. He worked in printing after school and summer vacations. When 17, he and his brother started a print shop in their garage. Bruce learned to run all of the few pieces of equipment they could afford to purchase.

Near the end of high school, Bruce decided he wanted to teach graphic arts in a high school. He made preparations to enter California State University, Los Angeles to major in Industrial Arts Teacher Training. During this period, Bruce

Fig. 2-2. Cutting and trimming a printed job is done to make all sheets exactly same size. This computer-controlled cutter automatically trims sheets to size. (Heidelberg U.S.A.)

Where The Jobs Are

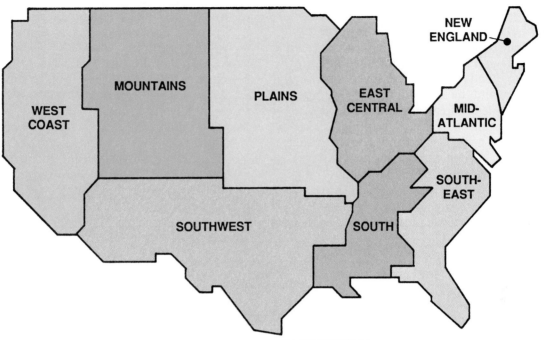

NEW ENGLAND

(Connecticut, Maine, Massachusetts, New Hampshire, Rhode Island)

Nearly 4,000 business establishments employ nearly 102,000 people. Boston has some 1,200 graphic communications companies.

MID-ATLANTIC

(Delaware, District of Columbia, Maryland, New Jersey, New York, Pennsylvania)

This area has the most people employed in graphic communications of any region in the country -- nearly 325,000 individuals working in nearly 13,000 businesses. The New York metropolitan area has more than 3,800 operations; nearby Newark, New Jersey, has another 1,500 and Philadelphia more than 1,000. Printing is the largest manufacturing industry in Washington, D.C.

SOUTHEAST

(Florida, Georgia, North and South Carolina, Virginia, West Virginia)

More than 6,000 businesses employ some 127,000 people.

SOUTH

(Alabama, Kentucky, Louisiana, Mississippi, Tennessee)

More than 3,100 businesses employ nearly 65,000 people.

EAST CENTRAL

(Illinois, Indiana, Michigan, Ohio, Wisconsin)

Nearly 13,000 businesses, the most in any region of the country, employ nearly 300,000 people. Chicago alone has nearly 3,500 operations. Detroit more than 1,000.

PLAINS

(Iowa, Kansas, Minnesota, Missouri, Nebraska, North and South Dakota)

Nearly 6,000 businesses employ more than 117,000 people in St. Louis, some 900 plans employ around 16,000 people, and in-plant operations employ an additional 4,000.

MOUNTAINS

(Colorado, Idaho, Montana, Nevada, Utah, Wyoming)

More than 2,000 plants employ more than 30,000 people.

SOUTHWEST

(Arizona, Arkansas, New Mexico, Oklahoma, Texas)

More than 5,500 plants employ nearly 100,000 people.

WEST COAST

(Alaska, California, Hawaii, Oregon, Washington)

More than 8,500 businesses employ more than 155,000 people. California has the most graphic communications firms of any state --- more than 6,800 and it ranks second only to New York in the number of people employed. Los Angeles has nearly 3,000 businesses in the field.

Fig. 2-3. Offset lithography industry is growing faster in South and West. California has most printing firms in any state. New York ranks first in number of people employed in printing and publishing.

and his brother continued to run the printing business. Bruce graduated from the university with a B.A. degree and a teaching credential. At age 21, Bruce was the youngest person in the State of California to have a designated subjects teaching credential. As it turned out, he was offered a position teaching graphic arts at his high school alma mater.

A busy teaching schedule consumed much of Bruce's time and energy. He loved teaching graphic arts and he got along well with the students. Besides his teaching duties, Bruce was responsible for all the district's printing production. This was a heavy burden and severely limited the time he was able to spend in his own business. Although he was quite happy with how things were going, still, there was something missing; the excitement of operating his own business. As Bruce grappled with a solution, it became evident

that most of what he liked to do was run the business. This eventually led him to resign his teaching position after four years and return to his business.

Today, Bruce is happy at IPF Printing doing quality printing in a shop which has expanded to 8 people. Bruce says he has a good group, including his mother and father. The company is growing and the future is bright. Bruce recently remodeled a 7,000 square foot building next door to his present location to expand his printing capabilities even more.

In addition to his day-to-day responsibilities of running the shop, Bruce still makes time to help customers while supervising the printing operation. At age 31, Bruce is one of the youngest owners in the printing industry and loving every second of it.

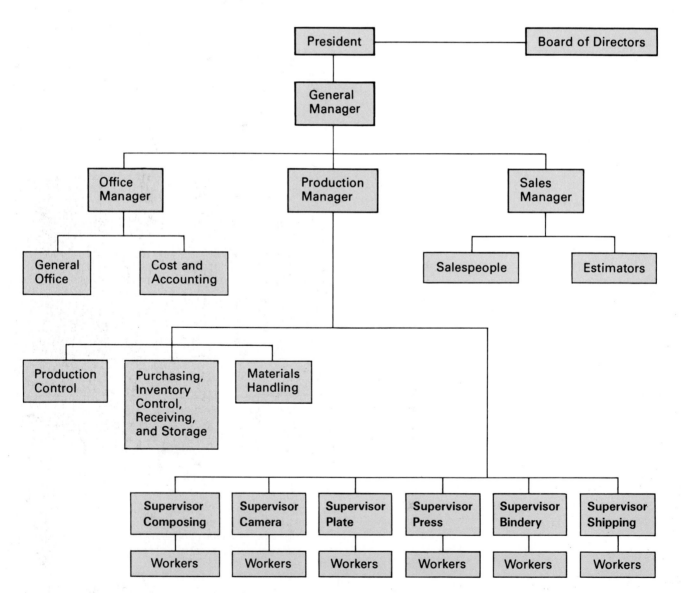

Fig. 2-4. Printing companies are categorized as small, medium, or large. This is an organizational chart for a medium size company.

Management, planning, and production control

Managers, estimators, and *office staff* are responsible for operating a printing plant efficiently and profitably. They handle business accounts, records and payroll. They are responsible for buying materials and managing the work flow. They also price the product, direct sales efforts, and are in charge of hiring personnel.

Creative jobs

Printed products are designed to deliver information. Information includes ideas, which are created by people. A printed product presents an idea or information to the person who needs it. The offset lithography process provides the equipment and techniques for printing this information.

A printed product, then, presents an image or printed information to the buyer. Those who develop the image and words have creative jobs. These people must be able to form visual images or write copy from verbal descriptions.

Creative jobs in offset lithography include the following tasks:

1. Once the ideas have been developed for a printed product, the message or the copy must be prepared.
2. Writers present an idea through the written word, Fig. 2-5. Thus, writing sometimes is the starting point for graphic arts products.
3. Illustrators or artists have the job of developing pictures that present ideas clearly, Fig. 2-6. Cartoons and drawings in newspapers, magazines, and books are examples of the illustrator's work.

The original copy is used by the artist to prepare the required illustrations. The final illustrations are generally referred to as *artwork.* Such artwork may consist of black-and-white inked drawings, tempera renderings, or charcoal drawings. Artwork may also include full-color photographs and transparencies. These are converted into color-separated negatives or positives.

In a small offset lithography printing firm, one person may perform a number of tasks related to artwork. Larger offset lithography plants generally employ a staff of artists. Each artist would have specific duties.

Photographers illustrate graphic arts products through continuous tone images (photographs). The camera is the photographer's main tool, Fig. 2-7.

Designers are artists who are responsible for the appearance of printed items, Fig. 2-8. Designers are sometimes called *layout artists.* They arrange or lay out the words and illustrations of printed products. That is, they decide where the copy, drawings, and photographs will be placed on the page.

Technical jobs

Technical workers and other skilled technicians in the offset lithography industry operate many different

Fig. 2-5. Writers prepare written material in form of manuscript or copy. Most graphic arts products start as a result of creatively written messages.   (Stephen Simms)

Fig. 2-6. Illustrators or artists prepare pictorial materials, such as cartoons and drawings, so that graphic ideas are presented clearly.   (Dynamic Graphics, Inc.)

Fig. 2-7. Commercial photographers create graphic images for many types of printed communications. Achieving desired outcome in photographic composition, and knowing how it will look in final printed form, are creative talents required of photographer. (Eastman Kodak)

Fig. 2-8. Graphic designers are responsible for planning printed pieces. Expressing ideas visually is essential requirement for this position.

types of equipment. Usually, this kind of knowledge is obtained through technical school or college courses. In several occupations, apprenticeship programs are available.

In some large offset lithography printing plants, technicians play a major role in the operation of the company, Fig. 2-9. For example, they work on ways to improve the handling of inks, plates, and paper. Equipment modifications, which improve production and quality, are also part of a technician's job.

A *phototypesetter,* or *compositor,* sets *copy* (words) on computerized machines called phototypesetting machines. These machines produce photographic type images on paper or film, Fig. 2-10. To set type, the operator types the copy on the keyboard of the phototypesetting machine. It is similar to that of a typewriter. The phototypesetter also should know the basics of photography. Since the operator may need to make minor adjustments or service the machine, a basic knowledge of the equipment is helpful.

After the type is composed, it is checked against the original manuscript by a *proofreader.* This person reads the typeset material and ensures that it is complete and accurately typeset.

The *paste-up artist* prepares a paste-up, or *mechanical.* The paste-up artist must be able to work accurately and quickly while following the layout artist's sketch, Fig. 2-11. The paste-up artist must deter-

32

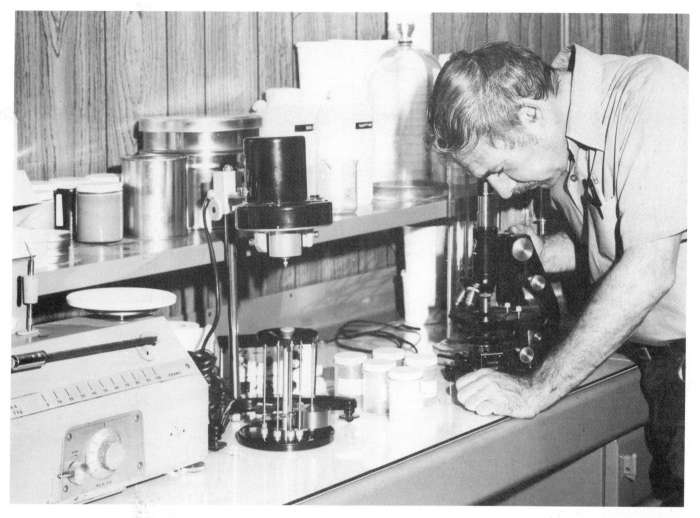

Fig. 2-9. Technicians are responsible for overall plant production and quality control. For example, technicians develop efficient ways to improve the handling of inks, plates, and paper. Technicians assure consistent quality and reduce spoilage and waste.   (General Printing Ink Co.)

Fig. 2-10. Typesetter inputs manuscript, called copy, on a phototypesetting machine. This person must also be able to use and maintain processing unit as well as make minor adjustments and service machine.

Fig. 2-11. Paste-up artist organizes all elements of a printed job on a base sheet. This is done according to what designer has indicated on comprehensive layout.   (Sir Speedy, Inc.)

mine the exact positions of the illustrations and typeset materials. The paste-up is prepared according to the original layout. The artwork and typeset material are cut apart and bonded to illustration board. This must be carefully done, since each space must be precisely measured. The material is then carefully pasted onto illustration board.

The *paste-up* is the master from which the offset lithography *printing plate* is prepared. It contains all the type and line artwork pasted in position on a piece of illustration board. In this format, it is ready for the process camera department. If type and artwork are NOT properly aligned, or if type is broken, these imperfections will be photographed and will then appear on the printing plate.

The graphic arts *process camera operator* photographs the paste-up or other images with a very large camera, Fig. 2-12. This is the first stage in making an offset printing plate of the job to be printed. The camera operator must be familiar with the films and lighting techniques used to photograph flat (two-dimensional) copy, such as the paste-up. A camera operator may specialize in either black-and-white or color photography.

If a color photograph is needed, an electronic color separation scanner must be used. These are complex machines. They combine the technologies of offset lithography and electronics. Their product is a set of color-corrected separations. The machines are operated by a *color separation scanner operator*. This person must have a good knowledge of color values and photography, Fig. 2-13.

Fig. 2-13. Electronic color separation scanner operator must have a good knowledge of color values and basic photography. Most scanners use a laser to create color separations on film. (Crosfield Electronics)

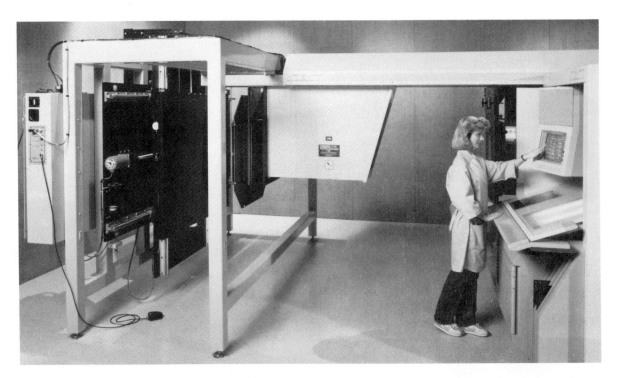

Fig. 2-12. Process camera operator must know exposure calculations, optics, lighting, chemicals, and processes for photographing different types of copy. (Opti-Copy, Inc.)

After the illustrations and typeset copy have been photographed, the *negatives* (reversed image) or *positives* (same image) are sent to a stripper, Fig. 2-14. The *stripper* positions the negative or positive on a special masking sheet in preparation for platemaking. A *masking sheet* is made of plastic or paper. It is the size of an offset printing press plate.

Fig. 2-15. Platemaker uses flat (negatives or positives taped on a masking sheet by a stripper) to place image on an offset plate (image carrier). The plate has a latent (invisible) image exposed on it when light passes through clear areas of the negatives.

Fig. 2-14. Stripper organizes and positions film negatives and positives on a special masking sheet in preparation for platemaking. Layout accuracy is important! (Allen McMakin)

The negative is taped to the masking sheet in the desired location. *Windows* (openings) are cut in the sheet to reveal the images. This makes up what is called a *flat*.

The stripper works over a light table to illuminate the negatives or positives. The room lighting is usually subdued for optimum visibility of the film material on the light table.

The sheet containing film negatives is passed along to the *platemaker,* Fig. 2-15. The offset lithography printing plate is generally made of thin, flexible metal. The plate is designed to wrap around a printing press cylinder. The process of transferring an image to the plate's surface is called *platemaking*. It is a photochemical process.

The plate may be made of aluminum, stainless steel, or a processed paper. Regardless of material, it is coated with a *light-sensitive* (photographic) chemical. Generally, the coating is applied during manufacture of the plate. The chemical reacts to high-intensity light and processing chemicals.

The flat is placed against the plate and exposed to light. The plate can be developed by hand or by an automatic plate processor. When more than one page of the same image is run on a single offset plate, an *automatic plate exposure* unit is used. This is called a *step-and-repeat* plate exposure machine.

The platemaker must know how to use various kinds of printing plates. The platemaker must also know the properties and uses of each type of plate being used in the shop.

The *press operator* prepares and tends the offset printing press, Fig. 2-16. This work may vary greatly,

Fig. 2-16. Press operator prepares and tends offset printing press. A mechanical aptitude and knowledge of paper, ink, and chemical solutions are important. (Sir Speedy, Inc.)

depending on the types and size of press. The press operator must be able to install the printing plate, adjust roller pressures, and mix inks and fountain solutions. Mechanical ability is important.

In large printing plants and on large presses, the press operator must be able to supervise press feeders and helpers. As might be expected, press operators must have excellent color perception. Much of the work printed today is in full-color.

The actual printing or reproduction operation is referred to as *presswork*. Presses range in size from small duplicators to large sheet-fed presses and giant web-fed presses. Most web-fed presses are capable of printing both sides of the sheet at the same time in several different colors of ink. Some can print an entire book in a single pass through the printing units.

Sometimes more than one printing process is used to reproduce a printed product. Often, this is necessary because of a special requirement, such as color or length of run. For example, a Sunday newspaper may be printed by the offset lithography process at one location. Its full-color magazine supplement may be printed by the gravure process at a different location.

The final operations in the manufacture of a printed product generally involve finishing and binding. These operations are performed by the *bindery operator,* Figs. 2-17 and 2-18.

Finishing operations include gathering, collating, sorting, wrapping, labeling, shipping, and similar processes.

Binding operations include the assembly and fastening of books, magazines, catalogs, business forms, and calendars.

Large offset lithography firms generally operate their own finishing and binding departments. Specialized finishing and binding firms are available to printers who do not have such facilities, Fig. 2-19. The finishing and binding operation is one of the most important phases in the manufacture of printed products, Fig. 2-20. An error at this stage can result in the loss of money, time, and materials, Fig. 2-21.

Shipping and maintenance

The *shipping department* packages and ships the product to the customer. The shipping department also receives supplies used in the printing operation.

Maintenance personnel are required to maintain the offset lithography plant. They ensure that equipment is serviced and functions properly. Maintenance personnel includes electricians, plumbers, and carpenters.

Industry trade services

As you can see, many operations and people are involved in offset lithography. Some printers specialize in certain phases of production. For example, a printer may do everything except finishing and binding operations. For these other operations, the printer relies on outside trade services. *Trade services* are firms that specialize in the various offset lithography and letterpress production processes. There are trade services that specialize in design and layout, typesetting, presswork, printing of forms, die cutting, and finishing and binding.

Quick printing

Several years ago, when quick printing was still new, it underwent a period of rapid expansion. Now it is

Fig. 2-17. Cutting and trimming paper stock is a function performed by bindery operator. Controls on many of the newer paper cutters are computerized for fast and efficient production. (Hart Bindery Services)

Fig. 2-18. Bindery workers often operate complex folding machines. Operators must have a knowledge of folding, slitting, scoring, and perforating techniques. (Hart Bindery Services)

Fig. 2-19. This gathering and collating machine requires operators and setup personnel who can produce accurate work and maintain equipment. (Hart Bindery Services)

Fig. 2-20. Since a wide variety of equipment is found in bindery, specialized finishing and binding firms are available to printers who do not have such facilities located in their plants. (Hart Bindery Services)

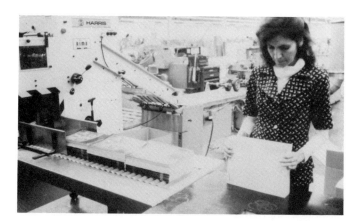

Fig. 2-21. Finishing and bindery operations require a high degree of mechanical aptitude and ability to work precisely. Men and women are employed in the bindery. (Hart Bindery Services)

considered a mature industry, having settled into a slower growth phase. But with more than 40,000 such firms in the United States, doing $7.8 billion in business every year, this is another excellent career area to consider, Fig. 2-22. Career opportunities in quick printing are plentiful in virtually every area which includes many of the following:

1. *Equipment operation*—You might run anything from a small offset duplicator to a large press, process camera, digital typesetting equipment, computerized cutters and folders.

2. *Art and design*—You might help clients prepare materials, assisting in type, graphics, color, paper, and layout selection.

3. *Marketing*—Your duties would include advertising and public relations.

4. *Research*—Looking at new equipment and processes and evaluating their potential; investigating possible new markets would be a few responsibilities.

5. *Quality control*—You would make sure that people and machines are turning out the best possible work and finding solutions to problems.

6. *Systems development*—You would handle duties

Fig. 2-22. Quick printing is considered a mature industry, having more than 40,000 firms in the United States. Career opportunities in quick printing are plentiful in all areas of the business. (Sir Speedy, Inc.)

from natural job-tracking to implementing sophisticated desktop publishing and communications networks.

7. *Technical coordination* — You would manage adapting new technologies and products into a shop.

8. *Sales* — Cold call selling, telemarketing, direct mail, franchise sales might be a requirement.

CAREER PROFILE

Shambala Simpson is supervisor of facsimile operations at USA TODAY in Arlington, Virginia. Translated into everyday English, that means she oversees the satellite transmission of newspaper pages to 30 national and two international printing sites. This is just one of the "daily miracles" behind the production of this colorful newspaper.

Beginning as a paste-up artist in the late 1970s for COCOA TODAY (now called FLORIDA TODAY), Shambala is still surprised that she plays such a crucial role in producing the most successful newspaper in recent history. "If you had told me years ago that I'd be doing this today," she jokes, "I'd have said you were nuts."

Starting at about 1:00 p.m. every weekday, Shambala begins sending up test signals. An hour later, the next day's advertising pages begin their 46,000 mile journey up to the satellite and back down to the remote site dish. From 5:00 p.m. to as late as 4:00 the next morning, editorial pages are beamed up. Using a personal computer, Shambala also keeps records on every page she receives and transmits, and troubleshoots equipment problems that develop at the sites.

Shambala's job evolution was simply a matter of taking advantage of the technological advances now changing the face of newspaper production. She had moved from paste-up artist to night manager of composing at COCOA TODAY when her supervisor moved to Washington, DC to help in the development of USA TODAY.

Summoned to Washington, DC in the summer of 1981 to paste up the first working model of the newspaper, Shambala returned to Florida, as Gannett, the parent company, continued to work to get the newspaper off the ground. Once they launched the newspaper, Shambala moved north permanently and joined them.

Aside from enjoying the "oohs and aahs" of visitors impressed with the futuristic hardware, Shambala has discovered some personal benefits from being on the forefront of printing technology. "I know people all over the country," she says. "I talk to them every day, not only about work, but about sports, weather, and everyday things."

Seeking employment

Your first choice of a job can influence your entire future career. Though most people change jobs at least three times during their working lives, many do not. Rather, they remain at the job for which they were first hired. They may, of course, be promoted. Nonetheless, they continue to work within the same general occupation in which they started.

Whether you plan to change jobs or not, it is important to begin your working life with a satisfying job, Fig. 2-23.

Refer to the United States Government OCCUPATIONAL OUTLOOK HANDBOOK for additional information regarding careers. It describes various occupational aspects of the printing industry with reference to offset printing. This handbook is updated periodically and is available at most public libraries.

A good job rewards an individual in several ways. Most obviously, there is the salary. This money enables you to provide the basic needs of life. It should be sufficient to allow you to save a small percentage each payday. A job also provides satisfaction, especially if the person has a keen interest in the job, Fig. 2-24. The

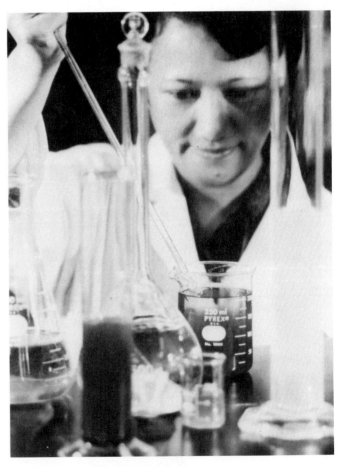

Fig. 2-23. This woman is a chemist for a large offset printing firm. Work is satisfing, challenging, and financially rewarding. A college or university degree in chemistry is generally a requirement for this position. (Van Son Holland Ink Corp.)

Fig. 2-24. This woman works in quality control department of an ink manufacturing company. The job requires constant sampling to assure consistent quality of inks.
(Van Son Holland Ink Corp.)

satisfaction and the interest can depend on the individual's ability to do the job. In thinking about a career, then, it is important to evaluate your skills, interests, and talents.

To find the employer and employment best suited for you, you may need to explore several employment opportunities. The following can provide possible leads to employment:

1. Personnel offices or business and industry.
2. Civil service announcements.
3. U.S. Government Printing Office.
4. Help wanted signs in printing firms.
5. Radio and television announcements.
6. Newspaper help wanted classified advertising.
7. School placement office.

Preparing for employment

Be prepared before you inquire about or apply for employment. Do the following two things:

1. Prepare a personal resumé.
2. Find out as much as you can about the company you are interested in working for.

By following this two-step procedure you will gain several advantages. You will acquire increased self-confidence and also find it easier to fill out applica-

tion forms and questionnaires. You will create a better impression on prospective employers and be more thoroughly prepared for personal interviews.

Personal resumé

Your *personal resumé* is a typed summary that represents you to people whom you have never met. It is a permanent outline of your education, employment history, and qualifications. A well-written resumé will go a long way toward helping you make a good first impression. It may lead to a job now or at a future date. In later years, it becomes increasingly important to keep your resumé up-to-date, particularly if you change jobs.

The contents and layouts of personal resumés vary as widely as the different individuals who apply for jobs, Fig. 2-25. You do not have to restrict the items on it to those shown in the sample. Add any information necessary in order to best acquaint the employer with your qualifications.

Your personal resumé should be mailed with a cover letter of application to a prospective employer. Then you may receive an application form and/or an appointment for an interview. You may also hand your resumé to a prospective employer first thing or very early during an interview.

In planning the layout and the writing of your resumé, remember these important points:

1. Keep the resumé simple and only one or two pages in length.
2. A resumé should be typed on good quality paper.
3. Your resumé must be neat and accurate in every detail.
4. Keep copies of the original resumé so you do not have to rewrite it completely at a later date.
5. Use some "white space" to create a pleasing impression of neatness and orderliness.

Investigating the company

Doing some investigation of the company with whom you may wish to seek employment will help you protect your own interests, Fig. 2-26. It may provide answers to questions you have about the company. It will give you something to talk about besides yourself during an interview. It will provide materials from which you can form questions you should ask during an interview. It will provide information to help you make decisions should you be considered for employment.

Information about a particular company can be located by contacting the school placement office, individuals in sales, contractors who are in touch with company personnel, persons employed by the company, and company brochures.

The following types of information are suggested as points you may want to know about a prospective employer:

1. Name and address of the company.

2. Where its plants and offices are located and what territory it covers.
3. How long it has been in business.
4. What its products or services are and the nature of the business.
5. What has been its growth.
6. How its prospects look for the future.

7. Is it a union or nonunion shop.
8. Name of owner or president.
9. Name and title of person in charge of hiring.
10. Name of relatives and friends employed by the firm.
11. Miscellaneous information about employee fringe benefits.

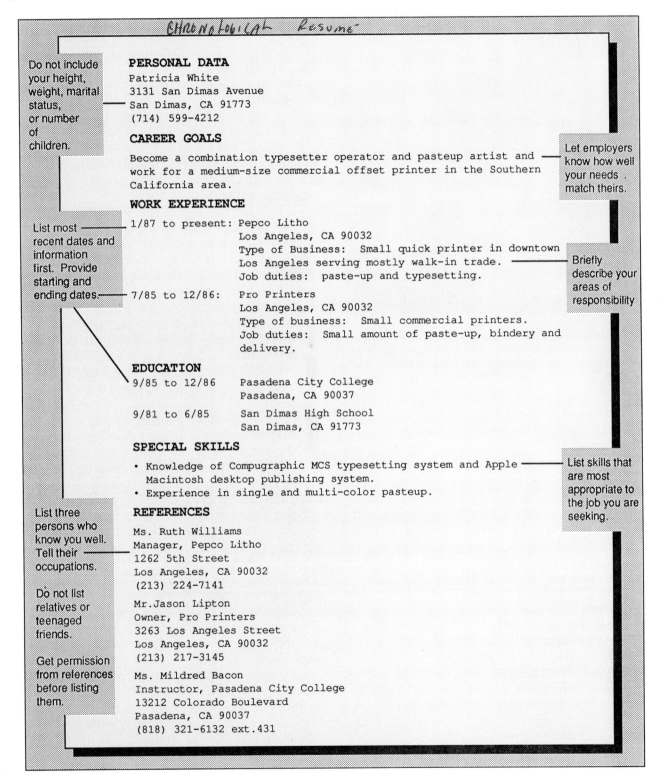

CHRONOLOGICAL Resume

Do not include your height, weight, marital status, or number of children.

PERSONAL DATA

Patricia White
3131 San Dimas Avenue
San Dimas, CA 91773
(714) 599-4212

CAREER GOALS

Become a combination typesetter operator and pasteup artist and work for a medium-size commercial offset printer in the Southern California area.

Let employers know how well your needs match theirs.

WORK EXPERIENCE

List most recent dates and information first. Provide starting and ending dates.

1/87 to present: Pepco Litho
Los Angeles, CA 90032
Type of Business: Small quick printer in downtown Los Angeles serving mostly walk-in trade.
Job duties: paste-up and typesetting.

Briefly describe your areas of responsibility

7/85 to 12/86: Pro Printers
Los Angeles, CA 90032
Type of business: Small commercial printers.
Job duties: Small amount of paste-up, bindery and delivery.

EDUCATION

9/85 to 12/86   Pasadena City College
Pasadena, CA 90037

9/81 to 6/85   San Dimas High School
San Dimas, CA 91773

SPECIAL SKILLS

- Knowledge of Compugraphic MCS typesetting system and Apple Macintosh desktop publishing system.
- Experience in single and multi-color pasteup.

List skills that are most appropriate to the job you are seeking.

List three persons who know you well. Tell their occupations.

Do not list relatives or teenaged friends.

Get permission from references before listing them.

REFERENCES

Ms. Ruth Williams
Manager, Pepco Litho
1262 5th Street
Los Angeles, CA 90032
(213) 224-7141

Mr.Jason Lipton
Owner, Pro Printers
3263 Los Angeles Street
Los Angeles, CA 90032
(213) 217-3145

Ms. Mildred Bacon
Instructor, Pasadena City College
13212 Colorado Boulevard
Pasadena, CA 90037
(818) 321-6132 ext.431

Fig. 2-25. Example of a typical one-page resumé. Information is intended to acquaint prospective employer with your work experience and education.

Applying for employment

Once you have collected and recorded the important information about yourself and the company, you may then contact your prospective employer. Methods for making the initial contact include mailing a letter of application and your personal resumé, inquiring in person about openings, and inquiring about openings by telephone. Your decision regarding which method to use is dependent upon how you learned of the position and how competitive the job market is for the job.

If you know only that the prospective employer hires individuals with qualifications similar to yours, you may prefer to compose a letter of application and mail it along with your resumé. If the employer is NOT hir-

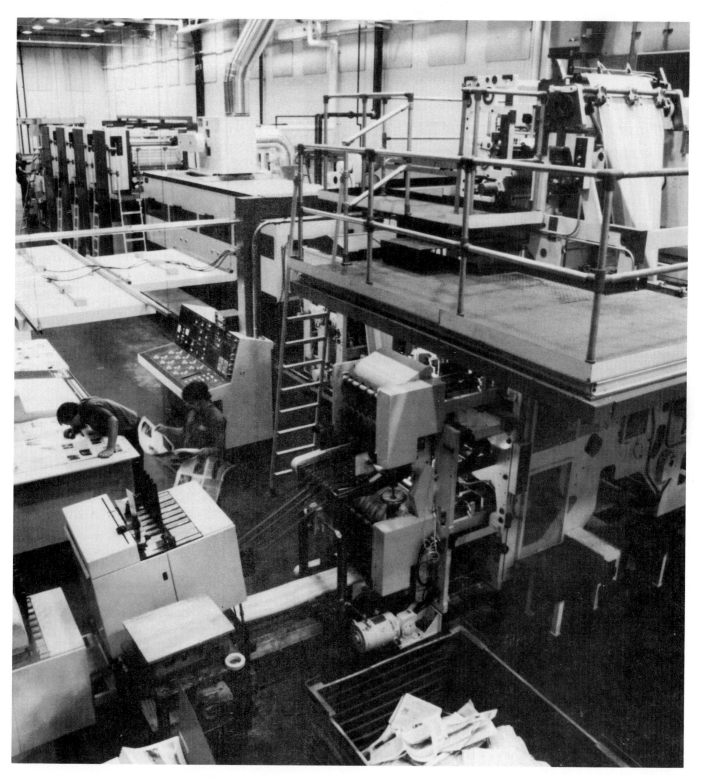

Fig. 2-26. Investigate the company with whom you may wish to seek employment. This will help you protect your own interests and may provide answers to questions you have about the company.

Fig. 2-27. When seeking a new job position, it is more impressive when you go to appropriate company office to make application. Prospective employers sometimes ''frown'' upon telephone calls regarding job openings.

ing, your letter might be ignored. In that case, you might elect to follow up on the possible position by inquiring in person or by telephoning. Many prospective employers "frown" upon telephone calls. Your interest and desire for a position is more impressive when you go to the appropriate company office to make application, Fig. 2-27.

CAREER PROFILE

Company names, such as Apple Computer, Memorex, Mattel, Lorimar, and Disney, are familiar to almost every consumer in the United States. Perhaps not as familiar is the name Ivy Hill Graphics and Packaging, with plants in New York, Terre Haute, and Los Angeles. Ivy Hill has the distinction of being one of the foremost manufacturers of printed packaging in the nation. The company designs and creates cartons, point-of-purchase displays, record album jackets, J-cards for cassette tapes, and life-size replicas of people and other types of objects.

As a Printing Management major at California State University, Los Angeles, Lisa Diaz became fascinated with the whole idea of the printing and publishing industry. Now she is employed at Ivy Hill as a production representative and estimator. Lisa took several "twists and turns" in her life before realizing that printing was really her "first love."

Lisa attended Sacred Heart of Mary High School and graduated in the top ten of the class. She then entered California State University, Los Angeles as an undeclared major. It was hard for her to find a niche. She took several business courses and after two years decided that it wasn't for her. Somehow it was not related to any one type of business endeavor such as printing. She then decided to take a few art classes which had always held a special interest for her. These included design, lettering, and by chance one graphic arts class in the Department of Technology. That's all it took! She was hooked on graphic arts from that moment on.

From that beginning, Lisa began to take a renewed interest in her education. She realized that she had finally found a career niche. With the encouragement of her printing teachers and the friendship of the students in the program, Lisa found herself with a strong feeling of accomplishment. During all this time at the university, Lisa supported herself by working as a waitress since she could schedule work around classes. She also held a job in a small "mom and pop" print shop which helped her gain some practical experience in printing.

After graduation, Lisa relaxed from the previous rigorous pace for three months. Then, as she began to explore the help wanted classified advertisements in the local newspaper, she came upon a printing position listed only with a telephone number. As it turned out, this was to be the Ivy Hill position. Although she did not have the two years experience asked for in the advertisement, she did possess enthusiasm, positive attitude, vigor, potential, and a Printing Management four-year degree. Needless to say, she was hired after interviewing with several department heads.

Lisa sums it all up by explaining that "Ivy Hill is a wonderful company to work for and I am looking forward to a bright future in the printing and packaging industry."

POINTS TO REMEMBER

1. Most people spend about one-third of their lives working for a living.
2. It is important to select a career that is not only interesting, but also challenging and profitable.
3. The demand for printed communications will continue to increase every year.
4. People with managerial, creative, and technical skills will be needed by the offset lithography industry in increasing numbers.
5. Jobs in offset lithography will be available in management, sales, copy writing, design, typesetting, photography, platemaking, press-

work, and finishing and binding operations.

6. According to U.S. Government statistics, the printing industry is growing fastest in the South and West. Those geographic areas now have a 44 percent share of the total industry, and more than 50 percent of the commercial printing sector.

7. The printing industry divides job descriptions into three general categories: management, creative, and technical.

8. *Management jobs* include estimators, managers, and office staff.

9. *Creative jobs* include writers, illustrators, photographers, and designers.

10. *Technical jobs* include such occupations as typesetting, paste-up, process camera, electronic color separation, stripping, plate-making, press operation, finishing and binding, shipping, and maintenance.

11. The printing industry has a number of trade services. These are firms that specialize in the various offset lithography and letterpress production processes.

12. In seeking and applying for employment, an applicant should make the necessary preparations by first thinking about a career and then evaluating his or her skills, interests, and talents.

13. Employment opportunities are found through Civil Service announcements, help wanted signs in printing firms, help wanted advertisements in newspapers, school placement office, and U.S. Government Printing Office.

14. The applicant for a job should prepare a concise one or two page resumé.

15. The applicant should investigate the company and find out as much as possible about its products and policies.

16. Information about a particular company can be located by contacting the school placement office, individuals in sales, contractors who are in touch with company personnel, persons employed by the company, and company brochures.

17. Methods of making the initial employment contact with a company include mailing a letter of application along with a personal resumé, inquiring in person about possible openings, and inquiring about openings by telephone.

KNOW THESE TERMS

Artwork, Phototypesetting machine, Paste-up. Step-and-repeat plate exposure machine, Press-work, Finishing and binding, Trade services, Resumé.

REVIEW QUESTIONS

1. Most people spend about _____ of their lives working for a living.

2. The offset lithography industry offers opportunities to work in managerial, creative, and _____ occupations.

3. Applied and pure sciences, such as chemistry and mathematics, are basic to offset lithography. True or false?

4. Which of the following is NOT a management job?
   a. Sales.
   b. Estimating.
   c. Writing.
   d. Office staff.

5. Writers, illustrators, photographers, and designers belong to the _____ occupational group of offset lithography.

6. Phototypesetting operators belong to the technical occupational group of offset lithography. True or false?

7. Why must the paste-up artist be extremely accurate in the preparation of camera-ready materials?

8. What is meant by two-dimensional copy?

9. A stripper generally works at a _____ under subdued room lighting.

10. A person who prepares offset plates for the press is NOT required to know the various properties and uses of each type of plate being used in the shop. True or false?

11. Why are press operators required to have excellent color perception?

12. Describe the general duties and essential qualifications of a press operator.

13. Explain why the finishing and binding operation is one of the most important phases in the manufacture of printed products.

14. What are trade services? Why are trade services used by some printers?

15. Describe the essential steps that should be followed when seeking employment in the offset lithography industry.

SUGGESTED ACTIVITIES

1. Prepare a one- to two-page report on one of the offset lithography occupations. Also, list what you would like and would NOT like about the job.

2. Study the classified advertisements in your daily newspaper under the help wanted heading. Prepare a list of the offset lithography jobs available, including qualifications and salaries. Discuss your findings with other

graphic arts students in your class.

3. Prepare a list of occupations that have been made possible directly or indirectly as a result of offset lithography.

4. Visit a commercial offset lithography firm. Talk with employees in the areas of production, management, and sales concerning the career opportunities, working conditions, salaries, and opportunities for advancement and promotion available.

5. Visit your local or school library and find a copy of the Occupational Outlook Handbook. Study the information relating to offset lithography. Summarize the information and share it with other graphic arts students in your class.

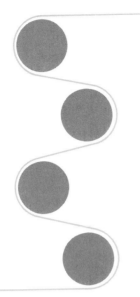

Chapter 3

Safety and Health in Offset Lithography

When you have completed the reading and assigned activities related to this chapter, you will be able to:

O List the general safety precautions that you must follow when working in industry or the school graphic arts shop.

O List the recommended procedures for storing and using chemicals, solvents, and flammable liquids.

O List the safety procedures to be followed when operating power equipment.

O List the safety procedures to be followed when using small hand tools.

O Define OSHA and describe its function.

O Explain the major concerns of a noise abatement program.

O List the major classifications of fires and recommended fire fighting treatment for each.

Safety and health are important considerations in the offset lithography printing industry. This is equally true of school graphic arts programs. Modern printing plants are complex operations. Many accidents are caused by lack of thought, carelessness, and ignorance. Most accidents can be prevented if all employees are *safety conscious*. This means taking the necessary precautions against all foreseen risks. Safety regulations and posters are frequently used to remind employees of potential hazards, Fig. 3-1.

PLANNING FOR SAFETY

Planning for safety must start with an organized program of accident prevention. Learning to work safely starts with a good program of safety education, whether it's in industry or school.

A new graphic arts class, or personnel new to the plant, should be taken on a *safety tour*. This should be done as soon as possible. The tour should include visitations to each department which might present the possibility of injury. In addition, items that are provided for cleanliness, first-aid, and safety, should be emphasized and explained in detail.

The safety tour should include information indicating inherent dangers, such as: carbon arc lights, offset presses, paper cutters, folders, stitchers, the darkroom environment, and chemistry. Stress the locations and use of master switches, emergency stop buttons, fire extinguishers, fire alarms, and first-aid kits. Point out that lights should be turned ON when working with machines, work stations, and upon entering stockrooms. This of course is NOT true where the process prohibits illumination, such as darkrooms.

PERSONAL CONDUCT

Although printing plant safety is the constant responsibility of EVERY PERSON, safety must start with a management commitment. Management must make sure that new employees are properly trained in all safety procedures and regulations. They must ensure that they develop the proper attitudes, dress safely, and use tools and equipment carefully.

In the school shop, the teacher must educate students in safety procedures and regulations. Graphic arts teachers should administer a written safety test to every student.

Unsafe conduct and work habits by a worker or student should be corrected immediately. The person should be informed that they could be permanently banned from the shop if the unsafe conduct/work

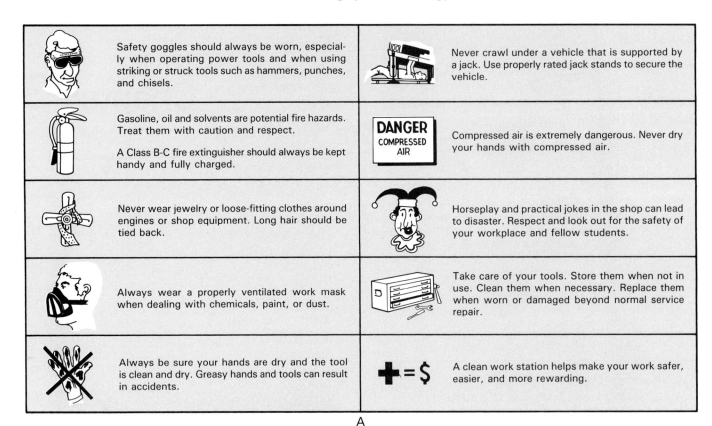

A

B

C

Fig. 3-1. Most accidents can be prevented if all workers are safety conscious. A—This poster serves as a ''common sense'' safety reminder. B—These posters act as constant visual reminders to those who work around machines. C—Some posters help remind workers about hazards of dangerous chemicals and other unsafe practices. (Lab Safety Supply)

habits continue. Horseplay of any kind is never allowed! Running in the shop is obviously an unsafe practice.

FEDERAL SAFETY REGULATIONS

The *Occupational Safety and Health Act* (OSHA) governs the safety and health of workers in the United States. Business and industry are required to comply with OSHA safety and health regulations. Otherwise, they face severe penalties. The regulations require that certain types of power equipment be equipped with guards and similar safety devices. Workers are required to wear proper clothing and eye protection when working around certain types of equipment and chemicals. In addition, OSHA regulations cover the facilities required for personal cleanliness and the comfort of the workers.

The following safety guidelines for each mechanical area in an offset lithography operation or school graphic arts shop should be followed. Careful adherence to regulations, good housekeeping, and a healthy attitude toward safety can make the facility a safe and pleasant place to work.

Copy preparation area safety

The modern copy preparation area is one of the safest in the entire printing plant. There are no presses, folders, or other heavy equipment. There is little noise. The loudest sound is probably that of an electronic keyboard of a phototypesetter.

However, remember the hazards of knives, razor blades, cutting boards, and flammable adhesives. Organize work areas, keeping each sharp tool in its proper place when NOT in use. Keep aisles clear of obstructions. Be sure that light levels are properly adjusted to allow glare-free work without eyestrain, Fig. 3-2.

Fig. 3-2. Light levels in all work spaces should be properly adjusted to allow glare-free work without eyestrain.

Stripping area safety

Like the copy preparation area, there are few hazards in the stripping department. If light levels are properly set, it will be relatively dark in the area to allow for comfortable work at light tables. Therefore, keep walkways clear so that people do not stumble over obstacles.

In the stripping area, store razor blades, knives, scissors, and cutting boards where they will be out of the way. Never place heavy materials or metal objects on the glass surface of a light table.

Use light pressure when cutting film or paper on the light table. This will help avoid scoring the glass. Organize the stripping area for convenience. Place light tables away from the traffic pattern. Keep materials, tools, and chemicals in a well-lighted area.

Camera and darkroom safety

Blinding lights, hot surfaces, corrosive chemicals, and hazardous voltages are potential problems in the camera and darkroom areas. Store chemicals out of the way and in covered containers to avoid fumes and accidental spillage. Mix darkroom chemicals only in approved containers placed in a sink at working level. Wear safety goggles and protective, spill-proof clothing. Make sure that the area is properly ventilated. It is most important that all chemicals are labeled correctly.

Exercise special care around camera lights to avoid burns and eye damage. Disconnect power plugs before changing lamps or working on electrical equipment. Inspect all wiring regularly for damage to insulation.

Platemaking area safety

Platemaking equipment varies widely. However, there are common hazards to guard against, particularly with plate exposure equipment, Fig. 3-3. Be careful when changing lamps or arc rods. Wait for the burned-up ends of lamps to cool before changing them. Be sure the power is disconnected before working on any electrical equipment.

When processing offset plates by hand, follow the manufacturer's recommended procedure. Splashed chemicals can get into your eyes or irritate your skin. Consider using the new water-developed offset plates. These eliminate chemicals for development. Be sure bottle caps are tight before shaking containers. Plate edges can be sharp and cause serious cuts.

When working around automatic plate processing equipment, do not wear loose clothing that could be drawn into the machine. Clean up spilled chemicals or oil immediately. Use an oil-absorbent cleanup product that will keep work area dry and slip-free without press contamination.

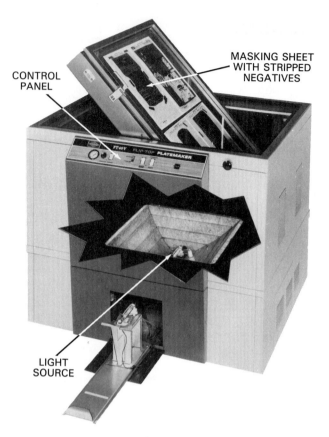

CONTROL PANEL

MASKING SHEET WITH STRIPPED NEGATIVES

LIGHT SOURCE

Fig. 3-3. Extreme care must be taken when changing lamps or arc rods in platemaker. Disconnect power before working on platemaker. (nuArc Co.)

Offset press area safety

Safety in the press area requires skill and alertness. Never reach into a moving piece of equipment to clear paper jams. Always shut down and safety lock the equipment before cleaning or making adjustments. Do not tamper with or disconnect electrical guard interlocks. Keep all guards and safety devices in place and functioning properly, Fig. 3-4.

Wear protective gloves when handling chemicals. Do not smoke around flammable cleaners, fountain solutions, or inks. Never wash the press rollers with gasoline (flammable and explosive), benzine and toulene (toxic and flammable), or turpentine (toxic).

Finally, the press area should be uncluttered. Busy press operators may stumble over a low skid of paper that extends into an aisle. Mark the edges of all walkways with bright safety tape or paint them using OSHA approved colors.

Bindery and finishing area safety

The bindery and finishing area has very powerful, high-speed equipment. Some have sharp blades. Often, the pace in this area is hurried. Never let speed get in the way of safety! Never bypass safety devices on mechanical or electrical

cutters. Keep the blades sharp and the equipment well lubricated. Place under the blade only the paper that you plan to cut. Do not test a blade for sharpness with your finger. Keep all extra blades bolted into safety covers.

Folders move faster than hands, Fig. 3-5. Be sure safety devices are in place on the folder. Keep your hands behind you unless the folder is shut off. Do not wear neckties or loose clothing around power machinery in the binding area. Be sure to keep shirttails tucked in.

To do a clean, fast job, paper drills should be sharp. Keep your hands away from such drills. Allow the bit to cool off before changing it. Keep hands away from the hold-down clamps. The same is true for stitchers. Make sure the throat guards are secure before operating a stitcher.

Shipping area safety

The shipping area of a printing plant can quickly become cluttered. One of the key maintenance-safety measures is to keep aisles clear. Never pile cartons too high. Keep all paper stock below the sprinkler levels. Be sure all racks are anchored firmly to the floor, Fig. 3-6.

Drive lift trucks slowly! Use warning devices to alert workers and pedestrians, particularly at blind intersections. Mark equipment movement paths. Consider the use of wide-angle mirrors on all 90 degree corners.

In lifting, use your legs instead of your back. Always lift from a squatting position.

Allow plenty of temporary storage space for incoming and outgoing materials. This space should be away from the work flow and traffic paths. Safety shoes should be worn where appropriate.

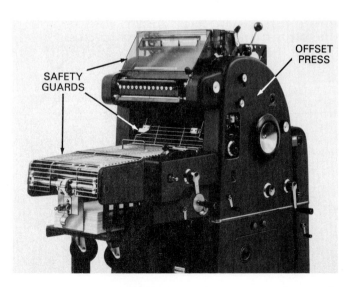

SAFETY GUARDS

OFFSET PRESS

Fig. 3-4. Always keep covers and guards in position when operating an offset press or other type of equipment. Do NOT tamper with, or disconnect, electrical guard interlocks. (A.B. Dick)

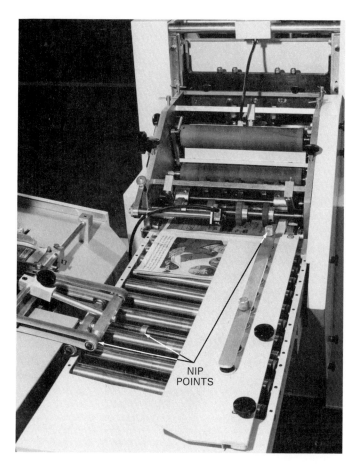

Fig. 3-5. Folding equipment has many rollers and nip points that produce tremendous pulling and crushing forces. Folding equipment can cause permanent disability to fingers, hands, and arms! (Harris Bindery Systems Div.)

Fig. 3-6. Shipping area of an offset printing plant can quickly become cluttered. Keep all materials in their proper storage locations and be sure all racks are anchored firmly to floor. (American Manufacturing Co.)

ELECTRICAL CONSIDERATIONS

The school graphic arts shop or commercial printing plant should be provided with a *main disconnect* (master) *switch*. This allows all power to machinery and small appliances to be turned off when the shop is not in use.

Emergency disconnect buttons or switches should be installed at every machine location. These allow anyone to turn off the power in case of emergency or malfunction. These buttons are sometimes referred to as *"panic buttons"* in the industry.

Frayed and damaged electrical cords, plugs, switches, etc. should be replaced. These items should be inspected periodically by a local electrical inspecting agency. All electrical components must comply with existing electrical codes. Outlet wiring and plugs should be three-prong. The extra prong and wire insures that a ground connection is automatically provided when the electrical device is plugged to power.

Do NOT locate electrical equipment close to sinks or water pipes. This could cause a serious shock or fatal electrocution.

TOXIC CHEMICAL PRECAUTIONS

There are several types of chemicals or products that are toxic and/or hazardous. The first can be classified as *chlorinated solvents*. Probably the most well-known of these types of solvents is methyl-ethylene chloride. In national toxicology studies performed by OSHA, *methyl-ethylene chloride* was shown to have the potential to present a serious health hazard. Some roller and blanket washes contain this substance.

The second hazardous family of products is *ethylene-glycol ethers* or EGBE. EGBE has been found to cause birth defects, liver damage, and serious blood disorders. Such problems can occur at relatively low concentration levels. However, many of the major products for the pressroom use this as an ingredient. People using these toxic products may not be adequately advised of the dangers.

United States federal law now demands that workers know about *material safety data sheets* (MSDS) listing ingredients, Fig. 3-7. They must have access to the MSDS, and the sheets have to be complete. The label on the product must indicate the potential hazards and give detailed handling instructions, Fig. 3-8.

The eye and face water wash unit is designed to furnish first aid to chemical splash victims in areas of a printing plant not accessible to plumbing. The eye and face wash unit is required to be used in incidents where a worker is accidentally subjected to toxic or hazar-

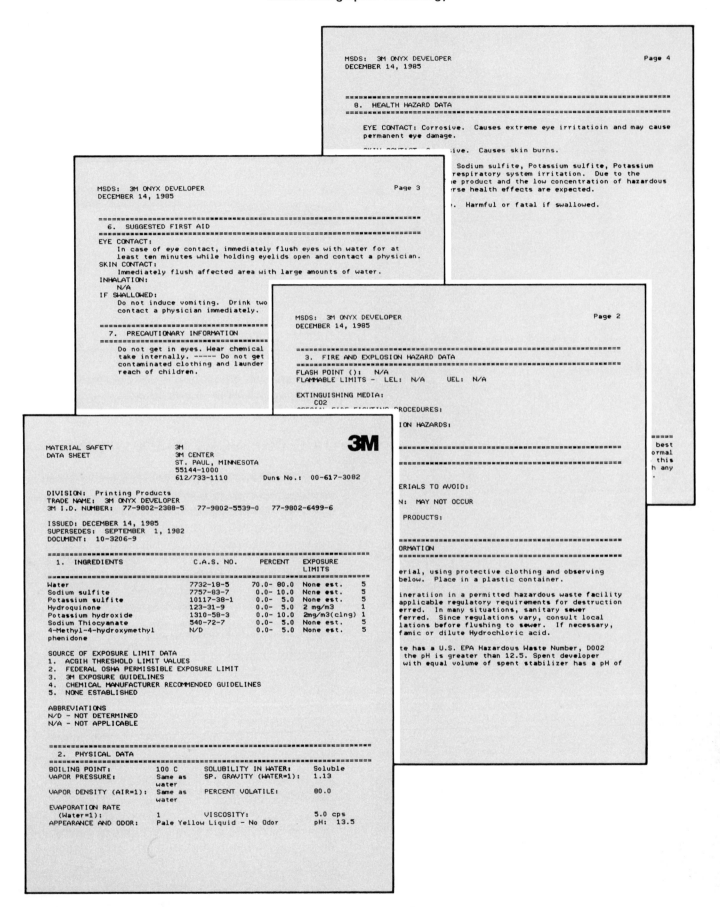

Fig. 3-7. Material Safety Data Sheets (MSDS), like this example, list ingredients of chemicals. These sheets contain information related to potential hazards of chemicals and give detailed handling instructions. All workers must have access to these safety sheets. (3M)

Fig. 3-8. Most chemicals used in offset lithography are potentially dangerous! Workers must be informed about these chemicals through use of Material Safety Data Sheets. Employers are required, by law, to keep MSDA sheets on file at all times so that workers have access to them. (Lab Safety Supply)

dous chemicals. Speed is of the essence in the administration of face and eye first aid. These units are generally located at key locations in the plant where chemicals are in general usage, Fig. 3-9.

Hazard communication laws

Federal law also administers the *Hazard Communication Laws*. These laws are directed at the "right-to-know" of all employees. The *right-to-know* is an extremely complex issue, but there are three major points or areas that are included:
1. Material safety data sheets.
2. Product labeling.
3. Employee education.

Material safety data sheets. Material safety data sheet forms must be filled out by employers regarding hazardous chemicals. The form is entitled OSHA 174 and it is very specific. The form serves as a preventive measure, but it can also be extremely useful if a medical emergency does occur. For treating symptoms, a doctor does not want to lose time calling the manufacturer. The doctor wants to know exactly what a product contains so the problem can be treated properly.

The material safety data sheets are of NO value unless they get into the hands of the right people. It is a supplier's responsibility to provide data sheets to dealers and make sure that the printers also receive copies. Printers should be aware that complete data sheets must be provided with chemicals.

Product labeling. The label of the product also plays an important role in safe handling. The *product label* must list all the ingredients and indicate if there are any safety issues involved. These might include flammability, reactivity, health and safety hazards, etc. These labels also include information that might be useful to firefighters. The whole idea behind the right-to-know regulation is to have a universal warning system.

Employee education. Employers who use chemicals in their plants are required to educate their employees as part of the Hazard Communication Law. This means that they must have a written educational program. Training must be done at regular intervals. Educational programs must include information on the potential hazards of the chemicals, how to use them, and what safety equipment is needed.

Using safe chemicals

There are many safe commercial solvents and other chemicals for removing ink from offset press rollers, blankets, ink fountains, and plates. Use only those solvents and chemicals which the local fire marshal approves.

Solvents can be purchased that are non-toxic, non-irritating to the skin, non-injurious to offset rollers and blankets, and non-flammable or have a high flash point of at least 100 or more.

Fig. 3-9. Eye and face wash units are required to be used in incidents where a worker is accidentally splashed with a toxic or hazardous chemical. (Lab Safety Supply)

The *flash point* of a solvent refers to the lowest temperature at which a substance will give off vapors that will support combustion. For example, say a solvent has a flash point of 0°F. This means that at a temperature as low as 0°F., the solvent's vapors are so highly combustible that they will ignite or support combustion. Safe solvents with a flash point of 100 or more are available and should be utilized.

NOISE CONTROL REGULATIONS

During the past few years, the hazards of noise have been recognized. Loud noise can harm the human ear.

For example, a jet airliner's noise on takeoff is annoying and causes fatigue. Aircraft are now required to adhere to strict noise abatement regulations at most airports.

Similarly, noise in a printing plant reduces the efficiency of its work force. It also interferes with clear communications. If extreme, it may cause permanent damage to the human ear. Generally, the louder the noise, and the longer the ear is exposed to it, the greater the potential hearing loss. Hearing loss caused by printing and binding equipment can be either temporary or permanent.

Recent court rulings have allowed individuals to collect Workman's Compensation for hearing loss allegedly caused by extreme noise at work. Many states have adopted noise control regulations. Most of these regulations call for the use of ear protection, Fig. 3-10.

For example, the United States Government requires printers with a government contract over $10,000 to adhere to certain noise control levels. Briefly, the regulation states the exact length of time employees may be exposed to continuous noise of certain levels. If the noise exceeds these levels, the hearing of employees must be protected, Fig. 3-11.

The permissible noise exposures are read (expressed) in *decibels*. A *decibel* is a unit of measure. It expresses the relative intensity of sounds on a scale. The scale runs from zero for the average, least perceptible sound to about 130 for the average pain level.

For example, 90 decibels of continuous noise exposure for 8 hours is the maximum allowable for an employee. The higher the decibel reading, the shorter must be the employee's exposure to it.

The National Safety Council measured typical sound levels in the printing industry, Fig. 3-12. Few areas have sound levels below 90 decibels (db). Unfortunately,

**PERMISSIBLE NOISE EXPOSURES
WALSH-HEALY ACT**

DURATION PER DAY	SOUND LEVEL "A" SCALE Standard Slow Meter Response
8 hours	90 db or less
6	92 db
4	95 db
3	97 db
2	100 db
1 1/2	102 db
1	105 db
1/2	110 db
1/4 or less	115 db
Exposure to impact noise should not exceed 140 db	

Fig. 3-11. Study table that shows typical permissible noise exposures. A db reading of 90 is maximum allowable limit for a worker in an eight-hour day.
(National Association of Photo Lithographers)

Fig. 3-10. Ear protection is often needed in the pressroom and bindery area of an offset lithography plant. Protective devices for ears include: ear muffs, moldable inserts, and ear plugs.

TYPICAL PRINTING INDUSTRY NOISE LEVELS IN DECIBELS

Pressroom	General area At specific machines	85-95 85-105
Sheet fed	Offset press	80-85
Web offset press	Between folder and packer Slitter	96 96
Dual folder	Operator's station	88-94
Offset press	Delivery end	90
Newspaper printing press		102
Bindery	Saddle wire stitcher Perfect binding and gluing Folding machine general area Stitcher-operator's station	85 85 85-90 90

Fig. 3-12. Study table that shows typical printing industry noise levels in decibels. Large printing presses and bindery equipment generally produce highest decibel readings.
(National Association of Photo Lithographers)

in some areas where noise exceeded 90 db, hearing protection was NOT used!

For comparison, you may be interested in typical sound intensities in your everyday activities. Some are given in Fig. 3-13. Try to become more aware of sound intensities. This will help you learn to protect yourself from hearing damage or loss.

FIRE PROTECTION PLAN

Printing industry employees must be alert to situations that can cause fire! Good housekeeping, preventive measures, and education are the key elements to a sound *fire protection plan*. The National Fire Protection Association lists the following as major problem areas in the printing industry:

1. Flammable debris left on floors and around equipment operating areas—safety trash containers should be used for this purpose. See Figs. 3-14 and 15.
2. Low flash-point cleaning solvents for presses—solvents with a flash point of over 100°F should be used. *Flash point* refers to the lowest temperature at which a substance gives off enough vapors to initiate combustion. For example, kerosene is relatively safe with a high flash point of about 140 degrees F. White gasoline has a low flash point of only 0 degrees F. This means that it will ignite or support combustion as low as zero degrees F. Solvent containers called *plunger cans* should be used to dampen wiping cloths, Fig. 3-16.

TYPICAL SOUND INTENSITIES MEASURED IN DECIBELS ON "A" SCALE

NORMAL HUMAN BREATHING	10 db
WHISPER	20 db
QUIET ROOM	40 db
NORMAL CONVERSATION	50 db
QUIET STREET	50 db
HEAVY TRAFFIC	70 db
SPORTS CAR or TRUCK	90 db
SHOUTED CONVERSATION	90 db
ELECTRIC BLENDER	93 db
PNEUMATIC JACK HAMMER	94 db
LOUD OUTBOARD MOTOR	102 db
LOUD POWER LAWN MOWER	107 db
JET PLANE AT PASSENGER RAMP	117 db
JET PLANE AT TAKE OFF	120 db
AT CLOSE RANGE	150 db
ROCK & ROLL BAND	138 db
PAIN THRESHOLD	140 db

Fig. 3-13. Compare this table of everyday sound intensities with those in typical printing plant or your school graphic arts facility. (National Association of Photo Lithographers)

Fig. 3-14. Flammable debris should be placed in a fire-safe trash container. The safety cans should be emptied daily! (Justrite Manufacturing Co.)

Fig. 3-15. Many types of fire-safe trash containers are available. Never store oily or ink-soaked rags in a container that does not meet fire safety requirements. (Lab Safety Supply)

Fig. 3-16. Solvent plunger is used to dampen wiping cloths around presses and other equipment in printing plant. Solvent-soaked rags, that are to be stored overnight, must be placed in a safety can. (Justrite Manufacturing Co.)

3. Static electricity accumulation on presses—*static eliminators* should be installed on all presses.
4. Oily cloths left outside cans with self-closing lids—cans with self-closing lids should be used for disposal of solvent-soaked and oily cloths. Refer to Fig. 3-17.
5. Flammable ink and solvent storage—ink and solvent cans should be stored in a safety cabinet when not in use, Fig. 3-18.
6. Paper dust from bindery and press areas—*dust collectors* should be installed in these areas.
7. Scrap paper accumulation—all scrap paper should be placed in bins.
8. Set-off spray powder accumulation, with possibility of dust cloud formation and ignition—avoid this condition.

Fire extinguishers and sprinklers

The maintenance of exterior fire department connections that serve sprinklers and hose systems should be examined regularly. Hydrants should be checked to make sure they operate. Firehoses

Fig. 3-18. Ink and solvent cans should be stored in a safety cabinet when not in use. Safety cabinets, like this one, must be approved for use by local fire department. (Justrite Manufacturing Co.)

and all emergency equipment must be available for immediate use. See Fig. 3-19.

Hose lines and valves should never be obstructed or blocked. Hoses should be checked regularly for rot, mildew, and other damage. The hoses should NEVER be used for anything but periodic testing and fire fighting.

Fire extinguishers should be located near potential hazard areas. They should be easily accessible. When an extinguisher is discharged in combating a fire, the fire department should be called.

Fig. 3-17. An enclosed safety can should be used to dispose of flammable rags or towels. Daily emptying of safety cans is a procedure generally followed by most printing plants. This procedure should also be followed in the school graphic arts facility. (Justrite Manufacturing Co.)

Fig. 3-19. Fire hose systems must be examined regularly by local fire department personnel. Hose lines and valves should never be obstructed or blocked. Indoor fire sprinkler systems must also meet local fire codes and be inspected on a regular basis. (Bob Bright)

Types of fires

You should learn the different classes of fires, since they must be controlled by different means. For example, water is ideal for putting out a paper fire, but it may ruin electrical equipment or cause short-circuits and electrocution. If water is used on gasoline fires, it may make the fire worse. There are three general types of fires, as shown in Fig. 3-20.

Class-A fires are fires in ordinary combustible materials such as clothing, wood, and paper. Water should be used to douse the fire. In a Class-A fire, embers or ashes will remain after the fire has been extinguished. These ashes must be cold before the fire is completely put out.

Class-B fires are fires in flammable liquids such as gasoline, oil, grease, ink, paint, and turpentine. Such liquids burn at the surface, where the vapors are given off. Smothering or blanketing the burning liquid is the best technique for extinguishing it. Foam and carbon dioxide (CO_2) are effective in smothering a Class-B fire. See Fig. 3-21.

Class-C fires are fires in electrical equipment. Here, the use of a "non-conducting" extinguishing agent is of first importance. A non-conducting agent will NOT transmit heat or electricity. In most electrical fires, it is necessary to de-energize the circuits before any progress can be made. Carbon dioxide (CO_2) is a non-conductor of electricity and will not damage electrical equipment.

SAFETY IS YOUR RESPONSIBILITY

Unsafe machines, work areas, and procedures are the cause of many accidents. Take time to look at the equipment and work area closely. A sam-

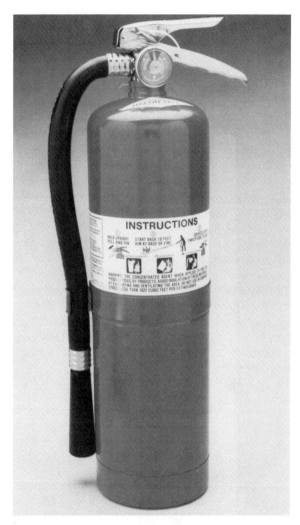

Fig. 3-21. Foam and carbon dioxide (CO_2) are effective in smothering a Class-B fire consisting of liquids, such as gasoline and oils. (Lab Safety Supply)

	Pressurized Water	Loaded Stream	CO²	Regular Dry Chemical	All Use Dry Chemical
CLASS-A FIRES Paper, wood, cloth, etc., where quenching by water or insulating by general purpose dry chemical is effective.	Yes, Excellent!	Yes, Excellent!	Small surface fires only.	Small surface fires only.	Yes, Excellent! Forms smothering, film, prevents reflash.
CLASS-B FIRES Burning liquids (gasoline, oils, cooking fats, etc.) where smothering action is required.	No! Water will spread fire.	Yes! Has limited capability.	Yes! Carbon dioxide has no residual effects on food or equipment.	Yes, Excellent! Chemical smothers fire.	Yes, Excellent! Smothers fire and prevents reflash.
CLASS-C FIRES Fire in live electrical equipment (motors, switches, appliances, etc.) where a non-conductive extinguishing agent is required.	No! Water is a conductor of electricity.	No! Water is a conductor of electricity.	Yes, Excellent! CO_2 is a non-conductor that leaves no residue.	Yes, Excellent! Non-conducting smothering film. Screens operator from heat.	Yes, Excellent! Non-conducting smothering film. Screens operator from heat.

Fig. 3-20. Learn three general classes of fires. Each type of fire requires a different procedure to extinguish it.

ple of a safety inspection form developed for use in the printing industry is shown in Fig. 3-22.

When unsafe conditions exist, or if any are brought to your attention, take immediate action to correct or eliminate them. All electrical repairs should be performed by qualified technicians. All work should be certified and guaranteed.

A copy of all accident reports should be filed. Schools and printing firms generally maintain accurate accident reports, including: dates, conditions, treatment, witnesses, and other pertinent data.

To:	OK	Not Right	Not Applicable
1. _____			
2. _____			
Electrical Equipment			
● Inspection—Current			
● Cords—In Good Condition			
● Cords—Away From Aisles			
● Tampering—No Evidence Of			
● Panel Boxes—Lock Prevent Exposed Fuses			
● Lighting—Adequate			
● Lighting—Emergency Adequate			
Machines—Equipment			
● Operators Trained			
● Guards—Adequate			
● Attire—Appropriate For Job			
● Jewelry—Appropriate For Job			
● Typewriters—Properly Placed			
● Space—Adequate			
Storage			
● Location Appropriate			
● Ladders & Stools—Adequate			
● Scissors—Blunted Or Scabbard			
● Heavy Objects—Near Floor Level			
● File Cabinets—Secured Or Weighted			
Open Drawers Don't Block Aisles			
One Drawer Open At A Time			
Drawer Handles—To Open—Close			
● Aisles—Not Blocked Or Cluttered			
Tripping—Slipping Hazards			
● Carpets & Mats—In Good Condition			
● Floors—Spill Free—Not Slippery			
● Clips—Rubber Bands—Staples—None On Floor			
Fire Safety			
● Waste Paper Disposal			
● Other Trash Disposal			
● Flammable Liquids—Proper Storage			
● Facilities For Smokers			
● Exits Not Blocked			
● Fire Extinghishers—Current Inspection			
● Other Fire Hazards—Explain*			

*Of area inspected

Fig. 3-22. This is a safety inspection checklist. Study checklist carefully and note items to look for when evaluating your graphic arts facility.   (Continued)

	OK	Not Right	Not Applicable
DATE _____			
AREA _____			
INSPECTORS (1) _____			
(2) _____			
Housekeeping			
● General Impression—Good			
● Desks—Tables—Orderly—Uncluttered			
● Waste Containers—Adequate			
● Storage Space—Adequate—Orderly			
Safety General			
● Chairs—In Good Condition			
Properly Used			
● Edges—Smooth—Not Sharp			
● Lifting—Proper Way			
● Staplers—In Good Condition			
Used Properly			
Remover Used			
● Desk Drawers Closed When Not In Use			
● Signs—Adequate—Obeyed			
● Safety Equipment—Where Needed			
Shoes—Gloves—Glasses—Goggles			
● Ladders—Inspection Current			
In Good Condition			
Office Security			
● Appears Adequate			
● Personal Valuables—Locked Up Or Attended			
● No Suspicious Happenings Noted			
● Locks—Appear Adequate			
● Other Observations*			

*Other Observations & Explanation Of Items Needing Attention _____

_____

_____

_____

_____

_____

_____

Fig. 3-22. Continued.

Safety inspection checklist

A successful safety program involves education of workers and inspection of facilities. The following industry safety rules should be followed in an offset lithography shop.

1. None of the following items should be worn while working around operating machinery: long sleeves (except if snug fitting), bulky sweaters, pants with cuffs, jewelry, medallions, necklaces, ties, watches, rings, bracelets, scarves, or other loose items!

2. Long hair should be tied back or tucked under a hair net to prevent it from getting caught in machinery!

3. Machinery should also be shut off and safety locked before being wiped down!

4. A rag or cloth that catches in moving machinery should never be held or pulled in an attempt to retrieve it!

5. The power to machinery should be locked out before lubrication or servicing!

6. Safety precautions should be used to ensure that others are clear of a press before it is started!

7. No one should operate a press or other power machinery when alone in the graphic arts shop!

8. All hazardous areas should be protected by machine guarding devices!

9. There should be specialized guards on all plate-to-blanket and blanket-to-impression cylinder nip points on offset presses!

10. There should be adequate mechanical exhaust ventilation and a supply of clean air available where platemaking chemicals are used or stored!

11. Food and beverages should NEVER be consumed or stored where platemaking and other press chemicals are used or stored!

12. When carbon arc lamps are used in vacuum frames and platemaking equipment, there should be adequate ventilation and replacement fresh air! This is necessary to carry away the ozone gas that results from the use of carbon arc lamps!

13. Saturated and soiled cloths and similar materials should be placed in a fire safety container with a self-closing lid!

14. The floor of the shop should be kept clean of liquids and debris at all times!

15. Hands should be washed before eating or drinking! This is necessary to remove the toxic products from various work assignments.

16. Heavy objects should be lifted using your legs and NOT your back!

17. First aid assistance should be readily available! Know where the first aid station is located. A first-aid kit should be stored in a prominent location in the shop.

18. Fire extinguishers should be visible! They should contain current content level certification.

19. Noise sources should be controlled whenever possible. Ear protection should be provided where needed.

20. Ink and spills should be cleaned up immediately!

POINTS TO REMEMBER

1. Safety and health are important considerations in the printing industry and school graphic arts shop.

2. Accidents can be prevented if you are safety-conscious at all times.

3. Learning to work safely starts with a good program of safety education in the printing industry and school graphic arts shop.

4. The *Occupational Safety and Health Act* (OSHA) governs the safety and health of workers in the United States. Business and industry (including the printing industry) are required to comply with OSHA safety and health regulations.

5. OSHA regulations cover such things as guards on equipment, safety clothing, eye protection, toxic materials and waste removal, personal cleanliness, and general comfort of employees.

6. General safety guidelines apply to each of the mechanical departments in a printing plant.

7. The copy preparation area is one of the safest areas in the printing plant. This is because there are no presses, folders, or other heavy equipment.

8. Care should be taken in the stripping area when working with razor blades, knives, scissors, and cutting boards.

9. Blinding lights, hot surfaces, corrosive chemicals, and hazardous voltages are potential problems in the camera and darkroom area.

10. Care should be taken in the stripping and platemaking areas when changing lamps or arc rods on platemakers.

11. Splashed platemaking chemicals can get into the eyes or irritate the skin.

12. Loose clothing should NOT be worn around automatic plate processing equipment.

13. All guards and safety devices must remain in place on an offset press.

14. Operators should never reach into a moving press to clear paper jams, lubricate, or make adjustments.

15. The bindery and finishing area has high-speed equipment that requires careful handling and set-up.

16. All electrical buttons, switches, cords, and plugs should be well maintained.

17. All electrical components must comply with existing electrical codes.

18. Outlet wiring and plugs should be of the three-prong variety.

19. Electrical equipment should NOT be located near sinks or water pipes.

20. Noise abatement is a factor in all printing plants and school graphic arts shops.

21. Hearing loss caused by printing equipment can be either temporary or permanent.

22. The maximum allowable noise exposure level for a worker is 90 decibels for eight hours.

23. All individuals in the school graphic arts shop

and those who work in the printing industry must be alert to situations that can cause fire.
24. Good housekeeping, preventive maintenance, and education are the key elements to a good fire protection plan.

KNOW THESE TERMS

Occupational Safety and Health Act (OSHA), Flash point, Decibel, Class-A fire, Class-B fire, Class-C fire.

REVIEW QUESTIONS

1. Whether in industry or school, most accidents can be prevented if all personnel are safety conscious. True or false?
2. Planning for safety must start with an organized program of _____.
3. The _____ _____ _____ _____ _____ governs the safety and health of employees within business and industry in the United States.
4. _____ _____ should never have heavy materials or metal objects placed on their glass surfaces.
5. The darkroom should be properly ventilated and all chemicals labeled correctly. True or false?
6. What should you do before working on any electrical equipment?
7. Why are three-pronged plugs recommended for all appliances?
8. A duplicator or offset press may be cleaned or adjusted while it is running. True or false?
9. Gloves should be worn when using photographic _____.
10. Bulky clothes, neckties, and shirt-tails should NOT be worn when operating bindery equipment. True or false?
11. Noise levels are expressed in:
    a. Nanoseconds.
    b. Decibels.
    c. Mach numbers.
    d. Degrees.

12. Define the term "flash point."
13. Solvent-soaked rags should be kept in _____ _____.
14. A fire consisting of ordinary combustible materials (clothing, wood, and paper) is a:
    a. Class-A fire.
    b. Class-B fire.
    c. Class-C fire.
    d. Unclassified fire.
15. An electrical fire is classified as a Class-____ fire.
16. A Class-B fire involving flammable liquids can be extinguished with _____ or _____.

SUGGESTED ACTIVITIES

1. Conduct a safety inspection in the graphic arts shop using the safety inspection form shown in Fig. 3-22.
2. Invite a local fire department official to discuss fire-drill procedures and fire prevention. Ask the official to demonstrate the proper use of fire extinguishers in the graphic arts shop.
3. Invite the school nurse or another health official to demonstrate first-aid procedures for each kind of accident that might occur in the graphic arts shop, such as: electrical shock, cuts, poisoning, burns, and chemical spill.
4. Design a bulletin board or other type of display with a "Safety In Graphic Arts" theme. Sources for visual aid materials include the Occupational Safety and Health Administration (Washington, DC) local Labor Department, school health office, and local printing plant safety representatives.
5. Ask the school nurse or another health official to assist the class in assembling a first-aid kit for the shop. Be sure that all required items are obtained and in good condition. Place the first-aid kit in a prominent location in the graphic arts shop.

Section I

Issues for Class Discussion

1. Economists use several measures to gauge the importance of an industry in a society's economy. These include such things as the gross national product and the value added or difference between the cost of raw materials and the final market price of a product. The higher the value-added figure, the more valuable the skills that go into the manufacture of the product.

   Discuss the significance of the increased use of computers, robotics, and specialization within the printing industry with implications for worker displacement, future economic importance, number of establishments, higher operating costs, and number of people employed.

2. It is important for you to understand that the printing industry consists of more than just manipulative skills and production techniques. The printing industry is part of the world's communications network. Thus, a broad view of the industry, as part of visual communications, is essential for those engaged in its practice.

   Discuss the present and future role of the printing industry in light of its impact on a society that relies heavily on visual communications for survival.

3. Offset lithography affects each one of us every day in everything we do. The industry is so diverse that the necessary skills, resources, and personnel often extend into other domains or areas of business and industry.

   To gain an overview of the career opportunities in offset lithography, discuss and compare present and future earning and working conditions of production and management workers with those of other kinds of manufacturing enterprises.

4. The typed resumé will give the printing company interviewer a considerable amount of information about yourself before you actually arrive for the interview. The resumé also provides a permanent record of yourself so the company interviewer will remember you.

   Describe how you should conduct yourself during the job interview and what steps you can take to prepare for, and successfully complete, the interview. Class mock interviews, and videotaping yourself and other class members, is an excellent means of seeing how you look to others.

5. Environmental pollution control and safety are of prime importance to most citizens in the United States and around the world. Local, state, and national environmental regulations apply to every industry which handles potential pollutants and other toxic materials. The United States Environmental Protection Agency (EPA) is responsible for safeguarding the health and safety of those inside and outside the workplace.

   Discuss the effects of industrial environmental pollution control programs in your community and in the state. Give opinions on what might result if such programs were nonexistent in your community and state. Relate this to the printing industry.

Section II
Estimating and Trade Customs

To produce a printed product, materials, machines, and people are required. The materials are passed through a number of manufacturing steps to produce the product. Before production is started, an estimator will determine a final cost for the product.

Estimating is part of selling a printed product. Preparation and presentation of an estimate to the customer or prospect is very important. Anyone who does estimating generally works from a complete list of standard prices for different types of work. Chapter 4 presents the various elements that must be considered when estimating the cost of printed products.

Chapter 5 covers the printing industry Trade Customs. The Trade Customs are an established method of dealing between the printer and customer. Printed Trade Customs are generally included with the job estimate or quotation. By doing so, the printer establishes some form of protection in the likelihood that a court of law becomes necessary.

Estimating requires close communication between the client and printer.

Chapter 4

Estimating Printing

When you have completed the reading and assigned activities related to this chapter, you will be able to:

O Identify and describe typical cost centers in an offset printing firm.
O Identify and describe typical cost rates in an offset printing firm.
O Describe the budgeted-hourly cost method of estimating offset printing.
O Describe the difference between fixed and variable costs.
O Describe the use and function of production standards in an offset printing firm.
O Describe how computers are used to estimate offset printing.
O List and describe typical management and accounting tasks that can be included with a computerized estimating system.
O List and describe the steps an estimator generally uses to price a printing job.
O Describe the function of an estimator's worksheet.
O Describe the function of an estimator's quotation sheet.

When it comes to pricing the cost of printing, printers cannot use guesswork or copy the costs of a competitor. Estimating is a serious process relating to the policies of each printing company.

Estimators are people within a company who determine job costs. The estimator must use careful cost analysis and proven methods of pricing printed products. Without valid information related to time and cost standards, it is virtually impossible to accurately estimate the cost of printing a job. The cost of the product must always be determined before production begins!

COST CENTERS

The estimator is a key person in the printing company. This person must know about all aspects of the plant. If all the data is NOT known, reliable estimates cannot be given to customers and prospects. The job of the estimator is crucial since the margin of profit or loss rests with the estimating calculations.

Cost centers must be determined. A *cost center* is a job operation (typesetting, press work, binding, etc.) that costs money to maintain. An example of a cost center is an offset press operation. During this phase, supplies are used and the person operating the press is being paid. Each of these elements makes up the press cost center.

Printing firms decide on cost centers based on their particular operation and kinds of equipment. All firms do not have the same equipment. Workers have different and special abilities within each firm. Some firms do not complete every operation in a printing job. Some steps are done by another specialty firm. This would also be considered a cost center.

The general cost centers include:
1. Design and layout.
2. Typesetting.
3. Image assembly.

4. Camera.
5. Stripping and platemaking.
6. Press.
7. Finishing and binding.

Within each of these general cost centers, there are several subcost centers. For example, typesetting may include the following subcost centers:

Markup, body type, display type, proofreading, corrections.

Cost rates

The cost centers must be identified. Then, a cost rate for each center must be set.

A *cost rate* is the cost of operating a specific cost center per hour. This sometimes is a very complex process. To find the cost rates for each cost center, the dollar operating cost of every piece of equipment is determined, Fig. 4-1. This is based on items such as a machine's share of the insurance, heating, lighting, maintenance, depreciation, and labor costs.

The total hourly operating cost, called *standard production time,* of the machine is determined based on these costs. This cost figure represents the sum of operating the machine for one hour. It also takes into consideration the total floor space (square footage) the machine occupies, Fig. 4-2.

ALL-INCLUSIVE HOURLY COST RATES

BASED ON PREVAILING WAGES AND A STANDARD 40 HOUR WORK WEEK

NAME OF COST CENTER	SIZE	APPROXIMATE INVESTMENT IN EQUIPMENT	RATED SHEETS PER HOUR	CREW	60%	70%	80%
DUPLICATORS & SMALL PRESSES							
Multi, Chief, chute del.	10 x 15	11600	9000	1	40.53	34.74	30.41
Multi, Chief, chain del.	10 x 15	13900	9000	1	40.97	35.11	30.74
Multi, Chief, chute del.	11 x 15	12000	9000	1	40.61	34.80	30.47
Multi, Chief, chain del.	11 x 15	14200	9000	1	41.02	35.16	30.78
Multi, Chief, chute del.	11 x 17	13600	9000	1	41.10	35.22	30.84
Multi, Chief, chain del.	11 x 17	15600	9000	1	41.48	35.55	31.12
Multi, Chief, 2/c chain del.	10 x 15	19600	9000	1	42.57	36.48	31.94
Multi, Chief, 2/c chain del.	11 x 17	20500	9000	1	42.93	36.79	32.21
Multi, Chief, chute del.	14 x 18	16700	8000	1	51.19	43.87	38.41
Multi, Chief, chain del.	14 x 18	18150	8000	1	51.47	44.11	38.62
Multi, Chief, 2/c chain del.	14 x 18	26300	8000	1	53.38	45.75	40.05
Ryobi 17	11 x 17	15000	9000	1	41.65	35.69	31.25
ATF Davidson 501	11 x 15	15000	8000	1	41.36	35.45	31.03
ATF Davidson 502P	11 x15	16200	8000	1	41.84	35.86	31.39
ATF Davidson 701	15 x 18	16900	8000	1	51.66	44.28	38.77
ATF Davidson 702P	15 x 18	27975	8000	1	53.88	46.18	40.43
ATF Davidson 901	15 x 20	19750	8000	1	52.58	45.06	39.45
Heidelberg TOM	11 x 15 1/2	18100	10000	1	42.11	36.09	31.60
Heidelberg TOK	11 x 15 1/2	19850	10000	1	42.44	36.38	31.85
Heidelberg GTO 52	14 x 20 1/2	39950	8000	1	56.19	48.16	42.16
Hamada 500 CDA	10 x 15	9600	9000	1	40.16	34.42	30.13
Hamada 600 CD	11 x 17	12700	9000	1	40.93	35.08	30.71
Hamada 700 CD	14 x 18	15200	9000	1	51.16	43.84	38.39
Hamada 770 CD	14 x 18	20900	9000	1	52.23	44.77	39.19
Hamada 800 DX	14 x 20	20300	9000	1	52.56	45.04	39.44
Hamada 880 DX	14 x 20	27200	9000	1	53.86	46.16	40.41
Hamada 880 CDX	14 x 20	29400	9000	1	54.28	46.52	40.72
Oliver 52	14 3/16 x 20 1/2	48500	12000	1	57.57	49.34	43.20
Imperial 2200 Maxim	12 x 18	13800	10000	1	50.58	43.35	37.95
Imperial 3200 Maxim	12 x 18	15400	10000	1	50.88	43.61	38.18
A.B. Dick 369	11 x 17	15800	9000	1	41.52	35.58	31.15
A.B. Dick 369 T	11 x 17	19200	9000	1	42.16	36.13	31.63
A.B. Dick 8810	11 x 17	10000	10000	1	40.42	34.64	30.33
A.B. Dick 9840	17 3/4 x 13 1/2	15200	9000	1	41.78	35.80	31.35
A.B. Dick 385	17 1/2 x 22 1/2	19200	8000	1	51.11	43.81	38.35
A.B. Dick 385 w/P head	17 1/2 x 22 1/2	25200	8000	1	53.61	45.95	40.22
Oliver 58	17 1/2 x 22 1/2	54000	12000	1	70.48	60.41	52.88

Fig. 4-1. Cost rates for each cost center are based on each machine's share of insurance, heating, lighting, maintenance, depreciation, and labor costs. (Continued)

ALL-INCLUSIVE HOURLY COST RATES

BASED ON PREVAILING WAGES AND A STANDARD 40 HOUR WORK WEEK

NAME OF COST CENTER	SIZE	APPROXIMATE INVESTMENT IN EQUIPMENT	RATED SHEETS PER HOUR	CREW	COST PER HOUR AT PRODUCTIVITY OF 60%	70%	80%
OFFSET PRESSES							
One Color Presses							
Solna 125/80	18 x 25 3/16	54500	8000	1	171.04	60.88	53.30
Heidelberg KORD	18 1/8 x 25 1/4	49950	6000	1	170.22	60.18	52.68
Oliver 68	19 x 26	68000	12000	1	173.66	63.13	55.27
Royal Zenith R2 26	19 x 26	59500	12000	1	172.11	61.80	54.11
Solna 164	19 x 26	64500	10000	1	173.06	62.62	54.82
Heidelberg MO	19 x 25 1/2	69200	10000	1	173.81	63.26	55.38
Heidelberg SORM	20 1/2 x 29 1/8	89500	10000	1	178.12	66.95	58.62
Miehle 28	20 1/2 x 28 3/8	75000	10000	1	175.57	64.77	56.70
Miller SC 29	20 1/2 x 29 1/8	75000	13000	1	175.57	64.77	56.70
Heidelberg SORD	24 1/4 x 36	112500	10000	1	185.99	73.70	64.52
Miehle 36	25 x 36	120000	10000	1	187.23	74.76	65.45
Miller SC 36	25 5/8 x 36 1/4	95000	12000	1	182.38	70.60	61.81
Miller SC 36 Special	25 5/8 x 37	97500	12000	1	182.91	71.06	62.21
Heidelberg SORS	28 3/8 x 40 1/8	121500	10000	1	189.18	76.43	66.91
Miehle 40 low pile	28 1/8 x 40 3/16	145000	10000	1	193.50	80.14	70.16
Miehle 40 high pile	28 1/8 x 40 3/16	165000	10000	1	197.28	83.37	72.99

OFFSET PRESSES

YOU MUST COMPUTE YOUR OWN COST FIGURES FOR TWO-, FOUR-, FIVE- AND SIX COLOR PRESSES. THE HOUR RATE FOR THE 2ND PRESS OPERATOR AND FEEDER MUST BE ADDED TO THE COST CENTER HOUR RATE WHERE APPLICABLE BEFORE COMPUTING THE HOURLY COST FOR MAKEREADY, IMPRESSIONS PER THOUSAND OR WASHUP.

Two Color Presses							
Heidelberg GTO ZP	14 x 20 1/2	74550	8000	1	74.87	64.17	56.18
Oliver 258	17 1/2 x 22 1/2	100000	10000	1	79.78	68.38	59.86
Solna 225	18 1/8 x 25 3/16	105000	8000	1	82.00	70.28	61.53
Fuji Perfector	19 x 25	222900	10000	1	104.35	89.43	78.30
Heidelberg MOZP	19 x 25 1/4	161900	10000	1	93.11	79.80	69.87
Royal Zenith 26	19 x 26	128500	12000	1	86.54	74.17	64.93
Royal Zenith 26-2CP	19 x 26	148500	12000	1	90.31	77.40	67.76
Solna 264	19 1/16 x 26	139500	10000	1	88.61	75.94	66.49
Komori	20 x 26	225000	13000	1	105.81	90.68	79.39
Akiyama 228	20 x 28	175000	12000	1	96.34	82.57	72.29
Miehle 28	20 1/2 x 28 3/8	165000	10000	1	94.79	81.24	71.12
Heidelberg SORMZ	20 1/2 x 29 1/8	151250	10000	1	92.26	79.07	69.22
OMSCA H226-NP 2	20 7/16 x 29 1/8	175000	12000	1	96.68	82.86	72.54
Miller TP 29S	20 1/2 x 29 1/8	180000	12000	1	98.34	84.28	73.79
Miller TP 29S Perfector	20 1/2 x 29 1/8	195000	12000	1	101.17	86.70	75.91
Miller TP 30S	24 x 30	170000	10000	1	101.05	86.60	75.82
Miller TP 30S Perfector	24 x 30	180000	10000	1	102.93	88.22	77.23
Miller TP 32S	24 x 32 1/4	205000	10000	1	108.15	92.69	81.15
Miller TP 32S Perfector	24 x 32 1/4	220000	10000	1	110.98	95.12	83.28
Miehle 36	25 x 36	210000	10000	1	109.35	93.72	82.05
Miller TP 36S	25 5/8 x 36 1/4	220000	10000	1	111.24	95.33	83.46
Miller TP 36S Perfector	25 5/8 x 36 1/4	235000	10000	1	114.07	97.76	85.59
Miller TP 36S Special	25 5/8 x 37	240000	10000	1	115.01	98.57	86.30
Heidelberg SORSZ	28 3/8 x 40 1/8	194850	10000	1	107.35	92.00	80.55
Komori 240 Lithrone	28 3/8 x 40 1/2	351000	13000	1	136.70	117.15	102.57
Akiyama 240	28 5/16 x 40 1/8	250000	13000	1	117.64	100.82	88.27
Miehle 40	28 1/8 x 40 3/16	245000	10000	1	117.38	100.60	88.07
Miller TP 104	28 3/4 x 41	305000	10000	1	129.15	110.69	96.91
Miller TP 104 Perfector	28 3/4 x 41	320000	10000	1	131.98	113.12	99.03
Miller TP 104 Special	29 x 42	330000	10000	1	133.87	114.73	100.45
Miehle 50	39 3/8 x 50	450000	10000	1	159.01	136.28	119.31
Miehle 63	43 5/16 x 63	530000	9000	1	177.36	152.01	133.08
Feeder					37.02	31.73	27.78
Second Press Operator					49.74	42.63	37.32

Fig. 4-1. Continued.

PRESSWORK SCHEDULE—OFFSET

(Time given represents machine hours on the decimal system)

PRESS NAME	SIZE	RATED SHEETS PER HOUR	HOUR RATE At 70% Productivity	PREPARATION/MAKE-READY TIME SIMPLE Time	AVERAGE Time	DIFFICULT Time	AVG. IMPRESSION PER RUN HR. SIMPLE	AVERAGE	DIFFICULT	WASH-UP Time
SMALL PRESSES										
Multi, Chief, chute del.	10 x 15	9000	34.74	0.20	0.25	0.30	6750	5400	4500	0.25
Multi, Chief, chain del.	10 x 15	9000	35.11	0.20	0.25	0.30	6750	5400	4500	0.25
Multi, Chief, chute del.	11 x 15	9000	34.80	0.20	0.25	0.30	6750	5400	4500	0.25
Multi, Chief, chain del.	11 x 15	9000	35.16	0.20	0.25	0.30	6750	5400	4500	0.25
Multi, Chief, chute del.	11 x 17	9000	35.22	0.20	0.25	0.30	6750	5400	4500	0.25
Multi, Chief, chain del.	11 x 17	9000	35.55	0.20	0.25	0.30	6750	5400	4500	0.25
Multi, Chief, 2/c chain del.	10 x 15	9000	36.48	0.30	0.40	0.50	6750	5400	4500	0.50
Multi, Chief, 2/c chain del.	11 x 17	9000	36.79	0.30	0.40	0.50	6750	5400	4500	0.50
Multi, Chief, chute del.	14 x 18	8000	43.87	0.20	0.25	0.30	6000	4800	4000	0.25
Multi, Chief, chain del.	14 x 18	8000	44.11	0.20	0.25	0.30	6000	4800	4000	0.25
Multi, Chief, 2/c chain del.	14 x 18	8000	45.75	0.30	0.40	0.50	6000	4800	4000	0.50
Ryobi 17	11 x 17	9000	35.69	0.30	0.40	0.50	6750	5400	4500	0.50
ATF Davidson 501	11 x 15	8000	35.45	0.30	0.40	0.50	6000	4800	4000	0.50
ATF Davidson 502P	11 x 15	8000	35.86	0.40	0.50	0.60	6000	4800	4000	0.40
ATF Davidson 701	15 x 18	8000	44.28	0.30	0.40	0.50	6000	4800	4000	0.50
ATF Davidson 702P	15 x 18	8000	46.18	0.40	0.50	0.60	6000	4800	4000	0.40
ATF Davidson 901	15 x 20	8000	45.06	0.30	0.40	0.50	6000	4800	4000	0.50
Heidelberg TOM	11 x 15 1/2	10000	36.09	0.30	0.40	0.50	7500	6000	5000	0.35
Heidelberg TOK	11 x 15 1/2	10000	36.38	0.30	0.40	0.50	7500	6000	5000	0.35
Heidelberg GTO 52	14 x 20 1/2	8000	48.16	0.30	0.40	0.50	6000	4800	4000	0.35
Hamada 500 CDA	10 x 15	9000	34.42	0.20	0.25	0.30	6750	5400	4500	0.25
Hamada 600 CD	11 x 17	9000	35.08	0.20	0.25	0.30	6750	5400	4500	0.25
Hamada 700 CD	14 x 18	9000	43.84	0.30	0.40	0.50	6750	5400	4500	0.35
Hamada 770 CD	14 x 18	9000	44.77	0.30	0.40	0.50	6750	5400	4500	0.35
Hamada 800 DX	14 x 20	9000	45.04	0.30	0.40	0.50	6750	5400	4500	0.35
Hamada 880 DX	14 x 20	9000	46.16	0.30	0.40	0.50	6750	5400	4500	0.35
Hamada 880 CDX	14 x 20	9000	46.52	0.30	0.40	0.50	6750	5400	4500	0.35
Oliver 52	14 3/16 x 20 1/2	12000	49.34	0.30	0.40	0.50	9000	7200	6000	0.35
Imperial 2200 Maxim	12 x 18	10000	43.35	0.20	0.25	0.30	7500	6000	5000	0.25
Imperial 3200 Maxim	12 x 18	10000	43.61	0.20	0.25	0.30	7500	6000	5000	0.25
A.B. Dick 369	11 x 17	9000	35.58	0.30	0.40	0.50	6750	5400	4500	0.35
A.B. Dick 369 T	11 x 17	9000	36.13	0.30	0.40	0.50	6750	5400	4500	0.35
A.B. Dick 8810	11 x 17	10000	34.64	0.20	0.25	0.30	7500	6000	5000	0.25
A.B. Dick 9840	17 3/4 x 13 1/2	9000	35.80	0.20	0.25	0.30	6750	5400	4500	0.25
A.B. Dick 385	17 1/2 x 22 1/2	8000	43.81	0.32	0.40	0.50	6000	4800	4000	0.35
A.B. Dick 385 w/P head	17 1/2 x 22 1/2	8000	45.95	0.40	0.50	0.60	6000	4800	4000	0.60
Oliver 58	17 1/2 x 22 1/2	12000	60.41	0.30	0.40	0.50	9000	7200	6000	0.35

Fig.4-2. In this example of cost rates, various offset presses are rated as to sheets-per-hour, using an average of 70 percent productivity level. Press preparation time, wash-up time, and average impressions-per-hour are also factors included in cost rates. (Continued)

PRESSWORK CLASSIFICATIONS

The succeeding pages have presswork schedules for letterpress and offset. The figures in these schedules are based upon good normal shop conditions; thus it is assumed that equipment, paper, and ink provide no unusual production problems and that the press personnel are competent operators.

Three basic classifications are allowed for in MAKEREADY as well as PRESS RUN. It is important that the job be thoroughly analyzed and proper classification be selected before judging time and determining the cost. Bear in mind that on the same job, makeready and running may fall under different classifications.

LETTERPRESS
(Starting with Locked up Form)

MAKEREADY*

SIMPLE	—Type and/or line cuts with minimum makeready —Simple register forms —Stock — News, Book, Bond or Coated
AVERAGE	—Type and/or square or silhouette halftones —Medium register forms —Stock — Book, Coated
DIFFICULT	—Type and/or vignetted halftones or process color work —Close register forms —Stock — Dull Coated or High Finish Coated —Heavy Coverage

For work of better grade than above add at least 25% to DIFFICULT Schedule.

PRESS RUN

SIMPLE	—Stock between .003 and .006 thickness
AVERAGE	—Stock between .002 and .003 or between .006 and .008
DIFFICULT	—Stock lighter than .002 or heavier than .008

The running classifications assume average quality and average forms.

WASH UP

AVERAGE	—Light ink to dark ink
DIFFICULT	—Dark ink to light ink

*NOTE: Line-up time which is included in makeready represents approximately 20%.

The makeready figures given are based upon utilizing press size to the best advantage. If only a portion of the press area is used reduce allowances accordingly.

OFFSET

MAKEREADY

SIMPLE	—Type and line work, one color, ink coverage up to 20%
AVERAGE	—Some halftone or simple two color work, ink coverage 20-70%
DIFFICULT	—Intricate register jobs, process work, ink coverage over 70%

For work of better grade than above add at least 25% to DIFFICULT Schedule. Add additional time for Perfecting Makeready.

PRESS RUN

SIMPLE	—Stock between .003 and .006 thickness —Type and line plate, one color, ink coverage up to 20%
AVERAGE	—Stock between .002 and .003 or between .006 and .008 —Some halftone or simple two color work, ink coverage 20-70%
DIFFICULT	—Stock lighter than .001 or heavier than .008 —Indicate register, solids, process work, ink coverage over 70%

Add on penalities if extra troublesome work.

WASH UP

AVERAGE	—Light ink to dark ink
DIFFICULT	—Dark ink to light ink

Fig. 4-2. Continued.

The estimator or accountant prepares a schedule showing the hourly cost for each piece of equipment in the plant. In addition, the amount of machine time for a given job is calculated. This means that a careful study is made of the machine time required for the job, or individual operations, in each of the departments.

For example, the study might determine that a certain operation takes thirty minutes in the camera department, one and one-half hours in the stripping department, and so on. The estimator or accountant must also rely on production staff and previous jobs when determining these figures.

It should be noted that machine time rates are averages of production output per unit of time. These calculations must be revised periodically because significant variations are common when workers and machines combine to produce a product.

Fig. 4-3 gives a listing of items used to set hourly costs. This listing can be used for nearly all of the subcost centers. Hourly costs are set for each subcost center to begin the estimates.

There are a variety of methods used by printers to estimate the cost of a job. The procedures used depend on the size and diversity of the firm.

For example, a quick printer may print on only one or two sheet sizes. In this case, costing might be based on units of one hundred. The quick printer's cost per one hundred sheets is fixed.

A large commercial printer may require a more complex costing system. This may include material and labor costs involving many operations. Computerized estimating is now in widespread use throughout the printing industry, Fig. 4-4. The one feature common to all successful estimating systems is that of accurate, consistent procedures.

Estimating methods

This chapter covers one popular estimating method used in the printing industry. It is called the budgeted-hourly cost rate. There are several other methods used to estimate printing. Some of these include competitive pricing, use of standard catalogs, price lists, and pricing based on past work. The information presented here is meant only as an introduction to the method. There are several outstanding estimating books available that cover the subject in detail.

Budgeted-hourly cost rate

Printing firms assign production or cost centers to each operation in the plant. Examples of cost centers include the process camera and typesetting. Some of these costs remain constant regardless of production output in the center. Therefore, these are termed *fixed costs* and include such items as: rent, depreciation, insurance, and taxes. Additional costs, called *variable costs*, include such items as the cost of labor and power. These costs are generated only when production actually occurs in the cost center.

ANNUAL HOURS @ 100% PRODUCTIVITY
NUMBER OF WORKERS / SHIFTS
INVESTMENT IN COST CENTERS
RATE OF DEPRECIATION
FLOOR SPACE OCCUPIED - (SQUARE FEET)
TOTAL HORSEPOWER OF MOTORS
FIXED CHARGES - (NON-VARIABLE)
DEPRECIATION
RENT & HEAT @ $1.50 PER SQUARE FOOT
FIRE & SPRINKLER INSURANCE
PERSONAL PROPERTY TAXES
VARIABLE CHARGES
DIRECT LABOR
INDIRECT LABOR - SUPERVISION @15% OF DIRECT LABOR
PAYROLL TAXES (F.I.C.A. AND UNEMPLOYMENT)
WELFARE BENEFITS
WORKERS' COMPENSATION INSURANCE
LIGHT & POWER
DIRECT SUPPLIES
REPAIRS TO EQUIPMENT
GENERAL FACTORY EXPENSES
SUB-TOTAL
DISTRIBUTION OF GEN'L FACTORY CLASSIFICATION
TOTAL MANUFACTURING COST
MANUFACTURING COST PER PRODUCTIVE HOUR,
BASED ON A PRODUCTIVITY FACTOR OF:
85%
75%
60%
ADMINISTRATIVE, SELLING & FINANCIAL EXPENSES
TOTAL ALL-INCLUSIVE COSTS
ALL-INCLUSIVE COST PER PRODUCTIVE HOUR,
BASED ON A PRODUCTIVITY FACTOR OF:
85%
75%
60%

Fig. 4-3. Items in this list are used to set hourly costs. It can also be used for nearly all of the subcost centers. (National Assoc. of Printers and Lithographers)

The *budgeted-hourly cost rate* is the cost determined for which all fixed and variable costs are derived based on the hourly operation of a particular cost or production center. Budgeted-hourly costs do NOT include materials such as film, plates, ink, and paper. These items are generally calculated and charged to the job separately.

The need for outside materials, supplies and services, called *buyouts,* is also determined. These are NOT always necessary, but when they are, the costs must be included in the estimate. Other kinds of buyouts include typesetting, artwork, color separations, and bindery.

Standard catalogs

There are several published *pricing guides* used as supplements to the method just described. These in-

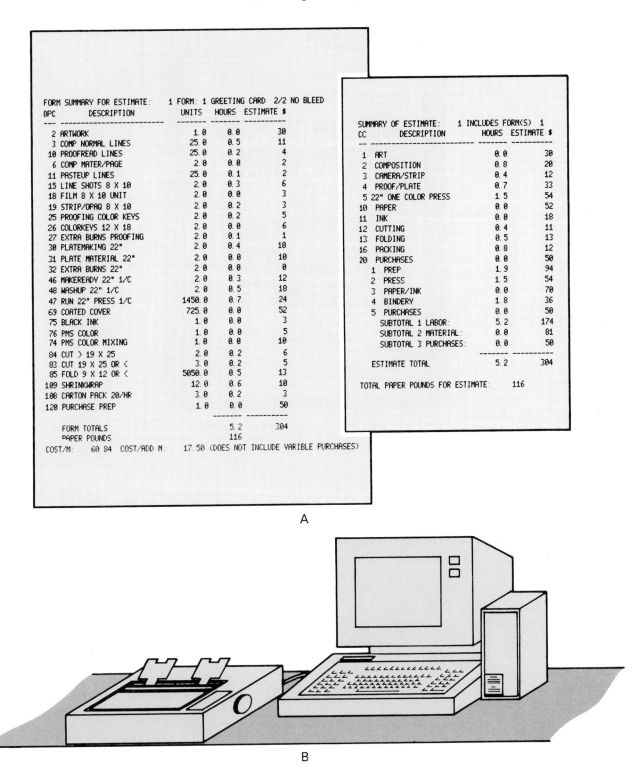

```
FORM SUMMARY FOR ESTIMATE:   1 FORM: 1 GREETING CARD  2/2 NO BLEED
OPC       DESCRIPTION          UNITS   HOURS  ESTIMATE $
---  -------------------       -----   -----  ----------
  2  ARTWORK                     1.0     0.0      30
  3  COMP NORMAL LINES          25.0     0.5      11
 10  PROOFREAD LINES            25.0     0.2       4
  6  COMP MATER/PAGE             2.0     0.0       2
 11  PASTEUP LINES              25.0     0.1       2
 15  LINE SHOTS 8 X 10           2.0     0.3       6
 18  FILM 8 X 10 UNIT            2.0     0.0       3
 19  STRIP/OPAQ 8 X 10           2.0     0.2       3
 25  PROOFING COLOR KEYS         2.0     0.2       5
 26  COLORKEYS 12 X 18           2.0     0.0       6
 27  EXTRA BURNS PROOFING        2.0     0.1       1
 30  PLATEMAKING 22"             2.0     0.4      10
 31  PLATE MATERIAL 22"          2.0     0.0      10
 32  EXTRA BURNS 22"             2.0     0.0       0
 46  MAKEREADY 22" 1/C           2.0     0.3      12
 48  WASHUP 22" 1/C              2.0     0.5      18
 47  RUN 22" PRESS 1/C        1450.0     0.7      24
 69  COATED COVER              725.0     0.0      52
 75  BLACK INK                   1.0     0.0       3
 76  PMS COLOR                   1.0     0.0       5
 74  PMS COLOR MIXING            1.0     0.0      10
 84  CUT > 19 X 25               2.0     0.2       6
 83  CUT 19 X 25 OR <            3.0     0.2       5
 85  FOLD 9 X 12 OR <         5050.0     0.5      13
109  SHRINKWRAP                 12.0     0.6      10
108  CARTON PACK 20/HR           3.0     0.2       3
120  PURCHASE PREP               1.0     0.0      50
                                        -----  ----------
     FORM TOTALS                         5.2     304
     PAPER POUNDS                        116
COST/M:   60.84  COST/ADD M:  17.50 (DOES NOT INCLUDE VARIBLE PURCHASES)
```

```
SUMMARY OF ESTIMATE:    1 INCLUDES FORM(S) 1
CC    DESCRIPTION          HOURS  ESTIMATE $
--  -----------------      -----  ----------
 1  ART                     0.0       30
 2  COMPOSITION             0.8       20
 3  CAMERA/STRIP            0.4       12
 4  PROOF/PLATE             0.7       33
 5  22" ONE COLOR PRESS     1.5       54
10  PAPER                   0.0       52
11  INK                     0.0       18
12  CUTTING                 0.4       11
13  FOLDING                 0.5       13
16  PACKING                 0.8       12
20  PURCHASES               0.0       50
 1    PREP                  1.9       94
 2    PRESS                 1.5       54
 3    PAPER/INK             0.0       70
 4    BINDERY               1.8       36
 5    PURCHASES             0.0       50
      SUBTOTAL 1 LABOR:     5.2      174
      SUBTOTAL 2 MATERIAL:  0.0       81
      SUBTOTAL 3 PURCHASES: 0.0       50
                           -----  ----------
      ESTIMATE TOTAL        5.2      304

TOTAL PAPER POUNDS FOR ESTIMATE:     116
```

A

B

Fig. 4-4. A—Automated estimating systems use special computer programs and a computer to calculate cost of a printed job. This results in a printout such as one shown. B—Computers used for estimating have television-like screens above keyboard. All calculations can be monitored on screen before a hardcopy printout is made. (Printer's Software, Inc.)

clude the PRINTING INDUSTRIES OF AMERICA ESTIMATING CATALOG, Fig. 4-5, and the FRANKLIN OFFSET CATALOG, Fig. 4-6. These references give the average costs for most of the operations and materials used by a printing firm. Step-by-step procedures for using either catalog are provided in the instruction section of each book.

Quick printers generally have *standardized cost sheets* for pricing their work. See Fig. 4-7.

Production standards

Depending upon the estimating method used in a printing firm, accurate production standards must be maintained.

ALL-INCLUSIVE HOURLY COST RATES

NAME OF COST CENTER	SIZE	APPROXIMATE INVESTMENT IN EQUIPMENT	CREW	TOP RATED SPEED	COST PER HOUR AT PRODUCTIVITY OF 80%	85%	90%
COMPOSITION							
Linotype Meteor Model 5 with 15 Fonts		$27,000	1		27.10	25.50	24.10
Mergenthaler V.I.P. System		42,850	1		30.90	29.05	27.45
***Compuwriter II System		10,658	1		18.35	17.25	16.30
Compugraphic ACM 9000 System		22,342	1		26.00	24.45	23.10
I.B.M. Magnetic Tape Selectric Composing System		21,000	1		26.85	25.25	23.85
A.M. 748 Photo Typesetter with Keyboard		25,365	1		27.95	26.25	24.80
*I.B.M. Selectric Typewriter		1,102	1		15.15	14.25	13.45
*I.B.M. Stand Alone Composer		5,506	1		15.75	14.85	14.05
***A.M. Varityper Model 1010 F-15 Fonts		6,672	1		17.25	16.25	15.35
Proofreading		300	1		21.10	19.85	18.75
					60%	70%	80%
Hand Composition		13,000	1		31.85	27.30	23.90
**Photo-Typositor		5,000	1		17.65	15.10	13.25
*Paste Up Art & Copy Preparation		500	1		20.50	17.60	15.40

*MPS Recommended Wage Rates of $5.50 per hour used in determining hour rate.
**MPS Recommended Wage Rates of $4.50 per hour used in determining hour rate.
***MPS Recommended Wage Rates of $6.00 per hour used in determining hour rate.

NAME OF COST CENTER	SIZE	APPROXIMATE INVESTMENT IN EQUIPMENT	CREW	TOP RATED SPEED	80%	85%	90%
LITHOGRAPHIC PREPARATORY							
NUARC SST Line & Halftone Camera	14 x 18	4,429	1		34.35	29.45	24.75
Robertson Line & Halftone Camera—Comet	24"	12,316	1		37.15	31.80	27.85
Litho Film Processor	24"	18,704	1		37.65	32.25	28.20
Robertson Line & Halftone Camera with Litho Film Processor	24"	31,020	1		43.35	37.15	32.50
Multi Scanner, Hell with Laser		250,000	1		99.15	85.00	74.35
Stripping		500	1		31.85	27.30	23.85
NUARC Flip Top Platemaker FT 40	30 x 40	2,122	1		33.95	29.10	25.45

Fig. 4-5. A—Note cover of PRINTING INDUSTRIES OF AMERICA ESTIMATING CATALOG referred to as the BLUE BOOK OF PRODUCTION STANDARDS AND COSTS FOR THE PRINTING INDUSTRIES. B—Note sample page from PRINTING IN-DUSTRIES OF AMERICA ESTIMATING CATALOG. In this example, hourly cost rates for type composition and preparatory cost centers are shown.   (PIA, Inc.)

Fig. 4-6. FRANKLIN OFFSET CATALOG is a pricing guide used by many printers. Step-by-step procedures for using catalog are provided in instruction section.

A *production standard* is an hourly value representing the average output of a particular operating area, producing under specified conditions. The average production output is generally measured as a quantity unit of material, such as pounds of paper, sheets of photographic film, or number of signatures. Therefore, the hourly value then translates into number of units produced. A good example of this would be the number of impressions produced on a given offset press.

Production standards are generally calculated on a weekly basis. However, some firms merely spot-check their production standards on a monthly or quarterly basis. Checking the accuracy of production standards ensures that the firm is within the cost limitations set for each machine operation.

Computerized estimating

Computerized estimating uses special computer programs and a computer to calculate the cost of a job. It has gained popularity over the past few years. The computer has a television-like screen above the keyboard and all calculations can be done on the screen *(softcopy)*. Look at Fig. 4-8.

POSTAL INSTANT PRESS
Price List
(Black Ink Only)

Number Copies	8 1/2 x 11 1 side 20 lb. white bond	8 1/2 x 11 2 sides white bond
10 (8¢)	$ .80	$ 1.60
25 (7¢)	$ 1.75	$ 3.50
50 (6¢)	$ 3.00	$ 6.00
75	$ 3.40	$ 6.40
100 (4¢)	$ 4.00	$ 8.00
150	$ 4.50	$ 8.75
200 (2 3/4¢)	$ 5.50	$10.40
250	$ 6.40	$12.10
300	$ 7.25	$13.50
350	$ 8.20	$15.15
400 (2 1/4¢)	$ 9.00	$16.75
450	$ 9.95	$18.40
500	$10.80	$20.00
	($1.75 per add'l 100)	($3.20 per add'l 100)

Note:
Add 22¢ per 100 for colored bond
Add 43¢ per 100 for No. 1 white watermarked bond
Add 81¢ per 100 for white 20 lb. 25% rag bond
Add 45¢ per 100 for 8 1/2 x 14 size

Fig. 4-7. Standardized cost sheets, similar to this example, are generally used by quick printers. Note that prices are for black ink only on one or two sides of 8 1/2 x 11 inch, 20-pound, white bond paper. Additional charges are made for any other variations on a job of printing.

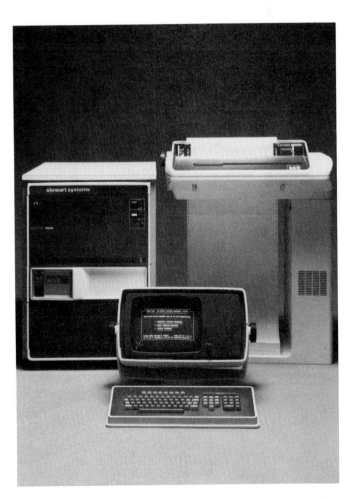

Fig. 4-8. Computerized estimating systems are popular throughout printing industry. With all job details entered in system, computer is capable of determining cost for each sub-part of job as well as cost for total job.   (Stewart Systems)

Special *software programs* are used to operate these machines for estimating. The various job details are placed into the computer's operating system. The computer then determines the answers for each part of a job. A total estimate is then output on a printout *(hardcopy)*. See Fig. 4-9.

Smaller versions of computerized estimating devices are widely used, Fig. 4-10. These machines are also used in the general operation of office business routines. Inventory control and mailing lists are typical tasks assigned to these computers.

The computer is capable of storing large amounts of *data* (information). It can be retrieved (called up) at any time. In addition to estimating, the computer has other business uses in the printing industry. For example, computers are used in production control, inventory, financial transactions, and to help keep sales records. Refer to Fig. 4-11.

Selecting the system

Computer systems should be selected on the basis of total plant integration, Fig. 4-12. In this way, each

```
        SUPER STEWY ESTIMATE

           ORDER SHEET SIZE
   INCHES   8THS    INCHES   8THS        POUNDS        SHEETS           1000 S
     23      0       35       0           1608         18980              1


           PRESS SHEET SIZE
   INCHES   8THS    INCHES   8THS     SIGS  PAGES/SIG  SHEETS           1000 S
     23      0       35       0         2      32      18980              1

   PRESS  NO. UP  NO. OUT  PLATES        FOLDER    PAGES/SIG   NO. UP   BIND UP
     1    1-SW      1        4             2          32         1         1


   DEPARTMENT      NO.       HRS.    LBR.$    UNITS   MAT.$  PURCH.$   TOT.$

   CAMERA         106       21.6    477.      1.     92.      0.      569.
   STRIP          107       52.9   1058.      1.      0.      0.     1058.
   PROOFS         108        1.7     34.      0.      0.      0.       34.
   PLATES         109        2.2     45.      4.     40.      0.       85.
   RDY&WASH         1        1.5     66.      0.      0.      0.       66.
   RDY-BACK         1        0.6     29.      0.      0.      0.       29.

   STOCK          113        0.0      0.   1608.    723.      0.      723.
   RUN&DRY          1        3.2    142.  18980.      0.      0.      142.
   INK            117        0.0      0.     16.     40.      0.       40.
   FOLD             2        6.9    167.      0.      0.      0.      167.

   SADDLEWIRE     125        2.6     84.      0.      0.      0.       84.
   PACKING        127        1.6     13.     29.      7.      0.       20.

       TOTAL      999       95.2   2118.      0.    904.      0.     3022.


   DEPARTMENT    HRS.    TOT.$

   CAMERA        21.6     569.
   STRIP         52.9    1058.
   PROOFS         1.7      34.
   PLATES         2.2      85.
   RDY&WASH       1.5      66.
   RDY-BACK       0.6      29.
   STOCK          0.0     723.
   RUN&DRY        3.2     142.
   INK            0.0      40.
   FOLD           6.9     167.
   SADDLEWIRE     2.6      84.
   PACKING        1.6      20.


   STDS. SET       QUANTITY       TOTAL $      $ ADDTL/M
     1              8500.          3022.        131.71
```

Fig. 4-9. Example of a total printing estimate printout (hardcopy) prepared on a computer estimating system. Computerized estimating is fast, efficient, and cost effective.

A

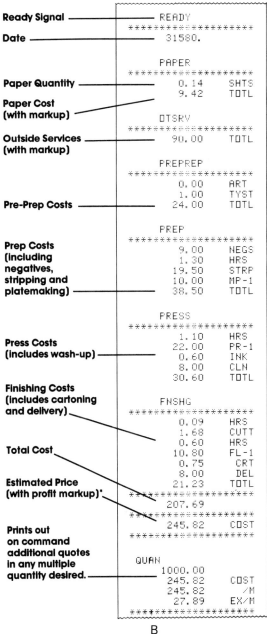

Ready Signal	READY	
	********************	
Date	31580.	
	PAPER	
	********************	
Paper Quantity	0.14	SHTS
	9.42	TOTL
Paper Cost (with markup)		
	OTSRV	
	********************	
Outside Services (with markup)	90.00	TOTL
	PREPREP	
	********************	
	0.00	ART
	1.00	TYST
Pre-Prep Costs	24.00	TOTL
	PREP	
	********************	
Prep Costs (including negatives, stripping and platemaking)	9.00	NEGS
	1.30	HRS
	19.50	STRP
	10.00	MP-1
	38.50	TOTL
	PRESS	
	********************	
	1.10	HRS
Press Costs (includes wash-up)	22.00	PR-1
	0.60	INK
	8.00	CLN
	30.60	TOTL
Finishing Costs (includes cartoning and delivery)	FNSHG	
	********************	
	0.09	HRS
	1.68	CUTT
	0.60	HRS
Total Cost	10.80	FL-1
	0.75	CRT
	8.00	DEL
	21.23	TOTL
Estimated Price (with profit markup)*	********************	
	207.69	
	********************	
	245.82	COST
	********************	
Prints out on command additional quotes in any multiple quantity desired.	QUAN	
	1000.00	
	245.82	COST
	245.82	/M
	27.89	EX/M
	********************	

B

Fig. 4-10. A—This small computerized estimating machine can also be used in general operation of office business routines. B—Hardcopy prepared on a small computerized estimating machine. This size computerized estimating machine is ideal for small and quick printers.   (M.P.Goodkin Co.)

application, as it is installed, will eliminate multiple entry of data. It will allow more data processing to be accomplished with the same number of office personnel. See Fig. 4-13.

In most printing firms, the applications suitable for computerization can be divided into two groups. These include: management applications and accounting applications. The various applications under these two categories are listed below:

A. MANAGEMENT APPLICATIONS
1. Quotation control.
2. Computer-assisted estimating.
3. Order entry.
4. Job costing.
5. Markup percentage rates.
6. Budgeted-hourly costs.
7. Raw material inventory.
8. Purchase order control.
9. Plant loading and scheduling.

B. ACCOUNTING APPLICATIONS
1. Accounts payable.
2. Accounts receivable.
3. Billing.
4. Finished goods inventory.
5. Payroll.
6. Financial statements.

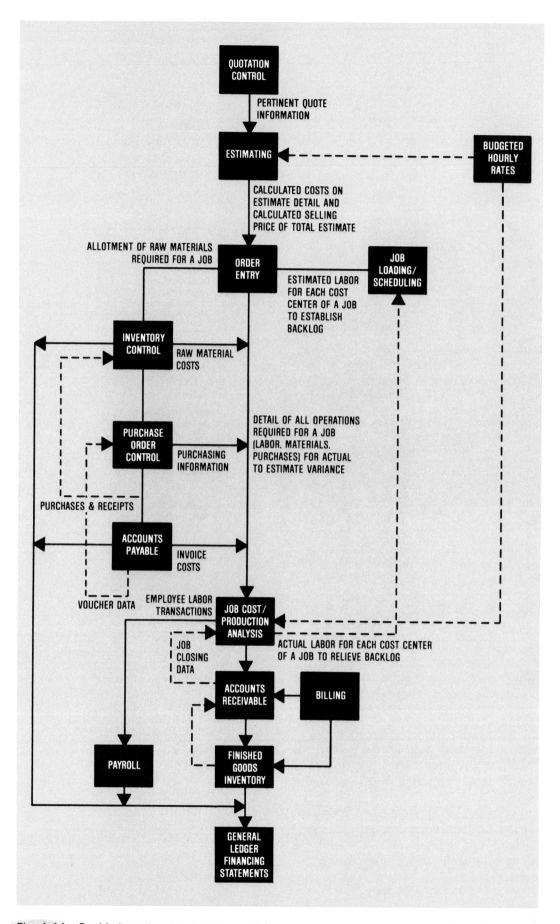

Fig. 4-11. Sophisticated computers are used for production control, inventory, financial transactions, and sales records. Most of these systems are used by large printing firms.

Ethernet is basically a coaxial cable that can be easily installed in a building through ceilings, walls or in existing ducts.

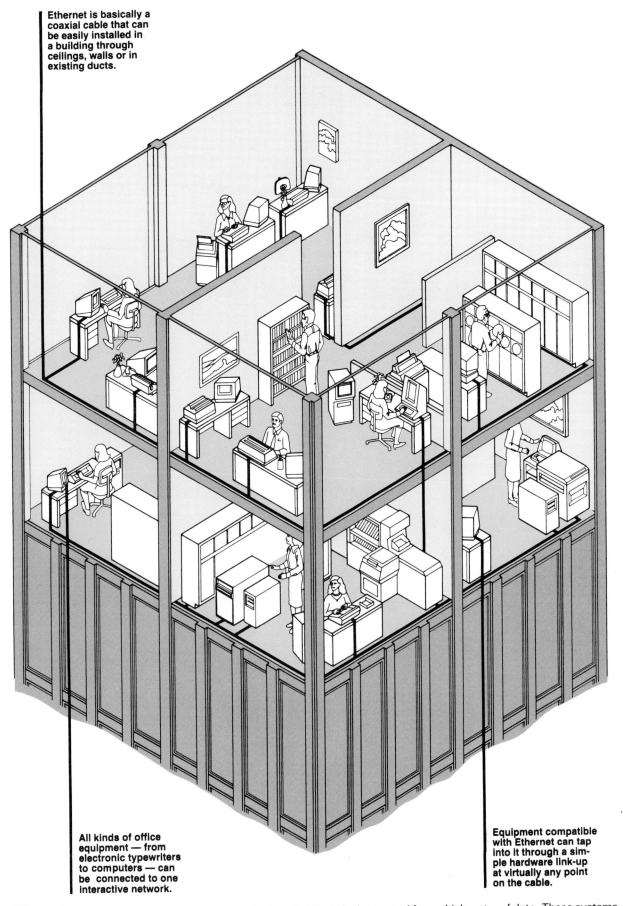

All kinds of office equipment — from electronic typewriters to computers — can be connected to one interactive network.

Equipment compatible with Ethernet can tap into it through a simple hardware link-up at virtually any point on the cable.

Fig. 4-12. Integrated computer systems can be installed that eliminate need for multiple entry of data. These systems are generally used by large commercial printing and publishing firms. (Ethernet)

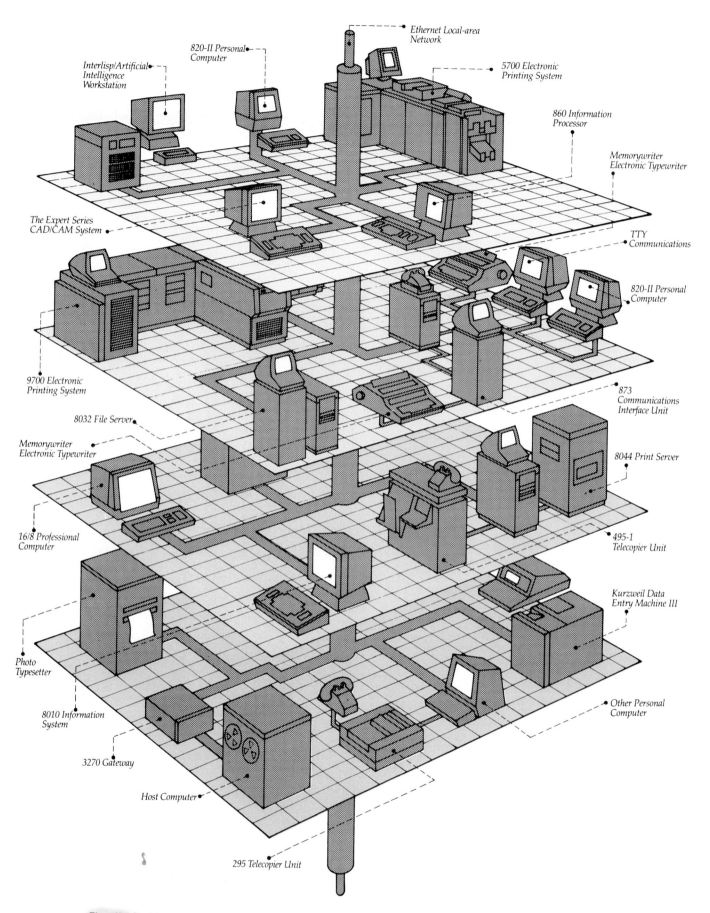

Fig. 4-13. Note example of a completely integrated computer system. Same office personnel can be used for multiple entry of data. (Ethernet)

These computer applications provide a solid foundation for any management information system. They represent a pyramid of modern computer applications for use in the printing industry. Estimating and quotation control are important components in the pricing of printing.

It should be noted that computers have NOT replaced printing plant personnel. Computer estimating, job costing, and accounting systems simply reduce the time and effort required in estimating and plant management. When it comes to judgment, people are still needed. Computers are sophisticated tools that only complement the people who use them!

ESTIMATING PROCEDURE

Regardless of how the estimator computes a job estimate, a systematic procedure must be followed to obtain the precise cost involved. The steps generally used to price a printing job include the following:

1. Secure accurate specifications for the proposed job. Prepare a layout/dummy.
2. Plan the sequence of operations involved in the job.
3. Determine the amount of material required and convert it to standard units.
4. Determine the time required to complete each operation.
5. Determine labor costs.
6. Determine fixed costs.
7. Total up the costs and add profit margin.
8. Prepare a quotation for the customer.

The estimator at work

Sales personnel in a printing firm are generally in close contact with the customer. Therefore, they are responsible for obtaining all information regarding the proposed job. All job specifications and special requirements must be noted. This ensures that all considerations are included in the final price quotation.

Estimator's worksheet

Estimators use a *worksheet* to write down all costs during the estimating process, Fig. 4-14. The estimator

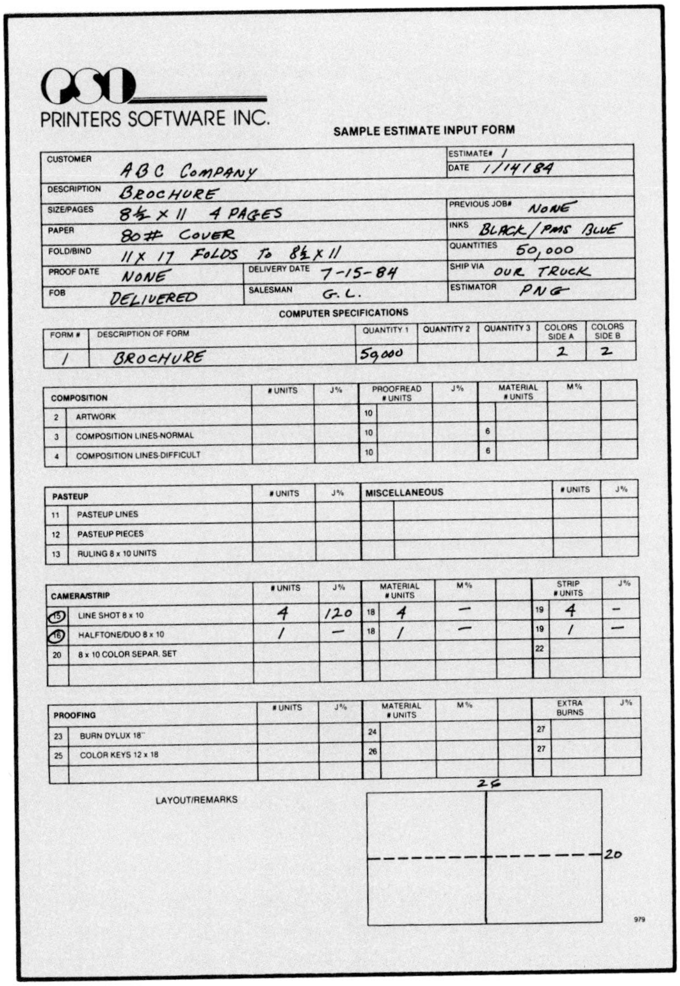

FRONT

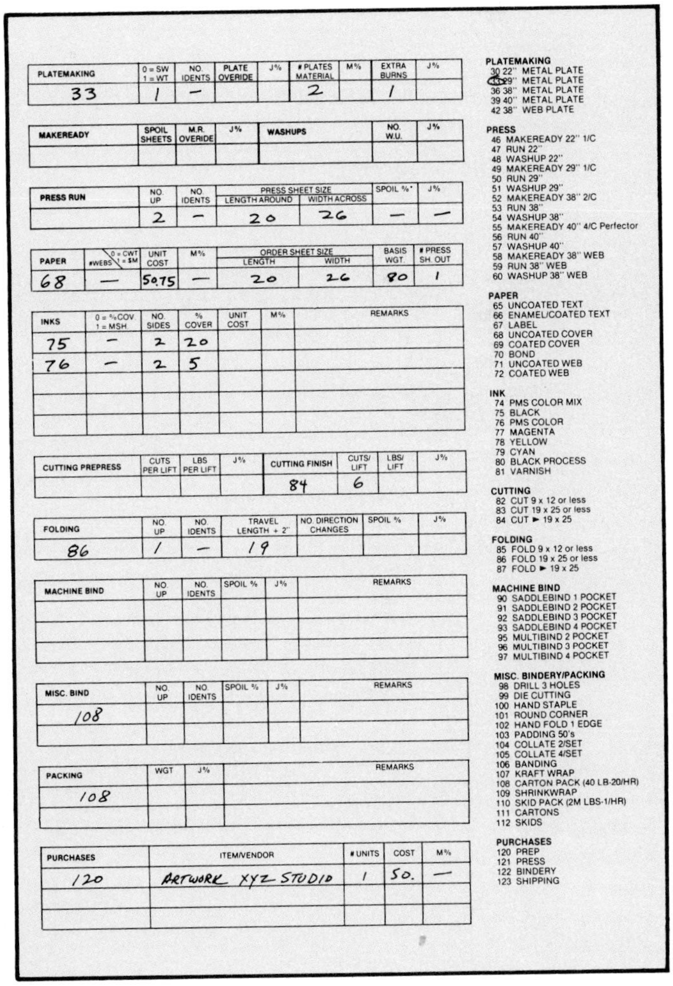

BACK

Fig. 4-14. Example of an estimator's worksheet. All costs are carefully noted during estimating process. Worksheets come in many formats and are generally prepared by individual printing firms to meet their particular needs. (Printer's Software, Inc.)

usually begins by outlining the requirements of a job. This is done to indicate the sequence of operations and to be certain that nothing has been missed.

Some estimators organize the worksheet around production departments, such as art, typesetting, presswork, etc. A layout or dummy is sometimes prepared to help the estimator visualize the job. Regardless of how the worksheet is set up, it should include all major production operations.

Once the requirements and sequence of operations have been determined, the quantities of needed materials are calculated. For example, the job may call for five pounds of blue ink. The prices are converted to unit amounts. Using a standard unit reference chart,

the estimator determines the amount of time necessary to complete all production operations.

Hourly labor costs are calculated for each operation. Overhead, or fixed costs, are added. These are generally based on the time of each operation. These figures are then totaled. A fixed percentage of profit is added to the total. The final sum represents the estimated cost of the job.

Quotation

Once the cost of the job has been submitted to the customer in the form of a *quotation,* the selling price is considered firm, Fig. 4-15. The quotation describes all the terms of the job, Fig. 4-16. The customer has

PRINTING ☐ **ESTIMATE**
☒ **QUOTATION**

FOR _ALASKA AIRLINES_ DATE _12-29-91_
ONTARIO INT'L AIRPORT QUOTE NO. _1141_
ONTARIO, CA 91773 ATTENTION _LISA WHITE_

WE ARE PLEASED TO SUBMIT OUR ESTIMATE FOR THE FOLLOWING:

JOB TITLE:	AIRLINE PRICE CHANGE FLYERS
QUANTITY:	10,000
NUMBER OF COLORS	1 - NAVY BLUE
PAGE SIZE NUMBER OF PAGES	8½" x 11" - 2 SIDES
STOCK/INSIDE: COVER:	8½" x 11 WAUSAU 24# WHITE BOND
PREPARATION CREATIVE ART: TYPE COMPOSITION: MECHANICAL ART:	2 LINE DRAWINGS - CUSTOMER SUPPLY TYPESET COPY PROVIDED
BINDERY & FINISHING	2 PARALLEL FOLDS
PACKAGING & SHIPPING	2 CARTONS
PRICE	$2137.95 - .213795 PER UNIT

We thank you for the opportunity to submit this estimate and we hope we may be of service to you.

PLEASE SIGN ONE COPY OF THIS FORM AND RETURN IT TO US OR SEND YOUR PURCHASE ORDER TO AUTHORIZE US TO BEGIN THIS JOB. PLEASE NOTE THAT ALL WORK IS ACCEPTED SUBJECT TO THE TRADE CUSTOMS SHOWN ON REVERSE SIDE

☐ PROCEED WITH JOB.
☐ PURCHASE ORDER ENCLOSED. Customer Signature _____

Fig. 4-15. Here is an example of a form that can be used as either an estimate or quotation. Most printers rely on one form for estimate and separate form for quotation. This particular form is used by a small printing company.

PRINTING TRADE CUSTOMS

Trade Customs have been in general use in the Printing Industry throughout the United States of America for more than 50 years.

1. QUOTATION A quotation not accepted within thirty (30) days is subject to review.

2. ORDERS Orders regularly entered, verbal or written, cannot be cancelled except upon terms that will compensate printer against loss.

3. EXPERIMENTAL WORK Experimental work performed at customer's request, such as sketches, drawings, composition, plates, presswork and materials will be charged for at current rates and may not be used without consent of the printer.

4. PREPARATORY WORK Sketches, copy, dummies and all preparatory work created or furnished by the printer, shall remain his exclusive property and no use of same shall be made, nor any ideas obtained therefrom be used, except upon compensation to be determined by the printer.

5. CONDITION OF COPY Estimates for typesetting are based on the receipt of original copy or manuscript clearly typed, double-spaced on 8½" x 11" uncoated stock, one side only. Condition of copy which deviates from this standard is subject to re-estimating and pricing review by printer at time of submission of copy, unless otherwise specified in estimate.

6. PREPARATORY MATERIALS Art work, type, plates, nega-

leakage insurance on all property belonging to the customer, while such property is in the printer's possession; printer's liability for such property shall not exceed the amount recoverable from such insurance.

13. DELIVERY Unless otherwise specified, the price quoted is for a single shipment, without storage, F.O.B. local customer's place of business or F.O.B. printer's platform for out-of-town customers. Proposals are based on continuous and uninterrupted delivery of complete order, unless specifications distinctly state otherwise. Charges related to delivery from customer to printer, or from customer's supplier to printer are not included in any quotations unless specified. Special priority pickup or delivery service will be provided at current rates upon customer's request. Materials delivered from customer or his suppliers are verified with delivery ticket as to cartons, packages or items shown only. The accuracy of quantities indicated on such tickets cannot be verified and printer cannot accept liability for shortage based on supplier's tickets. Title for finished work shall pass to the customer upon delivery, to carrier at shipping point or upon mailing of invoices for finished work, whichever occurs first.

14. PRODUCTION SCHEDULES Production schedules will be established and adhered to by customer and printer, provided that neither shall incur any liability or penalty for delays due to state of civil disorder, fire, strikes, accidents, action of Government ithority and acts of God or other causes beyond the control er or printer.

TOMER FURNISHED MATERIALS Paper stock, cam-film, color separations and other customer furnished ma-all be manufactured, packed and delivered to printer's ons. Additional cost due to delays or impaired production specification deficiencies shall be charged to the customer.

MS Payment shall be net cash thirty (30) days from date unless otherwise provided in writing. Claims for defects, or shortages must be made by the customer in writing period of thirty (30) days after delivery. Failure to make m within the stated period shall constitute irrevocable e and an admission that they fully comply with terms, s and specifications. Printer's liability shall be limited to ling price of any defective goods, and shall in no event pecial or consequential damages, including profits (or st). As security for payment of any sum due or to become r terms of any Agreement, printer shall have the right, if to retain possession of and shall have a lien on all cus-operty in printer's possession including work in process ned work. The extension of credit or the acceptance of de acceptances or guarantee of payment shall not affect rity interest and lien.

EMNIFICATION The customer shall indemnify and nless the printer from any and all loss, cost, expense and on account of any and all manner of claims, demands, nd proceedings that may be instituted against the printer ds alleging that the said printing violates any copyright or rietary right of any person, or that it contains any matter elous or scandalous, or invades any person's right to pri-ther personal rights, except to the extent that the printer ibuted to the matter. The customer agrees to, at the cus-wn expense, promptly defend and continue the defense of claim, demand, action or proceeding that may be brought e printer, provided that the printer shall promptly notify mer with respect thereto, and provided further that the hall give to the customer such reasonable time as the exi-f the situation may permit in which to undertake and con-defense thereof.

Jnited Typothetae of America, 1922. Revised nting Industries of America, Inc., 1945 & 1974.

IC COMMUNICATIONS CENTER GTON, VA. 22209

We are pleased to submit this proposal for your consideration:

DESCRIPTION

QUANTITY

TRIM SIZE

STOCK

ART WORK

NEGATIVES

COMPOSITION

PRESS WORK

BINDING

MISC.

PRICE

CHANGES AND ALTERATIONS WILL BE CHARGED AS AN EXTRA

This quotation good for_____days from above date.

☐ Estimate only

☐ Firm quotation

Subject to conditions printed on reverse side.

Accepted: _____   Approved: _____

BY   BY

Fig. 4-16. Note this form used by a quick-printer for an estimate or quotation. This dual-purpose format saves time and expense of printing and preparing two forms.

a fixed number of days to accept the quotation before it becomes void. The customer does NOT pay more than the quoted price. The estimator always reviews the quotation before submitting it to the customer.

POINTS TO REMEMBER

1. Estimating is a very important first step in the production of a printed product.
2. Estimators are people within a company who determine job costs.
3. The estimator must use careful cost analysis and proven methods of pricing individual printed products.
4. Without accurate information on time and cost standards, it is virtually impossible to estimate the cost of a job accurately.
5. In all cases, the cost of the printed product must be determined before production begins.
6. Cost centers are established by the estimator. A cost center is a job operation that costs money to maintain.
7. Printing firms decide on cost centers based on their particular operation and kinds of equipment.
8. Cost rates are also a part of cost centers. A cost rate is the expense of operating a specific cost center per hour.
9. One of the most common estimating procedures used in the printing industry is the budgeted-hourly cost method.
10. The computer is used extensively in the areas of estimating, management, accounting, and inventory control.
11. Many software programs are now available to assist printers, estimators, and managers with the tedious work of estimating, accounting, and inventory control.
12. Estimators generally use a worksheet to write down all costs during the estimating process.
13. When the cost of the printing job has been submitted to the customer in the form of a quotation, the selling price is considered firm.
14. The job of the estimator is a crucial one since the margin of profit or loss rests with the estimator's calculations.

KNOW THESE TERMS

Estimating, Estimator, Cost center, Fixed costs, Variable costs, Pricing guides, Standarized cost sheets, Production standards, Softcopy, Hardcopy, Data, Cost rate, Markup percentage, Budgeted-hourly cost, Buyout, Printing Industries of America, FRANKLIN OFFSET CATALOG, Software, Hardware, Computerized estimating, Quotation, Worksheet.

REVIEW QUESTIONS

1. A _____ _____ is the expense of operating a specific cost center per hour.
2. The _____ _____ method of estimating is often used by many printing firms to estimate the cost of printed jobs.
3. What do you call costs for outside materials, supplies, and services such as: typesetting, artwork, or bindery?
   a. Accounts.
   b. Buyouts.
   c. Markup.
   d. Contributions.
4. What is the FRANKLIN OFFSET CATALOG?
5. A computer program is called _____.
6. Computer systems can be adapted to printing management and accounting tasks in a printing firm. True or false?
7. Estimators use a _____ to write all costs when estimating a job.
8. The estimator generally prepares a _____ or _____ of the proposed job to help visualize the production steps.
9. A _____ describes all the terms of a printing job and serves as a contract between the printing firm and the customer.
10. A customer is often required to pay more than the quoted price for a job of printing. True or false?

SUGGESTED ACTIVITIES

1. Make a list of the cost centers in your graphic arts or printing facility. Describe the cost centers in a short written report.
2. With your teacher's assistance, use the budgeted-hourly cost method of estimating to find the cost of at least three jobs in your facility.
3. After discussing the proposal with your teacher, plan a class field trip to a printing firm that uses a computerized estimating system.
4. Prepare a worksheet and a quotation sheet for a substantial printing job in your facility. Have your teacher check the calculations and quoted price.
5. Examine a copy of the PRINTING INDUSTRIES OF AMERICA ESTIMATING CATALOG or the FRANKLIN OFFSET CATALOG. If these publications are not available at your facility, ask your teacher for assistance in locating a source for the books. Study the instruction section carefully to learn how the estimator uses the information contained in these books.

Chapter 5

Trade Customs and Legal Restrictions

When you have completed the reading and assigned activities related to this chapter, you will be able to:
O Describe the purpose of Printing Trade Customs practices.
O Explain the purpose of printing industry trade unions.
O Describe printing industry trade associations.
O Identify the three elements in a copyright notice.
O Describe those items that may be copyrighted.
O List items that may not be copyrighted.
O Name items that cannot be legally reproduced.

This chapter will summarize the most important information concerning Trade Customs, Trade Unions, Associations, and copyright laws. This knowledge is important to protect you when printing a job and to also protect your customer when agreeing to pay for a quoted job.

PRINTING TRADE CUSTOMS

The business practices used in the printing industry are called *Printing Trade Customs*. They were established at the annual convention of the United Typothetae of America in 1922. The *Typothetae* was a group of printers who banned together in order to support each other within the printing industry. Most printers today operate their businesses under these general guidelines.

The Printing Trade Customs are a useful and important tool to protect the printer from thoughtless customers and to also protect the customer from printing problems. For example, a customer may use one printer's creative ideas and then have the job printed by another printer. The Printing Trade Customs also cover such matters as proofs, alterations, press overruns, delivery, and terms of payment.

The most important factor in a business transaction is that the printer and customer be fully aware of the Printing Trade Customs. In addition, the contract for the work to be performed should be prepared in accordance with the Printing Trade Customs.

Many printers reproduce the Printing Trade Customs and submit them to a customer. This is especially important when a bid or quotation is given in written form. Printers find it convenient to print the Printing Trade Customs on the back of contract or quotation forms, invoice forms, and even letterheads. The Printing Trade Customs in general use in the printing industry are shown in Fig. 5-1.

PRINTING INDUSTRY LABOR UNIONS

The printing industry was the first to organize a labor union in the 1800s. Labor unions are organizations made up of many individuals. The primary purpose of a *labor union* is to improve the wages and working conditions of its members. The health and safety of workers is also a consideration of a union.

A company that hires only union members is called a *closed shop*. A nonunion company is called an *open shop*. In some cases, the workers in an open shop can be members of a labor union.

There are many good and bad points concerning labor unions. The ultimate decision of whether you

PRINTING TRADE CUSTOMS

Trade Customs have been in general use in the Printing Industry throughout the United States of America for more than 50 years.

1. **QUOTATION** A quotation not accepted within thirty (30) days is subject to review.

2. **ORDERS** Orders regularly entered, verbal or written, cannot be cancelled except upon terms that will compensate printer against loss.

3. **EXPERIMENTAL WORK** Experimental work performed at customer's request, such as sketches, drawings, composition, plates, press-work and material will be charged for at current rates and may not be used without consent of the printer.

4. **PREPARATORY WORK** Sketches, copy, dummies and all preparatory work created or furnished by the printer, shall remain his exclusive property and no use of same shall be made, nor any ideas obtained therefrom be used, except upon compensation to be determined by the printer.

5. **CONDITION OF COPY** Estimates for typesetting are based on the receipt of original copy or manuscript clearly typed, double-spaced on 8½" x 11" uncoated stock, one side only. Condition of copy which deviates from this standard is subject to re-estimating and pricing review by printer at time of submission of copy, unless otherwise specified in estimate.

6. **PREPARATORY MATERIALS** Art work, type, plates, negatives, positives and other items when supplied by the printer shall remain his exclusive property unless otherwise agreed in writing.

7. **ALTERATIONS** Alterations represent work performed in addition to the original specifications. Such additional work shall be charged at current rates and be supported with documentation upon request.

8. **PROOFS** Proofs shall be submitted with original copy. Corrections are to be made on "master set", returned marked "O.K." or "O.K. with corrections" and signed by customer. If revised proofs are desired, request must be made when proofs are returned. Printer regrets any errors that may occur through production undetected, but cannot be held responsible for errors if the work is printed per customer's O.K. or if changes are communicated verbally. Printer shall not be responsible for errors if the customer has not ordered or has refused to accept proofs or has failed to return proofs with indication of changes or has instructed printer to proceed without submission of proofs.

9. **PRESS PROOFS** Unless specifically provided in printer's quotation, press proofs will be charged for at current rates. An inspection sheet of any form can be submitted for customer approval, at no charge, provided customer is available at the press during the time of makeready. Any changes, corrections or lost press time due to customer's change of mind or delay will be charged for at current rates.

10. **COLOR PROOFING** Because of differences in equipment, paper, inks and other conditions between color proofing and production pressroom operations, a reasonable variation in color between color proofs and the completed job shall constitute acceptable delivery. Special inks and proofing stocks will be forwarded to customer's suppliers upon request at current rates.

11. **OVER RUNS OR UNDER RUNS** Over runs or under runs not to exceed 10% on quantities ordered up to 10,000 copies and/or the percentage agreed upon over or under quantities ordered above 10,000 copies shall constitute acceptable delivery. Printer will bill for actual quantity delivered within this tolerance. If customer requires guaranteed "no less than" delivery, percentage tolerance of overage must be doubled.

12. **CUSTOMER'S PROPERTY** The printer will maintain fire, extended coverage, vandalism, malicious mischief and sprinkler leakage insurance on all property belonging to the customer, while such property is in the printer's possession; printer's liability for such property shall not exceed the amount recoverable from such insurance.

13. **DELIVERY** Unless otherwise specified, the price quoted is for a single shipment, without storage, F.O.B. local customer's place of business or F.O.B. printer's platform for out-of-town customers. Proposals are based on continuous and uninterrupted delivery of complete order, unless specifications distinctly state otherwise. Charges related to delivery from customer to printer, or from customer's supplier to printer are not included in any quotations unless specified. Special priority pickup or delivery service will be provided at current rates upon customer's request. Materials delivered from customer or his suppliers are verified with delivery tickets as to cartons, packages or items shown only. The accuracy of quantities indicated on such tickets cannot be verified and printer cannot accept liability for shortage based on supplier's tickets. Title for finished work shall pass to the customer upon delivery, to carrier at shipping point or upon mailing of invoices for finished work, whichever occurs first.

14. **PRODUCTION SCHEDULES** Production schedules will be established and adhered to by customer and printer, provided that neither shall incur any liability or penalty for delays due to state of war, riot, civil disorder, fire, strikes, accidents, action of Government or civil authority and acts of God or other causes beyond the control of customer or printer.

15. **CUSTOMER FURNISHED MATERIALS** Paper stock, camera copy, film, color separations and other customer furnished materials shall be manufactured, packed and delivered to printer's specifications. Additional cost due to delays or impaired production caused by specification deficiencies shall be charged to the customer.

16. **TERMS** Payment shall be net cash thirty (30) days from date of invoice unless otherwise provided in writing. Claims for defects, damages or shortages must be made by the customer in writing within a period of thirty (30) days after delivery. Failure to make such claim within the stated period shall constitute irrevocable acceptance and an admission that they fully comply with terms, conditions and specifications. Printer's liability shall be limited to stated selling price of any defective goods, and shall in no event include special or consequential damages, including profits (or profits lost). As security for payment of any sum due or to become due under terms of any Agreement, printer shall have the right, if necessary, to retain possession of and shall have a lien on all customer property in printer's possession including work in process and finished work. The extension of credit or the acceptance of notes, trade acceptances or guarantee of payment shall not affect such security interest and lien.

17. **INDEMNIFICATION** The customer shall indemnify and hold harmless the printer from any and all loss, cost, expense and damages on account of any and all manner of claims, demands, actions and proceedings that may be instituted against the printer on grounds alleging that the said printing violates any copyright or any proprietary right of any person, or that it contains any matter that is libelous or scandalous, or invades any person's right to privacy or other personal rights, except to the extent that the printer has contributed to the matter. The customer agrees to, at the customer's own expense, promptly defend and continue the defense of any such claim, demand, action or proceeding that may be brought against the printer, provided that the printer shall promptly notify the customer with respect thereto, and provided further that the printer shall give to the customer such reasonable time as the exigencies of the situation may permit in which to undertake and continue the defense thereof.

Fig. 5-1. Printer and customer must be fully aware of Printing Trade Customs. The contract for work to be performed should be prepared in accordance with these understood rules. (Printing Industries of America, Inc.)

decide to work in a union or nonunion shop is a personal matter. Therefore, careful consideration should be given the matter.

INDUSTRY ASSOCIATIONS

There are a number of groups, called *associations,* that serve the people who work in the printing industry. Each association has its own operating structure and policies. The leaders of these organizations are either appointed or elected. These organizations generally conduct meetings at the local, regional, or national level.

The main purpose of *industry organizations* is to assist members and contribute to their welfare. Some of the services include technical expertise, new processes, and methods of operation.

The associations listed below serve people in the printing industry. There are many other organizations but these represent some linked to the offset printing industry. The list is intended to give you an idea of the types of organizations that serve the industry:

American Institute of Graphic Arts
New York, NY 10021

American Paper Institute
New York, NY 10016

Direct Mail/Marketing Association
New York, NY 10017

Graphic Arts Equipment and Supply Dealers
Association
Arlington, VA 22209

Graphic Arts Industries Association
Ottawa, Ontario, Canada KIP 5E7

Graphic Arts Technical Foundation
Pittsburgh, PA 15213

Graphic Communications Association
Arlington, VA 22209

In-Plant Printing Management Association
New Orleans, LA 70119

International Association of Photoplatemakers
South Holland, IL 60473

International Association of Printing House Craftsmen
Cincinnati, OH 45236

International Business Forms Industries
Arlington, VA 22209

International Graphic Arts Education Association
Pittsburgh, PA 15213

National Association of Lithographic Plate Manufacturers
Arlington, VA 22209

National Association of Printers and Lithographers
Harrison, NY 10528

National Association of Quick Printers
Chicago, IL 60601

National Business Forms Association
Alexandria, VA 22301

National Paper Trade Association
New York, NY 10017

National Printing Equipment and Supply Association
McLean, VA 22101

Paper Industry Management Association
Arlington Heights, IL 60005

LEGAL RESTRICTIONS

There are *legal restrictions* in the United States that regulate what may be printed. Printers must be alert to possible violations of the law. Printers who are in doubt about a printing job generally consult a lawyer. The most common legal problems relate to copyrights, counterfeiting, and obscene literature.

COPYRIGHTS

Copyrights protect the original owner of a literary, musical, or artistic work from having someone copy or sell their work. Printers have always been affected by copyright coverage. Increased protection of ownership by copyright has had a positive effect. This type of protection has increased printing production. For example, in the 19th century, the extension of copyright coverage to printed music and fine art prints stimulated their production on a wider scale. As the copyright law is extended in coverage, it will promote the sale of new media which use printing either in the product or in its promotion.

The following information is used as the standard when including a copyright notice in a printed piece.

Copyright notice
The copyright notice should contain the following three elements:
1. The word Copyright, the abbreviation Copr., or the symbol ©.
2. The name of the Copyright owner.
3. The year of publication.

These three elements should appear together on the copies. An example of a copyright notice is "Copyright 1991 by Kenneth F. Hird." For example, in the book you are now reading, refer to the notice on the page following the title page and note the copyright. In other documents, the copyright symbol may be included with the owner's initials, symbol, monogram, or mark if the name appears in another part of the document.

Expanded copyright coverage
The United States copyright law influences everyone involved in the production and distribution of infor-

mation. The United States Congress recently made extensive revisions to the U.S. Copyright law. Because it is impossible to give a complete and detailed description of this complex law, only general areas of copying are covered here.

The new copyright law provides increased protection of ownership by copyright. It includes expanded protection of the electronic media, including cable television, computers, and sound recordings. The new law includes the following significant changes.

1. There is now a single, national system of copyright administration. Previously, unpublished works were protected by common law (which varied from state to state) and published works were protected under federal law. The current law says that all copyrightable works, published or unpublished, are federally protected.

2. The period of time that a copyright is effective has been lengthened.

3. In the past, if a work was intended to be copyrighted, but was published without a copyright notice, any person was free to reprint, distribute, or market the work. This is no longer true. A copyright notice is still necessary for statutory (lawful) protection. However, innocently made mistakes (omissions or errors in names and dates) may be changed within five years of publication. In so doing, the owner does NOT lose the copyright privilege.

4. The ownership of a copyrighted work, and the persons liable for copyright infringement, have been more clearly defined.

5. The clause which banned certain types of printed works from U.S. copyright protection, if manufactured outside the country, no longer exists.

Copyrightable works

The United States Copyright Office defines copyrightable works as "original works of authorship fixed in any tangible medium of expression . . ." This includes the following categories: literary works, dramatic works, pantomimes and choreographic works, pictorial, graphic and sculptural works, motion pictures, other audiovisual works, and sound recordings.

Also copyrightable are compilation and derivative works, such as collections of articles previously published in a pre-existing work. Unpublished works, as well as published works, fall under protection of the law. However, protection does NOT extend to any idea, concept, principle, or discovery. For example, a specific article on nontoxic fountain solution developed for use with DiLitho presses can be copyrighted. However, the idea of nontoxic fountain solutions for use with DiLitho presses cannot be copyrighted.

Certain materials are simply NOT copyrightable and are free for anyone to use for any purpose. They include: titles, names, short phrases, slogans; common

symbols and designs; slight variation of typographic ornamentation; lettering or coloring; report forms; and any kind of work whose basis is common knowledge (calendars, tape measures and rulers, sporting event schedules, and tables derived from public sources), except those of unusual design. Many of these works such as calendars may use artwork that has been contracted for and is, therefore, copyrightable.

Length of copyright

The old term of copyright was 28 years plus an additional 28 years upon renewal. Under the present law, copyright for works currently covered by the old law has been extended an additional 17 years. Copyright on a work created on or after January 1, 1978, lasts for the lifetime of the designer, creator, or author and extends for 50 years after the person dies.

In the case of anonymous works, pseudonymous works, and works made for hire, the copyright lasts 75 years from the date of publication, or 100 years from the time of completion of the work, whichever is shorter.

A *"work made for hire"* is defined by the law as a work created by an employee or a work that has been specially ordered or commissioned, in writing, as a contribution to a collective work. These may include items such as audiovisual work, translation, supplementary work, compilation, instructional text, text, or an atlas.

The law defines *"supplementary works"* as introductions, conclusions, illustrations, charts, tables, bibliographies, appendices, and indexes.

Form of copyright notice

The *form of copyright notice* is the method of giving the copyright information. It is one of the areas of copyright that should be most familiar to printers. On most types of work, any of the following three forms is acceptable:

1. Copyright John Doe 1991
2. Copr. John Doe 1991
3. ©John Doe 1991

Exceptions to the forms listed above are those used on works of art, such as: lithographs, photoengravings, photographs, transparencies, pictorial illustrations, and labels. Because of design considerations, these works, and a number of others, are permitted a shorter version of the copyright notice. This may be displayed as ©, plus the initials, monogram, mark, or symbol of the copyright owner.

COUNTERFEITING

Counterfeit means to imitate or copy closely with the intent to deceive. There are a number of regulations which apply to counterfeiting and the most important ones are included here. You should remember that any request for printing or photographing, in part or whole, or the supplying of materials for any work which may seem to violate any part of the regulations quoted here, should be reported at once. The report should be made to the Department of the Treasury, United States Secret Service, Washington, DC 20223. If the proposed work is legally permissible, you will be so informed. If it is illegal, you will be far better off having had no part in it, Fig. 5-2.

Never proceed on the assumption that something similar to what is requested has been publicly distributed with permission. Prior violations may have gone unnoticed or unknown. They may also be under investigation at the time.

Fig. 5-3 gives some of the laws written to prevent counterfeiting.

OBSCENE LITERATURE

Obscene literature is material considered offensive to morality or virtue. The printing of articles, drawings, or photographs that might be classified as obscene requires careful legal advice. The printer's attorney should contact the proper governmental agency before any potentially obscene printing is undertaken.

The definition of what is obscene is sometimes a difficult matter to define. The United States Supreme Court in the 1960s declared that "to be *obscene* any printed matter must be totally without redeeming social value." As a result of that ruling, many pornographic publications were placed in circulation.

In 1973 the United States Supreme Court ruled that "the standards for judging obscenity could be drawn up by the people of each community." As a result, some communities have taken the initiative to control pornography.

The reputation of a printing company depends on how the people in a community view its operation.

Every company has a moral obligation to its employees and community.

RIGHTS OF THE PRINTER

Printers should also understand the importance of protecting their own works by copyright. Works created by artists, designers, and writers for the printer for personal distribution and sale should bear copyright notice. Examples include artwork on calendars and greeting cards.

Literature is available from the United States Copyright Office that clearly explains the copyright procedure. It can be useful in helping a printer answer basic customer questions on application for copyright, deposit of copies, and registration and fees. When discussing copyright law with customers, there is a point at which the printer should advise a customer to consult with an attorney regarding legal restrictions and obligations.

POINTS TO REMEMBER

1. The printing industry in the United States is governed by business practices called *Printing Trade Customs*. These guidelines were established in 1922.
2. The Printing Trade Customs protect the customer and the printer from unfair business practices.
3. The Printing Trade Customs also cover such matters as proofs, alterations, press overruns, delivery, and terms of payment.
4. Labor unions are also a part of the printing industry.
5. A company which hires only union members

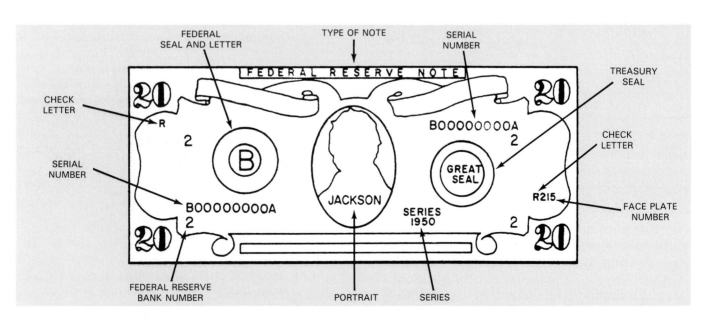

Fig. 5-2. Counterfeiting is considered a serious crime and is punishable by a prison sentence. This example shows various elements that form structure of a United States treasury bill.

U.S. CODE OF LAWS, TITLE 18, SEC. 474. PLATES OR STONES FOR COUNTERFEITING U.S. OBLIGATIONS OR SECURITIES

"Whoever, having control, custody or possession of any plate, stone or other thing, or any part thereof, from which has been printed, or which may be prepared by direction of the Secretary of the Treasury for the purpose of printing, any obligation or other security of the United States, uses such plate, stone, or other thing, or any part thereof, or knowingly suffers the same to be used for the purpose of printing any such or similar obligation or other security, or any part thereof, except as may be printed for the use of the United States by order of the proper officer thereof; . . . . .

"Shall be fined not more than $5,000 or imprisoned not more than fifteen years, or both."

U.S. CODE OF LAWS, TITLE 18, SECTION 475 (AS AMMENDED). IMITATING U.S. OBLIGATIONS OR SECURITIES; ADVERTISEMENTS

"Whoever designs, engraves, prints, makes or executes, or utters, issues, distributes, circulates or uses any business or professional card, notice, placard, circular, handbill or advertisement in the likeness or similitude of any obligation or security of the United States issued under or authorized by any Act of Congress or writes, prints or otherwise impresses upon or attaches to any such instrument, obligation or security, or any coin of the United States, any business or professional card, notice or advertisement, or any notice of advertisement whatever, shall be fined not more than $500."

U.S. CODE OF LAWS, TITLE 18, SECTION 481. PLATES OR STONES FOR COUNTERFEITING FOREIGN OBLIGATIONS OR SECURITIES

"Whoever within the United States except by lawful authority, controls, holds, or possesses any plate, stone, or other things, or any part thereof, from which has been printed or may be printed any counterfeit note, bond, obligation, or other security, in whole or in part, of any foreign government, bank, or corporation, or uses such plate, stone, or other thing, or knowingly permits or suffers the same to be used in counterfeiting such foreign obligation, or any part thereof; . . . .

"Shall be fined not more than $5,000 or imprisoned not more than five years, or both."

U.S. CODE OF LAWS, TITLE 18, SECTION 15. OBLIGATION OR OTHER SECURITY OF FOREIGN GOVERNMENT

"The term 'obligation or other security of any foreign government' includes, but is not limited to, uncanceled stamps, whether or not demonetized."

Fig. 5-3. Printers can be held responsible for printing produced illegally, even though they profess ignorance of the law for their offense.

is called a *closed shop*. A nonunion company is called an *open shop*.
6. Individuals must decide for themselves whether to work in a union or nonunion shop.
7. In addition to unions, there are a number of groups, called *associations,* which serve the people who work in the printing industry.
8. The main purpose of industry organizations is to assist members and contribute to their general welfare.
9. There are legal restrictions in the United States that regulate what may be printed. Printers must be alert to possible violations of the law.
10. The most common legal problems relate to copyrights, counterfeiting, and obscene literature.
11. Printers should understand the importance of protecting their own works by copyright.
12. Literature is available from the United States Copyright Office in Washington, DC that clearly explains the copyright procedure.
13. In some instances, printers should consult with an attorney concerning legal obligations and restrictions.

KNOW THESE TERMS

Closed shop, Copyright, Open shop, Printing Trade Customs, Obscene literature.

REVIEW QUESTIONS

1. Business practices enforced by law in the printing industry are called _____ _____ _____.
2. The printing industry was the first to organize a labor union in the:
   a. 1400s.
   b. 1500s.
   c. 1800s.
   d. 1900s.
3. A firm that hires only union members is called a _____ _____.
4. A nonunion firm is called a/an _____ _____.
5. Summarize 17 Printing Trade Customs.
6. The United States Copyright Law influences everyone involved in the production and distribution of information. True or false?
7. How long is a copyright in effect on a work created on or after January 1, 1978?
8. One of the following is NOT an acceptable copyright notice format:
   a. ©John Doe 1991.
   b. Copr. John Doe 1991.
   c. Copyright John Doe 1991.
   d. 1991 John Doe.
9. Counterfeiting violations come under the jurisdiction of the Department of the Treasury. True or false?
10. In 1983, the United State Supreme Court ruled that "The standards for judging obscenity could be drawn up by the _____ _____ _____ _____ in the United States."

SUGGESTED ACTIVITIES

1. Prepare a list of trade customs that could be used in the school graphic arts shop.
2. Determine if there are any printing associations near your town or city. Examples are the Printing House Craftsmen, Printing Industries Association, and Association of Quick Printers. With your teacher's assistance, contact these associations and find out who is eligible to attend these meetings and/or monthly dinner/speaker activities.

Various class members may want to attend such meetings, if possible.

3. Prepare a small display showing various types of copyrighted material. Label the samples for the benefit of those who will see the display. After getting permission, place the display in a busy part of the school.
4. Construct a small display showing various types of materials that are NOT copyrightable.
5. Write a historical research paper dealing with counterfeiting. Share this information with other members of the class.

Section II

Issues for Class Discussion

1. Once the need for a printed product has been determined, the customer needs an estimate and quotation. Selection of the printing company to do the work is generally made on the basis of price, quality, and service.

   Since it is often only possible to provide two of the three elements (price, quality, service), discuss, as a printing company owner, how and why you would determine their priority. Is there a class consensus on this important issue? If not, why?

2. The introduction of computer-assisted estimating has been the most important developoment in printing estimating in recent years. It has freed the estimator of nearly all manual calculations.

   Discuss the advantages and disadvantages of a computer-assisted estimating system and how it relates to small, medium, and large commercial printers, in-plant printers, and quick printers.

3. There has been, there still is, and there will always be problems with copyright infringement, pornography, counterfeiting, and unethical business practices.

   Discuss and summarize the rights and privileges of the Federal Copyright Act as it affects the commercial and quick printer.

4. The following scenario happens all too frequently in the real world of business. A printer completes and delivers a job valued at $5,200. The customer refuses to pay the bill, claiming that the price is higher than originally discussed. The customer did make several changes, along with adding new copy after proofs of the typesetting were supplied. A deposit for the job was NOT collected, nor were charges for the added copy provided the customer. The customer's comment was "I need these revisions and additions . . . I don't care about the cost."

   Discuss the scenario and provide possible solutions for the situation. As a printing firm owner, how would you have handled this customer?

Section III

Typography and Design and Layout

Hourly, hundreds of newly printed products enter the market. These products are carefully produced by printers. Most of these products all contain words and illustrations. This section of the textbook will assist you in developing a background and appreciation for our modern alphabet and the many typefaces that are in use today. In addition, you will learn about graphic design and layout. It will become evident to you that a printed product can only be as good as its typographic elements and original design format.

Chapter 6 is quite unique in that type and type styles or typefaces are thoroughly examined as being the core of all printed products. Certainly, without the alphabet and, in turn, without typefaces or typography, there would be little need for the printing industry. In this chapter, you learn an appreciation for the modern alphabet and the typefaces which have been categorized for recognition purposes.

In Chapter 7, design and layout is viewed as having to satisfy the needs of the customer and the audience for whom it is intended. Good design is typically created by someone called a design or layout artist. In order to design printing effectively, the design artist uses a knowledge of the product, the design process, and basic design principles. In this chapter, you will learn more about what you need to know to create successful designs for printed products.

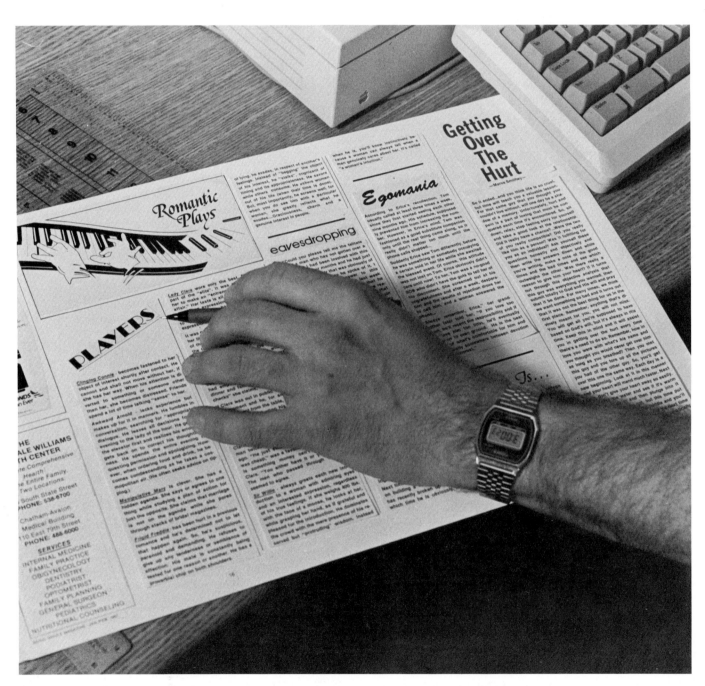

Type plays an important role in creating a ''feeling'' for a printed product.

Chapter 6

Typography

When you have completed the reading and assigned activities related to this chapter, you will be able to:

O Describe the major differences between hot type and cold type.
O Explain the principal features of the basic typeface classifications.
O Describe how type is measured and what basic units of measurement are used in printing.
O Explain how body and display size types differ, listing the usual point sizes included within each group.
O Define the terms ascender and descender and describe their relationship when measuring type point sizes.
O Distinguish among the several parts of a typeface.
O Compare the terms "font," "series," and "family" when used in relationship to type.
O List the main steps in copyfitting.
O Describe how the unit system is used in typography.

This chapter will review the most important information concerning typography. It will discuss how type is set, selected, and organized on a printed page. Knowing how to use type properly is a basic skill that you must understand to succeed in offset lithography.

TYPOGRAPHY

The term *typography* refers to the selection and arrangement of type elements included in the proposed design layout for a piece of printing. Skill in typography requires an understanding of type, typefaces, type measurement, and the various formats available for presenting a printed message, Fig. 6-1.

Type provides the means for including symbols (letters, numbers, and punctuation marks) in the printed message. Two basic kinds of type are available for this purpose: hot type and cold type.

Composition refers to the setting or placement of type for the print job. A *compositor* or *typesetter* follows the manuscript to compose the type.

Fig. 6-1. Typographic skill requires a thorough understanding of various types, typefaces, type measurement, and formats available for presenting a printed message.

Old, *hot type* composition methods used type made from molten metal. It is an older, outdated method of composition. Each piece of hot type consisted of one or more raised characters on a metal body, Fig. 6-2A. Hot type was composed by hand *(foundry type)* or by machine (Linotype, Ludlow, Monotype). After the desired words and sentences were composed, the type was inked and then printed directly on a press.

Sometimes the hot metal process was used to prepare printed sheets, called *proofs,* for checking the accuracy of the offset lithography process. After proofreading and correction of errors, a *reproduction proof* was prepared on dull, coated, white paper using black ink. The reproduction proof could then be photographed to produce a film negative (or positive) for the offset lithography process.

Cold type composition is a typesetting process that includes all other typesetting processes other than those using hot type or type made from molten metal. The copy required for present-day offset lithography is produced using one or more of the following methods: impact, dry transfer, phototypesetting, and laser.

Machines that place an image on paper using pressure action are called *impact machines,* Fig. 6-2B. A typewriter is an example of an impact machine.

A second method is *dry transfer,* Fig. 6-3. With this method, preprinted letters are positioned and adhered in place by hand.

Machines that put an image on paper photographically are called *phototypesetters,* Fig. 6-4A. One of the newest image-generating methods uses the laser printer in conjunction with a computerized desktop publishing system, Fig. 6-4B.

Most printers refer to letterpress type and letterpress relief plates as three-dimensional. The term cold type composition refers to the several methods of producing type composition on paper or film. The various forms of cold type composition can be called *two-dimensional type* because they all are on flat surfaces. Look at Fig. 6-5.

TYPEFACES

The term *typeface* refers to the unique shape or design of the characters included in a type font, type series, or type family.

A *font* is a complete assortment of characters of a SINGLE size and design. It includes *uppercase* (capital) letters, *lowercase* (small) letters, numerals, and punctuation marks, Fig. 6-6.

A *series* of individual types consists of ALL the sizes of one style or design, Fig. 6-7. For example, a series could include all the sizes available in Spartan bold. Another series might include all the sizes available in Optima medium.

A *family* is a group of related, but SLIGHTLY DIFFERENT type fonts and type series. Although family members all contain the same general design features,

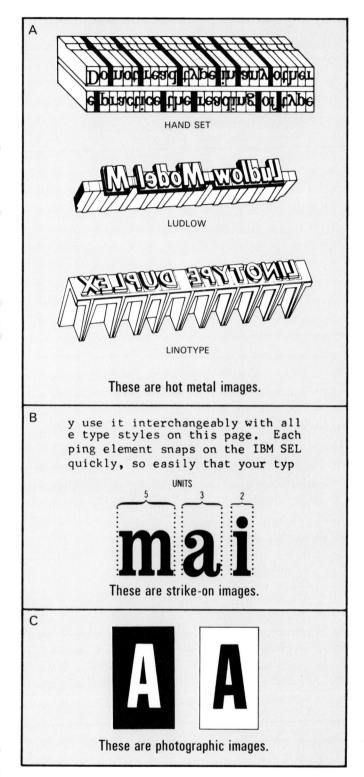

Fig. 6-2. A—Hot type composition methods have been replaced by computerized typesetting systems because they are faster and more versatile than the old metal type methods. Type examples shown above include hand-set type, Ludlow, and Linotype. B—A cold type composition method that uses a typewriter-like machine to produce type is referred to as impact composition. This method of typesetting is rarely used today since computer has replaced typewriter. C—High-speed machines used to put an image on paper photographically are called phototypesetters. These machines are extremely fast and can store large quantities of typeset material in electronic memory.

Fig. 6-3. A—Dry transfer lettering is available as either run-on or adhesive back. B—With this method, preprinted letters are positioned and adhered in place by hand. (Formatt Graphic Products, Corp.)

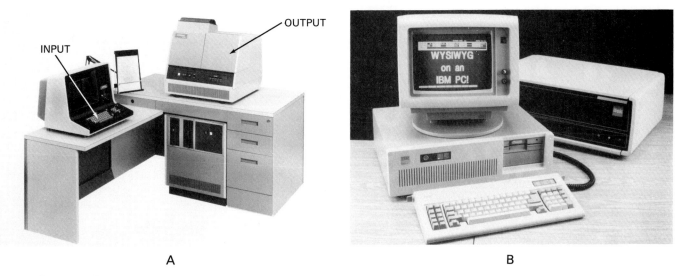

Fig. 6-4. A—Typeset images are prepared photographically on this phototypesetting machine. Input is on left (keyboard) and output is on right. B—This computerized desktop publishing system can be used to generate typeset material on a laser printer or digital typesetter, such as Itek Digitek shown here. An optional dot matrix printer can also be used for general office purposes or proofreading. (Linoterm and Itek)

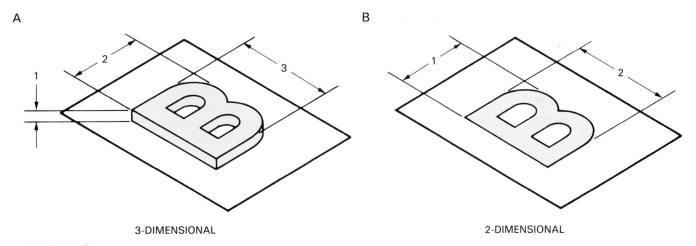

Fig. 6-5. In contrast to three-dimensional metal and wood types, two-dimensional cold types are on flat surfaces. Impact, dry transfer, phototypesetting, and electronic laser prints are two-dimensional type materials. A—Three-dimensional type. B—Two-dimensional type.

**Century Schoolbook Bold
ABCDEFGHIJKLMNOPQRS
TUVWXYZ&.,:;'"''!?()-abcdef
ghijklmnopqrstuvwxyz
$1234567890**

Fig. 6-6. Study a typical font which consists of a full assortment of characters for one size typeface.

6	ABCDEFGHIJKLMNOPQRSTUVWXYZABCDEFGH
8	ABCDEFGHIJKLMNOPQRSTUVWXYZA
10	ABCDEFGHIJKLMNOPQRSTUV
12	ABCDEFGHIJKLMNOPQR
14	ABCDEFGHIJKLMNOP
18	ABCDEFGHIJKLM
24	ABCDEFGHI
30	ABCDEFG
42	ABCDE
48	ABCD
60	ABC

Fig. 6-7. A series consists of all sizes of one style typeface. Not all typefaces are available in all sizes, however.

they may differ with respect to the width, weight, and/or slant (slope) of their individual characters. This is illustrated in Fig. 6-8.

The shapes of typefaces have persisted longer than any other artifacts in common use. Typeface designs more than 2,000 years old are still in service today. Many common typefaces are replicas of designs popular in the 15th and 16th centuries.

There are more than 6,000 typeface designs or styles in the Western world. Nevertheless, all can be categorized into SIX major styles called *type classifications*. Understanding the characteristics of each type classification will help you when selecting type for a particular job.

Garamond Light
Garamond Book
Garamond Bold
Garamond Ultra
Garamond Light Italic
Garamond Book Italic
Garamond Bold Italic
Garamond Ultra Italic
Garamond Light Condensed
Garamond Book Condensed
Garamond Bold Condensed
Garamond Ultra Condensed
Garamond Light Condensed Italic
Garamond Book Condensed Italic
Garamond Bold Condensed Italic
Garamond Ultra Condensed Italic

Fig. 6-8. A typeface family consists of the variations of one style of type. This is an example of the Garamond type family.

The six major typeface classifications include:
1. Roman.
2. Sans serif.
3. Square serif.
4. Textletter or blackletter.
5. Script and cursive.
6. Novelty.

Roman

The most important features of *Roman typefaces* are the numerous *serifs* (ends on character strokes) and the thin and thick strokes, Fig. 6-9. Serifs are rounded, angular, or rectangular, Fig. 6-10. Roman typefaces were patterned after the classical Roman letters. The characters are easily read because they are open, round, and wide. There also is a pleasing contrast between the thin and thick strokes. Roman types are further classified as Oldstyle, Transitional, and Modern. See Fig. 6-11.

Oldstyle — A typeface is classified as *Oldstyle* if it has little contrast between thick and thin lines. Serifs are pointed and generally slant or curve and extend outward at the top of the capital "T" and the bottom of the capital "E," Fig. 6-12.

Transitional — *Transitional Roman typefaces* have characteristics of BOTH Oldstyle and Modern

Typography Is the Selection of a Congruous Typeface, the Qua
Quality and Suitability for its
BASKERVILLE ITALIC TRUE CUT

Typography Is the Selection of a Congruous
Typeface, the Quality
BASKERVILLE BOLD

Fig. 6-9. Roman typefaces are most widely used. Characteristics commonly associated with Roman face is a pleasing contrast between thin and thick strokes.

Typography Is the Selection of a Congruous
Typeface, the Quality
CENTURY LIGHT

Typography Is the Selection of a Congruou
Typeface, the Quality
CENTURY LIGHT ITALIC

Fig. 6-12. Oldstyle Roman typefaces have little contrast between thick and thin strokes. With invention of printing press in 1450, Roman Oldstyle typefaces were used almost exclusively for book text matter.

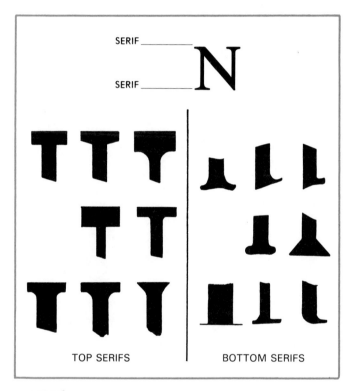

SERIF

SERIF

N

TOP SERIFS   BOTTOM SERIFS

Fig. 6-10. Serifs are rounded, angular, or rectangular strokes at tips of type characters. Sans serif means that type style does NOT have serifs. (Compugraphic Corp.)

typefaces, Fig. 6-13. Historically, Transitional typefaces were patterned after the Oldstyle designs, but prior to Modern typefaces. The contrast between thin and thick strokes is more evident in Transitional compared to Oldstyle. Also, in Transitional, the serifs are somewhat long and contain smooth, rounded curves. The characters are somewhat wider.

Typography Is the Selection of a Congruous Ty
Typeface, the Quality an
BODONI BOLD ITALIC NO. 2

Typography Is the Selection of a Co
Congruous Typef
BODONI EXTRABOLD NO. 2

Fig. 6-13. Transitional Roman typefaces are similar to Oldstyle but contrast between thick and thin strokes is more evident in Transitional. Transitional characters are also wider than Oldstyle.

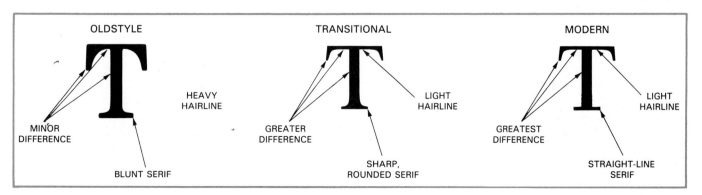

OLDSTYLE   TRANSITIONAL   MODERN

HEAVY HAIRLINE   LIGHT HAIRLINE   LIGHT HAIRLINE

MINOR DIFFERENCE   GREATER DIFFERENCE   GREATEST DIFFERENCE

BLUNT SERIF   SHARP, ROUNDED SERIF   STRAIGHT-LINE SERIF

Fig. 6-11. Three Roman typeface designs include: Oldstyle, Transitional, and Modern. Note weight variations in strokes and serifs.

Modern – Compared with Oldstyle, Modern typefaces show greater variation in stroke thickness, Fig. 6-14. The serifs are usually thin, straight, and somewhat rectangular. There is some rounding at the corners. There is a very strong contrast between the thick and thin lines of the characters.

Sans serif

Sans serif typefaces have NO serifs, Fig. 6-15. They also have little or no contrast in the thickness of letter strokes. Because the strokes have the same thickness,

the characters are cleanly designed, Fig. 6-16.

Sans serif typefaces are very popular today because of their simplicity, readability, and modern appearance. Sans serif typefaces are well suited to printing on smooth finished papers. They are used in magazines, books, and newspapers. They also are used for advertising, business cards, and stationery.

Square serif

Square serif typefaces are geometric in design, Fig. 6-17. The characters have square or blocked serifs and more or less uniform strokes. Square serif typefaces have limited usage as text matter. Square serif typefaces are appropriate for headlines and for short pieces of text matter. Square serif typefaces are used for display lines, in advertisements, newspaper headlines, and some letterheads.

Textletter

Textletter typefaces are very difficult to read, Fig. 6-18. Therefore, they are used very little today. These typefaces were patterned after the writing of the early scribes (writers). The serifs are pointed and the strokes are angular. Textletter is also called *Blackletter* and *Old English*. Because it is so difficult to read, textletter should never be composed in paragraph form or all uppercase. This typeface is most commonly used for certificates, diplomas, and religious programs.

HSMOUEAK
hsmoueakfg

Fig. 6-14. Modern Roman typefaces have a greater variation in stroke thickness. Serifs are usually thin, straight, and somewhat rectangular. (American Type Founders)

Typography Is the Selection of a Congru
Typeface, the Qualit

UNIVERS 53

Typography Is the Selection of a Congruous Typ
Quality and the Suitabilit

UNIVERS 55

Fig. 6-15. Sans serif typefaces tend to have a monotone (singular) look since they have NO serifs. Sans serif typefaces are popular today because of their simplicity, readability, and modern appearance.

Quality and Suitability for its Purpose of
Selection of a Congru

ITC LUBALIN GRAPH BOOK

Quality and Suitability for its Purpose
Selection of a Congr

ITC LUBALIN GRAPH MEDIUM

Fig. 6-17. These examples of Square serif typefaces exhibit a geometric design with square or blocked serifs and almost uniform strokes.

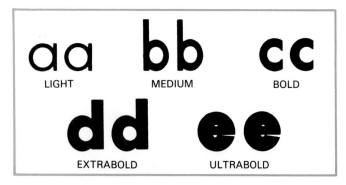

LIGHT MEDIUM BOLD

EXTRABOLD ULTRABOLD

Fig. 6-16. Sans serif characters are cleanly designed because strokes have same thickness. Sans serif typefaces are well suited to printing on smooth offset papers.

Typography Is the Selection of a Congruous
Typeface, the Quality an

OLD ENGLISH

Fig. 6-18. Textletter typeface resembles calligraphy (pen written) of German monks during Johann Gutenberg's era. This classification is also referred to as Blackletter and Old English.

Script

The *script*, also called *cursive*, typefaces are designed to resemble handwritten lettering, Fig. 6-19. Some authorities on type insist that SCRIPTS are letters resembling HANDWRITING and are connected while CURSIVE have a noticeable GAP between the letters.

Both scripts and cursives contain thin and thick strokes. These variations imitate the natural pressure variations exerted in handwriting. Scripts and cursives should NEVER be set in all uppercase. The letter combinations are awkward and difficult to read. It should be noted that scripts and cursives should NEVER be letterspaced. This would defeat the purpose for which they were intended — to imitate handwriting.

Decorative

The *decorative* typeface classification consists of those typefaces that do NOT fit any other classification, Fig. 6-20. Usually, these include hand-designed types and types designed for special effects. Decorative typefaces are commonly used for advertising purposes.

Typography Is the Selection of a Congruous Typeface, the Quality and

And Suitability for Its Purpose of

ORIGINAL SCRIPT

A

Typography Is the Selection of a Congruous Typeface

Quality and the Suitability

LISBON CURSIVE

B

Fig. 6-19. A—Script typefaces resemble handwriting and are joined. B—Cursive typefaces have a noticeable gap between characters.

TYPOGRAPHY IS THE SELECTION OF A CO

TYPEFACE THE QUALI

ITC NEON

TYPOGRAPHY IS THE SELECTION OF A CONGRUOUS TYPEFAC

THE QUALITY AND SUITABILI

NEW BOSTONIA

Fig. 6-20. Decorative typeface classification includes many styles that cannot be defined with specific features. Decorative typefaces are also referred to as Novelty or Occasional.

Typeface patents and licenses

There are a number of typefaces and phototypesetting machine trade names that are protected by patents and licenses as a result of original design and manufacturing contracts. Companies such as Linotype and Monotype secured original patents on many of the typefaces which they had cast in hot metal. These patents exist today, even though many of the original typefaces are used in impact typewriters, phototypesetters, and desktop publishing equipment. Some of the manufacturers of these machines have designed typefaces that appear as "look-alikes" to some of the original patented typefaces.

The following examples illustrate some of the popular typefaces that are protected by patents and licenses.

1. All typefaces whose names contain ITC are licensed by International Typeface Corporation.
2. Bembo, Emerson, Gill Sans, Perpetua, and Rockwell are licensed by the Monotype Corporation, Ltd.
3. Bauer Bodoni, Futura, and Schneider are licensed by Neufville.
4. The typeface names Caledonia, Helvetica, Optima, Palatino, Melior, and Spartan are registered trademarks of Allied Linotype.
5. Digitek, Quadritek and Itek Graphix are trademarks of, or under license to, Itek Graphix Corporation.

Note! Samples of many popular typefaces are included at the end of this chapter. The samples give the date of origin and company or person responsible for their design. The author would like to express gratitude to the Alphatype Corporation for giving permission to include the sample typefaces in this textbook. Such a list is a valuable tool for all typographers, designers, and printers.

TYPE MEASUREMENT

The *printer's point system* is used to designate type sizes and for measurements that relate to type composition. The basic units of the point system are the point and the pica.

A *point* is equal to abut 1/72 inch, and 12 points equal one *pica*. Six picas (72 points) equal 0.996 inch or 0.004 less than a full inch. Other units sometimes used in the point system are the nonpareil and the agate.

Fig. 6-21 illustrates the various units that comprise the point system and the inch.

1 POINT =	1/72 INCH
1 INCH =	72 POINTS
12 POINTS =	1 PICA
6 PICAS =	1 INCH
6 POINTS (1/2 PICA) =	1 NONPAREIL
5 1/2 POINTS =	1 AGATE

Fig. 6-21. Study this formula that illustrates relationships among various units that comprise point system and inch.

A *line gauge* is used to measure points and picas. The line gauge shown in Fig. 6-22 contains an inch scale and a nonpareil/pica scale. Some line gauges contain an agate scale as well.

Type is measured in *point sizes*. Common sizes range from six (6) through 72 points. Larger type is sized by the line, which equals 12 points. Therefore, a piece of 10-line type used for headlines measures 10 picas or 120 points in height.

The size of type is designated by referring to its point size, Fig. 6-23. Note that the size of a typeface is always smaller than its point-size designation. This is necessary to provide sufficient *white space* (nonprinting area) between lines when the type is *set solid* (without extra space added).

Display and text type sizes

Typefaces that are 12 points and SMALLER are referred to as *text type* or *body type*. Type sizes from 9 to 12 points are usually used for general reading literature, newspapers, and books such as you are now reading. Typefaces that are 14 points and LARGER are referred to as display type. These sizes of type are usually used as headlines or subheadings.

Type terminology

Some of the basic terms relating to type characters include:
1. *Serifs* are thin strokes at the ends of the main character strokes. Sans serif types do NOT have serifs.
2. *Stress* refers to the slant of a character.
3. *Stem* is the vertical stroke of a character, Fig. 6-24.

4. *Hairline stroke* is a thin line forming the elements of a character.
5. *Heavy stroke* is a heavy line forming the elements of a character.
6. *Set width* is the distance from the left to right sides of a character.
7. Cold *type sizes* are arrived at by measuring from the top of an ascender letter to the bottom of a descender character.
8. In lowercase characters, the upper stroke (as in b, d, h, or k) is called an *ascender,* Fig. 6-25.
9. The downstroke (as in g, p, or y) is called the *descender,* Fig. 6-25.
10. Type that slants (slopes) to the right is called *italic*.
11. *Uppercase* characters are capital or larger characters. They are also referred to as "caps."
12. *Lowercase* characters are small text or body size characters. Also referred to as "lc."
13. *Baseline* is the imaginary line along the bottom of the body height of characters. It is also referred to as the "x" height line, Fig. 6-25.
14. *Waist line* is the imaginary line along the top of the body height of characters. It is also referred to as "x" height line.
15. *Counter* refers to areas inside the loops (as in O) of type characters.

Leading

Inserting additional space between lines of type is called *leading* (pronounced ledding). The purpose of leading is to make type matter easier to read. *Long lines* (lines set to a wide measure) and large typefaces generally require MORE leading than *short lines* (lines

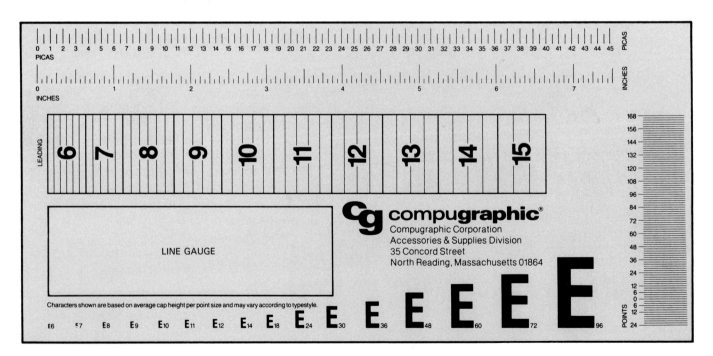

Fig. 6-22. Line gauge has inch scale and pica scale. The line gauge contains common point size comparisons, leading scale, and point scale. (Compugraphic Corp.)

72 Point

Roust DEN

60 Point

KING Invited

48 Point

Students DESKS

42 Point

HOMES Built Right

36 Point

Reach Elegant MODES

30 Point

**THOMAS HID
Cunning Crank**

12 Point

LIGHTENS THE BURDEN
Durable Type Faces Always
at Hand Provide Very Useful
Facilities for All 1234567890

24 Point

**MINES DEMAND
Enormous Output**

10 Point

SOUGHT ART OF VERS LIBRE
Literary man greatly disappointed
because in spite of all his studious
search he could find no free verse

18 Point

**PRINTING MACHINE
Lifting Higher Mankind**

8 Point

ANCESTRAL SEARCH INTERESTING
Perhaps it is due to their having no family
pride that so many persons do not concern
themselves about their ancestral descent

14 Point

**COTTON GOODS SECTION
Bargains in Ladies' Underwear**

6 Point

CHINESE AND INDIAN ARTCRAFT OBJECTS
Announcing a series of auction sales of an immense
consignment of freshly imported oriental makes of
wood metal and imitation china also textile articles

Fig. 6-23. This is an example of various common type sizes. Memorize these point sizes!

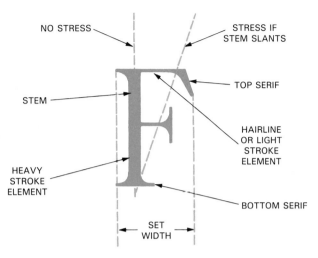

Fig. 6-24. Note elements of a character.

set to a narrow width) and small typefaces. Examples of type set solid and with various points of leading added are shown in Fig. 6-26.

In designating leading, if 9-point type is to be set with 1 point of leading, it is set 9 on 10, which is written as 9/10. In this example, the FIRST figure indicates the type size. The SECOND figure indicates the type size plus the leading desired. The text you are now reading is set in 11/12. Specifically, it is 11 point type with 1 point leading. Keep in mind that these terms apply to both older hot and present cold type composition.

Type formats

A variety of type formats are available for composing the printed message. The most common formats

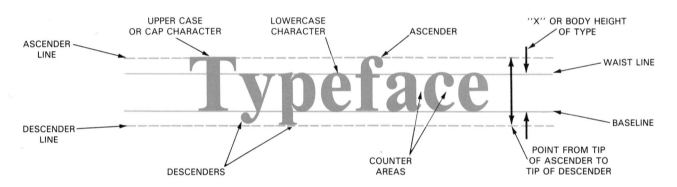

Fig. 6-25. Ascenders and descenders are strokes which go above or below common "x" height line of a type character. Note other names relating to type.

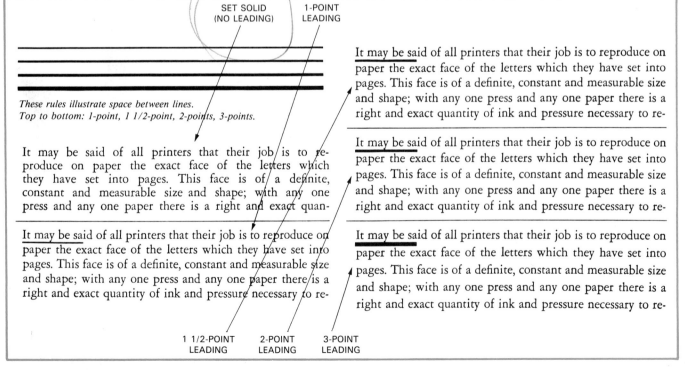

Fig. 6-26. Leading or linespacing controls how much space separates one line of type from another. In these examples, type has been set with no leading (solid), 1-point, 1 1/2 point, 2-points, and 3 points. (Westvaco Corp.)

are flush left, flush right, centered, and justified. These are shown in Fig. 6-27.

1. *Flush left* format has the type aligned evenly on the left side of the page and *ragged* (uneven) on the right side.
2. *Flush right* format is the opposite; it has the copy evenly aligned on the right and uneven on the left. This is not as commonly used as flush left.
3. *Centered* format has the copy centered on the page but it has uneven line lengths.
4. *Justified* format has the lines aligned on both the right and left sides of the page. This book is set with justified line length in the body copy. It is

very common. The words or characters are spaced differently to make the line workout even or justified.

An *initial character* is a larger first letter in a body of copy, Fig. 6-28. It is set in display type for emphasis. Initial characters are often used to begin a chapter of a novel. You rarely see initial characters used in the typesetting of textbooks. Instead the large letters have a decorative function. Thus, they are used in less serious or older style text. Many magazines use initial characters to break up the monotony of long, narrow columns of text matter, Fig. 6-29.

When initial characters are used, alignment must be carefully considered. For example, a wide gap between the initial character and the remainder of the word is NOT acceptable. In addition, the space around the initial should be in balance with the initial character. Initial character alignment spacing styles are shown in Fig. 6-30.

Using the right type style

Once the arrangement of type lines for a particular job has been determined, consistency must be maintained. This consistent use of one format is called *style*. The style for any given printed job is generally different

---

Since the printed word is intended primarily to be read, it is essential that the type should be of a size to produce maximum legibility. If the type is too small, it very quickly creates eye-strain and fatigue. If it is too large, it spreads out upon too great an area on the retina of the eye to be perceived quickly in as large groups as possible.

FLUSH LEFT

Since the printed word is intended primarily to be read, it is essential that the type should be of a size to produce maximum legibility. If the type is too small, it very quickly creates eye-strain and fatigue. If it is too large, it spreads out upon too great an area on the retina of the eye to be perceived quickly in as large groups as possible.

FLUSH RIGHT

Since the printed word is intended primarily to be read, it is essential that the type should be of a size to produce maximum legibility. If the type is too small, it very quickly creates eye-strain and fatigue. If it is too large, it spreads out upon too great an area on the retina of the eye to be perceived quickly in as large groups as possible.

CENTERED

Since the printed word is intended primarily to be read, it is essential that the type should be of a size to produce maximum legibility. If the type is too small, it very quickly creates eye-strain and fatigue. If it is too large, it spreads out upon too great an area on the retina of the eye to be perceived quickly in as large groups as possible.

JUSTIFIED

Fig. 6-27. Typesetting formats commonly used are flush left, flush right, centered, and justified.

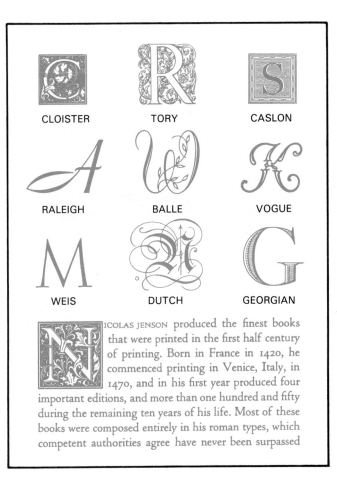

CLOISTER   TORY   CASLON

RALEIGH   BALLE   VOGUE

WEIS   DUTCH   GEORGIAN

NICOLAS JENSON produced the finest books that were printed in the first half century of printing. Born in France in 1420, he commenced printing in Venice, Italy, in 1470, and in his first year produced four important editions, and more than one hundred and fifty during the remaining ten years of his life. Most of these books were composed entirely in his roman types, which competent authorities agree have never been surpassed

Fig. 6-28. Initial characters are used for decorative emphasis. Example above shows effectiveness of an initial character used in conjunction with text matter.

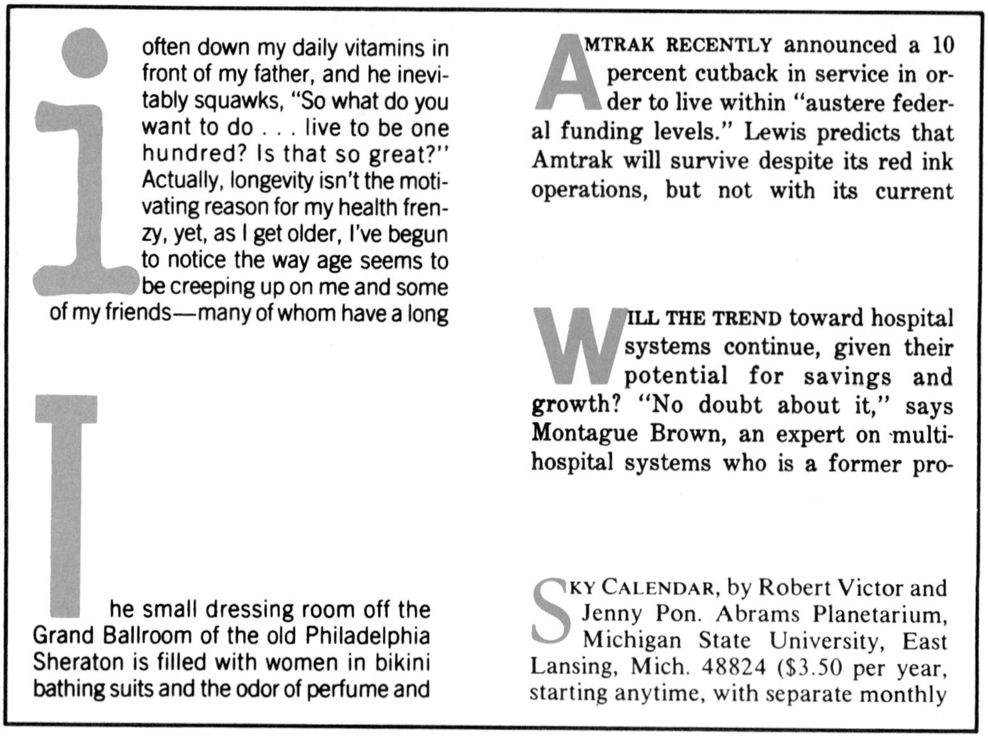

i often down my daily vitamins in front of my father, and he inevitably squawks, "So what do you want to do . . . live to be one hundred? Is that so great?" Actually, longevity isn't the motivating reason for my health frenzy, yet, as I get older, I've begun to notice the way age seems to be creeping up on me and some of my friends—many of whom have a long

T he small dressing room off the Grand Ballroom of the old Philadelphia Sheraton is filled with women in bikini bathing suits and the odor of perfume and

A MTRAK RECENTLY announced a 10 percent cutback in service in order to live within "austere federal funding levels." Lewis predicts that Amtrak will survive despite its red ink operations, but not with its current

W ILL THE TREND toward hospital systems continue, given their potential for savings and growth? "No doubt about it," says Montague Brown, an expert on multihospital systems who is a former pro-

S KY CALENDAR, by Robert Victor and Jenny Pon. Abrams Planetarium, Michigan State University, East Lansing, Mich. 48824 ($3.50 per year, starting anytime, with separate monthly

Fig. 6-29. Initial characters are often used in magazines to break up monotony of long, narrow columns of text matter.

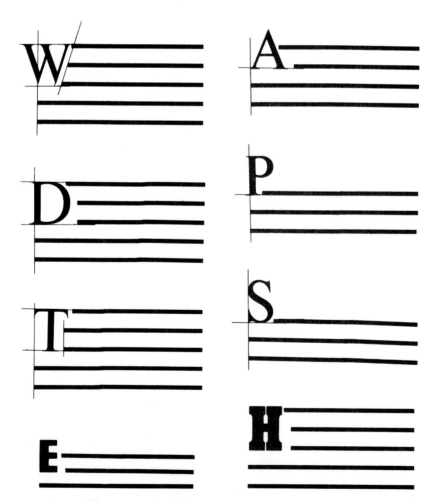

Fig. 6-30. Study these initial character alignment spacing styles. Space around initial should always be in balance with initial character.

from that for most other printed jobs. For example, the style of this textbook is somewhat different from those of other graphic arts textbooks with similar content.

Legibility

Legibility refers to the speed with which a type character can be identified by the reader. Some typefaces are more legible than others. For example, Roman type characters are generally easier to identify than square serif characters. This is because there is more difference between the shape of each character.

Readability

Readability refers to the ease that a printed page can be read. Some typefaces are "easy on the eyes." They can be read for long periods of time without tiring the reader's eyes. The readability of a page also depends upon the arrangement of type, line length, and the amount of white space on the page.

Appropriateness

The *appropriateness* of a typeface for a particular printed job is determined by the intended message and the intended audience. Some typefaces are said to "speak" in a normal tone; others "shout" their message. Some typefaces make us think of antiquity; others convey the impression of newness. In addition, characteristics such as femininity, masculinity, and formality can be suggested by a typeface.

Wordspacing

Spacing between words is called *wordspacing*. In three-dimensional typesetting, quads and spaces are pieces of metal cast less than type high to be used between words. These quads and spaces do not take ink and do not print when the type is put on the press. In two-dimensional typesetting, the wordspacing is achieved by pressing appropriate keys on the typesetting equipment. Depending upon the computer typesetting program being used, many different variations in wordspacing can be achieved either manually or automatically.

Words composed with narrow wordspacing appear to run together and are difficult to read. On the other hand, if the spacing is too wide, the line appears choppy and lacks unity. If the spacing between words is noticeably uneven, a displeasing effect is created. This interferes with the readability of the type.

Letterspacing

Spacing placed between individual characters is called *letterspacing*. When words are to be letterspaced, extra space must be allowed between the words so that they do not appear to run together. Letterspacing makes short words in titles or headings appear longer. The use of letterspacing usually improves the appearance of larger type sizes set in capitals. However, letterspacing reduces readability of words set in lowercase type.

Reverse type

Reverse type refers to type that drops out of the background and assumes the color of the paper. This technique is used to gain the reader's attention. It is also used to stress the importance of the message.

Magazines, newspapers and all types of advertising pieces use reverse type in certain advertisements for prominence. Since the reverse type is difficult to read, it should be used with extreme care. Excessive use of reverse type will cause a loss of readability.

UNIT SYSTEM

All typesetting systems are based upon a counting system that specifies type sizes and spacing. The point and pica system is used for setting cold type. As mentioned, 12 points equal 1 pica; 6 picas equal 1 inch.

A variety of *specifications* (measurements) are used to determine character width (set width) and spacing. All specifications are based upon the *unit system* of measurement. Remember that the size of the em quad depends upon the size of type. Thus, 10-point type uses em quads 10 points by 10 points square.

The unit system divides the em quad into eighteen uniform elements, Fig. 6-31. Each element is called a *unit*. The unit is used to define individual character set widths. The smaller the unit size, the more accurate the *wordspacing* (space between words) and *letterspacing* (space between letters) for most body (text) composition.

On most phototypesetting machines, the space between characters is adjustable. This means that the type can be set with regular, *loose* (more), or *tight* (less) letterspacing. It is also possible to letterspace selectively. This means kerning can be used to produce proper character spacing.

In *kerning*, it is possible to reduce space between certain characters and maintain normal spacing between the remaining characters, Fig. 6-32.

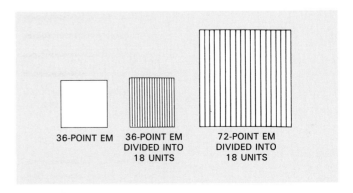

36-POINT EM   36-POINT EM DIVIDED INTO 18 UNITS   72-POINT EM DIVIDED INTO 18 UNITS

Fig. 6-31. Unit system divides em quad into eighteen uniform elements. Each element is called a unit and is used to define individual character set widths.

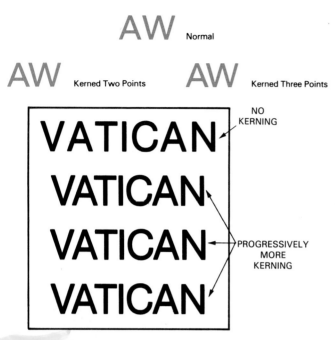

Fig. 6-32. Kerning allows spacing to be reduced between selected characters and maintain normal spacing between remaining characters. Study these examples to learn subtle differences in kerning techniques.

Preparing copy for typesetting

The term *copy* often refers to original typewritten manuscript. Copy is supplied to the printer who uses it to set type. The writer and typist must read and check manuscript carefully. This will help ensure its accuracy. It also will minimize errors that will need to be corrected on typeset copy.

The format for typewritten manuscript is important. Standard 8 1/2 x 11 inch bond paper is generally used. This allows the typist to make corrections and erasures. Margins should be 1 1/2 inch on the left, 1 inch on the right, 1 1/2 inch on the top, and 1 inch on the bottom. These margins provide space for typesetting instructions. The copy should be double-spaced, clean, and typed on only ONE side of the paper. Corrections should be neatly made in pencil, using standard *proofreaders' marks* (abbreviations indicating change in copy). Pages should be numbered consecutively at the top.

Markup of copy

The written information about copy parameters for typeface and column width is known as markup. Fig. 6-33. *Markup* involves writing the following *specifica-*

Instructions on a marked up rough include:

- The family, series, and size of all type faces.
- The length of all the lines of type, plus the leading between lines.
- The line setting, such as centered, flush left, or flush right.
- Any special instructions for the particular job in question.

Below are the most frequently used terms for indicating instructions on roughs:

CAPS	Set in all capital letters
clc	Set in capitals and lowercase letters
lc	Set in all lowercase
pt	Abbreviation for point size of type
BF	Set in boldface
8 on 10	(Also written as a fraction, 8/10.) The top number indicates the point size of the type and the lower number the leading between lines (in this case, 8-point type set on a 10-point base, or 8-point type leaded 2 points)
18	Set copy block 18 picas wide
8/10 × 18	Often the type size, leading, and line length are combined like this
⌐ ⌐	Center this line
⌐	Set flush left
⌐	Set flush right

A

Fig. 6-33. A—Markup involves writing down Linformation about copy (manuscript) parameters for typeface and column widths. Symbols used for markup are illustrated here. B—Study this chart containing a checklist for copy markup. Learn various symbols and their meanings. (Continued)

Symbol	Meaning
Helios	Kind of type
X 18	Line length in picas
9/11	Nine point type with two point leading
F-1	Font 1 (Helios light, for example)
F-2	Font 2 (Helios italic, for example)
F-3	Font 3 (Helios bold, for example)
QL	Set copy flush left on margin (quad left)
QR	Set copy flush right on margin (quad right)
QC	Set copy in center of margin (quad center)
⊡	Indent one em space
⧄	Indent one en space
———	Set in italic
═══	Set in small caps
≡≡≡	Set in regular caps
〰〰	Set in bold face
#	New paragraph
Just.	Set copy margin to margin

B

Fig. 6-33. Continued.

tions (measurements and style) on the layout and copy prior to typesetting:

1. Type size.
2. Leading.
3. Name of typeface.
4. Uppercase and lowercase, small capitals, etc.
5. Width of type lines.

For example, such information might be written as:

10/12–20 picas, Spartan bold, c. & l.c.

In this example, the type is to be set in 10 point with 2 points of line spacing (leading). It will be set 20 picas wide in Spartan bold using capitals and lowercase characters.

In some cases, the graphic designer who prepares the layout may also specify, or "spec," the type styles and sizes. In other instances, a person in the copy preparation area is assigned the task of markup.

Other specifications may be needed. These can include:

1. Amount of indentation for paragraphs.
2. Whether body copy is to be set in lightface or boldface.
3. Typeface and size for display lines.
4. Typeface and size for subheads.
5. Display line format (flush left, flush right, or centered).
6. Subhead format (flush left, flush right, or centered).

The detailed instructions for typesetting are written in the margins of the copy. Fig. 6-34 shows copy marked up for typesetting.

Copyfitting the manuscript

Before typewritten manuscript is set in high quality type, the typeface and type size must be determined. Sometimes, it is necessary to know the size of the area the type will fill.

A precise number of typeset characters, of any given point size, occupy a certain amount of space. Calculating the amount of space a given amount of copy will occupy when typeset is referred to as *copyfitting.*

By counting the number of characters in a typewritten manuscript, you can find out how much space the copy will take when it is typeset, Fig. 6-35. Follow the procedure outlined here when copyfitting typewritten manuscript.

1. Measure the width in inches of one average line of typewritten copy, Fig. 6-36.
2. Count the number of characters and spaces in this average line. (Note: pica typewriters reproduce ten characters per inch; elite typewriters reproduce 12 characters per inch.) Multiply the width of the average line by the number of characters and spaces in one inch. For example, a typewritten pica line six inches wide has 60 characters and spaces (10 characters per inch x 6 lines = 60 characters).
3. Multiply the number of typewritten characters per line by the total number of lines in the typewritten manuscript. You must then total the number of lines of typeset copy.
4. Use a line gauge to measure the width of the columns set aside on the layout for the copy. For ex-

WE ARE A GOOD NEIGHBOR) 18 Egmont Bold caps full 26 wide

Forty-one years ago when we were a young bank and Los Angeles was a growing city just about ready for its first pair of long pants, we decided there was no good reason why anyone should have to go traipsing all the way downtown to do his or her banking. So we inaugurated branch banking in Southern California--brought banking to you in your own community. Today, there are fifty convenient neighborhood Local Savings Bank offices in and around Los Angeles. Go to the nearest one the next time you have some banking to do. You'll find the officers and employees at any one of our fifty neighborhood offices eager to help you to be good neighbors.

14/24 Egmont Light 26 wide

LOCAL SAVINGS BANK) 24 Egmont Bold Italic caps full 26 wide

50 conveniently located neighborhood offices in and around Los Angeles. Member Federal Reserve System and Federal Deposit Insurance Corporation.

24 Egmont Light up

10/14 Egmont Light 26 wide

Copy "B"

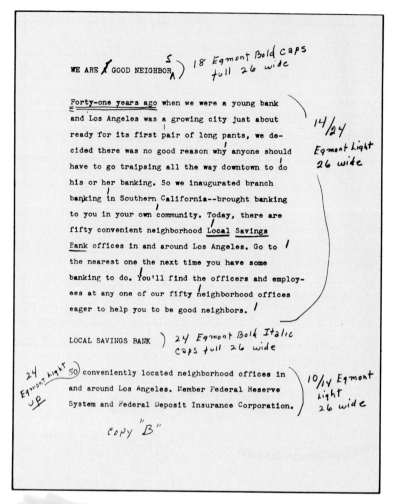

Fig. 6-34. This is copy marked up for typesetting. Detailed instructions are written in margins of copy.

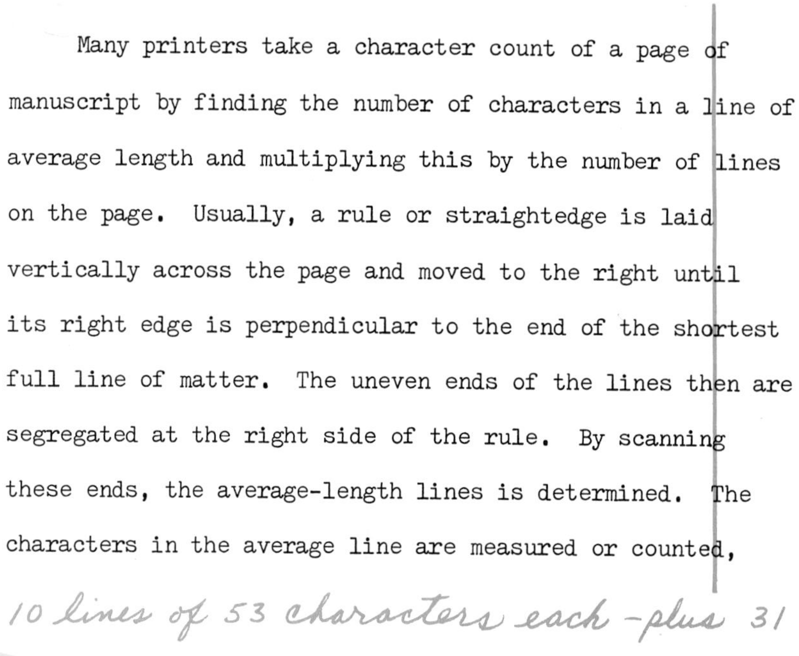

Many printers take a character count of a page of manuscript by finding the number of characters in a line of average length and multiplying this by the number of lines on the page. Usually, a rule or straightedge is laid vertically across the page and moved to the right until its right edge is perpendicular to the end of the shortest full line of matter. The uneven ends of the lines then are segregated at the right side of the rule. By scanning these ends, the average-length lines is determined. The characters in the average line are measured or counted,

10 lines of 53 characters each - plus 31

Fig. 6-35. Copyfitting typewritten manuscript is easily done by counting number of characters in manuscript. Simple mathematics is used to determine how much space copy will occupy when it is typeset.

← ———————————— 5 INCHES = 50 CHARACTERS ———————————— →

Graphic arts is defined by Webster as "the fine

and applied arts of representation, decoration, and

writing or printing on flat surfaces together with

Fig. 6-36. Number of typewritten characters per line is multiplied by total number of lines in manuscript. In this example, three lines of pica-size characters typed at a width of 5 inches equals 50 typewritten characters.

ample, a layout may call for columns to be set 18 picas wide.

5. Look in the *type specimen book* (book with type and copyfitting information) for the table showing the number of characters of the desired typeface size that will fit in one pica. A portion of a page from a type specimen book is shown in Fig. 6-37.

For example, 10-point, Cloister Bold has 2.8 characters per pica. To find the number of characters in a line 18 picas wide, multiply 2.8 × 18 picas. This gives a total of about 50 characters. A sample page is shown in Fig. 6-38.

6. Divide the total number of characters in the typewritten manuscript by the number of characters in one line of the typeface chosen. For example, if you are using 10-point, Cloister Bold and the typewritten copy has 2,000 characters and spaces, divide 2,000 by 50. This gives a total of 40 lines of copy set in 10-point, Cloister Bold.

In copyfitting, allowance must be made for leading when calculating the total depth of the type. For example, if 2 points of leading were added to each line of 8-point Spartan bold, each line would take 10 points of vertical spacing. For 6 lines, 60 points would be needed (10 × 6 = 60 points).

7. To convert points to picas, divide the 60 points by 12 points (number of points in one pica). The total vertical depth of the type in this example would be 5 picas (60 ÷ 12 = 5).

TYPOGRAPHY GUIDELINES

When dealing with the elements of typography, consider the following guidelines:

1. In a general way, Roman typefaces are preferred for body type because of their readability.
2. One and one-half times the width of the lowercase alphabet is considered the proper line width for easy reading.
3. Words crowded too closely together or spaced too far apart are difficult to read.
4. One or two points of space between the typeset lines is about right for text (body) matter.
5. As the width of the line is increased, the space between the lines should be increased.
6. Generally, the larger sizes of body types are the easiest to read. Sizes between 9- and 12-point for body type are acceptable.
7. Use a larger type size for *reverses* (words white on solid background for example). Reverses are more effective if the types are simple in design. Fine serifs should be avoided when planning reverses in a layout.

Typeface	6	8	10	12	14
Baskerville	3.8	3.2	2.6	2.3	2.0
Bodoni Bold	3.7	2.9	2.4	2.2	1.9
Bodoni Schoolbook	4.0	3.3	2.6	2.3	2.0
Caslon #540	3.5	3.2	2.8	2.2	1.9
Century Expanded	3.6	3.0	2.4	2.0	1.7
Century Bold	3.6	3.0	2.4	2.0	1.7
Cheltenham Wide	3.4	3.0	2.5	2.2	1.8
Cloister Bold	3.8	3.2	2.8	2.4	2.2
Futura Medium	3.8	3.7	2.9	2.5	2.2
Garamond	3.9	3.3	2.9	2.6	2.3
Garamond Bold	3.6	3.0	2.6	2.3	2.0
Memphis or Cairo	3.5	3.3	2.5	2.1	1.7
Metro Lite and Bold	3.6	3.2	2.3	2.1	1.8
Spartan	4.0	3.6	2.6	2.2	1.9

Fig. 6-37. This table shows a characters-per-pica chart from a type specimen book. How many characters-per-pica are contained in 8-point Futura Medium?

8. Type styles should be mixed with care. When two very different styles are used, one should dominate.
9. All-capital lines are difficult to read and their use should be limited.
10. Long amounts of body copy should be broken up for easier reading. Subheads, indented paragraphs, and other devices can help.
11. Avoid setting large quantities of body type in italics since it is difficult to read.
12. Ragged right does NOT seem to reduce readability, but ragged left does and should be used sparingly.

The primary objective of typography is to transform original manuscript information into readable, legible,

Times Roman, Century Expanded, and Similar Type

						Width in Picas						
	1	10	12	14	16	18	20	22	24	26	28	30
6 point	3.45	35	42	49	55	62	69	76	83	90	97	104
7 point	3.10	31	37	43	49	55	62	68	74	80	86	93
8 point	2.87	29	34	40	45	50	56	62	68	73	78	84
9 point	2.67	27	32	37	42	46	52	57	63	68	73	79
10 point	2.41	24	29	34	38	43	48	53	58	63	67	72
12 point	2.11	21	25	29	33	37	42	46	50	54	58	63
14 point	1.77	18	21	25	28	32	36	40	43	46	50	54
18 point	1.52	15	18	21	24	27	30	33	36	39	42	45
24 point	1.13	11	13	15	18	20	23	26	28	30	35	37

Century Schoolbook

						Width in Picas						
	1	10	12	14	16	18	20	22	24	26	28	30
6 point	3.71	37	44	52	59	67	74	82	89	96	104	112
7 point	3.20	32	38	45	51	58	64	70	77	83	90	96
8 point	2.89	29	34	41	46	52	58	64	69	75	81	87
9 point	2.69	27	32	38	43	48	54	59	65	70	75	81
10 point	2.50	25	30	35	40	45	50	55	60	65	70	75
12 point	2.14	21	26	30	34	39	43	47	51	56	60	64
14 point	1.79	18	21	25	29	32	36	39	43	47	50	54
18 point	1.50	15	18	21	24	27	30	33	36	39	42	45

Bodoni Book and Spartan

						Width in Picas						
	1	10	12	14	16	18	20	22	24	26	28	30
6 point	4.21	42	50	59	67	77	84	93	101	109	118	126
7 point	3.54	35	39	44	48	52	56	60	64	68	72	76
8 point	3.32	33	40	46	53	60	66	73	79	86	93	99
9 point	3.10	31	37	43	50	56	62	68	74	81	87	93
10 point	2.78	28	34	39	45	50	56	61	67	72	78	84
12 point	2.50	25	30	35	40	45	50	55	60	65	70	75
14 point	2.25	23	28	42	47	51	56	60	65	69	74	78
18 point	1.64	16	19	23	26	29	32	36	39	42	46	49
24 point	1.28	13	16	18	21	23	26	28	31	33	36	39

Garamond and Old Styles

						Width in Picas						
	1	10	12	14	16	18	20	22	24	26	28	30
6 point	3.80	38	46	53	61	68	76	84	91	99	106	114
7 point	3.64	36	43	51	58	65	72	80	87	94	102	109
8 point	3.35	34	40	46	53	60	67	74	80	87	94	100
9 point	3.05	30	36	42	48	54	61	67	73	79	85	91
10 point	2.89	29	35	40	46	52	58	64	69	75	81	87
12 point	2.57	26	31	36	41	46	51	57	62	67	72	77
14 point	2.15	22	26	30	34	39	43	47	52	56	60	65
18 point	1.77	18	21	25	28	32	35	39	42	46	50	53

Fig. 6-38. This type specimen page shows character-per-pica along with common pica widths. For other widths not listed, add or subtract one-pica figure as necessary. For example, typeset characters in 15 picas using 10-point Century Schoolbook is 37.50 (35 + 2.50 = 37.50).

and condensed blocks of type. All type must be symmetrically arranged for an aesthetic, functional effect. This means that the printed piece must be functional, attractive, pleasing, and inviting to the reader. If the type is too small, lines are too long or too short, and incongruent typefaces are used, it can cause eyestrain and fatigue. The organization of the printed type mass must permit a smooth flow of the reader's eye across a line, and from line to line.

Good typography results when the type images are arranged in a pleasing, attractive form and yet deliver the essence of the message clearly and concisely.

POINTS TO REMEMBER

1. Skill in typography requires an understanding of type, typefaces, type measurement, and the various formats that are available for presenting a printed message.
2. *Type* provides the means for including symbols (letters, numbers, and punctuation marks) in the printed message.
3. There are two basic kinds of type available for visual communications, hot type and cold type.
4. *Hot type* composition methods use type made from molten metal (three-dimensional).
5. *Cold type* composition has replaced hot type methods and is a non-metal typesetting process (two dimensional).
6. The term *typeface* refers to the unique shape or design of type characters.
7. Type is grouped into fonts, series, and families.
8. A *font* is a complete assortment of characters of a single size and design.
9. A *series* of individual types consists of all the sizes of one style or design.
10. A *family* is a group of related, but slightly different type fonts and type series.
11. There are more than 6,000 typeface designs or styles in the Western world.
12. Most typefaces are grouped into six major styles called *type classifications*. These include Roman, San serif, Square serif, Textletter, Script, and Decorative.
13. Typefaces are also grouped as to sizes. Type sizes from 9- to 12-point are called *text* or *body type*. Type sizes 14 points and larger are called *display type*.
14. The *printer's point system* is used to designate type sizes and for measurements relating to type composition.
15. Types are measured in *point sizes*.
16. Spacing placed between lines of type is called *leading* and is the amount of additional point space placed between individual lines for readability and appearance.
17. Once the arrangement of type lines for a particular job has been determined, consistency must be maintained. This is called *style*.
18. The style for any given printed job is generally different from that for all other printed jobs.
19. Typesetting is based upon a counting system that specifies type sizes and spacing.
20. The unit system divides the *em quad* (square of body size of type being used) into eighteen equal elements. Each element is called a *unit*, and the unit is used to define individual character set widths.
21. The term *copy* refers to typewritten manuscript. Copy is supplied to the printer who uses it to set type for the job.
22. The copy or manuscript is generally typewritten double-spaced on 8 1/2 × 11 inch bond paper.
23. The marking of copy for typeface styles and column widths is called markup.
24. *Markup* involves writing the specifications on the layout and copy or manuscript prior to typesetting. Markup information includes type size, leading (line spacing), name of typeface/s, uppercase and lowercase, small capitals, and width of lines in picas.
25. Calculating the amount of space a given amount of typewritten copy will occupy when typeset is referred to as *copyfitting*.
26. By counting the number of characters in a typewritten manuscript, it is possible to determine how much space the copy will occupy when typeset.

KNOW THESE TERMS

Typography, Hot type, Cold type, Impact machine, Dry transfer, Phototypesetter, Typeface, Font, Series, Family, Type classifications, Roman, Old Style Roman, Modern Roman, Sans serif, Square serif, Textletter, Script, Cursive, Decorative, Text type, Body type, Display type, Point, Pica, Nonpareil, Agate, Line gauge, Line, Ascender, Descender, Italic, Leading, Initial character, Style, Manuscript, Markup, Copyfitting, Type specimen book, Legibility, Readability, Appropriateness.

REVIEW QUESTIONS

1. Many of today's common typefaces are replicas of designs popular in the 15th and 16th centuries. True or false?
2. The selection and arrangement of type for a proposed design layout is called _____.
3. _____ type characters are prepared by hand lettering, art type alphabet sheets,

and photographically.

4. The sizes of type are always expressed in _____.
5. The fine end strokes of type characters are called _____.
6. Define the term "letterspacing."
7. Type set WITHOUT leading is called _____ matter.
8. Typefaces that are imitations of handwriting or hand lettering are called:
   a. Text typefaces.
   b. Sans serif typefaces.
   c. Oldstyle typefaces.
   d. Script typefaces.
9. The classification of type called _____ has characteristics of Oldstyle and Modern.
10. Typefaces that are straight and rectangular, without serifs, are called _____ _____ typefaces.
11. Square serif typefaces are frequently used for large areas of closely set type. True or false?
12. Letter forms that do NOT fit into any standard classification of typeface are referred to as _____.
13. The task of selecting typefaces, sizes of type, and width of lines for a job involves the process known as:
   a. Comping.
   b. Markup.
   c. Mitering.
   d. Typesetting.
14. Typewriter manuscript is referred to as _____.
15. Determining the size of the area that the type will fill on a printed page is called:
   a. Markup.
   b. Comping.
   c. Copyfitting.
   d. Justification.

SUGGESTED ACTIVITIES

1. Gather several good quality magazines and cut from them samples of the following typeface classifications: (a) Textletter, (b) Roman, (c) Sans serif, (d) Square serif, (e) Script, (f) Decorative. Carefully paste your samples to a sheet of 8 1/2 × 11 inch paper. Write the name of each typeface classification below the sample. Let your teacher examine the samples for accuracy.
2. Gather several good quality magazines and cut from them samples of initial letters. Carefully paste your samples to a sheet of 8 1/2 × 11 inch paper. Let your teacher examine the samples for accuracy.
3. Gather printed samples of approximately 20 different typefaces. Paste each sample to a separate index card. Write the type's classification on the reverse side of the card. Pick a partner from your class and use these as "flash cards" to test each other's typography knowledge.
4. Obtain a one-line printed sample of the following type categories: (a) all lowercase, (b) all capitals, (c) small capitals, (d) italic capitals, and (e) italic lowercase. Rank them as to which is easiest to read (number 1) and which is most difficult to read (number 5).
5. A thirty-page manuscript typed on an elite typewriter contains 22 lines per page. Each line contains about 83 characters. The manuscript will be composed in 9-point Times Roman type. Nine point Times Roman contains 3.0 characters per pica. Each line of type will be 20 picas wide. One point leading will be used between lines. How many typeset lines will there be altogether? What will their total depth be in picas? Show your calculations.

Alpha Gothic Light
Typography's function is to convey a visual me
1968 • ALPHATYPE • SIMILAR TO TRADE GOTHIC

Alpha Gothic Light Italic
Typography's function is to convey a visual me
1968 • ALPHATYPE • SIMILAR TO TRADE GOTHIC

Alpha Gothic
Typography's function is to convey a visual
1968 • ALPHATYPE • SIMILAR TO TRADE GOTHIC

Alpha Gothic Italic
Typography's function is to convey a visual
1968 • ALPHATYPE • SIMILAR TO TRADE GOTHIC

Alpha Gothic Bold
Typography's function is to convey a visual
1968 • ALPHATYPE • SIMILAR TO TRADE GOTHIC

Alpha Gothic Extended
Typography's function is to convey a
1968 • ALPHATYPE • SIMILAR TO TRADE GOTHIC

Alpha Gothic Bold Extended
Typography's function is to convey a
1968 • ALPHATYPE • SIMILAR TO TRADE GOTHIC

American Gothic Light
Typography's function is to convey a vi
1967 • VLADIMIR ANDRICH • ALPHATYPE/BERTHOLD EXCLUSIVE

American Gothic Light Italic
Typography's function is to convey a vi
1967 • VLADIMIR ANDRICH • ALPHATYPE/BERTHOLD EXCLUSIVE

American Gothic Medium
Typography's function is to convey a vi
1967 • VLADIMIR ANDRICH • ALPHATYPE/BERTHOLD EXCLUSIVE

American Gothic Medium Italic
Typography's function is to convey a vi
1967 • VLADIMIR ANDRICH • ALPHATYPE/BERTHOLD EXCLUSIVE

American Gothic Bold
Typography's function is to convey a vi
1967 • VLADIMIR ANDRICH • ALPHATYPE/BERTHOLD EXCLUSIVE

American Typewriter Light ITC
Typography's function is to convey a visu
1974 • KARDEN/STAN

American Typewriter Medium ITC
Typography's function is to convey a v
1974 • KARDEN/STAN

American Typewriter Bold ITC
Typography's function is to convey
1974 • KARDEN/STAN

American Typewriter Light Condensed ITC
Typography's function is to convey a visual message
1974 • KARDEN/STAN

American Typewriter Medium Condensed ITC
Typography's function is to convey a visual messa
1974 • KARDEN/STAN

American Typewriter Bold Condensed ITC
Typography's function is to convey a visual mess
1974 • KARDEN/STAN

Astro
Typography's function is to convey a visu
1968 • ALPHATYPE • SIMILAR TO ASTER

Astro Italic
Typography's function is to convey a visu
1968 • ALPHATYPE • SIMILAR TO ASTER

Astro Bold
Typography's function is to convey a visu
1968 • ALPHATYPE • SIMILAR TO ASTER

Avant Garde Extra Light ITC
Typography's function is to convey a vis
1970 • LUBALIN/CARNASE

Avant Garde Extra Light Oblique ITC
Typography's function is to convey a vis
1977 • GURTLER/MENGELT/GSCHWIND

Avant Garde Book ITC
Typography's function is to convey a visu
1970 • LUBALIN/CARNASE

Avant Garde Book Oblique ITC
Typography's function is to convey a vis
1977 • GURTLER/MENGELT/GSCHWIND

Avant Garde Medium ITC
Typography's function is to convey a vis
1970 • LUBALIN/CARNASE

Avant Garde Medium Oblique ITC
Typography's function is to convey a
1977 • GURTLER/MENGELT/GSCHWIND

Avant Garde Demibold ITC
Typography's function is to convey a vi
1970 • LUBALIN/CARNASE

Avant Garde Demibold Oblique ITC
Typography's function is to convey a
1977 • GURTLER/MENGELT/GSCHWIND

Avant Garde Bold ITC
Typography's function is to convey a
1970 • LUBALIN/CARNASE

Avant Garde Bold Oblique ITC
Typography's function is to convey
1977 • GURTLER/MENGELT/GSCHWIND

Avant Garde Book Condensed ITC
Typography's function is to convey a visual mes
1974 • EDWARD BENGUIAT

Avant Garde Medium Condensed ITC
Typography's function is to convey a visual me
1974 • EDWARD BENGUIAT

Avant Garde Demibold Condensed ITC
Typography's function is to convey a visual me
1974 • EDWARD BENGUIAT

Avant Garde Bold Condensed ITC
Typography's function is to convey a visual m
1974 • EDWARD BENGUIAT

NEW BASKERVILLE SMALL CAPS ITC
TYPOGRAPHY'S FUNCTION IS TO CONVEY A VISU
1982 • MERGENTHALER LINOTYPE

NEW BASKERVILLE SEMIBOLD SMALL CAPS ITC
TYPOGRAPHY'S FUNCTION IS TO CONVEY A VISU
1982 • MERGENTHALER LINOTYPE

New Baskerville ITC
Typography's function is to convey a visual
1982 • MERGENTHALER LINOTYPE

New Baskerville Italic ITC
Typography's function is to convey a visual messa
1982 • MERGENTHALER LINOTYPE

New Baskerville Semibold ITC
Typography's function is to convey a visual
1982 • MERGENTHALER LINOTYPE

New Baskerville Semibold Italic ITC
Typography's function is to convey a visual mes
1982 • MERGENTHALER LINOTYPE

New Baskerville Bold ITC
Typography's function is to convey a visua
1982 • MERGENTHALER LINOTYPE

New Baskerville Bold Italic ITC
Typography's function is to convey a visual me
1982 • MERGENTHALER LINOTYPE

New Baskerville Black ITC
Typography's function is to convey a vis
1982 • MERGENTHALER LINOTYPE

New Baskerville Black Italic ITC
Typography's function is to convey a visua
1982 • MERGENTHALER LINOTYPE

Bauhaus Light ITC
Typography's function is to convey a visual m
1975 • BENGUIAT/CARUSO

Bauhaus Medium ITC
Typography's function is to convey a visual
1975 • BENGUIAT/CARUSO

Bauhaus Demibold ITC
Typography's function is to convey a visual
1975 • BENGUIAT/CARUSO

Bauhaus Bold ITC
Typography's function is to convey a visu
1975 • BENGUIAT/CARUSO

Bauhaus Heavy ITC
Typography's function is to convey a vis
1975 • BENGUIAT/CARUSO

Belwe Light TSI
Typography's function is to convey a visual
1975 • TSI

Belwe Light Italic TSI
Typography's function is to convey a visual
1975 • TSI

Belwe Medium TSI
Typography's function is to convey a vi
1975 • TSI

Belwe Bold TSI
Typography's function is to convey a
1975 • TSI

Belwe Bold Condensed TSI
Typography's function is to convey a visual me
1975 • TSI

Benguiat™ Gothic Book ITC
Typography's function is to convey a vis
1979 • EDWARD BENGUIAT

Benguiat™ Gothic Book Italic ITC
Typography's function is to convey a vis
1979 • EDWARD BENGUIAT

Benguiat™ Gothic Medium ITC
Typography's function is to convey a vis
1979 • EDWARD BENGUIAT

Benguiat™ Gothic Medium Italic ITC
Typography's function is to convey a vi
1979 • EDWARD BENGUIAT

Benguiat™ Gothic Bold ITC
Typography's function is to convey a vis
1979 • EDWARD BENGUIAT

Benguiat™ Gothic Bold Italic ITC
Typography's function is to convey a vis
1979 • EDWARD BENGUIAT

Benguiat™ Gothic Heavy ITC
Typography's function is to convey a vi
1979 • EDWARD BENGUIAT

Benguiat™ Gothic Heavy Italic ITC
Typography's function is to convey a vi
1979 • EDWARD BENGUIAT

Benguiat™ Book ITC
Typography's function is to convey a vis
1978 • EDWARD BENGUIAT

Benguiat™ Book Italic ITC
Typography's function is to convey a vi
1978 • EDWARD BENGUIAT

Benguiat™ Medium ITC
Typography's function is to convey a vis
1978 • EDWARD BENGUIAT

Benguiat™ Medium Italic ITC
Typography's function is to convey a v
1978 • EDWARD BENGUIAT

Benguiat™ Bold ITC
Typography's function is to convey a
1978 • EDWARD BENGUIAT

Benguiat™ Bold Italic ITC
Typography's function is to convey
1978 • EDWARD BENGUIAT

Benguiat™ Book Condensed ITC
Typography's function is to convey a visual mes
1979 • EDWARD BENGUIAT

Benguiat™ Book Condensed Italic ITC
Typography's function is to convey a visual m
1979 • EDWARD BENGUIAT

Benguiat™ Medium Condensed ITC
Typography's function is to convey a visual me
1979 • EDWARD BENGUIAT

Benguiat™ Medium Condensed Italic ITC
Typography's function is to convey a visual
1979 • EDWARD BENGUIAT

Benguiat™ Bold Condensed ITC
Typography's function is to convey a visua
1979 • EDWARD BENGUIAT

Benguiat™ Bold Condensed Italic ITC
Typography's function is to convey a visu
1979 • EDWARD BENGUIAT

Bodoni
Typography's function is to convey a visual message
1964 • ALPHATYPE

Bodoni Italic
Typography's function is to convey a visual message
1964 • ALPHATYPE

Bodoni Book
Typography's function is to convey a visual m
1964 • ALPHATYPE

Bodoni Book Italic
Typography's function is to convey a visual mes
1964 • ALPHATYPE

Bodoni Bold
Typography's function is to convey a visual mes
1964 • ALPHATYPE

Bodoni Bold Italic
Typography's function is to convey a visual mess
1964 • ALPHATYPE

Bodoni Ultra
Typography's function is to convey a v
1964 • ALPHATYPE

Bodoni Ultra Italic
Typography's function is to convey a v
1964 • ALPHATYPE

Bookman
Typography's function is to convey a v
1970 • ALPHATYPE

Bookman Italic
Typography's function is to convey a v
1970 • ALPHATYPE

Bookman Light ITC
Typography's function is to convey a vi
1975 • EDWARD BENGUIAT

Bookman Light Italic ITC
Typography's function is to convey a v
1975 • EDWARD BENGUIAT

Bookman Medium ITC
Typography's function is to convey a
1975 • EDWARD BENGUIAT

Bookman Medium Italic ITC
Typography's function is to convey a
1975 • EDWARD BENGUIAT

Bookman Demibold ITC
Typography's function is to convey a
1975 • EDWARD BENGUIAT

Bookman Demibold Italic ITC
Typography's function is to convey
1975 • EDWARD BENGUIAT

Bookman Bold ITC
Typography's function is to convey
1975 • EDWARD BENGUIAT

Bookman Bold Italic ITC
Typography's function is to conv
1975 • EDWARD BENGUIAT

Boston Script
Typography's function is to convey a visual message quick
1967 • ALPHATYPE • SIMILAR TO PALACE SCRIPT

Bramley Light TSI
Typography's function is to convey a visual
1979 • ALAN MEEKS

Bramley Medium TSI
Typography's function is to convey a vis
1979 • ALAN MEEKS

Bramley Bold TSI
Typography's function is to convey a vi
1979 • ALAN MEEKS

Bramley Extra Bold TSI
Typography's function is to convey a v
1979 • ALAN MEEKS

Brewer Text Light
Typography's function is to convey a visual mess
1983 • ALPHATYPE • SIMILAR BAUER TEXT

Brewer Text Medium
Typography's function is to convey a visual me
1983 • ALPHATYPE • SIMILAR BAUER TEXT

Brewer Text Demibold
Typography's function is to convey a visual me
1983 • ALPHATYPE • SIMILAR BAUER TEXT

Brewer Text Bold
Typography's function is to convey a visual m
1983 • ALPHATYPE • SIMILAR BAUER TEXT

Bridal Script
Typography's function is to convey a visual message
1971 • ALPHATYPE • SIMILAR TO NUPTIAL SCRIPT

Brighton Light TSI
Typography's function is to convey a visual message
1981 • ALAN BRIGHT

Brighton Light Italic TSI
Typography's function is to convey a visual messa
1981 • ALAN BRIGHT

Brighton Medium TSI
Typography's function is to convey a visual mess
1981 • ALAN BRIGHT

Brighton Bold TSI
Typography's function is to convey a visua
1981 • ALAN BRIGHT

Cardinal Light
Typography's function is to convey a vis
1979 • ALPHATYPE • SIMILAR TO CARDIFF

Cardinal Light Italic
Typography's function is to convey a vis
1979 • ALPHATYPE • SIMILAR TO CARDIFF

Cardinal Medium
Typography's function is to convey a vis
1979 • ALPHATYPE • SIMILAR TO CARDIFF

Cardinal Medium Italic
Typography's function is to convey a vi
1979 • ALPHATYPE • SIMILAR TO CARDIFF

Cardinal Demibold
Typography's function is to convey a vi
1979 • ALPHATYPE • SIMILAR TO CARDIFF

Cardinal Bold
Typography's function is to convey a
1979 • ALPHATYPE • SIMILAR TO CARDIFF

CASLON 74 SMALL CAPS
TYPOGRAPHY'S FUNCTION IS TO CONVEY A VIS
1974 • ALPHATYPE

Caslon 74
Typography's function is to convey a visual
1974 • ALPHATYPE

Caslon 74 Italic
Typography's function is to convey a visual mess
1974 • ALPHATYPE

Caslon 74 Bold
Typography's function is to convey a vis
1974 • ALPHATYPE

Caslon 74 Bold Italic
Typography's function is to convey a vis
1974 • ALPHATYPE

Caslon 74 Bold Condensed
Typography's function is to convey a visual mes
1974 • ALPHATYPE

CASLON NO. 224™ BOOK SMALL CAPS ITC
TYPOGRAPHY'S FUNCTION IS TO CONVEY A VIS
1983 • EDWARD BENGUIAT

CASLON NO. 224™ MEDIUM SMALL CAPS ITC
TYPOGRAPHY'S FUNCTION IS TO CONVEY A VIS
1983 • EDWARD BENGUIAT

Caslon No. 224™ Book ITC
Typography's function is to convey a visual
1983 • EDWARD BENGUIAT

Caslon No. 224™ Book Italic ITC
Typography's function is to convey a visua
1983 • EDWARD BENGUIAT

Caslon No. 224™ Medium ITC
Typography's function is to convey a visu
1983 • EDWARD BENGUIAT

Caslon No. 224™ Medium Italic ITC
Typography's function is to convey a visu
1983 • EDWARD BENGUIAT

Caslon No. 224™ Bold ITC
Typography's function is to convey a vis
1983 • EDWARD BENGUIAT

Caslon No. 224™ Bold Italic ITC
Typography's function is to convey a vi
1983 • EDWARD BENGUIAT

Caslon No. 224™ Black ITC
Typography's function is to convey a vis
1983 • EDWARD BENGUIAT

Caslon No. 224™ Black Italic ITC
Typography's function is to convey a v
1983 • EDWARD BENGUIAT

Century Old Style Bold
Typography's function is to convey a vi
1976 • ALPHATYPE

CENTURY TEXT SMALL CAPS
TYPOGRAPHY'S FUNCTION IS TO CONVEY
1965 • ALPHATYPE • SIMILAR TO CENTURY SCHOOLBOOK

Century Text
Typography's function is to convey a visua
1965 • ALPHATYPE • SIMILAR TO CENTURY SCHOOLBOOK

CENTURY OLD STYLE SMALL CAPS
TYPOGRAPHY'S FUNCTION IS TO CONVEY A
1976 • ALPHATYPE

Century Old Style
Typography's function is to convey a visu
1976 • ALPHATYPE

Century Old Style Italic
Typography's function is to convey a visual
1976 • ALPHATYPE

Century Text Italic
Typography's function is to convey a visu
1965 • ALPHATYPE • SIMILAR TO CENTURY SCHOOLBOOK

Century Text Bold
Typography's function is to convey a visu
1965 • ALPHATYPE • SIMILAR TO CENTURY SCHOOLBOOK

Century Text Bold Italic
Typography's function is to convey a visu
1965 • ALPHATYPE • SIMILAR TO CENTURY SCHOOLBOOK

Century X
Typography's function is to convey a visu
1965 • ALPHATYPE • SIMILAR TO CENTURY EXPANDED

Century X Italic
Typography's function is to convey a visual
1965 • ALPHATYPE • SIMILAR TO CENTURY EXPANDED

Century X Bold
Typography's function is to convey a visu
1965 • ALPHATYPE • SIMILAR TO CENTURY EXPANDED

Century Light ITC
Typography's function is to convey a vis
1980 • TONY STAN

Century Light Italic ITC
Typography's function is to convey a vi
1980 • TONY STAN

Century Book ITC
Typography's function is to convey a vi
1980 • TONY STAN

Century Book Italic ITC
Typography's function is to convey a v
1980 • TONY STAN

Century Bold ITC
Typography's function is to convey a
1980 • TONY STAN

Century Bold Italic ITC
Typography's function is to convey a
1980 • TONY STAN

Century Ultra ITC
Typography's function is to con
1980 • TONY STAN

Century Ultra Italic ITC
Typography's function is to co
1980 • TONY STAN

Century Light Condensed ITC
Typography's function is to convey a visual mess
1980 • TONY STAN

Century Light Condensed Italic ITC
Typography's function is to convey a visual me
1980 • TONY STAN

Century Book Condensed ITC
Typography's function is to convey a visual me
1980 • TONY STAN

Century Book Condensed Italic ITC
Typography's function is to convey a visual m
1980 • TONY STAN

Century Bold Condensed ITC
Typography's function is to convey a visual
1980 • TONY STAN

Century Bold Condensed Italic ITC
Typography's function is to convey a visual
1980 • TONY STAN

Century Ultra Condensed ITC
Typography's function is to convey a
1980 • TONY STAN

Century Ultra Condensed Italic ITC
Typography's function is to convey
1980 • TONY STAN

Cheltenham Old Style
Typography's function is to convey a visual messag
1970 • ALPHATYPE

Cheltenham Old Style Italic
Typography's function is to convey a visual messag
1970 • ALPHATYPE

Cheltenham Old Style Condensed
Typography's function is to convey a visual message quickl
1970 • ALPHATYPE

Cheltenham Old Style Bold Condensed
Typography's function is to convey a visual message qu
1970 • ALPHATYPE

Cheltenham Medium
Typography's function is to convey a visual me
1970 • ALPHATYPE

Cheltenham Medium Italic
Typography's function is to convey a visual me
1970 • ALPHATYPE

Cheltenham Bold
Typography's function is to convey a visual me
1970 • ALPHATYPE

Cheltenham Light ITC
Typography's function is to convey a visual
1978 • TONY STAN

Cheltenham Light Italic ITC
Typography's function is to convey a visual
1978 • TONY STAN

Cheltenham Book ITC
Typography's function is to convey a visua
1978 • TONY STAN

Cheltenham Book Italic ITC
Typography's function is to convey a visual
1978 • TONY STAN

Cheltenham Bold ITC
Typography's function is to convey a vi
1978 • TONY STAN

Cheltenham Bold Italic ITC
Typography's function is to convey a vi
1978 • TONY STAN

Cheltenham Ultra ITC
Typography's function is to convey
1978 • TONY STAN

Cheltenham Ultra Italic ITC
Typography's function is to convey
1978 • TONY STAN

Cheltenham Light Condensed ITC
Typography's function is to convey a visual message q
1978 • TONY STAN

Cheltenham Light Condensed Italic ITC
Typography's function is to convey a visual message
1978 • TONY STAN

Cheltenham Book Condensed ITC
Typography's function is to convey a visual messa
1978 • TONY STAN

Cheltenham Book Condensed Italic ITC
Typography's function is to convey a visual messa
1978 • TONY STAN

Cheltenham Bold Condensed ITC
Typography's function is to convey a visual
1978 • TONY STAN

Cheltenham Bold Condensed Italic ITC
Typography's function is to convey a visual m
1978 • TONY STAN

Cheltenham Ultra Condensed ITC
Typography's function is to convey a visual
1978 • TONY STAN

Cheltenham Ultra Condensed Italic ITC
Typography's function is to convey a visual
1978 • TONY STAN

New Cheltenham Light
Typography's function is to convey a v
1974 • ALPHATYPE • SIMILAR TO MINI CHELTENHAM

New Cheltenham Light Italic
Typography's function is to convey a visu
1974 • ALPHATYPE • SIMILAR TO MINI CHELTENHAM

New Cheltenham Medium
Typography's function is to convey a vi
1974 • ALPHATYPE • SIMILAR TO MINI CHELTENHAM

New Cheltenham Medium Italic
Typography's function is to convey a vis
1974 • ALPHATYPE • SIMILAR TO MINI CHELTENHAM

New Cheltenham Bold
Typography's function is to convey a vi
1974 • ALPHATYPE • SIMILAR TO MINI CHELTENHAM

New Cheltenham Extra Bold
Typography's function is to convey a visu
1974 • ALPHATYPE • SIMILAR TO MINI CHELTENHAM

Clarendon
Typography's function is to convey a
1965 • ALPHATYPE

Clarendon Semibold
Typography's function is to convey a
1965 • ALPHATYPE

Clearface Bold
Typography's function is to convey a visual m
1977 • ALPHATYPE

Clearface Bold Italic
Typography's function is to convey a visual
1977 • ALPHATYPE

Clearface Extra Bold
Typography's function is to convey a v
1977 • ALPHATYPE

Clearface Extra Bold Italic
Typography's function is to convey a
1977 • ALPHATYPE

Clearface Regular ITC
Typography's function is to convey a visual m
1974 • VICTOR CARUSO

Clearface Regular Italic ITC
Typography's function is to convey a visual
1974 • VICTOR CARUSO

Clearface Bold ITC
Typography's function is to convey a visual m
1974 • VICTOR CARUSO

Clearface Bold Italic ITC
Typography's function is to convey a visual
1974 • VICTOR CARUSO

Clearface Heavy ITC
Typography's function is to convey a visua
1974 • VICTOR CARUSO

Clearface Heavy Italic ITC
Typography's function is to convey a vis
1974 • VICTOR CARUSO

Clearface Black ITC
Typography's function is to convey a vi
1974 • VICTOR CARUSO

Clearface Black Italic ITC
Typography's function is to convey a vi
1974 • VICTOR CARUSO

Clearface Outline ITC
Typography's function is to convey a visual
1979 • VICTOR CARUSO

Cloister Old Style
Typography's function is to convey a visual messag
1976 • ALPHATYPE

Cloister Old Style Italic
Typography's function is to convey a visual message q
1976 • ALPHATYPE

Cloister Bold
Typography's function is to convey a visual m
1976 • ALPHATYPE

Cloister Bold Italic
Typography's function is to convey a visual m
1976 • ALPHATYPE

Compacta Light TSI
Typography's function is to convey a visual message quickly and easily. The ty
1984 • FRED LAMBERT

Compacta TSI
Typography's function is to convey a visual message quickly and
1984 • FRED LAMBERT

Compacta Italic TSI
Typography's function is to convey a visual message quickly an
1984 • FRED LAMBERT

Compacta Bold TSI
Typography's function is to convey a visual message q
1984 • FRED LAMBERT

Compacta Bold Italic TSI
Typography's function is to convey a visual message q
1984 • FRED LAMBERT

117

Compacta Black TSI
Typography's function is to convey
1984 • FRED LAMBERT

Congress Regular FONTS
Typography's function is to convey a visual m
1982 • ADRIAN WILLIAMS

Congress Regular Italic FONTS
Typography's function is to convey a visual m
1982 • ADRIAN WILLIAMS

Congress Medium FONTS
Typography's function is to convey a visual m
1982 • ADRIAN WILLIAMS

Congress Bold FONTS
Typography's function is to convey a visual
1982 • ADRIAN WILLIAMS

Congress Heavy FONTS
Typography's function is to convey a visu
1982 • ADRIAN WILLIAMS

Contempo
Typography's function is to convey a visual
1967 • VLADIMIR ANDRICH • ALPHATYPE/BERTHOLD EXCLUSIVE

Contempo Bold
Typography's function is to convey a visual
1967 • VLADIMIR ANDRICH • ALPHATYPE/BERTHOLD EXCLUSIVE

COPPERPLATE GOTHIC
TYPOGRAPHY'S FUNCTION IS TO CONVEY A
1968 • ALPHATYPE

COPPERPLATE GOTHIC BOLD
TYPOGRAPHY'S FUNCTION IS TO CONVEY A
1968 • ALPHATYPE

COPPERPLATE GOTHIC CONDENSED
TYPOGRAPHY'S FUNCTION IS TO CONVEY A VISUAL
1968 • ALPHATYPE

COPPERPLATE GOTHIC BOLD CONDENSED
TYPOGRAPHY'S FUNCTION IS TO CONVEY A VISUAL
1968 • ALPHATYPE

Cooper Black HB
Typography's function is to con
1924 • OSWALD B. COOPER

Cooper Black Italic HB
Typography's function is to conve
1924 • OSWALD B. COOPER

Criterion Light TS
Typography's function is to convey a visual m
1982 • PHIL MARTIN

Criterion Light Italic TS
Typography's function is to convey a visual m
1982 • PHIL MARTIN

Criterion Book TS
Typography's function is to convey a visua
1982 • PHIL MARTIN

Criterion Book Italic TS
Typography's function is to convey a visua
1982 • PHIL MARTIN

Criterion Medium TS
Typography's function is to convey a visu
1982 • PHIL MARTIN

Criterion Bold TS
Typography's function is to convey a vis
1982 • PHIL MARTIN

CUSHING BOOK SMALL CAPS ITC
TYPOGRAPHY'S FUNCTION IS TO CONVEY A VISUAL
1982 • VINCENT PACELLA

CUSHING MEDIUM SMALL CAPS ITC
TYPOGRAPHY'S FUNCTION IS TO CONVEY A VISUA
1982 • VINCENT PACELLA

Cushing Book ITC
Typography's function is to convey a visual m
1982 • VINCENT PACELLA

Cushing Book Italic ITC
Typography's function is to convey a visual mes
1982 • VINCENT PACELLA

Cushing Medium ITC
Typography's function is to convey a visual
1982 • VINCENT PACELLA

Cushing Medium Italic ITC
Typography's function is to convey a visual mes
1982 • VINCENT PACELLA

Cushing Bold ITC
Typography's function is to convey a visua
1982 • VINCENT PACELLA

Cushing Bold Italic ITC
Typography's function is to convey a visual m
1982 • VINCENT PACELLA

Cushing Heavy ITC
Typography's function is to convey a vis
1982 • VINCENT PACELLA

Cushing Heavy Italic ITC
Typography's function is to convey a visua
1982 • VINCENT PACELLA

EDELWEISS SMALL CAPS
TYPOGRAPHY'S FUNCTION IS TO CONVEY A VISUAL
1974 • ALPHATYPE • SIMILAR TO WEISS ROMAN

Edelweiss
Typography's function is to convey a visual m
1974 • ALPHATYPE • SIMILAR TO WEISS ROMAN

Edelweiss Italic
Typography's function is to convey a visual message q
1974 • ALPHATYPE • SIMILAR TO WEISS ROMAN

Edelweiss Bold
Typography's function is to convey a visual
1974 • ALPHATYPE • SIMILAR TO WEISS ROMAN

Edelweiss Extra Bold
Typography's function is to convey a vis
1974 • ALPHATYPE • SIMILAR TO WEISS ROMAN

English
Typography's function is to convey a visua
1965 • ALPHATYPE • SIMILAR TO TIMES ROMAN

English Italic
Typography's function is to convey a visual
1965 • ALPHATYPE • SIMILAR TO TIMES ROMAN

English Bold
Typography's function is to convey a visua
1965 • ALPHATYPE • SIMILAR TO TIMES ROMAN

English Bold Italic
Typography's function is to convey a visua
1965 • ALPHATYPE • SIMILAR TO TIMES ROMAN

New English Bold
Typography's function is to convey a visual
1968 • ALPHATYPE • SIMILAR TO TIMES ROMAN

New English Bold Italic
Typography's function is to convey a visual m
1968 • ALPHATYPE • SIMILAR TO TIMES ROMAN

Eras™ Light ITC
Typography's function is to convey a visual m
1976 • BOTON/HOLLENSTEIN

Eras™ Book ITC
Typography's function is to convey a visual
1976 • BOTON/HOLLENSTEIN

Eras™ Medium ITC
Typography's function is to convey a visual
1976 • BOTON/HOLLENSTEIN

Eras™ Demibold ITC
Typography's function is to convey a vis
1976 • BOTON/HOLLENSTEIN

Eras™ Bold ITC
Typography's function is to convey
1976 • BOTON/HOLLENSTEIN

Eras™ Ultra ITC
Typography's function is to convey
1976 • BOTON/HOLLENSTEIN

Eurogothic
Typography's function is to convey a vis
1966 • ALPHATYPE • SIMILAR TO EUROSTILE

Eurogothic Bold
Typography's function is to convey a
1966 • ALPHATYPE • SIMILAR TO EUROSTILE

Eurogothic Condensed
Typography's function is to convey a visual message qui
1966 • ALPHATYPE • SIMILAR TO EUROSTILE

Eurogothic Bold Condensed
Typography's function is to convey a visual mes
1966 • ALPHATYPE • SIMILAR TO EUROSTILE

Eurogothic Extended
Typography's function is to con
1966 • ALPHATYPE • SIMILAR TO EUROSTILE

Eurogothic Bold Extended
Typography's function is to c
1966 • ALPHATYPE • SIMILAR TO EUROSTILE

FENICE LIGHT SMALL CAPS ITC
TYPOGRAPHY'S FUNCTION IS TO CONVEY A VISUA
1980 • ALDO NOVARESE

FENICE REGULAR SMALL CAPS ITC
TYPOGRAPHY'S FUNCTION IS TO CONVEY A VISU
1980 • ALDO NOVARESE

Fenice Light ITC
Typography's function is to convey a visual m
1980 • ALDO NOVARESE

Fenice Light Italic ITC
Typography's function is to convey a visual me
1980 • ALDO NOVARESE

Fenice Regular ITC
Typography's function is to convey a visua
1980 • ALDO NOVARESE

Fenice Regular Italic ITC
Typography's function is to convey a visua
1980 • ALDO NOVARESE

Fenice Bold ITC
Typography's function is to convey a vi
1980 • ALDO NOVARESE

Fenice Bold Italic ITC
Typography's function is to convey a vi
1980 • ALDO NOVARESE

Fenice Ultra ITC
Typography's function is to conve
1980 • ALDO NOVARESE

Fenice Ultra Italic ITC
Typography's function is to conve
1980 • ALDO NOVARESE

Folio Light HB
Typography's function is to convey a visual
1959 • K. F. BAUER, W. BAUM • FUNDICIÓN TIPOGRÁFICA NEUFVILLE

Folio Light Italic HB
Typography's function is to convey a visua
1959 • K. F. BAUER, W. BAUM • FUNDICIÓN TIPOGRÁFICA NEUFVILLE

Folio Book HB
Typography's function is to convey a visual
1965 • K. F. BAUER, W. BAUM • FUNDICIÓN TIPOGRÁFICA NEUFVILLE

Folio Medium HB
Typography's function is to convey a visua
1957 • K. F. BAUER, W. BAUM • FUNDICIÓN TIPOGRÁFICA NEUFVILLE

Folio Bold HB
Typography's function is to convey a
1959 • K. F. BAUER, W. BAUM • FUNDICIÓN TIPOGRÁFICA NEUFVILLE

Fotura Light AI
Typography's function is to convey a visual
1983 • PHIL MARTIN

Fotura Light Italic AI
Typography's function is to convey a visual
1983 • PHIL MARTIN

Fotura Medium AI
Typography's function is to convey a visu
1983 • PHIL MARTIN

Fotura Demibold AI
Typography's function is to convey a vi
1983 • PHIL MARTIN

Franklin-Antiqua HB
Typography's function is to convey a visual me
1976 • G. G. LANGE • BERTHOLD/ALPHATYPE EXCLUSIVE

Franklin Italic HB
Typography's function is to convey a visual me
1976 • G. G. LANGE • BERTHOLD/ALPHATYPE EXCLUSIVE

Franklin-Antiqua Medium HB
Typography's function is to convey a visual m
1976 • G. G. LANGE • BERTHOLD/ALPHATYPE EXCLUSIVE

Franklin Medium Italic HB
Typography's function is to convey a visual m
1976 • G. G. LANGE • BERTHOLD/ALPHATYPE EXCLUSIVE

Franklin-Antiqua Bold HB
Typography's function is to convey a visual
1976 • G. G. LANGE • BERTHOLD/ALPHATYPE EXCLUSIVE

Franklin Gothic
Typography's function is to convey a
1970 • ALPHATYPE

Franklin Gothic Italic
Typography's function is to convey a
1970 • ALPHATYPE

Franklin Gothic Condensed
Typography's function is to convey a visual me
1970 • ALPHATYPE

Franklin Gothic Condensed Italic
Typography's function is to convey a visual me
1970 • ALPHATYPE

Franklin Gothic Extra Condensed
Typography's function is to convey a visual message qu
1970 • ALPHATYPE

FRANKLIN GOTHIC BOOK SMALL CAPS ITC
TYPOGRAPHY'S FUNCTION IS TO CONVEY A VIS
1980 • VICTOR CARUSO

FRANKLIN GOTHIC MEDIUM SMALL CAPS ITC
TYPOGRAPHY'S FUNCTION IS TO CONVEY A VI
1980 • VICTOR CARUSO

Franklin Gothic Book ITC
Typography's function is to convey a visual
1980 • VICTOR CARUSO

Franklin Gothic Book Italic ITC
Typography's function is to convey a visual
1980 • VICTOR CARUSO

Franklin Gothic Medium ITC
Typography's function is to convey a visua
1980 • VICTOR CARUSO

Franklin Gothic Medium Italic ITC
Typography's function is to convey a visua
1980 • VICTOR CARUSO

Franklin Gothic Demibold ITC
Typography's function is to convey a visu
1980 • VICTOR CARUSO

Franklin Gothic Demibold Italic ITC
Typography's function is to convey a visu
1980 • VICTOR CARUSO

Franklin Gothic Heavy ITC
Typography's function is to convey a v
1980 • VICTOR CARUSO

Garamond Bold Italic
Typography's function is to convey a visual
1965 • ALPHATYPE

Franklin Gothic Heavy Italic ITC
Typography's function is to convey a v
1980 • VICTOR CARUSO

Garamond Light ITC
Typography's function is to convey a visual
1977 • TONY STAN

Frutiger 45 HB
Typography's function is to convey a visual mess
1976 • ADRIAN FRUTIGER

Garamond Light Italic ITC
Typography's function is to convey a visual
1977 • TONY STAN

Frutiger 46 HB
Typography's function is to convey a visual mess
1976 • ADRIAN FRUTIGER

Garamond Book ITC
Typography's function is to convey a visual
1977 • TONY STAN

Frutiger 55 HB
Typography's function is to convey a visual m
1976 • ADRIAN FRUTIGER

Garamond Book Italic ITC
Typography's function is to convey a visua
1977 • TONY STAN

Frutiger 56 HB
Typography's function is to convey a visual m
1976 • ADRIAN FRUTIGER

Garamond Bold ITC
Typography's function is to convey a vi
1977 • TONY STAN

Frutiger 65 HB
Typography's function is to convey a visu
1976 • ADRIAN FRUTIGER

Garamond Bold Italic ITC
Typography's function is to convey a vi
1977 • TONY STAN

Frutiger 66 HB
Typography's function is to convey a visua
1976 • ADRIAN FRUTIGER

Garamond Ultra ITC
Typography's function is to convey
1977 • TONY STAN

Frutiger 75 HB
Typography's function is to convey a vi
1976 • ADRIAN FRUTIGER

Garamond Ultra Italic ITC
Typography's function is to convey
1977 • TONY STAN

Frutiger 76 HB
Typography's function is to convey a v
1976 • ADRIAN FRUTIGER

Garamond Light Condensed ITC
Typography's function is to convey a visual message quic
1977 • TONY STAN

Futura Extra Bold HB
Typography's function is to convey
1930 • PAUL RENNER

Garamond Light Condensed Italic ITC
Typography's function is to convey a visual message qu
1977 • TONY STAN

Futura Extra Bold Condensed HB
Typography's function is to convey a visual m
1930 • PAUL RENNER

Garamond Book Condensed ITC
Typography's function is to convey a visual message
1977 • TONY STAN

Garamond
Typography's function is to convey a visual mes
1965 • ALPHATYPE

Garamond Book Condensed Italic ITC
Typography's function is to convey a visual messag
1977 • TONY STAN

Garamond Italic
Typography's function is to convey a visual mes
1965 • ALPHATYPE

Garamond Bold Condensed ITC
Typography's function is to convey a visual mes
1977 • TONY STAN

Garamond Bold
Typography's function is to convey a visual
1965 • ALPHATYPE

Garamond Bold Condensed Italic ITC
Typography's function is to convey a visual m
1977 • TONY STAN

Garamond Ultra Condensed ITC
Typography's function is to convey a visual mes
1977 • TONY STAN

Garamond Ultra Condensed Italic ITC
Typography's function is to convey a visual
1977 • TONY STAN

Glib Light
Typography's function is to convey a visual me
1974 • ALPHATYPE • SIMILAR TO GILL SANS

Glib Light Italic
Typography's function is to convey a visual messa
1974 • ALPHATYPE • SIMILAR TO GILL SANS

Glib Medium
Typography's function is to convey a visual m
1974 • ALPHATYPE • SIMILAR TO GILL SANS

Glib Medium Italic
Typography's function is to convey a visual messa
1974 • ALPHATYPE • SIMILAR TO GILL SANS

Glib Bold
Typography's function is to convey a visu
1974 • ALPHATYPE • SIMILAR TO GILL SANS

Glib Bold Italic
Typography's function is to convey a visual
1974 • ALPHATYPE • SIMILAR TO GILL SANS

Glib Extra Bold
Typography's function is to convey a v
1974 • ALPHATYPE • SIMILAR TO GILL SANS

Glib Ultra Bold
Typography's function is to conve
1974 • ALPHATYPE • SIMILAR TO GILL SANS

Goudy Old Style
Typography's function is to convey a visual m
1974 • ALPHATYPE

Goudy Old Style Italic
Typography's function is to convey a visùal mess
1974 • ALPHATYPE

Goudy Bold
Typography's function is to convey a visua
1968 • ALPHATYPE

Goudy Extra Bold
Typography's function is to convey a v
1968 • ALPHATYPE

Grotesque Light
Typography's function is to convey a vis
1967 • ALPHATYPE • SIMILAR TO GROTESQUE 216

Grotesque Light Italic
Typography's function is to convey a visu
1967 • ALPHATYPE • SIMILAR TO GROTESQUE 216

Heldustry TS
Typography's function is to convey a vi
1978 • PHIL MARTIN

Heldustry Italic TS
Typography's function is to convey a vis
1978 • PHIL MARTIN

Heldustry Medium TS
Typography's function is to convey a vi
1978 • PHIL MARTIN

Heldustry Medium Italic TS
Typography's function is to convey a vi
1978 • PHIL MARTIN

Heldustry Demibold TS
Typography's function is to convey a v
1978 • PHIL MARTIN

Heldustry Demibold Italic TS
Typography's function is to convey a vi
1978 • PHIL MARTIN

Helserif Light AI
Typography's function is to convey a vis
1983 • ED KELTON

Helserif Light Italic AI
Typography's function is to convey a vis
1983 • ED KELTON

Helserif Regular AI
Typography's function is to convey a v
1983 • ED KELTON

Helserif Medium AI
Typography's function is to convey a
1983 • ED KELTON

Helvetica Ultralight HB
Typography's function is to convey a visual
1970 • HAAS AG

Helvetica Light HB
Typography's function is to convey a visual
1965 • STEMPEL AG

Helvetica Light Italic HB
Typography's function is to convey a visual
1967 • STEMPEL AG

Helvetica HB
Typography's function is to convey a visu
1958 • MAX MIEDINGER • HAAS AG

Helvetica Italic HB
Typography's function is to convey a visual
1961 • MAX MIEDINGER • HAAS AG

Helvetica Medium HB
Typography's function is to convey a visu
1957 • MAX MIEDINGER • HAAS AG

Helvetica Medium Italic HB
Typography's function is to convey a visu
1957 • HAAS AG

Helvetica Bold HB
Typography's function is to convey a
1959 • HAAS AG

Helvetica Bold Italic HB
Typography's function is to convey a
1967 • HAAS AG

Helvetica Light Condensed HB
Typography's function is to convey a visual message
1963 • STEMPEL AG

Helvetica Medium Condensed HB
Typography's function is to convey a visual message quickly
1940 • HAAS AG

Helvetica Bold Condensed HB
Typography's function is to convey a visual m
1946 • HAAS AG

Helvetica Light Extended HB
Typography's function is to convey a
1964 • STEMPEL AG

Helvetica Medium Extended HB
Typography's function is to conv
1961 • STEMPEL AG

Helvetica Bold Extended HB
Typography's function is to
1961 • HAAS AG

Helvetica Inserat HB
Typography's function is to convey a visual mes
1969 • STEMPEL AG

Helvetica Packed
Typography's function is to convey a visual message qu
1983 • ALPHATYPE • SIMILAR TO HELVETICA COMPACT

Helvetica Pressed
Typography's function is to convey a visual messa
1983 • ALPHATYPE • SIMILAR TO HELVETICA COMPRESSED

Independence
Typography's function is to convey a
1983 • ALPHATYPE • SIMILAR TO AMERICANA

Independence Italic
Typography's function is to convey
1983 • ALPHATYPE • SIMILAR TO AMERICANA

Independence Bold
Typography's function is to conve
1983 • ALPHATYPE • SIMILAR TO AMERICANA

Independence Extrabold
Typography's function is to conve
1983 • ALPHATYPE • SIMILAR TO AMERICANA

Independence Black
Typography's function is to conv
1983 • ALPHATYPE • SIMILAR TO AMERICANA

Independence Bold Outline
Typography's function is to conv
1983 • ALPHATYPE • SIMILAR TO AMERICANA

Italia Book ITC
Typography's function is to convey a visual
1973 • COLIN BRIGNALL

Italia Medium ITC
Typography's function is to convey a visua
1973 • COLIN BRIGNALL

Italia Bold ITC
Typography's function is to convey a visu
1973 • COLIN BRIGNALL

Italian Script
Typography's function is to convey a visual message quickly and
1967 • ALPHATYPE • SIMILAR TO LORRAINE SCRIPT

Kabel Book ITC
Typography's function is to convey a visual m
1976 • PHOTO-LETTERING, INC.

Kabel Medium ITC
Typography's function is to convey a visual
1976 • PHOTO-LETTERING, INC.

Kabel Demibold ITC
Typography's function is to convey a visu
1976 • PHOTO-LETTERING, INC.

Kabel Bold ITC
Typography's function is to convey a visu
1976 • PHOTO-LETTERING, INC.

Kabel Ultra ITC
Typography's function is to convey a visu
1976 • PHOTO-LETTERING, INC.

Kaufman Bold
Typography's function is to convey a visual mess
1967 • ALPHATYPE

Kobel Light
Typography's function is to convey a visual mess
1975 • ALPHATYPE • SIMILAR TO KABEL

Kobel Light Italic
Typography's function is to convey a visual message
1975 • ALPHATYPE • SIMILAR TO KABEL

Kobel Demibold
Typography's function is to convey a visual messag
1975 • ALPHATYPE • SIMILAR TO KABEL

Kobel Bold
Typography's function is to convey a visua
1975 • ALPHATYPE • SIMILAR TO KABEL

Korinna ITC
Typography's function is to convey a visua
1974 • BENGUIAT/CARUSO

Korinna Kursiv ITC
Typography's function is to convey a visu
1977 • EDWARD BENGUIAT

Korinna Bold ITC
Typography's function is to convey a visu
1974 • BENGUIAT/CARUSO

Korinna Kursiv Bold ITC
Typography's function is to convey a vis
1977 • EDWARD BENGUIAT

Korinna Extra Bold ITC
Typography's function is to convey a v
1974 • BENGUIAT/CARUSO

Korinna Kursiv Extra Bold ITC
Typography's function is to convey a v
1977 • EDWARD BENGUIAT

Korinna Heavy ITC
Typography's function is to conve
1974 • BENGUIAT/CARUSO

Korinna Kursiv Heavy ITC
Typography's function is to convey
1977 • EDWARD BENGUIAT

Koronna
Typography's function is to convey a visua
1970 • VLADIMIR ANDRICH • ALPHATYPE/BERTHOLD EXCLUSIVE

Koronna Bold
Typography's function is to convey a visu
1970 • VLADIMIR ANDRICH • ALPHATYPE/BERTHOLD EXCLUSIVE

LEAWOOD™ BOOK SMALL CAPS ITC
TYPOGRAPHY'S FUNCTION IS TO CONVEY A VISUA
1985 • LES USHERWOOD

LEAWOOD™ MEDIUM SMALL CAPS ITC
TYPOGRAPHY'S FUNCTION IS TO CONVEY A VI
1985 • LES USHERWOOD

Leawood™ Book ITC
Typography's function is to convey a visual
1985 • LES USHERWOOD

Leawood™ Book Italic ITC
Typography's function is to convey a visual m
1985 • LES USHERWOOD

Leawood™ Medium ITC
Typography's function is to convey a visu
1985 • LES USHERWOOD

Leawood™ Medium Italic ITC
Typography's function is to convey a visual
1985 • LES USHERWOOD

Leawood™ Bold ITC
Typography's function is to convey a vi
1985 • LES USHERWOOD

Leawood™ Bold Italic ITC
Typography's function is to convey a vis
1985 • LES USHERWOOD

Leawood™ Black ITC
Typography's function is to convey a v
1985 • LES USHERWOOD

Leawood™ Black Italic ITC
Typography's function is to convey a vis
1985 • LES USHERWOOD

Libretto Light
typography's function is to convey a
1979 • GARY BRUNSELL • SIMILAR TO LIBRA

Libretto Bold
typography's function is to conve
1979 • GARY BRUNSELL • SIMILAR TO LIBRA

LIQUID CRYSTAL TSI
TYPOGRAPHY'S FUNCTION IS TO CONVEY A VISU
1974 • TSI

London Script
Typography's function is to convey a visual message quick
1967 • ALPHATYPE • SIMILAR TO EXCELSIOR

Lubalin Graph Extra Light ITC
Typography's function is to convey a v
1974 • HERB LUBALIN

Lubalin Graph Extra Light Oblique ITC
Typography's function is to convey a
1974 • HERB LUBALIN

Lubalin Graph Book ITC
Typography's function is to convey a vi
1974 • HERB LUBALIN

Lubalin Graph Book Oblique ITC
Typography's function is to convey a v
1974 • HERB LUBALIN

Lubalin Graph Medium ITC
Typography's function is to convey a
1974 • HERB LUBALIN

Lubalin Graph Medium Oblique ITC
Typography's function is to convey a
1974 • HERB LUBALIN

Lubalin Graph Demibold ITC
Typography's function is to convey a
1974 • HERB LUBALIN

Lubalin Graph Demibold Oblique ITC
Typography's function is to convey a
1974 • HERB LUBALIN

Lubalin Graph Bold ITC
Typography's function is to convey a
1974 • HERB LUBALIN

Lubalin Graph Bold Oblique ITC
Typography's function is to convey a
1974 • HERB LUBALIN

MAGNA CARTA SMALL CAPS
TYPOGRAPHY'S FUNCTION IS TO CONVEY A VIS
1974 • VLADIMIR ANDRICH • ALPHATYPE/BERTHOLD EXCLUSIVE

Magna Carta
Typography's function is to convey a visual me
1974 • VLADIMIR ANDRICH • ALPHATYPE/BERTHOLD EXCLUSIVE

Magna Carta Italic
Typography's function is to convey a visual messa
1974 • VLADIMIR ANDRICH • ALPHATYPE/BERTHOLD EXCLUSIVE

Magna Carta Demibold
Typography's function is to convey a visua
1974 • VLADIMIR ANDRICH • ALPHATYPE/BERTHOLD EXCLUSIVE

Magna Carta Bold
Typography's function is to convey a visua
1974 • VLADIMIR ANDRICH • ALPHATYPE/BERTHOLD EXCLUSIVE

Martin Gothic Light AI
Typography's function is to convey a visual m
1979 • PHIL MARTIN

Martin Gothic Light Italic AI
Typography's function is to convey a visual
1979 • PHIL MARTIN

Martin Gothic Medium AI
Typography's function is to convey a visu
1979 • PHIL MARTIN

Martin Gothic Medium Italic AI
Typography's function is to convey a vis
1979 • PHIL MARTIN

Martin Gothic Bold AI
Typography's function is to convey a vis
1979 • PHIL MARTIN

Martin Gothic Bold Italic AI
Typography's function is to convey a vis
1979 • PHIL MARTIN

MODERN NO. 216 LIGHT SMALL CAPS ITC
TYPOGRAPHY'S FUNCTION IS TO CONVEY A VIS
1983 • EDWARD BENGUIAT

MODERN NO. 216 MEDIUM SMALL CAPS ITC
TYPOGRAPHY'S FUNCTION IS TO CONVEY A VI
1983 • EDWARD BENGUIAT

Modern No. 216 Light ITC
Typography's function is to convey a visu
1983 • EDWARD BENGUIAT

Modern No. 216 Light Italic ITC
Typography's function is to convey a visu
1983 • EDWARD BENGUIAT

Modern No. 216 Medium ITC
Typography's function is to convey a vis
1983 • EDWARD BENGUIAT

Modern No. 216 Medium Italic ITC
Typography's function is to convey a vis
1983 • EDWARD BENGUIAT

Modern No. 216 Bold ITC
Typography's function is to convey a vi
1983 • EDWARD BENGUIAT

Modern No. 216 Bold Italic ITC
Typography's function is to convey a v
1983 • EDWARD BENGUIAT

Modern No. 216 Heavy ITC
Typography's function is to convey
1983 • EDWARD BENGUIAT

Modern No. 216 Heavy Italic ITC
Typography's function is to conve
1983 • EDWARD BENGUIAT

Monty Extra Light
Typography's function is to convey a vis
1984 • ALPHATYPE • SIMILAR TO SERIFA

Monty Extra Light Italic
Typography's function is to convey a vis
1984 • ALPHATYPE • SIMILAR TO SERIFA

Monty Light
Typography's function is to convey a vis
1984 • ALPHATYPE • SIMILAR TO SERIFA

Monty Light Italic
Typography's function is to convey a vi
1984 • ALPHATYPE • SIMILAR TO SERIFA

Monty Medium
Typography's function is to convey a
1984 • ALPHATYPE • SIMILAR TO SERIFA

Monty Medium Italic
Typography's function is to convey a
1984 • ALPHATYPE • SIMILAR TO SERIFA

Monty Bold
Typography's function is to convey
1984 • ALPHATYPE • SIMILAR TO SERIFA

Monty Extra Bold
Typography's function is to conv
1984 • ALPHATYPE • SIMILAR TO SERIFA

Musica
Typography's function is to convey a visual
1964 • ALPHATYPE • SIMILAR TO OPTIMA

Musica Italic
Typography's function is to convey a visual
1964 • ALPHATYPE • SIMILAR TO OPTIMA

Musica Semibold
Typography's function is to convey a visual
1964 • ALPHATYPE • SIMILAR TO OPTIMA

News Gothic
Typography's function is to convey a visu
1966 • ALPHATYPE

News Gothic Italic
Typography's function is to convey a visu
1966 • ALPHATYPE

News Gothic Bold
Typography's function is to convey a visu
1966 • ALPHATYPE

News Gothic Condensed
Typography's function is to convey a visual message q
1966 • ALPHATYPE

News Gothic Bold Condensed
Typography's function is to convey a visual message
1966 • ALPHATYPE

Newtext Book ITC
Typography's function is to co
1974 • RAY BAKER

Newtext Book Italic ITC
Typography's function is to con
1974 • RAY BAKER

Newtext Demibold ITC
Typography's function is to co
1974 • RAY BAKER

Newtext Demibold Italic ITC
Typography's function is to con
1974 • RAY BAKER

Novarese Book ITC
Typography's function is to convey a visua
1980 • ALDO NOVARESE

Novarese Book Italic ITC
Typography's function is to convey a visual mess
1980 • ALDO NOVARESE

Novarese Medium ITC
Typography's function is to convey a visua
1980 • ALDO NOVARESE

Novarese Medium Italic ITC
Typography's function is to convey a visual messa
1980 • ALDO NOVARESE

Novarese Bold ITC
Typography's function is to convey a v
1980 • ALDO NOVARESE

Novarese Bold Italic ITC
Typography's function is to convey a visu
1980 • ALDO NOVARESE

Novarese Ultra ITC
Typography's function is to conve
1980 • ALDO NOVARESE

Optima HB
Typography's function is to convey a visual m
1958 • HERMANN ZAPF • STEMPEL AG

Optima Italic HB
Typography's function is to convey a visual me
1958 • HERMANN ZAPF • STEMPEL AG

Optima Medium HB
Typography's function is to convey a visual
1969 • HERMANN ZAPF • STEMPEL AG

Optima Medium Italic HB
Typography's function is to convey a visual
1969 • HERMANN ZAPF • STEMPEL AG

Optima Bold HB

Typography's function is to convey a visual

1958 • HERMANN ZAPF • STEMPEL AG

Patina

Typography's function is to convey a visual

1965 • ALPHATYPE • SIMILAR TO PALATINO

Patina Italic

Typography's function is to convey a visual m

1965 • ALPHATYPE • SIMILAR TO PALATINO

Patina Semibold

Typography's function is to convey a visual

1965 • ALPHATYPE • SIMILAR TO PALATINO

Quorum™ Light ITC

Typography's function is to convey a visual me

1977 • RAY BAKER

Quorum™ Book ITC

Typography's function is to convey a visual me

1977 • RAY BAKER

Quorum™ Medium ITC

Typography's function is to convey a visual

1977 • RAY BAKER

Quorum™ Bold ITC

Typography's function is to convey a visua

1977 • RAY BAKER

Quorum™ Black ITC

Typography's function is to convey a visu

1977 • RAY BAKER

Rockwell Light HB

Typography's function is to convey a vis

1934 • MONOTYPE • SIMILAR TO STYMIE

Rockwell Light Italic HB

Typography's function is to convey a vis

1934 • MONOTYPE • SIMILAR TO STYMIE

Rockwell HB

Typography's function is to convey a vis

1934 • MONOTYPE • SIMILAR TO STYMIE

Rockwell Italic HB

Typography's function is to convey a vi

1934 • MONOTYPE • SIMILAR TO STYMIE

Rockwell Bold HB

Typography's function is to convey a

1934 • MONOTYPE • SIMILAR TO STYMIE

Rockwell Bold Italic HB

Typography's function is to convey a vi

1934 • MONOTYPE • SIMILAR TO STYMIE

Rockwell Extra Bold HB

Typography's function is to con

1934 • MONOTYPE • SIMILAR TO STYMIE

Rockwell Condensed HB

Typography's function is to convey a visual message quickly an

1934 • MONOTYPE • SIMILAR TO STYMIE

Rockwell Bold Condensed HB

Typography's function is to convey a visual mess

1934 • MONOTYPE • SIMILAR TO STYMIE

Romic Light TSI

Typography's function is to convey a vis

1979 • COLIN BRIGNALL

Romic Light Italic TSI

Typography's function is to convey a vi

1979 • COLIN BRIGNALL

Romic Medium TSI

Typography's function is to convey a vis

1979 • COLIN BRIGNALL

Romic Bold TSI

Typography's function is to convey a vi

1979 • COLIN BRIGNALL

Romic Extra Bold TSI

Typography's function is to convey a

1979 • COLIN BRIGNALL

Scenario Light AI

Typography's function is to convey a visu

1983 • GEORGE BRIAN

Scenario Light Italic AI

Typography's function is to convey a visu

1983 • GEORGE BRIAN

Scenario Demibold AI

Typography's function is to convey a v

1983 • GEORGE BRIAN

Scenario Bold AI

Typography's function is to convey

1983 • GEORGE BRIAN

Serif Gothic™ Light ITC

Typography's function is to convey a visua

1974 • LUBALIN/DISPIGNA

Serif Gothic™ ITC

Typography's function is to convey a visu

1974 • LUBALIN/DISPIGNA

Serif Gothic™ Bold ITC

Typography's function is to convey a visua

1974 • LUBALIN/DISPIGNA

Serif Gothic™ Extra Bold ITC
Typography's function is to convey a visu
1974 • LUBALIN/DISPIGNA

Serif Gothic™ Heavy ITC
Typography's function is to convey a vi
1974 • LUBALIN/DISPIGNA

Serif Gothic™ Black ITC
Typography's function is to convey a vi
1974 • LUBALIN/DISPIGNA

Slenderella
Typography's function is to convey a visual me
1970 • VLADIMIR ANDRICH • ALPHATYPE/BERTHOLD EXCLUSIVE

Slenderella Refined
Typography's function is to convey a visual mes
1970 • VLADIMIR ANDRICH • ALPHATYPE/BERTHOLD EXCLUSIVE

Souvenir™ Light ITC
Typography's function is to convey a visual
1970 • EDWARD BENGUIAT

Souvenir™ Light Italic ITC
Typography's function is to convey a visual
1970 • EDWARD BENGUIAT

Souvenir™ Medium ITC
Typography's function is to convey a visu
1970 • EDWARD BENGUIAT

Souvenir™ Medium Italic ITC
Typography's function is to convey a visu
1970 • EDWARD BENGUIAT

Souvenir™ Demibold ITC
Typography's function is to convey a vi
1970 • EDWARD BENGUIAT

Souvenir™ Demibold Italic ITC
Typography's function is to convey a vi
1970 • EDWARD BENGUIAT

Souvenir™ Bold ITC
Typography's function is to convey
1970 • EDWARD BENGUIAT

Souvenir™ Bold Italic ITC
Typography's function is to convey
1970 • EDWARD BENGUIAT

Souvenir Gothic TS
Typography's function is to convey a visual mess
1977 • GEORGE BRIAN

Souvenir Gothic Italic TS
Typography's function is to convey a visual mess
1977 • GEORGE BRIAN

Souvenir Gothic Medium TS
Typography's function is to convey a visual
1977 • GEORGE BRIAN

Souvenir Gothic Medium Italic TS
Typography's function is to convey a visual
1977 • GEORGE BRIAN

Souvenir Gothic Demibold TS
Typography's function is to convey a visu
1977 • GEORGE BRIAN

Souvenir Gothic Demibold Italic TS
Typography's function is to convey a visu
1977 • GEORGE BRIAN

Stymie Light
Typography's function is to convey a visual
1970 • ALPHATYPE • SIMILAR TO ROCKWELL

Stymie Light Italic
Typography's function is to convey a visual
1970 • ALPHATYPE • SIMILAR TO ROCKWELL

Stymie Medium
Typography's function is to convey a vis
1970 • ALPHATYPE • SIMILAR TO ROCKWELL

Stymie Medium Italic
Typography's function is to convey a vis
1970 • ALPHATYPE • SIMILAR TO ROCKWELL

Stymie Bold
Typography's function is to convey a v
1970 • ALPHATYPE • SIMILAR TO ROCKWELL

Stymie Bold Italic
Typography's function is to convey a vi
1970 • ALPHATYPE • SIMILAR TO ROCKWELL

Stymie Extra Bold
Typography's function is to convey
1970 • ALPHATYPE • SIMILAR TO ROCKWELL

Stymie Extra Bold Italic
Typography's function is to convey
1970 • ALPHATYPE • SIMILAR TO ROCKWELL

SYMBOL™ BOOK SMALL CAPS ITC
TYPOGRAPHY'S FUNCTION IS TO CONVEY A VISUA
1984 • ALDO NOVARESE

SYMBOL™ MEDIUM SMALL CAPS ITC
TYPOGRAPHY'S FUNCTION IS TO CONVEY A VIS
1984 • ALDO NOVARESE

Symbol™ Book ITC
Typography's function is to convey a visual
1984 • ALDO NOVARESE

Symbol™ Book Italic ITC
Typography's function is to convey a visual
1984 • ALDO NOVARESE

Symbol™ Medium ITC
Typography's function is to convey a visu
1984 • ALDO NOVARESE

Symbol™ Medium Italic ITC
Typography's function is to convey a visu
1984 • ALDO NOVARESE

Symbol™ Bold ITC
Typography's function is to convey a vi
1984 • ALDO NOVARESE

Symbol™ Bold Italic ITC
Typography's function is to convey a vi
1984 • ALDO NOVARESE

Symbol™ Black ITC
Typography's function is to convey
1984 • ALDO NOVARESE

Symbol™ Black Italic ITC
Typography's function is to convey
1984 • ALDO NOVARESE

Tiffany Light ITC
Typography's function is to convey a vi
1974 • EDWARD BENGUIAT

Tiffany Light Italic ITC
Typography's function is to convey a visu
1974 • EDWARD BENGUIAT

Tiffany Medium ITC
Typography's function is to convey a vi
1974 • EDWARD BENGUIAT

Tiffany Medium Italic ITC
Typography's function is to convey a visu
1974 • EDWARD BENGUIAT

Tiffany Demibold ITC
Typography's function is to convey a vi
1974 • EDWARD BENGUIAT

Tiffany Demibold Italic ITC
Typography's function is to convey a visu
1974 • EDWARD BENGUIAT

Tiffany Heavy ITC
Typography's function is to conve
1974 • EDWARD BENGUIAT

Tiffany Heavy Italic ITC
Typography's function is to conv
1974 • EDWARD BENGUIAT

Typewriter
Typography's function is to conve
1965 • ALPHATYPE

Univers 45 HB
Typography's function is to convey a visu
1957 • ADRIAN FRUTIGER • SIMILAR TO VERSATILE

Univers 46 HB
Typography's function is to convey a vis
1957 • ADRIAN FRUTIGER • SIMILAR TO VERSATILE

Univers 55 HB
Typography's function is to convey a vis
1957 • ADRIAN FRUTIGER • SIMILAR TO VERSATILE

Univers 56 HB
Typography's function is to convey a visu
1957 • ADRIAN FRUTIGER • SIMILAR TO VERSATILE

Univers 65 HB
Typography's function is to convey a
1957 • ADRIAN FRUTIGER • SIMILAR TO VERSATILE

Univers 66 HB
Typography's function is to convey a
1957 • ADRIAN FRUTIGER • SIMILAR TO VERSATILE

Univers 75 HB
Typography's function is to conve
1957 • ADRIAN FRUTIGER • SIMILAR TO VERSATILE

Univers 76 HB
Typography's function is to convey
1957 • ADRIAN FRUTIGER • SIMILAR TO VERSATILE

Univers 85 HB
Typography's function is to conve
1957 • ADRIAN FRUTIGER • SIMILAR TO VERSATILE

Univers 53 HB
Typography's function is to conv
1957 • ADRIAN FRUTIGER • SIMILAR TO VERSATILE

Univers 63 HB
Typography's function is to con
1957 • ADRIAN FRUTIGER • SIMILAR TO VERSATILE

Univers 73 HB
Typography's function is to co
1957 • ADRIAN FRUTIGER • SIMILAR TO VERSATILE

Univers 83 HB
Typography's function is to c
1957 • ADRIAN FRUTIGER • SIMILAR TO VERSATILE

University Light TSI
Typography's function is to convey a visual message quic
1982

University Bold TSI
Typography's function is to convey a visual message q
1982

Versatile 45
Typography's function is to convey a vis
1965 • ALPHATYPE • SIMILAR TO UNIVERS

Versatile 46
Typography's function is to convey a vis
1965 • ALPHATYPE • SIMILAR TO UNIVERS

Versatile 55
Typography's function is to convey a vis
1965 • ALPHATYPE • SIMILAR TO UNIVERS

Versatile 56
Typography's function is to convey a vis
1965 • ALPHATYPE • SIMILAR TO UNIVERS

Versatile 65
Typography's function is to convey a
1965 • ALPHATYPE • SIMILAR TO UNIVERS

Versatile 66
Typography's function is to convey a
1965 • ALPHATYPE • SIMILAR TO UNIVERS

Versatile 75
Typography's function is to conve
1965 • ALPHATYPE • SIMILAR TO UNIVERS

Vladimir Script
Typography's function is to convey a visual message quickly an
1982 • VLADIMIR ANDRICH • ALPHATYPE/BERTHOLD EXCLUSIVE

WEIDEMANN™ BOOK SMALL CAPS ITC
TYPOGRAPHY'S FUNCTION IS TO CONVEY A VISUAL
1983 • KURT WEIDEMANN

WEIDEMANN™ MEDIUM SMALL CAPS ITC
TYPOGRAPHY'S FUNCTION IS TO CONVEY A VISUA
1983 • KURT WEIDEMANN

Winslow Light
Typography's function is to convey a visual
1970 • ALPHATYPE • SIMILAR TO WINDSOR

Winslow Bold
Typography's function is to convey a
1970 • ALPHATYPE • SIMILAR TO WINDSOR

Zapf Book Light ITC
Typography's function is to convey a visu
1976 • HERMANN ZAPF

Zapf Book Light Italic ITC
Typography's function is to convey a visual
1976 • HERMANN ZAPF

Zapf Book Medium ITC
Typography's function is to convey a visu
1976 • HERMANN ZAPF

Zapf Book Medium Italic ITC
Typography's function is to convey a visua
1976 • HERMANN ZAPF

Zapf Book Demibold ITC
Typography's function is to convey a vi
1976 • HERMANN ZAPF

Zapf Book Demibold Italic ITC
Typography's function is to convey a vis
1976 • HERMANN ZAPF

Zapf Book Heavy ITC
Typography's function is to convey a v
1976 • HERMANN ZAPF

Zapf Book Heavy Italic ITC
Typography's function is to convey a vi
1976 • HERMANN ZAPF

Zapf Chancery Light ITC
Typography's function is to convey a visual messag
1979 • HERMANN ZAPF

Zapf Chancery Light Swash and Alternates ITC
Typography's function is to convey a visual messag
1979 • HERMANN ZAPF

Zapf Chancery Light Italic ITC
Typography's function is to convey a visual message q
1979 • HERMANN ZAPF

Zapf Chancery Light Italic Swash and Alternates ITC
Typography's function is to convey a visual message
1979 • HERMANN ZAPF

Zapf Chancery Medium ITC
Typography's function is to convey a visual messag
1979 • HERMANN ZAPF

Zapf Chancery Medium Swash and Alternates ITC
Typography's function is to convey a visual messa
1979 • HERMANN ZAPF

Zapf Chancery Medium Italic ITC
Typography's function is to convey a visual messag
1979 • HERMANN ZAPF

Zapf Chancery Medium Italic Swash and Alts ITC
Typography's function is to convey a visual messag
1979 • HERMANN ZAPF

Zapf Chancery Demi ITC
Typography's function is to convey a visual me
1979 • HERMANN ZAPF

Chapter 7

Design and Layout

When you have completed the reading and assigned activities related to this chapter, you will be able to:

O Identify, describe, and draw the general features of the design elements, such as: proportion, balance, contrast, rhythm, and unity.

O Describe the normal sequence of events in the preparation of a complete set of layouts.

O Draw thumbnail sketches, rough layouts, and comprehensive layouts when given copy requirements for a printing job.

O Describe the purpose of a mechanical and how it relates to a paste-up.

O Describe the importance of properly utilizing two-dimensional space.

O Recognize the several common page proportions used in design and layout.

O Describe and demonstrate the use of a proportion scale.

O Illustrate formal and informal balance.

O Describe the purpose of making up a layout and list the specifications usually included on the copy or layout.

O Describe the process of cropping and list the two methods used to indicate crop size.

O Crop artwork using the two methods described in this chapter.

O Figure the percentage of reduction or enlargement of a piece of art using the two methods described in this chapter.

O Distinguish among the several pigment colors known as primary, secondary, and intermediate.

O Design at least two multi-color printing product utilizing analogous and complementary harmonies.

O Apply the "psychology of color" in designing a two-dimensional multi-color record album jacket cover.

O Utilize and apply the basic design principles of page proportion, balance, contrast and unity.

O Establish page margins for a single-page and for two facing-pages in a book.

O Prepare a dummy (layout) for a booklet of eight or more pages.

Whether you are using a pen and T-square or a state of the art computer to do drawings and layout, you must understand the principles of planning, design, and paste-up. By knowing how to take an idea and develop it into an artistic, functional printed image, you will be better prepared to become a productive part of industry and society. This chapter will help you learn many rules for good design and layout.

DESIGN

Once the idea to produce a printed product has been formed, the first step is the planning, design, and preparation of a layout, Fig. 7-1.

The term *design* refers to the manner in which type and illustrations are selected and arranged on a printed page. The placing of the idea on paper refers to the preparations of a *layout*. These tasks are generally handled manually by a *graphic designer* or layout artist.

The graphic designer deals with a variety of visual

Fig. 7-1. Design refers to use of type and illustrations on a printed page. This advertisement demonstrates designer's skillful use of all design elements. The advertisement is both artistic and functional.

elements such as written copy, illustrations, and photographs. The design idea must be accomplished within a given amount of layout space. For example, a business card has a small area in which to project an image compared to that of a letterhead.

In the case of a business card design, the first printed impression a company or its representative makes on a potential client or customer is with a business card. That may be when the card is used as an introduction by a person making a business call, or when the card accompanies a letter. Therefore, the business card design should reflect two considerations. First, the card should provide continuity with the company's visual image in general, along with its letterhead and envelope. Second, the business card must be durable and easy to carry. Because the card is liable to be handled frequently, it should be printed on a heavy weight paper.

Fig. 7-2 illustrates variations in business cards and letterheads.

Principles of design

Good design is both a logical (purposeful) and *aesthetic* (artistic) function. The purpose of the printed piece will help determine its design. Good design is based on easily understood and identified principles.

The *principles of good design* include proportion, balance, harmony, contrast, rhythm, unity, symmetry,

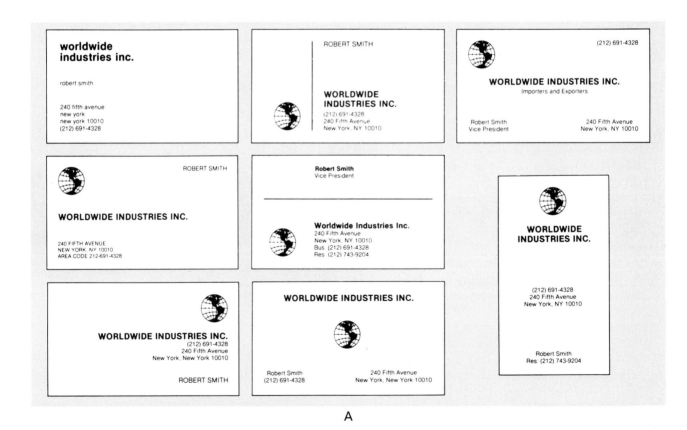

A

Fig. 7-2. A—Business cards are important form of graphic communication. These examples are all variations on the same theme. B—Designer must visually express an idea. The idea then becomes typeset material and is developed into final product. These letterheads contain unique expressions and constitute the designer's ideas about particular business or organization. (Regency Thermographers, Inc.)

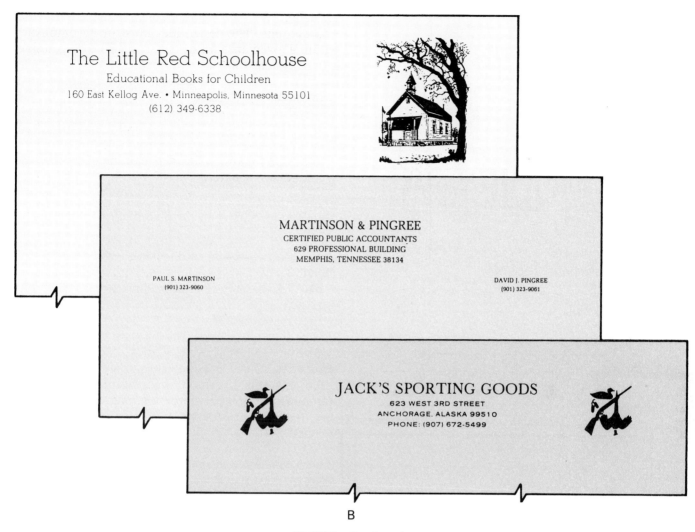

The Little Red Schoolhouse
Educational Books for Children
160 East Kellog Ave. • Minneapolis, Minnesota 55101
(612) 349-6338

MARTINSON & PINGREE
CERTIFIED PUBLIC ACCOUNTANTS
629 PROFESSIONAL BUILDING
MEMPHIS, TENNESSEE 38134

PAUL S. MARTINSON
(901) 323-9060

DAVID J. PINGREE
(901) 323-9061

JACK'S SPORTING GOODS
623 WEST 3RD STREET
ANCHORAGE, ALASKA 99510
PHONE: (907) 672-5499

B

Fig. 7-2. Continued.

variety, and action. A knowledge of these principles and how they interact will make any design task easier and more successful.

Proportion

In design, *proportion* refers to the pleasing relationship between the elements on the printed page. Proportion applies also to the general dimensions of the page itself. For example, a page can be too long and narrow to be pleasing. It can also be too square to be interesting. In pleasing page shapes, the dimensions are in a ratio of about 2 to 3, Fig. 7-3. Page sizes of 6 × 9 and 8 × 12 inches have this ratio.

When a single block or line of type is positioned in the exact CENTER of the page, the type will visually appear LOWER. This is an optical illusion, Fig. 7-4. Since the areas above and below the type are equal, the page will lack interest. Moving the type to the optical center of the page will make the type appear balanced on the page.

Optical center is a position approximately three units from the bottom and two units from the top of a page. Refer to Fig. 7-5.

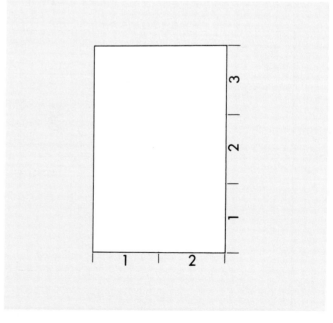

Fig. 7-3. Designers know that pleasing page shapes are essential to good communication. Effective design shapes are generally in a ratio of 2 to 3.

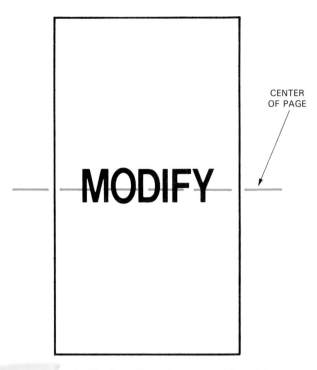

Fig. 7-4. A single block or line of type positioned in exact center of a page tends to appear lower, as illustrated. This visual phenomenon is an optical illusion.

Balance

A design is in *balance* when the different elements appear to be equalized. There are two kinds of balance: formal and informal.

In *formal balance,* the elements on the page are centered horizontally, Fig. 7-6. An equal amount of each major unit is positioned on either side of an imaginary centerline. This kind of balance suggests strength, order, and dignity.

Informal balance allows elements to be placed at different positions on the page, Fig. 7-7. This style is more modern than formal balance. Since it is less rigid, it allows more variety in the placement of the elements on the page.

Harmony

The placement of elements so that they do NOT clash with one another, or with the theme of the message, is called *harmony*. See Fig. 7-8. The typefaces, the format of the typeset copy, the ink, and the paper colors contribute to harmony. If a printed piece has proper harmony, its elements will appear organized.

For example, the traditional placement of the logo, company name, and address on a letterhead is in the

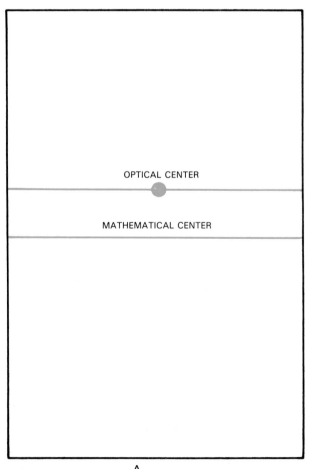

A

B

Fig. 7-5. A—Optical center is a position about three units from the bottom and two units from the top of a page. B—This advertisement illustrates where optical center is properly located.   (Dynamic Graphics, Inc.)

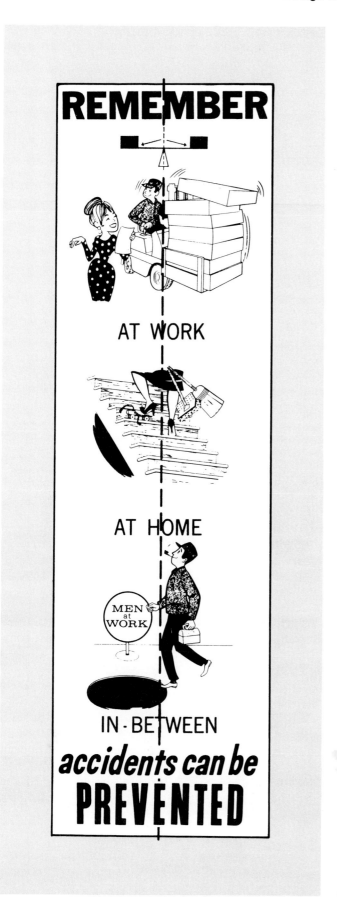

Fig. 7-6. This layout illustrates principle of formal balance. Various elements are placed symmetrically within design area. (Dynamic Graphics, Inc.)

Fig. 7-7. Informal balance is achieved by changing value, size, and location of unequal elements. Examples above contain variety and have a modernistic appeal.

Fig. 7-8. Harmony involves arrangement of image elements so that they flow smoothly and do not clash.

upper left-hand corner of the sheet. However, many design artists are placing that information elsewhere on the page, Fig. 7-9. Such placement can, in and of itself, suggest a non-traditional approach or a fresh look.

The important thing to remember about harmony is to let the design itself suggest the placement. A logo that suggests movement might look very different on the left or right side of the page. A strongly symmetrical design might look out of place anywhere but in the top center of the sheet.

Contrast

Contrast is achieved by using different sizes and weights (light and bold, etc.) of type and art. In design, contrast adds interest and emphasis to the printed page. Other ways of producing contrast include the use of a second ink color, underlining type, and shading backgrounds. Look at Fig. 7-10.

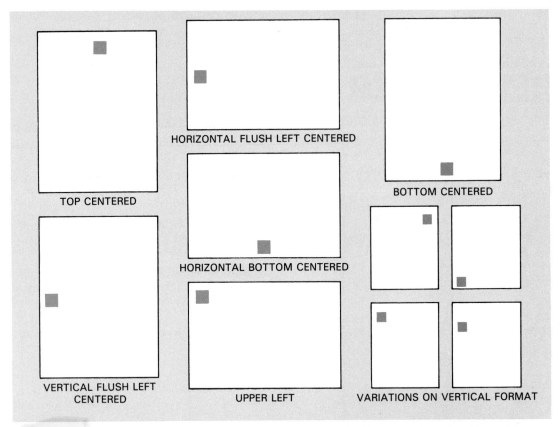

Fig. 7-9. Placement of a company logo, company name, and address on a letterhead is a critical factor for good communication. These examples suggest placement in locations other than traditional upper left-hand corner of sheet.

Fig. 7-10. Contrast can be achieved by using different sizes and weights of type. Different ink colors and shaded backgrounds also can add contrast. (Compugraphic Corp.)

Rhythm

Repetition of similar elements in a design creates *rhythm*, Fig. 7-11. Rhythm refers to the flow or movement of the individual elements in a design. Rhythm also can be achieved by selecting only one element in a design and repeating it several times. This element might be an attention spot or *focal point* such as a dot or snowflake. The use of definite shapes in text composition is another way of achieving rhythm. Rhythm contributes to the attractiveness of a design.

Unity

A design that leads the reader's eye easily from one element to another has *unity*. See Fig. 7-12. When all the type and illustrations in a design complement each other, unity is achieved. It is important that a design NOT appear crowded. Neither should any of the elements appear unnecessary. Each of the elements should support the others.

Symmetry

If a design contains parts or elements that are similar in size, shape, and location it is said to be *symmetrical*. An example is given in Fig. 7-13. When a design is symmetrical, the positions of the type and artwork are balanced on the page. Symmetry is similar to the principle of *formal balance*.

Variety

The design principle known as variety is the opposite of symmetry, Fig. 7-14. Designs having variety usually are NOT balanced.

Variety is used to raise the interest level of a design and to suggest action. Creating variety in a printed piece is generally more difficult than the design of a symmetrical layout.

Action

Most advertising uses the design principle known as action, Fig. 7-15. *Action* is achieved by placing type and artwork off-center and in unsymmetrical groupings.

For example, advertising pieces with action give the illusion of motion and activity. Some formal advertising may be quite pleasing to the eye. However, it usually contains no action. Some examples of ways in which a message can be designed to suggest motion and draw attention are shown in Fig. 7-16.

Color

Color contributes visual impact and interest to the printed page. Color lends a lifelike quality to photographic reproductions. Even printed pages with NO illustrations are more attractive if a second color is used along with black.

Color can be used on the printed page to create an overall mood. Sometimes, graphic designers use a second color for a single line of display type. Such a technique draws attention to the display line. The uses of color are varied depending on the product being designed.

Some colors are referred to as being either warm or

Fig. 7-11. Repeating one image in a design several times can contribute to rhythm. This could be an ornament, a trademark, or a certain word.

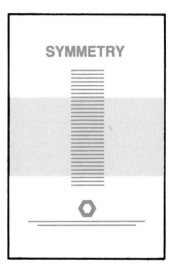

Fig. 7-13. A symmetrical design has type and illustrations occupying balanced positions. Symmetry is similar to principle of formal balance.

Fig. 7-12. A design is said to have unity if type and illustrations fit well together. Images should not appear to be crowded, but should have the look of belonging together.

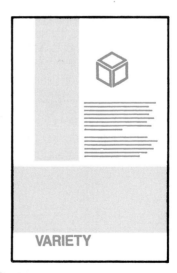

Fig. 7-14. Variety in design is opposite of symmetry. Variety raises interest level of the design. It adds a spirit of life, action, and freedom to the layout.

cool. Examples of *warm colors* include red, orange, and yellow. These colors give the feeling of warmth and closeness. *Cool colors* include violet, blue, and green. Violet suggests darkness; blue gives the impression of water; and green usually makes a person think of lush grass.

Certain combinations of color do NOT appeal to every reader's eye. For example, yellow, purple, and brown are NOT pleasing when used in combination. For this reason, the graphic designer must know how to select colors that are pleasing when used together. This practice is referred to as the *psychology of color*. Humans are influenced by different colors.

When printing on colored paper, the effect of the ink is pleasing if its color is a shade lighter or darker

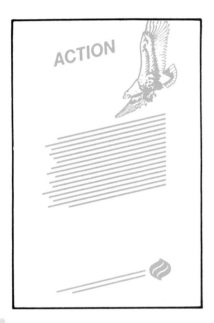

Fig. 7-15. Action plays an important role in modern advertising design. One way to create a feeling of motion and activity is through the use of illustrations containing action.

Fig. 7-16. Action can be achieved by designing out-of-center balances and unsymmetrical groupings of image elements. Feeling of motion and activity is essential to good design involving action.

than the color of the paper. For example, the use of dark blue ink on pale blue paper will result in a harmonious effect. Another example might be deep green images printed on pale green paper. This use of ink is called *monochromatic printing*. Monochromatic means of ONE COLOR.

Harmony can be achieved on white paper by using dark type, with the illustrations, borders, and ornaments in a lighter tone. Of all the elements, the type should normally be the darkest in tone. The tone of the other elements could be one-half as dark as the type. In this example, the paper has the lightest tone.

Monochromatic or one-color printing does NOT always consist of black images on white paper, although this is most common. A one-color printing job can be printed in red ink on tan paper. Obviously, the possible combinations are almost endless.

The use of more than one color refers to *multicolor*. Multicolor printing is used to increase reader interest in the printed products.

Color can highlight certain elements of a printed product. For example, in this textbook, color is used for the section and title page headings. This helps you identify major topics to be studied. The headings printed in color stand out from the black body type itself. A graphic designer is usually responsible for choosing the color combinations.

Process color printing reproduces an image in its natural colors. To do this, three basic *process inks* are used for such printing. *Process colors* are yellow, magenta, and cyan. A colored photograph, painting, or film transparency can be separated into its three component primary colors. This is done to allow the tones of each single primary color to be recorded as a separate halftone negative on monochromatic film. After these negatives are made, the remainder of the printing process is much like reproducing black-and-white halftones. However, colored inks are used and a much greater degree of quality control is maintained.

Carefully controlled tones of each of the three process colors are printed from separate plates in perfect *registration* (alignment). The result is a composite reproduction of the original color copy.

While these three primary colors theoretically can reproduce the full range of colors, usually a fourth plate, printed in BLACK INK, is also used. Using black accomplishes the following:

1. It increases the density range of the reproduction.
2. It improves *shadow* (darker areas) detail.
3. It makes control of the other three colors less critical in the area of ink balance.

In process color printing, *color separation negatives* are used to prepare printing plates for the offset press. This is a complex process. Knowledge of the process is NOT required for the graphic designer. However, it is important for the designer to know the basic steps involved. This is important so that the best colors can be chosen.

Note! Information about the color separation process is provided in Chapter 18.

PRINCIPLES OF LAYOUT

Before copy can be typeset for a printing job, there must be a clear *layout* or "blueprint" of its proposed format and style. This layout must show proportions, widths of lines, choice of typefaces, illustrations, and arrangement of the type. The graphic designer generally prepares layouts in three stages: thumbnail sketches, rough layout, and comprehensive layout.

Thumbnail sketches

The first step in designing a job is to make *thumbnail sketches* or crude hand renderings of layouts. Thumbnail sketches usually measure about 2 by 3 inches, Fig. 7-17. Though small, these have the shape and proportions of the piece that is to be printed. Several of these sketches are prepared in pencil. The best of them is then used to prepare a rough layout.

Rough layout

The most promising thumbnail sketch is drawn in pencil as a *rough layout*. See Fig. 7-18. The rough layout is usually the same size as the proposed job.

Space for type and the position of drawings and photographs are shown on the rough layout. Since the rough layout is drawn in pencil, changes are easily and quickly made.

Comprehensive layout

The *comprehensive layout* is a detailed drawing of how the final printed product will look, Fig. 7-19. A comprehensive layout is also referred to as a *"comp."* It shows the type styles, type sizes, drawings (illustrations), and position of photographs.

Fig. 7-18. Rough layouts are same size as printed job, showing space for type and illustrations. Since rough layout is drawn in pencil, changes are easy to make.

Fig. 7-17. Thumbnail sketches are small pencil drawings in same shape and proportion as job being designed. Designer selects most promising of these sketches for further development.

Comprehensive layouts are frequently drawn using colored pencils or felt pens. The use of color gives a better rendering of the printed piece.

The comprehensive layout is the final planning step. It is followed by typesetting and the preparation of the camera-ready mechanical or paste-up.

A MECHANICAL is the same as a PASTE-UP, Fig. 7-20. Mechanicals and paste-ups should NOT be confused with the comprehensive layout. Type is set and illustrations and photographs are prepared so that the elements can be pasted down to form a *mechanical*. It should be noted that mechanicals are also referred to as *camera-ready copy*.

Computerized makeup systems

Many graphic designers are now using computerized design, makeup, or layout systems for all of the steps described to this point.

A basic computerized page makeup system consists of the following:

1. SCANNER (electronic device for converting visual images into electronic or computer data).
2. DIGITIZING TABLET (electronic type drawing board for tracing existing images or drawing original images with electronic pen).
3. KEYBOARD (computer keypad for inputting information into computer).
4. COMPUTER (electronic means of processing, storing, and using electronic data from scanner, digitizing tablet, and computer keyboard).
5. SOFTWARE (programming disks for giving computer electronic information on how to operate).
6. VIDEO MONITOR (television type picture tube for viewing images being processed by computer).
7. OUTPUT DEVICE (electronic mechanism for outputting images: laser printer, phototypesetter, or platemaker).

Computerized page makeup systems vary in design and can be classified as low end or high end.

A *low end system* typically includes a personal com-

Fig. 7-19. Comprehensive layouts are detailed drawings showing how the printed product will look. Colors, if any, are included using colored pencils, felt pens, or tempera paints.

Fig. 7-20. Paste-up or mechanical is result of text, line drawings, and photographs being prepared for final assembly. Paste-up follows design format of comprehensive layout.

puter running on a pagemaking type program. The *PC* (personal computer) can use a laser printer for medium quality output or a phototypesetter for high quality output. A black and white system is fairly inexpensive and very common. However, a color system is costly and requires considerable more investment.

Often called *desktop publishing systems,* they can be used to produce line drawings, typeset copy, and do page layout. The line drawing and type output quality from these systems is dependent upon the output device (laser printer or phototypesetter).

Most systems will not scan and electronically paste down photographs without loss of image quality. Most desktop scanners only have low resolution of 300 to 400 dots per inch which is too coarse for many print jobs.

A *high end system* uses a more powerful "mainframe" type computer linked to a *high resolution* (over 1000 to 2000 dots per inch) four-color scanner and four-color output device. Look at Fig. 7-21. Color makeup systems or electronic pre-press systems attempt to perform all the steps from the original copy to the press plate in one network of integrated devices. These systems input text, black-and-white and color illustrations, and page layout and plate imposition instructions. They perform the functions of color correction, type composition, page makeup with all elements in correct position, *soft proofing* (no hardcopy on paper) on VDT (video display terminal) screens, and signature imposition of all pages. Design makeup systems output the digital information with high resolution electronic laser or CRT (cathode ray tube) dot generators onto film or high speed plates. Refer to Fig. 7-22.

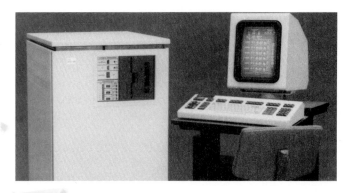

Fig. 7-22. Sophisticated color page makeup systems or electronic pre-press systems input text, black-and-white and color line art, and photographs to produce completed pages ready for platemaking. These systems perform color correction, type composition, and page makeup with all elements in correct position. (High Technology Solutions, Inc.)

Typical systems in use are produced by Scitex, Rudolph Hell, Crossfield, and Dainippon Screen. All systems are designed so that most functions can be done *off-line* (other parts of network still running), thus eliminating bottlenecks and enhancing flexibility. Some systems have input stations for graphic designers to input page layouts. Other systems interface with text, line illustrations, and scanners.

These high end systems are expensive and more productive than most printers need or can afford. Most of these systems are being installed in trade shops which makes the sophisticated services available to many printers.

Note! Electronic publishing systems are discussed in detail in Chapter 10.

Layout dummy

A *layout dummy* is an example set of page layouts, Fig. 7-23. A layout dummy is needed if the printed piece

Fig. 7-21. This computerized page makeup system can be used by a designer to input text, line art, and photographs.

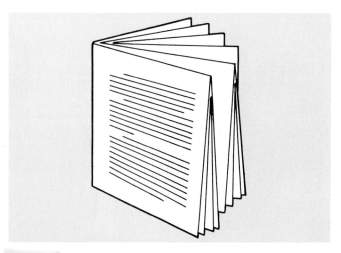

Fig. 7-23. A layout dummy is a visual preview of final printed product. It shows all details of job so that workers can gain a better understanding of what job will entail and how it goes together.

has more than one page. The dummy is a visual preview of the final printed product. It shows drawings, display and body type, and all page numbers. Page numbers in books and magazines are called *folios*.

A layout dummy is very useful since it shows clearly how the job fits together. It also contains much of the technical information and instructions needed to print the piece.

Most magazines and books are printed in sections called signatures. A *signature* consists of folded printed sheets in sequences of 4, 8, 16, 32, or 64 pages.

The master layout dummy provides production people with a quick reference for all critical printing requirements. The master dummy also indicates where color printing is to be used. Determining the number of signatures, printing sequence, and positioning of pages, refers to *printing impositions*. The subject of printing impositions is covered in detail later in this chapter.

ARTWORK

There are three kinds of original copy referred to as *artwork;* they are: line copy, continuous tone copy, and combination copy.

Line copy consists of solid lines on a white or contrasting background. It may include pen-and-ink drawings, diagrams, and type composition. See Fig. 7-24. Line copy is either black or white. It has no tone (image) variations except those made by the artist's pen. Look closely at the line drawing in Fig. 7-25. Notice how the different line widths provide tonal variations.

Continuous tone copy contains graduations (variations) of tone (image darkness). These vary from light to dark and are referred to as *intermediate tones of gray*. This is done by changing the amount of density.

A standard black and white photograph is the best example of continuous tone copy.

Look at Fig. 7-26. Even though this is no longer a continuous tone photo, notice the varying tones used to make the image. A magnifying glass will help you see the tiny dots. Notice that the dots vary in size but are all solid black (or ink color used to print).

Actual photos or continuous tones do NOT contain these dots. However, the printed reproductions are made to contain dots. This occurs because the printing press can only print solid ink image areas. Therefore, photographs are printed using the halftone process.

Chapter 16 will discuss halftones in detail. Besides photographs, other kinds of continuous tone copy include: wash drawings, pencil or charcoal drawings, and airbrush renderings, Fig. 7-27.

Fig. 7-24. Line copy consists of pen-and-ink drawings, pencil drawings, diagrams, type composition, and similar images. There are no tone (shade) variations. (NASA)

Fig. 7-25. Line drawing tonal variations are produced by different line widths drawn by artist. This example of a pen-and-ink line drawing has no tone variations except for those made by the artist's pen.

Combination copy is just what it suggests; it contains line copy and continuous tone copy. This is sometimes done to create an interesting effect, Fig. 7-28. Separate process camera operations are generally required to prepare combination copy offset negatives. A single photographic negative (or positive) can then be made for platemaking.

Fig. 7-26. A black and white photograph is an example of continuous tone copy. Printer converts continuous tone copy into halftones which contain dots. This is necessary because printing press can only print solid ink image areas.

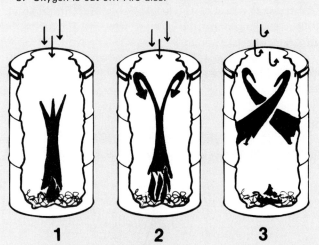

HOW "CEASE-FIRE" EXTINGUISHES FIRE
1. Fire starts and combustion vapors rise.
2. Baffle reverses flow.
3. Oxygen is cut off. Fire dies.

1   2   3

Fig. 7-27. Besides photographs, other kinds of continuous tone copy include: wash drawings (shown earlier), pencil or charcoal drawings, and airbrush renderings.

Fig. 7-28. Combination copy contains both line and continuous tone copy elements. This kind of copy arrangement creates interest. Separate camera operations must be performed to combine line and continuous tone copy. (Justrite Manufacturing Co.)

Handling artwork

Photographs and line art should be protected against dirt and fingerprints. The face of the illustration should be covered with a vellum paper overlay. The paper is fastened to the back of the illustration. The reproduction size (as a percentage of original size) should be written on the paper BEFORE it is attached to the artwork.

Writing on paper placed on top of an illustration may indent and leave the impression of the writing on the face of the photo or drawing. This might then be picked up by the camera. The vellum paper also allows space for other notations and instructions. Marks or indentations on the face of the artwork are NOT acceptable. They will reproduce when the illustration is photographed.

Paper clips should NEVER be used to attach notations to artwork, Instead, use draft tape or rubber cement. For example, a camera department work order is often rubber-cemented to the back of the artwork. See Fig. 7-29.

The desired reproduction size is usually hand printed on the vellum overlay or in the border of the photo.

Any other needed instructions are noted on the work order.

Cropping artwork

It is sometimes necessary or desirable to reproduce only part of an original photograph or drawing. Identifying that part of the illustration that is to be used is called *cropping*. Refer to Fig. 7-30.

Creative cropping of photographs allows the graphic designer to use photos that might not be usable uncropped. For example, one part of a photo might be blurred. If the blurred part is cropped out, the photo becomes usable. Cropping can also focus the reader's attention on one area of the photograph or to "blow up" (enlarge) one area of an image. Generally, artwork is cropped to fit the available space in the layout.

To prepare a photograph or drawing for cropping, you may have to mount it on white *illustration board* with rubber cement. This keeps the artwork rigid and smooth for camera operations.

Cropping marks are then placed in the margins, NOT in the image areas. These marks, normally made with ink, show only the area to be reproduced, Fig. 7-31.

If the margins cannot be marked, a sheet of vellum paper should be fastened to the back of the artwork and then folded over the front. Marks on the vellum

KEY _____ PAGE _____

PERCENTAGE _____ LINE SHOT ☐

DROPOUT ☐   DUOTONE ☐

SQ. HALFTONE ☐   Heavy Color ☐
                    Heavy Black ☐
SMALLER POS. ☐

Shoot Overlay ☐   TRI TONE ☐
For Color
                    SCREEN

FLOP ☐   110 ☐

POSTERIZE ☐   133 ☐

    Step 1 2 3 4   150 ☐
         ☐ ☐ ☐ ☐
                         175 ☐
       5 6 7 8 9
       ☐ ☐ ☐ ☐ ☐

EXPOSURE _____ FLASH _____
In Seconds   In Seconds

Fig. 7-29 Continuous tone copy and other types of illustrations must be handled carefully during the production phases. The form shown is used to give camera instructions. Form is pasted or rubber cemented to back of artwork.

Fig. 7-30. Cropping involves indicating what portion of an illustration is to be used or shown in final printing. In illustration, note that creative cropping allows for a number of layout options.

Fig. 7-31. A continuous tone photograph must be prepared for cropping. Crop marks are placed in white margin of print with a grease pencil or marker. Crop marks are placed on border to show area of image that will be illustrated.   (Minolta Corp.)

paper can then outline the part of the photo to be reproduced.

Sizing illustrations

Original drawings and photos can be enlarged or reduced. The operation of preparing illustrations for enlargement or reduction is known as *sizing*. Sizing also is called *scaling*. Enlargements tend to magnify defects in the original. Reductions usually reduce defects and sharpen the detail. For most purposes, artwork and photographs are reduced, rather than enlarged.

Enlargements and reductions are proportional. This means that a photograph or drawing will have its same proportions when reduced or enlarged. For example, a 3 × 5 inch photograph reduced 50 percent will then measure 1 1/2 × 2 1/2 inch. A reduction will reduce the width, as well as the depth. An enlargement will increase both the width and the depth. Fig. 7-32 shows examples.

When sizing a photograph or drawing, one of its two dimensions is called the *holding dimension* because it is the most important to the layout. For example, if a photograph is being reduced to fit a column that is 15 picas wide, the holding dimension is the width of the photograph, Fig. 7-33. Assume, however, that the depth of the column is the critical factor. Then, the depth becomes the holding dimension.

Several methods are used for calculating enlargements and reductions for artwork. These methods are not generally used for typeset copy. Phototypesetting machines are capable of producing the desired size type for both body and display type.

Proportion scale method

One way of calculating enlargements and reductions is the proportion scale method. A *proportion scale* is

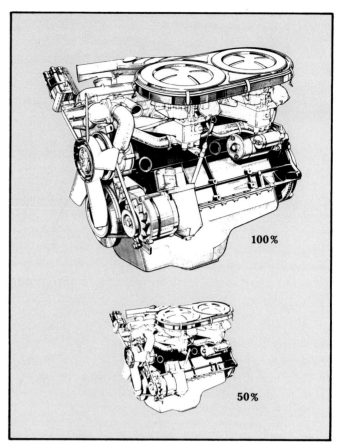

Fig. 7-32. When an image is reduced or enlarged, it will normally retain its same shape. Illustration above shows a 50 percent reduction of original artwork.   (Agfa-Gavaert,Inc.)

Fig. 7-33. A holding dimension on a photograph or illustration is dictated by critical dimension such as a pre-set column width or depth.   (Graphic Products Corp.)

a simple size calculator made up of a wheel that revolves on a circle. See Fig. 7-34.

The measurements on the smaller inside wheel, represent the dimensions of the original art. The measurements on the outside wheel refer to the dimensions of the art after scaling. This outside scale is known as the *reproduction scale.*

To size the art, line up the holding dimension on the inside scale with the desired size on the reproduction scale. The percentage of original size appears in the window opening. Assume that a photo is eight inches wide and ten inches deep. It must be scaled to fit a space six inches wide. The holding dimension is, then, the width of the photo, or eight inches.

Fig. 7-34. Proportion scale is commonly used to size images for layout. Scale will help calculate percent of reduction or enlargement of photographs and all types of illustrations.
(ANSCO Graphic Arts Products)

The proportional scale is used in the following manner:

1. Locate the original width of the artwork on the INNER wheel scale. As shown in Fig. 7-34, this would be eight (8) in our example.
2. Turn the wheel and match the original width with the desired new width, 6". This is found on the outer scale as six (6) in our example, Fig. 7-35.
3. Read the percentage figure under the arrow in the window opening of the scale. See Fig. 7-34 again.
4. To find the new depth of this artwork, locate ten inches (original depth) on the original-size scale. Read the new depth, which lines up opposite on the reproduction-size scale. This is 7 1/2 inches, Fig. 7-34.

Fig. 7-35. In this example, holding dimension's original size is 8 inches (A). It will be reduced to 6 inches. The 75 percent reproduction size (C) will reduce the remaining dimension from 10 inches to 7 1/2 inches (B).
(ANSCO Graphic Arts Products)

Diagonal line method

The diagonal of a rectangle or square can be used to see the final results of making an enlargement or reduction, Fig. 7-36. This is called the *diagonal line method.* Enlargements and reductions are proportional. This means that if a photograph or drawing is narrow in width and long in depth it will have these same proportions when it is reduced.

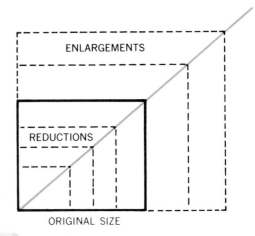

Fig. 7-36. Proportions of a reduction or enlargement remain same. Illustration shows how a diagonal line can be used to see final results of making a reduction or enlargement.

A reduction will reduce the width and the depth, as shown in Fig. 7-37. An enlargement will increase the width and the depth.

Fig. 7-38 illustrates how a diagonal line can be used to see the final results of making an enlargement or reduction. Follow the steps below when using the diagonal line method.

1. Determine the space allowed for the photograph or artwork on the comprehensive or rough layout.

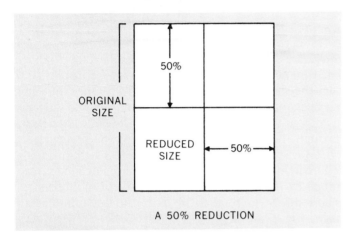

A 50% REDUCTION

Fig. 7-37. This example of a piece of copy shows how a reduction will reduce width and will also reduce depth proportionally.

REQUIRED SIZE: 4" WIDE

Fig. 7-38. Note diagonal line method to reduce an illustration. Diagonal line should always be lightly drawn on a tissue overlay over the illustration. Pencil mark indentations on face of photograph or illustration should be avoided since they can show up in final printed job. (A.B. Dick Co.)

2. Place a sheet of vellum paper over the photograph and hinge it to the back with tape.
3. Place a piece of transparent acetate film between the photo and the overlay sheet. This will protect the photo from damage.
4. Draw lines on the vellum sheet to show the image area of the photo to be reduced. This may be the entire photo or a cropped section. Use a T-square and triangle. The corners must be square.
5. Draw a diagonal line from the lower left corner to the top right corner. Use a straight edge.
6. On the baseline, measure and mark the desired "new width" on the photo.
7. Draw a vertical line from the baseline to the diagonal line. This shows the reduced photo width. This line must be exactly 90 degrees to the baseline.
8. Draw a horizontal line at the point where the new width, vertical line crosses the diagonal line. This line will then show the "new height." Use a T-square to make the line parallel with the baseline.

The same method can be used to enlarge a photo or artwork. Just extend the baseline to the right beyond the first photo width. Do the same to the diagonal line. Determine the new width or height. Draw the correct vertical and horizontal lines. They must intersect with the extended diagonal line.

The percentage of enlargement or reduction can also be found by mathematical calculations. To calculate the percentage, divide the desired reproduction size by the size of the original copy. The length of the sides can be stated in inches, millimeters, or picas. These calculations are shown in Fig. 7-39.

The width and depth of rectangular areas enlarge or reduce in direct proportion to each other. If the original copy was 3 inches wide and 6 inches deep, and

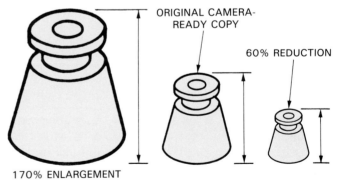

ORIGINAL CAMERA-READY COPY

60% REDUCTION

170% ENLARGEMENT

ENLARGEMENT CALCULATIONS

$$\frac{\text{LENGTH OF ENLARGED SIDE}}{\text{LENGTH OF SAME SIDE ON ORIGINAL}} = \frac{17}{10} = \frac{170\% \text{ OF}}{\text{ORIGINAL SIZE}}$$

REDUCTION CALCULATIONS

$$\frac{\text{LENGTH OF REDUCED SIDE}}{\text{LENGTH OF SAME SIDE ON ORIGINAL}} = \frac{6}{10} = \frac{60\% \text{ OF}}{\text{ORIGINAL SIZE}}$$

Fig. 7-39. Mathematical calculations can be used to determine a reduction or enlargement. To find percentage, divide desired reproduction size by size of original copy. Study examples and calculations.

the 3 inch width had to be reduced to 2 inches, how deep is the reduced copy? The following statement of proportion is used.

$$\frac{\text{Original width}}{\text{Desired width}} = \frac{\text{Original depth}}{\text{Desired depth}}$$

$$\frac{3''}{2''} = \frac{6''}{\text{depth}}$$

$$3 \times \text{Depth} = 12$$
$$\text{Depth} = 4''$$

PRINTING IMPOSITIONS

The arrangement of pages so that they will be in the proper sequence after the sheet is printed and folded is known as *imposition*. The matter of spacing and margins must be considered by the graphic designer in the case of magazines, books, catalogs, and programs. Magazines, books, catalogs, and programs are printed in units of several pages on a large sheet. A full sheet of paper is generally printed in units of 4, 8, 16, 32, or 64 pages. The task of determining printing impositions rests first with the graphic designer and later with the stripping department.

Mentioned briefly, folded printed sheets are called signatures, Fig. 7-40. This is a term used by the bindery and layout departments. A *signature* is a sheet that is printed on two sides and folded in a sequence of 4, 8, 16, 32, or 64 pages. Any number of these signatures when bound together will form a magazine, book, booklet, catalog, or program.

The textbook you are now reading is a good example of how signatures are bound together to form a complete volume. Look at the spine of this book and note how the signatures are bound together.

Imposition follows the same principle whether for offset, letterpress, or gravure printing. In offset lithography and gravure, a *stripper* arranges the pages in sequence. The page numbers in books and magazines are called *folios*. The stripper works on a light table. The film negatives or positives are arranged and taped on a sheet of masking paper, Fig. 7-41. Once the film has been taped in place, the stripper cuts window openings from the masking sheet. This allows light to pass through the film during exposure to the plate.

Certain fundamentals are more or less standard procedure in imposition. After the page size has been determined by the design artist, 1/8 inch to 1/4 inch extra is added for trim on the top, side, and bottom of each page, Fig. 7-42. *Trim* takes into consideration the amount of paper removed from the three outside edges after the signature is folded to make them even.

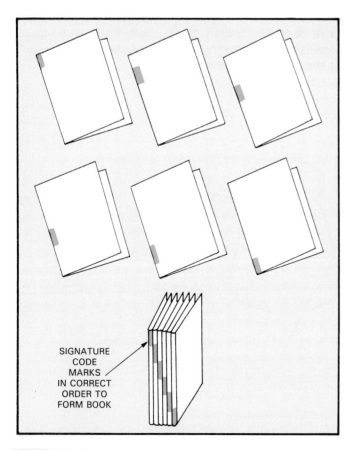

SIGNATURE CODE MARKS IN CORRECT ORDER TO FORM BOOK

Fig. 7-40. A signature is a large sheet that is printed on two sides and folded to make 4, 8, 16, 32, or 64 pages. Note how signatures are marked so that they form a book, booklet, or magazine.

Fig. 7-41. For purposes of imposition, film negatives or positives are arranged and taped on a sheet of masking paper. Careful attention to detail is necessary since a sequence error will result in all pages being out of order.

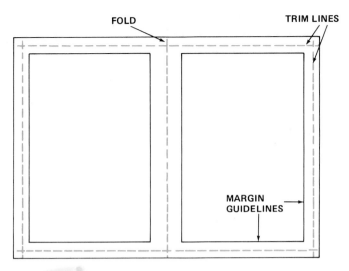

Fig. 7-42. Trim lines are indicated on layout. Trim takes into consideration amount of paper cut off from three outside edges after signature is folded to make them even.

In addition to trim allowance, provision must be made for creep. This allowance is made when planning a thick saddle-stitched magazine or book. *Creep* occurs when the signature is folded and the inside pages extend past the outside pages, Fig. 7-43. This means that the inside center pages will trim out smaller in size because they extend out farther than the outside signature pages.

Fig. 7-43. Creep often occurs in books and magazines when signatures are folded and inside pages extend past outside pages. Creep makes outside edges uneven. To overcome this tendency, provision is made for creep allowance when job is in layout stage.

Two-page imposition

A *two-page imposition* can be used for printing a single sheet on both sides. In this example, a double-sized sheet is printed on one side and then turned and printed with the same pages on the other side. The sheet is then cut in half, each full sheet making two complete copies. Two pages can be used to print one side of a four-page folder, the other side being printed with another two pages.

When a sheet is printed on both sides with the same pages and then cut in half, it is called a *work-and-turn job*. When it is printed on one side with one set of pages and then printed on the other side with a different set of pages, it is called a *sheetwise job*.

When four pages are printed at a time, work-and-turn, two different impositions are possible, Fig. 7-44. This is the smallest signature possible. On one side of the sheet are pages one and four; on the other side of the sheet are pages two and three.

Try this! Fold a sheet of paper once. Begin with the outside facing page and number them one through four. Open the sheet and lay it flat. This is how the pages are imposed to print in the correct sequence.

Eight-page imposition

Ordinary eight-page impositions are generally run either work-and-turn or sheetwise, Fig. 7-45. In addition to the space for margins, additional allowance must be made for trimming the pages after the signature is folded. An eight-page imposition is folded at the top or head and at the binding edge or spine. See Fig. 7-45.

Fold a sheet of paper twice. Number the pages consecutively one through eight. Open the sheet and lay

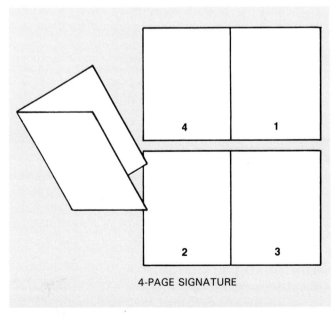

Fig. 7-44. Study layout for a four-page, work-and-turn imposition. One side of sheet contains pages one and four. Pages two and three are on other side of sheet.

149

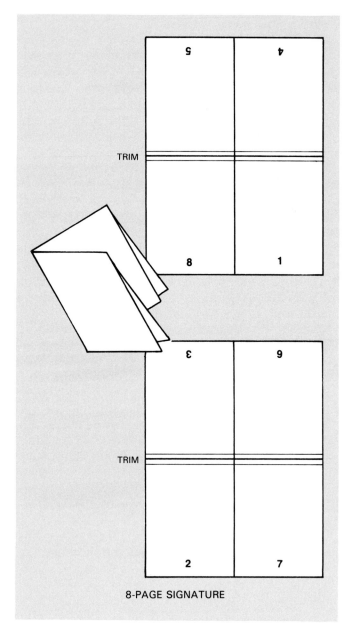

TRIM

TRIM

8-PAGE SIGNATURE

Fig. 7-45. Note layout for an eight-page, work-and-turn imposition. Trim allowance is made for trimming pages after signature is printed and folded.

it flat. This is the sequence for an eight-page signature. The numbers are NOT in order, and some of the pages are upside down.

A 16-page signature is usually printed work-and-turn, Fig. 7-46. This can be visualized by folding a sheet of paper three times and numbering the pages consecutively one through 16. The same procedure can be followed to make 32-page and 64-page signatures.

Determining imposition

It should be noted that imposition is determined at an early stage in the design and layout stage. The printer is in the best situation to determine imposition on the basis of the equipment to be used and the individual job requirements.

A safe procedure for checking imposition of pages is to fold and mark a sheet of paper. This is also one of the purposes of preparing a layout dummy. If more than one signature is involved, a rectangular "door" should be cut and numbered, as in Fig. 7-47. The top of the right reading number establishes the top (head) direction of each page. The top of a folded signature is called the *head*. The open layout clearly illustrates the correct page position. Look at Fig. 7-48.

To extablish the correct trim areas of the folded signature, a rectangular-shaped notch is cut on the top, right, and bottom sides of the layout, Fig. 7-49. The folded signature is then opened up flat and the trim lines are ruled in with a pen, Fig. 7-50. The folded signature is then trimmed along the trim lines, Fig. 7-51.

DESIGN AND LAYOUT CHECKLIST

When you begin the process of designing a printed piece, consider the following points:

1. Balance can be obtained through control of size, tone, and position of elements.
2. Balance can be upset by nonharmonizing typefaces and too many nonessential elements in a layout.
3. A unifying force should hold the layout together. White space, borders, and consistency in shape, size, and tone of elements can unify a layout.
4. Equal margins are monotonous. There should be more margin outside a rule than inside, but the rule should not crowd the type within.
5. Contrast adds interest to a layout. It can be achieved by varying widths of copy blocks, enlarging one in a group of pictures; and/or using italics or boldface sparingly.
6. Orderly repetition of some elements can provide motion to a layout.
7. Long horizontal or vertical elements will cause the reader to follow their direction.
8. The space within a layout should be broken up into pleasing proportions.
9. Simplicity is important for attractive layouts. It can be achieved by using a few type styles and reducing the number of shapes and sizes of artwork.
10. Use cool colors as the background for black type because they recede and do not detract from the type as much as hot colors.
11. When using colors, one color should dominate and the other should be used for accent or contrast.
12. Remember that color combinations can add to or detract from the legibility of the type. Black or yellow has high legibility; black on red is extremely difficult to read.
13. Use red sparingly, but remember that red can give a lift to the printed piece. Red is also a good background for white type, if the type is large.
14. Blue is an excellent background color. It is good

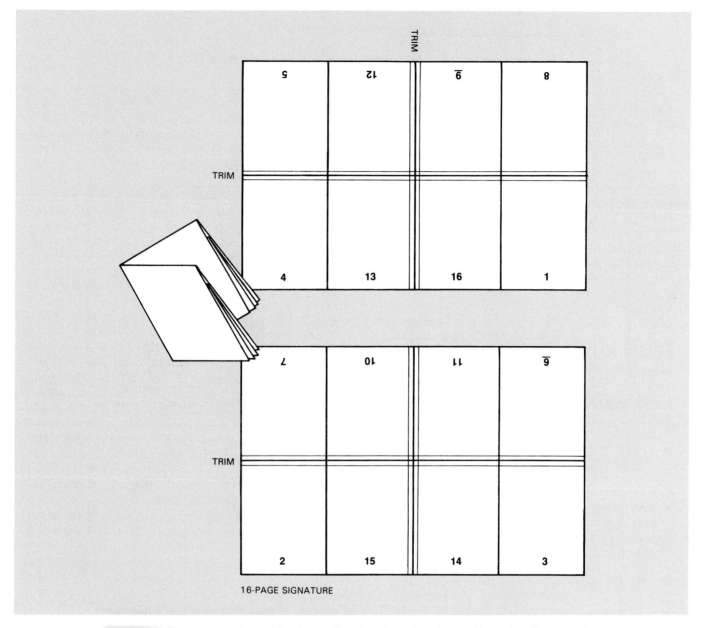

Fig. 7-46. Sixteen-page imposition is usually printed work-and-turn. Note trim allowance for head and sides

for tints behind black type and it is good for reverses if used full strength.

15. Brown prints well and has good legibility. It produces good tints.

16. In most instances, body copy should be printed in black ink or a tint or shade of black when printed on colored paper.

Although there are principles of design that can be helpful guides, the effective use of size, shape, and tone depend more on a sense of correctness and good taste than strict adherence to arbitrary rules. This sense is developed by experience and practice, and by carefully examining all the various printed pieces with which we come in daily contact. Your increased awareness of the principles of design and layout in printed pieces will gradually add to your own skill in this subject area.

Fig. 7-47. When more than one signature is involved in job, a rectangular "door" can be cut and numbered in layout. Top of the right-reading number establishes top (head) direction of each page.

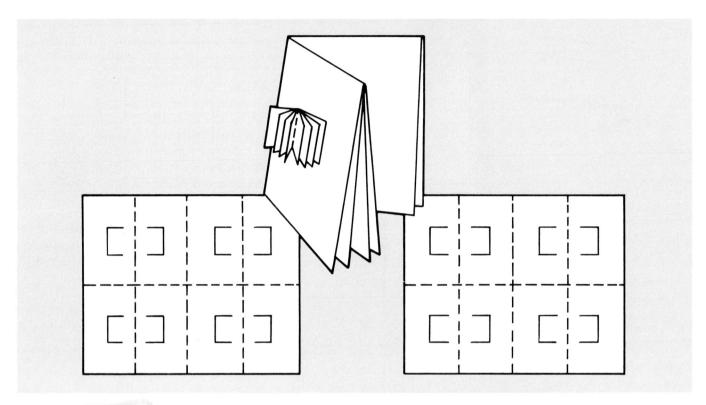

Fig. 7-48. Open layout, with doors cut and numbered, reveals correct page positions at a glance.

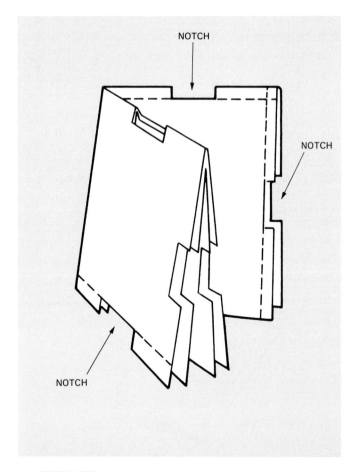

Fig. 7-49. Correct trim areas of folded signature layout are identified by cutting a rectangular-shaped notch on top, right, and bottom sides of layout.

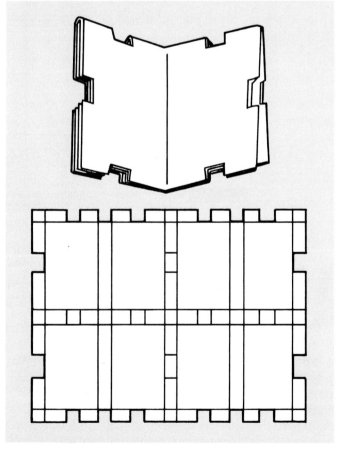

Fig. 7-50. After cutting trim notches in folded signature layout, it is opened up flat and trim lines are ruled in with a pen. Study this example!

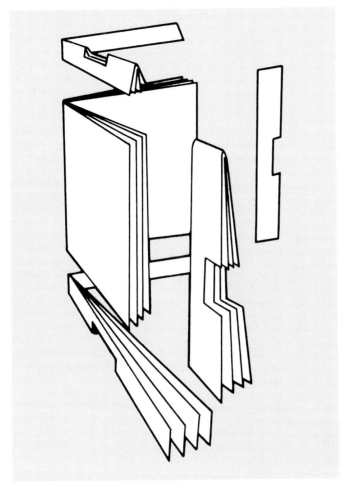

Fig. 7-51. Folded signature is trimmed along trim lines, resulting in an accurate layout dummy.

POINTS TO REMEMBER

1. Once the idea to produce a printed product has been formed, the first step is the planning, design, and preparation of a layout.
2. Design refers to the manner in which type and illustrations or photographs are arranged on a printed page.
3. The placing of the idea on paper refers to the preparations of a layout.
4. The principles of good design include proportion, balance, harmony, contrast, rhythm, unity, symmetry, variety, and action.
5. In good design, color contributes visual impact and interest to the printed page.
6. Colors are referred to as being either warm or cool. Red, orange, and yellow are examples of warm colors. Violet, blue, and green are cool colors.
7. Multicolor printing is used to add reader interest to printed products.
8. Process color printing reproduces an image in its natural colors. To do this, three basic inks, or process inks, are used.

9. The colors used to process color printing are black, yellow, magenta (red), and cyan (blue).
10. In process color printing, color separation negatives are used to prepare printing plates for the offset press.
11. In the layout process, the graphic designer generally prepares layouts in three stages. These include thumbnail sketches, rough layout, and comprehensive layout.
12. A layout dummy is needed if the printed piece has more than one page. A layout dummy is a set of page layouts.
13. Magazines and books are printed in sections called signatures. The page numbers in magazines and books are referred to as folios.
14. There are three kinds of original copy referred to as artwork. These include line copy, continuous tone copy, and combination copy.
15. Line copy includes pen-and-ink drawings, diagrams, and type composition.
16. Continuous tone copy contains gradations of tone or image, an original photo for example.
17. Combination copy contains some line as well as some continuous tone copy.
18. Artwork should always be handled with care. Marks or indentations on the face of the artwork are NOT acceptable.
19. Cropping artwork refers to the identification of the area of the illustration or photograph to be used in the final printed product. Generally, artwork is cropped to fit the available space in the layout and to show the most important content.
20. Sizing or scaling is the preparation of illustrations and photographs for enlargement or reduction. Enlargements and reductions are proportional.
21. Calculating enlargements and reductions can be done using either the proportional scale method or the diagonal line method.
22. Imposition is the term used for arranging pages so that they will be in the proper sequence after the sheet is printed and folded.
23. Imposition is necessary when printing such items as magazines, books, catalogs, and programs. A full sheet of paper is generally printed in units of 4, 8, 16, 32, or 64 pages.
24. In book and magazine printing, press sheets, once folded, become signatures. A signature is a sheet that is printed on two sides and folded into a sequence of 4, 8, 16, 32, or 64 pages.
25. Several signatures bound together form a magazine, book, booklet, etc. Careful attention must be given to establishing the correct trim areas of the finished signatures.

153

KNOW THESE TERMS

Design, Proportion, Formal balance, Informal balance, Harmony, Contrast, Rhythm, Unity, Symmetrical, Variety, Action, Warm colors, Cool colors, Monochromatic, Multicolor, Process color printing, Thumbnail sketch, Rough layout, Comprehensive layout, Mechanical, Paste-up, Camera-ready copy, Layout dummy, Folio, Signature, Artwork, Line copy, Continuous tone copy, Halftone, Combination copy, Cropping, Cropping marks, Sizing, Scaling, Holding dimension, Proportional scale method, Diagonal line method, Imposition, Signatures, Trim, Creep, Work-and-turn, Sheetwise.

REVIEW QUESTIONS

1. The relationship of sizes, widths, and depths of various parts on the printed page is called _____.
2. A pleasing page shape has dimensions with a ratio of about _____ to _____.
3. What happens visually if a single line of type is positioned exactly in the center of a vertical page?
4. Image elements on a page are centered horizontally when _____ balance is used.
5. When the image elements of a design appear to fit together and complement each other, they have:
   a. Harmony.
   b. Rhythm.
   c. Contrast.
   d. Unity.
6. Variety is included in a design by adding _____.
7. In design, _____ adds interest, life, and attraction to the printed page.
8. The repetition of similar elements in a printed job creates:
   a. Harmony.
   b. Rhythm.
   c. Contrast.
   d. Balance.
9. The first small layouts that are drawn when developing a printed piece are called _____ _____.
10. A detailed drawing showing the appearance of the final printed piece is called a _____ _____.
11. In your own words, explain the use of color on the printed page to create an overall mood.
12. The term _____ means one color.
13. Process color printing generally involves four-color reproduction. True or false?

14. The comprehensive layout is also referred to as the:
    a. Comp.
    b. Stat.
    c. Proof.
    d. Velox.
15. What is a computerized makeup system?
16. Computerized page makeup systems are classified as _____ end or _____ end.
17. A collection of design layouts which serves as a visual preview of a final printed product is called a _____ _____.
18. Most magazines and books are printed in sections called _____.
19. Reproducing only a part of a photograph or illustration is called _____.
20. What do enlargements of photographs tend to do to the image?
21. Reductions and enlargements can be calculated by using a _____ _____.
22. _____ _____ refers to the depth or width of a photograph and is used to calculate a reduction or enlagement.
23. Define the term "imposition."
24. Folded printed sheets are called:
    a. Folios.
    b. Sections.
    c. Signatures.
    d. Units.
25. _____ is the amount of paper removed from the three outside edges of a magazine after the signatures are _____ and _____ together.
26. What is a work-and-turn job?
27. The top of a folded signature is referred to as the _____.
28. One of the following is NOT a standard number of pages for an imposition:
    a. 4.
    b. 8.
    c. 12.
    d. 16.
29. A _____ dummy can be prepared to check for correct page imposition of a magazine or booklet.
30. _____ occurs when the signature is folded and the inside pages extend past the outside pages.

SUGGESTED ACTIVITIES

1. From good quality magazines, obtain one printed advertising example for each of the design principles: proportion, balance, harmony, contrast, rhythm, unity, symmetry, variety, and action. Share your examples with class members. Which of the nine design principles was most difficult to recognize? Why?

2. Prepare thumbnail sketches and rough and comprehensive layouts for a poster telling about the graphic arts program in your school. Ask your teacher to comment on the design selected.

3. Conduct a visual color experiment among the graphic arts class members. Obtain several different magazine advertisements or other printed products that contain substantial amounts of color printing. Ask members of the class to view the advertisements or printed products and give their responses as to how the colors influence their feelings about each. Compare responses of all the class members. What interesting results were revealed?

4. Find a printed sample of a poster or brochure that is used in your school library or counseling office. Prepare thumbnail sketches and rough and comprehensive layouts for what you think would be a more attractive design. Consult with other class members and see if your design has greater appeal than the original printed piece. If not, try to determine where your design might be improved.

5. Using the diagonal line method, calculate the following scaling problems:
   a. The resulting width when a 1 3/4 inch high by 3 1/4 inch wide illustration is increased in height to 2 1/2 inches.
   b. The resulting height when the same original is reduced to a width of 2 1/2 inches.

Section III

Issues for Class Discussion

1. As new typeface styles are introduced for desktop publishing systems in offices, they will become desirable for correspondence and other office documents. At first, the primary reasons for this may be novelty and fashion, but sounder, long-range considerations will probably prevail.

   Discuss important reasons why typefaces for business and office will need to be extremely functional.

2. Hundreds of typeface styles have been designed since the years of Johann Gutenberg. Many typeface variations are available, such as: inline, outline, shadow, contour, oblique, condensed, expanded, light, bold, and extra bold. All of these variations are obtainable in fonts in various degrees and combinations. In addition to the classic typefaces, more distinctive and fashion-oriented styles have become prominent.

   Discuss the primary human and machine factors that dictate the design of typefaces for use by printers and business office personnel.

3. Electronically created design work and artwork in full-color is a reality. Graphic designers and layout artists are using computers to create and communicate graphic ideas. The design artist can create full-color art with a keyboard and a cathode-ray tube. No paper, brushes, pencils, or pens are required. The artist can work faster and create more versions, to be stored in the computer or thrown away or produced at will.

   Discuss how the new electronic design devices will affect creative talent, job opportunities in design and layout, design choices, communication effectiveness, cost effectiveness, and speed.

4. The new computer design technologies will never replace human taste or judgment. These tools help graphic designers do what they must do faster and more precisely. Although specialized technicians will replace many traditional crafts people, the creative graphic designer who can work with the new technologies, such as keyboards, graphic tablets, electronic pens, and computer programs will be in greater demand than ever.

   Discuss how the new technologies will affect the design communications that will be used in newspapers, magazines, books, and all types of printing.

Section IV

Image Generation and Assembly

This section will discuss important aspects of the pre-press production sequence. In the printing industry, typesetting and image assembly is to the printed product as a foundation is to a completed building. The most important factor to consider in this section is the emphasis on producing high quality type composition and image assembly.

The concepts of typesetting and image assembly deal with very different and complex operations but they will be brought together in this section. With increased computer use, many of these operations occur together. For example, a computer-generated phototypesetter or desktop publishing system can generate the individual characters and assemble them into a full-page product in one complete operation. Line drawings and halftones can even be included in the computer-generated product.

Chapter 8 covers hand-mechanical methods of type composition. Hand-mechanical composition includes simple methods such as pen and ink drawings, transfer rub-off and adhesive materials, and hand lettering devices. Impact composition is also covered in Chapter 8. It is created by a relief form of a character (such as a typewriter) striking a ribbon. This transfers the shape of the character to a piece of paper. This is the least expensive method of generating acceptable quality type composition.

Chapter 9 covers phototypesetting. This method is two-dimensional and constitutes the most important area of typesetting and image assembly. Phototypesetting processes are categorized into five basic "generations." The first-generation machines were designed around hot metal composition machines in 1950. Second-generation machines produce an image from a master film matrix. Third-generation photographic typesetters create an image by using a character matrix grid and cathode ray tubes. Fourth-generation photographic typesetting creates an image from digital information stored in a computer's memory.

Chapter 10 covers electronic publishing, an important method of image generation and assembly.

Chapter 11 describes how visual elements are combined into final form called image assembly or paste-up. Provisions must be made on the paste-up for continuous tone copy and color copy. Therefore, paste-ups vary for single-color and multi-color work. During paste-up, provisions must be made for image variations, such as tints, bleeds, reverses, and overprints.

Chapter 12 covers the process of proofreading. A series of symbols have been developed to communicate changes in typeset composition and final assembled paste-ups. These symbols are called proofreaders' marks. The marks are generally placed in the margin of the page in blue pencil.

Hand-mechanical composition is still widely used in industry.   (Sir Speedy, Inc.)

Chapter 8

Hand-Mechanical and Impact Composition

When you have completed the reading and assigned activities related to this chapter, you will be able to:
- Recognize and use hand-lettering devices, such as pens, stencils, and templates.
- Recognize and use dry transfer type composition such as pressure-sensitive, tab, and adhesive types.
- Explain the characteristics of impact composition.
- Explain the advantages and disadvantages of impact composition.

Although computerized composition is replacing hand-mechanical and impact composition, you should still be familiar with these processes. Some shops still use these methods and they are basic to the industry.

Later chapters will expand on this information by discussing phototypesetting, computer generated art, and electronic page layout.

HAND-MECHANICAL COMPOSITION

Hand-mechanical composition can be drawn freehand, with mechanical instruments or it can be clipped or transferred from individual sheets. There are several variations of hand-mechanical composition.

Hand lettering

Hand lettering with pen and ink is a hand-mechanical composition process using special inking pens, stencils, and templates.

Freehand lettering is done on a good-quality white paper with an inking pen and India ink. A *technical fountain pen* is commonly used for freehand lettering, Fig. 8-1. These pens come with interchangeable point sizes for controlling line width, Fig. 8-2.

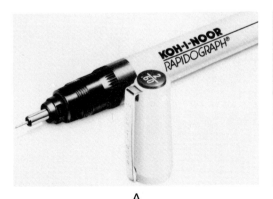

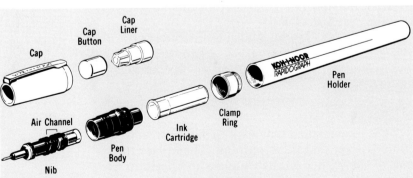

A

B

Fig. 8-1. A—Technical pens can be used to produce freehand lettering and line artwork. B—Technical pens contain several components that must be cleaned periodically for best results. (Koh-I-Noor Rapidograph, Inc.)

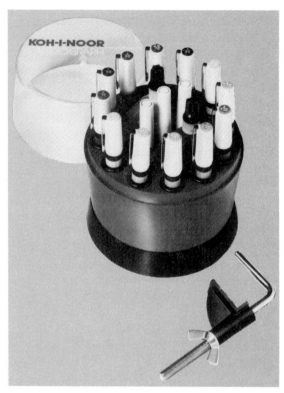

6x0	4x0	3x0	00	0	1	2	2½	3	3½	4	6	7
.13	.18	.25	.30	.35	.50	.60	.70	.80	1.00	1.20	1.40	2.00
.005 in.	.007 in.	.010 in.	.012 in.	.014 in.	.020 in.	.024 in.	.028 in.	.031 in.	.039 in.	.047 in.	.055 in.	.079 in.
.13 mm	.18 mm	.25 mm	.30 mm	.35 mm	.50 mm	.60 mm	.70 mm	.80 mm	1.00 mm	1.20 mm	1.40 mm	2.00 mm

B

A

Fig. 8-2. A—Some technical pens come with interchangeable points of different widths. A pen point holder is convenient. B—Note typical point widths available. (Koh-I-Noor Rapidograph, Inc.)

A set of interchangeable *Speedball® pen points* and a pen holder can also be used, Fig. 8-3. Each pen point is capable of producing a different weight line and configuration. Speedball pens in ordinary penholders are used for freehand lettering with special effects. Depending upon the skill of the user, some interesting artistic designs can be produced. Fig. 8-4 shows some examples.

Lettering and drawing guides

A *stencil* is generally used along with a technical fountain pen to produce letters and other symbols. *French curves* can also be used to make smooth, curved lines, Fig. 8-5. When using a stencil or French curve as a guide, hold the pen in a vertical position. Otherwise, the ink may run under the tool and blur or smear the line. See Fig. 8-6.

Devices called *templates* are used to create lettered type with pen and ink, Fig. 8-7. A common one, the *Leroy lettering set* consists of plastic templates, each with recessed letters, a pointed scriber, called a *stylus,* an inking pen, and interchangeable points of varying widths.

To letter with the Leroy set, fasten the pen and point in the scriber and fill it with ink. Then place the pin end of the scriber on the template and begin to trace one of the recessed letters. Because the pen point follows the same movement as the pin, it will create an inked duplicate of whatever letter the pin traces. See Fig. 8-8.

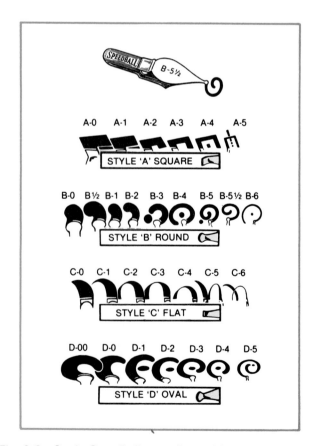

Fig. 8-3. Study Speedball pen nibs and lettering examples. Layout and paste-up artists frequently use these pens for lettering. (Hunt Manufacturing Co.)

A

B

Fig. 8-4. A—These are examples of lettering done with a C-5 pen point. B—Heavier Speedball pen points were used to letter these examples. (William Bockus)

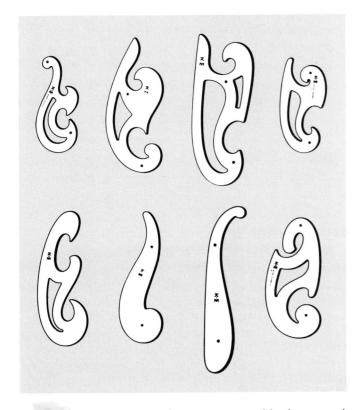

Fig. 8-5. Irregular or French curves are used by layout and paste-up artists to draw curved inked lines.

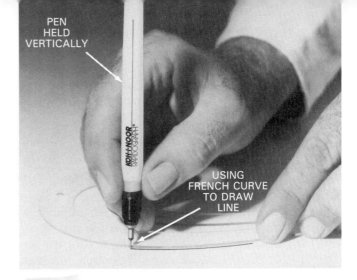

Fig. 8-6. Technical pen is held vertically against guide edge of a steel ruler, triangle, template, or T-square when drawing lines. Pen can also be used for hand lettering. (Koh-I-Noor Rapidograph, Inc.)

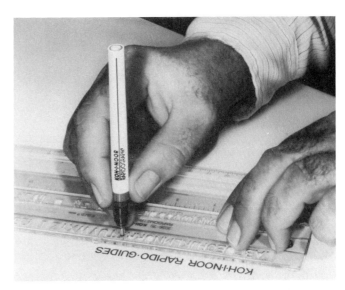

Fig. 8-7. Template and pen can be used to create simple lettering on a paste-up. (Koh-I-Noor Rapidograph, Inc.)

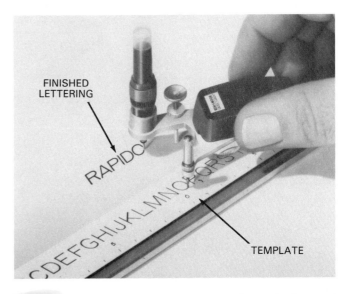

Fig. 8-8. Tracing a template produces an inked duplicate of lettering. (Koh-I-Noor Rapidograph, Inc.)

Templates may be used in combination with a technical pen for producing lettering symbols, or even sheets of music. Refer to Figs. 8-9 and 8-10.

Another versatile lettering device is the Varigraph Headwriter, Fig. 8-11. It works like the Leroy set in that a pointed stylus is used to trace characters engraved in a template. As the letters are traced, an inked duplicate is produced on paper. The Varigraph's height, width, and slant of the letters can be changed to produce thousands of different letter sizes and shapes. These are all done from a single template. Adjustment controls are provided on the Varigraph for this purpose.

Correcting inked errors

Errors can be corrected by painting over them with a *white opaque* or *China white paint* and then inking over the paint. Corrections can also be made by redrawing the copy on a separate sheet of the same paper and pasting it over the original error. All edges of the separate piece should be painted white to avoid shadows on the negative when it is photographically processed.

Preprinted type

Although the previously described methods are effective, they are slow and tedious. These methods also require some skill in forming the letter. Preprinted type, however, can be applied to a paste-up by a person who may be unskilled in lettering. Several kinds of preprinted letters are available for composing small quantities of hand-mechanical composition. These include dry-transfer type, pressure-sensitive type, paper type pads, and adhesive type.

MUSIC SYMBOLS

OUTLINE SHADOW

OLD ENGLISH

HEBREW

Fig. 8-10. These symbols were produced using a template in combination with a technical pen. (Keuffel and Esser Co.)

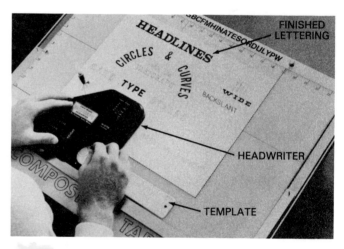

Fig. 8-11. This is a versatile lettering device. A pointed stylus is used to trace characters engraved in a template. (Varigraph, Inc.)

Dry transfer type

Dry transfer type consists of preprinted characters attached to a carrier. Dry transfer type comes in sheet form, Fig. 8-12. The characters are printed in reverse on the back of the carrier. The carrier is made from transparent plastic or translucent paper. It should be noted that the characters are right-reading (left to right) when viewed through the top of the carrier. Each carrier sheet usually contains a single size and style of type.

Shown in Fig. 8-13, follow the steps listed below to set dry-transfer type:

1. Draw a baseline on the paper with a light-blue pencil. Then position the base of the first letter to be

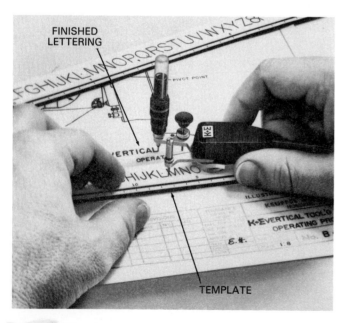

Fig. 8-9. Template is often used in combination with a technical pen to produce simple lettering. (Keuffel and Esser Co.)

162

A

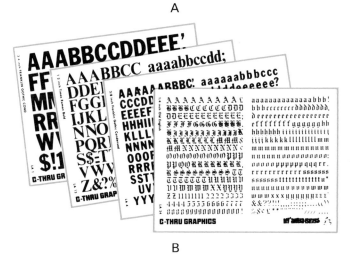

B

Fig. 8-12. A—Dry transfer type is available in sheet form. Individual characters are easily transferred to paste-up base. B—Many different styles and sizes of dry transfer type are available for paste-up work.

set on this baseline. Some brands of dry-transfer type have preprinted guidelines to aid in positioning the letter on the baseline, Fig. 8-14A.

2. The backs of the characters printed on the carrier sheet are coated with a waxlike adhesive. To transfer each one, hold the sheet firmly in position. Place the carrier side of the sheet up so that all characters are right-reading. Now rub lightly over the entire letter with a *burnisher* (tool with round, blunt end). Burnishing transfers the letter to the paper below, Fig. 8-14B.

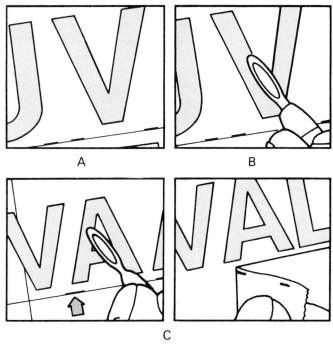

Fig. 8-14. A—This dry transfer type contains preprinted guidelines for easy positioning. B—Burnisher is used to adhere individual characters to base. C—Guidelines are carefully cut off and removed from bottom of each character.

1. Position letter.

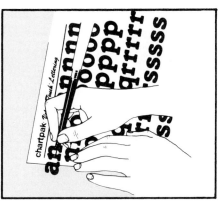

2. Rub over letter with burnisher—be sure to cover all areas and fine lines.

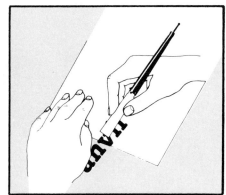

3. Remove sheet by carefully lifting from corner. For maximum adhesion, place a backing sheet over the letter and rub again with the flat end of the burnisher.

Fig. 8-13. To use dry transfer type on a paste-up or other artwork, follow these simple steps.   (Chartpak)

3. Carefully lift away the carrier sheet and position the next letter on the baseline. Then repeat the process described in Step 2 until the complete word or phrase has been set. Refer to Fig. 8-14C.

4. Letters that are improperly transferred or damaged during the transfer process may be removed in several ways. They can be scraped away with the edge of a razor blade, picked off the surface with the sticky side of adhesive tape, or erased. After removal, replace the letter as described previously.

5. Place the protective sheet that was supplied with the carrier over the lettering. Then rub firmly over the type with the burnisher. *Burnishing* serves to improve adhesion between the letters and paper.

6. Several light coatings of a *plastic spray* or *fixative* will protect the type from dust, dirt, and moisture.

Pressure-sensitive type

Each *pressure-sensitive type* character in a single alphabet is printed on the top surface of a clear plastic sheet, Fig. 8-15. The back of the sheet is coated with an adhesive and protected by a paper backing. Each sheet of pressure-sensitive type generally contains a single size and style of letters.

Use the following steps to set pressure-sensitive type:

1. Draw a baseline on the paper with a light-blue pencil. This baseline is needed to position the letters correctly as they are set.

2. Use a sharp knife to cut around the selected letter. Be sure to include the guideline that is printed below the letter in the cutout. Cut through the plastic sheet only. Do not cut into the protective paper backing. Then use the point of a knife to lift the cut out letter from the protective paper backing.

3. Place the guideline located at the bottom of the letter on the baseline drawn earlier. Rub lightly over the surface of the letter with a blunt tool, burnisher, or the handle of a knife. This adheres the letter to the paper below. Look at Fig. 8-16.

4. Repeat Steps 2 and 3 until the complete word or sentence has been set.

5. A damaged or improperly transferred letter is removed by carefully lifting it from the paper with a knife point. After removal, replace the letter.

6. After the letters are in final position, they are reburnished to improve adhesion. Burnish the type firmly this time. However, do NOT burnish the guidelines at the bottom of the letters. Use a knife to cut and peel the guidelines away.

Tab type

Preprinted pads of paper type, called *tab type,* are used to set small amounts of display type. Preprinted tab type pads are available in fonts, Fig. 8-17. Each of the padded sheets, or tabs, contains a single type character. Blank pad sheets are also supplied for spacing between words. Preprinted tab type pads are available in a wide range of styles and sizes of type.

Follow these steps to set tab type:

1. Preprinted tab type letters are assembled (composed) in a special composing stick provided for this purpose. See Fig. 8-18. Remove the first required letter from the pad containing the letter desired.

2. Align the letters carefully by placing the desired letter in the special composing stick.

3. Fasten the letters with transparent tape. Transparent tape is taped along the top of the composed line in the composing stick. The tape will

Fig. 8-15. Pressure-sensitive type sheets have a transparent plastic backing. Individual characters are cut out, removed, and adhered to paste-up base or other artwork.

Fig. 8-16. Note aligning and adhering transfer type. Type characters are aligned along their preprinted guidelines. Guidelines are then removed after final alignment.
(Graphic Products Corp.)

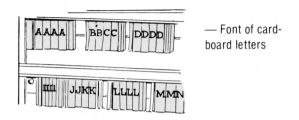

— Font of card-
board letters

Fig. 8-17. Preprinted tab type is used to set small amounts of display type. This method is seldom used today since faster typesetting systems are available.

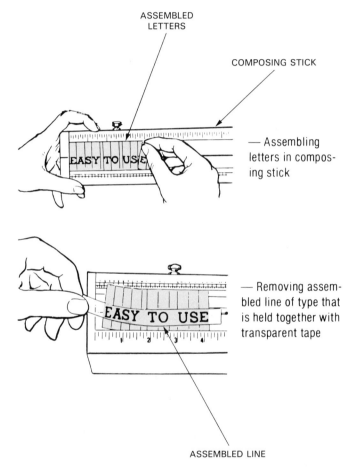

ASSEMBLED
LETTERS

COMPOSING STICK

— Assembling letters in composing stick

— Removing assembled line of type that is held together with transparent tape

ASSEMBLED LINE

Fig. 8-18. Preprinted tab type letters are assembled in a special composing stick. Line is then taped to hold it intact while transferring it to paste-up base or other artwork.

hold the line of letters securely. Since the dense black reproducing print of the letters is on the reverse side of the tabs, the taped line is turned over and it is mounted on the paste-up.

Adhesive type

The use of *adhesive type* is generally limited to small quantities of display type. The letters can be applied directly to the paste-up. This is done by cutting around the desired letter and lifting it from the plastic backing sheet.

Follow these steps to set adhesive type:
1. Draw a light guideline either on the paste-up or on a piece of white paper. A *nonreproducing* (will not show when photographed) *blue pencil* is generally used for this purpose.
2. Cut around the desired letter and lift it from the backing sheet. See Fig. 8-19. Note: the guideline under each letter should be lifted with the letter.
3. Then place the letter in the desired position by aligning the drawn guideline with the letter guideline, Fig. 8-20.
4. Burnish the letters, but not the guidelines, down and then cut off and remove the guidelines.

IMPACT COMPOSITION

Several kinds of typewriters can be used to compose type for offset lithography. This process of composition is referred to as *impact* or *strike-on composition*. The process is also known as *direct impression*. It

Fig. 8-19. Adhesive type is removed from the plastic backing sheet by cutting around desired character and lifting it from sheet. (Formatt Graphic Products Corp.)

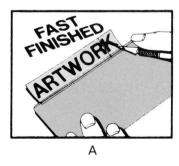

A

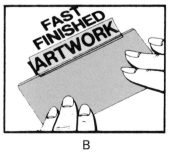

B

C

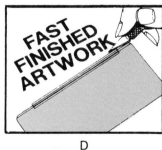

D

Fig. 8-20. A—Adhesive lettering is set and positioned on preprinted guidelines using a special transparent plastic carrier. B—Line of type is placed in correct position on paste-up board. C—Line of type is then burnished in position. D—Guidelines are removed from bottom of letters with an art knife. (Formatt Graphic Products Corp.)

should be noted that although many of the machines described here are still in general use, the major portion of typesetting produced today is accomplished on phototypesetting equipment.

Impact composition is produced on either standard office typewriters or specially designed type composition models. These typewriters are grouped into two general categories: standard typewriters and proportional spacing typewriters.

Standard typewriters

Office-type standard typewriters are sometimes adequate for simple type composition, Fig. 8-21. On most standard typewriters each of the characters and word spaces has the SAME WIDTH allotted. As a result, most type prepared on a standard office typewriter does not have a flush or *justified* right-hand margin. However, it is possible to justify type manually on a standard typewriter.

Each character typed on a standard typewriter occupies the same amount of space. For example, on an elite typewriter, every character requires 1/12 unit of horizontal space since any 12 characters will fill one

Fig. 8-21. A standard typewriter can be used to prepare simple type composition for a paste-up.   (A.B. Dick Co.)

inch of horizontal space. On pica typewriters every character typed requires a space that is 1/10 inch wide. Any ten characters will fill one inch of horizontal space. The horizontal space is the same for a capital "W" as it is for a period, lowercase "i," or lowercase "n." Because of this feature, standard typewriter composition lacks the appeal of type produced on *proportional* (varied according to character size) spacing typewriters.

Justifying margins on standard typewriters

Standard typewriters can be used to prepare composition with flush left and right margins. Flush left means beginning at the left margin. Flush right means going up to the right margin. This method of preparing composition is called justified. The appearance of the composition is similar to newspaper and magazine columns—flush left and right margins.

Follow the steps listed below to get quality copy:
1. Set margins.
2. Begin typing copy.
3. Stop with a complete or divided word as near the right-hand margin as possible.
4. If the line is short, type consecutive numbers until the margin is reached. If the line is over the right margin by more than three spaces, it may be difficult to reduce the space between the words.
5. Type the entire copy, following Step 4, Fig. 8-22.
6. Place a new ribbon on the typewriter or use a typewriter with a carbon ribbon to give a sharp, black image. A crisp, clear image is needed when a photographic negative is made from the copy.
7. Begin retyping the copy on clean paper.
8. On lines that are short, add as many extra word spaces as the highest number typed on the first typed copy. Add an extra space between words that have tall letters at the end of one word and at the beginning of the next one, example:
high   long.
9. Make up space on lines that went over the desired margin. Put half-spaces between words that have

```
    Fix up those/        Fix/up those/          Fix   up those
flat tires, boys     flat tires, boys      flat tires, boys
and girls, and//     and/girls,/and//      and  girls,  and
load your camera     load your camera      load your camera
with films--it's     with films--it's      with films--it's
our annual bi-//     our/annual/bi-//      our  annual  bi-
cycle trip back/     cycle/trip back/      cycle  trip back
to nature come//     to/nature/come//      to  nature  come
Saturday, May 17     Saturday, May 17      Saturday, May 17
        A                    B                     C
```

Fig. 8-22. Outdated method of using a standard typewriter to justify type composition required two typings. A—Lines were first typed with extra end-of-line spaces indicated by slant marks. B—Typed copy was then marked up to indicate where extra spacing in each line was needed. C—Final typed copy was then justified.

short letters at the end of one word and at the beginning of the next, example: little short.

10. Retype the entire copy, following Steps 8 and 9.

Copy for reproduction should be typed on smooth, dull-white paper. In most cases, a one-time carbon ribbon is used to produce sharp, dense, black images. The use of such a ribbon is necessary. The typewritten composition will become part of a paste-up. This, in turn, is photographed with a graphic arts process camera. Much of the standard typewriter composition produced is used directly as camera-ready copy on a photo-direct paper plate system.

Proportional spacing typewriters

Characters on a proportional spacing typewriter differ in width and therefore occupy varying amounts of space horizontally. This feature provides a choice of spacing widths between words. Each of the individual characters has varying widths.

Proportional typewriter composition is more pleasing than standard typewriter composition. Even so, proportional typewriter composition is still NOT equal to phototypesetting with respect to quality and proportional fit of the characters.

Phototypesetting is covered in Chapter 10.

Lettering machines

The *Kroy Lettering Machine* sets type in sizes up to 36 points by the strike-on process, Fig. 8-23. Characters are generated by pressing a raised-letter typedisc against an inked ribbon. The ink then transfers to a paper or plastic tape.

The Kroy machine does NOT have a keyboard. Instead, the typedisc is turned until the desired character is positioned over the tape. Then the printing mechanism is activated to transfer the image. Letter and word spacing is automatic and adjustable.

After setting, the adhesive-backed tape can be applied to the paste-up or other artwork and reproduced. See Fig. 8-24.

Some electronic or microprocessor systems can cut lines of type from 24 to 192 points in size, Fig. 8-25. The letters are cut from adhesive-backed vinyl material that can be applied to paste-ups or any smooth surface. The letters are pre-spaced and aligned automatically.

Type styles are changed by using different font cartridges. Lettering material is available in black, white, and a variety of other colors.

Electronic lettering machines can be used for such items as directory signs, nameplates, labels, posters, flipcharts, engineering drawings, reports and proposals, and all types of artwork and paste-ups.

Fig. 8-24. Adhesive-backed lettering is applied directly to paste-up base or other artwork.   (Kroy, Inc.)

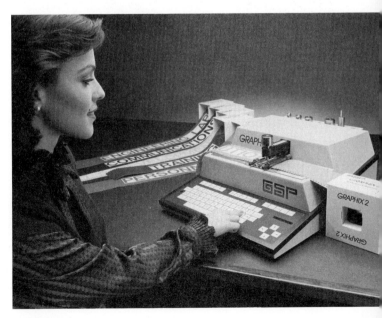

Fig. 8-25. This is another device that produces type on adhesive-backed vinyl material that can be applied to a paste-up or other artwork.   (Gerber Scientific Products, Inc.)

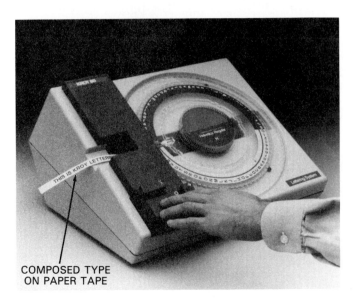

COMPOSED TYPE ON PAPER TAPE

Fig. 8-23. This letter machine produces strike-on type in sizes up to 36 points on paper or plastic tape.   (Kroy, Inc.)

POINTS TO REMEMBER

1. Some display composition is either drawn freehand or with mechanical instruments such as pens and templates.
2. Hand-mechanical composition refers to hand lettering with pen and ink. Special inking pens, stencils, and templates are available for this purpose.
3. Interchangeable Speedball pen points, along with a pen holder, can be used for artistic freehand lettering.
4. A stencil is generally used along with a technical fountain pen to produce letters and other symbols.
5. Templates, such as the Leroy lettering set, are used to create accurate lettered type with pen and ink.
6. The Varigraph Headwriter has a pointed stylus and is used to trace characters engraved in a template.
7. Preprinted type is slow and tedious. It can be used effectively for a few lines of type on a paste-up.
8. Preprinted type composition methods include dry-transfer type, pressure-sensitive type, tab type pads, and adhesive type.
9. When typewriters are used to compose type for offset lithography, the process is referred to as impact composition. Strike-on composition and direct impression composition are also terms applied to this process.
10. With standard typewriters, each character, whether a capital "W" or lowercase "i," occupies the same width of space.
11. Standard typewriters, if absolutely necessary, can be used to prepare composition with justified lines similar to the format used by newspapers, books, magazines, and similar publications.
12. Characters on a proportional spacing typewriter differ in width and therefore occupy varying amounts of space horizontally.
13. Proportional typewriter composition is more pleasing than standard typewriter composition because of its more precise letter fit and aesthetic effect.
14. Proportional typewriter composition is NOT equal in quality and proportional fit of the characters when compared to phototypesetting.

KNOW THESE TERMS

Hand-mechanical composition, Freehand lettering, Stencil, Template, Stylus, Preprinted type, Dry transfer type, Pressure-sensitive type, Tab type, Adhesive type, Impact composition, Strike-on composition, Direct impression composition, Standard typewriter, Flush left, Flush right, Justified, Proportional spacing typewriter.

REVIEW QUESTIONS

1. Describe the several different methods, tools, and materials used to prepare freehand lettering.
2. What type of pen is generally used along with stencils to produce hand lettering and other symbols?
3. Describe how type can be lettered using templates and special pens.
4. Preprinted type can be applied to a paste-up by a person who may be unskilled in lettering. True or false?
5. Dry transfer type is available in sheets and rolls. True or false?
6. Describe the procedure for using dry transfer type.
7. What is the difference between dry transfer type and tab-type?
8. How is a heading composed using tab-type letters?
9. Describe the procedure used to compose adhesive type.
10. One of the following preprinted typesetting processes requires the use of a composing stick:
    a. Adhesvie type.
    b. Transfer type.
    c. Tab-type.
    d. Hand lettering.
11. A strike-on typewriter ribbon that ensures dense, black images is known as a _____ ribbon.
    a. Nylon.
    b. Magnetic.
    c. Cotton.
    d. Carbon.
12. Describe how manual justification of lines was achieved on a standard typewriter.
13. Describe the term "proportional spacing" as it applies to a typewriter capable of justification.
14. All impact typewriters have proportional spacing. True or false?
15. What is a Kroy Lettering Machine?

SUGGESTED ACTIVITIES

1. Using the materials available to you, compose several lines of display size type using the dry transfer and tab-type methods of type-

setting. Compose the same words for each method and compare the quality of individual letter spacing, word spacing, and sharpness of letters.

2. Design a wall-size calendar for which preprinted type can be used.

3. Arrange to have a salesperson demonstrate to the class any or all of the preprinted type processes, including transfer type, adhesive type, and tab-type.

4. Plan to visit a small print shop or paste-up (ad) agency. Observe the various typesetting processes in use. Are any preprinted type processes being used? If so, why are these processes being used?

5. Using the materials available to you, design an 8 1/2 × 11 inch visual aid showing samples of the several preprinted type processes used in your school graphic arts shop. This can be used for a handout in the graphic arts classes.

6. Arrange a demonstration for the class of one of the proportional spacing typewriters described in this chapter.

7. Design a small project that involves some impact typewriter composition. Try to include lines that are flush left, centered, flush right, and fully justified (such as newspaper and magazine columns).

8. Redesign one or more forms used in the school office. Compose the type on an impact typewriter. Include some horizontal and vertical ruled lines.

9. Following your teacher's instruction, practice replacing the ribbons on the impact typewriters in the shop. In addition, check each printing element to be sure it is free of cracks, damaged letters, etc.

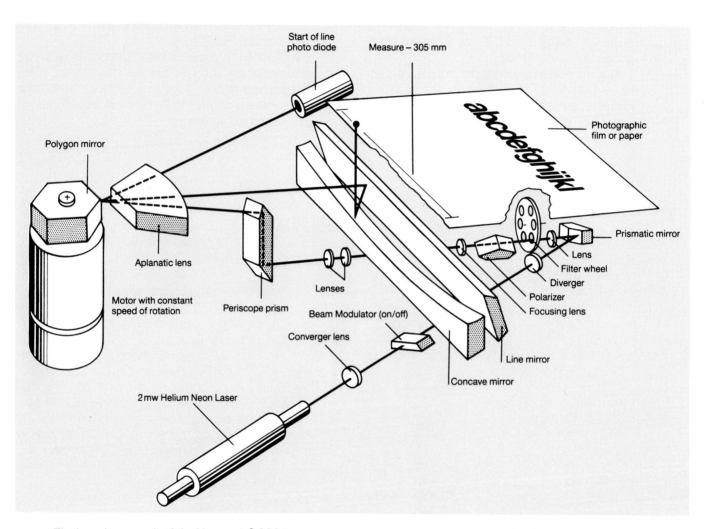

The laser beam path of the Linotronic® 300 imagesetter provides accuracy of exposure for both text and graphics. (Linotype Co.)

Chapter 9

Phototypesetting

When you have completed the reading and assigned activities related to this chapter, you will be able to:
O Describe photographic composition.
O Recognize each of the four generations of photographic typesetting.
O Distinguish between text (body) and display composition.
O Operate photographic composition machines.
O Process exposed phototypesetting paper.
O Describe how microcomputers are being used in the composition process.
O Describe the various components and operating systems of a typical phototypesetting system.

Typesetting has seen tremendous change in the past few years. Computers and electronics have taken over this area of production. This chapter will concentrate on discussing new electronic technology while reviewing traditional methods. This should give you a broad enough background to be well versed about the equipment and methods found in industry.

PHOTOTYPESETTING

The composition of type produced by means of photography is called *phototypesetting*. Some printers and professional typesetters refer to the process as *photocomposition*. Type characters are produced in the desired sequence in columns or entire pages on light sensitive or photographic paper or film. The image (copy) is placed on the special paper or film with light projection. The *latent* (invisible) image is then processed, as you would process or develop a photograph.

Study the basic parts of a typical phototypesetting system in Fig. 9-1.

The output of a phototypesetter is two-dimensional type images, called *phototype*. See Fig. 9-1. Phototype can be used directly as camera-ready copy or to complete image assembly (paste-up).

Note! Image assembly is covered in Chapter 11. Electronic or desktop publishing is covered in Chapter 10. Although not a photographic typesetting process, optional photographic printing devices can be used to produce phototypeset quality output with desktop systems. More often, the desktop publishing equipment is used in conjunction with a lower quality laser printer for final output.

Basic phototypesetting equipment

Although designs vary, there are several components typical to modern phototypesetting equipment. Shown in Fig. 9-2, these parts include:
1. KEYPAD (typewriter style keyboard with extra keys for programming in specifications and altering copy).
2. VIDEO DISPLAY TERMINAL (picture tube or monitor for showing words that have been keystroked and other copy information).
3. COMPUTER CONTROL UNIT (computer "brain" that temporarily stores data from keypad; also feeds data to, and controls disk drive, proof printer, and image generator).
4. DISK DRIVE (magnetically deposits data on "floppy" disks for storing typeset material).

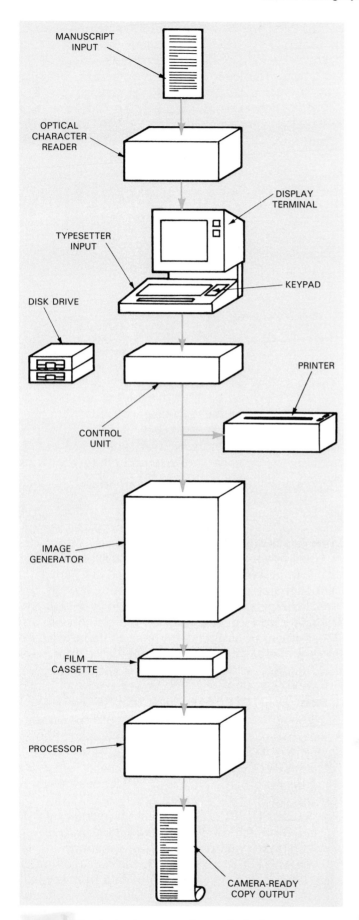

MANUSCRIPT INPUT

OPTICAL CHARACTER READER

DISPLAY TERMINAL

TYPESETTER INPUT

KEYPAD

DISK DRIVE

PRINTER

CONTROL UNIT

IMAGE GENERATOR

FILM CASSETTE

PROCESSOR

CAMERA-READY COPY OUTPUT

Fig. 9-1. Study basic parts of a typical phototypesetting system. Input is on top and output is on bottom.

Fig. 9-2. Phototype is used directly as camera-ready copy in most printing operations. Phototype can be used to complete image assembly as in paste-up shown above. (Joseph Gross)

5. PROOF PRINTER (low quality printer for making hard copy printouts for proofreading typeset material).
6. IMAGE GENERATOR (device that uses light source to deposit copy on photographic paper or film).
7. FILM CASSETTE (enclosed box or "portable darkroom" that holds photopaper or film before being processed).
8. PROCESSOR (mechanism for moving photographic paper or film through chemicals that develop copy image and make image permanent).
9. OPTICAL CHARACTER READER (sometimes used to convert typewriter copy into computer data, eliminates manual keystroking of copy by typesetter).

Phototype images

Phototype is usually produced as *right-reading images* (read characters normally). These images consist of positive phototype (black) characters on WHITE phototypesetting paper, Fig. 9-3. Sometimes, a transparent film base is used instead of a paper base.

Typeset material can also be composed with the characters in white (open or clear) on a BLACK background. This is called *reverse phototype* (characters are backwards), Fig. 9-4.

Basic phototypesetting process

A *phototypesetting machine* exposes type images on light sensitive paper or film. A large variety of phototypesetting machines are available for producing phototype.

Galley 1
Letter from President

It is with great pleasure that I welcome you to the 63rd annual conference of Western Society for Physical Education of College Women and to this beautiful setting where we will share much *conversation, thought, stimulation,* and *camraderie* over the next three days.

"Making Waves". Using the symbolism of the energy and power of waves, Jan Seaman and her committee have designed a program which will encourage each of us, separately and together, to develop, expand and channel our personal energy and power to create changes within our teaching, our professional lives and our personal lives. At the conclusion of this exploration, we hope you'll have picked up some "pearls of wisdom" which you can use to make changes in *your* environments.

Now, join me and the program committee and let's all make waves—together!!

Barbara J. Franklin

President

Fig. 9-3. Phototype is produced as black images on white phototypesetting paper, called photopaper. Column of type shown above is called a "galley."

Fig. 9-4. Typeset material with characters in white (open) on a black background is called reverse type.

The basic process inside a phototypesetting machine is illustrated in Fig. 9-5. The exposure light source transmits a light beam through a letter image of the type font. The letter image allows the light beam to pass through a shutter and through a sizing lens. The letter is then exposed directly onto the unexposed photographic paper or film by the light source. The previously described operations occur while the photographic paper or film from the unexposed cassette advances past the exposure point. Refer to Fig. 9-6.

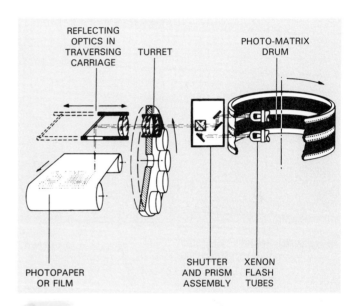

Fig. 9-5. Basic components of a phototypesetter. Light source is used to expose correct characters to form words and sentences. With character set, light shines through letter and is reflected onto photopaper or film.   (Compugraphic Corp.)

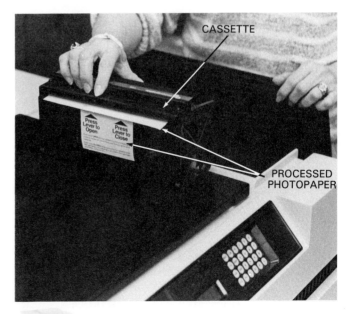

Fig. 9-6. A light-tight receiver cassette inside phototypesetter contains photopaper or film. Photopaper is exposed automatically after all copy has been set.

Generally, the exposed paper or film is fed into a receiver cassette. After the required amount of type has been set, the operator processes or develops the paper or film. This is done by cutting off the last of the exposed paper or film. The exposed paper or film in the cassette is then run through a separate *processor,* Fig. 9-7. This can also be done in the typesetting machine's own processor, if so equipped, Fig. 9-8.

Stabilization paper

There are two popular kinds of phototypesetting paper available. One is called stabilization (S) paper, and the other is resin-coated (RC) paper. *Stabilization paper* is used in a processor and once stabilized by a chemical bath, can be used for paste-up almost immediately, Fig. 9-9.

Resin-coated paper must be processed in the conventional way. This involves development, rinsing, fixing, and drying. A tray, tank, or mechanical processor must be used for this purpose, Fig. 9-10. Resin-coated photo papers lie flat and last a very long time.

Fig. 9-7. Exposed photopaper inside cassette is processed (developed) automatically in a separate processor.

Fig. 9-9. A galley phototype sheet is removed from processor. Latent image is now visible and permanent.

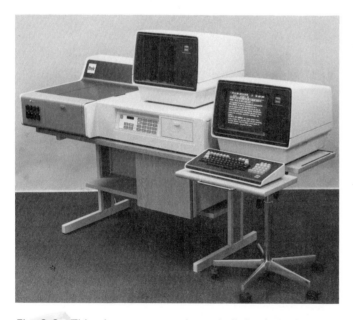

Fig. 9-8. This phototypesetter has a built-in phototype processor. (Itek Graphix Corp., Graphic Systems Div.)

For most operations, paper-base phototype is preferred over film-base phototype. This is because the paper-base phototype is used in the paste-up process. Under some conditions, paper-base phototype goes directly to the process camera in finished form. Film negatives or positives are prepared by the camera department.

Fig. 9-10. Typical compact, tabletop phototype processor used for processing photopaper or film. (Itek Graphix Corp., Graphic Systems Div.)

Display and text phototype

Phototype composition is divided into display and text. These terms describe the type sizes, purposes of the phototype composition, and kind of phototypesetting machines used to produce the type.

Display type, also called *photodisplay type,* ranges in sizes from 14 points up to 96 points and larger. Since display type is large, it is typically used for advertising, newspaper headlines, and posters.

Text type, also called *phototext type,* ranges in sizes from 5 1/2 points to about 12 or 13 points. This type, also referred to as *body type,* is used for large masses and columns of straight reading matter in newspapers, magazines, books, and brochures.

PHOTODISPLAY TYPESETTERS

Photodisplay type is generally composed on machines that set large type sizes up to 144 points, Fig. 9-11. Some of these machines are also capable of setting smaller text sizes of type. For example, the type sizes on individual machines may range from 14 to 72 points, 10 to 84 points, and 12 to 144 points.

Photodisplay typesetters operate either manually or automatically, Fig. 9-12. Many of the automated machines are driven through input such as a keyboard, tape, or on-line signal.

Manual photodisplay typesetters

The principle of operation among the manually-operated photodisplay typesetters is basically the same. These machines are prepared by installing the desired type font. This is usually a film negative such as a filmstrip or disk. The font contains transparent images in the form of characters.

The operator manually moves the font so that the character desired is in exposing position. A button is pressed and a light beam exposes the character onto a photosensitive paper or film. Exposure of the desired character is made through the film type font. Each desired character is positioned and exposed in the same manner until all the type is set. Spacing is accomplished

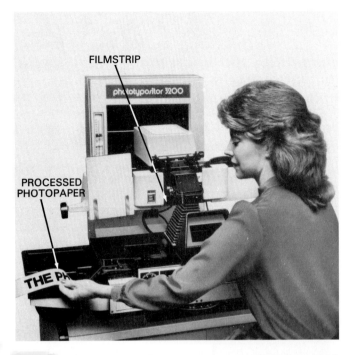

FILMSTRIP

PROCESSED PHOTOPAPER

Fig. 9-12. This photodisplay typesetter uses a filmstrip to hold character images. Strip is turned until correct character is positioned. Then, machine is activated to expose character onto photopaper or film. (Visual Graphics Corp.)

by a spacing device. Reference marks on the type font and machine are used for this purpose.

After the type has been set, the paper or film is cut, and then developed, fixed, and washed. This process is done in the machine on some models or outside in a separate processor on other models. The phototype is then ready for use on the paste-up or for further camera/darkroom operations.

It should be remembered that manually-operated photodisplay typesetters operate either by the contact principle or by the projection principle. Manually-operated photodisplay typesetters that use the *contact principle* (image and light sensitive materials in contact) are generally limited to producing same-size (one-to-one) images of the type font characters.

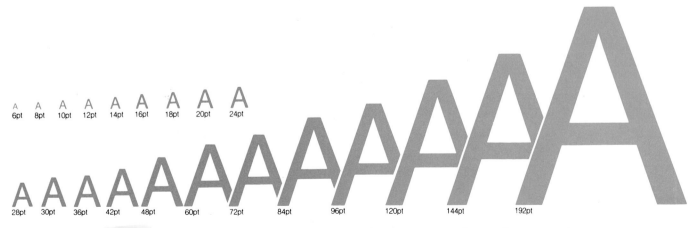

Fig. 9-11. Photodisplay type is composed on phototypesetters that set large type sizes.

Manually-operated photodisplay typesetters that operate on the *projection principle* (image and light sensitive materials have lens between them) increase the range of type sizes possible from any one type font. This is possible because of a built-in enlarging mechanism or lens arrangement, Fig. 9-13.

For special needs of the job, it is possible to photographically reduce or enlarge the finished phototype composition. This is generally done on a process camera.

Most photodisplay typesetters can be operated under normal room lighting conditions. With some machines, however, the operator must work under *safe light* (darkroom) conditions to avoid exposing the photosensitive paper to white light. See Fig. 9-14.

Fig. 9-15 shows a manually-operated projection photodisplay machine. It uses filmstrip type fonts. This unit will produce photodisplay characters from 3/16 to 11 inches high from each font on the filmstrip. Over 12,000 font styles are available. It has a projection mirror and several lenses which allow for type enlargement and reduction.

Fig. 9-16 shows how, with some machines, type can be composed in a variety of styles, such as: backslant, italic, condensed, staggered, etc.

Fig. 9-14. Most photodisplay typesetters can be operated under normal room lighting.   (Fotostar International, Inc.)

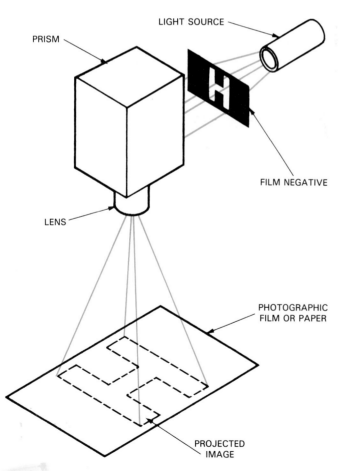

Fig. 9-13. This illustration shows basic principle of projection photocomposition. Light is projected through transparent image and onto light-sensitive photopaper or film.

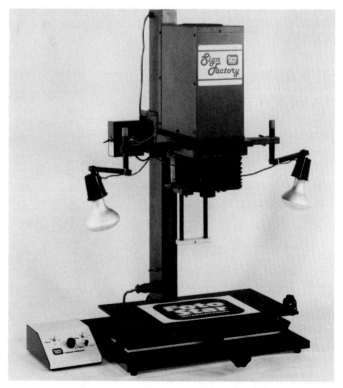

Fig. 9-15. A manually-operated projection photodisplay typesetter.   (Fotostar International, Inc.)

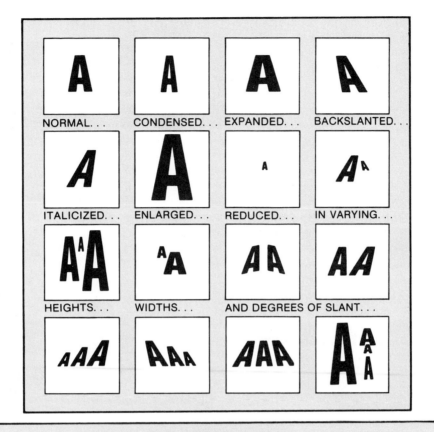

Fig. 9-16. Many different type styles and formats are available for use on photodisplay typesetters. (Visual Graphics Corp.)

Automated photodisplay typesetters

Several automated photodisplay typesetters are in general use. These include keyboard-operated machines. Some photodisplay typesetters are operated by tape or similar devices that produce a signal that becomes a machine command. Most of these machines function on the same principle as the phototext typesetters described later in this chapter.

Typeface selections on automated photodisplay typesetters are made by regulating a lens selection dial. Typefaces are available in many different sizes for each font. By adjusting the lens, it is possible to set type from about 14 through 72 points. These machines may also be equipped to set text and display sizes from 6 through 96 points.

A visual display unit allows the operator to view the characters. By pressing a button, the line is automatically exposed inside the typesetter. The exposed photographic paper is then transferred to a lightproof cassette for processing. The typeset material can then be cut apart and pasted into position on a paste-up.

Photodisplay typesetting can also be done using a composition and makeup (CAM) terminal. These machines are commonly used for daily newspaper advertising makeup and complete page makeup.

PHOTOTEXT TYPESETTERS

Phototext typesetters are generally capable of composing entire pages of fully justified text type. Type sizes range from about 6 through 36 points. It is possible on some of these machines to set larger type sizes. For example, some typesetter manufacturers designate their machine's low range (6 through 36 points) and high range (above 36 point). Width of lines generally runs to 45 picas or more.

Phototext typesetters have undergone major changes since the development of the first phototext typesetting machines, Fig. 9-17. Each of the major developments is classified by generation as follows:
1. First generation.
2. Second generation.
    a. Direct-entry.
    b. Indirect-entry.
3. Third generation.
4. Fourth generation.

First generation phototypesetters

The *first generation phototypesetters* were basically adaptations of earlier hot metal typesetting machines and an impact (typewriter) machine. The mechanical principles employed were the same. However, the metal casting matrix was replaced by a film negative matrix and the casting mechanism was replaced by the exposure unit.

Although these first generation machines were definitely photographic, they were limited by the mechanical principles of their operation. The matrices could not be moved into position quickly enough. The second generation phototypesetters came into existence in the search for increasing speed.

Second generation phototypesetters

The *second generation phototypesetters* employed greater sophistication in the exposure unit. For the first time, all the characters were contained on a disc or strip that was constantly spinning as each character was photographed. A high intensity, instantly rechargeable flash lamp was incorporated to photograph the moving character as if it were standing still. A computer was then added to control spacing of characters and words. These machines operated directly from a keyboard in which both the keyboard and phototypesetting components were one unit.

Some phototypesetters in use today are second generation machines. They offer a vast improvement in cost, speed, and convenience over first generation. However, they still contain many electro-mechanical parts and therefore require frequent servicing. They are slow when compared to the third generation phototypesetters. See Fig. 9-18.

It should be noted that second generation phototypesetters are either direct-entry or indirect-entry keyboard machines.

The *direct-entry keyboard* phototypesetters incorporate both the keyboard and phototypesetter unit. They are designed to be operated from a self-contained

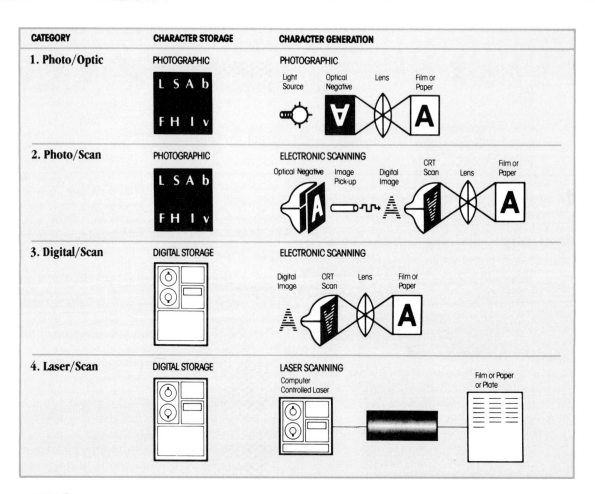

Fig. 9-17. Phototext typesetter development is classified according to generation. Four generations or developmental stages of typesetters are illustrated above.

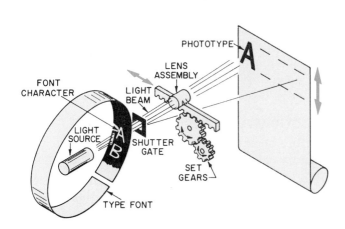

Fig. 9-18. Second generation phototypesetters are slow since they are operated electro-mechanically. (Compugraphic Corp.)

Fig. 9-20. This is a simple freestanding phototypesetter having a keyboard, computer (memory), and photographic unit. (Berthold Corp.)

or cable-connected keyboard. The direct-entry machine is a complete phototypesetting system and is available with a range of capabilities. See Fig. 9-19.

Group 1 phototypesetters are small, fairly simple machines, designed as a unit to fit on a table or desk top. These machines are at the lower end of the price range and aimed at the small company.

Group 2 typesetters are dimensionally larger than those in Group 1. They are usually designed as a single freestanding unit, Fig. 9-20. They are generally more

sophisticated and capable of setting complicated text matter. Work requiring both horizontal and vertical rules can be set to very accurate measurements.

Group 3 typesetters do NOT have a keyboard as part of the phototypesetter. The keyboard is linked by cable, sometimes referred to as the *umbilical cord,* to a separate freestanding phototypesetter unit. See Fig. 9-21.

The main advantage of this type of system is that the phototypesetter can be driven directly by the keyboard or by tape, floppy disks, or other devices.

Fig. 9-19. Direct-entry phototypesetter has limited use since it has no memory capabilities. (Compugraphic Corp.)

A

B

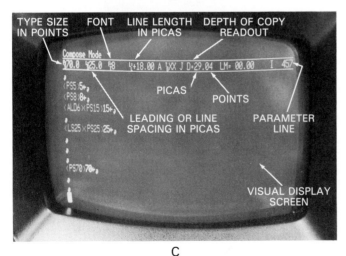

C

Fig. 9-21. A—Keyboard on this phototypesetter is separate. A cable, called umbilical cord, links keyboard to freestanding phototypesetter unit. B—Operator can see typeset material on a screen. C—Display screen on some advanced phototypesetters will give readout of programmed specifications on a parameter line. (High Technology Solutions, Inc.)

These machines allow other keyboards which produce input tape or disks to be added to expand the system. Because the keyboard is a completely separate unit, it can be purchased more cheaply than the direct-entry keyboard machines.

Direct-entry typesetters

Before beginning to compose the text matter, it is necessary to set up the phototypesetter with format and typographic data. This can be done manually with switches and dials or by using keyboard commands. For the input of text material, settings or *specifications,* called *parameters,* must be made. Such parameters include the following:

1. SIZE—set usually in points, using two figures, such as 08 or 10 representing 8 point or 10 point type sizes.
2. FACE—the typeface is selected manually or by special numbered or lettered keys.
3. LINE WIDTH (length)—this is usually indicated in picas or fractions.
4. FILM ADVANCE—this is usually indicated in points or fractions.
5. SETTING MODE—there are two variations that include:
   a. Justified or unjustified setting (range left).
   b. Fixed or variable work spacing. If variable spacing is set, the determined minimum and maximum word spacing must be allowed.

If formal changes are desired during typesetting, they can be done by the operator. Changes done by keystrokes are faster than those done manually.

Second generation indirect-entry phototypesetters generally have no built-in keyboard. These devices are intended to be part of a phototypesetting system. The machine is operated from either punched paper tape, magnetic tape or floppy disk, or by electronic signals produced by an operator on a separate keyboard, Fig. 9-22. Speed and efficiency are gained by this method.

Several keyboard operators can prepare tape, disks, or other signals. One indirect-entry phototypesetting machine will then accept and process all of the output produced by operators of one or more keyboard units or other signal-generating source.

A second generation indirect-entry phototypesetter usually includes:

1. READER—the *reader* converts the incoming tape, disk, or signal into a digital (on-off) electronic signal. This conversation process allows the tape, disk, or other signal to be read by the computer.
2. MEMORY—the *memory,* or *buffer,* acts as a storage depot that accepts and stores the incoming signal. The memory then transmits the stored signal to the computer.
3. COMPUTER UNIT—the *computer unit* contains a format (menu) program and memory. Its job is to process the unjustified tape, disk, or signal and make end-of-line decisions. The format program

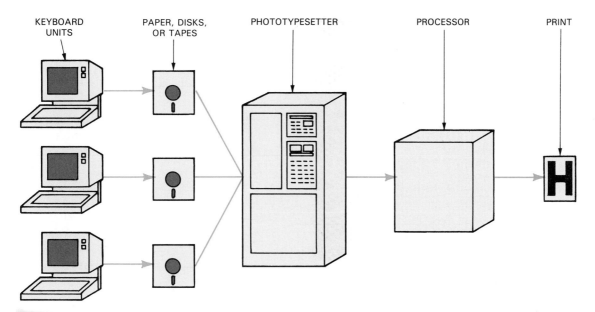

Fig. 9-22. Second generation indirect-entry phototypesetters operate from either punched tape, magnetic tape, floppy disk, or other electronic signal.

contains written specifications of typesetting requirements. It is written in a special language for the computer.

4. PHOTO UNIT—the output from phototypesetting comes from the photo unit. It is a combination of electronic and photographic components. The type is set using high-intensity light flashes through the character, projecting them onto photographic paper or film. Look at Fig. 9-23.

When the phototypesetting function is completed, the cassette containing the exposed material is removed from the photo unit, Fig. 9-24. The exposed paper or film is passed through a processor that automatically develops and fixes the image. Some photo units have built-in processing units for developing the photographic paper or film.

The production of phototype with a second generation indirect-entry phototypesetter consists of three phases. These include:

1. Signal input.
2. Preparation of the phototypesetter.
3. Operation of the phototypesetter.

The input unit is generally a separate keyboard machine used to prepare a tape, floppy disk, or other

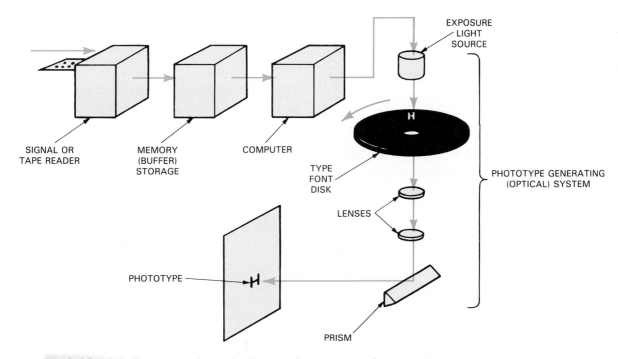

Fig. 9-23. This illustration shows basic steps in operation of a second generation phototypesetter.

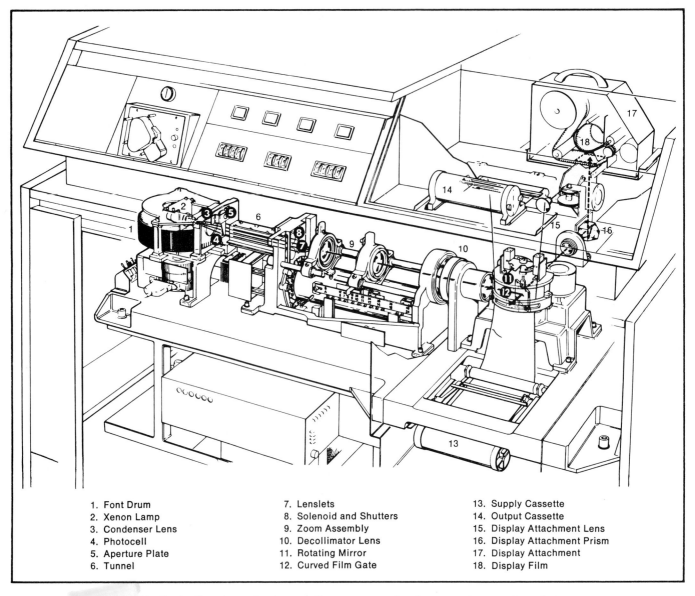

1. Font Drum	7. Lenslets	13. Supply Cassette
2. Xenon Lamp	8. Solenoid and Shutters	14. Output Cassette
3. Condenser Lens	9. Zoom Assembly	15. Display Attachment Lens
4. Photocell	10. Decollimator Lens	16. Display Attachment Prism
5. Aperture Plate	11. Rotating Mirror	17. Display Attachment
6. Tunnel	12. Curved Film Gate	18. Display Film

Fig. 9-24. This illustration shows basic operating components of a second generation phototypesetter. (Mergenthaller Linotype Co.)

signal, Fig. 9-25. Most current systems use a keyboard that is similar to a standard electric typewriter. There are two kinds of keyboards used to facilitate hyphenation and justification of lines.

With the older *counting keyboard* all end-of-line decisions are made by the operator as the copy is typed. The operator must know the typeface, point size, line width, wordspacing, and so on. In addition, the operator must have a good understanding of the machine's controls.

As the copy is being typed, the unit widths of the individual characters are totaled and displayed on a scale. As the minimum line width is approached, the machine informs the operator by a light or audio signal that an end-of-line decision must be made. After making the end-of-line decision, the operator then presses a key that instructs the unit to expand the wordspace to fill out the line to the desired width.

The *noncounting keyboard* system can be operated to make all end-of-line decisions. The operator must have the ability to type accurately at high speeds. The system can be adjusted to any desired line width, type style, and leading. All end-of-line decisions are made by the computer rather than by the keyboard operator.

TOTAL ELECTRONIC PHOTOTYPESETTING

There are now several total electronic indirect-entry second generation phototypesetting systems. These systems are versatile since they incorporate phototypesetting with word processing. It should be noted that the system described here is also available for third and fourth generation phototypesetters. A typical fully electronic newspaper system is illustrated in Fig. 9-26.

Phototypesetting

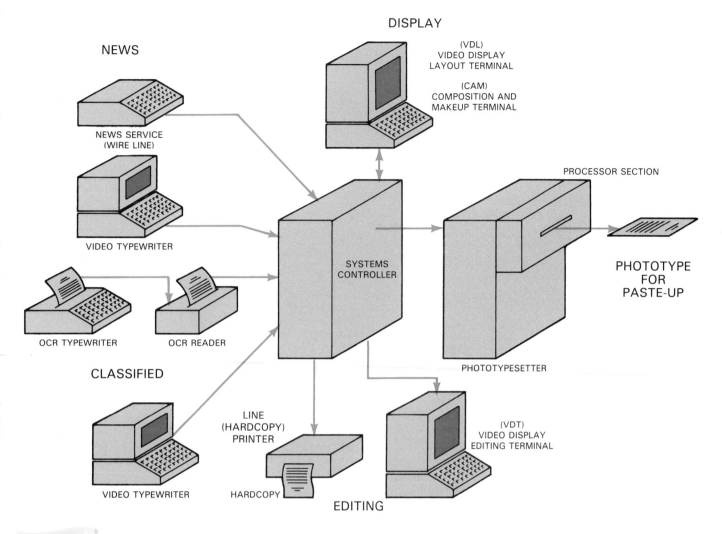

Fig. 9-25. Fully electronic phototypesetting systems are typical of second, third, and fourth generation phototypesetters.

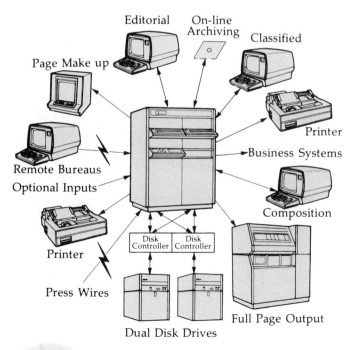

Fig. 9-26. This illustration shows a typical fully electronic, newspaper publishing, phototypesetting system. (Itek Corp.)

Input is generated from more than one location. As an example, a reporter's desk might contain a VDT (video display terminal) input and editing terminal, Fig. 9-26. Operators can also enter copy without keystroking or typing by using an *OCR* (optical-character-reader). The OCR converts typed matter into electronic data for the phototypesetter. Other staff members keyboard classified advertisements on the VDT. The main process and storage device looks like a computer installation, Fig. 9-27.

Copy for display advertisements can be entered by video keyboard terminal or by OCR units. The operator uses a CAM (composition and makeup terminal). CAM video devices show actual or simulated point sizes, type faces, line widths, leading, etc. These devices are used for display advertising and full-page makeup, Fig. 9-28. They can be passive or interactive.

Passive CAMs are often called *soft typesetters* or *previewers* since they only show how the typesetter will set the information in response to the commands.

Interactive CAMs let the operator change typographic format on the screen and instantly see what changes will take place on the page.

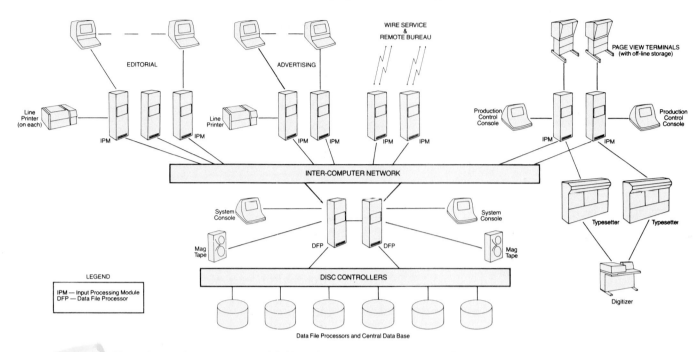

Fig. 9-27. Note electronic newspaper publishing phototypesetting system. Classified advertising and news stories can be entered by video keyboard terminal or by OCR units. (Dunn Technology, Inc.)

Fig. 9-28. This CAM (composition and makeup terminal) is used for display advertising and full-page newspaper and magazine makeup. (Camex)

CAMs can also be either systems oriented or stand alone. A *stand-alone CAM* has the width values for the type fonts to be used within its memory. Input may be accomplished directly into the CAM via an attached keyboard or from some recorded medium.

In some cases, the CAM may be connected to a larger computer typesetting system such as the one described here. This usually includes a minicomputer, rigid magnetic disk for storage, video editors, and input and output devices.

All of these and multiples of them, are interconnected. The CAM thus becomes a peripheral to this system.

Note! Electronic or desktop publishing systems are detailed in the next chapter.

Third generation typesetters

The *third generation phototypesetter,* because of electronic sophistication, offers a tremendous increase in typesetting speed (setting up to several thousand lines per minute) and low maintenance requirements. Look at Fig. 9-29.

Instead of exposing type from a film negative, third generation machines reproduce type electronically on a cathode ray tube (CRT) by very high speed scanning of an electron beam. The photographic paper is then exposed by the light emanating from the television tube-like screen. This is shown in Fig. 9-30.

Analog and digital storage

Third generation phototypesetters can be subdivided into two basic categories depending upon the manner in which the original characters or images are stored. The two types include: analog and digital.

Fig. 9-29. This is an example of a third generation, high speed phototypesetting machine.

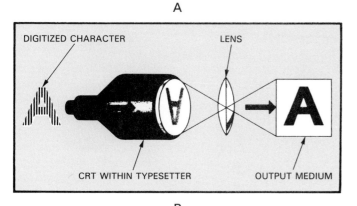

Shaping up. Digital technology "paints" characters, stroke by stroke, in blue-white light on the CRT tube. By calling a character's outline from font memory, the imaging system knows where to begin and end each stroke. Digitized fonts, pioneered by Mergenthaler, have eliminated the need for font replacements.

A

DIGITIZED CHARACTER   LENS

CRT WITHIN TYPESETTER   OUTPUT MEDIUM

B

Fig. 9-30. A—In third generation phototypesetting, digitized characters are "painted," stroke by stroke, in blue-white light on the CRT tube. B—Digitized character is output to photopaper or film medium. (Mergenthaller Linotype Co. and Autologic, Inc.)

An *analog unit* creates the image on the face of the CRT by first scanning a photographic master, usually held on a grid. When the original is scanned, its characteristics are converted into electronic information. This information is then transmitted through a photomultiplier and a logic box (computer) where sizing, spacing, italicizing, and similar functions are performed. The electronic information then goes to a printout CRT where the image is recreated by exciting phosphors (substance that glows when excited by radiation) on the surface of the CRT. These phosphors emit a light that travels through a lens and exposes the photographic paper or film.

The *digital unit* does NOT use photographic masters. The scanning step has been done at a previous station by the manufacturer. The unit retrieves the digitized information necessary to reproduce a character from a magnetic disk. This data is then sent to a main computer where information is added for enlarging, reducing, italicizing, and positioning the type on the line. The information, in electronic form, then activates phosphors on the screen of a CRT and stroke-by-stroke reproduces that character on the screen. The light emitted by the tube then travels through a lens and exposes the photographic material.

The digital third generation machine has eliminated many moving parts. The type font film negatives and flash are replaced by digital storage. The intricate lens system and prisms are replaced by a computer. With few moving parts, the machine is extremely fast and easy to maintain. These machines require a very high speed photographic paper or film. A typical third generation digital phototypesetter is shown in Fig. 9-31.

Fourth generation typesetters

Fourth generation phototypesetters employ lasers as the light source to scan copy and expose photopaper. See Fig. 9-32.

Fig. 9-31. This is an example of a third generation, digital phototypesetter. (Itek Corp.)

Some Advantages of Digital Typesetters

You can condense, expand, slant and produce small
caps of correct weight by keyboard command. ★ For rush jobs simulate condensed, expanded, italic and
small caps fonts not in your library. ★ Produce new font variations to create interesting and original type
faces. ★ Copyfit by condensing instead of kerning, to keep inter-character space open. Can you find the
18 point Helvetica line below which has been condensed?

Philip Kinsel
set in Times Bold with a normal width

Philip Kinsel
expanded to 140% of normal width by keyboard command

Philip Kinsel
condensed to 60% of normal width

PHILIP KINSEL
height reduced by keyboard command for a small cap of consistent weight

This is a Helvetica Roman *slanted* by keyboard command. This is Helvetica Italic *slanted* by the artist who drew the face.

Fig. 9-32. Fourth generation phototypesetters use a laser as a light source to scan copy and expose photopaper. Many typographic variations are possible with laser light source.

A laser is basically a tube of glowing glass. The word LASER is an acronym formed from the phrase, "Light Amplification by Stimulated Emission of Radiation." The laser produces a powerful, sharp beam of light (energy) that can be precisely controlled. The laser generates type characters dot by dot. The laser beam is extremely bright, very pure, and consistent in color. Refer to Fig. 9-33.

Information can be input into the laser phototypesetter either by tape or disk. All the information necessary to produce a given character is stored as digitized information. The laser phototypesetter employs a stationary helium-neon laser that generates a pure, constant, red, light beam. After leaving the laser, the light travels to a modulator that deflects the beam so that it enters the optical path only when necessary. From there, it passes through a neutral density filter, which ensures correct light intensity. The beam is directed to the photographic material by a constantly rotating mirror.

A *modulator* deflects the beam on or off, causing

Fig. 9-33. Examples of fourth generation, digital laser phototypesetters. Information (copy) can be input into these phototypesetters either by tape or disk.

the photographic material to be exposed in such a way that all characters on one line are composed at the same time. The advanced computer system within this machine not only resizes the type, but justifies and sorts all the information so the machine exposes across a line rather than exposing one character at a time.

The basic operation of a laser phototypesetter is illustrated in Fig. 9-34. Laser phototypesetters can produce type sizes from 5 to 256 points. They offer many type selections on the machine at one time. Fig. 9-35 gives some examples.

PHOTOTYPESETTING TRENDS

Two trends have characterized modern phototypesetting development. The first is full *page makeup* (type,

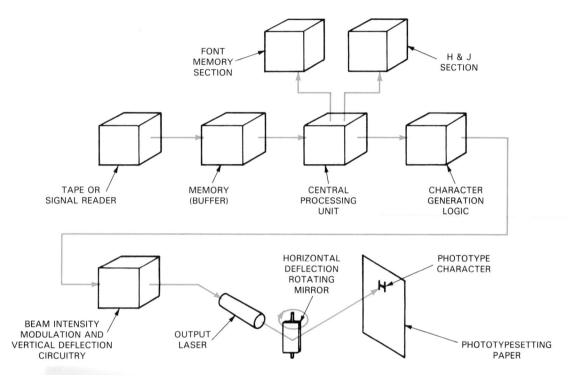

Fig. 9-34. This illustration shows basic operating steps of a digital laser phototypesetter.

Fig. 9-35. These are some examples of many type variations possible on a fourth generation digital laser phototypesetter. (Mergenthaller: Mycro-Tek Production Div.)

halftones, and drawings output at once). The second is *dry output* (no chemical processing). Lasers will play an important role in both areas.

The Associated Press LaserPhoto which is used to transmit halftones over telephone lines is an example of the technology used to create complete pages. These trends are moving the typesetting process closer to platemaking in the printing industry and copying in the office environment.

Advances in digitization and scanning now permit the reproduction of any visual item (photographs, illustrations, even real objects) to a scrics of digital signals. High resolution video cameras can capture these images. Computer Assisted Makeup equipment (CAMs) will eventually handle type and other visual elements routinely.

The implications of electronic page makeup will lead to the plate and reproduction steps of the offset printing process. Once a page is in electronic form, it can be stored, transmitted, and output to a typesetting machine, platemaker, or electronic reproduction system. In some cases, it may be transmitted by earth-orbiting satellites to remote printing locations.

Computerized equipment for writing, editing, and publishing is becoming more common. The new laser composing systems can produce both type, line drawings, and halftones on photographic paper, film, or printing plates. This reduces the time needed by authors, editors, and publishers to create materials quickly and efficiently.

Computer-aided publishing (CAP) is one of the fastest growing segments of typesetting. Computer-aided publishing is an outgrowth of the automation trend the publishing industry has been experiencing for several years. The most powerful systems are those that can:

1. Accept input from several entry devices.
2. Merge and process elements of text, graphics, and imagery.
3. Produce formatted and paginated (pages in correct sequence) output.
4. Output to printers, platemakers, and photo-typesetters.

Typesetting with personal computers

The *personal computer* or *microcomputer* is finding increasing use within the typesetting field, Fig. 9-36. This procedure eliminates the need for a second keyboarding (typing) since the copy is stored on a diskette.

A *diskette* or *disk* is a thin material with a magnetic surface capable of storing digitized information.

The keyboarded copy on the diskette is "read" by the typesetting machine. This is done by coding the copy being set on the microcomputer. These codes are necessary to alert the typesetting machine to certain elements in the copy. Codes are inserted for typeface, width of line, leading, paragraph indention,

etc. Other codes are used to indicate boldface, italic, or underlined words. Customers who supply the typesetting firm or printer with keyboarded copy in the form of a diskette do not always insert the codes. The typesetting operator will often insert the codes under direction from the customer.

It should be noted that some computer software programs do NOT require the insertion of codes. This type of computer software program is referred to as *codeless*. The advent of personal computers has created a whole new array of terms. You should learn these basic terms since they will also apply to the area of desktop publishing described in the next chapter.

Keyboarded diskettes are often delivered directly to the typesetting firm or printer. However, copy can be transmitted to the typesetting firm or printer by use of a modem.

The term *modem* means modulator-demodulator. This means that the modem alters the signals of a computer into signals that can be transmitted over telephone lines. The signal is received and reconverted by another modem attached to the typesetting machine. The machine identifies the codes and transforms the codes into commands that instruct the typesetting machine to set the copy. The actual typesetting output can be accomplished on a laser printer or phototypesetting processor machine.

Typesetting is usually one of the most expensive parts of any printing job. Therefore, the use of a personal computer saves time and expense. The reason for this is that the typesetting operator's charge for keyboarding the type is bypassed. This will reduce the typesetting expense by up to 50%-60%. Many companies who require typesetting, send copy, via a modem, since this is the least expensive method.

Fig. 9-36. This personal typesetting workstation (PTW) is coupled with a personal computer and a digital phototypesetter or optional dot matrix printer.
(Itek Graphix Corp., Graphic Systems Div.)

By using computerized equipment, companies that need to produce publications like catalogs, service manuals, product documentation, technical specifications, or transit schedules can realize a greater degree of productivity over conventional methods.

POINTS TO REMEMBER

1. Phototypesetting is the term given to all type composition produced by means of photography.
2. The process of phototypesetting results in two-dimensional type images called phototype. Phototype is usually produced as right-reading black images on a white background.
3. The basic phototypesetting process is a result of projecting a light source through a letter image of the type font. The letter image allows the light beam to pass through a shutter and lens.
4. In phototypesetting, the letter image is projected directly onto the unexposed photographic paper or film and the paper or film is then processed.
5. For most operations, paper-base phototype is preferred over film-base phototype.
6. There are two popular kinds of phototypesetting paper available: stabilization paper (S) and resin-coated paper (RC).
7. Phototype composition is divided into larger display and smaller text types. Display type is called photodisplay type. Text type is called phototext type.
8. Photodisplay type is generally composed on machines that set large type sizes up to 144 points.
9. Some phototypesetting machines set type in ranges from 14 to 72 points, 10 to 84 points, and 12 to 144 points.
10. Manually-operated photodisplay typesetters operate either by the contact printing principle or by the projection printing principle.
11. Most photodisplay typesetters can be operated under normal room lighting conditions.
12. Several automated photodisplay typesetters are in general use. These include keyboard-operated machines. Some operate by tape, disk, or similar devices that produce a signal that becomes a machine command.
13. Most of the photodisplay machines operate on the same principle as the more sophisticated phototext typesetters.
14. Phototext typesetters are generally capable of composing entire pages of fully justified text type.
15. Phototext typesetters have undergone several major changes since the development of the first phototext typesetting machines.
16. Each of the major developments in phototext typesetters is classified by generation. These include: first generation, second generation, third generation, and fourth generation. These machines span a period of time from around 1950 to the present.
17. With most fourth generation phototypesetters, a laser is used to produce photographic characters directly onto photographic paper or film.
18. Lasers play an important role in modern phototypesetting development.
19. Advances in digitization and scanning now permit the reduction of any visual image (pictures, illustrations, etc.) to a series of digital signals.
20. The implications of electronic page makeup will lead to the plate and reproduction steps of the printing process.
21. Personal computers, or microcomputers, have found increasing use within the typesetting field. These machines help eliminate the need for a second keyboarding of typeset material.
22. Personal computers use codes to indicate typeface selection, width of line, leading, paragraph indention, etc. Other codes are used to indicate boldface, italic or underlined words.
23. Using a computer, keyboarded copy is formed on diskettes which are used with a laser printer or phototypesetting unit for true typeset quality.

KNOW THESE TERMS

Stabilization paper, Resin-coated paper, Photodisplay type, Phototext type, Contact printing, Projection printing, Filmstrip, First generation phototypesetter, Second generation phototypesetter, Direct-entry, Indirect-entry, Reader, Memory, Buffer, Computer unit, Photo unit, Counting keyboard, Justified, Noncounting keyboard, Passive CAM, Interactive CAM, Analog, Digital, Third generation phototypesetter, Raster line, Fourth generation phototypesetter, CRT, CRT raster line, Laser.

REVIEW QUESTIONS

1. Type composition produced by means of photography is referred to as _____.
2. Two popular kinds of phototypesetting paper are _____ paper and _____-_____ paper.
3. Display type ranges in sizes from _____ points to _____ points and larger.

4. _____ size type ranges in sizes from 5 1/2 points to about 13 points.
5. Photodisplay typesetting generally refers to type sizes ranging to 144 points. True or false?
6. Photodisplay typesetters that operate on the projection printing principle are limited to producing same-size images. True or false?
7. Second generation phototypesetting machines are electro-mechanical in operation. True or false?
8. Describe the principle of operation of a second generation phototypesetter.
9. Explain the differences between second generation direct-entry and indirect-entry phototypesetting machines.
10. Describe the differences between a counting keyboard and a noncounting keyboard as used on indirect-entry phototypesetting equipment.
11. What are the principles and major operating features of a total electronic indirect-entry second generation phototypesetter?
12. Describe the duties of a CAM operator.
13. Describe the purpose of a VDT terminal when used in a newspaper publishing system arrangement.
14. Describe the differences between a passive CAM and interactive CAM.
15. Describe the components and operation of a third generation phototext typesetter.
16. Describe how a laser is used in fourth generation phototext typesetting.
17. What are two trends that have characterized modern phototypesetting development?
18. How can a microcomputer be used to facilitate typesetting operations?
19. A _____ alters the signals of a computer into signals that can be transmitted over telephone lines.
20. Typesetting is usually one of the least expensive parts of any printing job. True or false?

SUGGESTED ACTIVITIES

1. Using the equipment available to you, set one or two lines of text and display type using some or all of the phototypesetting methods described in this chapter. Compare such features as quality of letter fit, word spacing, hyphenation, leading, and quality of images.
2. Plan a field trip to a local phototypesetting firm. Prepare a brief description or outline of the equipment and processes you observed. Report your findings to the class.
3. Analyze the phototypesetting equipment capabilities in your shop. Plan the area that would include newer equipment, more work stations, and a systems approach for future growth. Check with local distributors to determine the cost of your plan.
4. Prepare a bulletin board display in the shop that illustrates your plan for present and future phototypesetting needs in the graphic arts program.
5. Invite a phototypesetting salesperson to talk about typesetting equipment with the class. Coordinate the visit with your instructor. Prepare for a question-answer period to follow the presentation. Ask the salesperson to provide printed brochures and other literature related to the phototypesetting equipment.

Chapter 10

Electronic Publishing

When you have completed the reading and assigned activities related to this chapter, you will be able to:

O Define electronic publishing.
O Explain the basic hardware of a desktop publishing system.
O Describe electronic publishing software.
O Summarize the use of a word processing program.
O Summarize the use of drawing and paint programs.
O Describe the use of art modification programs.
O Summarize the use of a page layout program.
O Compare different electronic publishing systems.
O Explain the difference between a low- and high-resolution system.

Electronic publishing, sometimes called *desktop publishing*, uses a computer to generate and combine TEXT and ART on a display screen before being output as hardcopy or printing plates. It eliminates the need to draw art with a technical ink pen, and paste down columns of copy by hand. Instead, you can use a personal computer to type the manuscript into a word processing program, draw the art in a drawing program, import existing artwork with a scanner, and then combine everything in a page layout program. Photos can be stripped in later or they can be scanned in with a high-resolution scanner. This decreases the amount of time required for the publishing process because many manual steps are eliminated and done electronically with a computer.

The first section of this chapter discusses the *hardware* (equipment) and the *software* (computer programs) that are used with smaller publishing systems. You will learn the basics of word processing, drawing on a computer, and using computer page layout to develop a completed publication.

In the second section of the chapter, you will learn about more complex electronic publishing systems that let you go from a computer screen directly to the printing plate. These systems eliminate the need for layout sheets or flats. The electronic data in the computer is used to run an electronic output device for making the printing plate directly.

The printing industry is experiencing an "electronic revolution." The capabilities of small printing shops, and their customers, has increased tremendously in the last few years due to the development of powerful personal computers. This chapter summarizes the most important information relating to this exciting new facet of printing. See Fig. 10-1.

ELECTRONIC PUBLISHING HARDWARE

Electronic publishing hardware refers to the physical equipment or components that make up the computer system. This commonly includes a computer, printer, monitor, and sometimes other devices. Electronic publishing is often referred to as *desktop publishing*. The term "desktop" generally implies a smaller, less expensive system that fits on top of a desk, Fig. 10-2. More advanced publishing systems might also include a slide scanner, graphics input pad, and a video input device. See Fig. 10-3.

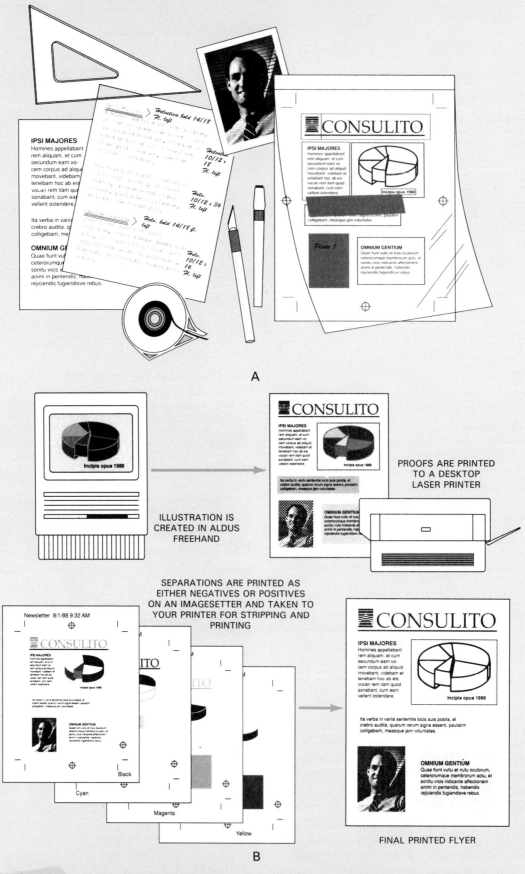

Fig. 10-1. A—Traditional image assembly procedure involves typesetting, and proofing, designing artwork, pasting up boards, and creating overlays for color, and photographing negatives of the boards. B—With computerized desktop publishing system, all of these tasks can be carried out on computer. (Aldus FreeHand, used with permission from Aldus Corporation)

Fig. 10-2. As a general-purpose computer, Macintosh™ SE suits a wide range of needs and can expand as those needs grow with communications. (Apple Computer)

Fig. 10-3. The DuPont HighLight Monochrome computerized scanner-recorder can be coupled with raster image processor for desktop publishing. (DuPont Imaging Systems)

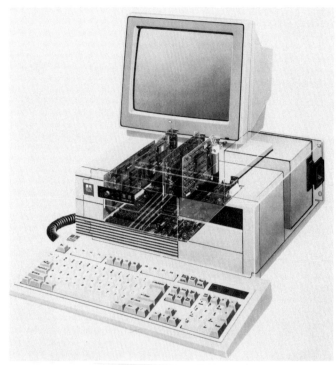

Fig. 10-4. A personal computer is a small computer typically consisting of a keyboard, mouse, monitor, electronic processor, and disk drive. (DTK Computer, Inc.)

Fig. 10-5. Most computer keyboards use a conventional QWERTY key configuration.

PERSONAL COMPUTER

A *personal computer*, or *PC,* is a small computer typically consisting of a keyboard, mouse, monitor, processor, and disk drive. These parts allow you to organize your thoughts and work electronically, Fig. 10-4.

Computer keyboard

A *computer keyboard* is the primary input device for a computer. You can input words, numbers, and commands to control the operation of the computer. Most keyboards use a conventional *QWERTY* key configuration, where the upper left letters form the term "QWERTY." See Fig. 10-5.

The *cursor keys* on a keyboard allow you to move the cursor on the computer screen. A *cursor* (arrow, square, or other location-indicating icon) is used to show where you are inputting data. When typing a letter, such as "R," the cursor shows you the exact placement of the character you are inputting. When drawing with a computer, the cursor shows where you are working on the screen. You can move the cursor up, down, left, and right using either the cursor keys or the mouse.

Function keys allow you to change the "function" of the keyboard or mouse. Pressing one of the function keys, for example, may change the typed letters into bold or italic typeface. They are usually the top row of keys on the keyboard.

Mouse

A *mouse* is a small, hand-operated input device. It is commonly used in conjunction with the keyboard to work more efficiently, Fig. 10-6. A mouse is helpful

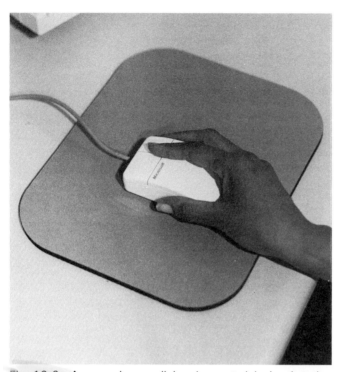

Fig. 10-6. A mouse is a small, hand-operated device that also allows input into computer.

when doing word processing, drawing, or placing elements of your layout into place on the monitor.

When a mouse is moved on the work surface, movement sensors output electrical signals to the computer, Fig. 10-7. These signals are used to move the cursor on the monitor. A button or buttons on the top of the mouse also allow you to select objects shown on the monitor.

A *mechanical mouse* generally uses a small rubber ball to turn metal wheels inside the mouse. The metal wheels trigger the movement sensors, thereby, sending the electrical signals to the processor. A mechanical mouse is shown in Fig. 10-8A.

A *puck,* or *electronic mouse,* often uses electrical sensors to read a grid pattern on a digitizing tablet or special mouse pad. The tablet receives signals from the puck or mouse and determines direction and distance of mouse motion. Refer to Fig. 10-8B.

Dragging refers to pressing the mouse button while moving the mouse. This allows you to quickly select or modify objects on the monitor. Some word processing programs allow you to press the mouse button while dragging over words to select them. You can then delete, modify, or move the words at one time.

Fig. 10-7. When the mouse is moved on mouse pad or work surface, movement sensors output electrical signals to computer. (Apple Macintosh)

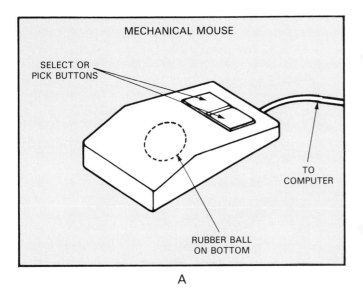

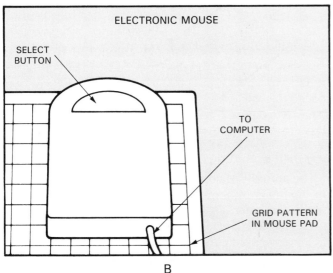

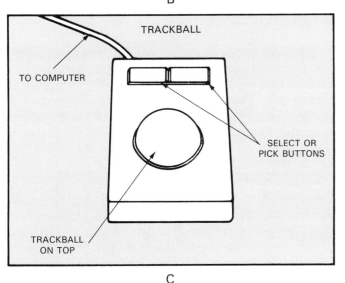

Fig. 10-8. A—Mechanical mouse uses a small rubber ball to turn metal wheels. Metal wheels trigger movement sensors. B—Electronic mouse uses optical sensors to read a grid pattern on a special mouse pad. C—Trackball is an upside-down mouse. Motion of ball turns small wheels to input data for moving cursor.

Trackball

A *trackball* is an upside-down mouse. You simply use your fingers to spin a large ball, instead of moving the mouse over the work surface. The motion of the ball turns small wheels to input data for moving the cursor. It also has buttons for selecting objects and performing other operations. A trackball has the advantage of requiring little space on the work surface, Fig. 10-8C.

Monitor

The *monitor,* or *display screen,* is a television-type picture tube or CRT (cathode ray tube) used to show the information being processed on the computer. The user views the monitor while working. Images are generated on the facc of the monitor by an electron beam directed onto a phosphor coating on the inside surface of the tube. Refer to Fig. 10-9.

The electron beam is automatically turned on and off and scans horizontally and vertically across the screen very quickly. The phosphor glows whenever the electron beam hits the inner surface of the monitor. It glows long enough and the beam moves fast enough that the human eye perceives a full screen image without seeing the movement of the electron beam.

A *monochrome monitor* only shows one color. It is sufficient for some desktop publishing jobs. A *color monitor* displays in full color. A color monitor is used to do color separations and overlays on the computer. Calibration devices are available to help adjust the color monitor so that it will match the final color of the final printed product. See Fig. 10-10.

Fig. 10-9. A monitor or display screen is a television-type picture tube or CRT (cathode ray tube) for showing information being processed on computer.
(BARCO Calibrator, BARCO, Inc., Kennesaw, GA)

Fig. 10-10. Linotype Series 1000 desktop publishing system utilizes Apple Macintosh® personal computer to bring text and graphics together on-screen as composition workstation. (Courtesy of Linotype Company, Hauppauge, NY)

Monitor resolution refers to the number of dots or pixels visible on the computer screen. A low-resolution monitor has fewer pixels and a lower quality image. A typical low-resolution monitor might have 640 pixels horizontally and 350 pixels vertically. A low-resolution monitor produces a problem called *jaggies* or *stair-stepping* that make angled lines ragged or rough along their edges. A high-resolution monitor has more, smaller pixels and produces a better image. A high-resolution monitor might have 1024 by 1024 resolution. The images appear smoother and more realistic.

Another way of denoting monitor resolution is *dots per inch,* or *dpi.* A monitor might have 72 dpi. However, each dot can carry a different amount of information. A high-quality, photo realistic, 4-color computer image might have 32 bits of information per pixel. A low-quality, monochrome monitor might have only one bit of information per pixel.

The term *WYSIWYG* means "What You See Is What You Get." The image shown on the monitor is similar to what will be produced when printed. The resolution will normally be lower on screen than the printed product, but the font and other values will be comparable. A WYSIWYG screen image is very useful for electronic page layout. It lets you check style more accurately and is much easier to read, thus reducing eye strain and errors.

Processor

The *processor* is the "electronic brain" of a personal computer. It has complex electronic logic circuits capable of converting and storing inputs as *digital* (numerical) data. Any input (copy, art, etc.) can be altered quickly and easily when it is in digital form.

The term *RAM,* or *random access memory,* refers to the amount of memory of the computer's electronic circuits or "chips." It generally refers to how much information the computer can hold and process in its in-tegrated circuit (IC) chips. The amount of RAM in a personal computer ranges from 512K (512,000 bytes of data) to 32M (32,000,000 bytes of data). One to two megabytes of RAM is usually the least amount of memory required to operate a basic monochrome desktop publishing system. Four or more megabytes of RAM may be required for color page layout programs. See Fig. 10-11.

The amount of computer memory required depends on the program and type of artwork being processed on the computer. Graphics programs are more memory intensive than word processing programs. A single, high-resolution color illustration may require over one megabyte of memory to be displayed, Fig. 10-12.

STORAGE DEVICES

A *storage device* provides a means of keeping information intact when the computer is turned off. The information kept in RAM is lost when a computer is turned off. Therefore, it must be transferred to a storage device for safekeeping until the information is needed again.

Disk drives

A *disk drive* is the most common storage device for computer data. It places tiny electrical charges on the surface of a magnetic disk to remember any information input by a computer. The disk drive is used to store programs, characters of a word processing program, line drawings, and halftones. See Fig. 10-13.

A *hard drive* is generally used for storage of programs and other large amounts of information. An *internal hard drive* is mounted inside the computer housing. An *external hard drive* is mounted on the outside of the computer in its own enclosure. Although a hard drive is not essential, several programs and hundreds of *files* (letters, drawings, and layouts) can be

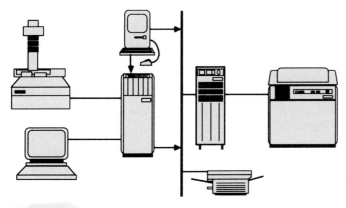

Fig. 10-11. Color page layout programs generally require four or more megabytes of circuit memory. (Courtesy of Linotype Company, Hauppauge, NY)

Fig. 10-12. A raster image processor converts PostScript page description information received from a workstation into electronic data for output on a laser imagesetter. (Courtesy of Linotype Company, Hauppauge, NY)

Fig. 10-13. A disk drive is most common storage device for computer data. It uses tiny electrical charges on surface of a magnetic disk to remember information input by a computer.

kept on hard disk for maximum work efficiency. Hard drive storage typically ranges from 20 megabytes to over 100 megabytes of data.

A *removable hard drive* is an external unit with large, hard disk cartridges that can be removed and replaced easily. This type of drive is useful when large amounts of data must be stored and retrieved very quickly. It has unlimited storage capabilities because you can buy more disk cartridges as needed. It is also much faster than a tape drive. A single removable hard drive can typically store 45 megabytes of data.

The term *microdisk drive* refers to a 3.5-inch internal or external drive for storage of moderate amounts of data. Programs and files of information are handled using microdisks and a microdisk drive. A 3.5-inch microdisk refers to the overall dimensions of the disk that inserts into the drive. High-density, double-sided, 3.5-inch floppy disks can store over one megabyte of data.

Floppy disk drive refers to the drive that uses the flexible 5 1/4-inch disks. The drive can be either internal or external. This type of drive is slowly being replaced by the 3.5-inch drives. Note that the term "floppy" is also used to denote a 3.5-inch microdisk, even though the disk is not flexible or floppy, Fig. 10-14.

Tape drives

A *tape drive* is a storage medium that uses cassette-type tapes to store data. It *reads* (inputs) and *writes* (outputs) data much slower than a disk drive but can hold larger amounts of information. This type of drive is very useful for archival purposes. See Fig. 10-15.

Optical disk drives

An *optical disk drive* is presently the fastest and most efficient drive for storing computer data. It is ideal for electronic publishing because the artwork in computer page layout requires huge amounts of electronic storage and it can recall images quickly. An optical disk typically hold 650 megabytes of data per disk. It can retrieve data as fast as a hard drive and can hold much more data than a tape drive. The optical disk drive offers almost instant access to requested information, unlike tape drives that take considerable time to do a linear search for information. This capability increases the speed of changes or revision of your electronic publications.

A laser is used to write data to or retrieve data from a light-sensitive disk. This is the newest medium for reading and writing computer data. The laser deposits data using an intense light source rather than using a magnetic head to write data.

Some optical disk drives are *write once, read many (WORM)* drives. This means data can be stored once, but read many times. Newer optical drives are *erasable* which means the data can be removed from the disk so that new information can be written in its place.

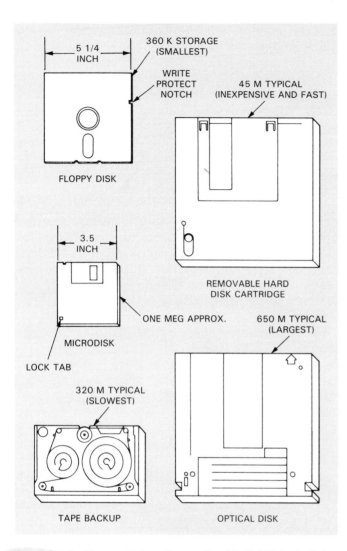

FLOPPY DISK

5 1/4 INCH

360 K STORAGE (SMALLEST)

WRITE PROTECT NOTCH

45 M TYPICAL (INEXPENSIVE AND FAST)

REMOVABLE HARD DISK CARTRIDGE

3.5 INCH

MICRODISK

LOCK TAB

ONE MEG APPROX.

650 M TYPICAL (LARGEST)

320 M TYPICAL (SLOWEST)

TAPE BACKUP

OPTICAL DISK

Fig. 10-14. Floppy disks refers to flexible 5 1/4-inch disks. Drive can be either internal or external. This type drive is slowly being replaced by smaller 3 1/2-inch disks. A removable hard drive is an external unit that has large, hard disk, cartridges that can be removed and replaced easily. Optical disks are capable of storing the largest amounts of data.

Fig. 10-15. A tape drive is a storage medium that uses audio cassette-type tapes to store data. It inputs and outputs data much slower than a disk drive but holds larger amounts of information.

Other optical disk drives are *read only,* which means the computer can only pull existing data off of the disk.

The *erasable optical drive* is desirable for electronic publishing because it lets you alter data as needed with revisions of manuscript and artwork. New information can be written an unlimited number of times and the data is very stable and safe. The laser does not write data on the surface of the disk, as with tape and magnetic disks. It writes the data inside the disk, below a clear, protective layer of plastic. The electronic data is buried safely inside the disk, away from dust, fingerprints, wear, and other sources of data damage. You can handle the disk, store it in a drawer, even mail it without danger of data loss, Fig. 10-16.

Two lasers are used during operation of an erasable optical drive. A high-power laser writes data into the disk and also erases data. It heats a tiny area in the disk to deposit or remove the digital information. A low-power laser is then used to read data from the disk.

Fig. 10-16. Optical disk drives are presently fastest and most efficient drive for storing large amounts of computer data. (Pinnacle Micro)

SCANNERS

A *scanner* is a device for changing hard copy images into digital computer data, Fig. 10-17. A scanner can be used to transfer a line drawing or photograph into computer memory. The image can then be modified or placed into an electronic page layout system, Fig. 10-18.

A *grayscale scanner* inputs various densities of black and white. A *color scanner* inputs full color illustrations. They are more expensive than a grayscale scanner.

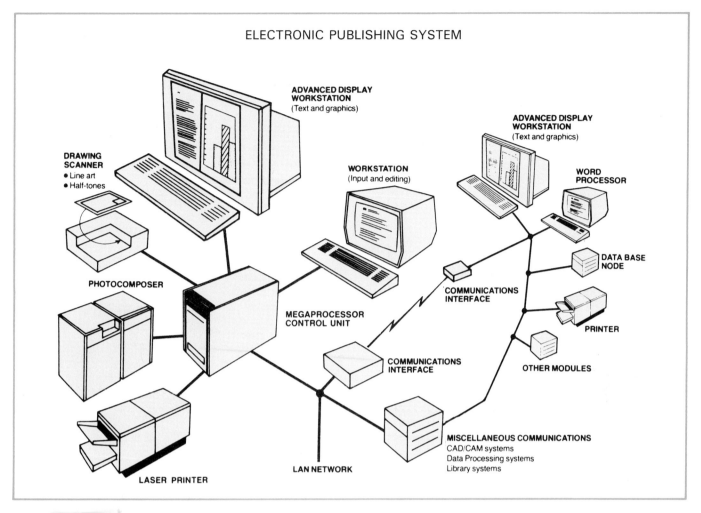

ELECTRONIC PUBLISHING SYSTEM

Fig. 10-17. A scanner can be used for changing hardcopy images into digital computer data. It can transfer a line drawing or photograph into computer memory. (Xyvision, Inc.)

Fig. 10-18. Once in scanner, image can then be modified by computer or placed into an electronic page layout system. (Hewlett Packard, Photography Copyright 1990, Jim Cambon)

A two-dimensional image is mounted on a table-like surface when using a *flatbed scanner*. This offers the advantage of mounting stiff or rigid copy on the scanner without bending it around the scanner drum. This prevents damage to the original.

A typical scanner has a resolution of 300 dpi, which is equal to the output of most laser printers. This is adequate quality for many print jobs. However, more expensive high-resolution scanners have over 1000 dpi and produce high-quality results without retouching or redrawing the scanned image.

Note! Large color flatbed scanners operate like conventional rotary scanners and small scanners discussed in this chapter. Refer to the index for the location of more information on scanners.

Optical character reader

An *optical character reader (OCR)* is a specialized input device for reading typed pages into the computer automatically. Typed pages are fed through the OCR and input as digital data. Spell checking programs are often used to help proof the scanned copy.

OCR programs are also available that allow a common 300 dpi scanner to read and import typed copy into a computer. These are discussed in the next section. Refer to Fig. 10-19.

OUTPUT DEVICES

An *output device* is used to place computer-generated images on a substrate (paper, photopaper, film, or printing plate itself with direct-to-plate systems). The output device converts the *softcopy* (image on computer screen) into *hardcopy* (images on paper or other carrier).

Dot matrix printers

A *dot matrix printer* is a low-quality output device that prints dots to form an image. It is commonly used to make proof copies of manuscript. It is not suitable as a hardcopy device for professional offset printing. A dot matrix printer uses a typewriter action to strike a carbon ribbon that places the images on the paper, Fig. 10-20.

Laser printers

A *laser printer* is a medium- to high-quality output device that can be used for many types of print jobs. A typical laser printer has a resolution of 300-400 dots per inch. The computer directs a laser beam across the surface of a spinning drum. The drum then rotates through a tray of *toner* (fine, black powder substance). The toner adheres to the areas of the drum hit by the laser beam. The toner is then deposited on the paper to produce the printed image. See Fig. 10-21.

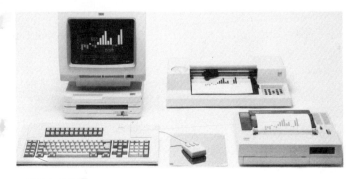

Fig. 10-20. A dot matrix printer is a low-quality output device that prints large dots to form an image. It is not suitable as a hardcopy device for professional offset printing. (International Business Machines Corp.)

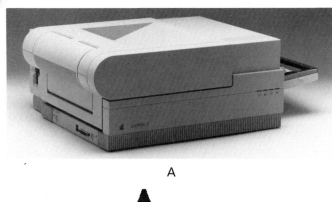

A

B

Fig. 10-21. A—A laser printer is a medium-quality output device that can be used for camera-ready copy. It has a resolution of 300-400 dots per inch. B—Laser printer output. (Apple Computers, AGFA Compugraphic Division)

Photo imagesetter

A *photo imagesetter* is a high-quality output device. The computer layout can be sent to this type of unit, producing a high-resolution image on light-sensitive paper or film. A photo imagesetter uses a CRT or laser beam to print at a resolution of 1200 dpi and higher. Since an imagesetter is very expensive to purchase and maintain, it is common to send jobs to a printing company having a computer and a photo imagesetter when the job requires superior print quality. See Fig. 10-22.

PostScript

PostScript® is a common computer format, or *language,* used to output information to a printer or photo imagesetter. It allows the computer to output high-quality images. Other computer output formats are available, but are not as common with electronic publishing. Refer to Fig. 10-23.

Fig. 10-19. Operator is using optical character reader (OCR). It will automatically typeset a full page in less than 30 seconds which is several times faster than best typewriter. It will also read almost any typewriter face. Problems are coded so data can be checked and corrected easily. (CompuScan)

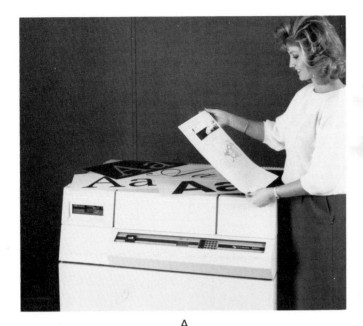

A

Aa
B

Fig. 10-22. A—A photo imagesetter stores up to 1000 typestyles on-line and can set type in sizes from 4 to 1000 points in 1/2-point increments. B—Photo imagesetter output. (Courtesy of AGFA Compugraphic Division)

ELECTRONIC PUBLISHING SOFTWARE

Electronic publishing software refers to the computer program designed to electronically produce page layouts. The programs are usually available on 3.5-inch or 5 1/4-inch program disks. *Program disks* give the computer the instructions needed to operate and perform specific tasks. Various kinds of programs are used during electronic publishing.

A *start-up program* provides initial information to the system so the computer can *"boot up."* The start-up, or boot up, program gives the computer the data needed to read signals from the keyboard, mouse, and to output information. The start-up program is usually kept on the hard drive. However, lower-quality systems require that a floppy disk be inserted every time the computer is turned on.

A program manual is generally supplied with the electronic publishing software. The *program manual* is a booklet that explains how to use the specific program. It should be studied thoroughly before attempting to work on a desktop publishing system.

WORD PROCESSING PROGRAMS

A *word processing program* allows you to input manuscript quickly and easily. Personal computers with word processing programs are starting to replace dedicated typesetting machines because of their add-

A

B

Fig. 10-23. A—Adobe Systems Incorporated originally developed PostScript® for Apple Macintosh Computers. Program is now available for IBM computers. B—Hewlett Packard-Apple desktop publishing system. (Courtesy of International Business Machines: Hewlett Packard, Photography Copyright 1990, Jim Cambon)

ed capabilities. Manuscript is input into a word processing program as you would type it on a typewriter, Fig. 10-24.

A word processing program provides easy modification of your typed manuscript. Refer to Fig. 10-25. A menu is often provided for selecting the font, its size, and style, Fig. 10-26. The program generally allows simple selection of margins: flush right, flush left, ragged right, centered, etc. You can modify leading, tabs, and other attributes at the press of a key or "click" of a mouse. See Fig. 10-27.

Word processing programs may also include features such as sorting or alphabetizing, indexing, and outlining. Some word processing programs are small page makeup programs. They can be used to produce single-page publications, Fig. 10-28. You can import artwork, wrap text around art, and alter style easily.

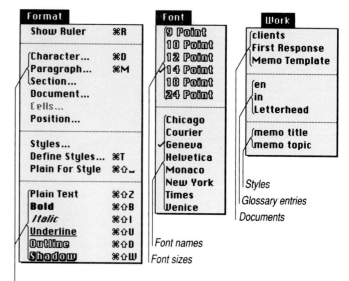

Character, paragraph, and section formats

A

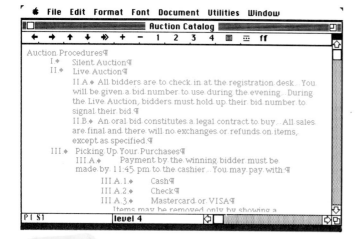

Fig. 10-24. An example of a word processing program. (Microsoft Word)

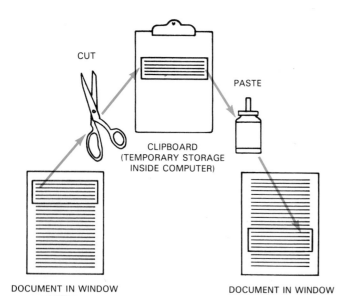

DOCUMENT IN WINDOW   DOCUMENT IN WINDOW

Fig. 10-25. On a computer, moving text is a process of cutting and pasting, as if you were using scissors and rubber cement. (Microsoft Word)

B

Fig. 10-26. A—A computer menu shows items such as type styles, font sizes, and kind of font. B—With some programs, illustrations can be combined with words to achieve interesting logos and designs. (Microsoft Word; Aldus FreeHand, used with permission from Aldus Corporation)

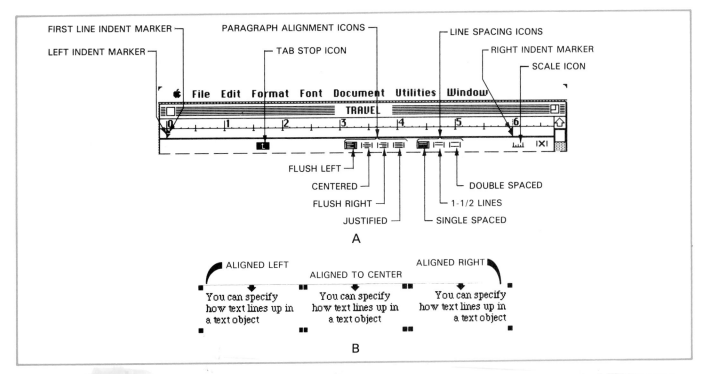

Fig. 10-27. A—An example of word processing program that allows for selection of flush left, flush right, ragged right, centered, etc. Headings, tabs, and other features can also be selected at press of a key or click of the mouse. B—This shows how text alignment appears as flush left, centered, and flush right. (Microsoft Word; Superpaint)

Fig. 10-28. This example illustrates a word processing program that can be used to import artwork, wrap text around artwork, and alter style easily. (Microsoft Word)

A *spell checking program* automatically detects words not stored in a *computer dictionary* (thousands of words loaded into memory). It is often a part of the word processing program, but it can also be an add-on program. When the spell checker is activated, the computer scans through all of the words that have been input. It stops on any word not in the computer dictionary. The spell checker then suggests alternate words for you. Pressing enter replaces the incorrect word with the correctly spelled word. This is a time-saving feature that can help reduce typographical errors. An example is shown in Fig. 10-29.

Grammar checking programs are capable of scanning manuscript and checking for unusual word and phrase usage. This can be helpful when a strict style of writing is needed.

A *thesaurus* is capable of suggesting alternate words with the same general meaning. This is useful when writing new copy. The computer shows other words that can be used in its place. Pressing enter automatically installs the new word into the copy.

OCR PROGRAMS

An *optical character recognition (OCR) program* allows a common 300 dpi scanner to read typed pages into computer memory. This saves the time of having to manually keystroke copy. Modern OCR programs have a high level of accuracy and can read various fonts and type sizes. They analyze difficult-to-decipher words using spell checking and grammar checking technology.

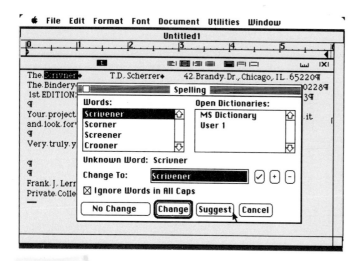

Fig. 10-29. A spell checking program automatically detects words not stored in a computer dictionary. When activated, spell checker scans all of the text. It stops on any word not in the computer dictionary and suggests possible correct words. (Microsoft Word)

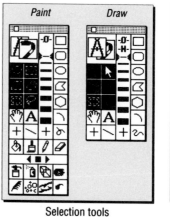

Selection tools

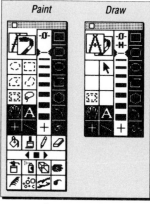

Common tools

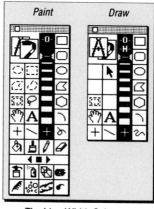

The Line Width Selector

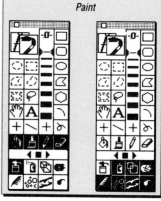

Built-in Paint tools   Plug-in Paint tools

Fig. 10-30. Drawing programs have replaced technical ink pen and hand drawing in most printing facilities and design studios. Study drawing and painting tool icons. Each purpose is described by visual image of icon, making program operation very intuitive. (Silicon Beach Software)

Many programs mark an unrecognizable word so that it can be visually checked later when proofing or rewriting the copy.

DRAWING AND PAINT PROGRAMS

Drawing programs are used to generate high-resolution line drawings. The drawings are *object oriented,* meaning that each line or shape is a mathematical description of the figure. You cannot edit objects one dot at a time, but you can resize, reshape, fill, split, and change them in other ways.

Drawing programs have replaced the technical ink pen and hand drawing in many modern facilities, Fig. 10-30. Most drawings can be completed more quickly and accurately using a computer. Drawings can also be stored on disk and used later or modified as needed for other jobs, Fig. 10-31. Once you draw a basic shape, it never has to be drawn again. You can start building a library of images for future publications.

A *paint program* produces a lower quality drawing when output because it is bitmapped, Fig. 10-32. *Bitmaps* consist of tiny dots displayed as pixels on the computer screen. Since the image is a series of dots, you can easily erase or modify small parts or individual dots that make up the drawing. However, the image is coarse when printed. See Fig. 10-33.

Note! The image quality of draw and paint programs may look the same on the monitor. Typically, this is because the monitor only shows 72 dpi, even though the object may be output at high resolution. A paint object that is output only has 72 dpi. This is the same resolution as the object shown on most monitors. Therefore, drawings created with paint programs are primarily used to make renderings for a high-quality printed product. Refer to Fig. 10-34.

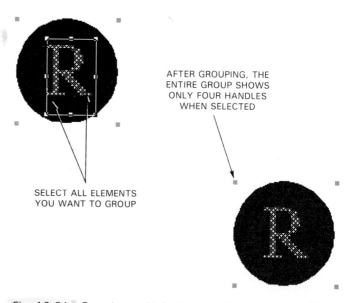

AFTER GROUPING, THE ENTIRE GROUP SHOWS ONLY FOUR HANDLES WHEN SELECTED

SELECT ALL ELEMENTS YOU WANT TO GROUP

Fig. 10-31. Grouping multiple elements freezes their positions relative to one another so you can manipulate them as a single element. Group can be resized, copied, or moved to desired position easily.
(Aldus FreeHand, used with permission from Aldus Corporation)

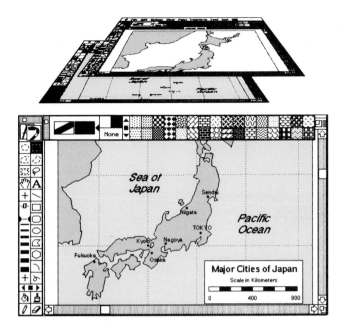

Fig. 10-32. In this paint-draw program, operator works with two layers. One is a draw layer and other a paint layer. Layers are superimposed on screen. (Silicon Beach Software)

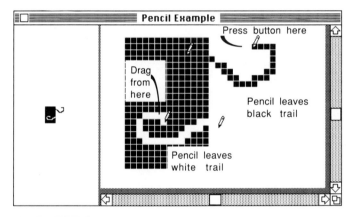

Fig. 10-33. In paint program, tiny dots are displayed as pixels on computer screen. Since image is a series of tiny dots, operator can easily erase or modify small parts or individual dots that make up image. (Superpaint)

Fig. 10-34. Image quality of paint objects can vary. Left: apple as a 300 dots per inch object. Right: apple at 72 dots per inch before it was edited. (Silicon Beach Software)

The *drawing and editing tools* in a drawing program allow you to create various shapes, Fig. 10-35. Each tool performs a specific function. The number and types of tools vary with each program. Some of the more common drawing tools are:

1. **Select tool** — The select tool might be an arrow, dotted box, or lasso-shaped icon. It is used to select drawing elements to be moved, modified, copied, or erased. An object is selected by moving the selection tool over the object and pressing the mouse button or enter key. The selected object flickers or otherwise indicates that it is selected.

2. **Shape tools** — Shape tools, as the name implies, are used to draw specific shapes: lines, circles, ellipses, squares, and other geometric shapes. The shape of the icon usually indicates its purpose. A variety of these tools may be provided depending upon the specific program. Sometimes, one tool can perform several functions by pressing additional command keys.

3. **Freeform tools** — These tools are used to draw any freeform shape: lines, smooth curves, and corners. They provide an accurate way of drawing complex, irregular shapes, Fig. 10-36.

4. **Trace tool** — This tool is used to convert bitmapped objects and grayscale images into draw images. The computer detects the shape of the image and

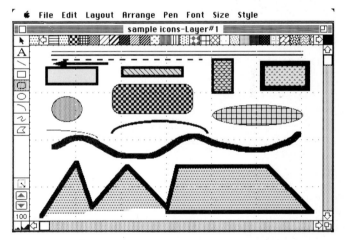

Fig. 10-35. Drawing tools in drawing program allow operator to make various shapes quickly. (Apple Macintosh)

Fig. 10-36. Freeform tools are used to draw any shape including freehand lines, smooth curves, and corners. These tools provide an accurate way of drawing complex, irregular shapes. (Silicon Beach Software)

automatically generates lines that represent the original image. The trace tool can be set to produce different effects. The drawing can then be output as a high-resolution line drawing. See Figs. 10-37 and 38.

5. **Editing tools** — These tools can be used to rotate, mirror, scale, skew, or smooth images. The *rotate tool* is used to position images or words at various angles. The *mirror tool* is used to generate an opposite image, as if you were looking at it in a mirror. See Fig. 10-39. The number and type of editing tools will vary with the specific program.

6. **Break tool** — The break tool is used to divide a continuous path into two or more separate paths. A continuous circle, for example, can be divided into two semicircles with this tool or command.

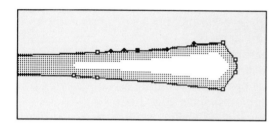

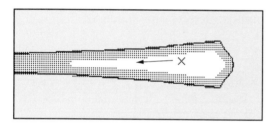

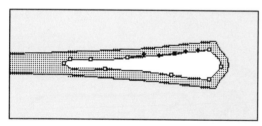

Fig. 10-37. Trace tool can be used to convert paint (bitmapped) objects and grayscale images into high-resolution, draw-type images. (Adobe Illustrator)

ORIGINAL BITMAPPED IMAGE   AUTOTRACED IMAGE

Fig. 10-38. Trace tool command lets operator take bitmapped images that have been created, scanned, or imported and convert them into draw objects. This can save hours of manual tracing or drawing when an object-oriented version of bitmap art is desired. (Silicon Beach Software)

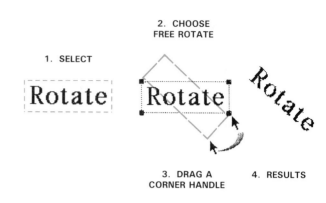

1. SELECT   2. CHOOSE FREE ROTATE   3. DRAG A CORNER HANDLE   4. RESULTS

Fig. 10-39. Transformation tool can be used to rotate, mirror, scale, skew, or smooth images. (Superpaint)

7. **Magnification tool** — This tool is used to zoom in on a section of a drawing. Some programs have several levels of magnification for doing detailed work. This allows you to zoom in to carefully work on parts of your drawing.

8. **Text tool** — The text tool is used to add copy to your drawings. You can enter words and numbers into your drawing with the keyboard after selecting the tool or pressing the correct command key.

3-D PROGRAMS

A *3-D program* is used to draw very precise and accurate illustrations that show the three dimensions — height, width, and depth. Most drawing programs are designed for two-dimensional images, but can be used to manually draw in perspective or in three dimensions. Some 3-D programs automatically calculate the shape of the third dimension after you draw two, two-dimensional views of the image. See Figs. 10-40 and 41.

The cost and capabilities of 3-D programs vary. Some 3-D programs only output as bitmapped images, which are acceptable for some effects and publications. Others output as object-oriented images.

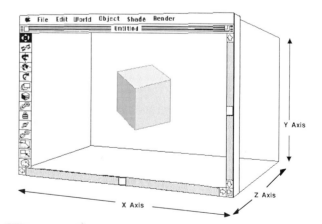

Fig. 10-40. 3-D program can be used to make very precise and accurate illustrations that show three dimensions: height, width, and depth. (Paracomp Swivel 3D)

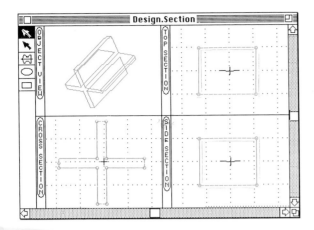

Fig. 10-41. Designing an object in this 3-D program involves drawing and editing three sectional views that define shape of object. Counterclockwise, four views shown are object view, cross-section, side section, and the top section.
(Paracomp Swivel 3D)

Drawing program terms

There are some typical terms that you should understand before drawing with any computer program.

1. **Layer** — Refers to the plane on which a drawing is placed. Layers are similar to transparent overlay. See Figs. 10-42 and 10-43.

2. **Duplicate** — Used to make an exact copy of an image. If you are drawing a top view of a car, draw one tire and then duplicate it three more times as shown in Fig. 10-44.

3. **Skew** — To slant or distort an image. After selecting the object, you can use the skew command while dragging to change the shape or angle of the object, Fig. 10-45.

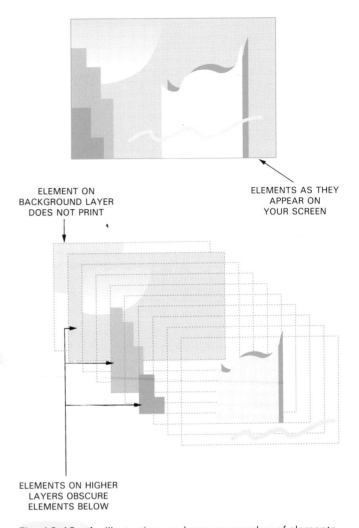

ELEMENT ON
BACKGROUND LAYER
DOES NOT PRINT

ELEMENTS AS THEY
APPEAR ON
YOUR SCREEN

ELEMENTS ON HIGHER
LAYERS OBSCURE
ELEMENTS BELOW

Fig. 10-43. An illustration can have any number of elements. Some programs provide 200 printing layers and one nonprinting layer where operator can distribute elements.
(Aldus FreeHand, used with permission from Aldus Corporation)

Fig. 10-42. Layering refers to drawn objects placed in front of one another, although they do not necessarily overlap.   (Adobe Illustrator)

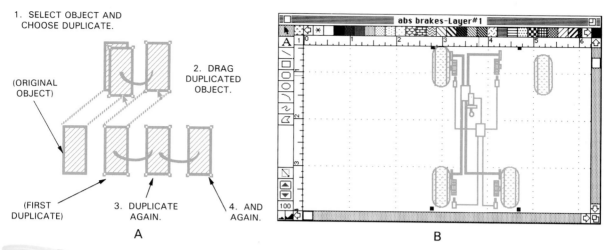

Fig. 10-44. A—Duplicating is done to make an exact copy, or clone, of an object. This illustration shows how three duplicates were quickly made from original object. B—In drawing top view of an automobile, operator draws one tire and then duplicates three more tires with Duplicate command.   (Silicon Beach Software)

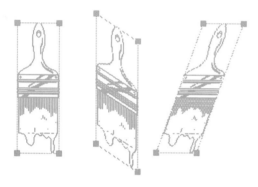

Fig. 10-45. Slant command is used to skew or distort an object. Slant command allows operator to change shape or angle of object.   (Silicon Beach Software)

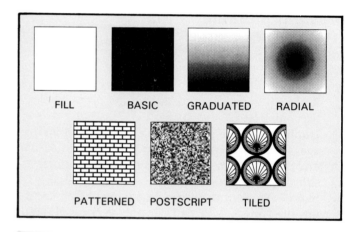

Fig. 10-46. Fill means to add a texture or color to inside of a closed path or shape.
(Aldus FreeHand, used with permission from Aldus Corporation)

4. **Save** — Refers to storing an electronic image to disk. This makes the image data permanent, even if the computer is shut down.
5. **Fill** — To add a texture or color to the inside of a closed shape. It can be used to produce a variety of effects. See Fig. 10-46.
6. **Template** — A template is a bitmapped or scanned image used as a drawing guide, Fig. 10-47. If a line drawing is scanned with a low-resolution scanner, it can be used as a template. A *template* provides a guide so that you can quickly and accurately redraw the image at high resolution. Refer to Fig. 10-48.
7. **Zoom** — Refers to a command that enlarges a drawing for doing detailed work.
8. **Toolbox** — The area with icons for selecting tools or commands that are used for your drawing.
9. **Blend** — Used to draw a series of images that progressively change. This might be a change from one line drawing shape to another, or a change from a pattern-filled area to a solid-filled area. See Fig. 10-49.

Fig. 10-47. Template is a bitmapped or scanned image used as a drawing guide. The template appears less distinct than the image being created. Templates are created by scanning an existing image or by converting a drawing created with another application.   (Adobe Illustrator)

TEMPLATE ONLY   ARTWORK AND TEMPLATE   ARTWORK ONLY   PREVIEW ILLUSTRATION

Fig. 10-48. A template provides a guide so that the operator can quickly and accurately redraw the image at high resolution.　(Adobe Illustrator)

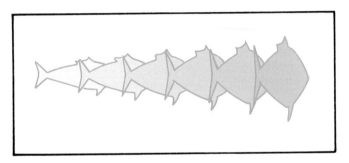

Fig. 10-49. Blending allows the operator to create a series of intermediate shapes between two objects. Depending on the way the operator paints the objects being blended, airbrush effects can be created such as shading, highlighting, or different line weights and shades of gray.　(Adobe Illustrator)

10. **Grid** — A grid is a drawing guide similar to graph paper.

11. **Preview** — This view shows exactly how the drawing will look when printed. The preview mode is needed so you can see the appearance of the image before it is output to a printer or imagesetter.

12. **Group** — Refers to making several individual images or objects into one complex object. This allows you to move, drag, or reshape the objects together.

13. **Ungroup** — As the name implies, ungroup can be used to make a grouped set of objects independent.

14. **Modifiers** — Keys on the keyboard that change the function of a command key. See Fig. 10-50.

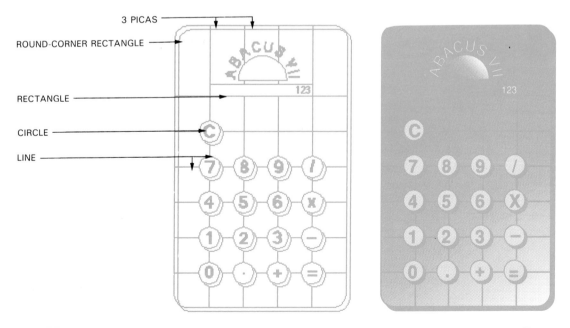

Fig. 10-50. Special basic shape tools, called modifiers, can be used for round-corners, squares, ellipses, and lines.　(Aldus FreeHand, used with permission from Aldus Corporation)

Electronic clip art

Electronic clip art consists of generic images stored on computer disk. In most cases, the drawings can be modified and then used as needed. The quality and usefulness of clip art can vary considerably. Review samples of the printed clip art before purchasing a clip art package. See Fig. 10-51.

Electronic halftones

An *electronic halftone* is a continuous tone photograph that has been scanned into a computer for manipulation and layout in the page layout program. This technology has improved tremendously.

Image processing programs are available for retouching the halftones and line drawings electronically. They can be used to enhance, retouch, and compose scanned and other bitmapped images. Once the photo is scanned into the computer, each pixel or group of pixels can be modified in any way to enhance the image. The halftone can be output on a laser printer for proofing or may be used for an average-quality publication. It can also be output on a photo imagesetter for high-quality results.

Fig. 10-51. Electronic clip art consists of images stored on computer disk. The quality and usefulness of clip art can vary considerably. (3G Graphics)

Halftone processing programs allow you to set the screen value (coarseness) before generating hard copy or film. You must set the screen size before generating the image. Then, the printer can use the film or photopaper image of the completed layout at the same size for good printing results. You can also use your electronic layout with halftones to go directly to plates, bypassing any photographic screening or stripping stages.

PAGE LAYOUT PROGRAMS

A *page layout program* is designed to quickly combine electronically stored manuscript, line art, and halftones on the computer screen to produce a layout. This eliminates the need for manually drawing, cutting, and bonding down galleys of copy and line drawings. You can import and integrate this material electronically. The computer and its accessories replace the drawing table, T-square, art knife, stat camera, and other conventional tools of the artist. This can save considerable time, effort, and expense if done properly.

The computer operator eliminates or combines many of the conventional steps of document preparation. These steps typically include:

1. Re-keystroking copy at the typesetting machine.
2. Physically cutting and bonding materials to a layout sheet.
3. Setting up and pasting running heads on each page.
4. Creating line drawings with a technical ink pen.
5. Using a proportional scale to size art.
6. Photographing line art to size with a stat camera.

The capabilities and procedures for using page layout programs varies considerably. Some programs are more user-friendly than others. Some programs use a working area that resembles a conventional drawing board and layout sheet with icons that resemble layout tools. Other programs are more keyboard and word oriented.

ELECTRONIC PAGE LAYOUT

Electronic page layout is the end result of using a page layout program on a computer. It is either hard copy on paper, photopaper, film, or plates generated and output using the electronic data from a computer. This section of the chapter summarizes how to do page layout on a computer. Keep in mind that computers and their programs are designed with different operating systems. You must always refer to the operating manuals for the specific computer and programs.

Fig. 10-52 illustrates the basic function of electronic page layout. A dedicated word processing program stores the manuscript. A drawing program contains the artwork. The data in each of these programs can be imported into the page layout program for organization and manipulation. Fig. 10-53 shows how other devices can be used to import even more images into the computer for electronic layout.

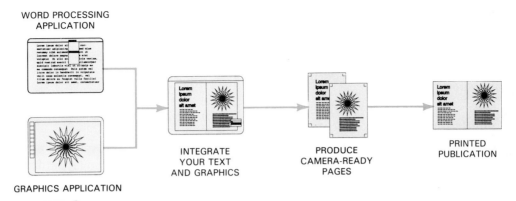

WORD PROCESSING
APPLICATION

GRAPHICS APPLICATION

INTEGRATE
YOUR TEXT
AND GRAPHICS

PRODUCE
CAMERA-READY
PAGES

PRINTED
PUBLICATION

Fig. 10-52. Page makeup programs allow the operator to combine text and artwork into a planned publication design. The operator can resize, trim, and reposition drawings, charts, and graphs, and use drawing tools to create lines, boxes, and circles, or other special design effects. (Aldus PageMaker, used with permission from Aldus Corporation)

Importing data

Importing data refers to moving manuscript, style information, and artwork into the page layout program from other programs or applications. Images can be imported from a scanner, draw program, video camera, paint program, or other electronic input device. See Fig. 10-53.

File format is the type of language used to write the information into computer memory or storage. The file format must be the same as the page layout program being used. It is possible to convert some formats over to others for importation. A few of the most common file formats are:

1. **ASCII** – This is the most common text file format. ASCII is the abbreviation for American Standard Code Information International. The text data is stored in the same order as written. ASCII is often called a print or text file.

2. **CCITT** – Consultative Committee International Telegraph And Telephone is a series of formats for facsimile (FAX) transmission over telephone lines.

3. **CGM** – Computer Graphics Metafile is a device-oriented file format that records the vector and raster commands that compose an image. *Vector* is a method of computer graphic coding that codes a series of LINES to produce an image. *Raster* is a computer graphic coding technique that uses individual image ELEMENTS to produce the image.

4. **CUT** – This is a popular file format supported by graphic adapters for IBM PC and compatible computers.

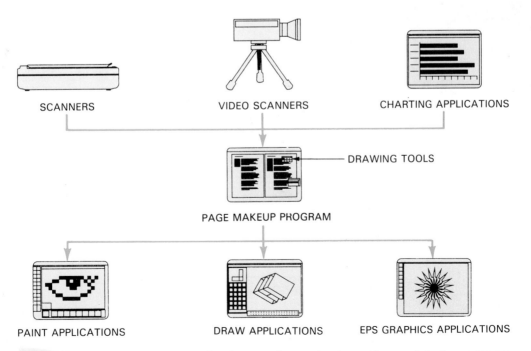

SCANNERS

VIDEO SCANNERS

CHARTING APPLICATIONS

DRAWING TOOLS

PAGE MAKEUP PROGRAM

PAINT APPLICATIONS

DRAW APPLICATIONS

EPS GRAPHICS APPLICATIONS

Fig. 10-53. Separate graphics applications can be used to import images into the computer for electronic page layout. (Aldus PageMaker, used with permission from Aldus Corporation)

5. **EPS®** — The Encapsulated PostScript® file format is used with IBM PC and compatible computers to allow them to print on PostScript® devices.
6. **GIF** — This stands for Graphic Interchange Format. It is an independent raster image compression and encoding standard.
7. **PICT** — An Apple computer file format consisting of both raster and vector Quickdraw® commands for imaging graphics and text on the computer screen. PICT is automatically converted to PostScript® when output to a compatible laser printer or photo imagesetter.
8. **POSTSCRIPT®** — This was the first commercially available device-oriented file format or page description language. It is very common with desktop publishing.
9. **RIFF** — This is a raster file format for storing images in either a compressed or uncompressed state.
10. **TIFF** — The Tag Image File Format is a very flexible file format used in many applications.

Note! There are many other file formats. Refer to the equipment and program manuals for more information.

Computer layout area

Computer layout area refers to the section on the monitor for combining your manuscript and artwork into an organized layout. The style of the computer layout area varies with the specific program. Fig. 10-54 shows a common, intuitive page layout area. Note its arrangement. Fig. 10-55 shows a menu for selecting commands.

Computer layout terminology

Some of the terms used with page layout programs, other than the ones already explained in this chapter, include:

1. **Actual size** — This is a view in an electronic window that shows the image at its correct, printed size. *Preview size* is usually a reduced size that shows how the document appears on one or more printed pages; for example, it might display two consecutive pages of a layout for checking general organization of images. The preview size is also needed on smaller monitors that cannot display a full page at actual size. Refer to Fig. 10-56.
2. **Alignment** — This refers to how the manuscript aligns on the page: flush right, flush left, ragged right, ragged left, or centered.
3. **Anchored** — This means an element is fixed or attached to a location on the page. If you try to drag a line drawing to make it larger or smaller on the computer screen for copyfitting, one corner of the drawing remains anchored in place.
4. **Application default** — A specific setting or style that is automatically loaded by the program when it is opened.
5. **Archive** — This refers to updating files with any corrections or changes as you lay out the document.
6. **Caption** — This is special text commonly used to explain an illustration. It is separate from the body copy, but can be part of an illustration.
7. **Bounding box** or **frame** — This is a rectangular box that defines a space on your electronic page layout. You can often create a bounding box to define the

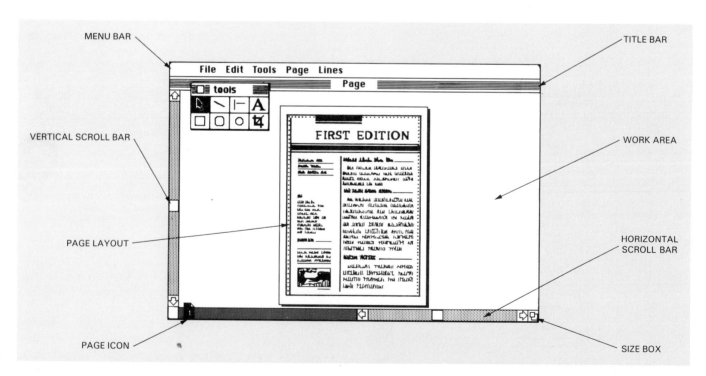

Fig. 10-54. Note basic terms used to describe images on the screen of this particular desktop publishing system. The terms and locations of areas can vary with the type of system.

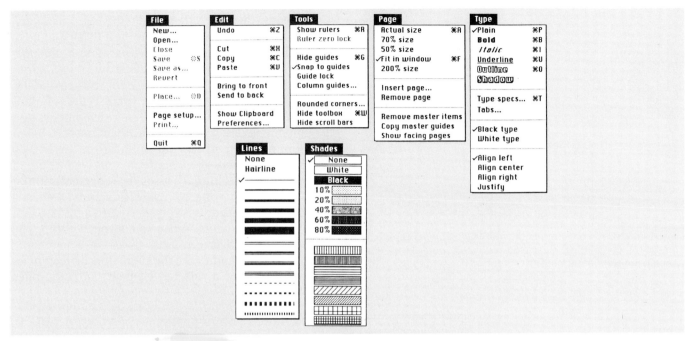

Fig. 10-55. This is an example of a menu for selecting commands.
(AldusPageMaker, used with permission from Aldus Corporation)

space for text or a graphic by clicking and dragging your mouse.

8. **Cancel** — To halt a selection.

9. **Clipboard** — This is a temporary, invisible storage place for copy or artwork. Elements may be copied to the clipboard in one application or program and then can be put in position in another application or program. This is a very useful feature available on some platforms.

10. **Close** — A command that exits the current document, but does not quit the application or program.

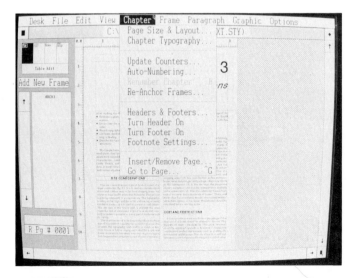

Fig. 10-56. Actual size on the computer screen shows the page image at its correct printed size. Preview size is usually a reduced size that shows how the document will look on one or more printed pages.

11. **Column guides** — These are the dotted, nonprinting, vertical lines that denote the left and right sides of a column in a publication.

12. **Comp** — This is an abbreviation for "comprehensive." It refers to a complete, detailed dummy showing how the finished product will look when printed.

13. **Crop** — In electronic publishing, this means to trim a graphic or illustration using the computer.

14. **Default** — The settings for the program are set automatically to commonly used values or styles.

15. **Design element** — A graphic image that you design into the publication to make it more attractive or informative.

16. **Desktop** — The working area on your computer screen for completing your electronic page layout. The specific program determines the organization and tools on your desktop. An example is shown in Fig. 10-57.

17. **Dialogue box** — A square or rectangle used to present choices you can make about specific functions, Fig. 10-58.

18. **Downloadable font** — An extra font that can be purchased separately and installed into your computer system.

19. **Edit** — This refers to a change in your electronic document. You can edit graphics, text, and specifications for your publication.

20. **Export filter** — A program that changes how the computer information is organized so it can be imported into another program using a different format.

21. **Folder** or **directory** — It is a place where you keep information similar to a conventional file folder

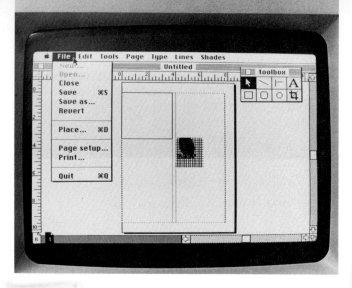

Fig. 10-57. Desktop refers to the working area on a computer screen for completing electronic layout or publishing. The specific program determines the organization and tools on the desktop.

or phone directory. It can contain a document, artwork, or a publication. See Fig. 10-59.

22. **Formatting**—This is the type, paragraph, and artwork specifications for the electronic publication. Also refers to initializing a disk for storage of data.

23. **Guides**—These are one of three types of nonprinting lines—margin guides, ruler guides, and column guides. You can create them to speed alignment of elements in your publication. Fig. 10-60 shows some types of guides.

24. **Halftone screening**—This is the representation of a continuous tone by using tiny dots. When a photo is scanned, it is converted into a bitmap that contains small dots. When you output a photo, you must assign a screen size.

25. **Handles**—Handles are the small dots or rectangles that surround a selected element. In some page

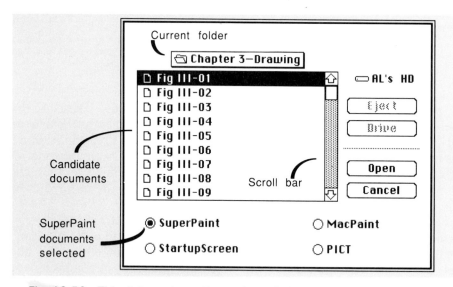

Fig. 10-58. This dialogue box gives various choices that may be selected by the operator. (Superpaint)

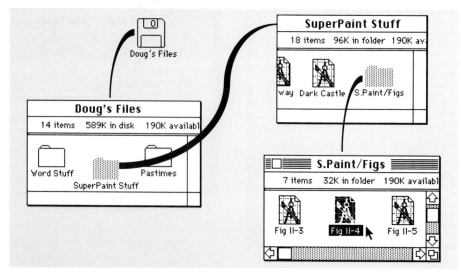

Fig. 10.59. A folder or directory is similar to a conventional file folder or phone directory. It is used to store information such as artwork, document, publication, or other items. (Superpaint)

214

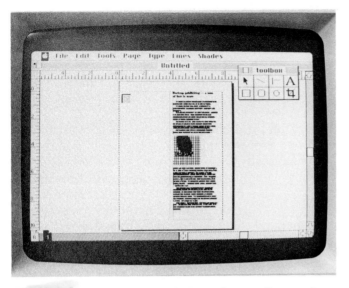

Fig. 10-60. Guides are nonprinting reference lines and are available as margin guides, ruler guides, and column guides. They are used to speed alignment of elements in a publication.

layout programs, you can use handles for changing the shape and size of elements to rapidly copyfit your layout.

26. **I-beam** or **text cursor** — An icon used when selecting or working with manuscript in your electronic document.

27. **Import filter** — Part of a program that converts data into another file format so that the data can be understood by the program being used. An import filter might convert a line drawing into another file format so that it can be positioned on your electronic layout sheet.

28. **Key combination** — Two or more command keys or a command key and character key are pressed at the same time to perform a different function.

29. **Layout grid** — An underlying design or layout plan for a publication.

30. **List box** — An area in a dialogue box that provides more options.

31. **Manual text flow** — A text flow option that allows you to assemble text on the page until it meets an illustration or the bottom of the column.

32. **Master items** — Items on the master pages, such as running heads, page numbers, guides, etc.

33. **Master page** — A page in your publication that has defined styles, such as column widths, running head locations, etc.

34. **Page description language** — A computer language that includes commands for creating text and graphics on the electronic layout page.

35. **Page view** — Refers to the size of the electronic layout sheet, as it will appear in the publication window: full page view, close-up, or zoom in or out. See Fig. 10-61.

36. **Palette** — A small window on the screen with a list of styles, colors, tools, and other attributes.

37. **Print area** — The area on the electronic layout sheet that will print when output to hardcopy.

38. **Printer driver** — The program information that allows your electronic layout system to be accepted and printed by the output device.

39. **Replace** — This command allows you to replace a specific image block with a new image while retaining the same form.

40. **Ruler guides** — Horizontal and vertical, non-printing, page lines that are extensions of the tick marks on the rulers.

41. **Rulers** — Horizontal and vertical rules used as guides for layout work on the computer screen. They can be scaled in inches, millimeters, or picas.

42. **Rules** — Refers to printed lines that are added to a page to enhance the design, organization, and aesthetics of the publication.

43. **Scaling** — Resizing of a graphic or a page during electronic layout.

44. **Scroll bar** — A feature that allows you to move around in your layout by dragging along or clicking on horizontal and vertical bars on each side of the page layout.

45. **Spooler** — A program that allows you to continue working on an electronic layout while printing publications. This reduces the time you need to wait for the printing to be completed. See Fig. 10-62.

46. **Stacking order** — This refers to how the text and artwork overlap or "stack" on the page. It affects what part of each image will show in some situations.

47. **Standoff** — Refers to the distance between a boundary and a graphic.

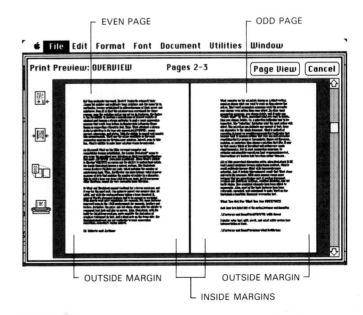

Fig. 10-61. Page view refers to the size of the electronic layout sheet as it will appear in the publication window. This example illustrates full page views of two pages of a publication. (Microsoft Word)

Fig. 10-62. By using spooler applications, the operator can continue electronic layout while printing out publication pages. This greatly speeds work since the operator does not have to wait until printing is completed.

48. **Submenu** — Provides additional selections other than those in the main menu. Refer to Fig. 10-63.
49. **Supplemental dictionary** — An additional dictionary of hyphenation, technical words, etc., that can be added to some page layout programs.
50. **Template** — Any image that is used as a tracing guide.
51. **Tag** — A term used in some programs for assignment of a format to a block of text or other image.
52. **Title bar** — A name usually at the top of your display that gives the name of the publication being produced.

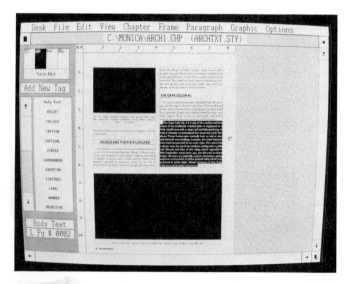

Fig. 10-63. In addition to a main computer program menu, submenus can be selected that give additional function selections.

Using a page layout program

The specific instructions for using a page layout program vary with the program and type of computer. Review the operator's manual of any page layout program and computer before working on any document or job. Generally, you will use the same principles needed for conventional or manual layout.

First, set the desired parameters or style for the job in the word processing or page layout program: font, type size, heads, leading, column width, and other attributes. Then, input the text in the word processing program. Use the spell checking program to check for typographical errors. Proofread the copy to check for grammar and context errors that are not caught by the spell checking program. The computer thesaurus is also helpful for selecting new words.

Use your drawing program to generate any high-resolution line art for the job. Add callouts and leaders to the drawing, if required, while using the drawing program. Some drawing programs allow you to automatically cut color screens or overlays for the artwork. If you have a complex piece of art and want to cut a screen to surround it with color, use the drawing program to create the screen. This could save hours of trying to use an art knife to manually cut the screen.

If you have access to a high-resolution scanner, photographs can also be imported into an image processing program and readied for the layout. A screen ruling can then be assigned to the photographs, and the photographs can be retouched as necessary. Since a high-resolution scanner is expensive, many people use conventional methods of stripping photos into their jobs after doing the layout electronically.

When all files to be used in the page layout program are ready, you may begin the electronic layout. First, set the style or parameters for your layout sheet or master page. Assign the number of columns, column widths, page number location and size, and other attributes. You can then import text and graphics from the other programs.

In some jobs, it might be desirable to first import all of the manuscript into the layout sheet columns. You can then import the artwork and place it in the proper position in the layout. The text will automatically be readjusted. This is useful if you want to place artwork in the vicinity of the figure references. In other jobs, you might want to import the artwork first and place it on your electronic layout sheet in its general location. The text can then be positioned around the artwork. Another method of layout provided by some programs allows you to manually flow text until you reach a graphic. The artwork is then placed in position and the text flows again until the next piece of artwork.

Save your work frequently when doing electronic page layout. When your work is not saved, it remains resident in the computer's integrated circuit chips and is very vulnerable. Power surges, computer bugs, and other problems can erase your information and waste hours of work. Make back-up copies of your disks or tapes after each layout session to further protect your work.

Remember that procedures vary with the program and type of computer you are using. Some programs require you to tag all captions, artwork, and text. In other instances, you can type the captions in the word processing program. This lets you import the captions with the text. This saves considerable time because you do not have to tag and position every caption separately.

SELECT YOUR PROGRAMS AND COMPUTER CAREFULLY

The procedures for using word processing programs, drawing programs, paint programs, page layout programs and computers are very different. Select your programs and computer carefully. Some computers are easier to learn and keep running than others. Take the time to research different brands of software and hardware and compare their strengths and weaknesses, Fig. 10-64.

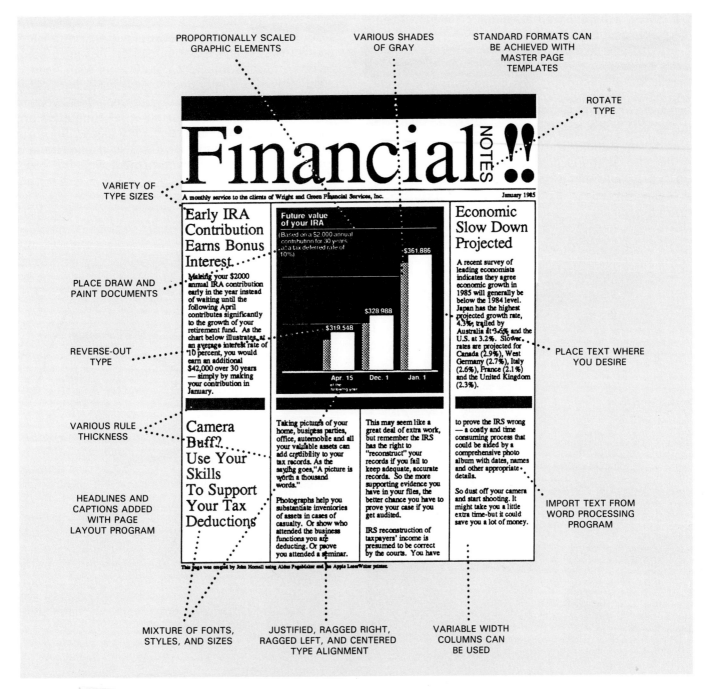

Fig. 10-64. Before purchasing new computer software and hardware, research the different brands and compare their features. (Aldus PageMaker, used with permission from Aldus Corporation)

Make a wise investment in a computer and its programs by finding and studying existing electronic publishing systems in industry. The opinion of end-users is generally more reliable than that of a sales clerk in a computer store. Talk to the people that are successfully running electronic publishing systems in industry. Learn from their experience and you will be able to make a more intelligent decision on your electronic publishing hardware and software.

Determine the amount of time it takes to learn the programs. Research the programs that are needed now and the programs that will be needed in the future as your electronic publishing program grows. Purchasing user-friendly programs is critical. You will need to use several programs for maximum efficiency during electronic publishing.

The trend in program development is to use *intuitive icons* or small, self-explanatory images. Drop-down menus are also becoming more common. This trend is important since your electronic publishing program will be similar to your word processing program, drawing program, OCR program, and other programs.

Program standardization makes it possible for you to become computer literate more efficiently. Try to select programs that use similar commands and icons.

If you learn the standard commands, menu locations, and icons for one program, you will already understand how to manage and quickly utilize another standardized program. You won't have to "start from scratch" looking for new commands and memorizing key combinations. Many computer manufacturers and program developers are now realizing the importance of standardization.

HIGH-END SYSTEMS

The term *high-end system* generally refers to a powerful computer-processing system. It is an expensive version of a desktop publishing system, but with more capabilities. The gap between personal computers and high-end systems is closing rapidly. PCs are becoming much more competent. See Fig. 10-65.

The term *front end* basically describes the equipment used to process a job before it goes to the press department, Fig. 10-66. Now that personal computers are more capable, "front end" generally means all of the electronic devices that process the electronic text and images for a publication before entering the press room.

Fig. 10-65. This is a powerful high-end computer processing system. (Xerox)

Fig. 10-66. This front end system is used to process a job before it goes to the press department. Personal computers are now capable of processing electronic text and images for publication before entering the press room.
(High Technology Solutions)

POINTS TO REMEMBER

1. Electronic publishing, sometimes called desktop publishing, uses a computer to generate and combine text and art on a display screen before being output as hardcopy or printing plates.
2. Desktop publishing hardware refers to the physical equipment or components that make up the computer system.
3. A personal computer (PC) is a small computer, typically consisting of a keyboard, processor, disk drive, monitor, and a mouse.
4. A computer keyboard is the primary input device for a computer. You can input words, numbers, and commands on a keyboard to control the operation of the computer.
5. A mouse is a small, hand-operated device that allows input into the computer.
6. Dragging refers to pressing the mouse button while moving the mouse.
7. The display screen or monitor is a television-type picture tube or CRT (cathode ray tube) for showing the information being processed on the computer.
8. A monochrome monitor only displays one color. It is sufficient for some desktop publishing jobs. A color monitor displays in full color.
9. Monitor resolution refers to the number of dots or pixels visible on the computer screen.
10. The term "WYSIWYG" means "What You See Is What You Get." The image on the computer monitor is similar to what will be produced when printed.
11. The processor is the "electronic brain" of a personal computer. It has complex electronic logic circuits capable of converting and storing input as digital (numerical) data.
12. The term "RAM" means "Random Access Memory." It refers to the amount of memory inside the computer's electronic circuits or chips.
13. A disk drive is the most common storage device for computer data. It uses tiny electrical charges on the surface of a magnetic

disk to remember any information input by a computer.

14. A tape drive is another storage medium that uses cassette-type tapes to store data.

15. An optical disk drive is presently the fastest and most efficient drive for storing computer data. It is ideal for electronic publishing because the art in computer layout requires huge amounts of electronic storage.

16. A scanner is a device for changing hard copy images into digital computer data.

17. A dot matrix printer is a low-quality output device that prints large dots to form an image.

18. A laser printer is an average-quality output device that can be used for many print jobs.

19. A photo imagesetter is a high-quality output device. The computer layout can be sent into this type of unit and a high-resolution image will be formed on light-sensitive paper or film.

20. PostScript® is a common computer format, or language, used to output information to a printer or photo imagesetter.

21. Program disks are used to give the computer the information needed to operate and perform specific tasks. Various kinds of programs are used during electronic publishing.

22. A word processing program allows you to input manuscript quickly and easily. In fact, computers are starting to replace dedicated typesetting machines because of their added capabilities.

23. A spell checking program will automatically detect words not stored in a computer dictionary.

24. A computer thesaurus is capable of suggesting alternate words with the same general meaning.

25. An optical character recognition (OCR) program allows a common 300 dpi flatbed scanner to read typed pages into computer memory.

26. Drawing programs can be used to efficiently generate high-resolution line drawings because they are object oriented.

27. A paint program produces a low-quality drawing when output because it is bitmapped. Bitmaps consist of tiny dots displayed as pixels on the computer screen.

28. A 3-D program can be used to draw very precise and accurate illustrations that show three dimensions—height, width, and depth.

29. Electronic clip art consists of images stored on computer disk. They are drawings that can be purchased on disk and then modified and used as needed.

30. An electronic halftone is a continuous tone photo that has been scanned into a computer

for manipulation and layout in a page layout program.

31. Image processing programs are available for electronically retouching the halftones and line drawings.

32. A page layout program is designed to quickly combine electronically stored manuscript, line art, and sometimes, halftones on the computer screen to produce a layout.

33. Importing data refers to moving manuscript, style information, and artwork into a page layout program from other programs or applications.

34. File format is the type of language used to write the information into computer memory or storage.

KNOW THESE TERMS

Electronic publishing, Desktop publishing, Hardware, Software, Personal computer (PC), Keyboard, Cursor keys, Cursor, Function keys, Mouse, Mechanical mouse, Puck, Electronic mouse, Dragging, Trackball, Monitor, Monochrome monitor, Color monitor, Monitor resolution, Jaggies, Stair stepping, Dots per inch (dpi), What you see is what you get (WYSIWYG), Processor, Digital, Random access memory (RAM), Storage device, Disk drive, Hard drive, Internal hard drive, External hard drive, Files, Removable hard drive, Microdisk drive, Floppy disk drive, Tape drive, Read, Write, Optical disk drive, Write once—read many (WORM), Scanner, Grayscale scanner, Color scanner, Flatbed scanner, Optical character reader (OCR), Output device, Softcopy, Hardcopy, Dot matrix printer, Laser printer, Toner, Photo imagesetter, PostScript®, Language, Program disk, Start-up program, Boot up, Program manual, Word processing program, Spell checking program, Computer dictionary, Grammar checking program, Computer thesaurus, Optical character reader program, Drawing program, Object oriented, Paint program, Bitmap, Drawing and editing tools, Select tool, Shape tool, Freeform tools, Trace tool, Editing tools, Rotate tool, Mirror tool, Break tool, Magnification tool, Text tool, 3-D program, Layer, Duplicate, Skew, Save, Fill, Template, Zoom, Toolbox, Blend, Grid, Preview, Group, Ungroup, Modifier, Electronic clip art, Electronic halftone, Image processing program, Page layout program, Import, File format, American Standard Code Information International (ASCII), Consultative Committee International Telegraph And Telephone (CCITT), Computer Graphics Metafile (CGM), Vector,

Raster, CUT, Encapsulated PostScript® (EPS®), Graphic Interchange Format (GIF), PICT, RIFF, TIFF, Computer layout area, Actual size, Preview size, Alignment, Anchored, Application default, Archive, Caption, Bounding box, Frame, Cancel, Clipboard, Close, Column guides, Comp, Crop, Cutout, Design element, Dialogue box, Export filter, Import filter, Folder, Dictionary, Formatting, Halftone screening, Handles, I-beam, Text cursor, Key combination, Layout grid, List box, Master items, Master page, Page description language, Page view, Palette, Print area, Printer driver, Replace, Ruler guides, Rulers, Rules, Scaling, Scroll bar, Spooler, Stacking order, Standoff, Submenu, Supplemental dictionary, Template, Tag, Title bar, Intuitive icon, High-end system, Front end.

REVIEW QUESTIONS

1. Electronic publishing is referred to as _____ _____.
2. Hardware refers to computer physical equipment. True or false?
3. The primary input device for a computer is the _____.
4. A _____ is used to show the computer operator the point on the screen at which data is being input.
5. A function key can be used to change typed letters into bold or italic typefaces. True or false?
6. Besides using the keyboard for input, a _____ can be used in conjunction with the keyboard.
7. The computer display screen is referred to as
   a. Frame.
   b. Grid.
   c. Monitor.
   d. Pallette.
8. The number of dots or _____ refers to a computer screen's resolution.
9. The term WYSIWYG means "what you see is what you get." True or false?
10. The _____ is the electronic brain of a personal computer.
11. The term RAM means _____ _____ _____.
12. The most common storage device for computer data is a:
   a. Directory.
   b. Scrapbook.
   c. Layout grid.
   d. Disk drive.
13. Computer hard drives typically store from 20 to over 100 megabytes of data. True or false?
14. A _____ _____ is a storage system that uses audio cassette-type tapes to store data.
15. The fastest and most efficient drive for storing computer data is:
   a. Optical.
   b. Tape.
   c. Floppy.
   d. Digital.
16. A digitizer is the same as an optical disk drive. True or false?
17. A _____ is used to change hardcopy images into digital computer data.
18. An _____ _____ _____ is an input device used for automatically reading typed pages into the computer.
19. The type of printer used for medium-quality output is a:
   a. Photo imagesetter.
   b. Laser.
   c. Dot matrix.
   d. High end.
20. The term dpi refers to _____ _____ _____.
21. PostScript® is a common computer format language that is used to output information to a printer or photo imagesetter. True or false?
22. A _____ _____ is used to give the computer the information needed to operate and do specific tasks.
23. A _____ _____ automatically detects words not stored in a computer word dictionary.
24. Computer drawing and paint programs can be used to generate line drawings. True or false?
25. Converting paint objects and grayscale images into draw type images is done with a:
   a. Break tool.
   b. Bezier curve.
   c. Trace tool.
   d. Transformation tool.
26. A _____ program can be used to draw precise and accurate illustrations that show height, width, and depth.
27. An _____ _____ is a continuous tone photo that has been scanned into a computer for use in page layout.
28. The hand method of pasteup has been largely replaced by electronic page layout systems. True or false?
29. The type of language used to write information into computer memory or storage is referred to as _____ _____.
30. When doing electronic layout, the document should always be stored on a hard disk, tape, or optical drive. True or false?

SUGGESTED ACTIVITIES

1. Visit a daily or weekly newspaper publishing facility and ask to view the operation of an electronic pagination system.
2. List the type of publications that you feel are especially suited to being produced by several desktop systems.
3. Use the manufacturer's manual to operate a computer system in your school facility. Compose several types of jobs including those with text and graphics.
4. Make a listing of page description software used with each desktop unit and identify its compatibility with other systems.
5. Visit a computer store and identify the equipment commonly associated with each desktop publishing system.

Chapter 11

Image Assembly

When you have completed the reading and assigned activities related to this chapter, you will be able to:

O Define the term "image assembly."

O Describe camera paste-up copy and its major elements.

O List and describe the three kinds of camera paste-up copy.

O Describe the primary differences between line and continuous tone copy.

O Explain what screening is and why it is necessary.

O Describe the three kinds of register and their applications.

O Describe the four methods used to locate continuous tone elements on the paste-up.

O Explain how the Pantone Matching System (PMS) is used.

O Assemble a single-color paste-up from a layout.

O Assemble a two-color paste-up from a layout.

O Describe the red keyline technique in paste-up.

O Summarize the information usually included on a paste-up tissue overlay.

Chapters 8, 9, and 10 described how to create or generate type matter and various other visual elements. These elements are brought together in a final format called *image assembly*.

As you have seen, most of the type and related visual elements are assembled on the screens of phototypeset-ters, desktop publishing systems, and other kinds of computers. In other instances, type, borders, and artwork are *assembled* (called image assembly) mechanically or by hand on an artboard called a *paste-up* or *mechanical,* Fig. 11-1.

Note! The term *paste-up* is also used to identify the actual procedure of locating and attaching image elements, such as type and illustrations, on the artboard.

All kinds of printed products are prepared as paste-ups, including: newspapers, magazines, brochures, letterheads, and business forms. The final printed product will only be as good as the accuracy and quality of the paste-up. Cleanliness and sharp paste-up image elements are essential for utmost quality in printing.

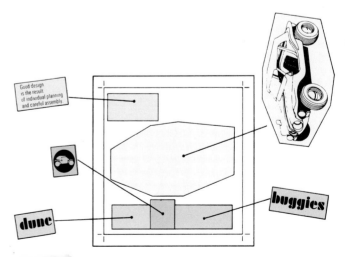

Fig. 11-1. Image assembly is a process of adhering type elements, borders, and artwork by hand on an artboard, called a paste-up or mechanical. (nuArc Company, Inc.)

PASTE-UP TOOLS

In preparing a paste-up, several kinds of small tools and accessories are required. Fig. 11-2 illustrates some of the basic tools needed to perform the tasks involved in paste-up. Among these are items such as the following:

1. Drawing table or drawing board.
2. T-square.
3. Triangles (45° and 30° − 60°).
4. Measuring scale, such as a line gauge.
5. Centering scale.
6. Curves, irregular, and French.
7. Compass.

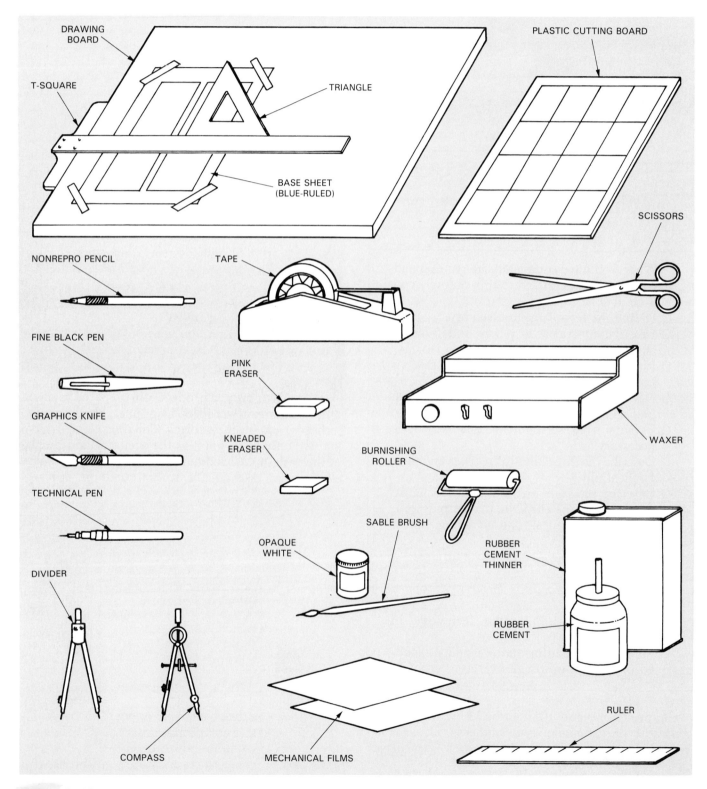

Fig. 11-2. Many tools are needed for paste-up. Tools should be kept in good condition and stored in a convenient location.

8. Dividers.
9. Ruling pen or technical drawing pen.
10. Opaquing brushes.
11. Nonreproducing blue pen or pencil.
12. Knife, such as an X-acto® No. 1 holder with No. 11 blade.
13. Scissors.
14. Burnishing roller or bone.
15. Proportional scale for reductions and enlargements.
16. Loop or 12X magnifier.
17. Percentage screen tint guide.
18. Pantone Matching System (PMS)® color swatch book.
19. Black India drawing ink.
20. Opaque white-out (deletion) fluid.
21. Paper sample catalogs.
22. Ink sample catalogs.

Depending on the requirements of the job, there are many other tools and materials that may be required for paste-up. For purposes of this textbook, only the most common tools are covered.

The basics of paste-up

A *drawing table* or *drawing board* is needed as a work surface. The working edge of the table or drawing board (left side as you sit) must be absolutely square. Some table tops and most light tables have a steel insert on the working edge. This provides a better wearing surface.

The paste-up artist should be comfortable at the table. Some of the newly-designed tables allow for a low profile seating arrangement. This normal seating posture helps to overcome much of the strain and tension of sitting on a high stool. Look at Fig. 11-3.

Fig. 11-3. A drawing table is frequently used for paste-up. Paste-up artist should be comfortable at table, assuming a normal seating posture at all times.
(Smith System Manufacturing Co.)

Base material for paste-ups

As a *carrier* or *base material* for the paste-up elements, hot press illustration board is recommended. This type of board consists of a smooth, white surface that is ideal for the paste-up process. For some applications, a lighter weight board, such as a 110-pound index bristol, may be used. However, the surface of index bristol separates easily when image elements must be repositioned.

The base material should be cut approximately 2 to 3 inches (50 to 75 mm) larger on all four sides than the finished size of the printed piece being pasted up. This gives enough space for marginal information, crop marks, center marks, and register devices. See Fig. 11-4.

Fig. 11-4. Paste-up base material should contain approximately 2 to 3 inches of border on all four sides. This will provide adequate space for marginal notes, crop marks, center marks, and register marks.

To start a paste-up, the top edge of the base material is aligned with a *T-square* to assure horizontal alignment. After the base material has been aligned, it is taped to the drawing surface with drafting or masking tape at all four corners, Fig. 11-5. The preparation of an actual paste-up is detailed at the end of this chapter.

Guidelines

Guidelines on the layout sheet or board show the image and non-image areas for the layout artist. They are drawn or printed in blue so that they will not be seen when photographed to make printing plates. Layout sheets with guidelines can be preprinted or drawn by hand for custom jobs.

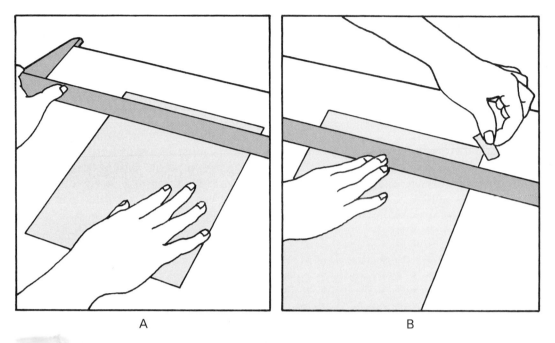

Fig. 11-5. A—The paste-up base material is aligned with a T-square to assure horizontal alignment.
B—The base material is taped to drawing surface with masking tape at all four corners.

If not preprinted, the next step in preparing a paste-up might be to draw guidelines with a nonreproducing blue pen or pencil. Guidelines should include vertical and horizontal centerlines as well as finish size lines and image area lines, Fig. 11-6. Blue guidelines should also be drawn to the size of the elements to be pasted down on the base material. These lines assist in aligning material along the margins.

The black inked crop marks are ruled approximately one-half inch (13 mm) long at right angles in the corners. These marks show the actual finish size of the job. Study Figs. 11-7 and 11-8.

When an image area of the paste-up is planned to extend off the edge(s) of the final printed page, it is called a *bleed*. Bleed is designed to eliminate white spaces between the printing and the trimmed edges of the printed job, Fig. 11-9. If one side of a printed page is to be bound or folded, it is not necessary to indicate bleed on that side. Normal bleed allowance is 1/8 to 1/4 inch (3 to 6 mm).

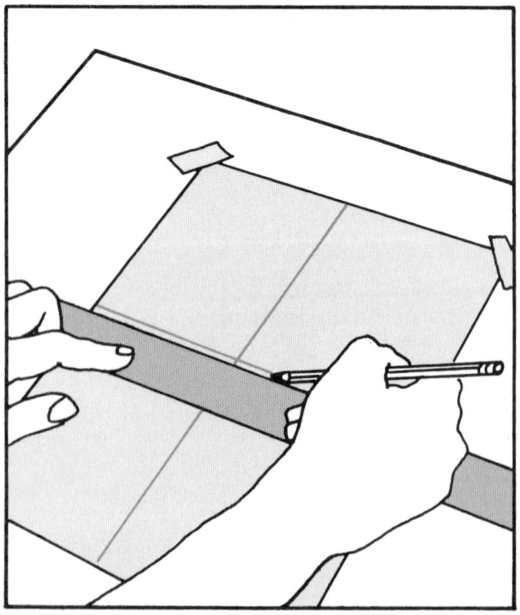

Dividing space into equal parts

Sometimes, it will be necessary to divide certain spaces on the paste-up into equal parts. Fig. 11-10 illustrates how a given space of 16 picas is divided into three equal parts.

Since 16 is not divisible by three, take the next larger number above 16 that is divisible by three. That number is 18. Tilt the line gauge so that 18 is now on the line. Three goes into 18 six times. Put a dot at increments of six and drop a vertical line from those dots. If necessary, use a nonreproducing blue line to extend the right vertical line, so that you may tilt the line gauge at the proper angle to place the dots at the correct increments. This method will give you accurate equal spaces using any number. It can also be used to divide horizontally.

Centering scales

Paste-up artists sometimes use a *centering scale* to locate the center of illustrations, type matter, and paste-up boards. The middle section of the top scale of a centering scale is generally at half scale, Fig. 11-11. It

Fig. 11-6. Paste-up guidelines are drawn in nonreproducing blue pen or pencil. Guidelines include vertical and horizontal centerlines, finish size, and image area lines.

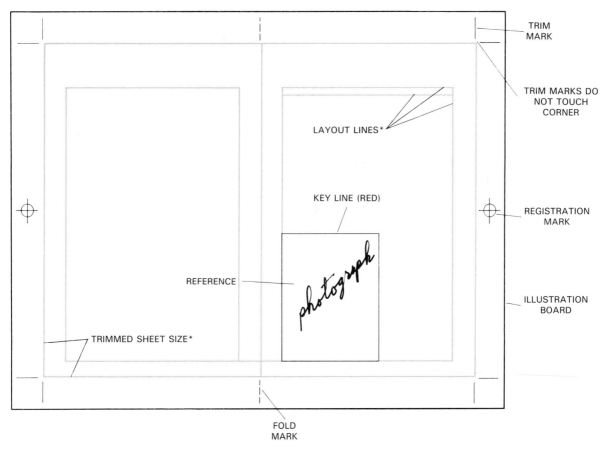

*All layout lines are drawn in non-repro blue and will not reproduce.

Fig. 11-7. Crop marks, trim marks, and fold marks are drawn as thin black lines.

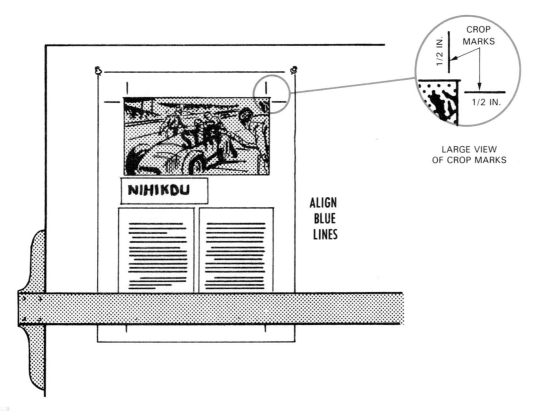

Fig. 11-8. Crop marks are drawn in black ink approximately 1/2 inch long on paste-up base. These marks show actual finish size of job. (nuArc Company, Inc.)

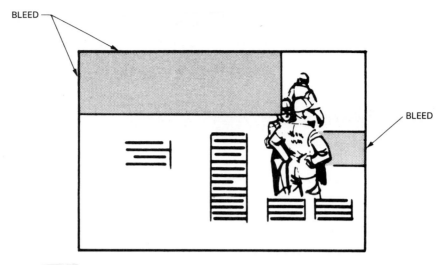

Fig. 11.9. Bleed refers to an image area of paste-up that is planned to extend off edge(s) of final printed page. (nuArc Company, Inc.)

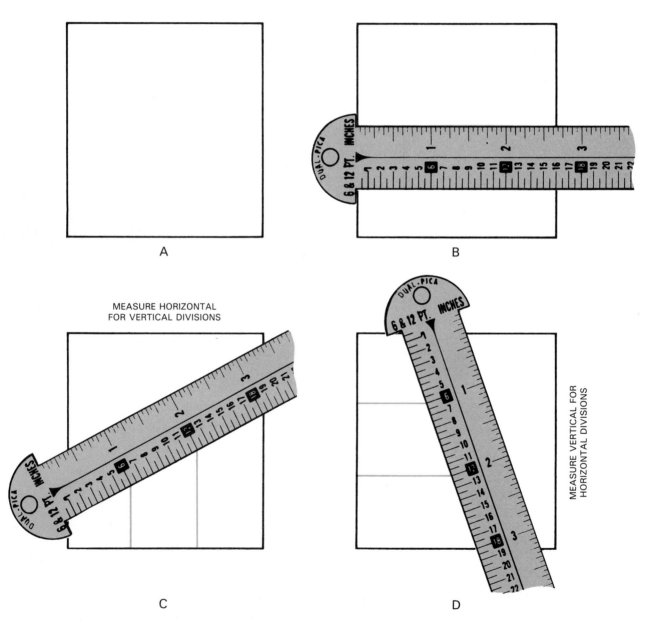

A

B

MEASURE HORIZONTAL FOR VERTICAL DIVISIONS

C

MEASURE VERTICAL FOR HORIZONTAL DIVISIONS

D

Fig. 11-10. To divide a space on paste-up into equal parts, follow procedure illustrated above.

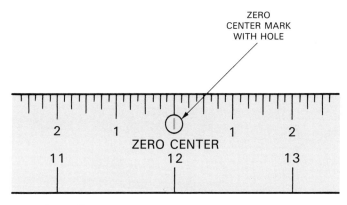

Fig. 11-11. This centering scale is frequently used in paste-up process. It has markings running left and right of zero center mark or hole.

has markings running left and right of the zero center mark or hole. The bottom half of the scale is generally in full inches.

To find and mark the center of some type matter, place the centering scale over a line in the column of type. Adjust the scale, as in Fig. 11-12, so that the top scale reads the same on both sides of the type line. Mark the center of the line at the center mark or hole.

By placing the centering scale over a layout, you can mark a width so it will be centered on a centerline. To do this, align the center of the scale with the centerline of the layout. If you want a total actual centered width of two inches, mark your layout at two on each side of the center mark of the top scale. If you want a total actual centered width of one and three quarter inches, mark at one and three-quarter on each side of the center mark of the top scale.

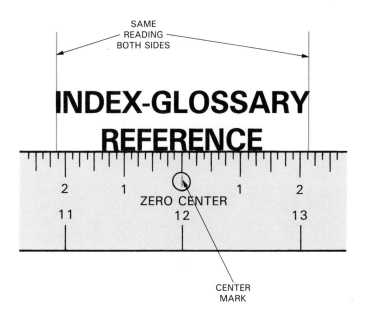

Fig. 11-12. Centering scale is being used to find center of a type line. To do this, scale is placed over a line in column of type. Scale is then adjusted so that top scale reads same on both sides of type line. Mark center of line at center mark or hole.

Nonreproducing blue lines

Paste-up artists use nonreproducing blue pens and pencils to make notations on the base material or layout sheet so they will not reproduce. These generally include instructions for the camera and platemaking departments.

A nonreproducing blue is seen as a white image when photographed with the usual kind of orthochromatic film in the process camera and material used in direct camera/platemakers. This is an advantage for the paste-up artist because the guidelines and other notations do NOT require erasing since they do not photograph.

The image elements

In most instances, type elements (typeset composition) for paste-up comes as phototypeset material, Fig. 11-13. Sometimes type for the paste-up is provided in the form of a reproduction proof.

Reproduction proofs are high-quality prints made from hand-set and machine-set type. The term *reproduction proof* is sometimes loosely used to also designate typeset material generated on a phototypesetter or impact typewriter.

Any line illustrations required for paste-up are provided as inked drawings, Fig. 11-14.

Example: (typewritten copy)

```
                      Just. X21
(10)-1 [1]  The entire appearance of a printed      /10
            piece can be altered by the selection
            of a typeface.  Many characteristics
            can be suggested by the typeface used.
            Delicacy, formality, feminity, mas-
            culinity, can all be suggested by the
            proper style.  Remember that type is
            designed for easy reading.  Also im-
            portant to the appearance of a printed
            piece is proper line space.
```

Example (typeset copy)

The entire appearance of a printed piece can be altered by the selection of a typeface. Many characteristics can be suggested by the typeface used. Delicacy, formality, femininity, masculinity, can all be suggested by the proper style. Remember that type is designed for easy reading. Also important to the appearance of a printed piece is proper line space.

Fig. 11-13. Most typeset material used on a paste-up consists of phototypeset composition. (Compugraphic Corp.)

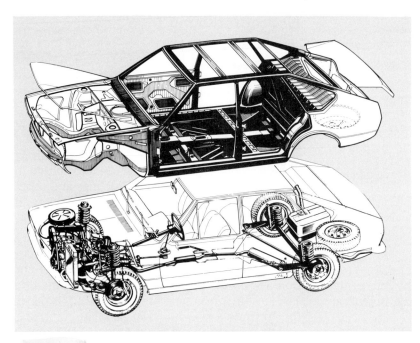

Fig. 11-14.   Inked drawings are prepared in black ink. Drawings are positioned on paste-up board according to layout.   (Volkswagen of America, Inc. and Koh-I-Noor Rapidograph, Inc.)

In addition, transfer lettering, borders, and ornaments are often used in paste-up, Fig. 11-15 and Fig. 11-16. These materials are either of the rub-off or adhesive variety.

Transfer lettering is easily applied to the paste-up by first locating the exact position of the character and then placing the transfer lettering sheet in position over the spot. A special *burnishing tool* is used to rub the character onto the paste-up, Fig. 11-17. The sheet is then lifted and moved into position for the next character.

Adhesive lettering is handled somewhat differently from transfer lettering. The desired adhesive character is cut from the lettering sheet and then positioned on the paste-up along a pre-drawn light blue line. A special tool is available that can be used to position individual characters. This makes letterspacing and wordspacing easier.

The procedure for positioning, aligning and burnishing characters on a paste-up is illustrated in Figs. 11-18 through 11-20.

Borders and *rules* are decorative lines which are easily applied to the paste-up, Fig. 11-21. Special applicators are also available for use in applying adhesive borders, Fig. 11-22.

Borders are cut to the desired length with a knife. The technique used to *miter* (square) a corner using border material is shown in Fig. 11-23.

It should be noted that typographically, *borders* are classified as DECORATIVE (fancy) in design. *Rules* are generally classified as merely STRAIGHT LINES in various thicknesses in either single or parallel form.

Ornaments include various designs which are generally available in adhesive form. These materials

Fig. 11-15. Transfer lettering is often used directly on paste-up board when a line or two of display type is required. (Letraset, Inc.)

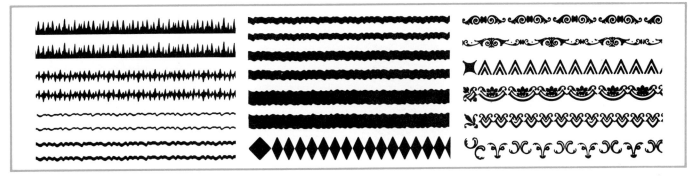

Fig. 11-16. Borders and ornaments are available in many sizes and designs. These images offer paste-up artist additional versatility in layout and design effects. (Letraset, Inc.)

   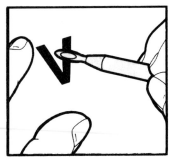

A—Remove protective blue backing sheet and place lettering sheet over a pre-drawn guideline. Letters transfer most easily when a Letraset burnisher is used. Position lines under letters on your pre-drawn guideline. This ensures both vertical and horizontal alignment.

B-Transfer letter by rubbing lightly with burnisher. Work diagonally from the top left of the letter to the bottom right. The letter will appear to turn gray as it transfers.

C—When letter is completely "grayed" out," peel sheet back from top. If the letter hasn't transferred completely, you can easily let it fall back into position by keeping it firmly in place with your hand.

D—Once entire word or image is transferred, burnish by covering it with blue backing sheet and rubbing over it with burnisher. The more thorough the burnishing, the more the letters will resist abrasion.

Fig. 11-17. Burnishing is done to make image elements adhere properly to paste-up base. Stick or pencil burnishing tool (shown above) is for small elements. Roller burnishing tool is used for large section of image elements. (Letraset, Inc.)

Fig. 11-18. Positioning adhesive lettering elements on paste-up board requires perfect alignment and good letter fit.

Fig. 11-19. Cutting away pre-drawn line after lettering elements are positioned.

BURNISHING STICK

Fig. 11-20. Paste-up artist uses a burnishing stick to bond adhesive lettering elements to paste-up board. A roller burnishing tool can also be used for this purpose.

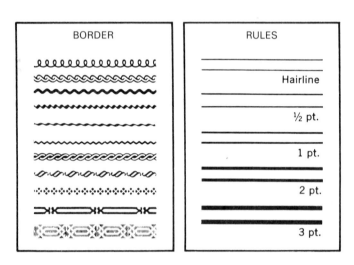

BORDER	RULES
	Hairline
	½ pt.
	1 pt.
	2 pt.
	3 pt.

Fig. 11-21. Adhesive borders and rules are decorative designs and straight lines which come in rolls or flat sheets. (Letraset, Inc.)

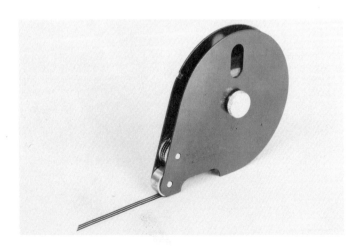

Fig. 11-22. This applicator is used to apply adhesive borders to paste-up base. (Chartpak, Inc.)

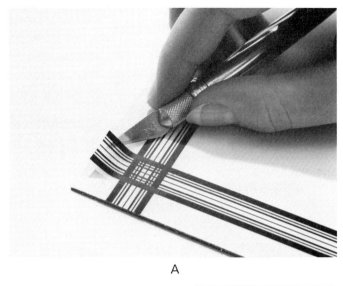

A

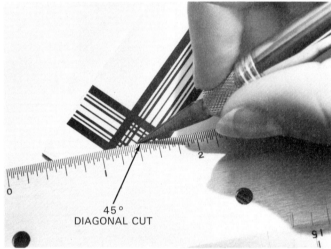

45° DIAGONAL CUT

B

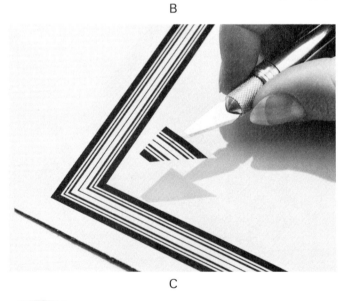

C

Fig. 11-23. A—Adhesive borders are commonly mitered at corners to provide an accurate fit. This is done by first overlapping two border tapes at corner. B—Steel straight edge is used, along with an art knife, to cut diagonally through overlapped areas. C—Two overlapping end pieces of border tape are gently removed. This leaves a perfect miter (45 degree cut) at corner. (Formatt Graphics Products Corp.)

are applied directly to the paste-up in much the same way as border materials.

In working with transfer materials, the paste-up artist generally indicates the exact position on the paste-up by drawing blue guidelines. These lines act as a reference mark for the actual application of the image elements.

Line drawings and hand lettering can be drawn directly on the paste-up base. It is also possible to draw these images on another sheet of white paper. The sheet is then attached to the main paste-up base. Sometimes, it is necessary to enlarge or reduce this type of artwork to fit the proportions of the job. The camera is used to prepare a photostat or other diffusion-transfer material.

Shading sheets

Shading sheets are shading patterns on transparent acetate. They are generally used on pen and ink illustrations.

First, the pen and ink drawing is drawn in outline and the solid black areas are inked. Some shading sheets have an adhesive back. They are adhered to the area of the drawing where shading is desired. With a sharp knife, they are cut to the shape required. The material is then rubbed gently into place. See Fig. 11-24.

Shading sheets should be used in the following manner. The artwork should be clean and free of pencil marks, fingerprints, and surface dirt. Cut a piece of shading material from the backing sheet to cover the area of original art requiring the shading. Position the shading material over the copy to be shaded. Rub the shading material onto the artwork with the adhesive side down. Use slight pressure to burnish the material. Start at the bottom and rub from left to right, working upward as the pattern adheres.

Excess or unwanted shading is removed by cutting the shading material with an X-acto knife, Fig. 11-25. Light pressure is all that is necessary. Do not scratch the original artwork. Peel off any unwanted pieces of the shading material.

Rub the shading material onto the artwork until the pattern is completely smooth and perfectly adhered. Use a burnisher for this procedure.

Cut away or paint out with white opaque areas that require highlighting. Use a brush or pen.

Halftones

All continuous tone copy must be photographed separately from line copy when making the films for the printing plates. This is because continuous tone copy must be separated into a series of fine dots of

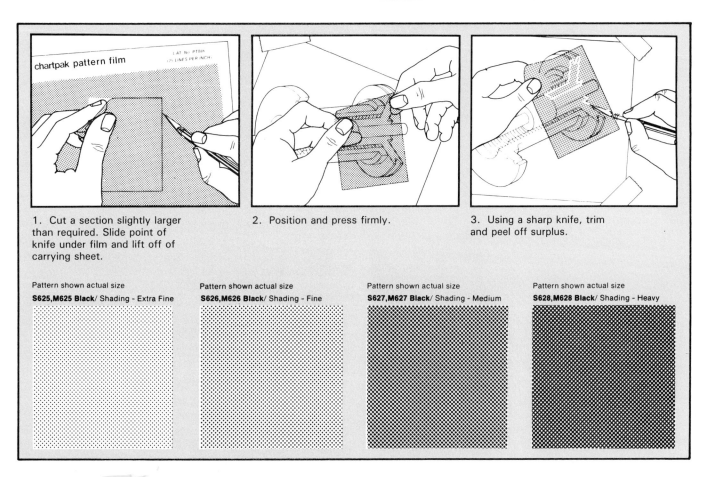

1. Cut a section slightly larger than required. Slide point of knife under film and lift off of carrying sheet.

2. Position and press firmly.

3. Using a sharp knife, trim and peel off surplus.

Pattern shown actual size
S625, M625 Black/ Shading - Extra Fine

Pattern shown actual size
S626, M626 Black/ Shading - Fine

Pattern shown actual size
S627, M627 Black/ Shading - Medium

Pattern shown actual size
S628, M628 Black/ Shading - Heavy

Fig. 11-24. Adhesive shading material can be applied to an area of a line drawing where shading is desired. Follow these steps.   (Chartpak, Inc.)

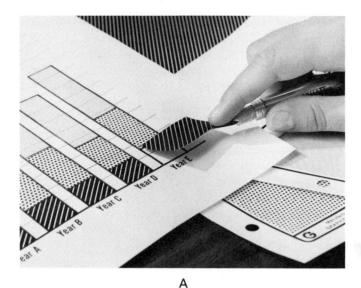

A

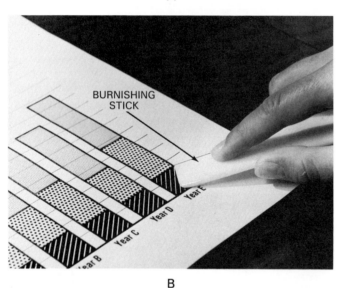

BURNISHING
STICK

B

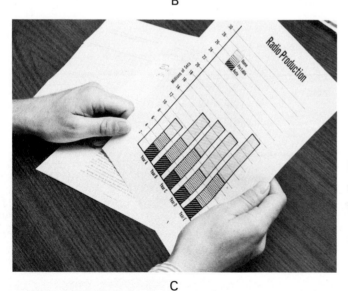

C

Fig. 11-25. A—Adhesive shading material can be used for many different paste-up applications. B—Once material is positioned, it is burnished thoroughly. C—Professional quality paste-ups can be obtained by using shading materials appropriately.

various sizes, called *halftones.* The result is a tonal effect resembling the original continuous tone copy. Because of this, original photographs are NOT placed directly on the paste-up base. Instead, three methods are commonly used to locate the position of photographs on the paste-up base. These include:

1. Halftone outline.
2. Halftone block.
3. Screened photo-mechanical print.

Note! The halftone process is covered in Chapter 16.

In using the *halftone outline method,* the location of the photograph is marked on the paste-up base by four thin black or red (key) lines, Fig. 11-26. Separate negatives are made of the paste-up copy and continuous tone copy. When the two negatives are prepared by the stripper, a window opening is cut into the paste-up line negative where the lines were drawn on the paste-up. The halftone negative is then taped behind the window on the line negative.

Ruby masking film or black construction paper is used in the *halftone block method,* Fig. 11-27. It is cut to the desired size of the reproduction and adhered with rubber cement to the paste-up base. The masking film or black paper is cut to the final desired size of the continuous tone copy.

As in the outline method, separate negatives are made of the paste-up copy and continuous tone copy. The masking film or black paper on the paste-up becomes an open window (transparent) in the negative when photographed and processed. The halftone negative of the continuous tone copy is then taped behind the transparent window opening of the line negative.

The *screened photo-mechanical print method,* also called the *halftone print method,* starts with the preparation of a halftone print from the original continuous tone copy. The diffusion-transfer process is the most popular method used to prepare halftone prints for use directly on the paste-up base. Refer to Fig. 11-27B.

Note! The photo-mechanical transfer (PMT) process is described in Chapter 17.

Screened prints can also be prepared by using a halftone negative made from the original continuous tone copy. A photographic contact print is made from the halftone negative. The screened print or Velox® is pasted directly on the paste-up base. The entire paste-up is photographed as line copy because the screened print is made up of dots and the film negative records them. A screened print or Velox® of finer than 120-line screen ruling should be avoided because the fine dots will be lost in the final preparation of the line negative.

The major disadvantage of the screened print method is that halftone quality is reduced because of the two extra steps involved. In preparing a conventional halftone, the procedure is to go from continuous tone copy to a halftone negative. In preparing screened prints, the procedure is to go from continuous tone

234

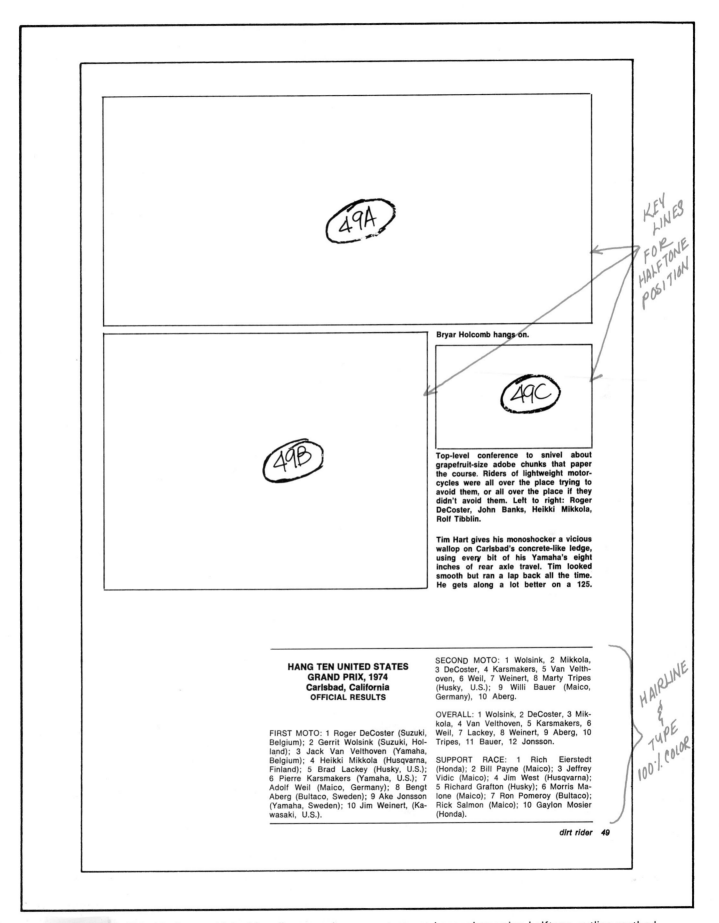

49A

Bryar Holcomb hangs on.

49B

49C

KEY LINES FOR HALFTONE POSITION

Top-level conference to snivel about grapefruit-size adobe chunks that paper the course. Riders of lightweight motorcycles were all over the place trying to avoid them, or all over the place if they didn't avoid them. Left to right: Roger DeCoster, John Banks, Heikki Mikkola, Rolf Tibblin.

Tim Hart gives his monoshocker a vicious wallop on Carlsbad's concrete-like ledge, using every bit of his Yamaha's eight inches of rear axle travel. Tim looked smooth but ran a lap back all the time. He gets along a lot better on a 125.

HANG TEN UNITED STATES GRAND PRIX, 1974
Carlsbad, California
OFFICIAL RESULTS

FIRST MOTO: 1 Roger DeCoster (Suzuki, Belgium); 2 Gerrit Wolsink (Suzuki, Holland); 3 Jack Van Velthoven (Yamaha, Belgium); 4 Heikki Mikkola (Husqvarna, Finland); 5 Brad Lackey (Husky, U.S.); 6 Pierre Karsmakers (Yamaha, U.S.); 7 Adolf Weil (Maico, Germany); 8 Bengt Aberg (Bultaco, Sweden); 9 Ake Jonsson (Yamaha, Sweden); 10 Jim Weinert, (Kawasaki, U.S.).

SECOND MOTO: 1 Wolsink, 2 Mikkola, 3 DeCoster, 4 Karsmakers, 5 Van Velthoven, 6 Weil, 7 Weinert, 8 Marty Tripes (Husky, U.S.); 9 Willi Bauer (Maico, Germany), 10 Aberg.

OVERALL: 1 Wolsink, 2 DeCoster, 3 Mikkola, 4 Van Velthoven, 5 Karsmakers, 6 Weil, 7 Lackey, 8 Weinert, 9 Aberg, 10 Tripes, 11 Bauer, 12 Jonsson.

SUPPORT RACE: 1 Rich Eierstedt (Honda); 2 Bill Payne (Maico); 3 Jeffrey Vidic (Maico); 4 Jim West (Husqvarna); 5 Richard Grafton (Husky); 6 Morris Malone (Maico); 7 Ron Pomeroy (Bultaco); Rick Salmon (Maico); 10 Gaylon Mosier (Honda).

dirt rider 49

HAIRLINE & TYPE 100% COLOR

Fig. 11-26. Thin black or red inked key lines are drawn on paste-up base when using halftone outline method for location of photographs. (Dirt Rider Magazine)

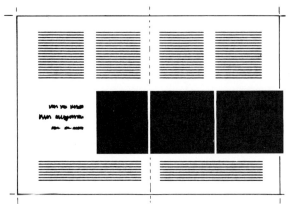

All line elements (type and line drawings) are in position on this mechanical prepared for two pages of a book. The black squares and rectangles show the size and position of halftones. These shapes will form clear windows in the line negative. Halftone negatives, shot separately, will be stripped into the line negative.

A

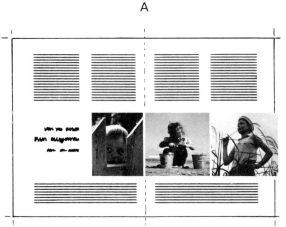

In this mechanical, the halftone illustrations to appear on the page were prepared as halftone paper prints in reproduction size. These halftone prints were pasted into position on the mechanical. This mechanical can be photographed completely as a line negative, without the need for stripping in other elements.

B

Fig. 11-27. A—Ruby masking film or black construction paper is used to prepare image areas for halftone block method. B— Screened photo-mechanical print is adhered directly on paste-up base. (Eastman Kodak Co.)

copy to a halftone negative to a screened print and then back to a line negative of the already screened copy. Slight amounts of quality are lost at each stage in the screened print process.

The screened diffusion-transfer method mentioned earlier is the preferred method of preparing halftone prints to be used on a paste-up base. A special 85- to 120-line PMT halftone contact screen is commonly used. Processing of a halftone is exactly the same as line work for most diffusion-transfer applications.

Rescreening a halftone

Sometimes it is necessary to rephotograph an already printed halftone. This process is called *rescreening*. This

technique may be necessary where a printed halftone has a poor screen pattern, a screen pattern is too fine, or a large reduction in size is required. The closer the printed halftone is to the original photograph the better will be the quality.

When a halftone must be rescreened from a newspaper or magazine, a screen having a ruling either 50 lines finer or 50 lines coarser than the screening on the original (at its new size) should be used. When rescreening, the copy or screen is angled so that the result is exactly 30° more than the original angle. This is necessary to minimize the possibility of producing a disturbing pattern on the negative, called *moiré*. Look at Figs. 11-28 and 11-29.

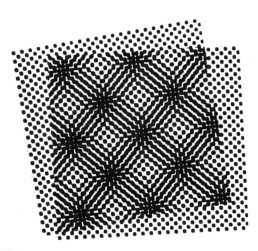

Fig. 11-28. Rescreening halftones can produce a disturbing pattern, called moiré. This occurs because two screen angles (original halftone screen angle of print and new rescreening screen angle) are not set exactly 30 degrees apart.

Fig. 11-29. Objectionable moiré pattern can destroy entire effect of an otherwise good halftone. Example above illustrates moiré pattern. (International Business Machines)

The final reproduction of a rescreened printed halftone is slightly inferior to that obtained by the other methods just described. Rescreening is covered in Chapter 16.

Adhering image elements

There are three basic methods of adhering image elements to the paste-up base. These include waxing, rubber cement, and spray mount.

Waxing uses melted wax to hold the elements on the paste-up base. A waxing machine or hand-held waxer applies the hot wax to the back of the image element, Fig. 11-30.

Waxing is the most popular method of adhering image elements to the paste-up base. This is because of the speed of preparation, cleanliness of the finished paste-up, and advantage of easily moving the waxed piece if necessary. A special kind of wax is used in the waxing machines, Fig. 11-31.

The image elements can easily be removed from the surface of the paste-up base if changes are required. The elements to be waxed are completely coated with a thin film of wax applied to the back of the typeset sheet. Some waxing equipment applies stripes instead of a solid coating of wax.

The simplest procedure is to wax the back of the entire sheet and then place it face up on a piece of stiff cardboard. Special plastic cutting surfaces are also available and are ideal for this purpose. The individual image elements are cut out as needed with a knife and lifted into position on the paste-up base, Fig. 11-32. This keeps small pieces of type from being lost and acts as a check to make certain that all of the image elements have been transferred to the paste-up base.

Using a T-square or straightedge ruler, the paste-up artist places the individual image elements in final position and checks for squareness before pressing the elements to the base material with a burnisher.

One of the most familiar adhesives used in paste-up is *rubber cement*, Fig. 11-33. It is applied to both surfaces and allowed to partially dry before being joined. Material placed on the paste-up while the rubber cement is still wet has a tendency to shift and may become stained because the trapped rubber cement does NOT always dry properly. When dry, the typeset sheet or other copy element is placed face up on a sheet of stiff cardboard, called a *cutting board*. The blocks of image elements are cut out with a knife and a T-square as needed. The dried rubber cement will hold the elements on the cutting board until they are ready to be lifted to the paste-up board.

Available in an aerosol spray container, *spray mount* has become popular among paste-up artists, Fig. 11-34. It is easy to use and provides a simple way in which to apply adhesive to paste-up elements. The elements to be sprayed are completely coated with a thin film of spray mount to the back of the sheet. After the back of the sheet has been sprayed, it is placed face up on

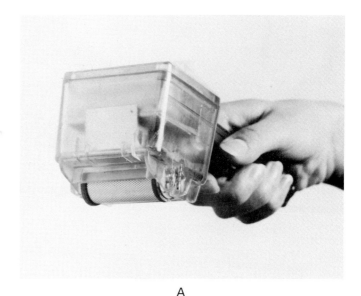

A

B

Fig. 11-30. A—This hand-held waxer is convenient for smaller paste-up projects. B—Large waxing machine is required where production is extensive. (Daige Products, Inc.)

Fig. 11-31. Special wax is used in waxing machines. (Daige Products, Inc.)

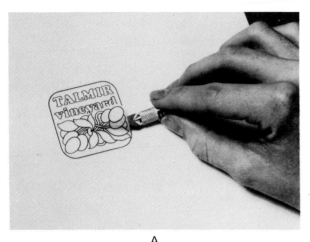 

A   B

Fig. 11-32. A—Individual image element is lifted with a knife into position on paste-up base. B—Element is then checked for position and alignment before burnishing. (Formatt Graphic Products Corp.)

Fig. 11-33. Rubber cement is a familiar adhesive used in the paste-up process.

Fig. 11-34. Spray mount is an aerosol spray adhesive available in artist stores. It is easy to use and provides a simple way to apply adhesive on paste-up elements. (Grumbacher, Inc.)

a cutting board. The individual image elements are cut out as needed with a knife and lifted into position on the paste-up base.

Checking paste-up alignment

A T-square and triangle are used to obtain perfect alignment of all elements on the paste-up base, Fig. 11-35. When the image elements on the paste-up base have been positioned and checked for alignment, a sheet of clean white paper is placed on top of the elements. A hand burnishing roller or other type of burnisher is then used over the sheet of white paper to burnish the elements securely.

Sometimes typeset material is not absolutely square even when phototypesetting equipment is used to produce the type matter. This must be carefully checked and, if necessary, the problem lines must be cut apart and positioned to make certain that they are aligned accurately. The paste-up artist is responsible for checking and repairing smudged or broken type elements. Obvious cases of damaged type should be returned to the typesetting department. Even though the type has been proofread, the paste-up artist should be alert for errors.

Flat color paste-up

Much of the printing done today is single color. *Single color* means that it is printed in only one color whether black, green, red, blue, etc. In the printing industry, black is considered a color. The trend to print more in color grows each day. One way this is being achieved is through the use of flat color.

Flat color is any color the layout artist or customer chooses for reproduction in the initial layout stage. In this case, each color requires a separate printing plate. Many presses are equipped to run two or more colors at a time, Fig. 11-36.

Flat color is NOT the same as color separation (process color). *Process color* is used to reproduce full-color

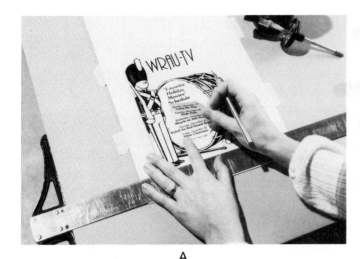

A

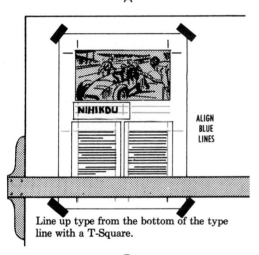

ALIGN
BLUE
LINES

Line up type from the bottom of the type
line with a T-Square.

B

Fig. 11-35. A—T-square is used to obtain perfect alignment
of all elements on paste-up base. B—Type composition should
be aligned from bottom of type lines with T-square.
(Dynamic Graphics, Inc. and nuArc Company, Inc.)

continuous tone copy such as color prints, paintings, and transparencies. The subject of process color is described in more detail in Chapters 18 and 19.

The addition of a second color provides a number of design variations. Not only can the two colors be used independently of each other, but they can also be combined as solids and screened tints to produce a variety of other colors and tones. The use of additional colors on a printed job increases the range of design possibilities and aesthetic effects. Most flat color work uses from one to four ink colors. Two-color printing is the most popular choice.

Specifying flat color

Two methods are used to specify flat color for reproduction purposes. The first method is to ask the printer to mix a color that is part of a color-matching system. The second method is to ask the printer to match a color that is not part of a color-matching system. The first method is the most practical because it allows the printer to mix inks using preestablished instructions for every color. The second method is more difficult for the printer, since the color requested is mixed through a less scientific technique.

There are several ink color-matching systems in use throughout the printing industry. The most widely accepted method used by layout and paste-up artists is the Pantone Matching System® referred to as PMS. The Pantone Matching System® consists of ten Pantone colors which include eight basic colors plus black and transparent white. These colors are mixed in varying amounts to achieve a desired ink color. Over 900 possible colors can be mixed with this system. The colors are numbered and arranged in an ink color book, called a *swatchbook,* Fig. 11-37. These swatchbooks are available from printers' supply houses.

Fig. 11-36. This sheet-fed offset press is capable of printing two colors at one time.
(Graphic Systems Division, Rockwell International)

Fig. 11-37. Pantone Matching System color swatchbook has over 900 possible ink colors that can be mixed. (Pantone, Inc.)

To specify an ink color using the PMS formula, the layout artist refers to the swatchbook. The desired color(s) is chosen and its number is indicated on the initial layout and finally on the paste-up board. This prevents any possibility of error when it comes time to print the job. The swatchbook usually contains sheets of numbered tear-out samples for this purpose, Fig. 11-38. A numbered tear-out sample of the color ink desired can be attached to the paste-up base or tissue overlay.

Fig. 11-38. Numbered tear-out samples in Pantone Matching System swatchbook are used by layout and paste-up artists to specify ink colors. (Pantone, Inc.)

Use of register

In preparing a paste-up where two or more colors of ink are to be printed, *register* (alignment) of the different color images on the paste-up is critical, Fig. 11-39. There are three kinds of register in common use. These include commercial, hairline, and nonregister.

The use of *commercial register* allows for slight variations in color images. This means that adjacent colors can vary approximately plus or minus 1/64 inch (0.4 mm).

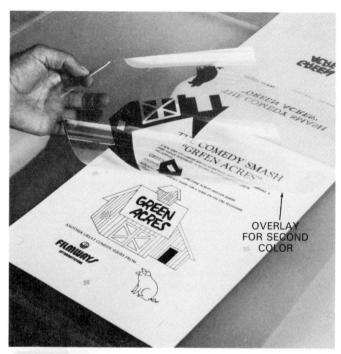

Fig. 11-39. When two or more colors of ink are to be printed, register of different color images on paste-up is critical. This example shows an overlay to be used for a second color on job. (Stephen B. Simms)

When image elements must be extremely precise where colors meet, *hairline register* is necessary. No overlap or white space is allowed to show where elements touch.

The least exacting flat-color registration is that of *nonregister*. Allowance is made for several different colors to be completely independent of each other. For example, the comic pages of a Sunday newspaper are printed under nonregister conditions. In this instance, exact register is not necessary.

Key-line art

When the registration of two or more colors is critical, a slight overlapping of colors is planned. The overlapping of color is referred to as *trap*.

Where two or more colors are to touch, the original line art is prepared with a *key line,* Fig. 11-40. The width of the key line is the amount of trap or overlap-

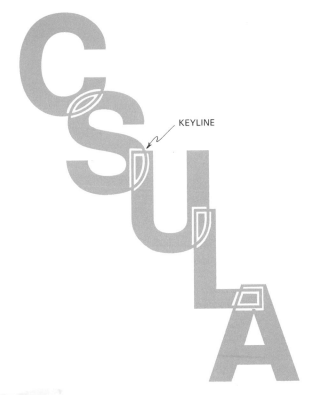

KEYLINE

Fig. 11-40. Where two or more colors are to touch, original line art is prepared with a key line.

ping of the colors. During camera operations, a negative is made from the paste-up for each of the colors to be printed. Each negative is then stripped with only the area open for the color to be printed.

Two colors that appear to join actually should overlap slightly. This procedure allows some variation in register without showing an unsightly gap between colors. Such an overlap is difficult to draw precisely on an overlay. Instead, key-line drawings are prepared.

A single drawing is made of artwork or type in black ink. A fine line, the width of the desired overlap or trap, is drawn where the two colors join. The actual areas to be printed are not inked close to the line. The edge of the solid inking is sometimes left ragged to show it is not complete.

Two negatives of the same artwork or type are given to the stripper, along with a drawing showing the final appearance of the job. The stripper then opaques up to one side of the line and scrapes away unwanted emulsion on the other side. In the example shown, the negative contains a color key line (guideline) with a narrow, black area surrounding it. The width of the key line is the amount of overlap or trap of the two colors. Each negative is stripped with only the area open for the color to be printed.

Overlay method

The simplest kind of paste-up prepared for flat multi-color printing involves a design layout in which the body type prints in one color and the display type or

line drawings print in another color. Since no two image elements touch, there is no problem of hairline register. It is a simple task to paste all of the type and/or line drawings that are to print in one color on the paste-up base. The remainder of elements to print in the second color are positioned in register on an overlay, Fig. 11-41.

The overlay is attached at the top of the paste-up base. The *overlay* is a transparent sheet (similar to acetate) to which second-color image elements are attached.

REGISTER MARKS

Fig. 11-41. Elements to print in a second color are positioned in register on an overlay attached to paste-up base. (Stephen B. Simms)

Register marks

The overlay must be in register with the material on the paste-up base. *Register* consists of providing markings that allow the overlay to be removed and then replaced in the identical location. This is done with the use of *register marks,* Fig. 11-42. Register marks are first drawn or positioned on the paste-up base on at least three sides. These positions should include the top, left, and bottom. Register marks are NEVER placed in a planned bleed area of the paste-up.

Register marks are drawn or positioned on the overlay exactly over those on the paste-up base. If more than one overlay is being planned for a job, the register marks on each overlay should be slightly shorter than the ones on the paste-up base or overlay below. This allows the paste-up artist to make certain that the marks are directly on top of each other. The thinner the marks, the more exact the register.

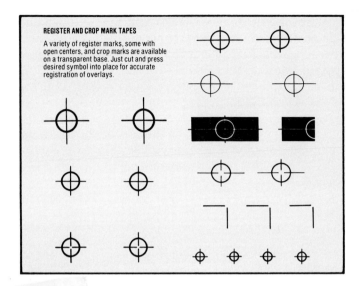

REGISTER AND CROP MARK TAPES

A variety of register marks, some with open centers, and crop marks are available on a transparent base. Just cut and press desired symbol into place for accurate registration of overlays.

Fig. 11-42. Preprinted register marks on transparent adhesive tape are convenient and easy to apply to most surfaces. (Chartpak, Inc.)

An alternative to drawing register marks on the paste-up base and overlay are preprinted register marks on transparent adhesive tape, Fig. 11-42. This kind of register mark is convenient to use and easy to apply to most surfaces, Fig. 11-43. A 12X magnifier or loop is needed when applying adhesive register marks to overlays.

Dust and dirt should be avoided when applying adhesive register marks since these particles will interfere with register and show up on the negative. Some printers prefer register marks pasted only on the artboard. They then cut holes in the overlays and shoot all overlays using only the register marks on the board. This technique assures perfect registration.

One-piece copy

There are times when overlays are not necessary in the preparation of a job. For example, in cases where colors do NOT overlap, the color breaks (separations) can be indicated on a single tissue overlay. This is called *one-piece copy.*

Two identical film negatives are made from the original paste-up. On the first negative, everything which is not to print in the first color is opaqued out or left covered by the stripper's masking sheet material. On the second negative, the parts of the paste-up which were NOT opaqued on the first negative are opaqued or left covered with masking material.

In this way, the paste-up is separated into its two colors. The nonopaqued parts represent an individual negative for each printing plate. It is important to understand that the negatives are NOT colored. They are black and identical in appearance to any other film negative. In this case, the two component color areas of the paste-up are disassembled on negative photographic film.

A plate is made from each negative in the usual manner. Obviously, both plates do not impress the paper at the same instant. The offset press has a separate plate cylinder for each plate. The press feeding mechanism is arranged so that the paper comes in contact with each plate in perfect register. Each plate is inked in any color desired. It is possible to use a one-color offset press by printing the first color on all of the sheets. The second offset plate is then used to print the second color. This is an extremely inefficient method when printing long runs. Presses capable of printing more than one color simultaneously are commonly used for this purpose.

Pin register systems

The use of pin register systems in copy preparation departments is almost universal. Multiple overlays can be created in original artwork or paste-ups in precise register. This is done by prepunching the overlay sheets and pasting them up in proper sequence, Fig. 11-44. Punched overlays eliminate the time required to accurately draw register marks or align pressure sensitive marks. Punched overlays also provide the advantage of consistently precise register. This eliminates the human error inherent in the previously described techniques.

The prepunched paste-up with overlays can be accurately and consistently placed in the process camera copyboard. This is done by attaching short pins to the copyboard through the use of a register tab. Use of the register system in this manner assures consistent placement of originals and eliminates the need to shift original copy in order to place it properly in the camera copyboard.

Multicolor overlay system

The multicolor overlay system is a more recent preparation procedure. It no longer limits the paste-up artist to the use of black, red, or amber copy elements when preparing paste-up copy. With the

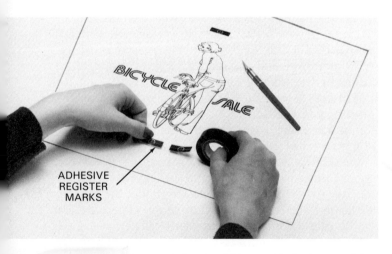

ADHESIVE REGISTER MARKS

Fig. 11-43. Dust and dirt should be avoided when applying adhesive register marks to paste-up base. (Chartpak, Inc.)

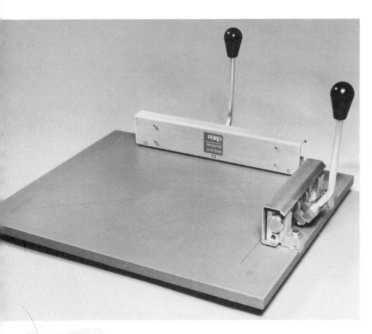

Fig. 11-44. Multiple overlays can be created in original artwork or paste-ups by use of a pin register system. This is done by prepunching overlay sheets and pasting them up in proper sequence, one on top of another. (Pako Corp.)

multicolor overlay system, paste-ups consisting of overlays can be constructed in the colors that are going to be produced on the press. The same applies to the colors that appear on the final layout or comprehensive. This system coordinates the paste-up with the printed press sheet colors. It provides the layout artist, paste-up artist, and printer with a close approximation of the printed job before it actually goes to press.

The multicolor overlay system is a sophisticated copy preparation method. It may not be suitable for every graphic arts paste-up application. The color coordination is possible because a number of printers' suppliers have introduced transparent and opaque block-out films. These materials match many of the printing inks on the market as well as the felt tip pens used by layout artists. It is even possible to obtain color block-out films which relate to specific ink numbers of the major ink manufacturers. This allows the layout artist to standardize design, paste-up copy, and printing ink colors.

For the multicolor overlay system, the usual paste-up routine is followed. The only change occurs in the photographic stage where the process camera operator must use a filter to photograph each color as black. The camera operator may have to use panchromatic film depending on the colors of the copy. More opaquing and outlining may be required with panchromatic film than with orthochromatic film. Many layout artists feel that the advantages of having copy in reproduction colors outweighs the disadvantages related to panchromatic film. This is generally NOT true in the case of the small print shop.

When the multicolor overlay system is used, paste-up copy serves as a stripping guide for the color breaks. It is also used as a prepress guide to help color matching during the press run.

Use of solids

Solid or screened image elements planned for a multicolor job are usually indicated on the tissue overlay of the paste-up, Fig. 11-45. This is done in the same way as it would be for a single-color paste-up. Red key lines are used to indicate the size, shape, and position of all halftones, tint, and color areas on the paste-up base.

Fig. 11-45. Solid or screened image elements planned for a multicolor job are usually indicated on tissue overlay of paste-up.

Red lines drawn on the paste-up base appear on the film negative as transparent openings. Later these areas are removed along the lines in the negative to accept the illustration that is taped in behind the window cutout of the negative.

Hairline register of solid color areas is done with ruby masking film material, Fig. 11-46. It is possible to cut a more accurate edge on masking film than can be ruled or drawn with a pen. If a panel or shape for artwork is cut slightly out of register, it can be easily picked up and placed in position. The masking film can be trimmed or patched to produce more accurate register. Some masking films are manufactured so that the color can be peeled off the acetate base.

When using masking film, a piece should be cut that is approximately two to three inches wider on all four dimensions than the finished size of the job. The film

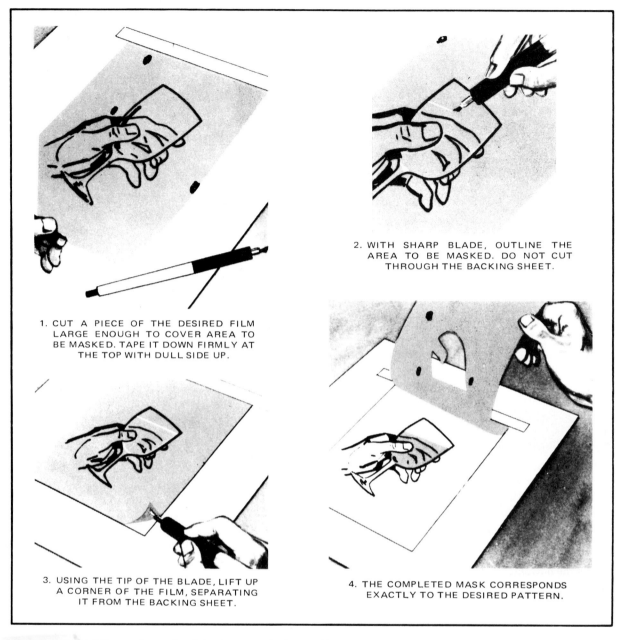

1. CUT A PIECE OF THE DESIRED FILM LARGE ENOUGH TO COVER AREA TO BE MASKED. TAPE IT DOWN FIRMLY AT THE TOP WITH DULL SIDE UP.

2. WITH SHARP BLADE, OUTLINE THE AREA TO BE MASKED. DO NOT CUT THROUGH THE BACKING SHEET.

3. USING THE TIP OF THE BLADE, LIFT UP A CORNER OF THE FILM, SEPARATING IT FROM THE BACKING SHEET.

4. THE COMPLETED MASK CORRESPONDS EXACTLY TO THE DESIRED PATTERN.

Fig. 11-46. Hairline register of solid color areas is done with ruby masking film material on paste-up base. Steps in cutting a ruby masking film overlay are illustrated. (nuArc Company, Inc.)

is attached to the paste-up base at the top with a single piece of masking tape. The *emulsion* (dull side) of the masking film must face upward. The design area is then outlined with a knife and the unwanted or nonimage areas of film are removed, Fig. 11-47. When using these films, care must be taken to avoid scratching or damaging the emulsion.

Paste-up cleanliness

All excess wax, rubber cement, pencil lines (other than light blue), eraser residue, and dirt MUST be removed from the surface of the paste-up. It is easy to see unremoved rubber cement if the paste-up is held at an angle to the light. This part of cleanup is important since unwanted elements, if left for the camera department, will be photographed along with the image elements, Fig. 11-48.

Check to see that all image elements are secure and have not moved. A quick visual check will reveal any missing elements. A cleanup eraser similar to the type used in technical drafting should be used to remove any leftover debris.

Final paste-up assembly

Before the paste-up is sent to the camera department, it must be prepared for conversion to negative form. Continuous tone photographs provided with the paste-up are marked to designate their location on the page. For example, the page number and letter may be marked as "1-A," "1-B," "1-C," etc., for identification.

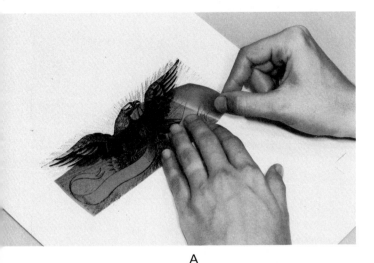

A

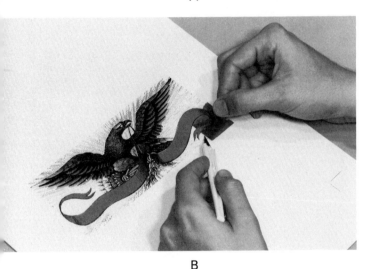

B

Fig. 11-47. A—Ruby masking film is placed in position. B—Then, remove the unwanted or nonimage area of film by peeling it away. (Chartpak, Inc.)

Fig. 11-48. Paste-up artist is responsible for checking paste-up cleanliness before releasing it to camera department. All excess wax, rubber cement, pencil lines (other than light blue), eraser residue, and dirt must be removed from surface of paste-up. (Dynamic Graphics, Inc.)

This process is known as *keying*. The halftones are keyed to the paste-up in the exact spaces that the halftones will occupy. Refer to Fig. 11-48.

If the outline method is used, a letter is usually written in the space provided for the halftone or line illustration. If the block method is used, the letter is normally written on top of the black paper square. In the case of ruby masking film, the letter is written underneath the ruby material in the square on the paste-up base. This allows it to be picked up later by the stripping department. In addition, the same letter is written on the reverse side of the photograph or illustration to match its position on the paste-up.

Keying each continuous tone photograph and line illustration is the only way these elements can be correctly combined after the negatives have been prepared. The following steps are recommended when preparing continuous tone photographs and line illustrations for reproduction.

1. Use cropmarks to indicate which part of the illustration you want to print, or what size you want the illustration to print.
2. Put cropmarks in the borders of the illustration rather than on the face itself. The only possible exception to this would be if you are using a grease pencil which easily wipes off with your finger or a rag. This should only be attempted on glossy photographs and never on line illustrations.
3. Mount illustrations with rubber cement or dry mount them on illustration board whenever possible to help keep illustrations from being damaged during handling.
4. Put a tissue flap or cover-flap over the illustration to protect it. The flap is used for writing instructions and drawing cropmarks. Use a felt tip pen or grease pencil.
5. Never use paper clips on continuous tone photographs for any purpose. Paper clips will scratch the photograph's surface and will then reproduce in the halftone.
6. Do not write on the back of continuous tone photographs with a sharp, hard pointed writing tool. If the photograph is mounted on illustration board, you may write on the back of the illustration board without damaging the photograph.
7. Do not leave paste-ups in an automobile on a hot day. The heat will cause the wax to melt. The elements on the board will pop off or move.

A blank sheet of tissue or vellum paper should be attached to the paste-up base. This provides a surface for marking corrections and prevents the temptation to mark on the paste-up base. It is also used to show the reproduction size desired. The paste-up should also be protected with a paper flap cover.

Fig. 11-49 shows a paste-up checklist that should be followed by the paste-up artist. The items on the checklist also require attention prior to proofing the paste-up for customer approval.

Checkpoints on preparing camera-ready artwork

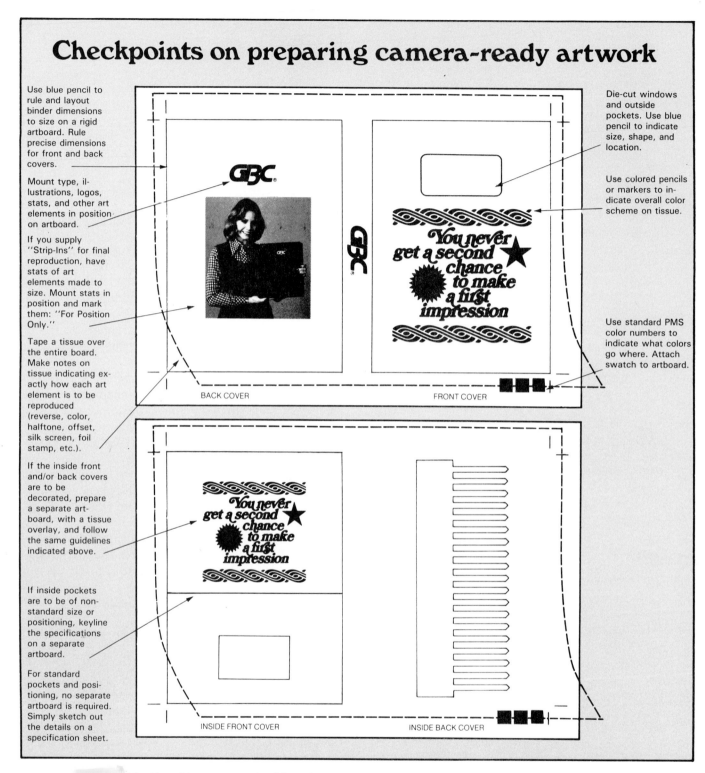

Use blue pencil to rule and layout binder dimensions to size on a rigid artboard. Rule precise dimensions for front and back covers.

Mount type, illustrations, logos, stats, and other art elements in position on artboard.

If you supply "Strip-Ins" for final reproduction, have stats of art elements made to size. Mount stats in position and mark them: "For Position Only."

Tape a tissue over the entire board. Make notes on tissue indicating exactly how each art element is to be reproduced (reverse, color, halftone, offset, silk screen, foil stamp, etc.).

If the inside front and/or back covers are to be decorated, prepare a separate artboard, with a tissue overlay, and follow the same guidelines indicated above.

If inside pockets are to be of nonstandard size or positioning, keyline the specifications on a separate artboard.

For standard pockets and positioning, no separate artboard is required. Simply sketch out the details on a specification sheet.

Die-cut windows and outside pockets. Use blue pencil to indicate size, shape, and location.

Use colored pencils or markers to indicate overall color scheme on tissue.

Use standard PMS color numbers to indicate what colors go where. Attach swatch to artboard.

GBC

You never get a second chance to make a first impression

BACK COVER   FRONT COVER

INSIDE FRONT COVER   INSIDE BACK COVER

Fig. 11-49. Use this paste-up checklist when preparing camera-ready copy.   (General Binding Corp.)

Proofing the paste-up

Once the paste-up has been prepared, it should be proofed and presented for final approval. Some type of electrostatic copying equipment is generally used for the proofing of paste-ups. Use a copier that does not pull the original material through the exposure area. This will ensure that none of the paste-up elements are damaged or pulled off.

After errors have been marked on the tissue overlay, they must be corrected. Image elements may need repositioning or resetting. The image elements that are replaced or corrected must be a perfect match. All corrections should be undetectable. This requires diligent attention during typesetting, processing, and layout.

Show all corrected proofs to the customer. Be certain to supply the rough or comprehensive layout along with original text copy. This makes it easy to compare the paste-up with the approved layout. The customer should note any errors directly on the proof sheet and date and sign the job in the margin of the proof sheet.

Note! Proofreading is described in Chapter 12.

Paste-up checklist

When you begin the process of paste-up, consider the following points.

1. Border tapes used with care can add a pleasing element to the paste-up.
2. Cut border corners at a 45° angle for neater, more professional paste-ups.
3. Shade artwork for a more realistic effect. Use shading films for shading or for color overlays.
4. Use creativity and customize borders on paste-ups.
5. All copy should be proofread.
6. Type should be checked for correct sizes.
7. The paste-up should be clean. Excess wax, rubber cement, dirt, fingerprints, and pencil lines should be carefully removed.
8. Line paste-up copy should be prepared in black, regardless of the color of ink that will be used to print the job. The black should be black and NOT gray.
9. Broken type should be repaired or replaced. Smudged type should be replaced.
10. Elements should be squared on the paste-up by using a T-square and triangle. Type should be aligned against both square edges.
11. Any overlay material should be accurately registered.
12. Separate continuous tone photographs and illustrations should be keyed for position identification. The required enlargement or reduction of every piece of separate artwork should be indicated.
13. Any windows for halftones should be of correct size and in the proper position.
14. Screen tint areas should be marked with the tint percentage and the color and number of ink desired.
15. Instructions for all desired operations should be written on the base paste-up and separate artwork. All instructions should be written in light blue pencil.
16. The paste-up should be covered with a protective tissue or vellum overlay along with a sheet of text or book paper.

POINTS TO REMEMBER

1. The goal of image assembly is to bring all the different image elements together into a final form that meets the job specifications.
2. Printers generally refer to camera-ready copy as a paste-up or mechanical.
3. Copy for photographic reproduction is prepared in several ways. Each element can be prepared and photographed separately. The negatives are brought together later during the stripping stages or preparation for making the plate.
4. The usual method of paste-up is to place as many elements as possible on a base and photograph them at the same time. This is referred to as the paste-up technique.
5. In paste-up, the elements to be printed are attached in the proper location on a suitable base material. The paste-up is then used as the copy to make the photographic negative.
6. Flat color is any color the layout artist or customer chooses for reproduction of the printed job.
7. Flat color can be handled as an overlay(s) on the paste-up or as one-piece copy.
8. Register is a critical alignment factor when preparing multicolor overlays.
9. Once the paste-up has been prepared, it should be proofed and presented to the customer for final approval.

KNOW THESE TERMS

Copy preparation, Base material, Bleed, Commercial register, Crop mark, Finish size, Guideline, Hairline register, Halftone block method, Halftone outline method, Illustration board, Image area, Image assembly, Keying, Keyline, Multicolor overlay system, Nonregister, Ornament, Overlay, Pantone Matching System®, Paste-up, Pin register system, Register mark, Reproduction proof, Rescreen, Shading material, Swatchbook, Transfer type.

REVIEW QUESTIONS

1. The copy elements used in _____ _____ are prepared for the single purpose of being photographed.
2. A curve can be used for drawing _____ and _____.
3. A ruling pen or technical drawing pen is used for inking straight lines and non-circular curves. True or false?
4. Nonreproducing pens and pencils are light _____ in color.
5. Cold press illustration board is used as a carrier or base material for paste-up. True or false?
6. Most of the typeset material used in paste-up comes in the form of phototype. True or false?
7. Adhesive lettering is applied to the paste-up base with a special _____ _____.
8. What is it called when an image area extends

off the edges of the final printed page?

9. One of the following is NOT a common method used to locate the position of photographs on the paste-up base.
   a. Outline.
   b. Register line.
   c. Block.
   d. Screened photo print.
10. _____ sheets are artistic patterns on transparent acetate.
11. A Velox® is a _____ print.
12. The process of rephotographing previously printed halftones is called _____.
13. What is the Pantone Matching System®?
14. When different color overlays on a paste-up align exactly, they are in _____.
15. Hairline register allows for image elements to be completely independent of each other. True or false?
16. The most exacting kind of registration is called:
   a. Hairline register.
   b. Commercial register.
   c. Nonregister.
   d. Step register.
17. An overlay is attached at the top of the paste-up base. True or false?
18. _____ marks are thin 90 degree cross lines used on color overlays.
19. The _____ register system is popular in the preparation of the paste-up, especially where color overlays are being prepared.
20. Red keylines are drawn on the paste-up to indicate the size, shape, and position of all halftones, tints, and color areas. True or false?

SUGGESTED ACTIVITIES

1. Gather several examples of printed material such as a brochure, booklet, newsletter, magazine, and poster. With the help of your classmates, try to determine how the materials were assembled at the copy preparation stage. For example, some materials may have been assembled by hand-mechanical paste-up while others were done on computerized phototypesetting equipment.
2. Prepare three kinds of paste-up copy: a simple one-piece single color job, a two-color job with overlay, and a hand-ruled business form.
3. Use the keyline procedure to prepare a two-color job involving flat color.
4. Gather at least two examples each of printed materials containing the three kinds of register: commercial, hairline, and nonregister. Label each sample and describe its primary use in printing.
5. Prepare a bulletin board display in the classroom that illustrates the several categories of paste-up copy described in the textbook. Obtain some of the samples from commercial sources and the school shop files.
6. Prepare a rough layout for a two-color radio station bumper sticker. The copy should include the call letters of the station in addition to a suitable slogan. The size of the sticker is 3 1/2 x 10 inches. Obtain the type and illustration, if desired, to complete the paste-up with an overlay. The competition for the liveliest and most unique radio station bumper sticker in your class will be an exciting feature of this activity.

Chapter 12

Proofreading

When you have completed the reading and assigned activities for this chapter, you will be able to:
O List the responsibilities of the proofreader.
O Describe the importance of quality proofing and proofreading in relationship to the final printed product.
O Demonstrate a basic knowledge of the commonly used proofreaders' marks.
O Recognize and describe an electronic proofing and editing system.
O Perform the tasks associated with proofing and editing phototypesetting material.

A unique procedure of detecting errors is used in the printing industry to make certain that typeset material is correct before it becomes a printed product. The procedure is called *proofreading*. See Fig. 12-1.

Proofreading is both an art and a skill, surviving in a time of high technology in the printing industry. The scribes used proofreaders to check for errors in their handwritten books. Proofreading became even more important after the invention of movable type in 1450. The practice of proofreading has survived because it is a means of quality control vital to the success of a printed product, Fig. 12-2.

Proofreading has changed very little over the years since the scribes. Even the computerized equipment of today still requires the skill of a proofreader. Look at Fig. 12-3.

PROOFREADERS' MARKS

Symbols called *proofreaders' marks* are a standard set of special codes used by the proofreader to eliminate writing out explanations to the typesetter. Like Morse code, proofreaders' marks are a universal language by

℘	Delete Take it Out	⊃	Close Up Print as Single Word	℘	Delete and Close Up	∧	CARET Insert *extra* Word Here	#	Insert space	stet	Leave Marked Word In	tr	Transpose
[	Move to the Left	]	Move to the Right	✕	Imperfect Or Broken Character	¶	Begin New Paragraph	sp	Spell out (10 lbs.) as Ten Pounds	cap	Set in Capitals	lc	Set in Lowercase
ital	Set in Italic	bf	Set In Boldface	∧	Insert a Comma	⌄	Insert Quotation Marks	⌄	Insert Apostrophe	⊙	Insert a Period	:	Insert a Colon
;	Insert a Semicolon	][	Center the Copy	wf	Wrong Font	⌐	Start a New Line	OK/?	Query to Author: Is it set as Intended?	⊃	Turn Over Inverted Letter	⊨	Insert a Hyphen

Fig. 12-1. Proofreading is an important step in preparing type composition. Before work of typesetting can be used to reproduce printed product, output must be proofread with care.

PROOF SYMBOLS USED TO CORRECT ERRORS IN LINES OF TYPE	THE LINES AFTER CORRECTIONS HAVE BEEN MADE
Change this to a period, It marks the	Change this to a period. It marks the
A comma, inserted here will make more	A comma, inserted here, will make more
Machinecast is a compound word when	Machine-cast is a compound word when
The printers work is most interesting	The printer's work is most interesting
Place a colon here, this is a place to	Place a colon here: this is a place to
Semi-colon, one should be inserted here	Semi-colon; one should be inserted here
Quotation marks are called "quotes by	Quotation marks are called "quotes" by
Delete, or take out, a characters marked	Delete, or take out, a character marked
The meeting was held in Columbus, (O.)	The meeting was held in Columbus, Ohio

Fig. 12-2. Proofreading of type composition is a means of quality control vital to success of a printed product.

which accurate and definite instructions are transmitted. Refer to Figs. 12-4 and 12-5.

For example, typeset material can be proofread in New York and its corrections accurately made in Los Angeles.

An example of the use of proofreaders' marks is illustrated in Fig. 12-6. Typesetters do NOT generally make this many errors. Can you find any errors which have not been marked and corrected?

Every proofreader should have access to a good dictionary. This is an essential tool for people who work as proofreaders. Many proofreaders use more than one dictionary. It may also be useful to have a dictionary designed especially as a spelling and hyphenation reference. This kind of dictionary is smaller and faster to use than a dictionary containing complete definitions.

The following is a summary of the most common proofreading or editing marks:

1. The *period symbol* is a dot with a circle around it. The circle is needed so that the period is easier to identify. Just a dot could be missed by the typesetter.
2. The *caret* is a right-side up "V" or an inverted "V" and it shows where something must be added to the copy.
3. *Punctuation symbols,* such as commas, semi-colons, apostrophes, etc., are denoted with a caret over or under the punctuation mark. If the punctuation goes near the bottom of the line, the mark (comma, for example) is placed under the line and vice versa.
4. The *hyphen symbol* is usually a small line with a caret under it. Some readers use two small lines to represent a hyphen. Also, a small line with an "N" above or below it could be used for hyphen and a small line with an "M" above it is used as a *dash symbol.*
5. A *delete symbol,* for removing letters or words, is a distorted capital "S." This symbol varies slightly from person to person but consistency is helpful.
6. A *close-up symbol* is two small curves above and below the letters or words and it means remove space. For example, it would be used if the typesetter accidentally hit the space bar in the middle of a word.

Fig. 12-3. High-speed computerized typesetting requires proofreading, which may be accomplished on screen as operator sets type composition. A hard copy or print of material is always provided for proofreading phase.

Proofreading

PROOFREADERS' MARKS

⊙	Insert period	*tr*	Transpose [1]—used in margin	⌐	Move left
⌒	Insert comma	∼	Transpose [2]—used in text	⌐	Move up
:	Insert colon	*sp*	Spell out	⌐	Move down
;	Insert semicolon	*ital*	Italic—used in margin	‖	Aline vertically
?	Insert question mark	___	Italic—used in text	=	Aline horizontally
!	Insert exclamation mark	*b.f.*	Boldface—used in margin	⊐⊏	Center horizontally
=/	Insert hyphen	∼∼∼	Boldface—used in text		Center vertically
∨	Insert apostrophe	*s.c.*	Small caps—used in margin	∪	Push down space
⌄⌄	Insert quotation marks	≡	Small caps—used in text	⌒	Use ligature
⊥/N	Insert 1-en dash	*rom.*	Roman type	*eq.#*	Equalize space—used in margin
⊥/M	Insert 1-em dash	*caps.*	Caps—used in margin	∨∨∨	Equalize space—used in text
#	Insert space	≣	Caps—used in text	*stet.*	Let it stand—used in margin
ld>	Insert lead	*c + sc*	Caps & small caps—used in margin	....	Let it stand—used in text
shill	Insert virgule	≣	Caps & small caps—used in text	⊗	Dirty or broken letter
∨	Superior	*l.c.*	Lowercase—used in margin	*run over*	Carry over to next line
∧	Inferior	/	Used in text to show deletion or substitution	*run back*	Carry back to preceding line
(/)	Parentheses	*w.f.*	Wrong font	*out, see copy*	Something omitted—see copy
[/]	Brackets	⌒	Close up	*♀/?*	Question to author to delete [3]
☐	Indent 1 em	ᵍ	Delete	∧	Caret—General indicator used to mark exact position of error in text.
☐☐	Indent 2 ems	ᵍ	Close up and delete		
¶	Paragraph	⊙	Correct the position		
no ¶	No paragraph	⊐	Move right		

Fig. 12-4. Proofreaders' marks are a universal language used by printers to transmit accurate instructions to person doing typesetting.

Abbreviations of States, Territories, and Possessions of the United States

AL	Alabama	Ala.	MT	Montana	Mont.
AK	Alaska	. . .	NE	Nebraska	Nebr.
AZ	Arizona	Ariz.	NV	Nevada	Nev.
AR	Arkansas	Ark.	NH	New Hampshire	N.H.
CA	California	Calif.	NJ	New Jersey	N.J.
CO	Colorado	Colo.	NM	New Mexico	N. Mex.
CT	Connecticut	Conn.	NY	New York	N.Y.
DE	Delaware	Del.	NC	North Carolina	N.C.
DC	District of Columbia	D.C.	ND	North Dakota	N. Dak.
FL	Florida	Fla.	OH	Ohio	. . .
GA	Georgia	Ga.	OK	Oklahoma	Okla.
GU	Guam	. . .	OR	Oregon	Oreg.
HI	Hawaii	. . .	PA	Pennsylvania	Pa.
ID	Idaho	. . .	PR	Puerto Rico	P.R.
IL	Illinois	Ill.	RI	Rhode Island	R.I.
IN	Indiana	Ind.	SC	South Carolina	S.C.
IA	Iowa	. . .	SD	South Dakota	S. Dak.
KS	Kansas	Kans.	TN	Tennessee	Tenn.
KY	Kentucky	Ky.	TX	Texas	Tex.
LA	Louisiana	La.	UT	Utah	. . .
ME	Maine	. . .	VT	Vermont	Vt.
MD	Maryland	Md.	VI	Virgin Islands	V.I.
MA	Massachusetts	Mass.	VA	Virginia	Va.
MI	Michigan	Mich.	WA	Washington	Wash.
MN	Minnesota	Minn.	WV	West Virginia	W. Va.
MS	Mississippi	Miss.	WI	Wisconsin	Wis.
MO	Missouri	Mo.	WY	Wyoming	Wyo.

Use the two-letter abbreviations on the left when abbreviating state names in addresses. In any other situation that calls for abbreviations of state names, use the abbreviations on the right.

Fig. 12-5. Proofreaders use various reference materials for handy access to vital information not generally available in any other form.

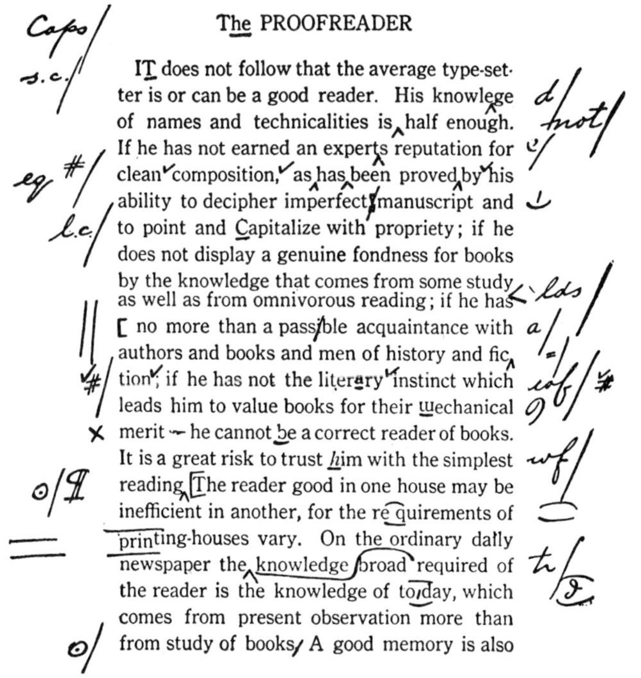

The PROOFREADER

IT does not follow that the average type-setter is or can be a good reader. His knowledge of names and technicalities is half enough. If he has not earned an experts reputation for clean composition, as has been proved by his ability to decipher imperfect manuscript and to point and Capitalize with propriety; if he does not display a genuine fondness for books by the knowledge that comes from some study as well as from omnivorous reading; if he has no more than a passible acquaintance with authors and books and men of history and fiction; if he has not the literary instinct which leads him to value books for their mechanical merit — he cannot be a correct reader of books. It is a great risk to trust him with the simplest reading. The reader good in one house may be inefficient in another, for the requirements of printing-houses vary. On the ordinary daily newspaper the knowledge broad required of the reader is the knowledge of to-day, which comes from present observation more than from study of books. A good memory is also

Fig. 12-6. All proofreaders' marks made on a proof must be legible so that person doing the typesetting can see clearly what to do.

7. A *stet symbol* means to "let copy stand" or ignore any denoted change in copy. The word "stet" (written in lowercase letters) denotes correction is unnecessary. Dots may be placed under the letters or words to denote the stet symbol.

8. A *paragraph symbol* is two lines with a curve at the top. It is used to show where a sentence should start and new paragraph is to be indented.

9. A *flush symbol* is a lowercase "fl" and means copy should be shifted to margin.

10. A *move* or *shift symbol* is a square with one side missing. Copy can be moved up or down and right or left with this symbol.

11. A *center symbol* means to center copy and it is denoted with half squares on each side of the copy or a lowercase "ctr."

12. A *transpose symbol* means to reverse the sequence of letters or words. It is shown as a sideways or lazy "S," wavy line, or as the abbreviation "tr." An elongated "S" would be used to transpose words.

13. The *insert space symbol* is like a small tick-tack-toe board or numeral sign and it means separate or add space. This symbol might be needed if the typesetter failed to hit the space bar between two words.

14. A *lowercase symbol* means change letters from capital to lowercase and is given by the letters "lc."

15. The *capital symbol* means change to capital letters and is usually denoted by placing three lines under the letters. The letters "cap" in the border can also show capitalization.

16. The *italic symbol* is usually a single line under the characters.

17. The *broken type symbol* is an "X" and it means that the typeset material is defective or has been damaged in layout.

18. A *push down symbol* is an inverted "T" and it denotes that copy should be moved down.

19. A *see layout symbol* means something is wrong and you should compare the manuscript or rough with the layout. It is given by writing "see l/o."

20. A *query symbol* denotes that the editor or proof-reader is not sure if something is correct. This may refer to content of material, sequence, etc. A circled question mark and the word "query" would be used to denote a possible problem.

The use of these and other proofreading marks will vary. It is important, however, that everyone in the same facility use the same marks. This consistency will reduce errors and increase communication as a printed product moves through each step of production.

PROOFS

A *proof* is a preliminary copy of typeset material used for finding and correcting errors. A proof is also used for making final editorial changes. In the case of paste-ups and phototypesetting materials, proofs are prepared from the originals using some form of duplicating equipment. It is important to remember that proofreaders' marks are seldom indicated on original paste-ups or original typeset materials.

Proofs are read by a proofreader or by two people in cases where complicated text matter is involved, such as highly technical publications. When two individuals read proof, one is called *copy holder* and the other a *reader*. Original copy must be compared word for word to assure accuracy. Each proof is marked with the proofreaders' marks to indicate the kind of error or omission.

Most of the typesetting used for offset lithography printing comes from hand-mechanical and phototypesetting methods. There are generally four kinds of proofs. These include:

1. Galley proof.
2. Revise proof.
3. Page proof.
4. Press proof.

Galley proof

A *galley proof* is a preliminary or first proof of typeset material. The term comes from galley trays, made of metal in which metal letterpress type was collected and stored. The same term has evolved over the years and is used for proofs from hand-mechanical and phototypesetting sources. These processes produce type in black images on white paper and phototypesetting paper respectively, Fig. 12-7.

The Craftsman's Creed

THIS IS MY CREED: The kind of work to be done should be that which is of real use. One of the best tests of whether work is useful or not is that people are willing to pay money for it.

● I believe the joy of good work well done is the highest form of satisfaction, and that the most dependable kind of happiness is that which is a by-product of work. No pay is worth while unless it refreshes and restores the power and disposition to work. No rest is sweet unless it is earned by work.

● I believe that no man can do good work unless he submits to training and practice. All work done in love, and following training and practice, will be beautiful. The perfectly useful is always beautiful. Whatever is beautiful is useful.

● I believe that no man lives unto himself nor can do his best work by himself; for he must learn of his masters, he must co-operate with his fellow Craftsmen and he must produce something that shall be valuable to the people.

Every human being was born to do some kind of good work. In doing it he finds his best excuse for living, and the most intelligent answer to the question, "Why was I born?"

by Philip Kinsel

Fig. 12-7. A copy machine is generally used to prepare a galley proof of phototypesetting composition. (Mergenthaller Linotype Co.)

To make galley proofs from these materials, copying equipment is generally used. Galley proofs show individual text and display type reproduced in single-column widths. Wide margins on the proof paper are included for use by proofreaders to mark corrections. The term *galley* also implies that proofs are uncorrected. The type is shown in the exact form as it came from the typesetting operation.

Revise proof

A *revise proof* is the same as a galley proof except that a first set of corrections has been made. Revise proofs are used to check the corrections made on the galley proof. It should be noted that in typesetting, the greatest number of errors occur during the correction phase. This means that typesetters generally produce more errors, line for line, in making corrections than they do on the original typesetting.

When a revise proof is being proofread, it should be carefully compared with the previous proof. Particular attention should be given to the corrections. However, it is necessary to make a careful check of the entire proof. Attention should be given to lost, damaged, or improperly spaced material during the process of proofreading.

The following checklist is especially important when proofreading revise proofs.

1. Read lines containing corrections and make sure that they are properly placed in the text.
2. Make sure that none of the lines have become *transposed* (positioned wrong) during correction.
3. Check against any possible lost type.
4. Check ends of lines for *dropouts* (missing characters) or unaligned characters.
5. Check all type, including illustrations, for any damage.
6. Make sure the original spacing and positioning have NOT been altered.
7. Carry over any notations from the previous proof(s).
8. File or destroy the previous proof(s) as directed.

Page proof

A *page proof* shows type and illustrations made up into a page ready for printing, Fig. 12-8. In some printing firms, page proofs are used for checking galley corrections as well as for making sure that copy fits.

When phototypesetting is the only equipment being used, page proofs may be the only kind of proof available.

Desktop publishing composition is generally output in full pages. Therefore, page proofs are usually made on a laser printer to take advantage of the cost savings over output on a phototypesetter.

The proofreading of page proofs is generally done to check for major errors. It is expensive and time-con-

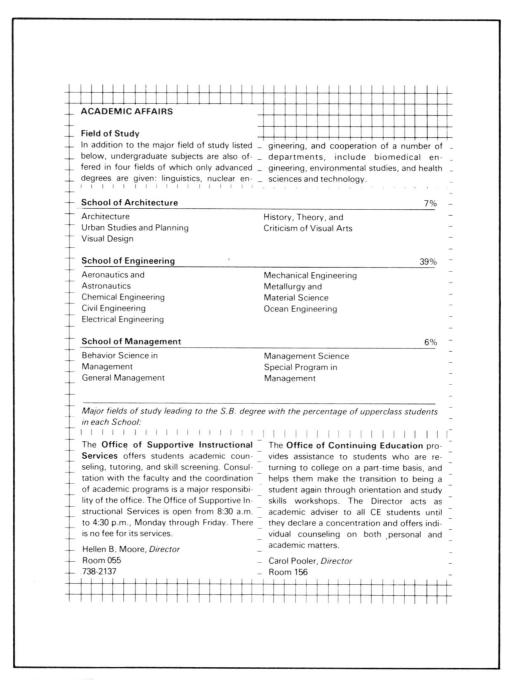

Fig. 12-8. A page proof shows type and illustrations made up ready for printing.

suming to make changes on page proofs which should have been caught in the galley proof stage. Therefore, it is an established rule that proofreaders do NOT edit or rewrite on page proofs.

A proofreader should never let mistakes get into print if they are caught in time to make corrections. This means that it is considered poor practice to make copy changes on page proofs primarily because of cost factors. However, it is worse yet to let errors get through.

The following checklist is especially important when proofreading page proofs:
1. Follow the specifications in the layout.
2. Check type sizes and width of lines.
3. Check for correct typefaces.
4. Proof for uniform line spacing in type groups.
5. Check spacing between units and in margins.
6. Proof the proper letter spacing and word spacing.
7. Check that illustrations are inserted and positioned correctly.
8. Are captions and/or legends correct?
9. Are rules, borders, and ornaments positioned correctly?
10. If type matter is for a book or otherwise paged, check that *folios* (page numbers, etc.) are correct and in sequence and that figure and footnote numbers are correct.
11. Eliminate widows at the top or bottom of pages. A *widow* is a single word in a line by itself at the end of a paragraph. Widows are frowned upon in good typographic layout.
12. Eliminate rivers of whitespace (contoured space resembling a stream).
13. Check ends of lines for missing or unaligned characters.
14. If a galley proof is being read, be sure that the head and foot of the type connects properly with type of preceding and following pages.

15. Check assembly of individual units of a border.
16. Check for defective rule, border, or other characters outside the body of the text matter.
17. Check positions and alignment of initial letters, captions, and all items which require exact positioning.

Press proof

A press proof is NOT generally used for checking the accuracy of typesetting. Instead, a *press proof* is used to check for proper page locations, illustration placement, and print quality. Making corrections at this stage would be like redoing the entire job.

The following checklist is especially important when proofreading a press proof.
1. Check position of type and illustrations on the sheet, the margins, and the gripper and guide edges (referring to orientation of sheet passing through offset press).
2. If the type and illustrations consist of a multiple of pages, check folios and the page sequence.
3. If the type and illustrations back up a previous page, check backing of pages, position, and register.

How proofs are marked

Corrections are generally marked in the margins of proofs, Fig. 12-9. Care should be taken to keep the typeset area as clean as possible. All writing should be done in the outer margins and kept free of typeset areas. Only the necessary indication at the point of each error should be made in the body of the text (actual typeset material). These include the diagonal stroke, caret, underline, circle, or transposition sign, Fig. 12-10. Proofreaders' marks should seldom be written directly on a paste-up. Instead, a vellum overlay should be used for this purpose. The safest and most efficient

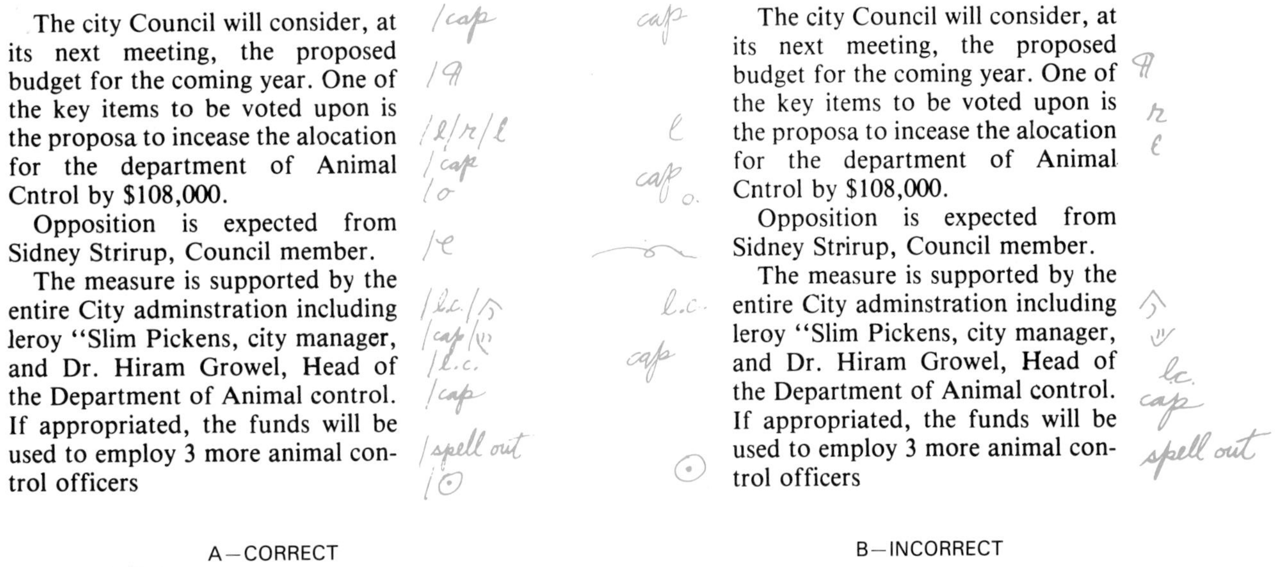

The city Council will consider, at its next meeting, the proposed budget for the coming year. One of the key items to be voted upon is the proposa to incease the alocation for the department of Animal Cntrol by $108,000.

Opposition is expected from Sidney Strirup, Council member.

The measure is supported by the entire City adminstration including leroy "Slim Pickens, city manager, and Dr. Hiram Growel, Head of the Department of Animal control. If appropriated, the funds will be used to employ 3 more animal control officers

A—CORRECT

The city Council will consider, at its next meeting, the proposed budget for the coming year. One of the key items to be voted upon is the proposa to incease the alocation for the department of Animal Cntrol by $108,000.

Opposition is expected from Sidney Strirup, Council member.

The measure is supported by the entire City adminstration including leroy "Slim Pickens, city manager, and Dr. Hiram Growel, Head of the Department of Animal control. If appropriated, the funds will be used to employ 3 more animal control officers

B—INCORRECT

Fig. 12-9. Marked-up proofs. A—Correct method. B—Incorrect method is confusing and difficult to decipher.

method is to make a photocopy of the paste-up for proofreading purposes.

In marking a galley proof, each marginal symbol indicating a correction should be placed as near to the error as possible. This will be to the left or right end of the line in which the error occurs. Errors in the left half of the line should be marked in the left margin. Errors in the right half of the line should be marked in the right margin.

Responsibility for editing the original manuscript is usually assigned to the editor. However, when obvious errors of fact or form are encountered, it is customary for the proofreader to question them. Whenever an editorial correction is marked or questioned, the proofreader should state the source of authority for the recommended change. It is important to remember that the proofreader's duty is to question—NOT to change. The editor and/or author is responsible for the editing.

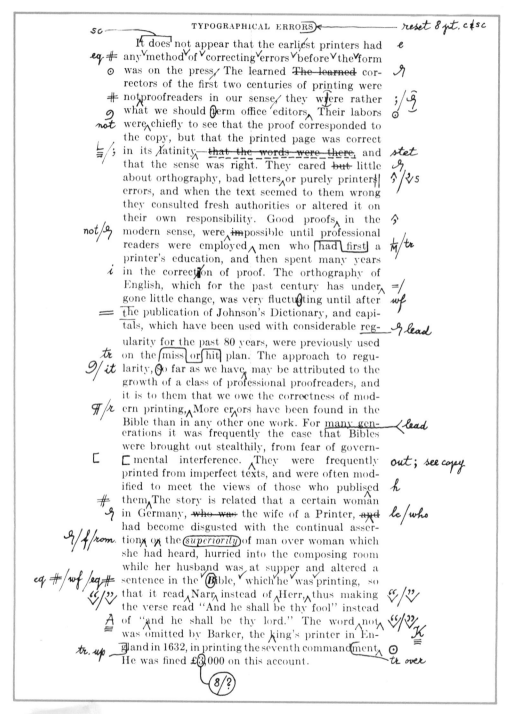

Fig. 12-10. All proofreaders' marks should be placed in outer margins of proof, balanced equally on both sides of an imaginary vertical centerline through typeset material. Only necessary indication at point of each error should be made in body of typeset material.

Proofs should always be signed or initialed by the proofreader. Some proofreaders are instructed to include the date and job number as further identification of the proof. If a proof happens to be free of all errors, the proofreader marks it OK and initials it.

CORRECTING TYPESET MATERIAL

The actual correcting of typeset material after it has been proofread is time-consuming and expensive. The typeset material must be made free of all errors after the proofs are read by the proofreader. There are special correcting techniques for each of the typesetting processes used in offset lithography.

Correcting dry transfer lettering

Correcting dry transfer letters on a paste-up or sheet is relatively simple. The letters are removed from the base with a knife. A new letter is selected and positioned on the base. This method of typesetting is extremely slow and tedious. Therefore, it is only used for small quantities of display type. See Fig. 12-11.

If you wish to remove letters or Spacematic bars, it can easily be done before burnishing. Cover the letter with a low tack adhesive tape e.g. masking tape. Rub over the letter lightly with the finger and then lift off.

Fig. 12-11. Correcting dry transfer letters on a paste-up or other camera-ready copy requires removal of unwanted letter and careful positioning of desired letter.

Strike-on type

Type composed with a strike-on (impact) typewriter is handled separately. To change one character or word, a complete new line is prepared on the typewriter. The corrected line is pasted over the incorrect line.

Some paste-up artists remove the incorrect word or line from the base with a knife. Then the corrected material is cut and pasted in the vacant space. This method eliminates any possibility of photographic *"shadow effect"* caused by one line being higher than another. This can cause an unwanted shadow line in the negative close to the type itself.

Extreme care should be exercised when retyping correction lines to ensure that the new type lines are of the same density (blackness) and weight as the original typeset material. Type of varying densities within the text matter is unsightly and in poor taste.

Non-computer-assisted phototypesetting

In a non-computer-assisted phototypesetting system, corrections must be keyboarded into a machine-readable form. They are then processed through the phototypesetting system. The corrections must be cut apart and either pasted over incorrect lines or cut into the original output.

There is always the possibility that the corrected type may not perfectly match the original in density. This could result in some type appearing darker than the other type. Rigid control of processing procedures will help to minimize or eliminate this problem.

Computer-assisted phototypesetting

In a computer-assisted phototypesetting system, the machines are capable of greater speeds than non-computer assisted systems. They can also create complete pages of text, display, and illustrations in position. The characters are stored as either master images or as digital information. They are then translated into characters on the cathode ray tube (CRT) screen of the phototypesetter. Corrections are made by having computer memory display previously typeset material. Corrections can be made by adding, deleting, replacing, and moving text matter.

The video display terminal (VDT) is one of the more useful tools in today's phototypesetting departments. The operator can edit keyboarded copy and do full-page makeup directly on the screen. The material is stored on magnetic tapes or disks. Editing and corrections are made directly on the screen, Fig. 12-12. The

Fig. 12-12. Operator of a video display terminal (VDT) can edit and correct full pages of typeset material quickly and easily.

operator can add, delete, replace, move sections of text matter around, etc. This allows flexibility during the editing and correction cycles.

Improved proofing and editing can be done when a preview editing terminal is linked directly with the phototypesetter. These units give an operator considerable flexibility in designing copy in electronic form, according to the prepared layout. Spelling, punctuation, word and line spacing, type styles and sizes, alignment of columns, and many other typographic functions can be seen on the screen and adjusted.

PROOFING PROGRAMS

A *proofing program,* also called *spell checker program, spelling program,* etc., will automatically check the spelling of words typed into the computer system. This is a very handy feature that reduces the work load on proofreaders and editors. The typed copy can be quickly checked for typos and spelling errors by activating the computer spelling program.

There are two basic types of proofing or spelling programs: running spell checkers and batch spell checkers.

Some programs, termed *running spell checkers,* inspect each word as it is typed into the computer. The program will make the computer produce an audible "beep" if a word NOT in the program is entered. Similar types of programs, called *batch spell checkers,* review spelling after all the copy has been input by the typesetter.

Batch spell checkers are more common than running spell checkers because the typesetter is NOT distracted by frequent beep sounds. The typesetter can concentrate on keystroking speed and accuracy. He or she can check all copy for errors after typesetting. Some spell checking programs will function as both running and batch spell checkers.

The proofing program typically has around 80 to 100,000 words loaded into disk memory. These words are called the *program dictionary* and they can contain as many words as a small, paperback dictionary. Many programs allow you to program new or specialized words into the dictionary. This is handy for specialized types of publishing where technical words are commonly used and are not included in the standard program dictionary. Look at Fig. 12-13.

Proofing programs can be very helpful to production speed and quality. They can greatly reduce typos and make the correction cycle much easier. The proofreader must still go over the copy, but with less typing errors to mark, more effort can be placed on style, illustrations, meaning, consistency, etc.

A proofing program will NOT check the meaning of words. For example, the word usage mistakes in this sentence—"My SUN went TOO school FOUR a long time TWO . . . ." would NOT be caught by a proofing program. None of the words are misspelled even though several of their meanings are incorrect. Keep this in mind when you use a spell checking program.

Note! Programs are available that will check word usage and sentence structure. Called *grammar checkers,* they will help you with style, word usage, incomplete sentences, incorrect homonyms, possessives, etc., and will offer suggestions for correcting the sentence structure.

A *thesaurus program* can be used to find synonyms or words that have the same meaning but are spelled differently than the word being checked. This is more

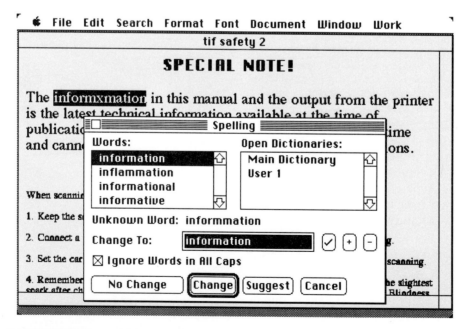

Fig. 12-13. A computer proofing program generally has between 80 and 100,000 words in its disk memory. Most computer software programs allow the operator to program new or specialized words into system.

handy for a writer or editor than for a proofreader or typesetter. The writer or editor can use the program to quickly find another word to use that has a similar meaning. For example, if the manuscript uses the same word over and over, the thesaurus program could be used to insert another word to prevent word repetition. See Fig. 12-14.

SOFT AND HARD PROOFS

The term *soft proof* refers to checking the copy while still on the video display terminal or computer screen. There is no hard copy just electronic data in the computer that is displayed on the CRT. All copy is read and corrected while still in the computer or phototypesetting machine. This system is used when production speed is the most important concern, as with a Sunday newspaper.

A *hard proof* is copy that has been output onto paper with a printer or other device. The proofreader can then read the copy off of paper to check for errors. When accuracy is more important than turnaround time, a hard copy is used, as with this textbook.

A hard copy proof is better than trying to check soft copy on the computer screen. The monitor has very low resolution and the characters are NOT as defined as they are on hard copy. It is much easier for mistakes to slip through when trying to proof on a display screen. Therefore, most facilities output onto paper or hard copy for proofing.

Impact printers can be interfaced with phototypesetters and computers to print out hard copy rapidly for the purpose of proofreading. Again, proofs on paper are easier to see than images on the video screen of the typesetter. Printed proofs let the proofreader compare the keyboarded copy with the original copy more accurately. The proofreader can also study the proofs longer without tying up the phototypesetter or editing terminal.

Hard copy impact printers can be located on-line or off-line. An *on-line printer* is electronically and directly connected to a phototypesetter or computer. An *off-line printer* is NOT tied directly to a phototypesetting machine or computer. The keyboarded input comes from a separate memory—floppy disk, laser or optical disk, paper tape, etc.

Hard copy printers produce copy rapidly and accurately. Some will print out the body and display copy in black ink and the typesetter command codes in red ink. This gives the proofreader instant identification of all the material printed on the proof. Some hard copy impact printers can produce 80 type characters per second. Up to four duplicate copies can be printed at once. This lets several people read the proofs at the same time.

Desktop publishing systems, such as on Macintosh or IBM computers are generally interfaced with a laser

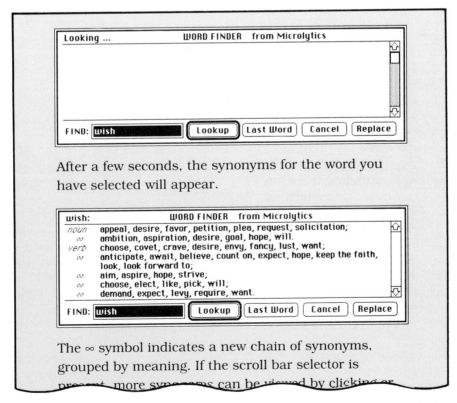

Fig. 12-14. A computer thesaurus program is used to find synonyms or words that have same meaning but are spelled differently than word being checked. Program is used more by writers and authors than proofreaders and typesetters.

printer and/or phototypesetter. For proofreading purposes and cost effectiveness, these systems can also be interfaced with a dot matrix printer.

POINTS TO REMEMBER

1. Proofreading is a procedure used in the printing industry to make certain that typeset material is correct before it becomes a printed product.
2. A proof is a preliminary copy of typeset material used for finding and correcting errors and omissions.
3. To proofread paste-ups and phototypeset materials, proofs are initially prepared from the originals using a photo copier.
4. Proofreaders' marks are never indicated on original paste-ups or original typeset materials.
5. Proofs can be read by one proofreader or by two individuals in cases where complicated text matter is involved.
6. When two individuals read proof, one is called the copy holder and the other a reader.
7. Symbols, called proofreaders' marks, are a universal language by which accurate and definite instructions are transmitted.
8. Proofreaders' marks are used to indicate errors, changes, additions, deletions, etc., on a proof.
9. There are four kinds of proofs. These include the galley proof, revise proof, page proof, and press proof.
10. A galley proof is a preliminary or first proof of typeset material.
11. A revise proof is the same as a galley proof except that a first set of corrections have been made.
12. A page proof shows type and illustrations made up into a page ready for printing.
13. The purpose of a press proof is to check for proper page locations, illustration placement, and print quality. Major corrections or changes are seldom made at the press proof stage.
14. Proofs are generally marked in the margins to show corrections. All writing should be done in the outer margins and kept free of the typeset or image areas.
15. Avoid placing proofreaders' marks directly on a paste-up. A vellum overlay should be used for this purpose.
16. Proofs should always be signed or initialed by the proofreader.
17. The actual correcting of typeset material after it has been proofread is time-consuming and expensive.
18. There are special correcting techniques for

each of the typesetting processes used in offset lithography, including dry transfer lettering, strike-on, non-computer-assisted phototypesetting, and computer-assisted phototypesetting.
19. Impact printers can be interfaced with phototypesetters to print out keyboarded copy for the purpose of proofreading. This is called a hard copy.
20. Desktop publishing systems are generally interfaced with a dot matrix printer for use in proofreading.

KNOW THESE TERMS

Galley proof, Page proof, Press proof, Proof, Proofreading, Proofreaders' marks, Review proof, Widow, Hardcopy, Proofing program, Running spell checker, Batch spell checker, Thesaurus program, Grammar checking program.

REVIEW QUESTIONS

1. _____ are preliminary copies of typeset materials used for finding and correcting errors or for making final editorial changes.
2. When two people are assigned to a proofreading job, one is called a _____ _____ and the other is the _____.
3. If a proofreader underlines a character or word on the proof, the character or word should be:
   a. Deleted.
   b. Set in boldface.
   c. Set in capitals.
   d. Set in italic.
4. If a proofreader writes "stet" on the proof, it means:
   a. Delete.
   b. Leave as is.
   c. Refer to manuscript.
   d. Set in bold face.
5. A _____ proof is a preliminary or first proof.
6. Revise proofs are used to check the corrections made on press proofs. True or false?
7. Proofreaders do NOT generally edit or rewrite copy of page proofs. True or false?
8. A _____ is a single word in a line ending a paragraph; it is frowned upon in good typography.
9. A _____ proof is used primarily for checking proper page locations, illustration placement, and print quality.
10. If a proof is free of all errors, the proofreader writes _____ and _____ the proof.

11. What is a spell checking program?

12. What is a computerized thesaurus?

13. A _____ _____ is a copy that has been output onto paper with a computer printer or similar device.

14. _____ printers can be interfaced with phototypesetters and computers to print out copies for proofreading.

15. An on-line printer is directly connected to a phototypesetter or computer. True or false?

SUGGESTED ACTIVITIES

1. Prepare a chart (poster) containing the proofreaders' marks. Use the visual aid to memorize as many symbols as possible. Share the chart with other members in the class.

2. Ask your teacher to give you a floppy disc that contains keyboarded manuscript for several typeset jobs. Load the disk in the computer and make the needed corrections. Prepare a corrected reading proof and make the necessary corrections on the proof.

3. Gather a quantity of newspaper and magazine articles and advertisements containing typographical errors. Rubber cement each article and advertisement onto a larger piece of white paper or cardboard. Mark the errors using the correct proofreaders' marks. Arrange an interesting bulletin board display in the classroom with an appropriate theme.

4. Visit a daily newspaper plant. Observe how reporters use the CRT and VDT screens to prepare copy, proofread, and edit material.

5. If your shop does not already have one, design a proofreading form that can be used as a customer approval sign-off. The form should be used for every job requiring some form of proofreading and/or approval for printing. Ask your teacher for assistance in identifying content details for the form.

Section IV

Issues for Class Discussion

1. With electronic typesetting, marking copy properly is an important control phase. This should be done prior to the initial keyboarding stage. However, in some shops, it is not done prior to keyboarding, thus decreasing production efficiency. The person responsible for the format and markup of copy must know the language of the system and what the system can and cannot do. The markup instructions, in coded form, are keyboarded as input, with the text. In large firms, a markup specialist may figure out command codes, so that the operator can concentrate on keyboarding.

   Discuss the several ways in which an organized program of copy markup and formatting can contribute to increased typesetting productivity.

2. Desktop publishing is a reality for in-office reproduction centers. The low-cost, easy to operate, compact, highly capable electronic typesetting equipment is helping to complete the production chain of in-office input, editing, typesetting, and copier services. As in-office desktop publishing and high speed copying operations take shape, one can expect them to grow in size, equipment sophistication, and complexity of copy processed.

   Discuss how this will affect the commercial and quick printing markets.

3. The processes that encompass the broad spectrum of typesetting are changing quickly and dramatically. There is a clear contrast between typesetting and how a foundation is prepared for a building. The carpenter constructs forms for a foundation of a residential unit. Concrete is poured and the foundation is ready for the next phase of construction. As construction proceeds, the foundation is permanently covered with the necessary framework and outside covering. Thus, the foundation is totally out of sight to all who view the structure. The typesetter's work, however, is in full view of the consumer and open to critical review.

   Discuss the implications of this contrast with regard to the several consumer impacts directly related to these important technologies.

4. The past 500 years have exemplified tremendous change in typesetting technology. Many people in the printing industry have been affected by the changes since the invention of movable type.

   Discuss some of the important technological and human implications that have evolved in typesetting since the advent of phototypesetting.

5. Desktop publishing has emerged as a viable means for many printers to enter the typesetting arena.

   Research and discuss factors that are important to printers who might be contemplating acquisition of a desktop publishing system.

6. Several nationally recognized communication experts have predicted that the nation, perhaps the world, will soon reach a point of becoming a "paperless society." This means that most of the printing and copying that we take for granted today would be replaced by other communication technologies.

   Discuss the practicality and implications of such a turn of events. Is it really possible that a paperless society could evolve in the near future?

Section V

Image Conversion

Image conversion includes processes used to change camera copy to negative or positive film materials. The negative or positive is then used to produce an offset printing plate.

The basic elements of the photographic process are light and light-sensitive materials. Chapter 13 introduces the process camera and darkroom layouts. The basic types of process cameras include horizontal and vertical. Darkroom layouts are designed to accommodate each type of camera and the production needs of the individual operation.

Chapter 14 covers film processing techniques, including the various tools and equipment necessary.

Chapter 15 introduces the most common of the photographic processes, called line photography. In line photography, high-contrast black and white copy is converted or changed from a positive image on paper to a photographic negative or to a positive film image.

Chapter 16 covers the conversion of original black and white copy that has continuous tone or tonal qualities into halftones. The continuous tone original is converted into dot patterns. This dot pattern consists of image and nonimage areas. The converted halftone print is an optical illusion of the tone of the original copy.

In Chapter 17, special effects photographic processes are covered. Besides the regular black and white halftone and process color photography, additional techniques have been developed to enhance or change the visual effect of the final reproduction. These "special effects" can assist in achieving unusual printed design effects.

Full-color reproduction color process photography, one of the most exciting areas of offset lithography, is covered in Chapter 18. As with black and white continuous tone copy, continuous tone color copy must also be converted to a dot pattern. However, in full-color reproduction, four halftone film negatives must be produced using color filters. This process is often referred to as color separation. The four film negatives represent the colors yellow, magenta, cyan, and black.

Chapter 19 covers one of the most popular methods used for producing full-color separation negatives and positives. The color separation scanner is an electronic device that accomplishes the intricate tasks of color separation easily and rapidly. Compared to the process camera, the color separation scanner is more versatile, efficient, and faster. This chapter also discusses how computers can be used to modify images electronically.

An electronic horizontal camera.   (nuArc Company, Inc.)

Chapter 13

Process Cameras and Darkroom Layouts

When you have completed the reading and assigned activities for this chapter, you will be able to:

O Identify the various parts of a process camera.
O Describe the purpose of a process camera.
O Distinguish between horizontal and vertical process cameras.
O List advantages and disadvantages of both the horizontal and vertical process cameras.
O Prepare the camera for a same-size reproduction.
O Prepare the camera for an enlargement and a reduction.
O Determine the basic exposure for a process camera.
O Determine the best f-stop that will give the sharpest image on a process camera.
O Reduce and enlarge copy using the constant time exposure method and the constant aperture method.
O Name common darkroom equipment and related accessories.
O Describe several darkroom arrangements, listing advantages and disadvantages of both.
O Identify the essential elements of darkroom cleanliness and housekeeping.
O Practice safe procedures while working in the darkroom.

The camera used to convert images in offset lithography is called a *process camera*. A process camera differs from an ordinary snapshot camera. It is a large piece of equipment with special features to make it suited for use in offset lithography, Fig. 13-1. For most of the work in offset lithography, the process camera converts copy into sharp high-contrast film negatives. For some applications in offset lithography, film positives are used instead of film negatives. A process camera is equipped with a special *process lens*. This means that the lens is constructed to give optimum results with flat or two-dimensional copy. The lens on a snapshot camera is constructed for three-dimensional subject matter.

TYPES OF PROCESS CAMERAS

Process cameras are installed and used in one of two ways. These include: darkroom camera or gallery camera.

A *darkroom camera* may be installed in a light-tight room. The usual way is to have the lens, lights, and copyboard outside the darkroom, with the camera back built into the wall of the darkroom, Fig. 13-2. This allows the operator to position and process the film in the dark.

A *gallery camera* is installed entirely in a lighted room. This arrangement requires that a light-tight film holder be loaded in a darkroom and then taken to the camera. To avoid this inconvenience, the camera, along with processing sink or processor unit, can be located within the darkroom. This arrangement requires safelights and regular white room lights, Fig. 13-3.

There are two basic process camera designs. They include: horizontal and vertical. Both types of cameras consist of essentially the same basic components and operate on the same principles.

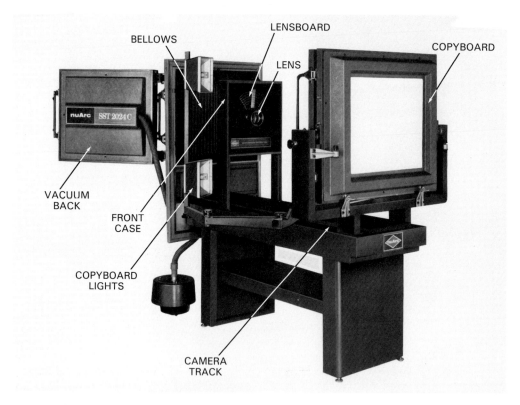

Fig. 13-1. Process camera is used to reproduce two-dimensional copy for offset lithography printing process. (nuArc Company, Inc.)

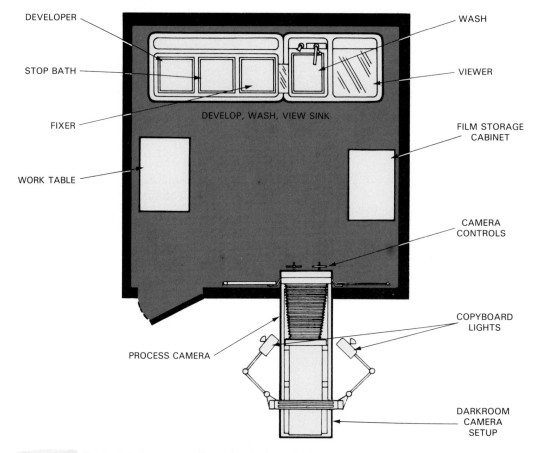

Fig. 13-2. Typical darkroom configuration is for a darkroom-type process camera. It should be noted that most darkrooms now use automatic film processors instead of hand-developing method.   (nuArc Company, Inc.)

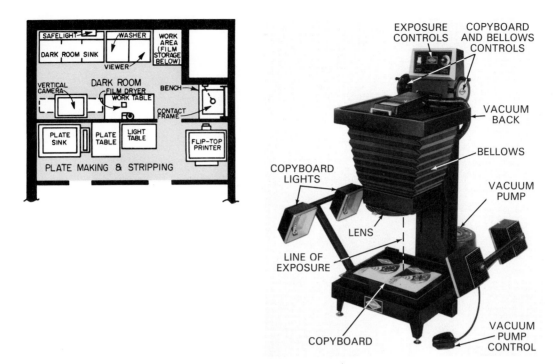

Fig. 13-3. Gallery process camera is installed entirely in a lighted room (see darkroom diagram). This vertical camera is compact and requires less space than horizontal process camera. (nuArc Company, Inc.)

Horizontal process camera

A *horizontal process camera* is constructed in a horizontal line or with the lens facing parallel with the floor, Fig. 13-4. The parts of a horizontal process camera are illustrated in Fig. 13-5.

Most horizontal cameras are installed through a wall and are used as a darkroom camera. The part of the camera that holds the film is in the darkroom. The major portion of the camera remains in a lighted, adjacent room. This arrangement makes it possible to load and unload film within the darkroom. As a result, film processing can be carried on without interruption.

When very large copy must be photographed, an overhead horizontal camera is often used, Fig. 13-6.

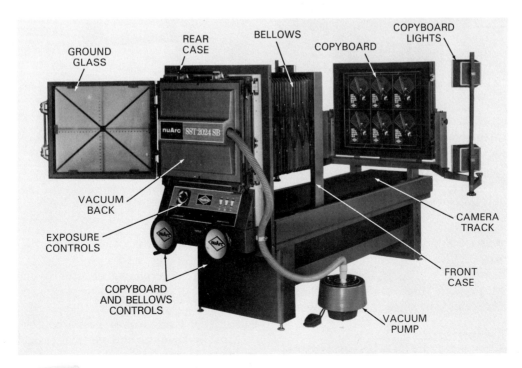

Fig. 13-4. Note rear view of a typical horizontal process camera. (nuArc Company, Inc.)

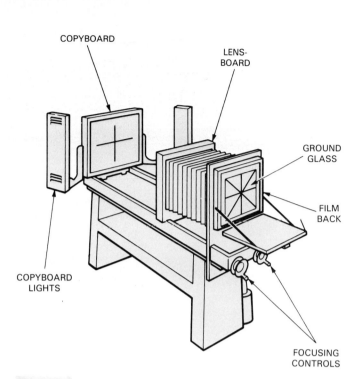

Fig. 13-5. Study main components of a typical horizontal process camera. (A.B. Dick Co.)

Fig. 13-7. A vertical process camera requires less space than a horizontal process camera. However, its physical size limits size of copy enlargements and reductions. (Kenro)

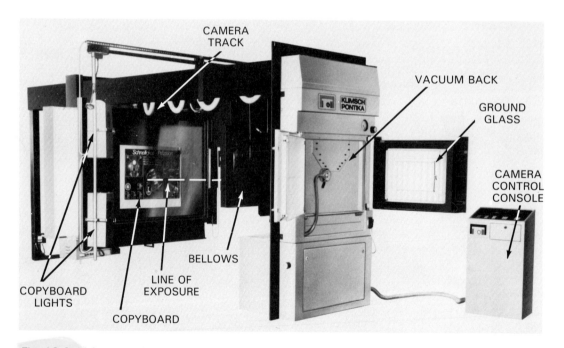

Fig. 13-6. A large overhead horizontal process camera is used for exceptionally large pieces of camera copy. Note computerized camera control console at far right. (Heidelberg U.S.A.)

This type of camera can handle copy up to 60 x 84 inches (152 x 213 cm). It also has a vacuum film holder of about 56 x 56 inches (142 x 142 cm). Process cameras of this type and size are used for poster work, mechanical drawings, and architectural drawings.

Vertical process camera

A *vertical process camera,* like the name implies, is constructed in a vertical line or the lens faces the floor,

Fig. 13-7. The parts of a vertical process camera are illustrated in Fig. 13-8.

Vertical cameras can be used in the darkroom or gallery. They are useful as back-up cameras or in areas where space is restricted. The newer models have a wide latitude of reduction and enlargement capability.

Most horizontal and vertical process cameras are made to accept single sheets of film. However, larger horizontal roll-film process cameras are in common use

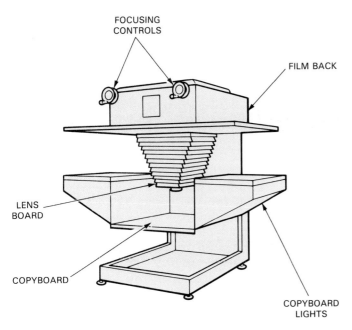

Fig. 13-8. Note main components of a vertical process camera. (A.B. Dick Co.)

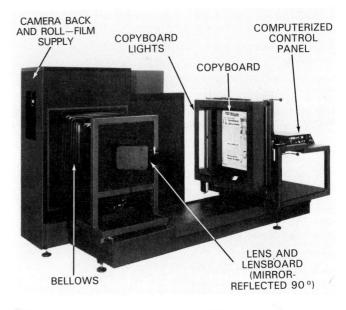

Fig. 13-9. High volume horizontal roll-film process camera is often used for daily newspaper operations. (Chemco Photoproducts Co.)

where high volumes of film processing are required. Refer to Fig. 13-9.

Fully automated camera systems are available that include components such as roll-film feed, camera-to-processor film transport, automatic film processing, and automatic negative conveyor belts, Fig. 13-10.

Flat-field scanners

Sometimes referred to as *electronic cameras, flat-field scanners* may eventually replace conventional process cameras and film for many applications. These scanners combine optical imaging with electronic image manipulation.

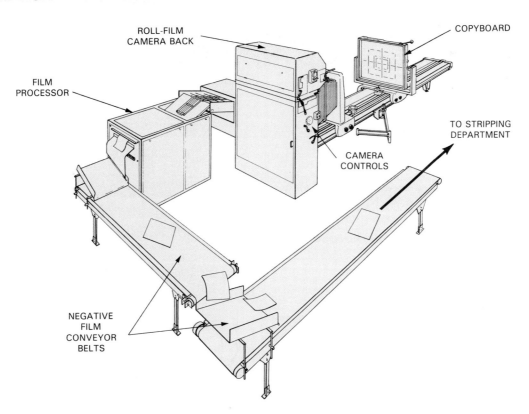

Fig. 13-10. With a fully automated roll-film camera system, features typically include camera-to-processor film transport, automatic film processing, and automatic negative conveyor belts. (Chemco Photoproducts Co.)

With a flat-field scanner, an original illustration is scanned by a light source, usually a laser. The information is transmitted to a computer which digitizes and manipulates it electronically. It is output through another optical system to a display medium like paper or film. In this series of steps, the image can be reproduced at any size in either or BOTH directions *(anamorphic enlargement)*. It can be slanted, outlined, bordered, or reversed (laterally, or from positive to negative).

Copy detail can be increased or sharpened by techniques known as *unsharp masking* and *edge enhancement*. Tone reproduction can be adjusted to suit variable printing conditions. The copy can be reproduced as continuous tone or halftone in a range of screen rulings and with special dot configurations like a double dot, chain dot, random grain, etc.

These systems are very cost effective. They relieve production in the film assembly department. They provide accurate films for 4-color process printing. The systems provide high levels of mechanical accuracy for fully imposed film flats. The systems are especially effective for complex magazine and catalog pages with multicolor halftone images, screen tint backgrounds, and facing page crossover images.

PROCESS CAMERA COMPONENTS

Regardless of camera size or design, the method and function of the controls are basically the same. There are approximately seven components which are common to all process cameras. These include:

1. Copyboard.
2. Copyboard lights.
3. Lensboard.
4. Bellows extension.
5. Ground glass.
6. Vacuum back.
7. Focusing controls.

Copyboard

The *copyboard* is the part of the process camera on which the copy to be photographed is positioned, Fig. 13-11. It has a large flat surface with a hinged glass cover to hold the copy during exposure. The copyboard is mounted on a track so it can be moved forward and backward for reductions and enlargements.

Copyboard lights

The *copyboard lights* act as a source of illumination of the copy, Fig. 13-12. On most process cameras, the lights are mounted in relation with the copyboard. This allows for the same distance between lights and copyboard regardless of reduction or enlargement settings. The lights are generally positioned at a 45° angle to the copyboard. See Fig. 13-13.

Carbon arcs, tungsten lamps, pulsed xenon, quartz-iodine, and mercury lamps are in common use. The

illumination from each of these light sources varies in *intensity* (amount of light) and *kind* of light (degree of visible spectrum).

Never touch lamp bulbs, especially the tubular quartz-iodine kind. Your fingerprints may cause the bulbs to darken. When one lamp fails, all lamps should be replaced at the same time. This will maintain equal illumination of the copyboard. Follow the manufacturer's instructions when installing camera lamps. Keep camera light reflectors free of dust.

Process cameras require special lighting arrangements to expose film in the camera back. The illumination from each of the light sources previously mentioned varies. To better understand lighting and how it affects film sensitivity, you should comprehend that part of the visible radiant energy called light.

The sun emits or radiates energy. A common light bulb also radiates energy. Radiation is measured by the length of the carrier wave on which it travels. The *full spectrum* (a series of waves arranged by length) ranges from short gamma waves to long radio waves. The visible spectrum consists of a narrow band of waves. The waves in the visible spectrum are made up of pure colors. Each color relates to a certain spectral wavelength.

The *wavelengths* (specified in angstrom units) for six common colors in light are:

1. Violet (4100A).
2. Blue (4700A).
3. Green (5200A).
4. Yellow (5800A).
5. Orange (6000A).
6. Red (6500A).

The wavelengths are from about 400 to 700 nanometers in length. A *nanometer* is one billionth of

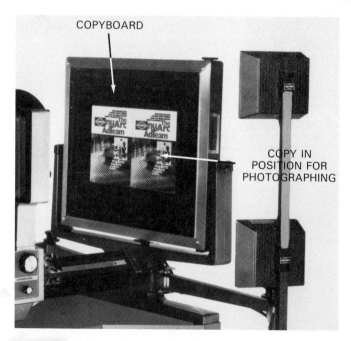

Fig. 13-11. Copy to be photographed is held in place in process camera copyboard. (nuArc Company, Inc.)

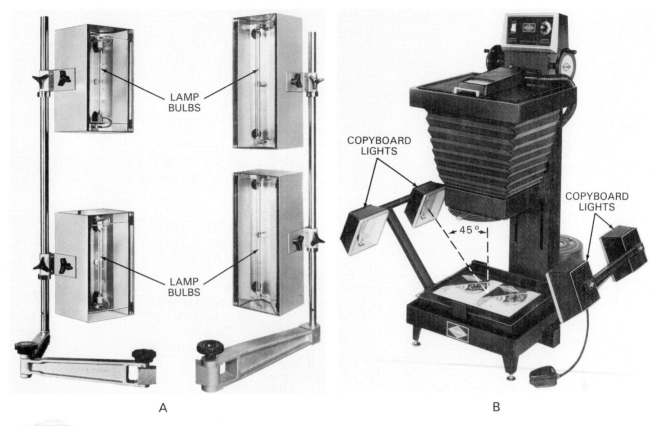

Fig. 13-12. A—Copyboard lights are usually held on copyboard frame of process camera. Examples above are solid state pulsed xenon lights. B—Copyboard lights are generally positioned at a 45° angle to copyboard as in this vertical process camera set-up.   (nuArc Company, Inc.)

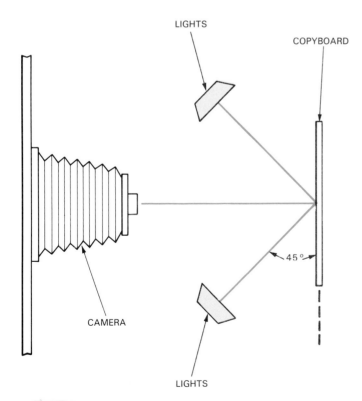

Fig. 13-13. Copyboard lights are generally positioned at a 45° angle to the copyboard on both horizontal and vertical process cameras. This angle gives even illumination at copy plane.   (nuArc Company, Inc.)

a meter. One nanometer is equal to 10 angstrom units (A). An *angstrom unit* is a wavelength of light equal to one ten-billionth of a meter. In suitable proportions, the total of ALL the colors (wavelengths) is seen by the human eye as *white light*.

Spectral distribution of light

Incandescent light sources, including tungsten and carbon arc, produce a continuous output of wavelengths. The emission of wavelengths is throughout the visible spectrum. The intensity of wavelengths within the spectrum depends on the temperature at which these lamps burn, Fig. 13-14.

Gas discharge light sources (mercury vapor and xenon arc) have separated spectral wavelengths. This type of light source is called a *discontinuous,* or line, spectrum. In this case, the nature of the gas, rather than temperature, determines the spectral distribution. Look at Fig. 13-15.

The spectral distribution of light for the various common graphic arts process camera light sources differ significantly.

Carbon arc—The spectral distribution of light for the carbon arc is shown in Fig. 13-16. The light intensity is high, allowing short exposures. The light covers the entire spectrum. These lights are adequate for color photography. Carbon arcs are NOT widely used to-

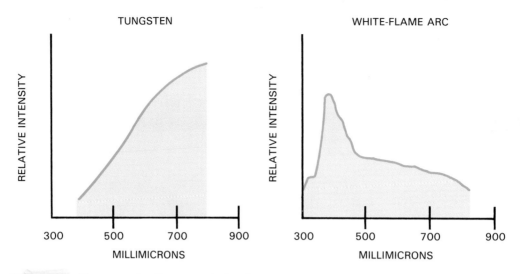

Fig. 13-14. These graphs illustrate relative light intensity of two kinds of incandescent light sources: tungsten and carbon arc. Tungsten burns brighter with smoother overall intensity and little or no drop-off as compared to carbon arc. (Eastman Kodak Co.)

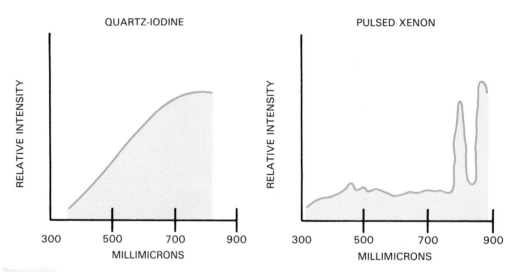

Fig. 13-15. Spectral distribution of gas discharge light sources are determined by nature of gas, rather than temperature. (Eastman Kodak Co.)

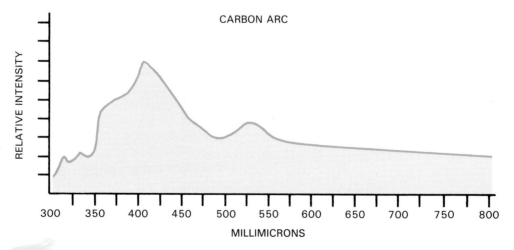

Fig. 13-16. Carbon arc light intensity is high, allowing short exposures. Carbon arcs are not widely used today because they are environmentally unhealthy and better light sources are now available.

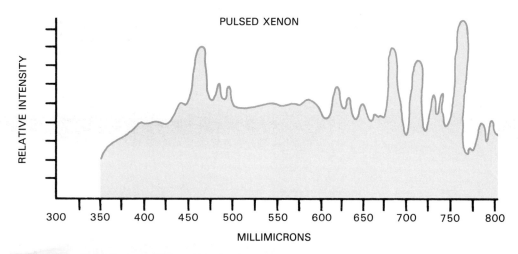

Fig. 13-17. Pulsed xenon light source is about same as carbon arc. However, pulsed xenon lamps are clean and are not affected by voltage fluctuations. They provide a constant level of light.

day because the smoke and ash are environmentally unhealthy.

Pulsed xenon — Pulsed xenon is similar to the electronic flash tube used in photography. The pulsed xenon lamp is designed to light, or pulse, with each half cycle of alternating current applied. Carbon arcs and pulsed xenon lamps are about equal in film exposure time. The xenon lamp has a spectral distribution across the visible light range, Fig. 13-17.

Pulsed xenon lamps are clean and are NOT affected by voltage fluctuations. The level of light is constant. No warm-up time is required. The main disadvantage of pulsed xenon lighting is the high initial cost of installation.

Tungsten — A tungsten filament is used in regular light bulbs. The same metal is used in some lamps for the process camera. The quartz-iodine lamp uses a tungsten filament inside a quartz glass envelope. The envelope is the glass shell over the filament. Tungsten, which makes an incandescent source of light, has a con-

tinuous spectral distribution. Much of the energy is dispersed along the red area of the spectrum, Fig. 13-18. Tungsten filament lighting systems are generally inexpensive. They are clean and easy to operate.

Lensboard

The lensboard is mounted on a track so it can be moved forward and backward, Fig. 13-19. This is necessary to maintain the proper distance between copyboard and lensboard during reductions and enlargements. The primary function of the *lensboard* is to act as a carrier of the lens. Look at Fig. 13-20.

The most important part of the camera is the lens. It is composed of several curved and polished glass circles mounted in a cylinder called a *lens barrel*. The *lens* gathers reflected or transmitted light from the copy. The light is then projected onto the camera back where the film is positioned. Curved surface of lens causes light to *converge* (focus) on film, Fig. 13-21.

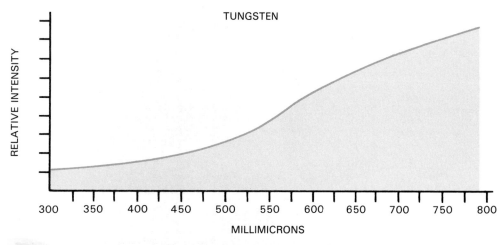

Fig. 13-18. Tungsten lamps, such as quarts-iodine, have a continuous spectral distribution. These lamps are inexpensive, clean, and easy to operate.

273

Fig. 13-19. Process camera lensboard is mounted on a track so it can be moved forward and backward for reductions and enlargements.

COLLAR   F-STOPS

Rodenstock Apo-Rodagon

LENS BARREL

Fig. 13-20. The lens is the most precise and costly part of a process camera. Quality of reproduction depends largely on the quality of the lens.   (Rodenstock Precision Optics, Inc.)

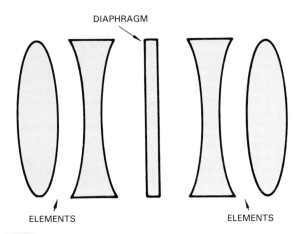

DIAPHRAGM

ELEMENTS   ELEMENTS

Fig. 13-21. Curved optical elements of lens causes light to converge or focus on film.   (Goerz Optical Co.)

Spherical aberration is the loss of focus, when the outer edges of the lens transmit light, Fig. 13-22. Distortion and curvature are caused by the difference in lens curvature and thickness. See Figs. 13-23 and 13-24.

Modern process camera lenses are coated to reduce flare. *Flare* is stray light reflected through the lens to the film, reducing contrast of the negative. The lenses are ground to avoid spherical aberration (fuzzy image from outer lens perimeter). These lenses are made "flat" so that they have little depth of field.

Depth of field refers to the distance in front of and behind the image that is in focus. Since process camera lenses are used for flat copy, they do not have to record a third dimension. Process camera lenses have relatively long focal lengths and are therefore "slow," compared to a short focal length, high-speed lens or a home camera. Refer to Fig. 13-25.

The *focal length* of a lens is the distance from the lens' optical center (nodal point) to the film when the image is in focus. The *optical center* is the point where the light rays cross.

The lens barrel generally contains the *iris diaphragm*. This is a device for restricting light gradually from the outer edges of the ends toward the center. The closure of the iris is measured in focal ratios called *f-stops*.

The *f-stop number* expresses the diameter of the diaphragm opening as a factor of the focal length of

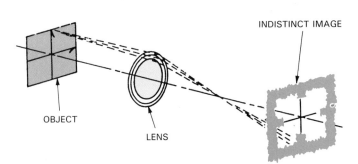

INDISTINCT IMAGE

OBJECT   LENS

Fig. 13-22. When outer edges of a lens transmit light, there is a loss of focus. This unwanted action of lens is referred to as spherical aberration.   (Robertson Photo-Mechanix, Inc.)

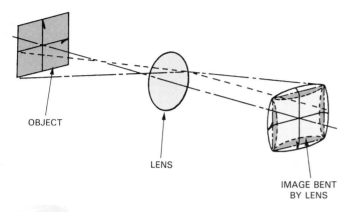

Fig. 13-23. This example of spherical aberration illustrates how an image can be bent or distorted by a lens. (Robertson Photo-Mechanix, Inc.)

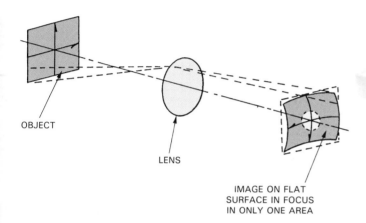

Fig. 13-24. This example of spherical aberration illustrates how an image on a flat surface may be in focus in only one small area. (Robertson Photo-Mechanix, Inc.)

the lens. For example, a 16 inch FL lens which is set at f/32 has an opening (aperture) of 1/2 inch, Fig. 13-26. The formula appears as follows:

$$\frac{FL}{f\text{-number}} = \text{diameter of aperture}$$

$$\frac{16}{32} = 1/2 \text{ inch}$$

To find the diameter of the aperture in the above example, divide FL (16 inch) into the f-number 32.

The larger the f-stop number, the smaller the opening of the iris. Each larger f-stop number allows one-half of the previous amount of light to go through the lens. Each larger f-stop number also doubles the exposure time. For example, an 8-second exposure at f/11 would require a 16-second exposure at f/16, or a 32-second exposure at f/22.

Lens speed is determined by the largest opening of the iris, which admits the maximum amount of light. Most graphic arts process cameras have f-stops of f/8, f/11, f/16, f/22, f/32, f/45, Fig. 13-27.

Fig. 13-26. The f-stop number expresses diameter of diaphragm opening as a factor of focal length of lens. (A.B. Dick Co.)

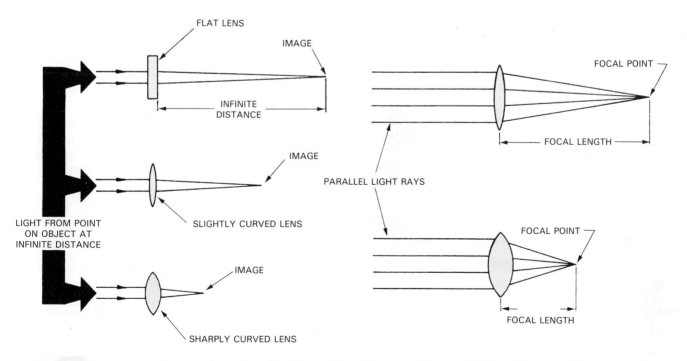

Fig. 13-25. Lens manufacturer determines focal length by aiming parallel rays of light at a lens and then measuring distance from center of lens to a spot where light rays converge. (Robertson Photo-Mechanix, Inc.)

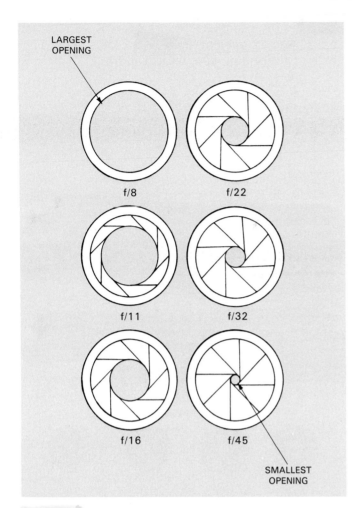

Fig. 13-27. Lens speed is determined by largest opening of iris, which admits maximum amount of light.   (A.B. Dick Co.)

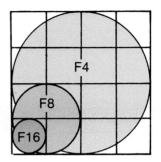

Fig. 13-28. Each higher f-stop number (smaller aperture) doubles exposure time. Small f-stop openings should be avoided since they cause light to bend or diffract around edges of iris, causing slight distortion to edges of image.
(Robertson Photo-Mechanix, Inc.)

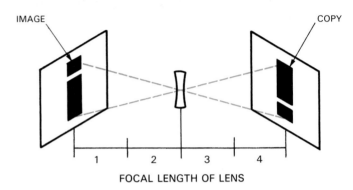

FOCAL LENGTH OF LENS

Fig. 13-29. Focal length of a lens is one-fourth distance between copy and same-size focused image on ground glass of process camera.

The lenses are ground so that the sharpest focus and most effective working f-stop is TWO F-STOPS from the lens' largest aperture (opening). Remember that the largest aperture corresponds to the smallest f-stop number.

Light from the copyboard is collected by the lens and restricted by the aperture of the iris. When the opening of the iris is small, say f/32, it causes the light to bend or diffract around the edges of the iris. The image is slightly distorted on its outer edges. As a result, this fuzziness may be transferred to the film in the camera. As pointed out earlier, each higher f-stop number (smaller aperture) doubles the exposure time and increases the diffraction of light, Fig. 13-28. Long exposure times at small apertures should be AVOIDED.

The longer the focal length of the lens, the larger will be the image. For example, at two focal lengths from the lens, the focused image is the same size as the copy. The focal length of a lens is one-fourth the distance between copy and same-size focused image on the ground glass, Fig. 13-29. For example, if the total distance from the object (copy) to the image on the ground glass is 40", the camera has a 10 inch FL lens.

This is determined by dividing the copy to image distance:

$$FL = 40 \text{ inches} \div 4 = 10$$

In another example, a horizontal process camera measures 60 inches from copy to ground glass. What is the FL? Divide the copy to image distance:

$$FL = 60 \text{ inches} \div 4 = 15$$

Enlargements and reductions of the focused image are determined by moving the lensboard and copyboard. These movements should be made in focal length ratios to each other if the image is to remain in focus.

Fig. 13-30 illustrates the ratio of focal length distances between the image, lens, and copy which determine enlargement or reduction of the image. When the lens is moved closer to the film, the image size is REDUCED. When the lens is moved away from the film, the image size is ENLARGED. See Fig. 13-31.

The focal length of the lens on the camera determines the fixed positions of focus and enlargement and reduction (definition), Fig. 13-32. These positions are preset on the camera. Indicators are provided on the lens barrel. Flexible marked tapes and control wheels are often used to set the copyboard and lensboard.

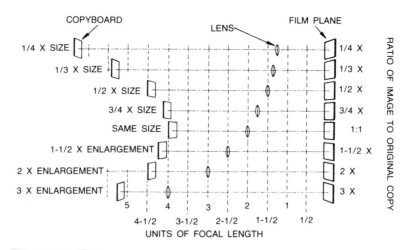

Fig. 13-30. This example illustrates ratio of focal length distances between image, lens, and copy which determine enlargement or reduction of image. (Robertson Photo-Mechanix, Inc.)

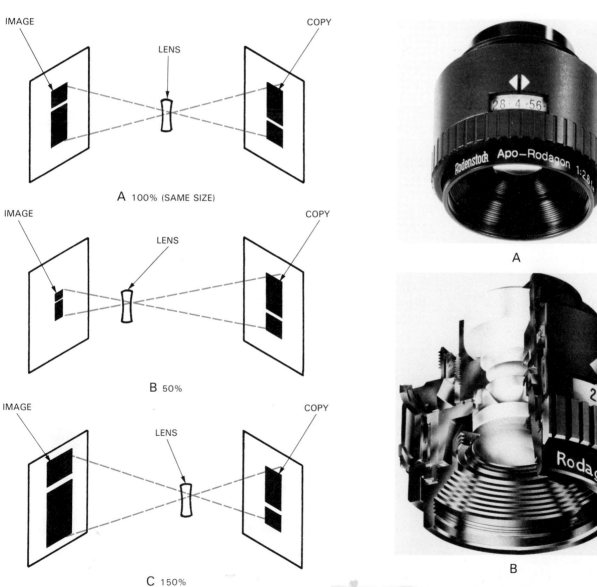

A 100% (SAME SIZE)

B 50%

C 150%

Fig. 13-31. When lens is moved closer to film, image size is reduced. When lens is moved away from film, image size is enlarged.

Fig. 13-32. A—Focal length of lens on camera determines fixed positions of focus, enlargement, and reduction. B—Focal length of a lens is provided on lens barrel. (Rodenstock Precision Optics, Inc.)

Bellows extension

The *bellows extension* is an accordion-shaped component that forms a light tunnel from the lens to the film plane (surface). It is attached to the lensboard, Fig. 13-33. The accordion arrangement of the bellows is designed to allow the lensboard to be moved for reductions and enlargements.

The bellows should be checked periodically for cracks and breaks. These can cause light leaks that, in turn, can cause flare. *Flare* is nonimage light that reaches the photographic emulsion during camera exposure. Its source may be any stray light falling directly on, or reflected to, the lens, or by internal lens reflections of image light. Periodically, the bellows extension should also be cleaned internally with a hand-held vacuum cleaner to remove dust and dirt.

Ground glass

Most process cameras are equipped with a *ground glass* to assist the camera operator in positioning and focusing the image, Fig. 13-34. The ground glass is attached to the rear camera case. It is usually mounted in a hinged frame so that it can be swung out of the way when not in use.

When in the viewing position, the ground glass surface is on the same plane (surface) as the film during exposure. Some cameras have a separate ground glass attachment stored in a compartment below the camera back. To use this type of glass, the camera operator removes the glass from the compartment and places it in position on the camera back for viewing purposes.

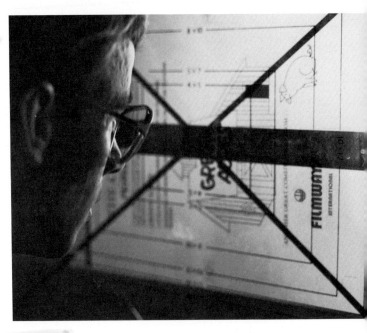

Fig. 13-34. Process camera operator can check position and focus of an image on ground glass.

Vacuum back

The *vacuum back,* like the ground glass, is hinged to the rear case of the camera, Fig. 13-35. The dull, black-colored vacuum back has markings on it to show the proper position for standard film sizes. A series of holes or narrow channels in the vacuum back connect to a vacuum chamber. With the vacuum pump turned on, the photographic film is held in position during the exposure. The vacuum back may also be fitted with register pins which accommodate pre-punched film register systems.

Focusing controls

The lensboard and copyboard focusing controls are located at the rear camera case, Fig. 13-36. Each control consists of a dial that runs a sliding tape with percentages printed on them, Fig. 13-37. When reducing or enlarging copy size on a process camera, the copyboard and lensboard tapes must be set at the same percentage size. For a REDUCTION, the lensboard moves closer to the vacuum back to reduce the image size. For an ENLARGEMENT, the lensboard moves away from the vacuum back to enlarge the image size.

CAMERA ACCESSORIES

There are a number of optional *camera accessories* which contribute to the ease and efficiency of camera work. The choice of this equipment is generally based on specific requirements of the camera department.

Light integrator

The goal of every camera operator is to produce accurate tonal reproduction of halftones. This means get-

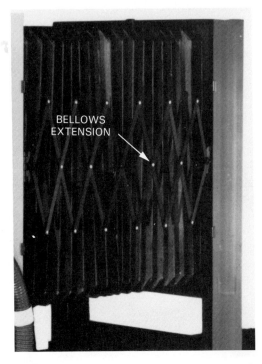

BELLOWS EXTENSION

Fig. 13-33. Bellows extension is an accordion-shaped component of process camera that forms a light tunnel from lens to film plane. (nuArc Company, Inc.)

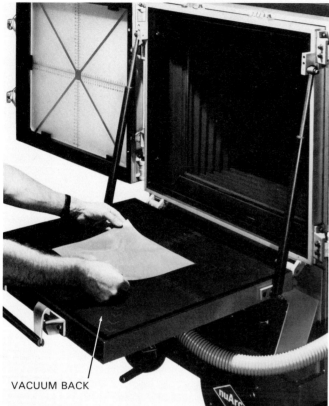

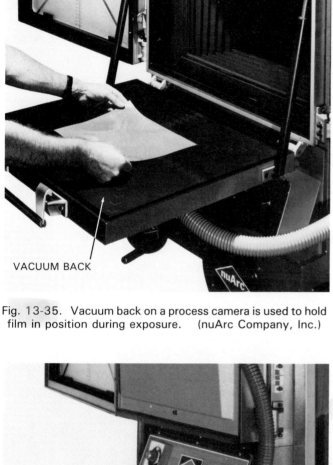

VACUUM BACK

Fig. 13-35. Vacuum back on a process camera is used to hold film in position during exposure. (nuArc Company, Inc.)

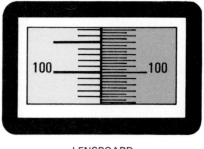

LENSBOARD
TAPE

TAPES SET AT 100%

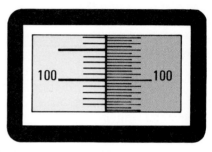

COPYBOARD
TAPE

Fig. 13-37. Process camera lensboard controls and copyboard controls consist of a dial that operates a sliding tape with percentages printed on them. Each tape is set at same percentage size whether for same-size, reduction, or enlargement. (Eastman Kodak Co.)

ting the exact quality of the original copy reproduced in the screened halftone. The basic tool for doing this is the light integrator, Fig. 13-38.

A *light integrator* is similar to a light meter. A light integrator is capable of making certain that the exposure is correct for originals of any given density. This is necessary because of the large number of variables involved such as film, type of illumination, distance from copy to lens, ambient light, and voltage fluctuations. In addition, a light integrator assures that if a

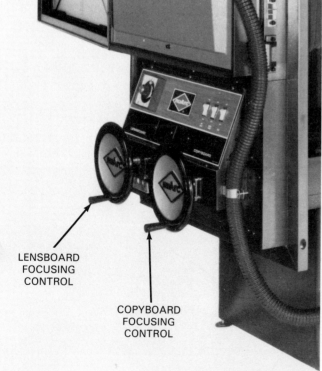

LENSBOARD
FOCUSING
CONTROL

COPYBOARD
FOCUSING
CONTROL

Fig. 13-36. Process camera copyboard and lensboard focusing controls are located at rear of camera case. (nuArc Company, Inc.)

Fig. 13-38. Computerized process camera light integrator is used for obtaining correct exposure for original copy of any given density. (Agfa-Gevaert, Inc.)

halftone negative must be reshot, consistent results can be obtained because these variables can be quantified. Light integrators are either built into the camera or are a stand-alone unit.

A light integrator has a programmable memory because it is often partly computerized. This allows certain combinations of exposures to be stored if they are used frequently. It also allows for color balance and precise exposure times and light amplitude.

Densitometer

A *densitometer* is a unit that allows for the tonal evaluation of copy of all kinds. A densitometer is NOT truly a part of the process camera, Fig. 13-39. However, the densitometer permits the operator to achieve more even illumination across the film plane. This is necessary because, although light may appear at the copyboard to provide even illumination, the light on the film plane does not. In most cases, the edges are more dense than the center.

Using the densitometer, white paper in the copyboard, and clear glass in place of the ground glass, density readings can be taken of the image. The results are recorded for various lens settings. Illumination can then be adjusted accordingly to achieve zero tonal variation.

Automatic focusing

Since focus is a mathematical function, it can be determined for each degree of reduction or enlargement. *Scale focusing* refers to a camera in which a tape or screw-drive revolution counter is used to determine focus for reductions and enlargements. This type of device has been used for many years. However, automatic focusing and scaling devices are becoming common. Using computer electronics, cameras with this feature can be moved automatically to the necessary position for the desired degree of reduction or enlargement. The unit then focuses itself. All that is required of the operator is to dial or key in the percentage size desired.

Lenses

In addition to high-quality same size, reduction, and enlargement lenses, a number of specialized lenses are available. One that is used by newspaper operations is the *variable-squeeze lens,* Fig. 13-40. It permits the reduction of copy in one dimension while allowing the other dimension to remain unchanged.

Camera processor

A *camera processor* is a stand alone daylight-operating unit containing an automatic processor and dryer, Fig. 13-41. These units are usually designed for in-plant and quick printers. They feature simplicity of operation, minimum operator training, built-in focusing and exposure, and fast operation.

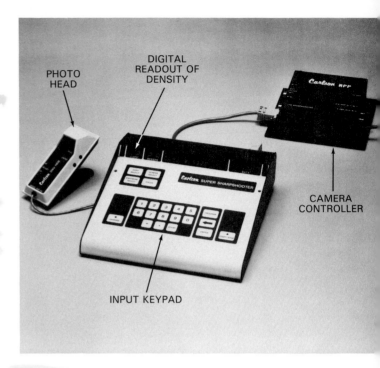

Fig. 13-39. An electronic densitometer accurately measures densities of photographic materials. (Chesley F. Carlson Co.)

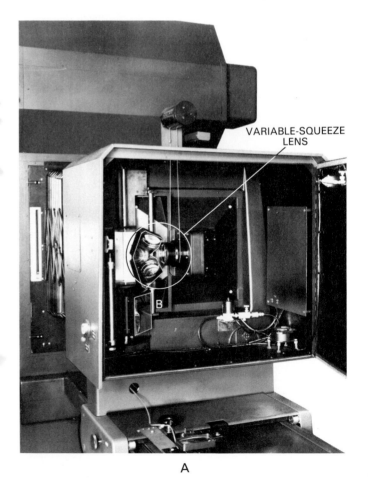

A

Fig. 13-40. A—This specialized lens, called a variable-squeeze lens, permits reduction of copy in one dimension while allowing other dimension to remain unchanged. B—Note mirror components of variable-squeeze lens.

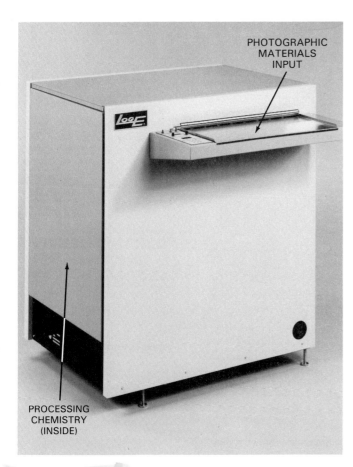

Fig. 13-41. Automatic camera processor units are capable of producing film negatives and positives, halftones, stats, typesetting materials, and in some instances, offset plates. (LogEtronics, Inc.)

The original versions of these units were limited to film and photopaper reproductions. The most sophisticated camera processors now produce stats, film positives or negatives in continuous-tone or halftone transparency enlargements, special effects, typesetting material, and small offset press plates.

A camera processor is NOT used in the same way as a process camera. It is NOT used to make large quantities of film negatives for stripping and platemaking. The reason for this is because a camera processor lacks critical exposure controls, size range, and versatility. However, the camera processor is an inexpensive, extremely convenient way of producing small stats, duplicates, transparencies, and small quantities of halftone prints. For this reason, camera processors are generally found in in-plant operations, quick printers, and advertising agencies.

DARKROOM LAYOUTS

A *darkroom* consists of a specially lighted room for processing light sensitive photographic materials. Most darkrooms have basically the same types of layouts, equipment, and accessories. The processing of

photographic materials requires development, stopping, fixing, and washing. Since all the operations in a darkroom involve the use of photographic materials, the darkroom must be light-tight.

The essential element of a good darkroom layout is utility. Typical darkroom layouts are in Fig. 13-42.

Most darkrooms are planned for movement of work from left to right. The major components of a darkroom generally include:

1. Light-tight entrance.
2. Processing sink.
3. Processing trays.
4. Film storage cabinets.
5. Safelights.
6. Darkroom utensils.
7. Timer.
8. Contact vacuum frame.
9. Inspection light.
10. Film processor.
11. Film dryer.
12. Ventilation equipment.
13. Electrical wiring.

Light-tight entrance

To protect light sensitive paper and film, it is essential to be able to enter and leave a darkroom without admitting light. This is made possible by the construction of double doors or light locks. In addition, the walls of entrances and passageways to the darkroom are painted with a special paint color to prevent reflection of light around the entrance.

There are three basic types of darkroom entrances. These include:

1. Open-passage.
2. Double door.
3. Revolving door.

Open-passage — The *open-passage entrance* is considered best because it provides both easy access and good ventilation. Easy access is important when people are carrying packages of film and plates, Fig. 13-43. The exact size of these entrances depends on the size of the photographic materials used and space allocations.

Double door — When floor space is at a premium, it may be necessary to use a double door arrangement, Fig. 13-44. With this opening, one or both of the doors can be replaced by heavy single or double curtains. If two solid doors are used, it is necessary to place a light-trapped vent in the wall of the passageway. This relieves the changes in air pressure caused by opening and closing the doors.

Revolving door — A revolving door is quite effective as a light lock, Fig. 13-45. The door consists of two cylinders, the smaller diameter one fitting into the larger one, Fig. 13-46. Revolving doors are generally used where space is limited. They may also be used to connect other darkrooms and contacting rooms to form a central facility.

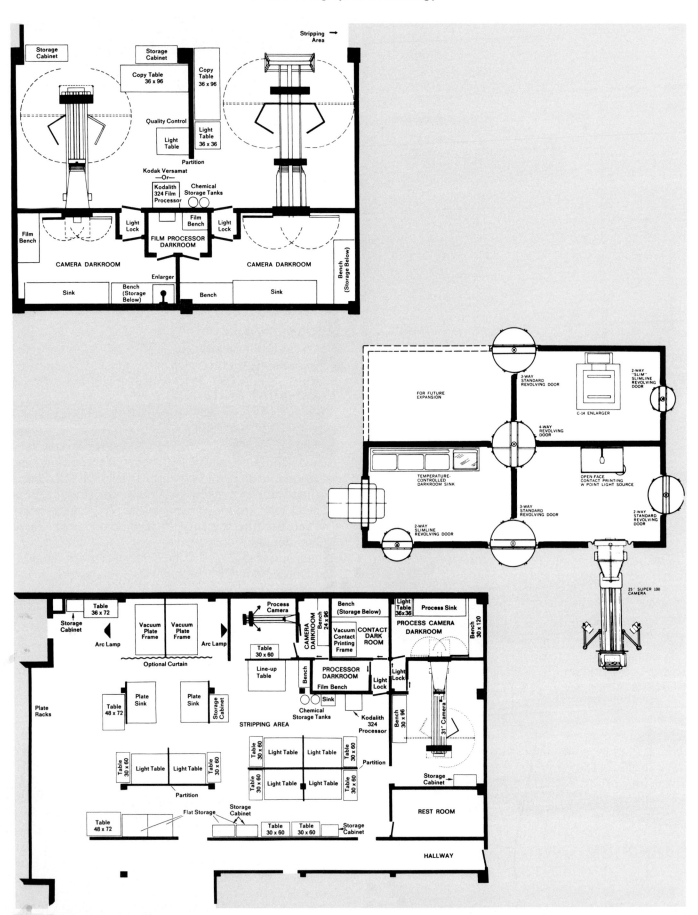

Fig. 13-42. Most darkrooms have basically the same types of layouts, equipment, and accessories. Three darkroom configurations are illustrated. (Eastman Kodak Co., Consolidated International, Inc., Eastman Kodak Co.)

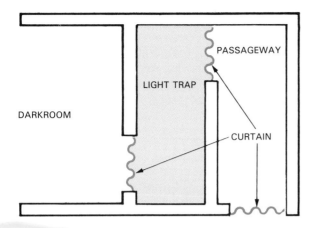

Fig. 13-43. Example of an open-passage darkroom entrance. This type of entrance provides easy access and good ventilation.

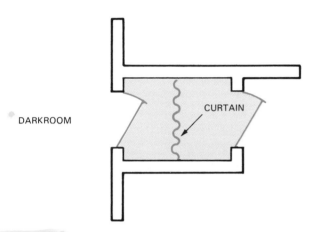

Fig. 13-44. The double door darkroom entrance arrangement is generally used when floor space is at a premium.

Fig. 13-45. Revolving door darkroom entrance is one of the most widely used arrangements today. (Consolidated International, Inc.)

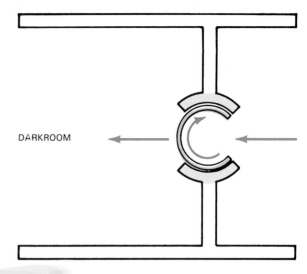

Fig. 13-46. A revolving door consists of two cylinders, smaller diameter one fitting into larger one.

Processing sink

The *processing sink* provides a place to hold the developing trays. It is generally fitted with a source of fresh running water and a means to dispose of used chemicals. Processing sinks are made primarily of fiberglass or stainless steel, as illustrated in Figs. 13-47 and 13-48.

Many factors must be considered in selecting a sink. These include resistance to the corrosive action of the solutions, size, mechanical durability, ease of cleaning, and appearance.

Automatic, thermostatically controlled *mixing valves* provide for the control of water, Fig. 13-49. These valves operate by mixing warm and cold water to obtain the desired temperature. They are capable of mixing water to temperatures accurate to within ±1/2 degree F. The mixing valve shown in Fig. 13-50 is accurate to within ±1/4 degree F. *Mixing valves* are used to provide a supply of temperature controlled, circulating water around the processing trays.

Several styles of *temperature-controlled processing sinks* are available. There are two basic types. One type

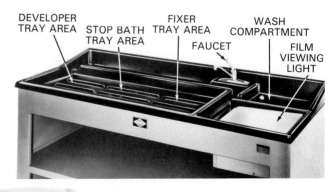

Fig. 13-47. This photographic darkroom processing sink is made to accommodate developing, stop bath, and fixing tray areas. A built-in wash compartment and lighted viewing area are included. (nuArc Company, Inc.)

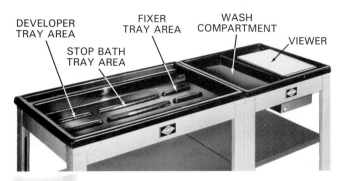

Fig. 13-48. Space-saving photographic darkroom processing sink includes developing, wash, and fixer tray areas. Unit also includes wash compartment and lighted viewing area. (nuArc Company, Inc.)

Fig. 13-49. Automatic, thermostatically controlled mixing valves provide for control of water temperature in darkroom. (Leedal, Inc.)

Fig. 13-50. This darkroom water mixing valve is accurate to within ±one-quarter degree F. (Leedal, Inc.)

has a built-in refrigeration unit with cooling coils under the sink, Fig. 13-51. This type maintains a shallow level of water at a constant temperature. The other type has no coils directly under the sink. This type continuously circulates water through a built-in mechanical cooler at the required temperature.

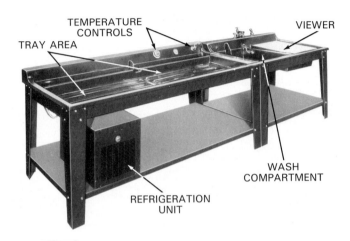

Fig. 13-51. This is a temperature-controlled photographic processing sink with built-in refrigeration unit. (Leedal, Inc.)

Processing trays

The *processing trays* hold the chemicals during the development cycle. At least three trays are required. These include:

1. Developer.
2. Stop bath.
3. Fixer.

Trays are made from plastic, fiberglass, stainless steel, and hard rubber. Trays should be stored in a vertical arrangement. This permits easy access and, at the same time, allows wet trays to drain. Refer to Figs. 13-52 and 13-53.

Fig. 13-52. Darkroom processing trays are available in several sizes and are made from plastic, fiberglass, stainless steel, and hard rubber. (nuArc Company, Inc.)

Fig. 13-53. This is a stainless steel darkroom processing tray. (Leedal, Inc.)

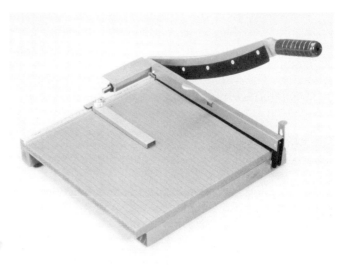

Film storage cabinets

Most darkrooms are equipped with at least one *film cabinet,* Fig. 13-54. The drawers of the cabinet can be used to store various sizes of film, photographic paper, and halftone contact screens. The top of the cabinet is convenient for holding a *print trimmer,* which is used to cut photographic materials. See Fig. 13-55.

Safelights

The handling of photosensitive materials must be done under *safelight* conditions or, in some cases, in

Fig. 13-55. A print trimmer can be used to cut various photographic film materials to desired working sizes.

total darkness. Safelights should be located where most of the developing activity occurs. This is generally over the sink and near the film storage area.

Safelights are available in many styles, Fig. 13-56. They may be plugged directly into a wall, attached to a wall, or hung from the ceiling. Safelights with correct wattage bulbs should be used in the darkroom.

Caution! Ruby red bulbs should NEVER be used, because even though they look red, they often transmit light to which orthochromatic films are sensitive.

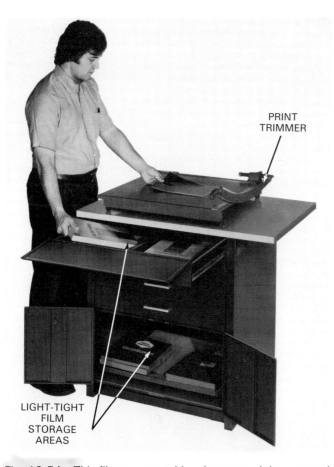

PRINT
TRIMMER

LIGHT-TIGHT
FILM
STORAGE
AREAS

Fig. 13-54. This film storage cabinet has several drawers and a compartment for storing various photographic materials. It is also equipped with a print trimmer to facilitate cutting materials to required sizes. (nuArc Company, Inc.)

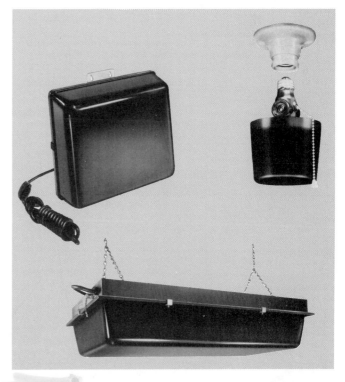

Fig. 13-56. Safety lights are available in many styles. They can be plugged directly into a wall outlet, attached to a wall, or hung from ceiling. (nuArc Company, Inc.)

It is usually necessary to have several safelights and safelight filters for the different types of photographic materials. The chart in Fig. 13-57 gives the correct safelight filters for various types of photographic materials. The correct safelight filters are recommended in the manufacturers' directions or data sheets included in each film package or box.

Any photographic material will *fog* (expose) if left too long under safelight illumination. This is because noncolor-sensitive materials have some sensitivity to green, yellow, and red light. For example, all Kodak safelight filters, when used with the recommended bulb and at the recommended distance, are safe for at least 30 seconds. This assumes that dry photographic materials are being used. The safe time is longer for materials in the developing tray.

Darkroom utensils

Darkroom utensils are containers used for mixing, pouring, and measuring chemicals. Utensils commonly found in a darkroom are graduates, beakers, pails, funnels, and storage tanks. Utensils are made of plastic, glass, and stainless steel.

Chemicals and water are measured in *graduates,* Fig. 13-58. They are usually calibrated in both cubic centimeters and ounces.

Fig. 13-58. A graduate is used to measure chemicals and water. (Leedal, Inc.)

Beakers and *pails* are containers generally used for mixing chemicals. Beakers have a SMALLER capacity than pails, Fig. 13-59. Pails are used to mix large quantities of chemicals, Fig. 13-60.

Chemicals are poured into storage containers using a *funnel,* Fig. 13-61. Funnels are available in many sizes to fit the requirements of every size of container.

Large quantities of chemicals are stored in *storage tanks,* Fig. 13-62. The kind and size of tank depends on the quantity of chemicals used in the darkroom. Tanks should be clearly labeled to describe the contents. Storage tanks should be constructed so that they prevent air and light from reaching the chemicals.

KODAK SAFELIGHT FILTERS

Wratten Series No.	Color	For Use With:
OA	Greenish Yellow	Contact printing and enlarging papers.
OC	Light Amber	High-speed enlarging papers, including Kodak Polycontrast Papers.
1	Red	Blue-sensitive films and plates, such as Kodak Commercial Film and Kodak Lantern Slide Plates; Kodagraph Projection Paper.
1A	Light Red	Kodalith Ortho materials and Kodagraph Contact Papers.
2	Dark Red	Orthochromatic films and plates, Kodagraph Fast Projection Paper, and green-sensitive film for photoradiography.
3	Dark Green	Panchromatic films and plates.
6B	Amber	X-ray film and blue-sensitive film for photoradiography.
7	Green	Infrared-sensitive films and plates. Not safe for orthochromatic materials.
8	Dark Yellow	Eastman Color Print Film, Type 5382.
10	Dark Amber	Kodak Ektacolor Paper; Kodak Ektacolor Print Film; Kodak Panalure Paper.

Fig. 13-57. A safelight filter chart gives correct safelight filters for various types of photographic materials. (Eastman Kodak Co.)

Fig. 13-59. A beaker is used for mixing chemicals in darkroom. (Leedal, Inc.)

Fig. 13-60. A pail is used for mixing large quantities of chemicals in darkroom. (Leedal, Inc.)

Fig. 13-61. A funnel is used to pour quantities of chemicals into storage containers. (Leedal, Inc.)

Fig. 13-62. Storage tanks are used to store large quantities of darkroom chemicals. Tanks should be clearly labeled to identify contents. (Leedal, Inc.)

Timer

A *timer* is used to determine how long the film is held in the processing solutions, Fig. 13-63. The luminous face is easily seen in the dark. There is a buzzer to indicate the elapsed time. Some timers are made to operate or control darkroom equipment. This type of timer resets itself after the specific time has elapsed.

Contact vacuum frame

A *contact vacuum frame* is used for making prints, negatives, positives, spreads, and chokes, Fig. 13-64. Contact printers are used to make same-size reproductions on film or photographic paper. The negative or positive is held in vacuum contact with the photographic material being made. A point source light is used to expose the material, Fig. 13-65. The lamp is located over or under the contact frame. A transformer is used to adjust the *intensity* (brightness) of the light.

Fig. 13-63. A timer is used to determine elapsed time of various processing tasks in darkroom. Some timers are made to operate or control various darkroom accessories and equipment.

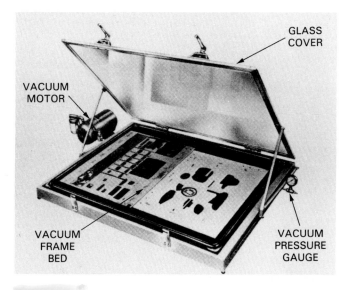

Fig. 13-64. A contact vacuum frame is used for making same-size prints, negatives, positives, spreads, and chokes. (nuArc Company, Inc.)

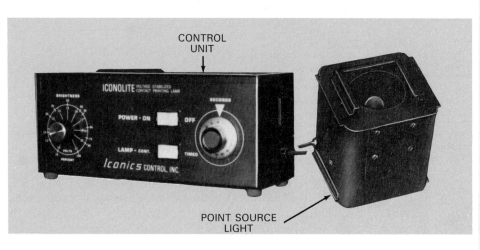

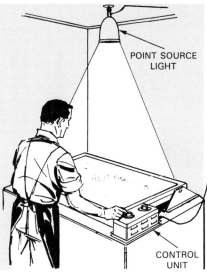

Fig. 13-65. This point source light, with control unit, is operated as illustrated on right. (Iconics Control, Inc.)

Inspection light

The darkroom is generally equipped with an inspection light. This is used to examine negatives and film positives during the development process. Most inspection lights are wall-mounted and are equipped with ortho-safelights and white lights.

Film processor

A *film processor* is used for fast, automatic processing of film materials, Fig. 13-66. It eliminates the need

Fig. 13-66. A photographic film processor is used for automatic processing of film materials. It eliminates the need for a processing sink, trays, and related utensils. (LogEtronics, Inc.)

for a processing sink, trays, and related utensils. These machines allow for the control of all variables and reduce the chance of human error. Film processors accept the film after it has been exposed.

The schematic diagram shown in Fig. 13-67 illustrates the operation of a typical film processor. The light-sensitive material enters the processor and travels through the chemistry and dryer. The material emerges ready for the stripping department. Film processors are available as through-the-wall and stand alone models, or as part of the camera itself. The affordable price of small processors now makes them feasible for most low to medium volume camera departments.

Processor advantages—Automatic processors have several advantages over manual processing. These can be grouped in the areas of time, cost, space, and quality.

Generally, processors can complete a sheet of film in four to six minutes. Additionally, processors will continuously process film by the hour. The processing rate is normally faster than the manual rate. Also, the operator is free from tray tending and can utilize this time for other work. Rapid processing and continuous volume save time.

In tray work, the depleted developer is discarded, and a new batch is mixed after developing several pieces of film. A mechanical processor automatically replenishes the developer as required.

The enclosed system retards exhaustion of developer, since air contact is less than in tray development. Filtering systems are provided in mechanical processors to recirculate clean developer and provide more active chemistry.

Most processors use less water than in tray processing, since the film is rinsed for a shorter period of time. Tempered water usage for controlling chemistry temperature is reduced during stand-by periods. A reduction in chemistry and water costs are the result

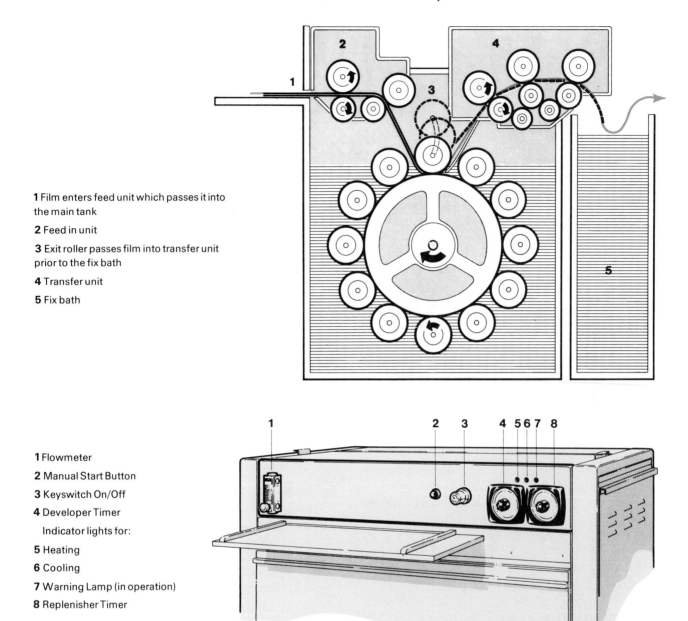

1 Film enters feed unit which passes it into the main tank

2 Feed in unit

3 Exit roller passes film into transfer unit prior to the fix bath

4 Transfer unit

5 Fix bath

1 Flowmeter

2 Manual Start Button

3 Keyswitch On/Off

4 Developer Timer

  Indicator lights for:

5 Heating

6 Cooling

7 Warning Lamp (in operation)

8 Replenisher Timer

Fig. 13-67. Schematic diagram shows details of how a photographic film processor operates. Light-sensitive material enters processor and travels through chemistry and dry units, emerging ready to use. (Crosfield Graphic Equipment, London, England)

of the enclosed design and automatic systems of a mechanical processor.

Stand alone models of processors are designed to save space since they are generally located within the darkroom. These units are self-contained and require no external plumbing. Stand alone processors are especially useful in shops where floor space is minimal in the darkroom area.

The quality and consistency of machine-processed negatives are the result of control of variables in processing. In machine processing, the time, temperature, agitation, chemical strength, and drying are controlled. Negatives of consistent quality can be produced. This quality is evident as the negative is used in related work.

Pinholes in negatives are caused mainly by dust. Machine-processed negatives have fewer pinholes, due to developer flow. As a result, less *opaquing* (touchup of pinholes) is required in the stripping department.

Heat control in drying produces better control in maintaining film size and reduces streaking. Film registration is aided by consistent size. Duplicates can be processed to the same size consistently.

With the control of variables, film density can be more consistent. *Exhausted developer* makes soft, mushy dots; sharp, hard dots come from *fresh developer*. Hard dots make good plates. Remember, film products with consistent density lead to consistent quality in plates.

Film dryer

For tray processing, a *film dryer* is useful, Fig. 13-68. The majority of film dryers are of the *forced-air type*. This means they operate much like a hair dryer in blowing hot air over the film surface. Film drying time is reduced to a matter of seconds. The film should be fed into the dryer emulsion-side up. In this way, the film base is protected from scratches as it is carried through the unit.

If a film dryer is NOT available, the film should be hung to dry on a line in the darkroom. Use a squeegee with a soft blade to remove the excess water from the film. The film is then hung to dry using film clips or clothespins on the two top corners. The drying area should be dust-free.

Darkroom ventilation

The proper control of darkroom ventilation is important for several reasons. The health and efficiency of the camera department personnel must be considered. Temperature and humidity must be controlled to avoid adverse effects on film materials. If the air is too dry, film has a tendency to attract static accumulations.

For example, excessive humidity causes the body to perspire, and damp fingers will easily leave marks on dry films. Lack of proper humidity control can cause nasal passages of workers to become dry and the skin to become chapped.

Incoming air should pass through filters to remove dust particles. The air flow should be in sufficient volume to change the air in the darkroom six to ten times an hour. Air should be pumped in rather than out. This will prevent dust entering through windows and doors.

Electrical wiring

The electrical wiring and equipment in the darkroom must be safe. This comes under the regulations of the National Board of Fire Underwriters. These regulations were established to safeguard the worker and the facility.

Voltages of 110 volts or less can be fatal if electric contact with the body are made on moist skin. Care should be taken to avoid a situation in which the body becomes part of an electric current.

All exposed non-current carrying metal parts of both fixed and portable equipment must be grounded. All outlets, switches, sockets, and similar items should be composed of insulating materials. Foot switches eliminate the need for the use of hands in operating electric fixtures. However, the precautions which apply to the grounding of other fixtures are even more important when foot switches are installed. This is because the floor may be damp or, on occasion, even wet.

The placement and circuit wiring of the various outlets should be planned for the convenience of the worker. All darkroom circuits, white lights, safelights, and outlets should be controlled by a master switch.

A *red signal light* should be located near the door outside the darkroom. This light should glow to indicate that the darkroom is in use. This will prevent accidental opening of the door while films are being processed.

Another switch should be placed below the master switch to control the white ceiling light in the darkroom. It is also advisable to have a *switch lock* in the form of a flat plate placed on the white light switch. This will prevent the white light from being turned on accidentally with light sensitive material exposed.

Darkroom walls should be painted light green, light gray, or white. These colors are recommended by the film manufacturers. Darkroom walls are easily cleaned when these colors are used. In addition, the illumination in the darkroom is increased by the use of these colors and provides safe working conditions for all personnel. A special pink darkroom paint is available that increases darkroom illumination. However, this color is not as pleasant as those already mentioned.

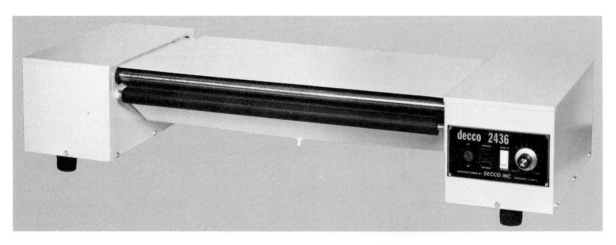

Fig. 13-68. A film dryer can be useful when tray processing photographic film materials. This compact unit operates much like a hair dryer in blowing hot air over film surface. (Decco, Inc.)

POINTS TO REMEMBER

1. The camera used to convert two-dimensional copy for offset lithography is called a process camera.
2. A process camera is equipped with a lens of the process type. This means that the lens is made to give optimum results with flat or two-dimensional copy.
3. Process cameras are installed and used in two different ways, including: darkroom camera and gallery camera.
4. A darkroom camera is contained in two rooms. The lens, lights and copyboard portion of the camera are in one room and the camera back is in the adjoining room, called the darkroom.
5. A gallery camera is installed in a lighted room. This arrangement requires that a light-tight film holder be loaded in a darkroom and then taken to the camera.
6. A process camera along with all necessary processing sinks or a film processor can be located entirely within the darkroom.
7. There are two basic types of process cameras: horizontal and vertical. Regardless of design, these two types of cameras operate essentially the same.
8. The components common to process cameras include the copyboard, copyboard lights, lensboard, bellows extension, ground glass, vacuum back, and focusing controls.
9. Process cameras require special lighting arrangements to expose film in the camera back. Tungsten, carbon arc, mercury, quartz-iodine, and pulsed xenon lamps are in common use.
10. The illumination from each of the common light sources varies in amount of light (intensity) and kind of light (extent of visible spectrum).
11. The various process camera light sources play an important role in camera lighting and its affect on film sensitivity.
12. The lens of a process camera is its most important part. The lens is composed of several curved and polished glass circles mounted in a cylinder, called a lens barrel.
13. The process camera lens gathers reflected light from the copy. The light is then projected onto the camera back where it exposes the film.
14. The copyboard and camera back must be parallel so that light passing through the lens does NOT form a distorted image.
15. The focal length of a process camera lens is the distance from the lens' optical center (nodal point) to the film when the image is in focus and the camera set at 100%.
16. The f-stop number expresses the diameter of the diaphragm opening as a fraction of the focal length of the lens.
17. There are a number of optional camera accessories which contribute to the ease and efficiency of process camera work. Examples of these accessories include the light integrator, densitometer, automatic focusing, special lenses, and film processor.
18. There are a number of darkroom layouts used for process camera operations. However, all darkrooms have basically the same types of layouts, equipment, and accessories.
19. The major components of a darkroom generally include a light-tight entrance, processing sink, processing trays, film storage cabinets, safelights, utensils, timer, contact vacuum frame, inspection light, film processor, film dryer, ventilation system, and electrical wiring.
20. Darkroom walls are generally painted light green, light gray, or white.

KNOW THESE TERMS

Bellows extension, Camera/processor, Copyboard, Copyboard lights, Copyboard tape, Darkroom camera, Densitometer, Diaphragm, Flare, F-stop, Gallery camera, Ground glass, Horizontal process camera, Lens barrel, Lensboard, Lensboard tape, Light integrator, Vacuum back, Vertical process camera.

REVIEW QUESTIONS

1. A process camera lens is made to give best results with flat or _____ copy.
2. A process camera installed with the film holder end built into the wall is referred to as a _____ camera.
3. A _____ process camera is located entirely in a lighted room and has a removable film holder.
4. Roll-film process cameras are in common use where high volumes of film processing are required. True or false?
5. The large flat surface, with a hinged glass cover that holds the paste-up during the exposure, is called the _____.
6. The _____ lights act as a source of illumination of the copy on a process camera.
7. The process camera lens is made of several optical glass elements which are assembled into a lens _____.

8. The size of opening produced by the process camera diaphragm is referred to as the f-stop number. True or false?

9. The metal blade arrangement at the rear of the camera lens that allows light to enter is called the:
   a. Lens barrel.
   b. Diaphragm.
   c. Ground glass.
   d. Bellows extension.

10. The accordion-shaped tunnel that channels light from the lens to the film plane is called the _____ _____.

11. Unwanted light reflected from objects or bellows light leaks is called _____.

12. Most process cameras are equipped with a _____ _____ to assist the camera operator in positioning and focusing the image.

13. The dial controls that adjust the percentages of reduction or enlargement, by moving the copyboard and lensboard back and forth, are called the _____ controls.

14. Why is a light integrator often used on a process camera?

15. Density readings of original copy can be measured with a(an):
    a. Integrator.
    b. Processor.
    c. f-stop.
    d. Densitometer.

16. The three basic types of darkroom entrances include:
    a. Revolving door.
    b. Double door.
    c. Single door.
    d. Open-passage.

17. A processing sink provides a place to hold the _____ _____.

18. There are two basic types of temperature-controlled sinks. True or false?

19. A darkroom contact vacuum frame is used for making reductions and enlargements. True or false?

20. Proper control of darkroom ventilation is important because it affects personnel and _____ _____.

21. The ventilating system in a darkroom should be capable of changing the air:
    a. Two to four times an hour.
    b. Six to ten times an hour.
    c. Ten to fifteen times an hour.
    d. At least twenty times an hour.

22. All exposed noncurrent carrying metal parts of both fixed and portable darkroom equipment must be _____.

SUGGESTED ACTIVITIES

1. Examine the graphic arts process camera in your shop. Identify all of the major parts on the camera as described in this chapter.

2. Visit a printing facility that has a process camera. Talk to the camera operator about the production methods used to produce film negatives on the camera.

3. List the basic advantages of a vertical process camera.

4. Describe the term focal length (FL) and illustrate the way in which focal length of a lens is determined.

5. Illustrate how enlargements and reductions are made by changing the bellows (lensboard) and copyboard extensions on the process camera.

6. Prepare a scale drawing of a darkroom layout for a medium-size offset firm and determine the cost of the camera and utensils required for the darkroom facility.

Chapter 14

Film Processing

When you have completed the reading and assigned activities related to this chapter, you will be able to:
- ○ List the various graphic arts films and describe their uses.
- ○ Describe the various parts of a piece of graphic arts lithographic film.
- ○ Distinguish between orthochromatic, panchromatic and silverless film.
- ○ Explain the correct procedures for handling and storing film.
- ○ Describe the primary parts and operation of a film processor.
- ○ Identify the common orthochromatic chemicals and mix the required quantities for film processing.
- ○ Identify common methods of contact printing and methods of application.
- ○ Name typical kinds of equipment and materials used in contact printing.
- ○ Make contact negatives and positives on film and paper using the equipment and materials available.

In the previous chapter, you learned about the organization and use of process cameras and darkrooms. This chapter will expand upon this information by detailing film construction, types, and processing. Film processing is very important to the production of a printed product. By understanding the nature of film and how its *latent* (invisible) image is made permanent and visible, you will be much better prepared to become employed in the printing field.

FILM

The term *film,* when used in graphic arts photography, refers to a transparent or translucent acetate or plastic base with a light sensitive coating. The *light sensitive coating* is actually a photographic emulsion containing gelatin and silver salts. The sensitivity of the *silver salts* to light is the basis of the photographic process. See Fig. 14-1.

Photographic paper is similar to film but it is *opaque* (not transparent). It is normally used for producing high resolution copy (words) and line art. Photographic paper is usually processed in an automatic processing machine. This was discussed in Chapter 13. Automatic film processors are also explained in that chapter.

Fig. 14-1. Photographic process relys on sensitivity of silver salts to light.

293

Structure of film

Whatever their chemical differences, all films are constructed to include the same general components. These basic components are shown in Fig. 14-2.

Graphic arts films are generally made from either cellulose-based, estar-based, or polystyrene-based material. The *gelatin* compound containing the silver halide (silver salts) is bonded to the base material by an adhesive substratum. The *substratum* can be compared to a primer paint necessary when painting metal. The *primer* adheres to the metal and the final coat of paint adheres to the primer.

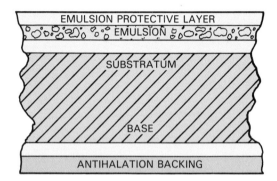

Fig. 14-2. Study cross section of a piece of graphic arts film. (A.B. Dick Co.)

Once the substratum is applied to both sides of the film, the emulsion is bonded on one side of the base. The other side of the film is covered with an antihalation coating. The *antihalation coating* prevents light from reflecting back to the emulsion side of the film during exposure.

Classification of film materials

Light sensitive materials, such as graphic arts film, can be classified by three main variables. These include:
1. Color sensitivity.
2. Contrast.
3. Film speed.

Color sensitivity

The *color sensitivity* of a light sensitive material describes the type of light that will expose the emulsion. It refers to the area of the visible electro-magnetic spectrum that will cause a chemical change in a particular silver halide emulsion. A *wedge spectrogram* is used to illustrate a film's reaction to light across the visible spectrum, Fig. 14-3. A *spectrogram* is similar to a diagram used to describe a light source output.

Fig. 14-3A shows a wedge spectrogram to describe the relative sensitivity of the human eye to the visible spectrum.

The idea of color sensitivity can be explained by example. X rays, which are usually considered a form of light (technically a form of energy), cannot be seen by the human eye. However, they can be recorded on certain types of films. The human eye is also blind to infrared rays or ultra-violet rays. These bands of light can be recorded on special films. Just as the eye has its limitations as to the type of light it can perceive, the same is true of light sensitive materials.

There are basically three types of light sensitive emulsions. These include:
1. Blue sensitive.
2. Orthochromatic.
3. Panchromatic.

Blue sensitive materials are known as color-blind or

COLOR SENSITIVITY

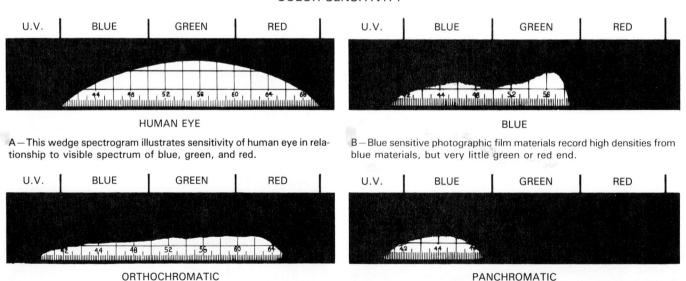

A—This wedge spectrogram illustrates sensitivity of human eye in relationship to visible spectrum of blue, green, and red.

B—Blue sensitive photographic film materials record high densities from blue materials, but very little green or red end.

C—Orthochromatic photographic film materials are sensitive to all colors except blue.

D—Panchromatic photographic film materials come closest to sensitivity of the human eye since they are sensitive to all visible colors.

Fig. 14-3. Study information relating to photographic materials.

monochromatic, Fig. 14-3B. On a negative, they record high densities from blue materials, but they record very little from the green or red end. This characteristic is useful because the film can be used outside the darkroom. It is ideal for copying black and white photographs.

Since blue sensitive materials are NOT sensitive to green and red light, the film sees these colors as black. If this page you are now reading were photographed with a blue sensitive film, it would make no difference in the finished negative whether the black type were green, red, or black. This is because the film sees green and red as black. All light sensitive emulsions that contain silver are sensitive to blue. This group of films is usually limited to contact printing, although there are some uses for it in camera work.

Orthochromatic materials are NOT blue sensitive. However, orthochromatic materials are sensitive to all other areas of the visible spectrum, Fig. 14-3C. This characteristic increases the range of possible applications. These include black and white originals, use of a magenta contact screen, and some filter work. Orthochromatic materials are much faster acting than blue sensitive materials. Since this type of emulsion is NOT sensitive to red light, red safelights can be used for darkroom illumination.

The **panchromatic films,** nicknamed *pan films,* come closest to the sensitivity of the human eye, Fig. 14-3D. They are sensitive to ALL visible colors. Natural tonal variations can be reproduced with pan films. This type of film is used to reproduce color originals in monochromatic form. Filters may be used with this film. Panchromatic film must be processed in total darkness.

Film manufacturers provide a *wedge spectrograph* for evaluating each of their films, Fig. 14-4. The most accurate photographic negatives are obtained when the peak sensitivity of a film's spectrograph corresponds with the peak output of a light source. For example, Fig. 14-5 illustrates how tungsten light has its lowest output in the range that orthochromatic film has its peak sensitivity. Pulsed xenon light source would produce better results.

Contrast

The difference between the lightest and darkest portions of an original photographic image creates contrast on the film. If this difference is great, the contrast

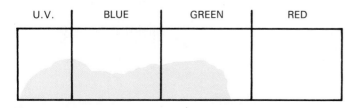

Fig. 14-4. Film manufacturers provide a wedge spectrograph for evaluating each of their films. (Eastman Kodak Co.)

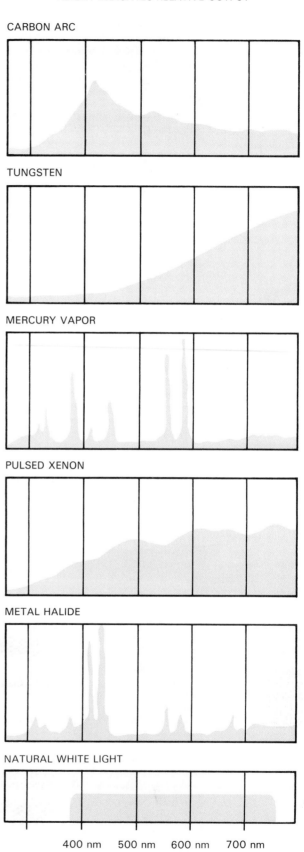

HEIGHT INDICATES RELATIVE OUTPUT

CARBON ARC

TUNGSTEN

MERCURY VAPOR

PULSED XENON

METAL HALIDE

NATURAL WHITE LIGHT

400 nm   500 nm   600 nm   700 nm

Fig. 14-5. Peak color sensitivity of a film can be compared to peak output of a light source. Notice how tungsten light has its lowest output in range that orthochromatic film has its peak sensitivity.

295

is considered high. For example, snow compared to coal is high in contrast, just as black type on white paper is high contrast copy. To reproduce the original, the proper contrast film must be used. Line copy which is high in contrast requires a high contrast film. Halftones are reproduced by photographing the continuous tone original through a halftone contact screen onto a high contrast film.

Contrast is described by a *film's characteristic curve,* also called a *Log E curve.* A simple characteristic curve can be drawn for any film by photographing a step scale. The density of each step of the original and film negative are measured. The results are plotted on a graph similar to the one in Fig. 14-6.

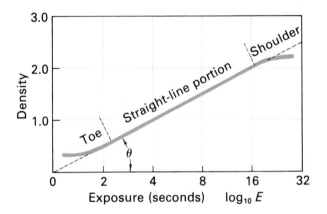

Fig. 14-6. Typical characteristic Log E curve has toe, straight line, and shoulder for sensitive material. (Eastman Kodak Co.)

Film manufacturers provide characteristic curve data for each of their films. Fig. 14-7 illustrates the curve for one brand of orthochromatic film. This is a high contrast film. The curve reveals that a very slight change in exposure time will result in a rapid jump in density of the film.

Film speed

All materials require a different amount of light to cause a chemical change in the silver halide emulsion. This is referred to as *film speed.* Emulsions that require LITTLE LIGHT are called *fast.* Emulsions that require STRONG LIGHT are called *slow.* These terms become almost meaningless because there are so many different materials, each requiring a different amount of light.

For this reason, an *exposure index* is used to classify the speed for each film manufactured. A *film speed number,* called *ASA* is assigned to each film—the HIGHER the number, the FASTER the film. For example, a film with an ASA rating of 25 will require twice as much light to create an image as does a film with an ASA of 50.

The exposure index of a particular film is assigned as a function of the kind of light source used to expose it.

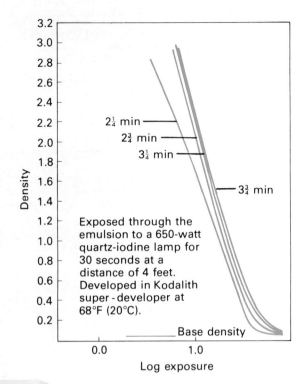

Fig. 14-7. Example of a film manufacturer's characteristic curve data sheet for use with an orthochromatic film. (Eastman Kodak Co.)

FILM PROCESSING

Film processing by the tray method, discussed briefly in the previous chapter, generally requires the use of four separate solutions. These include: developer, stop bath, fixer, and a wash, Fig. 14-8. These solutions are usually placed in trays in the darkroom sink so that processing is handled from left to right. This procedure eliminates confusion in the subdued safelight illumination of the darkroom.

With some types of processing sinks, the running water wash area is built in. With this type of sink, only three trays are required. All trays should be large enough so that the film can lie flat during processing.

Processing sinks with temperature control use a system of running water. The trays are actually surrounded in circulating temperature-controlled water. The water temperature is maintained at 68 degrees F (20 degrees C) using water mixing valves.

Film processing theory

Exposure of the film on a process camera causes a latent image in the emulsion. A *latent image* is an invisible image caused by a chemical change in the emulsion of the film. To make this image visible, it must be processed in *developer.*

The entire procedure of *film processing* involves developing, stopping, fixing, and washing. *Tray processing* is the basic hand method of developing film and it will help you grasp automatic machine processing.

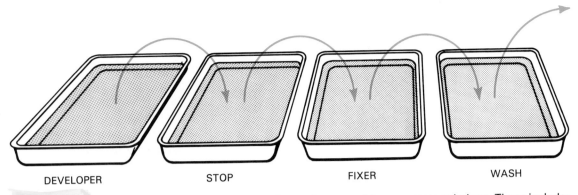

DEVELOPER STOP FIXER WASH

Fig. 14-8. Film processing by tray method generally requires use of four separate solutions. These include: developer, stop bath, fixer, and wash. (A.B. Dick Co.)

Developing is a process that basically dissolves the light exposed area of the film. The real action of the developer is somewhat more complex, however. It involves a chemical reduction of the silver salts in the emulsion. The white areas of the camera copy reflect light through the lens of the camera and expose the film. The black areas of the copy absorb (do not reflect) the light, thereby producing no exposure on the film.

The exposed areas of the film react with the developer to form black metallic silver. The developer does NOT affect the areas which are NOT exposed to light. During development, the antihalation coating starts to dissolve.

Developing time for most line work is usually 2 1/2 to 2 3/4 minutes.

Stop bath is used to halt the developing action. With development completed, the film is placed in a stop bath solution for about 10 seconds. This bath consists of a mild solution of acetic acid and water. Since the bath is acid, it has the effect of immediately stopping the action of the developer.

Fixing bath is basically used to make the developed image permanent or stable. After stopping the developing action in the stop bath, the film is then placed in a fixing bath.

Technically, the fixing bath serves three purposes:
1. Dissolving the remaining emulsion in the unexposed areas, which are black areas of the original.
2. Hardening the emulsion so that it resists scuffing and scratching when it is dry.
3. Dissolving any unused dye remaining in the antihalation coating.

Fixing chemicals are available in both liquid concentrate and dry powder form. Most films used in graphic arts photography fix in about 1 - 3 minutes.

Washing bath is used to remove any remaining processing chemicals. After fixing the film, the film is washed in running water for about 10 - 30 minutes. When washing is complete, the film is squeegeed to eliminate excess water. It is then hung to dry in a dust-free area of the darkroom. Automatic film dryers are also used for this purpose.

MIXING TRAY CHEMISTRY

The chemistry used in processing should be compatible with the type of film. Film processing chemistry is sold in *concentrate form*. This means that it must be diluted before it can be used in the trays. This is called a *working solution*. Follow the manufacturer's directions when mixing chemistry. In general, the procedure is as follows.

Mixing developer

Developer is generally purchased in concentrate liquid or powder form. It consists of two separate containers—Part A and Part B. Each part is prepared separately to form working solutions. Mixing should be done according to the manufacturer's instructions. When measuring liquids, use a graduate beaker on a level surface. In this way, an exact amount of each solution can be measured. Accuracy is very important when mixing solutions for film processing.

Solutions A and B are stored in separate containers marked DEV "A" and DEV "B." To use these solutions, pour equal amounts of A and B into the developer tray and stir thoroughly. The developer is generally discarded after use.

Mixing stop bath

The stop bath consists of water and a mild acetic acid. Stop bath is prepared by pouring about one gallon of water into a tray. To this, add 6 to 8 ounces (177 to 236 mL) of 28% glacial acetic acid. The stop bath is generally discarded after use.

SAFETY TIP! Do NOT use *concentrated* glacial acetic acid. It can cause serious burns, irritation, and even blindness if allowed to come in contact with skin, eyes, or clothes.

Mixing fixer

Fixer is generally purchased in concentrate liquid or powder form. Mixing of the concentrate should be done according to the manufacturer's instructions. The working solution is stored in a container marked FIX. To use the working solution, pour enough into the tray to give about 1 inch (25 mm) depth.

A pre-mixed liquid hardener is sometimes used in the tray with the fixer. The *hardener* aids in hardening the emultion of the film, making it more resistant to scratches.

Fixer can be reused, but should be discarded when it fails to clear film within 2 to 3 minutes. Spilled fixer should be wiped up with a damp cloth immediately. This is because the solution leaves white spots on objects when dry.

Water bath conditions

As mentioned, the water bath serves to wash away all traces of the processing chemicals. If a tray is used for the wash, it should be provided with some means of continuously changing the water. A tray siphon is generally used for this purpose, Fig. 14-9. If the water wash area is built into the sink, a temperature-controlled faucet can be used to run water into the sink. The temperature should be set at 68 degrees F (20 degrees C).

SAFETY TIP! When your hands have been in chemical solutions, always rinse them in clean water before wiping them on a towel or before touching your skin or eyes.

Temperature control

The control of solution temperatures during film processing is critical. The first requirement for temperature control is a thermometer. Make sure the precision of the thermometer is at least as good as the precision you expect from your process.

Do NOT expect a thermometer that reads to the nearest degree Fahrenheit to be good enough to control a process in which temperatures must be controlled within ONE-HALF (1/2) DEGREE. Remember that the ideal processing temperature is 68 degrees F (20 degrees C).

The degree Celsius (C) measures temperature and in some cases replaces the Fahrenheit (F) thermometer in darkroom operations. At freezing, the Celsius thermometer registers 0 degrees (0°C) while the Fahrenheit thermometer registers 32 degrees (32°F). These comparisons are illustrated in Fig. 14-10.

PROCESSING PROCEDURE

Exposed film is processed in the darkroom under safelight conditions. A red safelight can be used with orthochromatic film since it is sensitive only to blue and green light. Safelights should be no closer than four feet from the area where the film is handled. Overhead white or yellow lights should NOT be turned on until the film is in the wash bath.

After exposure of the film on the process camera, with safelights on, open the camera back. Turn off the vacuum pump and remove the film by the edges.

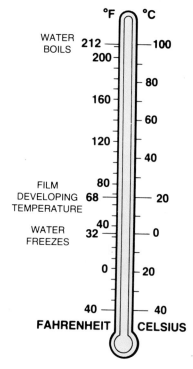

Fig. 14-10. This darkroom thermometer measures degrees Fahrenheit and Celsius.

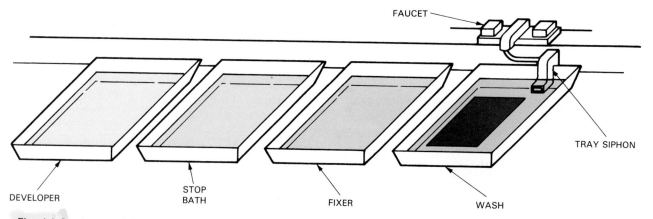

Fig. 14-9. A tray siphon should be used for film washing so that contaminated water is exchanged for fresh water throughout processing period.

Developing the film

Place the piece of exposed film in the developer with the emulsion side down. Immerse the film in the developer by pressing it firmly downward. Then quickly turn the piece of film over so that the emulsion side is up. Agitate the developer solution during the developing step by alternately raising and lowering the near edge of the tray about one inch. See Fig. 14-11.

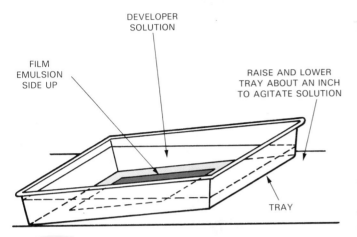

Fig. 14-11. Exposed film is dipped in the developer solution with emulsion side down. Piece of film is then quickly turned over (emulsion up) for remaining developing/agitation cycle.

As development continues, check the increasing density of the gray scale. For normal line work, a solid step four on the gray scale is generally used as the aim point. Continue agitation until the desired results are obtained.

The negative may be lifted out of the developer to check density. To do this, lift the negative by one edge, letting it drip for a moment. Hold the negative up to the inspection safelight. If more development is needed, return the negative to the developing solution.

When viewed before the inspection safelight, a properly developed negative will appear dense black in the background with clear, transparent images. Refer to Fig. 14-12.

Another method used to develop film in the tray involves an agitation cycle. After placing the piece of film in the tray with the emulsion side down, drag the film through the developer and then quickly turn it over so the emulsion side is up. Agitate the developer continuously.

Raise the left side of the tray about one inch. Lower the tray smoothly, and then immediately raise and lower the far side. Next, raise and lower the right side and then the near side. These four operations form the basis of the agitation cycle, which requires a total time of about eight seconds.

It should be noted that there are three generally ac-

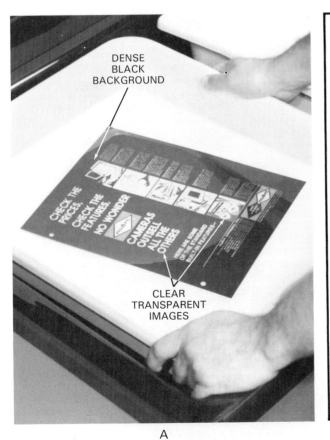

A

CORRECT EXPOSURE
This segment was exposed correctly. Negative areas are either clearly transparent or densely opaque. Edges are sharp, and detail proportions are true to the original.

UNDEREXPOSURE
This segment was underexposed. Although transparent areas are clear, the dark areas have low density. A positive made from a negative of this type shows thickening of all detail.

OVEREXPOSURE
This segment was overexposed. Although dense areas are opaque, density appears in some areas which should be clear. A positive made from a negative of this type shows loss of fine detail.

B

Fig. 14-12. A—Properly developed negative appears dense black in background with clear, transparent images. B—Study these examples of negatives that illustrate correct exposure, underexposure, and overexposure. (Eastman Kodak Co.)

cepted methods of controlling the film during the developing step. These include:

1. Inspection.
2. Time-and-temperature.
3. Gray scale.

In the **inspection method,** the operator waits until development is nearly complete. The negative is then lifted briefly from the developer and viewed in front of an inspection safelight. A magnifier can be used to inspect fine detail. By observing the density of the black background, a judgment is made when development should be stopped. Developing by inspection permits the operator to compensate for some exposure error. This is done by stopping short of the recommended developing time or by developing a little longer, whichever is indicated by inspection.

In the **time-and-temperature method,** the film is developed for the exact length of time and at a solution temperture recommended by the film manufacturer. This method is difficult because of the many variables involved. These include such things as exposure time, solution temperature, agitation, and freshness of the solution. The time-and-temperture method is usually combined to some degree with the inspection method.

In the **gray scale method,** control of development can easily be monitored. In addition, this method corrects for slight variables in exposure, temperature, and strength of developer. This method is sometimes used with ortho or graphic arts high contrast film.

Since the gray scale is photographed with the copy, it develops with the film. The operator observes the development stages of the gray scale on the film being developed in the tray. Each step of the gray scale, beginning with step one, develops up in progression, turning solid black. When the desired step turns black, development is stopped by placing the film in the stop bath. See Fig. 14-13.

Using stop bath

When the film has been completely developed, lift it up and let it drain for a few seconds. Transfer the film to the stop bath tray. Allow the film to remain in the stop bath for at least ten seconds with continuous agitation.

Using fixing bath

Next, transfer the film from the stop bath to the fixing bath. The film should remain in the fixer for about three minutes, or until the images clear completely. Do NOT turn on the white lights until the film appears to be clear in the unexposed, light areas. In general, film should be fixed for twice the length of time it takes to clear.

Washing the film

The film is then transferred from the fixing bath to the washing bath. The film should be washed in running water for 10 to 30 minutes. The washing cycle removes all traces of the chemicals. If washing is NOT thorough, the film may discolor.

After washing, wipe the film gently with a soft rubber squeegee and hang it up to dry. The film dries more quickly and uniformly if both of its sides are squeegeed on a rigid surface such as glass. Look at Fig. 14-14.

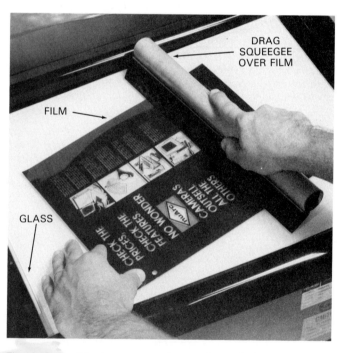

Fig. 14-14. Film is squeegeed on a piece of glass to remove excess water after wash cycle. The squeegee should be pulled gently across the acetate side of film to protect emulsion side from scratching.

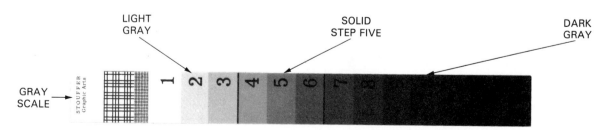

Fig. 14-13. Developing by inspection method permits operator to compensate for some exposure error. This is done by stopping short of recommended developing time or by developing a little longer, whichever is indicated by inspection.

FILM PROCESSORS

When medium to large quantities of film are handled, *automatic film processors* are used to develop, stop, fix, and rinse the film. Processors maintain quality control and are useful in operations where medium to high volume film processing is done. One is shown in Fig. 14-15(B).

Service conditions must be met before a processor can be operated. The machine should always be operated according to the manufacturer's directions. Temperature, water flow, and transport speed must conform to specifications. Machine control is basically the control of developer activity.

Development decreases as films are processed. To prevent this, *developer replenisher* is added. The amount added is related to the square inches of film developed. Replenishment balances developing agents for uniform activity over a period of time.

Control of the processor is achieved with the use of process-control strips. A *control strip* is a sheet of film about four by ten inches having a pre-exposed gray scale image on it. The strips are pre-exposed at the factory under very rigid quality control conditions. They are shipped to the customer in boxes of 50 or 100, depending on the manufacturer.

The control strips are used to give the operator an accurate indication of the condition of the chemistry in the processor. This is usually done the first thing in the morning and every two hours thereafter. All strips are exposed exactly the same. Any strip show-ing a difference after processing will be the result of a difference in chemistry strength or activity.

In addition to the gray scale image, the control strip carries a scale with a series of numbers. When the completely processed strip is read with a densitometer, the number on the scale opposite the area being read is logged on a processor control chart. The operator can easily see if the chemistry in the machine has experienced a change in strength or activity. This may be a result of an increase or decrease in the workload.

Once the automatic film processor is set up and operating properly, it must be kept that way. This can only be achieved through proper and regular maintenance. It is important to use the same products which performed best with the original installation conditions.

SILVER RECOVERY SYSTEMS

Since large amounts of silver are still used in making photographic films, silver recovery systems are widely used in the printing and publishing industry, Fig. 14-16. A standard silver recovery unit uses the *electrolytic principle* (electricity causes chemical change in solution). The silver is actually recovered from the liquid fixing solution in the darkroom.

In operation, the electrolytic unit has a stainless steel cathode. The cathode has a negative charge. The carbon anode has a positive charge. Electrical charges which are unlike attract. The positively charged silver ions are repelled by the anode. They are attracted by

A

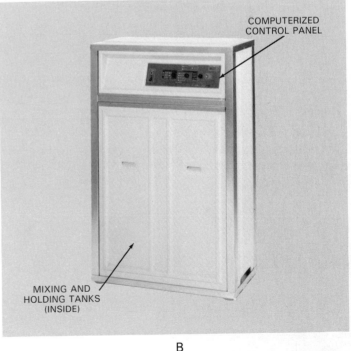

COMPUTERIZED CONTROL PANEL

MIXING AND HOLDING TANKS (INSIDE)

B

Fig. 14-15. A—Automatic film processors are used when medium to large quantities of film are handled. B—This automatic film processor-blender mixes necessary processing chemicals from concentrate and feeds solutions directly to processor as required. System eliminates mixing errors and reduces chemical handling. (LogEtronics, Inc., Kreonite, Inc.)

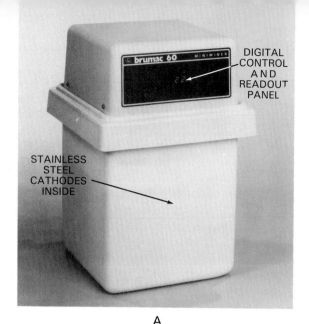

DIGITAL CONTROL AND READOUT PANEL

STAINLESS STEEL CATHODES INSIDE

A

ACCUMULATED SILVER ON STAINLESS STEEL CATHODE

B

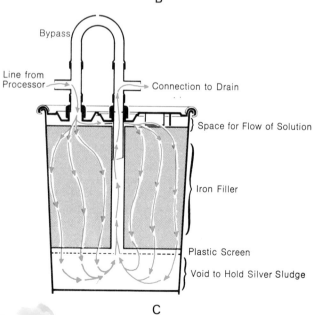

Bypass

Line from Processor

Connection to Drain

Space for Flow of Solution

Iron Filler

Plastic Screen

Void to Hold Silver Sludge

C

Fig. 14-16. A—This is a compact darkroom silver recovery unit for use in darkroom. B—Stainless steel cathode inside silver recovery unit attracts silver and is later removed. C—Typical silver recovery system connects to a film processor.

the stainless steel cathode. The recovered silver is then removed from the cathode. It can be sold and reused in making film. Exhausted fixer contains up to one ounce of silver per gallon.

CONTACT PRINTING OPERATIONS

Same-size contact prints on film and paper are often made with the use of a *vacuum contact frame,* Fig. 14-17. Reductions or enlargements are prepared by projection printing with the use of an enlarger, Fig. 14-18.

Same-size contact printing is done by placing a film negative or film positive in direct contact over a sheet of film or photoprint paper in a vacuum contact frame. A point-source light is used to expose the material. The exposed film or paper is then processed in the usual manner. Contact printing is a photomechanical process and there are a number of uses for it.

1. Making positives from negatives—a *film positive* is the reverse of a negative. The image is black and the non-image area is transparent, Fig. 14-19.
2. Making negatives from positives.
3. Changing the image orientation from *right-reading* (read normally) to *wrong-reading* (as if viewed backwards in a mirror), and vice versa.
4. Making photoprint proofs of negatives and positives.
5. Making screened halftone photoprints for use in paste-up.
6. Using tint screens to make tint blocks of various percentages.
7. Making spreads and chokes for multiple color printing. Spreads and chokes are covered later in this chapter.
8. Making duplicate negatives or positives for multiple printing on one printing plate.
9. Making negatives and positives for color separation printing.

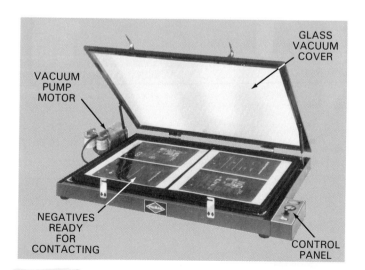

GLASS VACUUM COVER

VACUUM PUMP MOTOR

NEGATIVES READY FOR CONTACTING

CONTROL PANEL

Fig. 14-17. A vacuum contact frame is used to make same-size contact prints on film or paper. (nuArc Company, Inc.)

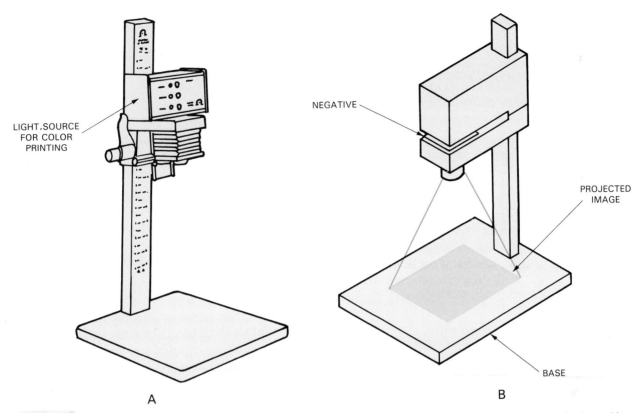

Fig. 14-18. A—Reductions and enlargements can be prepared by projection printing with enlarger. B—Projected image spreads out as negative moves away from base of enlarger.

Fig. 14-19. A positive is made from a negative. Image is black and non-image area is transparent. (Kenro Photoproducts Co.)

Contact printing equipment

Contact printing involves the use of three pieces of equipment. These include: vacuum frame, point source light, and transformer.

The *transformer* includes a light intensity (brightness) control, timer, and on-off switch.

The contact printing equipment is generally set up in a separate darkroom adjacent to the camera darkroom. This arrangement makes it possible for personnel in both darkroom areas to work independently. This avoids accidental exposure of light sensitive materials.

The vacuum frame is set up so that the lamp will illuminate the entire glass of the frame. The lamp is hung directly overhead at a distance of four to five feet from the frame. Exposure is determined by the amount (intensity) of light at the frame and the time it is on.

The *Inverse Square Law of Light* shows that the distance between the light and the film affects intensity, Fig. 14-20. To achieve proper exposure, the light source should be the prescribed distance from the frame and regulated by a variable voltage transformer. This gives longer and more accurate exposures which are easier to control.

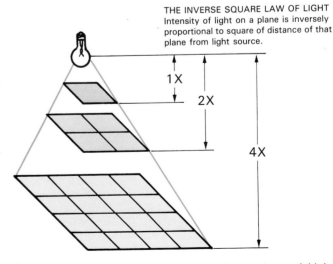

THE INVERSE SQUARE LAW OF LIGHT
Intensity of light on a plane is inversely proportional to square of distance of that plane from light source.

Fig. 14-20. This example of Inverse Square Law of Light demonstrates that distance between light and film affects intensity. (E.I. DuPont de Nemours and Co.)

303

Contact printing films

Two general types of reproducing materials are used for contacting. One type reproduces OPPOSITE values, such as negative to positive or positive to negative. The other type of material reproduces LIKE the original, that is, positive to positive or negative to negative. This type makes duplicate images of the original material.

Film or photographic paper that produces an opposite image is often called *contact reversal.* The reverse image is similar to the action of regular negative-producing film used in a camera.

Contact film is made especially for contact printing. This film is used for *indirect contacting.* Contact film is blue sensitive and the film emulsion is photographically slow. It can withstand a large amount of safelight illumination. A greenish-yellow safelight can be used in the contacting room. This makes it easier to see than with the red safelight generally used with orthochromatic film.

Duplicating film is another type of contacting film that is blue sensitive. It is used to make negatives from negatives and positives from positives. This is called *direct contacting.* Duplicate negatives or positives are required when stripping for multiple images on one printing plate.

The film emulsions used for most contact printing are slower than those used in process photography. This means that more light is needed to expose the film. Contact printing films have good emulsion density and are available in several thicknesses.

Photoprint papers are available in a variety of finishes and contrasts for various purposes including contact printing.

Contact printing procedure

Before you start contact printing operations, prepare the vacuum frame and point source light. Make certain that the glass on the vacuum frame is clean. Use the correct safelight illumination for the film or photoprint paper. Have the correct processing chemicals prepared and readily available.

The preparation of a contact print is done by using either a film negative or film positive. If a *film negative* is used, the resulting image will be positive. If a *film positive* is used, the resulting image will be negative.

In making a **positive print,** a piece of photographic paper (such as Velox, Azo, or ortho) is placed emulsion-side up on the bed of the contacting frame, Fig. 14-21. Place a piece of dark, smooth plastic on top of the vacuum frame blanket. This is called a *draw-down sheet.* It should be slightly larger than the sheet of photographic paper. The plastic provides a smooth surface so better contact can be made. The film negative is placed emulsion-side down over the paper.

Exposure to a point source light is made according to the time recommended for the kind of film or paper.

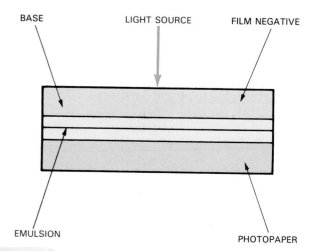

Fig. 14-21. Contacting a film negative to photoprint paper results in a positive contact print. (Eastman Kodak Co.)

The photoprint paper is then processed according to the manufacturer's specifications. For example, an ortho paper may be exposed to a point source light five feet away for seven seconds using the lowest number on the voltage regulator. The paper is developed in a developer for about one minute. It would then be stopped, fixed, and washed. The print is hung to dry or placed in a print dryer.

To speed the development cycle, a stabilization paper may be used. For example, one type paper is exposed for 15 seconds using the lowest number controlling the point source light. The paper is processed in the recommended processor, using the right activator and stabilizer solution. The processing time can be less than ten seconds for an 8 x 10 inch size print. If the print is to be saved for later use, it should be fixed and washed.

In preparing a **reverse print,** a film positive is required, Fig. 14-22. The procedure is exactly the same as that for preparing a positive print. The only difference is a film positive is used instead of a film negative. In a reverse print, as in a film negative, the

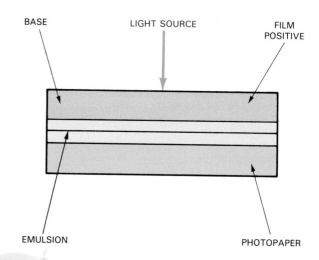

Fig. 14-22. Contacting a film positive to photoprint paper results in a reverse contact print.

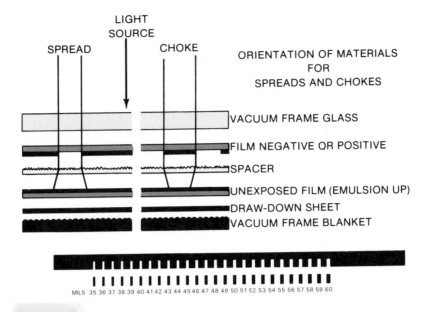

Fig. 14-26. Study this illustration to learn procedure for making spreads and chokes. (E.I. DuPont de Nemours and Co.)

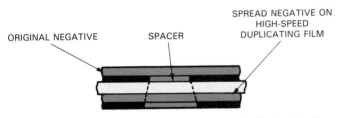

A—Note method of preparing a spread negative on duplicating film from an original negative.

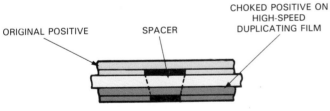

B—Original positive is used to prepare choke positive on duplicating film.

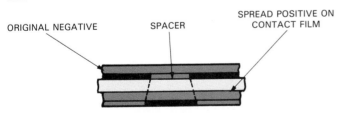

C—Original negative is used to prepare a spread positive.

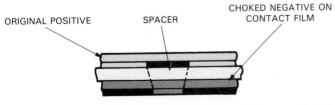

D—A choked negative on contact film is prepared from an original positive.

Fig. 14-27. Study these four useful contacting operations. (Eastman Kodak Co.)

5. Place the negative over the spacer, emulsion against the spacer.
6. Expose according to specifications, Fig. 14-28.
7. If a spread positive is required, use contact film in place of the duplicating film.

Making a choke

1. Make a contact positive from the original negative of the line copy.
2. Set up the contact frame as shown in Fig. 14-26.
3. Place duplicating film emulsion side down against the contact frame back.
4. Put on spacer. Use 0.004 inch thickness for a medium choke, a 0.008 inch thickness for a larger choke.
5. Place the positive on top of the spacer, emulsion side toward the spacer.
6. Expose according to specifications, Fig. 14-28.
7. If a choke positive is needed, use contact film instead of duplicating film.

To place line work over a window (clear, unexposed area) in a background, make a spread on duplicating film of the line work to be superimposed. If the line work cannot be spread, then the background can be choked. Make a positive of the background and choke it, using duplicating film.

Making outlines

Outlining is the procedure of exposing only the outer edges of the image. The outlined image is larger than the original when a spread is used. It is smaller than the original when a choke is used.

Fig. 14-29 shows the procedure for a spread outline. Fig. 14-30 shows the procedure for a choke outline.

1. Contact a positive from the original negative.
2. For a spread outline, use the original negative or

307

EXPOSURE TABLE

FILM	VOLTAGE SETTING	LAMP DISTANCE	FILTER	EXPOSURE
Contact Film 2571 (ESTAR Base)	20 volts	5 feet	KODAK WRATTEN Neutral Density Filter No. 96 (0.60 density)	20 seconds
High-Speed Duplicating Film 2575 (ESTAR Base)	20 volts	5 feet	None	40 seconds

Fig. 14-28. Contacting exposure tables give important information relative to various films, duplicating papers, voltage setups, lamp distances, filters, and exposure times. (Eastman Kodak Co.)

OUTLINE **OUTSIDE** THE ORIGINAL LETTER

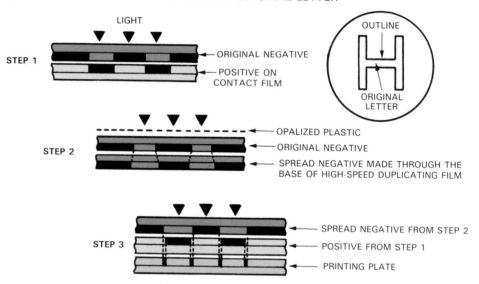

Fig. 14-29. To prepare a spread outline, study procedure in this illustration. (Eastman Kodak Co.)

OUTLINE **INSIDE** THE ORIGINAL LETTER

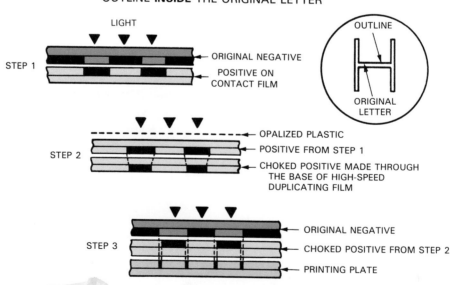

Fig. 14-30. To prepare a choke outline, follow this procedure. (Eastman Kodak Co.)

spread negative with duplicating film. When exposing the offset plate, register the spread negative over the positive, emulsion to emulsion, to produce an outline exposure.

3. For a choke outline, make a choked positive by exposing the positive from the first procedure to duplicating film. Register the original negative over the choked positive, emulsion to emulsion, to expose a choke outline on the offset plate.

The amount of spacer material used, the exposure time, and amount of diffusion determine the extent of spread or choke. A record of these factors should be maintained in order to repeat previous results.

A reference guide to make chokes and spreads is shown in Fig. 14-31.

PROBLEMS IN THE DARKROOM

There are common problems in the darkroom, especially with line and halftone work. A system should be established and maintained that takes exposure, temperature, processing, handling techniques, and product types into consideration. This also applies to the products being used.

Eliminate as many variables as possible. If this rule is adhered to, problems that arise resulting from poor or unusual copy can be overcome with a minimum of changes in the system. When changes must be made to meet the demands of the copy, only one change should be put into practice at a time. In this way, you can easily identify which change produced the desired result.

To assist you in identifying problems in both line and halftone photography, use the troubleshooting chart shown in Fig. 14-32.

After a little practice, you will become proficient in the handling of line and halftone copy. The key to success is adherence to an established system of controlling all variables.

POINTS TO REMEMBER

1. The term film, when used in graphic arts photography, refers to a transparent or translucent acetate or plastic base which has a light sensitive coating.
2. Graphic arts films are generally made from either cellulose-based, ESTAR-based, or polystyrene-based materials.
3. The gelatin compound containing the silver halide (silver salts) in film is bonded to the base material by an adhesive substratum. The other side of the film is covered with an antihalation coating.
4. The film's antihalation coating prevents light from reflecting back to the emulsion side of the film during exposure.
5. Light sensitive materials are classified with respect to three main variables, including color sensitivity, contrast, and film speed.

SPREADS AND CHOKES

ORIGINAL	RESULT DESIRED	PROCEDURE
Film Negative	Choked Negative	1. Make a normal contact positive on contact film. 2. Make a choke on contact film.
Film Negative	Choked Positive	1. Make a normal contact positive on contact film. 2. Make a choke on high-speed duplicating film.
Film Negative	Spread Negative	Make a spread on high-speed duplicating film.
Film Negative	Spread Positive	Make a spread on contact film.
Film Positive	Choked Negative	Make a choke on contact film.
Film Positive	Choked Positive	Make a spread on high-speed duplicating film.
Film Positive	Spread Negative	1. Make a normal contact negative on contact film. 2. Make a spread on high-speed duplicating film.
Film Positive	Spread Positive	1. Make a normal contact negative on contact film. 2. Make a spread on contact film.

Fig. 14-31. This handy reference guide can be used to prepare spreads and chokes. Guide should be displayed near contacting area.   (Eastman Kodak Co.)

LINE SHOTS

PROBLEM	CAUSE	REMEDY
1. Image does not develop.	A. Material not exposed.	Remove lens cap. Inspect lens, shutter, and camera lights for proper operation.
	B. Developer exhausted.	Mix new chemistry.
	C. Developer too cold.	Warm to 68-70 degrees F by placing in pan of warm water.
2. Image comes up too slow.	A. Material underexposed.	Check the lens opening. Check setting on timer. Be sure to allow for filter or halftone screen in determining exposure.
	B. Developer too cold.	Warm to 68-70 degrees F.
	C. Developer too old or oxidized.	Dump and change chemistry.
3. Image develops too quickly.	A. Material overexposed.	Check lens opening, timer setting, and lamp position. Correct discrepancy or reduce exposure.
	B. Developer too warm.	Reduce temperature by placing in pan of cold water or dilute the developer. Underexpose the material.
4. Film clears too slowly in fixer or fails to clear at all.	A. Fixer exhausted.	Dump and replace fixer.
5. Negative fogged or image filled-in.	A. Light leaks.	Correct any light leaks around doors, walls, and ceilings. Also look for pin holes in camera bellows.
	B. Material overexposed.	Reduce exposure to hold image.
	C. Over-development.	Check temperature of chemistry. Reduce development time and/or agitation of chemistry.
	D. Prolonged development due to old chemistry.	Replace developer.
	E. Flare.	Turn out room lights in vicinity of lens.
	F. Dirty lens surface.	Clean with lens tissue.
	G. Smoke or dust in camera room.	Ventilate the area.
	H. Old, outdated material.	Replace the material.
	I. Material not stored properly.	Find more suitable storage area free of humidity and heat.
6. Negative develops unevenly.	A. Improper agitation of developer.	Immerse negative emulsion down, agitate tray in smooth, regular manner. Do not let film lay motionless in tray.
	B. Developer ingredients not fully mixed together.	Agitate tray to completely mix all ingredients. Take bite off fresh chemistry with scrap of exposed film.
7. Film lacks density.	A. Not enough exposure and/or development time.	See item 2 above.
8. Negative thin in corners and edges.	A. Uneven illumination.	Move lights back from copy. Stop lens down one F-stop and double exposure. Check for uneven white surface in copyboard.
	B. Improper developer agitation.	Decrease developer activity. Agitate tray very gently and change film position in tray occasionally.
9. Blurred image.	A. Camera out of focus.	Check image on ground glass, and re-check camera settings.
	B. Vibration.	Look for equipment nearby causing camera to vibrate. Check location of vacuum pump.
10. Pinholes.	A. Copyboard film, or halftone screen dusty.	Check each area for cleanliness before re-shooting.
	B. Developer cold.	Increase temperature to 68-70 degrees F.
	C. Developer too old.	Replace chemistry.
	D. Material underexposed.	Refer to item 2 above.
11. Film scratches easily.	A. Not enough hardener in fix.	Add small amount of hardener to fixer and agitate tray.
	B. Fixer exhausted.	Dump and re-mix according to instructions on package, being sure to include hardener.

Continued

Fig. 14-32. Troubleshooting reference charts for line and halftone work are useful when displayed near camera and darkroom area. (Eastman Kodak Co.)

HALFTONES

PROBLEM	CAUSE	REMEDY
1. Image develops slowly.	A. Material underexposed.	Check lens opening, lamp position, and exposure setting. When determining exposure allow for halftone screen. If necessary, use gray rather than magenta screen to reduce length of exposure.
2. Highlight dots plugged-up.	A. Material overexposed.	Reduce main exposure.
	B. Too much bump.	Reduce or eliminate the no-screen exposure.
	C. Developer too warm.	Cool to 68 degrees F.
	D. Too much agitation of developer.	Agitate less vigorously.
3. Highlight dots too open.	A. Underexposure.	Increase main exposure. Try a no-screen (bump) exposure for 5% of the main exposure to pinch-up highlights.
	B. Inactive chemistry.	Increase agitation of developer.
	C. Developer cold.	Warm to 68 degrees F.
4. No shadow dot.	A. Not enough flash.	Increase flash exposure.
	B. Not enough main exposure.	Increase main exposure, (this will also affect highlights).
	C. Too much developer activity.	Slow down developer agitation to give shadow dots a chance to come up. If necessary, try to still develop for a third of the developing time.
	D. Exhausted developer.	Dump and replace with fresh chemistry.
	E. Developer too cold.	Warm to 68 degrees F.
5. Shadow dots weak in areas adjacent to highlights.	A. Developer losing strength.	Dump and replace with fresh developer.
6. Fingerprints.	A. Chemistry impregnated in fingers of operator.	Make sure hands are fully dry before handling new film. Use powder or rubber gloves.
	B. Fingerprints in original photo.	Try to clean original with film cleaner. Change angle of copy in copyboard where lights will strike from different angle.

Fig. 14-32. Continued.

6. Color sensitivity describes the area of the visible electromagnetic spectrum that will cause a chemical change in a particular silver halide emulsion.

7. Film materials are described as either blue sensitive (color blind), orthochromatic (sensitive to all other areas of visible spectrum), or panchromatic (sensitive to all visible colors).

8. Contrast refers to the lightest and darkest portions of an original photographic image and how it creates contrast on the film.

9. Each film material has a characteristic curve which can also be referred to as the Log E curve.

10. All film materials require a different amount of light to cause a chemical change in the silver halide emulsion. This is referred to as film speed. For this reason, an exposure index is used for each film manufactured.

11. Films are assigned ASA numbers to indicate their speed. The higher the number, the faster the film's speed.

12. Film processing by the tray method generally requires the use of four separate solutions, including developer, stop bath, fixer, and running water.

13. Temperature-controlled sinks are popular since precise temperature control of the chemicals provides accuracy in film processing.

14. After film has been exposed on a graphic arts process camera, the processing procedure follows. This involves development, stop bath, fixing bath, and washing bath.

15. Each of the chemicals used in film processing must be mixed with accuracy. Any variation from the recommended mixing procedure will result in processing difficulties.

16. There are three generally accepted methods to control the film during the developing step, including inspection, time-and-temperature, and gray scale.

17. The time-and-temperature method is usually combined to some degree with the inspection method.

18. The four steps used in tray development must be performed carefully in order to overcome each of the variables and achieve optimum results.

19. Silver recovery systems are used in an effort to recover the unused silver in graphic arts film during processing.
20. Silver recovery equipment is available for almost any size darkroom operation.
21. Recovered silver can be re-used by film manufacturers to produce new film.
22. Contact printing is an important part of graphic arts photography.
23. Contact printing involves several operations that use a contact vacuum frame, point source light, and a transformer for controlling light intensity and length of exposure.
24. Contact printing operations include the following: making positives from negatives, negatives from positives, photoprint proofs of negatives and positives, screened halftone photoprints, spreads and chokes, duplicates, and outline type.

KNOW THESE TERMS

Antihalation coating, Blue sensitive, Characteristic curve, Choke, Color sensitivity, Control strip, Developer, Exposure index, Film speed, Fixer, Latent image, Negative, Orthochromatic film, Panchromatic film, Positive, Spread, Stop bath, Wedge spectrogram.

REVIEW QUESTIONS

1. What is the difference between film and photopaper?
2. A processing sink is used to hold film developing trays. True or false?
3. _____ _____ are used to control the temperature of water around the processing trays.
4. Chemicals and water are measured in a container called a:
   a. Funnel.
   b. Tank.
   c. Basin.
   d. Graduate.
5. Contact printers are used to make _____ reproductions on film or photographic paper.
6. When are automatic film processors used?
7. The _____ sensitivity of a light-sensitive material describes how it will be affected by light in a particular silver halide emulsion.
8. Color-blind light-sensitive materials are known as _____ sensitive.
9. Red safelights can be used with panchromatic film. True or false?
10. The characteristic curve of a particular film describes its _____.
11. Film speed refers to the specific amount of light that will cause a chemical change in the silver halide emulsion of a particular film. True or false?
12. The ASA number assigned to a given film designates its:
    a. Speed.
    b. Size.
    c. Color.
    d. Thickness.
13. After development is completed, the film is placed in a _____ _____ for about ten seconds.
14. Film should remain in the _____ _____ for at least ten minutes.
15. Developer is available as either a _____ _____ or _____.
16. A tray siphon is generally used in the developer tray. True or false?
17. One of the following is NOT a technique used to control film development:
    a. Inspection.
    b. Touch.
    c. Gray scale.
    d. Time-and-temperature.
18. Describe the basic procedure of film processing by the tray method.
19. A film processor _____ _____ is used to determine the condition of the machine's chemistry.
20. In contact printing operations, a film negative produces a positive, and a film positive produces a negative. True or false?
21. The procedure of making a same-size duplicate of a film negative or positive is called _____ printing.
22. A positive print is made from a film _____.
23. A reverse print is made from a film _____.
24. A spread is contacted on film from a _____.
25. Spreads and chokes are made in a vacuum contacting frame using a point light source. True or false?

SUGGESTED ACTIVITIES

1. Draw a cross section of a piece of lith ortho film and label the various parts.
2. Make a wall chart showing the order of trays for hand processing film in the darkroom.
3. Prepare a chart illustrating the color sensitivity of orthochromatic and panchromatic films.
4. Make a film positive from a film negative.
5. Make a spread and a choke from copy supplied by your teacher.
6. Make a photographic paper contact print from a film negative.
7. Prepare a line of outline type.

Chapter 15

Line Photography

After you have completed the reading and assigned activities related to this chapter, you will be able to:

O List and describe the various classifications of line copy.

O Describe the purpose and use of camera filters.

O Establish camera standards for making line negatives through a series of tests.

O Describe the purpose and use of the camera gray scale as related to line photography.

O Describe the exposure procedure known as constant aperture method.

O Describe the exposure procedure known as constant time method.

O Prepare the darkroom for making a line negative.

O Expose and develop (process) a line negative to acceptable standards.

During *line photography,* original single color or single tone camera copy (images) are reproduced on film. There are NO continuous tones (gradations of tone) in line photography. These images do not contain shades of gray as in continuous tone photography. Continuous tone photography is covered in Chapter 16.

Examples of line copy include typeset composition, inked drawings, scratch board, block prints, reversals, screened tints, and screened prints. The film negatives produced by line photography are called *line negatives.*

Kinds of line copy

All kinds of **typeset composition** and similar symbols, such as those you are now reading, are classified as *line copy.* Typeset composition is prepared by any of the methods described in Chapters 8, 9, and 10.

The **inked drawings,** shown in Figs. 15-1 and 15-2, are examples of line copy. These drawings are prepared by drawing on white paper or illustration board with

Fig. 15-1. Inked drawings in these examples are known as line drawings. Original drawings were prepared in pen and ink. (Letraset® Inc.)

313

Fig. 15-2. Artist uses pen and ink to render a line drawing. Exceptional detail can be achieved in such drawings like this example. (Koh-I-Noor Rapidograph, Inc.)

black ink. They can also be drawn with a computer driving a pen plotter.

A specially-treated, black-coated drawing board is used for **scratch board renderings,** Fig. 15-3. The artist scratches lines in the surface of the board and exposes the white surface below.

Some drawings, classified as line drawings, are made from original wood blocks called **block prints,** Fig. 15-4. The design is cut from a piece of wood so that the image part to be printed is higher than the remaining part of the block. Various effects can be achieved by printing the block on a rough-textured paper. Block prints are a form of letterpress plates.

A **reversal** has the same general characteristics as any other kind of line copy. The difference is that selected image areas are reproduced in white instead of black (when printing black ink on white paper). Fig. 15-5 illustrates how type characters have been reproduced as white instead of black images.

Images, called **screened tints,** are backgrounds composed of many evenly spaced dots, Fig. 15-6. Other patterns are also used in this category of line copy.

Dry transfer screen tint material can be applied directly to the paste-up before it is photographed on a process camera. Screen tints on acetate film differ from adhesive dry transfer materials. Screen tints on acetate are positioned between the negative and plate before making the offset plate exposure. Screen tints can also be generated and placed on art using a computer and drawing program.

A **screened print** is made from an original continuous tone photograph. These prints are made with a special contact screen in much the same manner as a halftone

Fig. 15-3. This is an example of a scratch board rendering. It is done on a specially-treated, black-coated drawing material. (Kimberly-Clark Corp.)

Fig. 15-4. Block prints are generally made from hand-cut wooden blocks. Block prints are a form of letterpress printing plate and print best on rough-textured paper.

Fig. 15-5. A reverse consists of image areas that reproduce in white, or color of paper on which printing is done, instead of black ink. (Kenro Photoproducts Co.)

Fig. 15-6. Screened tints are images containing backgrounds composed of many evenly spaced dots. Screened tints are classified as line copy. (Zipatone, Inc.)

negative. Because a screened print is made on photographic paper, a diffusion-transfer process is used to make this type print.

Since a screened print consists entirely of dots of various sizes, it can be handled as line copy on the process camera along with other line copy on a paste-up. This process often saves time and allows the paste-up artist to combine screened prints with other type elements directly on the paste-up base.

Line copy theory

Line copy is generally positioned upside down on the camera copyboard. This allows the operator to check the copy on the ground glass since it will be right-reading on the glass. The film is placed in the camera back with the emulsion side facing the camera lens, Fig. 15-7.

When the camera lights are turned on, light rays are reflected from the white areas of the copy. The reflected rays pass through the camera lens and are projected onto the film. The image is formed on the film by the contrast between the white base and the black type characters and inked drawings of the copy, Fig. 15-8. The white areas of the copy reflect light through the lens of the camera. The light then strikes the film emulsion to expose it. The result is a line negative, Fig. 15-9.

When processed, the negative is opaque (dark) in the areas that were white on the original copy. The negative is transparent (light) in the areas that were black on

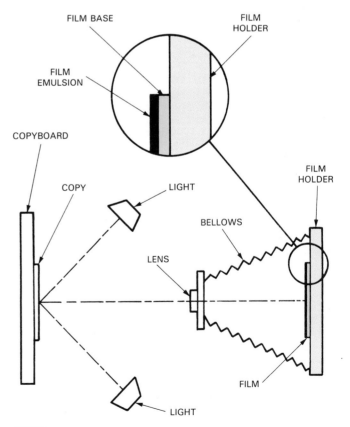

Fig. 15-7. Photographic film is loaded in process camera back with emulsion side facing camera lens.

Fig. 15-8. Note example of line copy used for preparing a line negative. Image is formed on film by contrast between white base and black type characters and inked drawings of copy.   (Eastman Kodak Co.)

Fig. 15-9. When fully processed, a line negative appears opaque (dark) in areas that were white on original copy. Negative is transparent in areas that were black on original copy.   (Eastman Kodak Co.)

the original copy. The areas of the film that were struck by reflected light rays were intensified in the developing process. These are the areas that show as dense, black areas on the negative. The areas of the film that received NO reflected light are washed away as a result of the processing. They become the transparent (image) areas of the negative.

Line photography test shot

Good line photography results from accurate camera focus and care in film processing. The following test exposure procedure is recommended to achieve the highest quality line negatives. First, refer to the details of copyboard preparation and focusing described in Chapter 13.

1. Select a piece of line copy containing various line images, such as in Fig. 15-13. This will be referred to as a line copy target.
2. Clean both sides of the copyboard glass. Use a graphic arts glass cleaner.
3. Place the line copy target upside down on the camera copyboard. A *sensitivity guide* (gray scale) should be placed next to the copy, Fig. 15-10. This guide is an aid to be used in film processing.
4. Set the camera for same-size (100%) reproduction.
5. Set the lens for the best f-stop. Each film manufacturer recommends a setting for optimum results. An aperture of f/16 or f/22 is common. For purposes of this test, set the lens at f/16.

   Note! The process of determining best f-stop number is described at the end of this test shot sequence.

6. The control arm setting on the aperture scale should match the percentage of enlargement or reduction of the copy. In this test, the percentage size is 100%.
7. Position the camera lights. The proper angle of the lights is very important. Make sure the camera lights are pointing toward the center of the copyboard.
8. As a starting point, use recommended exposure time procedure in step 14.
9. Focus the copy on the ground glass. Refer to the procedure described in Chapter 13.
10. Place a sheet of orthochromatic film on the camera back. Turn on the vacuum and make sure the film is smooth, flat, and dust-free. The DULL SIDE of the film is the emulsion side and should face the lens when the camera back is closed. Handle the film by the edges to avoid fingerprints in the image area.
11. Cut a piece of black construction paper that measures a little wider than, and at least as long as, the sheet of film. This is called a *mask*.
12. Place the mask over the film so that there is an uncovered one-inch strip across the top of the film. Then, close the camera back and make the exposure. All exposures in the test series should be made at the same f-stop setting at same size (100%).
13. Open the camera back, leaving the vacuum pump on, and move the mask down about one inch. Close the camera back and make the next test exposure in the series.

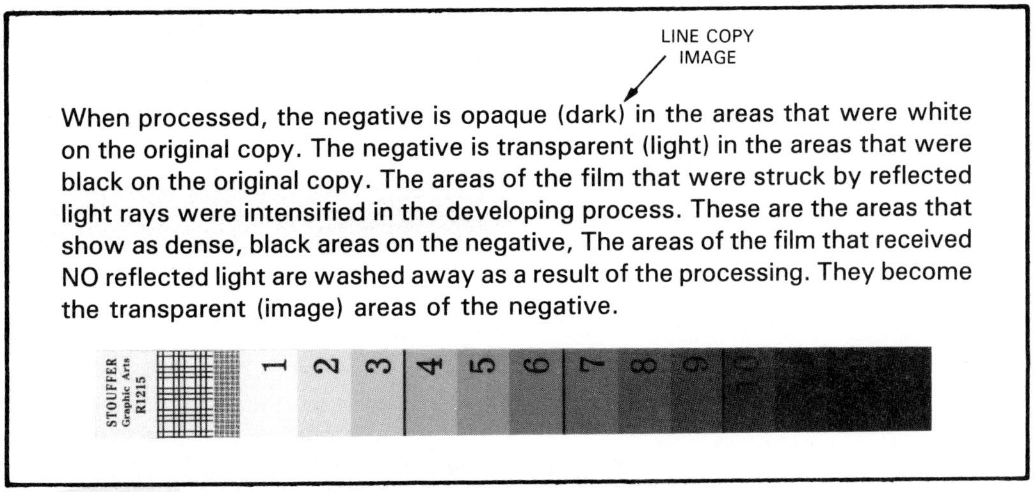

LINE COPY
IMAGE

When processed, the negative is opaque (dark) in the areas that were white on the original copy. The negative is transparent (light) in the areas that were black on the original copy. The areas of the film that were struck by reflected light rays were intensified in the developing process. These are the areas that show as dense, black areas on the negative. The areas of the film that received NO reflected light are washed away as a result of the processing. They become the transparent (image) areas of the negative.

STOUFFER Graphic Arts R1215

1 2 3 4 5 6 7 8 9 10

Fig. 15-10. A sensitivity guide (gray scale) is placed alongside line copy originals on copyboard of process camera.

14. Repeat this procedure until about seven exposures have been made. Fig 15-11 illustrates the mask exposure procedure.

Note! Select a series of exposure times in which each successive exposure represents a uniform increase over the last exposure. For example, 3, 6, 9, 12, and 15 seconds. This is easier to time than a more complicated logarithmic series, like 5, 8, 12, 18.

15. Turn off the vacuum pump and remove the film from the camera back.

16. Develop and fix the test film using the procedure described in Chapter 14. Develop to a solid step 4. After a short wash, squeegee the film on a piece of glass in the darkroom. Examine the film over a light table using a magnifier of about 10-power magnification, Fig. 15-12.

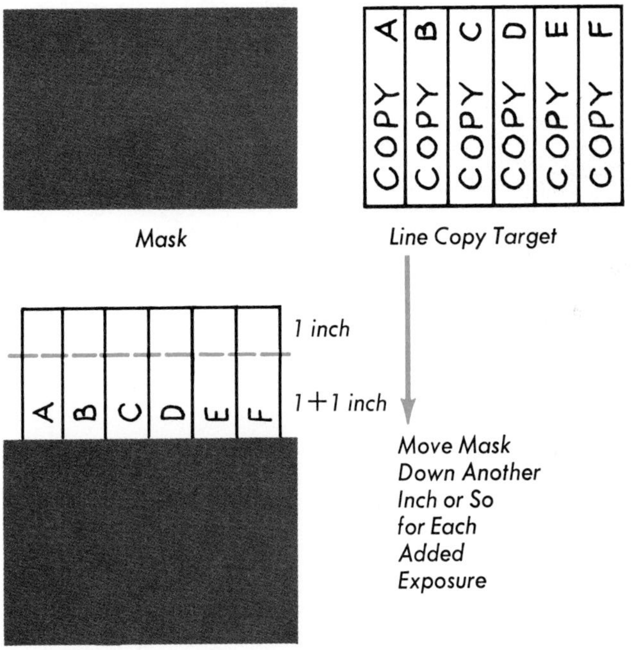

You have now produced a line test negative. Select the exposure that produced the best image. Here is what to look for during evaluation.

1. The black areas should be dense enough so that light can barely be seen through the area.
2. There should be no pinholes in the solid black areas.
3. The transparent areas should be clean and clear.
4. The edges of the image should be as sharp as in the original copy.
5. Line width and shape of small image areas should be as sharp as in the original copy, especially in the serif areas of the Roman type. See Fig. 15-13.

Remember that appearance similar to that of the underexposed segment can result even after correct ex-

Mask

COPY A COPY B COPY C COPY D COPY E COPY F

Line Copy Target

1 inch

1+1 inch

A B C D E F

Move Mask Down Another Inch or So for Each Added Exposure

Fig. 15-11. For process camera exposure tests, a mask procedure is recommended for best results and accurate line copy exposure times. (Eastman Kodak Co.)

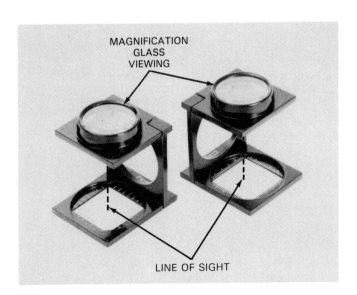

MAGNIFICATION GLASS VIEWING

LINE OF SIGHT

Fig. 15-12. Magnifier is used to check detail of a line test negative. (Direct Image Corp.)

posure if the developer is too cold or is improperly mixed. It may also result if the development time is too short.

Fogging

Appearance similar to that of the overexposed segment can result from *fogging*. This is caused by flare from a dirty lens, from use of an unsafe safelight in the darkroom, or from nonimage room light or reflected light in either the camera room or darkroom. Exhausted fixer, or failure to use a stop bath, can also lead to fogging.

Pinholes

If the camera room or darkroom is dusty, and if the relative humidity is too low (air too dry), the film will attract dust particles, just as a magnet attracts iron filings. These dust particles cast tiny shadows during exposure, which leave unexposed pinholes in the negative.

Pinholes are unwanted tiny openings in the negative. Pinholes can also be caused by dirt on the copyboard glass and by specks in the copy. Such unwanted spots (holes) can be eliminated from the negative by painting with *film opaque*.

Final negative

By selecting the best resolution strip in the test negative, you have now determined the correct exposure time. You can now expose a full piece of film to the copy. Since you will have eliminated all the variables, good line negatives should result each time if the same conditions are duplicated.

Whenever you get a negative as you want it, record the details of how it was made. The details include type of film, exposure time, developer, and development time. By duplicating the conditions and procedures of the test, you can get good negatives of any line copy

without further exposure testing. The gray scale will show the correct development with the same camera, lens aperture, film, and conditions.

Occasionally, a large piece of line copy may require a different light position for even illumination. If the lights are moved farther away, the exposure time must be increased. Similarly, the exposure must be adjusted for enlargements and reductions, as shown in Fig. 15-14.

Determining best f-stop number

The previous test determined the correct exposure time for a line shot. You must also determine the best f-stop number for the camera. Lenses, due to manufacture, are sharpest at one f-stop number. This f-stop should always be used for line shots. To determine the sharpest f-stop, a series of test exposures can be made.

1. Set the camera for same-size (100%) reproduction.
2. Position the copy used in the previous test on the copyboard.
3. Place film in camera back. Use a large sheet. Position film for exposure in upper right corner. Mask the remainder of the film. Set exposure time for ten seconds.
4. Expose the film with the largest (opening) f-stop number.
5. Reposition the film to expose in a clockwise direction around the sheet of film.
6. Reduce f-stop number for each exposure. DO NOT change the exposure time.
7. Process the film by the tray method. Select the sharpest image by using the magnifier. Always use the f-stop that produced the sharpest image.

LENS FILTERS

Line copy in a color other than the usual black on white is often photographed on a process camera. In

Fig. 15-13. Line test negative should reveal line width and shape of small image areas as sharp as in original copy. (Eastman Kodak Co.)

Reproduction Size	Suggested Exposure-Time Factor	
	Lights Attached to Copyboard	Copyboard Moves, Lights Do Not
200%	2.40	2.00
175%	2.00	1.70
150%	1.60	1.40
125%	1.30	1.20
100%	1.00	1.00
90%	.92	.92
80%	.85	.84
70%	.75	.75
60%	.68	.67
50%	.60	.60
40%	.52	.55
30%	.44	.50
20%	.38	.50

Fig. 15-14. Camera reductions and enlargements require adjusted exposure times for best results. This information should be placed near process camera.
(Stouffer Graphic Arts Equipment Co.)

some instances, the line copy is NOT always printed on a white background paper. These kinds of copy can be photographed to appear as though they were black and white line copy. This is done by using a specific *lens filter* and film.

Various combinations of filters and films used to photograph certain colors of original line copy are shown in Fig. 15-15. For example, original copy consisting of black ink printed on a yellow paper can be photographed as though it were black ink on white paper. For example, this can be done by using a Kodak Wratten Filter No. 16 and orthochromatic film. If more than one filter is recommended, the first one is preferred.

There may be other shades and combinations of colors encountered. In this case, the camera operator should prepare a chart showing satisfactory results to be used for future reference. If possible, the chart should show examples of the original copy, numbers of the filters, kind of film used, and length of exposure or filter factors.

Types of filters

There are two kinds of filters in common use. These include the gelatin film and gelatin cemented between two sheets of optical glass. Sizes of these filters are either 2 x 2 inches or 3 x 3 inches, depending on the camera's lens diameter. Fig. 15-16 shows a filter used in a process camera.

In operation, a filter is inserted into the filter slot (Waterhouse Stop) of the lens barrel. A *Waterhouse Stop* is a diaphragm for photographic lenses devised

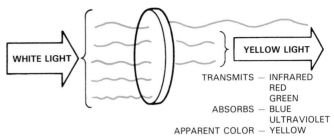

TRANSMITS — INFRARED
RED
GREEN
ABSORBS — BLUE
ULTRAVIOLET
APPARENT COLOR — YELLOW

YELLOW FILTER
KODAK WRATTEN NO. 8 (K2)

Fig. 15-16. A filter is used to photograph line copy in a color other than usual black on white. This illustration shows how a filter is used in a process camera. (Eastman Kodak Co.)

by John Waterhouse in 1856. The aperture takes the form of a thin sheet of metal or paper with circular or irregular openings of a definite size. It is intended for insertion in a slot cut into the barrel of a lens. A filter can also be inserted into a filter holder mounted on the front of the lens barrel.

Camera focus must be checked carefully when photographing copy using a filter.

Filter factors

When a lens filter is used during the exposure, the filter absorbs part of the light from the exposure. This absorption means that a GREATER exposure time is needed with a filter. The number of times an exposure must be increased for each specific filter and film combination is called a *filter factor*. See Fig. 15-17.

TO PHOTOGRAPH AS BLACK (to hold a color) use the film and filter° suggested below:			Color of Copy Being Photo-graphed	TO PHOTOGRAPH AS WHITE (to drop a color) use the film and filter° suggested below:		
Blue-sensitive Film	Orthochromatic Film	Panchromatic Film		Blue-sensitive Film	Orthochromatic Film	Panchromatic Film
Not recommended	Orange (16) Yellow (15, 12) Green (61, 58) Yellow (9, 8)	Green (61, 58)	Magenta (process red)	No filter needed°°	Can try:°° Blue (47B) Magenta (30) Blue (47)	Red (25, 29)°° Magenta (30) Blue (47B)
No filter needed	Blue (47B) Magenta (30) Blue (47)	Blue (47B, 47)	Red or Orange	Not recommended	Not recommended	Red (29, 25, 23A)
No filter needed	Blue (47B) Magenta (30) Blue (47)	Blue (47B, 47)	Yellow	Not recommended	Orange (16) Yellow (15, 12) Green (61, 58) Yellow (9, 8)	Red (29, 25, 23A) Orange (16) Yel (15, 12, 9, 8) Green (61, 58)
No filter needed	Blue (47B) Magenta (30) Blue (47)	Magenta (30) Red (25) Blue (47B, 47)	Green	Not recommended	Orange (16) Yellow (15, 12) Green (61, 58) Yellow (9, 8)	Green (58)
Not recommended	Not recommended	Red (25)	Cyan (process blue)	No filter needed	No filter needed	Blue (47, 47B)
Not recommended	Orange (16) Yellow (15, 12) Green (61, 58) Yellow (9, 8)	Green (58) Red (25)	Blue or Violet	No filter needed	Magenta (30) Blue (47B, 47)	Blue (47B, 47)

Fig. 15-15. This filter and film information is useful when photographing line copy in a color other than usual black on white. (Eastman Kodak Co.)

Suggested Film and Filter Factors

Film	Light Source	6 Light Yellow	8 Yellow	9 Deep Yellow	12 Deep Yellow	15 Deep Yellow	16 Yellow Orange	23A Light Red	25 Red	29 Deep Red	30 Magenta	47 Blue	47B Deep Blue	58 Green	61 Deep Green
							Wratten Filter Number and Color								
Orthochromatic	Pulsed-Xenon	1.5	2		5			This film not sensitive to red light.			6		12	4	
	Tungsten/Quartz	1.4	1.5		3						10		18	3	
	Carbon-Arc	1.5	2		6						5		10	5	
Panchromatic	Pulsed-Xenon		2.5						5				20	10	
	Tungsten/Quartz		2							3.5			38	13	
	Carbon Arc		2.5						5				15	15	

Because of variations in reflectors, copy (image and paper) colors, lamp age, and other conditions, these filter factors are approximate. Accurate factors for individual conditions may be arrived at experimentally, using the above figures as a guide. Keep a record of the results obtained through trials.

Fig. 15-17. Use of filters and increased camera exposure times for various colored line originals is referred to as filter factor. This chart gives filter factors for orthochromatic and panchromatic films. (Eastman Kodak Co.)

For example, a filter factor of 1.5 indicates that the normal or standard exposure for that film and setup should be multiplied by 1.5 because of the filter interference.

To illustrate the use of a filter, consider that your normal exposure time with quartz lighting using ortho line film is ten seconds. You plan to use a No. 12 (deep yellow) filter, which has a filter factor of three (3). To find the new exposure time, simply multiply the basic exposure of ten seconds by the filter factor of three (3) (10 x 3 = 30). The new filtered exposure time is 30 seconds.

It is also possible to use two or more filters in combination. In this case, you must multiply all the individual filter factors together to obtain the combined filter factor. To illustrate, suppose you want to use a No. 30 (magenta) filter and a No. 12 (deep yellow) filter in combination. Your camera is equipped with quartz lighting. Referring to Fig. 15-17, you find that the filter factors are ten (10) and three (3), respectively. To find the filter factor for this combination, multiply 10 x 3 to get a combined filter factor of 30. To find the new exposure time, multiply the factor 30 by the basic exposure time of ten seconds to get the new exposure time of 300 seconds.

Film manufacturers include filter factor information in their packages. Some filter factors will require experimentation. Keep records of all work performed with filters for future reference.

SENSITIVITY GUIDE

A *sensitivity guide* or gray scale should be included along with the copy for every line shot, Fig. 15-18. The *gray scale* is actually an aiming device used during the development stage. The gray scale represents a visual description of the logarithmic density scale. Each step on the scale equals a specific measurable density.

Gray scales are commercially available in several forms. The number of steps is arbitrary. Most camera operators use the 12 or 12-step scales. The opposed gray scale, in Fig. 15-19, is a newer addition to this area.

Stouffer gray scale

The most popular kind of gray scale is the *Stouffer 12-step gray scale*. This scale contains 12 steps in blocks ranging from a neutral white to a dense black. As the exposed negative begins to develop, the number one step fills in (blackens). Within seconds, the number two step fills in and so forth. Generally, the number FOUR STEP is used as an *aim point* for normal black and white line copy development. Look at Fig. 15-20.

Opposed gray scale

The *opposed gray scale* is another device used for quick visual reference for checking exposure/development changes.

The term *"break"* is commonly used to describe the point at which a line negative, containing a gray scale,

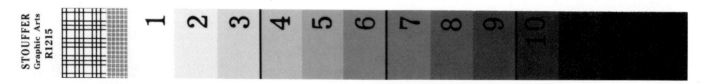

Fig. 15-18. A sensitivity guide or gray scale is included alongside copy for every line shot. Typical opposed gray scale is shown here. (Eastman Kodak Co.)

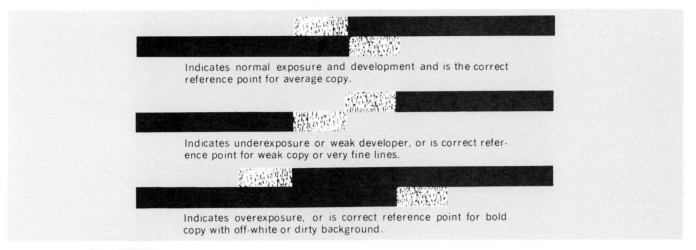

Fig. 15-19. Opposed gray scale gives accurate indication as to normal, underexposed, or overexposed line negatives. (Eastman Kodak Co.)

appears to have lost most of the separation between the steps. It shows a major shift in density. Therefore, the opposed gray scale will look different at the break as exposure is adjusted to suit the kind of copy.

The test exposure that gives the best negative from good line copy will be the proper aim point for a fairly wide range of copy conditions. Because of differences in shop conditions and operations, experience in using the opposed gray scale will soon establish the visual reference that is most suitable. Once the best reference point on the guide has been established for different types of copy and conditions, results can be interpreted quickly, with repeatability and consistency.

The guide should be checked against moderately overexposed or underexposed films during development. This will result in more reliable corrections in development time. When development progresses to the point where the guide shows the proper break, the film should then have a good image. This is because you will have adjusted the development time to compensate for the exposure error. Extreme overexposure or underexposure of the film will NOT respond well to development compensation and the exposure will need to be adjusted.

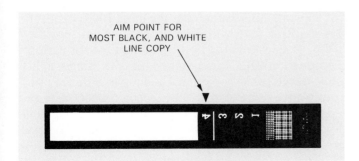

AIM POINT FOR MOST BLACK, AND WHITE LINE COPY

Fig. 15-20. This 12-step gray scale is a popular reference guide used to determine proper development of line negatives. Gray scale aim point for normal black on white line copy is step-4.

Using a gray scale

For normal black and white line copy, the aim point, on a 12-step scale, is generally a step-4, which leaves step-5 gray or with only a few black specks, Fig. 15-20. This is assuming that the previous test procedure was followed carefully, including fresh developer, temperature at 68 degrees F (20 degrees C), 2 3/4 minutes development time, with continuous agitation. Chapter 14 covers the details relating to film processing.

With every exposure, the camera operator places the gray scale on the copyboard next to the material being photographed. It is best placed in an open area near the center of the copy. This portion of the negative will be covered over with masking paper in the stripping operation. If necessary, the gray scale may be included at the edge of the copy to be photographed.

As the film is processed, both the copy image and the gray scale will become visible. As the gray scale becomes darker, it serves as a visual check that indicates the stage of development. After washing, the image of the gray scale serves as a means of judging the film's usability. For normal black and white line copy, a solid step-4 on the gray scale is recommended.

The gray scale can also be used for another special problem area in line photography. It is possible to correct for both the density of the copy and for any reduction or enlargement by varying the development time. The time is just enough to darken the negative to some other specific step on the gray scale.

The chart in Fig. 15-21 illustrates this procedure. First, determine the column best describing the copy density. Locate the row giving the desired reproduction size. This indicates the step to which the film should be developed.

Keep in mind that normal black and white line copy shot at 100% is generally developed to step-4 on a 12-step gray scale. However, a good pencil drawing being reduced in size to 50% of original size would be developed only to step-3. In another example, if bold

CRITICAL STEP CHART

DENSITY OF COPY	SIZE OF COPY		
	10-40%	41-120%	121-400%
Extra Heavy Copy Black Bold type Etching Proofs Photo Proofs	4 Black	5 Black	6 Black
Normal Copy Good black type proofs with fine serifs Pen and Ink Drawings Printed Forms	3 Black	4 Black	5 Black
Light Copy Gray Copy Ordinary Typewritten Sheets Printed Forms Light Lines Good Pencil Drawings	2 Black	3 Black	4 Black
Extra Light Copy Extra Fine Lines* Pencil Drawings Extra light gray copy	1-2 Black	2 Black	3 Black

*Difficult fine line copy and fine line reductions can us ally be improved by fine line development, or still development (not agitated) in regular developer (refer to manufacturer's instructions).

Fig. 15-21. This chart gives information relative to correcting for both density of line copy and for any reduction or enlargement by varying developing time. (Stouffer Graphic Arts Equipment Co.)

type is being enlarged to 200% it will require development to about a step-6. It can be seen that without these exposure corrections, the second example given here would take an extremely long development time.

Adjusted exposure time

It is generally recommended that basic exposure time be decreased for development to less than step-4 (on a 12-step scale), and increased for development to more than step-4. For example, if a step-3 development is recommended, multiply the basic exposure time by .71 (20 x .71 = 14.2 or 14 seconds). This is the adjusted exposure time for the copy. It will be close to that recommended for normal line copy at 100%. This will be true even though the negative is being developed to a different step from the normal step-4.

As in all darkroom processing operations, procedure is the most important factor. Once you have established a procedure for your various kinds of line copy, do NOT deviate. The procedure includes exposure, chemistry temperature, processing, and handling techniques. When certain problems do arise, they can generally be corrected with a minimum of changes in the procedure.

Reductions and enlargements

It is often necessary to reduce or enlarge line copy to make it fit a given area. There are two basic methods

used for setting the camera to make reductions and enlargements. These include: constant aperture exposure method, and constant time exposure method.

Constant aperture exposure method

A high degree of control can be achieved by using the *constant aperture exposure method* because the length of exposure time is varied to fit the desired reproduction size. The lens aperture, called f-number, is held constant for any desired reproduction size. See Fig. 15-22.

To use this method, determine basic exposure time at 100% for the camera you are using. The procedure was described earlier in this chapter. For purposes of this explanation, assume that the basic 100% exposure was made at f/16 with an exposure of ten seconds.

Fig. 15-23 shows the new exposure time for various reproduction sizes. It is assumed that the camera lights are attached to the copyboard and move with it. In this example, assume that a 70 percent reduction is required. Read across the chart from the 70 percent reproduction size to the column for ten seconds basic exposure. The new exposure time is shown as 7 1/4 seconds.

Constant time exposure method

The manual diaphragm control, with its percentage bands, is frequently used with the *constant time ex-*

Fig. 15-22. Normal black and white line copy photographed at 100 percent is generally developed to step-4 on a 12-step gray scale. Pencil drawings and reductions and enlargements require special handling procedures as demonstrated in this illustration.

EXPOSURE TIMES FOR REDUCTIONS AND ENLARGEMENTS USING CONSTANT APERTURE EXPOSURE METHOD

| | | Same-Size Basic Exposure Time | | |
		10 Seconds	20 Seconds	30 Seconds
	300%	40	80	120
	275%	35	70	105½
	250%	30½	61¼	92
	225%	26½	53	79
	200%	22½	45	67½
	175%	19	38	57
	150%	15½	31	47
	125%	12½	25	38
	100%	10	20	30
	95%	9½	19	28½
Reproduction Size Required	90%	9	18	27
	85%	8½	17	25½
	80%	8	16¼	24¼
	75%	7½	15	23
	70%	7¼	14½	21½
	65%	6¾	13½	20½
	60%	6½	12¾	19¼
	55%	6	12	18
	50%	5¾	11¼	17
	45%	5¼	10½	15¾
	40%	5	10	15
	35%	4½	9	13½
	30%	4¼	8½	12¾
	25%	4	8	12
	20%	3½	7¼	10¾

Fig. 15-23. Exposure times for reductions and enlargements using constant aperture exposure method.   (Eastman Kodak Co.)

posure method for making film negatives. Every exposure is made for the same length of time. The lens aperture (f-number) is changed to admit MORE light for an enlargement or LESS light for a reduction.

To use the constant time method, you must know the basic exposure time and best f-number for 100% reproductions. The procedure for determining these figures was described earlier.

The following example is given to illustrate the constant time exposure method. Assume that the basic exposure time is ten seconds at the f/16 diaphragm control percentage band. We also assume that the process camera has a manual diaphragm control and percentage calibrated tapes. The copy is 4 x 8 inches. A reduction of 50 percent is required. The procedure is as follows:

1. Determine the percentage of reproduction. The percentage is generally indicated on the copy. You may be required to determine the percentage using the method described in Chapter 11.
2. The next step requires that you make the correct camera settings for this exposure. Set the timer for ten seconds (basic length of exposure). Set the camera tapes for lensboard and copyboard extensions at 50 percent reproduction size. Set the diaphragm control pointer at 50 percent on f/16.
3. The reproduction size of 50 percent at f/16 will automatically reduce the lens aperture from f/16 to f/22, as shown in Fig. 15-24. The 50 percent reproduction will actually be made at an aperture of f/22. However, the length of exposure will remain constant at ten seconds. An exposure of ten seconds at f/22 admits less light through the lens than an exposure of ten seconds at f/16. As you can see by this example, it takes less light for the 50 percent reproduction.

DIFFICULT COPY

There will be times when you must work with camera copy that is poor quality or in which fine detail is difficult to record. The following types of copy problems require special handling in the camera department.

1. Sometimes, the original copy will have grayed blacks or poor contrast. At normal exposure, clear lines in the negative will be *veiled* (slightly covered). To adjust for or correct this problem, try using slightly less exposure.
2. A problem can occur when the original paper is yellowish. To correct for the yellow, place a wratten filter No. 8 over the lens. Give 2 1/2 times the usual exposure when using arc lights, or 1 1/2 times the exposure when using tungsten lights. These multipliers are called *filter factors*.
3. Fine detail can be difficult to duplicate. To pull up fine detail, try using still development. In still development, the film is allowed to lie perfectly still in the bottom of the tray for about one minute after a normal two minute agitation. Any movement of the film or developer during the still period may produce streaks in the film.
4. Dirty copy can be an annoying problem. Clean up all paste-up copy where possible. This will result in cleaner negatives requiring less opaquing. Erase penciled guidelines and dirt smudges. Cover bad ink spots with white opaque. Never do any of these

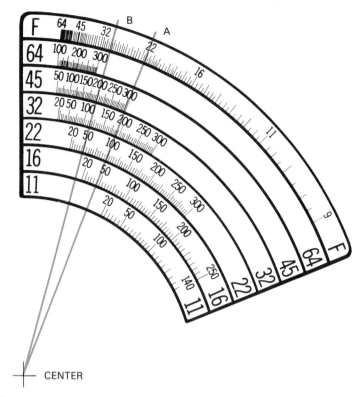

Fig. 15-24. Percentage of reproduction size is determined by setting diaphragm control pointer (shown here), timer, and lensboard and copyboard extensions.

things without permission from the paste-up artist, however.

CAMERA SET-UP AND OPERATION

The following general procedure applies to almost any kind of process camera when preparing to reproduce copy. In daily practice, you should plan ahead as far as possible. Organize all line shots that are to be made, then all halftone shots that are to be made. In addition, you should arrange all the shots in sequence by size. They will include those that are to be shot the same size, reduced, and enlarged.

1. Check the copy and the copyboard glass for dirt, rubber cement, and other debris. Use an approved glass cleaner.
2. Place the gray scale along a trim or non-image area of the copy.
3. Set the lens and copyboard for the desired percentage of reproduction.
4. Set the aperture for correct value, or adjust the diaphragm arm for the percentage of reproduction using the correct aperture scale.
5. Set the timer for the basic exposure, or adjust the basic exposure if the standard aperture method is being used.
6. Set up the processing sink in the following order:
   1. Developer.
   2. Stop bath.
   3. Fixer.
   4. Running water wash.

7. Prepare litho developer, if necessary, by mixing equal parts of A and B. Follow instructions on the chemical containers. If the developer is already mixed, make sure the developer is NOT brown. This indicates that the developer is depleted and needs to be replaced.
8. Prepare the stop bath, if necessary, by mixing one ounce (29.5 mL) of 28 percent glacial acetic acid with each 32 ounces (944 mL) of water.

   SAFETY NOTE! Always wear safety glasses when preparing stop bath from stock acetic acid solution. Always add the acid to the water, never the reverse.
9. Prepare fixer, if necessary, by mixing stock solution, water, and hardener. Follow the instructions on the chemical container.
10. Set up the running water wash.
11. Turn off the white overhead lights and turn on the correct type safelight. Never open boxes of light-sensitive film with white lights on. Instructions in each box of film give safelight requirements.
12. Remove a sheet of ortho film. Rewrap the film, and close the container. Cut the film if necessary so that it is at least one inch (25.4 mm) larger in each dimension than the copy being photographed.
13. Center the film, emulsion-side up, on the vacuum back of the camera. Turn the vacuum pump on.
14. Close the camera back and make the exposure.
15. Remove the exposed film from the vacuum back and place the emulsion-side down in the developer. Push it firmly under the surface. Quickly turn the

piece of film so that the emulsion side is up. Begin agitating the film by lifting or tilting one edge of the tray up approximately two inches (50 mm) every five seconds. Develop the film until a solid black step-4 appears (a partial five will be visible). The developing time should be around 2 3/4 minutes.

16. Remove the film, draining over the developer for two seconds. Place it in the stop bath for ten seconds. Agitate the film for the entire ten seconds.
17. Remove the film, draining over the stop bath for two seconds, and place it in the fixer. Fix for twice as long as it takes for the milky color to clear, usually between two to four minutes with some agitation.
18. Wash the film in the running water bath. The film should remain in the wash for a minimum of ten minutes. Remove the film and squeegee off excess water. Hang the film up to dry or place it in a film dryer.
19. Return film and gray scale to their storage areas.
20. Close the camera back and reset for 100% and turn off the power.
21. Discard the developer if it is depleted. Discard the stop bath. Save the fixer.
22. Clean the sink and trays. Replace all utensils and graduate beakers.
23. Clean up the scrap film and paper towels.

POINTS TO REMEMBER

1. Line photography is a process in which original camera copy composed of a single color or tone is copied on film.
2. There are no gradations of tone in line photography. Line copy images contain no shades of gray as in continuous tone photography.
3. Examples of line copy include typeset composition, inked drawings, block prints, and screened prints.
4. Line copy is generally positioned upside-down on the camera copyboard. This allows the operator to check the copy on the ground glass since it will be right-reading.
5. Film is placed in the camera back with the emulsion side facing the camera lens.
6. When the camera lights are turned on, light rays are reflected from the white areas of the copy. The reflected rays pass through the camera lens and are projected onto the film. The image is formed on the film by the contrast between the white background of the copy and the black characters and inked drawings.
7. White areas of the copy reflect light through the lens of the camera and the light then strikes the film emulsion to expose it. This process results in a line negative.
8. When processed, a negative is dark or opaque in the areas that were white on the original copy. The negative is light or transparent in the areas that were black on the original copy.
9. Sometimes it is necessary to photograph copy in a color other than the usual black on white. In some instances, the line copy is NOT printed on a white background paper. This type of copy calls for a lens filter.
10. When a lens filter is used during the exposure of problem copy, the filter absorbs part of the light from the exposure. As a result, longer exposure times are required when using lens filters.
11. Filter factors are used to determine the exact exposure time required when using various filters.
12. Good line photography results from accurate camera focus and care in film processing.
13. Test exposures are performed to achieve the highest quality line negatives. The tests must be performed under controlled conditions to eliminate variables.
14. It is often necessary to reduce or enlarge line copy to make it fit a given area of the layout.
15. There are two basic methods for setting the camera to make reductions and enlargements. These include the constant aperture exposure method and the constant time exposure method.
16. Gray scales commonly used for line camera work include the Stouffer 12-step gray scale and the opposed gray scale. These aides, when used correctly, assist the camera operator in achieving repeatable and consistent line photography results.

KNOW THESE TERMS

Block print, Filter, Filter factor, Inked drawing, Opposed gray scale, Line copy target, Line photography, Reversal, Scratch board, Screened tint, Sensitivity guide, Stouffer gray scale.

REVIEW QUESTIONS

1. The copying of original images composed entirely of dots, lines, and areas of a single tone is called __________ __________.
2. One of the following is *not* an example of line copy.
   a. Photograph.
   b. Scratch board.
   c. Inked drawing.
   d. Typeset composition.

3. Images, called _____ tints, are backgrounds composed of many evenly spaced dots.
4. A screened paper print is made from an original continuous tone photograph. True or false?
5. How can colored line copy be reproduced on a process camera?
6. Some process camera lens filters are made of gelatin film. True or false?
7. The correct exposure time for a given filter and film combination is known as a:
    a. Filter multiplier.
    b. Filter denominator.
    c. Filter factor.
    d. Filter target.
8. Good line photography results from accruate camera _____, _____ time, and care in film _____.
9. A camera _____ guide is an aiming device used during the developing of film.
10. For normal black and white line copy, the aim point on a 12-step gray scale is generally step _____.

SUGGESTED ACTIVITIES

1. Make a chart using examples of the various kinds of line copy. Label each example and place the chart near the process camera area.
2. Make a chart giving common filter factors. Place the chart near the process camera area.
3. Prepare the required developer, stop bath, fixer, and running water wash for processing line negatives.
4. Expose and process a test negative to determine the correct exposure time on your process camera.
5. Expose and process a line negative using copy supplied by your teacher.

Chapter 16

Halftone Photography

When you have completed the reading and assigned activities related to this chapter, you will be able to:
○ Define what is meant by the term "halftone."
○ Describe the terms highlight, middletone, and shadow—both in negative and positive (printed) form.
○ Identify original copy that should be prepared as a halftone.
○ Compute halftone negative exposures.
○ Calibrate a halftone negative computer.
○ Use a halftone negative computer to determine exposure times for continuous tone copy.
○ Make halftone negatives to faithfully reproduce a variety of continuous tone copy.

Halftone photography enables the printer to print a continuous tone image or photo with a single, uniform density of ink. It is a special photographic procedure used almost exclusively by the printing and publishing industry. The printed image looks as though it had been printed with intermediate tones (densities) ranging from white to black. See Fig. 16-1.

HALFTONE THEORY

Camera copy that has intermediate densities or tones of gray is referred to as *continuous tone copy*. Photographs and wash drawings are examples of continuous tone copy, Fig. 16-2. Continuous tone photographs have variations of tones of gray that blend together to form a picture.

To print a reproduction of a continuous tone photograph, the printer must break the image into *halftone dots* that bear the ink. These include large dots covering more area and representing *dark tones,* small dots covering less area and representing *light tones,* and medium-sized dots of all sizes representing other *middle tones*. Although the dot sizes are different, all the dots have the same density so that the printer can print them on an offset press. When the continuous tone copy is printed, the dots and the white (or light) paper

Fig. 16-1. In a printed halftone, dots of varying sizes and white paper fuse visually to give illusion of a continuous array of tones. (Interlake, Inc.)

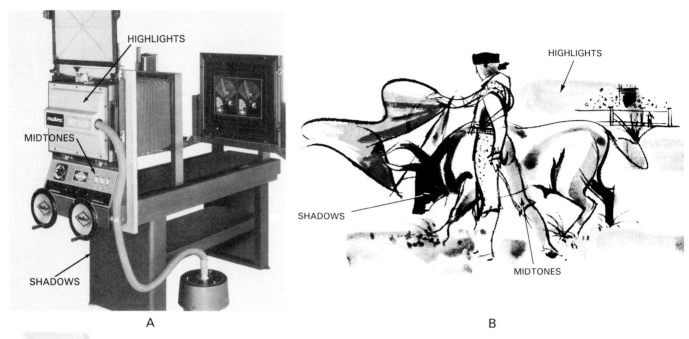

Fig. 16-2. A—A photograph is an example of continuous tone copy. B—A wash drawing is handled as continuous tone copy. (nuArc Company, Inc.; Eastman Kodak Co.)

fuse visually and give the illusion of a continuous array of tones.

Look at the halftone print in Fig. 16-3. It has been enlarged to show you the highlights, middletones, and shadow areas. Middletones are equal-size dots with the background, or mid-size dots. Shadow dots are the largest black dots in and around the eye. In making a negative of such copy, the camera operator must produce a film image made up of dots of various sizes. This is called a *halftone image.*

Halftone photography makes the printing of continuous tone photographs, paintings, and drawings possible. This is done by converting the continuous tone copy into a pattern of clearly defined dots of varying sizes. This is necessary because the offset press cannot lay down varying tones of ink. However, the printing plate can distribute ink onto paper as varying areas. If the areas or dots are small enough, the eye will not detect the dot pattern but will see the mixture of black ink dots and unprinted specks of white paper as gray tones, Fig. 16-4. In this example, it is assumed that the images are printed in black ink on white paper.

HALFTONE SCREENS

The major difference between line photography and halftone photography is the use of the halftone screen. When a halftone negative is made from continuous tone copy, a *halftone screen* is placed in the light path between the process camera lens and a sheet of film in a process camera. A continuous tone image can also be screened into a halftone on a scanner.

There are two classifications of halftone screens. They include: glass screens and contact screens. *Glass screens* are used primarily in photoengraving and,

therefore, are not covered in this book. Contact screens are used in offset lithography. For now, it is sufficient to say that these screens have the same basic function. *Screens* convert the tones of the original continuous tone copy into solid dots of equal density but varying sizes.

As you will recall from Chapter 15, when a line negative is made, the black and white line copy is exposed directly to a sheet of film in a process camera WITHOUT a screen. The white areas of the copy reflect the most light through the camera lens to the film. This produces solid black areas when the negative is processed. The dark areas of the copy reflect little light to the film, producing clear areas on the processed negative.

In similar fashion, the whitest or highlight areas of the continuous tone copy reflect the most light to, and through, the halftone screen. This produces broad highlight dots covering most of the area. This is illustrated in Fig. 16-5. The dark or shadow areas of the continuous tone copy reflect the least light and produce the small shadow dots shown as five percent dots. As the intensity of the reflected light increases from shadows to highlights, the intermediate tones of the copy produce intermediate dot sizes.

It should be pointed out that dot percentage always refers to the size of the black dots. An area in the negative containing 90 percent dots (90 percent of area black and 10 percent clear) is a highlight area. This area produces a ten percent black dot in the positive or print.

Contact screens

A contact screen is a precision pattern of vignetted (diffused) dots on a flexible support such as polyester.

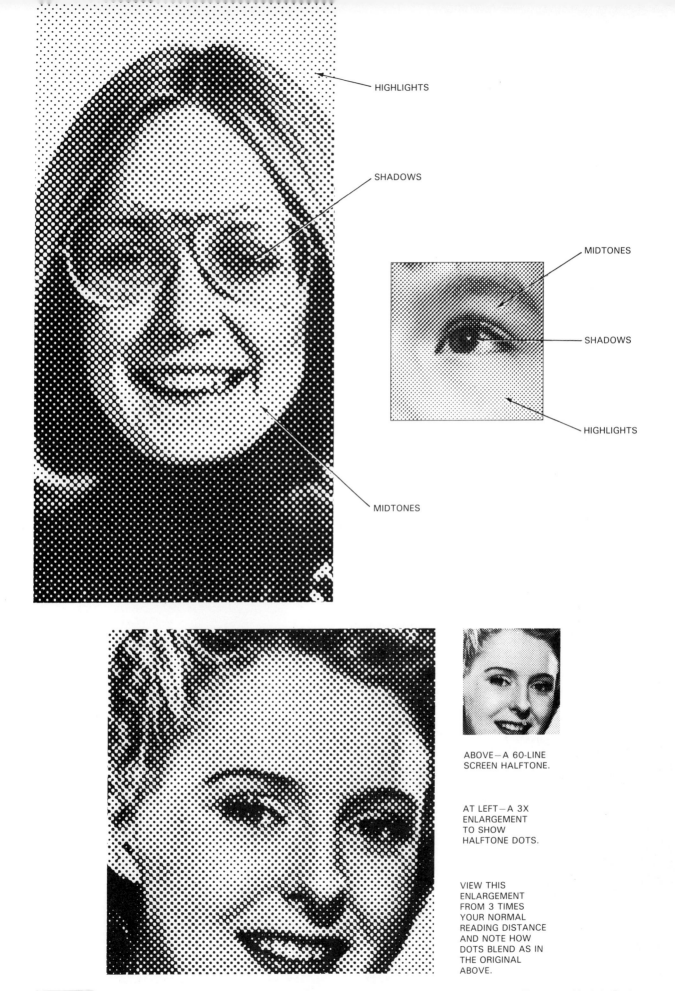

HIGHLIGHTS

SHADOWS

MIDTONES

SHADOWS

HIGHLIGHTS

MIDTONES

ABOVE—A 60-LINE SCREEN HALFTONE.

AT LEFT—A 3X ENLARGEMENT TO SHOW HALFTONE DOTS.

VIEW THIS ENLARGEMENT FROM 3 TIMES YOUR NORMAL READING DISTANCE AND NOTE HOW DOTS BLEND AS IN THE ORIGINAL ABOVE.

Fig. 16-3. Note halftone reproduction showing highlight, midtone, and shadow areas.   (Eastman Kodak Co.)

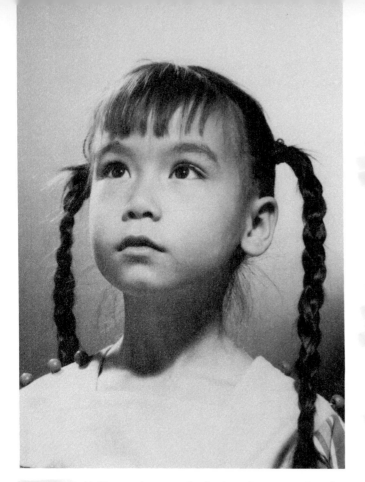

Fig. 16-4. Halftone photography is done by converting the continuous tone copy into a pattern of clearly defined dots of varying sizes.

Looking at the screen with a magnifier, you would see a soft, out-of-focus pattern of diffused dots and corresponding spaces. Look at Fig. 16-6.

As the name implies, a *contact screen* is used in direct contact with the light sensitive film, Fig. 16-7. Because it is essential that this be the closest possible contact, a vacuum set-up is used, Fig. 16-8. To provide a good vacuum seal, the screen must be at least one-half inch (12.7 mm) larger on all four edges than the film being exposed.

Types of contact screens

Contact screens are used for making halftone negatives on a process camera for black and white reproduction from reflection copy. They are also used for making halftone positives for indirect color separation photography. This is done on a process camera or in a contact printing frame. Contact screens are also used for making direct-screen color separation halftone negatives on a camera, in a contact printing frame, or with an enlarger. Contact screens can also be mounted on a scanner to produce halftones.

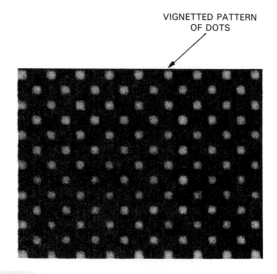

Fig. 16-6. Contact screen is a precision pattern of vignetted or diffused dots on a flexible support, such as polyester. Vignetted area appears as a soft, out-of-focus pattern of diffused dots and corresponding spaces. (Eastman Kodak Co.)

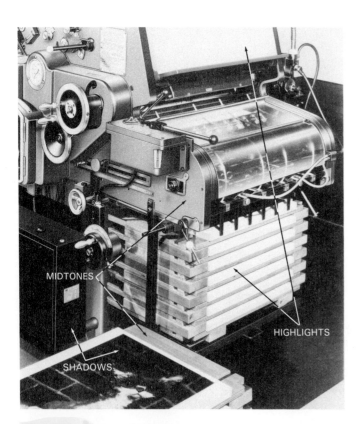

Fig. 16-5. Whitest or lightest areas of a continuous tone original reflect most light to, and through, halftone screen. This produces broad highlight dots covering most of the print. (Heidelberg U.S.A.)

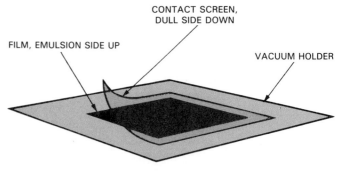

Fig. 16-7. A contact screen is placed in direct contact with light sensitive film. (Eastman Kodak Co.)

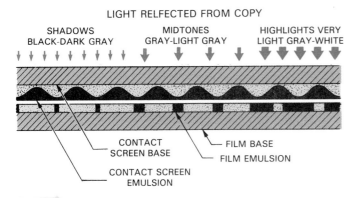

LIGHT RELFECTED FROM COPY

SHADOWS
BLACK-DARK GRAY

MIDTONES
GRAY-LIGHT GRAY

HIGHLIGHTS VERY
LIGHT GRAY-WHITE

CONTACT
SCREEN BASE

FILM BASE

FILM EMULSION

CONTACT SCREEN
EMULSION

Fig. 16-8. This shows set-up for a contact screen and unexposed film in direct contact on process camera back.

There are many types of contact screens used for making halftone negatives in offset lithography. Each type of screen fits some requirements better than others. The types of screens covered here are available from most graphic arts supply firms.

Elliptical and square dot screens

The primary difference between an elliptical dot screen and a conventional square dot screen is found in the middletones of a halftone reproduction.

The *elliptical dot screen* produces diamond-shaped midtone dots that join only two opposite corners as the dots reach 50 percent size. Refer to Fig. 16-9.

A *conventional dot screen* produces square midtone dots that join all four corners as the dots reach 50 percent size. See Fig. 16-10. When the four corners of the

Fig. 16-9. Elliptical dot halftone screen produces diamond-shaped midtone dots. (Eastman Kodak Co.)

Fig. 16-10. Conventional square dot halftone screen produces square midtone dots. (Eastman Kodak Co.)

square dot join, the density of that midtone area jumps slightly, resulting in a small tone break across the 50 percent dot.

A *tone break* refers to an uneven blend of the dots in the 50 percent region. Because only two corners join between elliptical dots, the tone gradation across the 50 percent dots is smoother.

Fig. 16-11 illustrates a reproduction of halftones made with a square dot screen and elliptical dot screen.

The elliptical dot screen should be used for copy with SOFT midtone vignetting, such as flesh tones. In the absence of soft midtone gradation, either the square dot or elliptical dot contact screens work well.

Magenta screens

As the name implies, the *magenta screen* is a purplish shade of red. Since the screen is colored, it acts as a filter on the film. The filtering action usually limits its use to black and white original copy. An enlarged section of a magenta screen showing the dot pattern is illustrated in Fig. 16-12.

SQUARE DOT

ELLIPTICAL DOT

Fig. 16-11. Examples of square dot and elliptical dot halftone prints. Magnification of dot shapes shows that elliptical dots allows a much smoother tonal gradation than square dot in midtone areas. (Chemco, Inc.)

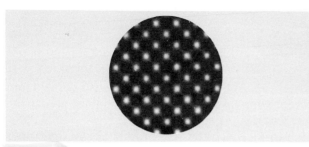

Fig. 16-12. Enlarged section of a magenta halftone contact screen showing detailed dot pattern. (Eastman Kodak Co.)

Magenta screens are used in the case of indirect color separation since it requires separate steps. Magenta and yellow compensating filters can be used to improve tone reproduction.

Gray screens

The *gray screen* is used in tone reproduction processes where a magenta dye offers no advantage or if it offers a disadvantage. Being neutral gray in color, it can be used to photograph colored original copy.

For direct screen color separation, the halftone screening takes place at the same time as the actual color separating. In addition, color filters are used to separate the printing colors. Any further color added to the process by the contact screen will only distort the effect of the color separation filters. Therefore, gray contact screens are used for direct screen color separation.

For making halftone negatives from black and white copy for black and white reproduction, either magenta or gray contact screens work well. It should be noted that the gray screen does NOT have the range of tonal contrast of a magenta screen.

PMT gray contact screen

The *Photo Mechanical Transfer* (PMT) *gray contact screen* is used for making screened paper prints. The screened prints are often combined with reflection line copy for paste-up on a mechanical. This screen has an extra-long screen range to maintain tonal separation and detail in the shadow areas of a screened print.

The PMT gray contact screen comes in coarse screen rulings (65, 85, and 100 lines per inch) which allows for halftone printing on uncoated and rough-textured papers. A finer screen, with 133 lines per inch, is available for printing halftones on smoother grades of paper.

Negative and positive screens

Either negative or positive contact screens can be used to make halftones.

Negative screens (magenta or gray), are most often used to make halftone negatives. Negative screens are made with a built-in highlight exposure. This assures good tonal separation with normal continuous tone copy.

Positive screens are required to make halftone positives for indirect color separation. They are also required when making halftones in the diffusion transfer process.

CARE OF CONTACT SCREENS

Contact screens must be protected from scratches, dirt, and liquid stains. Reasonable care will prolong the useful life of a contact screen. Follow these guidelines when working with contact screens:

1. Keep the screen flat and in its protective folder when it is NOT in use!
2. Handle the screen by the edges only! One way to protect the screen is to mount it on a larger size piece of plastic sheeting, Fig. 16-13. This plastic sheet should be about two inches (50 mm) wider and longer than the screen itself. Cut a window in the sheet. Insert the screen, and tape all four edges to prevent air from leaking out around the edges. You can now handle the screen without fear of fingerprinting its surface.

   Each side of the screen can be identified for easy handling. Information such as emulsion side, date purchased, screen ruling, type of dot, kind of screen, etc. This kind of information is especially useful for the camera operator.
3. Dust the screen when necessary by wiping it lightly with a clean, dry chamois. Do not use a camel-hair brush on the screen at any time. It will cause tiny scratches that can affect reproduction quality.
4. Do NOT allow the screen to come in contact with water or any chemical solution. If necessary, the screen can be cleaned with a soft fabric pad lightly moistened with film cleaner. After applying the cleaner, and before it is completely dry, wipe the screen surface lightly with a soft, dry cloth.
5. If the screen becomes spotted with water-soluble substances which cannot be removed with film cleaner, immerse the screen in a tray of distilled water at about 80 degrees F (27 degrees C). Sponge it with a wad of clean cotton.

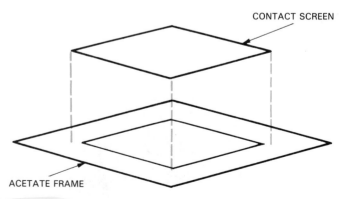

CONTACT SCREEN

ACETATE FRAME

Fig. 16-13. Halftone screen can be protected from damage through constant handling by mounting screen on a larger size piece of plastic sheeting. (Eastman Kodak Co.)

Be extremely careful with magenta screens. If any contaminants get into the water or if you wash the screen more than 15 minutes, some of the magenta dye may be removed from the screen.

Following the water wash, immerse the screen in a dilute photo-flo solution to prevent water spotting. Hang the screen to dry in dust-free air at room temperature.

HALFTONE NEGATIVE CHARACTERISTICS

There are three main characteristics to be considered when evaluating a halftone negative. These include: highlights, middletones, and shadow areas.

Highlights

The *highlights* are the darkest part of the negative and consist of small, clear openings in an otherwise solid area. See Fig. 16-14.

The highlights on the negative should be approximately 95 percent dots in order to produce five percent highlight dots on the printed sheet. The exact size that the highlight dot must attain cannot be specified. It will vary with individual shop conditions and with specific paper stocks. What is wanted is the smallest dot that will be solid enough to be carried through platemaking and printing.

The printed image of the highlights will show the area as small, black ten percent dots in a large white area. Look at Fig. 16-15.

Midtones

The halftone dots in the range from 30 to 70 percent comprise the *midtones* of the negative. Fig. 16-16. This is the area where the small black dots are increasing in size until they connect to form the 50 percent dots. At this point, the clear, open areas are equal in size to the black dots and form a checkerboard pattern. Mentioned briefly, the 50 percent area is less obvious with an elliptical dot screen. Between 50 and 70

percent, the open areas decrease in size as the connected black dots increase in size.

The reproduction of light to medium gray tones represent the *middle range*. It is important that the dots here change smoothly and gradually from one dot size to another. In the midtones, minor changes in dot size cause obvious changes in the appearance of the printed reproduction. This change is more obvious than equivalent variations in the highlight or shadow areas. Tone reproduction in the midtones is critical.

Sometimes, the continuous tone copy will have a tone in the 50 percent range, blending smoothly into another tone. Here it is likely that the four corners of each 50 percent dot will join all along the line of blend. This may cause a sudden jump in density instead of a smooth transition. It was this problem which brought about the development of the elliptical dot screen. A 50 percent elliptical dot joins only two corners instead of four, Fig. 16-17. This reduces the harshness of the

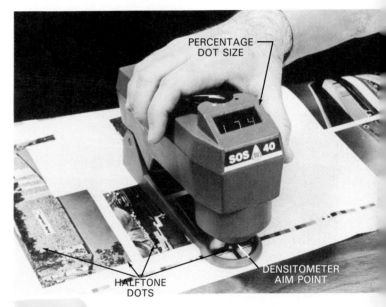

Fig. 16-15. Printed halftone dots are measured in size using a reflection densitometer. Digital readout measures dot size in percentage. (Cosar Corp.)

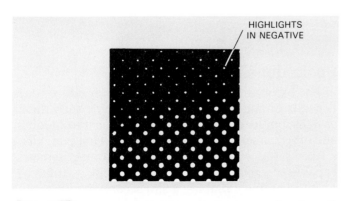

Fig. 16-14. Highlights in a halftone negative consist of small, clear openings in an otherwise solid black area. (Eastman Kodak Co.)

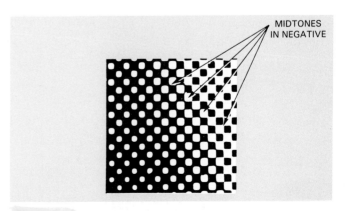

Fig. 16-16. Midtones in a halftone negative consist of 30 to 70 percent dots. (Eastman Kodak Co.)

333

Fig. 16-17. In an elliptical dot screen, 50 percent dots join at two corners instead of four as in conventional square dot screen. This reduces harshness of break in tonal value. Note 50 percent dots in this magnified section of a halftone print. (Eastman Kodak Co.)

break in tonal value. An elliptical dot screen is recommended for use with copy that has subtle blends in the midtones.

Shadows

The images of least density in the halftone negative are the *shadow areas*. Shadow dots range from tiny, pinpoint black dots to a dot value of about 30 percent, Fig. 16-18. Shadow dots print as 70 percent to 99 percent in the reproduction image. As in the case of the highlight dot, it is impossible to specify the exact size that the shadow dot should attain. Dot size is dependent on the individual printing conditions and the paper stock.

The blackest shadows of the reproduction, where little detail is necessary, are often allowed to print as solids. Areas that are slightly less dense, when detail is desired, should have the smallest dot that can be held open on the printed sheet. The size of this open dot on the printed sheet determines the size of the shadow dot needed on the halftone negative.

To determine the *just-printable dot,* make a series of test halftone negatives of several dark steps of a gray scale. Prepare a printing plate of these steps, and proof

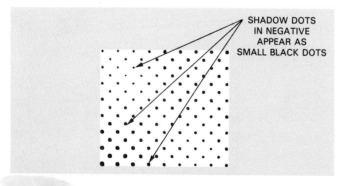

Fig. 16-18. Shadow dots in a halftone negative range in size from tiny, pinpoint black dots to a dot value of about 30 percent. Shadow dots print as 70 percent to 99 percent in reproduction image. (Eastman Kodak Co.)

it on both coated and uncoated paper. The proof will show the smallest dot that can be held on your equipment.

Tonal range

Original copy for halftone reproduction represents a *tonal range* from the whitest highlights through varied grays, to the black of the shadows. A good black and white photograph has a range of densities from paper white to a 1.7 density black or slightly darker. The entire tonal range of the original can seldom be reproduced on the printed sheet. Solid printing press black ink yields a density of about 1.40 to 1.60 (compared to a solid 2.0 black) on coated paper and usually less on uncoated papers.

The loss of some of the tones of the original may occur in the highlights, in the shadows, or as a fairly uniform compression (reduction) of the tonal range.

The usual alternative is to reproduce the highlights through middletones as accurately as possible and accept the necessary loss of detail in the shadows. In the process, the highlight, middletone, and shadow areas are controlled by the use of multiple exposures. A *main* (detail) *exposure* is always used. A *shadow* (flash) *exposure* is USUALLY required. A *no-screen* (bump) *exposure* is used where additional highlight contrast is desired.

Tone control of contact screens

Different types of contact screens vary in the way they control the tonal range of the negative. Therefore, a more detailed discussion of screen characteristics should prove helpful.

It is common to discuss highlight, midtone, and shadow areas of a halftone as though they were interdependent of each other. This mistake is often made in thinking that each of these elements can be manipulated without affecting each other. This is NOT true.

The flash and no-screen exposures affect one part of a tone scale more than another part. However, the whole scale is affected, with the relative amount of evfect being determined by the characteristics of the contact screen. A contact screen is limited as to the range or amount of the original continuous tone copy it can convert to tones.

Basic density range (BDR)

The ability of a contact screen to reproduce tones depends on many factors. These include such things as light source, exposure and development techniques, and other camera-darkroom variables. However, given a fixed set of working conditions, a contact screen is capable of reproducing a specific range of tones.

Disregarding variables, a single camera exposure of a gray scale through the contact screen to a sheet of film will result in a measure of the screen's specific range of tones. The tonal range capability of the contact screen is called the *basic density range* (BDR), or *screen range*.

The following procedure is recommended to determine basic density range of a contact screen:

1. Place a calibrated gray scale in the camera copyboard, Fig. 16-19. Set the controls for making a same-size (100%) reproduction.
2. Cover a sheet of orthochromatic film with a selected contact screen, emulsion to emulsion.
3. Expose a halftone negative of the gray scale. Try an exposure of 30 seconds at f/16 if two 1500-watt, pulsed-xenon lamps are set four feet from the copy.
4. Process the film and evaluate the resulting test negative. Refer to Chapter 14 for film processing procedure.

On the test negative, locate the halftone dots that represent typical highlight and shadow dots as you would want them in the highlights and shadows of your halftone negatives. Fig. 16-20 illustrates a test negative of a reflection density guide. If this is similar to your negative, you might select the dots in the 0.20 step as typical highlight dots, and the dots in the 1.30 step as typical shadow dots.

Using this test and these numbers as examples, the BDR of that contact screen is 1.30 minus 0.20, or a 1.10 range. Had the gray scale been a continuous tone photograph, all tones lighter than 0.20 density in the photo would have filled in solid. All the tones above 1.30 in density would have gone clear.

To carry the example one step further, say the continuous tone photo had been a full-range photo, with a highlight density of 0.05 and a shadow density of 1.80 (total copy density range of 1.75). This would result in the equivalent of 0.65 missing copy density range (1.75 minus 1.10 BDR = 0.65). For all of the copy density range to be reproduced, the basic density range must be extended until it equals the copy density range. This is done by the use of a *flash exposure* or by using a longer range screen.

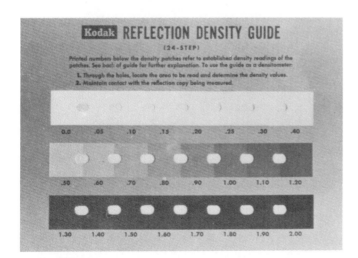

Fig. 16-19. A reflection density guide is a calibrated gray scale used for halftone photography. (Eastman Kodak Co.)

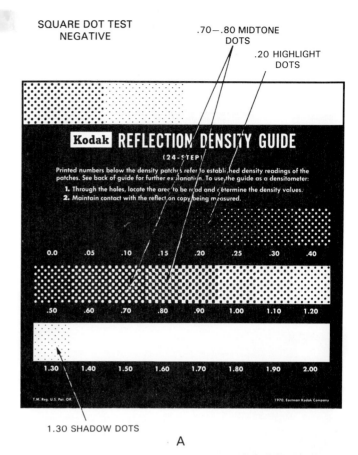

A

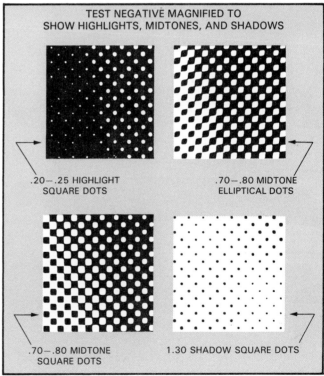

B

Fig. 16-20. A—Example of a test negative of a reflection density guide, indicating desired dot sizes for highlight, middletone, and shadow dots. B—Magnified portions of test negative of a reflection density guide comparing square dot and elliptical dot middletone structure. (Eastman Kodak Co.)

There is a basic rule applied in the previous and subsequent examples. For the halftone negative to represent all of the tones in the copy, the basic density range must be modified to equal the density range of the copy. To lengthen the BDR, use a flash exposure; to shorten it, add a no-screen exposure. In most cases, the BDR falls short of the copy range, requiring a flash exposure.

Following this rule, if the BDR is exactly equal to the copy density range (CDR), you can reproduce the copy with a main exposure alone. The example below illustrates this rule:

$$\begin{array}{l} 1.10 \;\; \text{BDR} \\ \underline{1.10} \;\; \text{Copy Density Range (CDR)} \\ 0 \end{array}$$

The screen can accommodate the range of the copy without any additional exposures in the example above.

If the range of the contact screen is actually longer than the copy range, then you must modify the BDR downward until it equals or is lower than the copy range. The screen range actually exceeds the copy range. The BDR must be reduced if the screen is able to reproduce all tones in the copy. This is done with a *no-screen* (bump) *exposure.*

The no-screen exposure needs only to be long enough to reduce the BDR, but you need not equalize the BDR and CDR exactly. The flash exposure increases the BDR, and is much more easily controlled than the no-screen exposure. Using the previous example:

$$\begin{array}{l} 1.10 \;\; \text{BDR} \\ \underline{1.00} \;\; \text{CDR} \\ 0.10 \;\; \text{(excess BDR)} \end{array}$$

The remedy for this problem is a no-screen (bump) exposure, which drops the BDR below the copy range.

$$\begin{array}{l} 0.90 \;\; \text{new BDR (after bump)} \\ \underline{1.00} \;\; \text{CDR} \\ 0.10 \;\; \text{(excess CDR)} \end{array}$$

The solution for this is a short flash exposure, which increases the BDR to equal the copy range.

Using a halftone calculator

A device used for calculating halftone exposures is the halftone calculator, Fig. 16-21. The basic principle of this calculator applies only to reproducing copy as is, or as you see it. The calculator is extremely accurate if calibrations and procedures are done with control and if exposures and processing are handled carefully. The greater the control of variables, the greater the accuracy of the calculator.

The steps in setting up a typical halftone calculator are illustrated in Fig. 16-22. For this procedure, use orthochromatic film and fresh chemistry. Refer to Chapter 14 for film processing instructions.

When the steps in Fig. 16-22 have been completed, record the following information:

a. The density of the original gray scale that produced a printable highlight dot with main exposure only.

b. The density of the original gray scale that produced a printable shadow dot with main exposure only.

c. The density of the original gray scale that produced a printable highlight dot with no-screen plus main exposure.

d. The exposure time that produced a printable shadow dot on the stepped-off flash test negative (same size as dot in b).

e. The exposure time for the main exposure.

Setting the calculator

Pick up the calculator and practice with the dials to see how they function, Fig. 16-23. For a quick start, leave 0.0 at about twelve o'clock; leave A pointing at one o'clock; leave B pointing at eight o'clock; leave C at about four o'clock; and leave D at about five o'clock.

The steps in setting the halftone calculator are illustrated in Fig. 16-24. When you have finished calibrating the calculator, it is ready for production. This calibration will remain unchanged as long as your production conditions remain unchanged. However, if you make a change (new contact screen, different film, new lights on camera, etc.), you must recalibrate the calculator.

Using the calculator for production

When using the calculator for production, read all halftone copy with a densitometer, Fig. 16-25. This assures correct highlight and shadow densities. If you

Fig. 16-21. A halftone calculator can be used to determine halftone exposures on a process camera. Calculator is accurate if calibrations, procedures, and processing are done with care and consistency. (Eastman Kodak Co.)

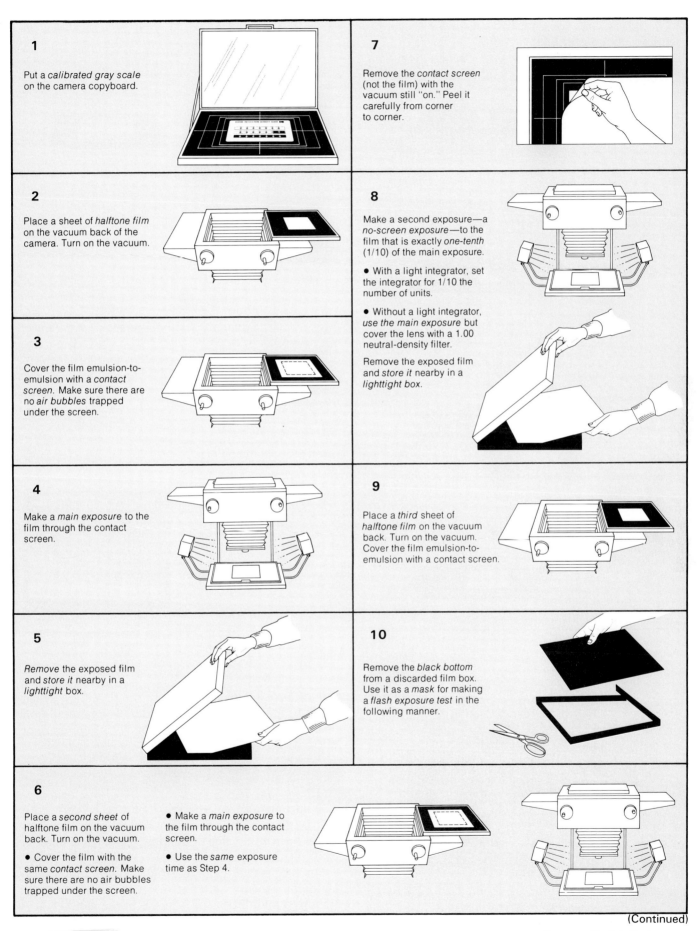

1

Put a *calibrated gray scale* on the camera copyboard.

2

Place a sheet of *halftone film* on the vacuum back of the camera. Turn on the vacuum.

3

Cover the film emulsion-to-emulsion with a *contact screen*. Make sure there are no *air bubbles* trapped under the screen.

4

Make a *main exposure* to the film through the contact screen.

5

Remove the exposed film and *store it* nearby in a *lighttight* box.

6

Place a *second sheet* of halftone film on the vacuum back. Turn on the vacuum.

• Cover the film with the same *contact screen*. Make sure there are no air bubbles trapped under the screen.

• Make a *main exposure* to the film through the contact screen.

• Use the *same* exposure time as Step 4.

7

Remove the *contact screen* (not the film) with the vacuum still "on." Peel it carefully from corner to corner.

8

Make a second exposure—a *no-screen exposure*—to the film that is exactly *one-tenth* (1/10) of the main exposure.

• With a light integrator, set the integrator for 1/10 the number of units.

• Without a light integrator, *use the main exposure* but cover the lens with a 1.00 neutral-density filter.

Remove the exposed film and *store it* nearby in a *lighttight box.*

9

Place a *third* sheet of *halftone film* on the vacuum back. Turn on the vacuum. Cover the film emulsion-to-emulsion with a contact screen.

10

Remove the *black bottom* from a discarded film box. Use it as a *mask* for making a *flash exposure test* in the following manner.

(Continued)

Fig. 16-22. Study steps for setting up a halftone calculator using orthochromatic film. (Eastman Kodak Co.)

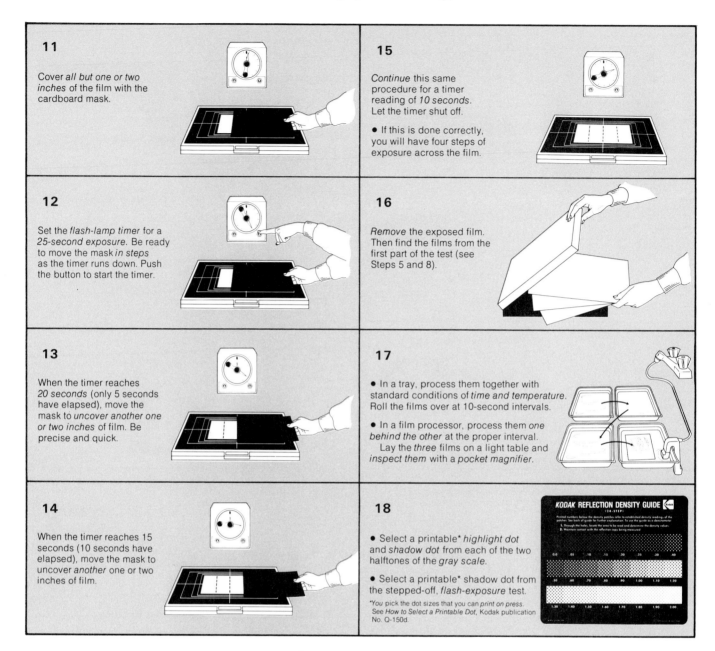

11

Cover *all but one or two inches* of the film with the cardboard mask.

12

Set the *flash-lamp timer* for a *25-second exposure*. Be ready to move the mask *in steps* as the timer runs down. Push the button to start the timer.

13

When the timer reaches *20 seconds* (only 5 seconds have elapsed), move the mask to *uncover another one or two inches* of film. Be precise and quick.

14

When the timer reaches 15 seconds (10 seconds have elapsed), move the mask to uncover *another* one or two inches of film.

15

Continue this same procedure for a timer reading of *10 seconds*. Let the timer shut off.

• If this is done correctly, you will have four steps of exposure across the film.

16

Remove the exposed film. Then find the films from the first part of the test (see Steps 5 and 8).

17

• In a tray, process them together with standard conditions of *time and temperature*. Roll the films over at 10-second intervals.

• In a film processor, process them *one behind the other* at the proper interval. Lay the *three* films on a light table and *inspect them* with a *pocket magnifier*.

18

• Select a printable* *highlight dot* and *shadow dot* from each of the two halftones of the *gray scale*.

• Select a printable* shadow dot from the stepped-off, *flash-exposure* test.

*You pick the dot sizes that you can *print on press*. See *How to Select a Printable Dot*, Kodak publication No. Q-150d.

KODAK REFLECTION DENSITY GUIDE

Fig. 16-22. Continued.

Fig. 16-23. A halftone calculator must be calibrated before using it for production work. Calibration will remain unchanged as long as production conditions remain unchanged. (Eastman Kodak Co.)

do not have a densitometer, use the reflection density guide to measure or read densities. It has small holes drilled in the density patches for visual comparison.

Mark the density readings on the back or outside border of each piece of continuous tone copy. Do NOT mark in the copy area or indent the copy on the back side. See Fig. 16-26.

When positioning the copy on the camera copyboard, place a reflection gray scale next to the copy. It will be helpful when evaluating the halftone negative. Look at Fig. 16-27.

On the calculator, Fig. 16-28, set A at the copy highlight density. Set D at the copy shadow density. Required main exposure appears in the window at A. Required flash exposure appears behind the hairline of D.

STEP 1
MOVE POINT A TO DENSITY OF THE
HIGHLIGHT DOT (DENSITY OF ORIG-
INAL GRAY SCALE THAT PRODUCED
A PRINTABLE HIGHLIGHT DOT WITH
MAIN EXPOSURE ONLY).

STEP 2
MOVE TAB B SO THAT THE EX-
POSURE TIME (FROM EXPOSURE
TIME FOR MAIN EXPOSURE) APPEARS
IN WINDOW AT A.

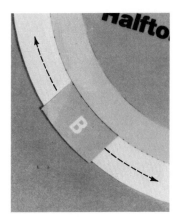

STEP 3
ROTATE C TO DENSITY OF SHADOW
DOT (DENSITY OF ORIGINAL GRAY
SCALE THAT PRODUCED A PRINT-
ABLE SHADOW DOT WITH MAIN
EXPOSURE ONLY).

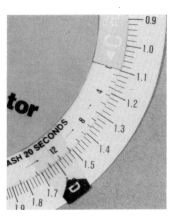

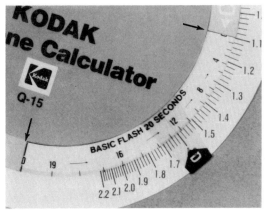

STEP 4
ROTATE BASIC FLASH DIAL (CLEAR
DIAL) UNTIL FLASH EXPOSURE
TIME (EXPOSURE TIME THAT
PRODUCED A PRINTABLE SHADOW
DOT ON STEPPED-OFF "FLASH"
NEGATIVE) IS CENTERED IN SLOT.

STEP 5
CHECK ALL SETTINGS TO MAKE SURE
THAT THEY HAVE CORRECT IN-
FORMATION. TAPE ALL THREE DIALS
TOGETHER AT POINTS SHOWN HERE.
IN ADDITION, TAPE TAB B TO
YELLOW BASE.

STEP 6
ON YELLOW DIAL, LOCATE DENSITY
OF HIGHLIGHT DOT (DENSITY OF
ORIGINAL GRAY SCALE THAT
PRODUCED A PRINTABLE HIGH-
LIGHT DOT WITH NO-SCREEN AND
MAIN EXPOSURE). MAKE A MARK
ON RED DIAL OPPOSITE THAT
DENSITY. LABEL THE MARK 10%.

Fig. 16-24. Learn six steps in setting up a halftone calculator for production work.   (Eastman Kodak Co.)

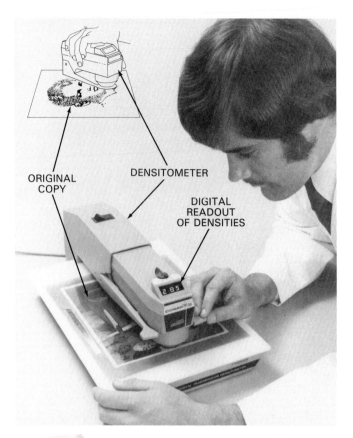

Fig. 16-25. Densities of original halftone copy are read with a densitometer. This assures correct highlight and shadow densities. (Cosar Corp.)

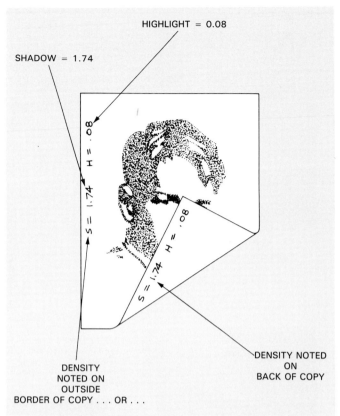

SHADOW = 1.74

HIGHLIGHT = 0.08

DENSITY NOTED ON OUTSIDE BORDER OF COPY . . . OR . . .

DENSITY NOTED ON BACK OF COPY

Fig. 16-26. Density readings should be noted on back or outside border of each piece of continuous tone copy. (Eastman Kodak Co.)

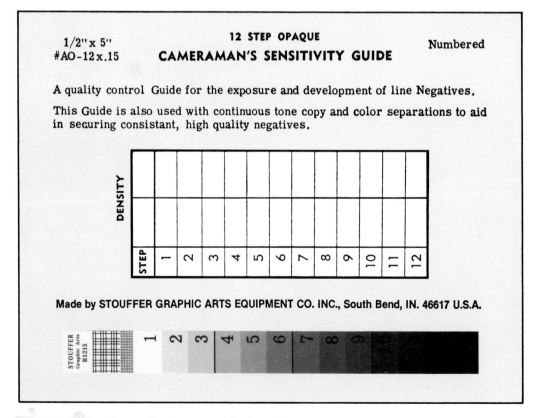

Fig. 16-27. Position a reflection gray scale alongside continuous tone copy on camera copyboard. (Stouffer Graphics Arts Equipment Co.)

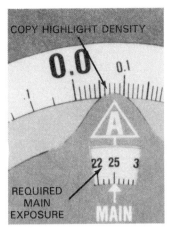 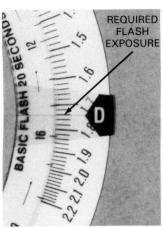

Fig. 16-28. In using calculator for halftone reproduction, set "A" at copy highlight density. Set "D" at copy shadow density. Required main exposure appears in window at "A." Flash exposure appears behind hairline of "D." (Eastman Kodak Co.)

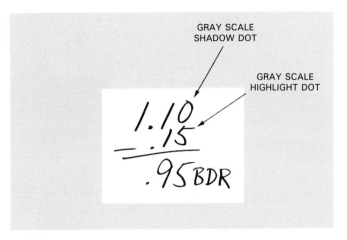

Fig. 16-30. Example shows how BDR is calculated.

If the hairline of D, Fig. 16-29, does not appear in the window, then use the 10 percent mark instead of the pointer for A. Add a ten percent no-screen (bump) exposure to the main exposure sequence. Refer to step 8, Fig. 16-22. You will now be able to use D.

Go back to the point where you recorded the computations from steps 1 through 18. Subtract the density of the original gray scale highlight dot (a) from the density of the original gray scale shadow dot (b). The result is the basic density range (BDR). This is illustrated in the example in Fig. 16-30.

If the BDR is between 1.00 and 1.40, the calculator will work for all halftones without further manipulation. However, if the BDR is between 0.80 and 1.00, use the small red arrows in the basic flash slot.

After the calculator is set for the exposure, move the hairline of D to the point of the nearest arrow, Fig. 16-31. Rotate the calculator counterclockwise the length

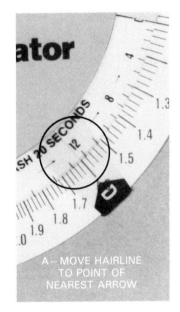 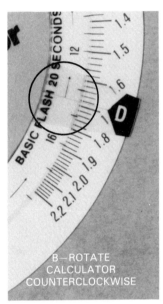

A—MOVE HAIRLINE TO POINT OF NEAREST ARROW

B—ROTATE CALCULATOR COUNTERCLOCKWISE

 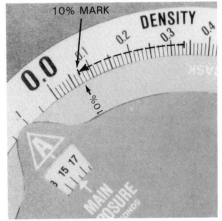

Fig. 16-29. A no-screen (bump) exposure may be required on some copy. This is usually calculated at ten percent of main exposure time. Refer to step 8 in Fig. 16-22. (Eastman Kodak Co.)

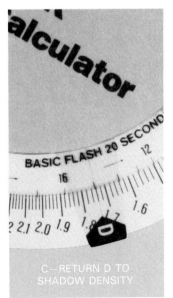 

C—RETURN D TO SHADOW DENSITY

D—READ NEW MAIN AND FLASH EXPOSURES AT A

Fig. 16-31. Study procedure used for calculating main and flash exposure times when BDR is between 0.80 and 1.00. (Eastman Kodak Co.)

of the arrow. Return D to the copy shadow density. Read and use the new main and flash exposures at A and D.

If the BDR is above 1.40 you will need to add the no-screen exposure to almost all copy. Remember that no-screen exposures shorten the range. This can be done by subtracting the copy highlight density from copy shadow density. If the difference is less than the BDR, you must add the 10 percent no-screen exposure. To do this, use the 10 percent mark as a substitute for A, Fig. 16-32, and add the 10 percent no-screen exposure to your exposure sequence. Use the correction arrows, as shown in Fig. 16-31. Refer to the no-screen calibration table, Fig. 16-33. Read down the left column to the density difference from Fig. 16-30. Then read across. These numbers are the no-screen percentages available for your use. In this example, the readings are: 2-4-7-10-15-20.

If the BDR is below 0.80, your ability to make good halftones from a variety of copy will be limited. *Contrasty* (extreme darks and lights) or very *gray copy* (not much contrast) will be almost impossible to reproduce. A contact screen with a longer screen range is recommended.

PRODUCTION HINT—once the calculator has been calibrated, remake the main exposure negative so that the highlight dot will fall between the 0.0 and 0.10 density steps. Then, reset the calculator to the new information. It will be easier to use the calculator, and the procedure will be less subject to flare and other variables. To do this, set the calculator to the original calibration, and rotate A to 0.0, Fig. 16-34. Remake the calibration at the new main exposure time in the window at A. The halftone calculator is now ready for production.

The sequence of exposures can be either main/flash/no-screen, or no-screen/main/flash. Choose the sequence you prefer best and stick with it.

NO-SCREEN (BUMP) CALIBRATION TABLE

Density Difference

DENSITY DIFFERENCE	.05	.10	.15	.20	.25	.30	.35	.40	.45	.50
.10	3	10	20							
.15	2	5	10	18						
.20	2	4	7	10	15	20				
.25	2	3	5	8	10	13	16	20		
.30	1	3	5	7	8	10	12	15	18	20
.35	1	2	4	5	7	8	10	12	15	18
.40	1	2	4	5	6	7	8	10	12	15
.45	1	2	3	4	5	6	7	8	10	12
.50	0	1	2	3	4	5	6	7	8	10

(DENSITY DIFFERENCE .19 arrow pointing to .20 row)

Fig. 16-33. Note example of a no-screen (bump) calibration table. Find density difference by reading down left column, then read across for no-screen percentages available for use. (Eastman Kodak Co.)

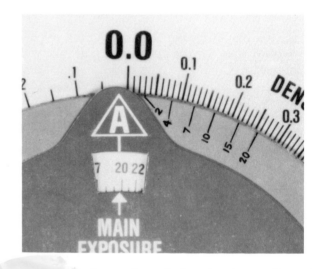

Fig. 16-34. Final step after calculator has been calibrated is to adjust it so that highlight dot will fall between 0.0 and 0.10 density steps. (Eastman Kodak Co.)

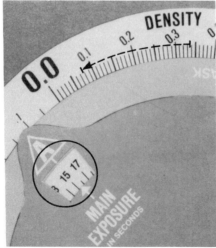

Fig. 16-32. To add a ten percent bump exposure, use the ten percent mark on red dial instead of "A." Read main exposure in window, and add a ten percent bump to sequence. Use correction arrows described in Fig. 16-31. (Eastman Kodak Co.)

The steps in making a halftone negative are illustrated and described at the end of this chapter. For purposes of assisting you in the task of making a halftone negative, the following points should be kept in mind.

FLASH AND NO-SCREEN EXPOSURES

Flash exposures affect shadow areas more than highlight areas. No-screen (bump) exposures affect highlight areas more than shadows. Contrast in highlight and shadow areas shifts with the use of auxiliary exposures, and midtone contrast will move noticeably. When you reproduce the copy as is, the natural shifts in contrast are completely acceptable. In fact, they are needed. If the copy needs modification of highlight, midtone, or shadow contrast to meet a specific requirement, you can produce a change by using a different type of contact screen. For example, use gray instead of magenta, magenta instead of gray, or interchange short, medium, or long-range screens.

You may also alter the BDR of the screen you are using. For example, suppose you have copy with a range of 1.70, and your contact screen has a BDR of 1.40. To reproduce the copy as is, you would use a flash exposure to account for the missing 0.30 in density range.

Suppose the copy lacks highlight contrast (highlights blend together with little separation of tones). To improve highlight contrast, use a no-screen (bump) exposure. The bump exposure will reduce the BDR of the screen to below 1.40, will affect the highlights more than the shadows, and will create a need for more flash. The main and flash exposures alone can reproduce the copy as is without a bump exposure. However, the addition of a bump exposure will change the main and flash exposure ratios while improving highlight contrast.

The same result can be obtained by using a contact screen with a shorter BDR. Screens with shorter BDRs react the same way because they have extra highlighting built into their design. Most negative gray contact screens have a built-in highlighting effect. Only a main exposure and a shadow flash exposure are required in most instances. While the flash affects the shadow areas most, the midtones are also affected to a considerable degree, and the highlights to a smaller degree. This often requires a slight reduction in the main exposure in cases where the original needs more than an average flash exposure. Without this reduction, the midtones tend to be overexposed.

Each piece of copy requires a different flash exposure to match the range of tones on the original. It obviously becomes essential for the camera operator to establish a basic flash exposure.

Fig. 16-35 is a flash exposure table which shows the difference in exposure times once the basic time is established. As every piece of continuous tone copy varies, the difference between the copy and screen ranges will fluctuate and so should the flash exposure. The camera operator should post a flash exposure table next to the camera.

The darkroom flashing lamp should be equipped with a filter holder that will accept ND (neutral density) filters, Fig. 16-36. *Neutral density filters* complement the flashing lamp's normal yellow filter. Neutral density filters can be used to shorten the flash exposure time. This allows better control of the exposure time and makes it easier to duplicate exposure times accurately.

Basic Flash	0.1	0.2	0.3	0.4	0.5
10	2	3¾	5	6	6¾
11	2¼	4	5½	6½	7½
12	2½	4½	6	7¼	8¼
13	2¾	4¾	6½	7¾	8¾
14	3	5¼	7	8½	9½
15	3¼	5½	7½	9	10¼
16	3¼	6	8	9½	11
17	3½	6¼	8½	10¼	11½
18	3¾	6¾	9	10¾	12¼
19	4	7	9½	11½	13
20	4	7½	11	12	13½

Fig. 16-35. A flash exposure table shows difference in exposure times for various copy and screen ranges once basic time has been established. This table should be posted next to process camera.

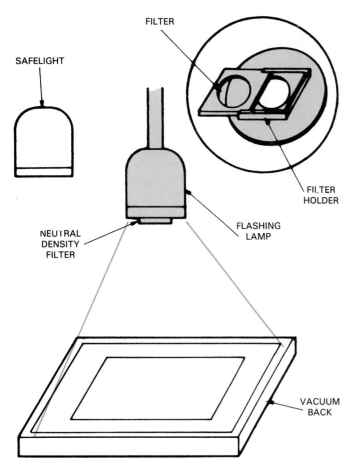

Fig. 16-36. Darkroom flashing lamp is equipped with a filter holder that will hold neutral density filters. These filters can be used to shorten flash exposure time.

Note! *Newton's rings* (concentric bands of light) may become a problem with the gray contact screen when using a neutral density filter. If Newton's rings become a problem, try flashing without the filter.

The negative magenta contact screen differs as to highlighting effect. It also has a built-in highlighting effect. However, the main and shadow flash exposures are more independent of each other than when the gray screen is used. One area of the tonal scale can be adjusted considerably with less effect on the other areas.

Variables to consider

Flash and bump exposures are methods of altering the screen range to better suit the copy requirements. Other variables also influence the screen range. Although you can control them as you control exposures, they are slight and often overlooked. Some of the variables to be aware of include: development rate, change of screens, laterally reversing a screen, changes in film, flare, and light sources.

Development rate—When using trays for processing, an INCREASE in the agitation rate may SHORTEN the BDR. In a similar way, the still development technique (a decrease in agitation rate) may lengthen the BDR substantially.

Change of screens—Halftone contact screens are made to very rigid tolerances. However, slight differences may be noticeable between a four year old screen and a new screen due to changes in manufacturing methods. A magenta screen which has been washed several times may show a slightly lower BDR. The original magenta screens for offset lithography had long BDRs. Comparing an original magenta screen with a new magenta screen, the differences in BDR may be great.

Laterally reversing a screen—Positioning a contact screen with its emulsion side away from the film will lower the BDR of that screen. This is mentioned as a caution. Purposeful reversing of a screen is NOT recommended because the emulsion is then exposed to more damage, and the degree of highlighting is highly questionable and is difficult to control. Use bump exposures for controlling the highlighting of a screen.

Changes in film—A change from one film product to another will result in a change of the BDR. It is important to calibrate the BDR for each film separately.

Flare—Nonimage light, referred to as *flare,* will lengthen the BDR of a screen. If you clean or change a lens on a process camera, you may be altering an accepted level of flare. This may cause a substantial change in BDR. Although the effect may be minor for most halftone work, the BDR should be recalibrated with a new or cleaned lens.

Light source—The negative gray contact screen and negative magenta contact screen are usable with most commercial light sources. However, if tungsten light is used, the negative magenta screen will produce a lower highlight contrast and a little longer BDR than the negative gray screen. The longer BDR requires less flash exposure.

Pulsed xenon light sources will also show a similar effect, but to a much smaller degree. Higher contrast can be obtained by using a Wratten gelatin filter No. 80B. In this case, double the exposure that would be made with no filter. This filter technique, used with tungsten or pulsed xenon light sources and a negative magenta contact screen, will produce shorter BDRs than nonfiltered halftone techniques.

Manipulating tone values

Different exposing and developing techniques produce variations in halftone negatives. When making a halftone exposure, include a typical gray scale. After processing the negative, examine the image of the gray scale to find which step has the 50 percent (checkerboard) dot. For example, when you use a negative gray contact screen with normal copy and give only a main and flash exposure, the checkerboard dot will be between 0.5 and 0.7 steps. This is an average condition among many possible tone-scale adjustments. It will yield high quality reproductions for most work.

If necessary, you can make the dot pattern in the 0.5 step more nearly 50 percent by increasing the highlighting exposure. The gray screen has a highlight effect built into it, but more effect may be desired. In this case, a no-screen (bump) exposure can be added. This adjustment may be desirable when the work is to be printed on coated or high-reflectance papers. For this purpose, the 50 percent dot can be moved toward the 0.7 step, making the 0.7 step more nearly a checkerboard dot. This may be desirable when you are printing on uncoated, low-reflectance papers. It is achieved by a modification of the BDR. Changing conditions so that the BDR is lengthened will tend to move the 50 percent dot to a higher density.

Identifying the 50 percent dot

Identifying the 50 percent dot is a relatively simple matter with a screen that has a conventional square dot at the 50 percent point, Fig. 16-37. It is easily seen.

Identifying the 50 percent dot on an elliptical dot halftone is another matter. The dots join on two diagonally opposite corners at less than 50 percent and on the other two corners at more than 50 percent, Fig. 16-38. The 50 percent dot will have two corners well joined and two corners not quite joined. The width of the joined corners will be approximately equal to the separation of the unjoined corners.

Until you are accustomed to recognizing the patterns, the best method is to make a quick check with a densitometer. From the table in Fig. 16-39, you will note that the 50 percent dot area should have a density

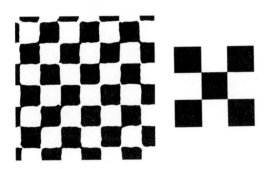

Fig. 16-37. The 50 percent dot from a conventional square dot screen can be identified as a checkerboard pattern. (Eastman Kodak Co.)

Fig. 16-38. The 50 percent dot from an elliptical dot screen is joined on two opposite corners. (Eastman Kodak Co.)

Conversion from Density Readings* to Percent Dot Area			
Integrated Halftone Density	Percent Dot Areas	Integrated Halftone Density	Percent Dot Areas
.004	1	.310	51
.009	2	.319	52
.013	3	.328	53
.018	4	.337	54
.022	5	.347	55
.027	6	.357	56
.032	7	.366	57
.036	8	.377	58
.041	9	.387	59
.046	10	.398	60
.051	11	.409	61
.056	12	.420	62
.061	13	.432	63
.066	14	.444	64
.071	15	.456	65
.076	16	.468	66
.081	17	.482	67
.086	18	.495	68
.092	19	.509	69
.097	20	.523	70
.102	21	.538	71
.108	22	.553	72
.114	23	.569	73
.119	24	.585	74
.125	25	.602	75
.131	26	.620	76
.137	27	.638	77
.143	28	.658	78
.149	29	.678	79
.155	30	.699	80
.161	31	.721	81
.168	32	.745	82
.174	33	.770	83
.181	34	.796	84
.187	35	.824	85
.194	36	.854	86
.201	37	.886	87
.208	38	.921	88
.215	39	.959	89
.222	40	1.000	90
.229	41	1.046	91
.237	42	1.097	92
.244	43	1.155	93
.252	44	1.222	94
.260	45	1.301	95
.268	46	1.398	96
.276	47	1.522	97
.284	48	1.699	98
.292	49	2.000	99
.301	50		

*Integrated halftone densities are given to three decimal places only as an aid to interpolation.

Fig. 16-39. Chart shows percent of dot areas converted to density readings. For example, 90 percent dot area should have a density reading of 1.000. (Eastman Kodak Co.)

reading of 0.30. This is true with either an elliptical dot or a square dot pattern.

Adjusting the BDR

The purpose of a change in the basic density range or BDR is to alter the tone scale to adjust for the type of copy. It is also necessary to satisfy printing requirements resulting from changes in the type of print stock and the degree of dot gain. Changing the BDR causes the 50 percent dot to move within the tone scale.

Local printing conditions will determine the most useful values. The table in Fig. 16-40 will serve as a rough guide. The modified basic density range is shown at the top of the table. Underneath each value is a rough guide to its effect on the tone scale when printed on various papers. In the table, "HR" represents high-reflectance papers, such as clay-coated papers. "NR" indicates normal-reflectance papers, such as offset paper. "LR" indicates low-reflectance papers, such as newsprint.

In general, the following effects will be noted:
1. The LONGER the modified basic density range, the better the shadow detail and the more loss in the highlights.
2. The SHORTER the modified basic density range,

the better the highlight detail and the more loss in the shadows.

The shorter modified basic density ranges are obtained by adding a no-screen (bump) exposure to the main exposure. To work out a set of values for your conditions, first determine the basic density range with the main exposure only. If you have followed the directions for calibrating the halftone calculator covered earlier, you already know this value.

Next, repeat a similar exposure of the reflection density guide. However, this time add a five percent no-screen exposure. Examine the negative and select the steps having the desired highlight and shadow dots. Subtract to get the difference between the marked density values of the two steps. This will be a modified (and shorter) basic density range. Repeat the procedure with a ten percent no-screen exposure. This will give you a set of values that can be used as a reference.

PRODUCTION HINT—To find out how each modification affects the actual "Printability" of the negatives, make plates for the offset press. Make press proofs of the test negatives on the paper normally used. Compare the press proofs instead of the negatives.

To get the longer modified basic density ranges, the method previously described must fit the particular

EFFECT OF MODIFIED BASIC DENSITY RANGE ON TONE REPRODUCTION

Modified Basic Density Range	.75 and below	.80—.90	.95—1.10	1.15—1.30	1.35 and higher
Effect on tone reproduction.	Greatly accentuates highlight detail on all stocks.	1. Tends to accentuate highlight detail on HR stock. 2. Greatly accentuates highlight detail on NR and LR stocks.	1. Satisfactory highlight detail on HR and NR stocks. 2. Accentuates highlight detail on LR stock.	1. Accentuates shadow detail on HR stock. 2. Satisfactory highlight detail on NR and LR stocks.	1. Greatly accentuates shadow detail on all stocks.
Approximate placement of 50 percent dot on gray scale (when the highlight dot is in the .00 step)	.40 and below	.40—.50	.50—.60	.60—.70	.70 and higher

HR=High Reflectance, NR=Normal Reflectance, LR=Low Reflectance.

A

The 50 percent dot will appear at about the 0.70 step when the modified BDR is about 1.25 to 1.35.

The 50 percent dot will appear at about the 0.50 step when the modified BDR is about 0.90 to 1.00.

B

Fig. 16-40. A—Modifying basic density range (BDR) of various kinds of copy is sometimes necessary under certain conditions. Type of paper, ink, press, and purpose of printed piece are considerations used to modify density range of copy. B—The 50 percent dot is used to illustrate modified BDR effect. (Eastman Kodak Co.)

screen being used. With a negative magenta contact screen, exposing with a Wratten gelatin filter No. 4 (yellow) will result in a modified BDR of about 1.40 when using orthochromatic films processed in a film processor. The range will be somewhat shorter with tray development.

With a negative gray contact screen the resulting BDR will be about 1.00 when using orthochromatic films processed in a film processor. If the screen is used with tray processing, it will be difficult to obtain longer modified basic density ranges without using still-development techniques.

Applying the modified BDR

Assume that you have worked out how to obtain the several modified basic density ranges under your conditions. You will have recorded the procedure for each BDR. Proceed as follows:

1. Determine the paper on which the printing is to be done: HR, NR, or LR.

2. Study the photograph being reproduced to determine if one portion of the tone scale needs improvement. From the table in Fig. 16-40, select the modified BDR that indicates that particular improvement on your paper.
3. Expose the negative, using the contact screen and technique which gave you the indicated BDR. Add the flash required.
4. You will find that a large number of your negatives call for normal exposures. You will have to apply extreme measures in relatively few cases. This is the result of the design of the negative magenta and gray contact screens.

MAKING A HALFTONE NEGATIVE

The steps in making a halftone negative are illustrated in Fig. 16-41. This procedure assumes that the original copy is normal or near average. However, many photographs are poorly made. The photograph

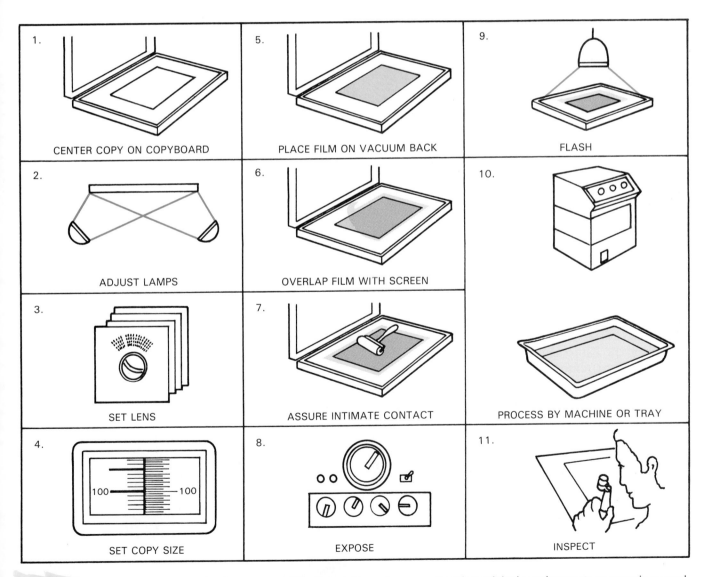

1.	5.	9.
CENTER COPY ON COPYBOARD	PLACE FILM ON VACUUM BACK	FLASH
2. ADJUST LAMPS	6. OVERLAP FILM WITH SCREEN	10. PROCESS BY MACHINE OR TRAY
3. SET LENS	7. ASSURE INTIMATE CONTACT	
4. SET COPY SIZE	8. EXPOSE	11. INSPECT

Fig. 16-41. Major halftone reproduction steps are illustrated. Procedure assumes that original continuous tone copy is normal or average. (Chemco Photoproducts Co.)

may have no highlight detail or may be dark and muddy. The photograph may have a density range shorter than that of the contact screen.

Before making a halftone negative, let's review what you will need in the way of equipment and supplies.
1. Darkroom.
2. Process camera (horizontal or vertical).
3. Flash lamp (if not built into camera).
4. Contact screen (magenta or gray, 133 line).
5. Lith ortho film.
6. Processing trays and sink or film processor.
7. Processing chemicals.
8. Reflection density guide, Q-16.
9. Halftone calculator, Q-15.
10. 10X or 12X magnifier (loop).

To make a halftone negative, apply what you have just learned by following these steps. Before you begin, determine the correct main and flash exposure times.
1. Clean the copyboard glass on both sides.
2. Place a piece of black and white continuous tone copy upside down in the copyboard with a 12-step gray scale beside it.
3. Set the bellows extension and copyboard extension for same-size (100%) reproduction.
4. Set the lens for the same f-stop number as used in line work at 100%.
5. Adjust the lights for proper angles.
6. Set the main exposure time.
7. Turn off the overhead white lights and turn on the red safelights.
8. Select the correct lith ortho film box and remove a sheet. Cut the sheet to proper size if required.
9. Open the camera back and position the piece of film.
10. Position the contact screen over the film, emulsion side to emulsion side. The contact screen must be at least one inch larger on all four sides than the film, Fig. 16-42. This is to ensure good vacuum contact.
11. Turn on the vacuum pump.
12. Close the vacuum back.
13. Press the exposure button and expose the film.
14. Open the camera back but do not turn off the vacuum or remove the contact screen.
15. Set the flash exposure timer for the required time.
16. Make the flash exposure with yellow light, Fig. 16-43.
17. Turn off the vacuum pump and remove the contact screen and film from the camera back. Replace the contact screen in its container.
18. Set the darkroom timer for 2 3/4 minutes in readiness for film development.
19. Process the exposed film in the same way as described for line negative work. The developing solutions should be fresh and maintained at 68 degrees F (20 degrees C). Move the film back and forth during the developing cycle.
20. After the negative has been stopped, fixed, and

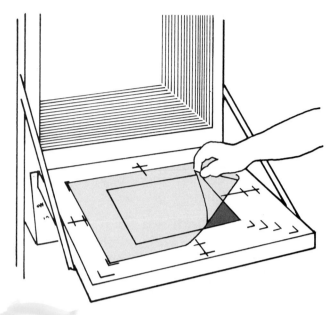

Fig. 16-42. Halftone contact screen is positioned over unexposed film, emulsion to emulsion. A contact screen must be at least one inch larger on all four sides than film. (nuArc Company, Inc.)

Fig. 16-43. Flash exposure is made with yellow light. Halftone screen remains in position over film during exposure. (nuArc Company, Inc.)

washed, squeegee the acetate side. Then hang the film to dry in a dust-free area of the darkroom or process it in a film dryer.

21. Examine the halftone negative on a light table. An acceptable negative should give a 90 percent dot in the highlights and a 10 percent dot in the shadows. It may be necessary to adjust exposure times slightly and try again.

RESCREENING HALFTONES

When you must prepare a halftone, it is always best to begin with a continuous tone photographic print. However, occasions arise when you do not have access to an original print. In such instances, you must start with an image that is already a screened halftone. If you try to photograph the printed halftone using a conventional halftone screen and following conventional reproduction methods, your final image may contain a moiré pattern. See Fig. 16-44.

A *moiré pattern* is an undesirable wavelike or checkerboard pattern visible across the halftone or screened image. Moiré patterns are usually caused by a directional misalignment of the dot pattern of the original halftone print and the dot pattern of the halftone screen used to rephotograph the original.

To avoid a moiré pattern in the rephotographed halftone, you should use a special technique, called *rescreening,* to reproduce a previously screened original. It will give the rephotographed halftone print a professional look. The process involves very few special requirements. You will need the following tools and materials.

1. A *screen-angle indicator* to help you find the angle of the original screen in the halftone print. Two good indicators are the T&E Center/RIT Screen-Angle Selector, and the GATF Screen Angle Guide. The GATF Screen Angle Guide is shown in Fig. 16-45.
2. The contact screen must be either a different angle than the original screen by 30 degrees or large enough to be rotated to a different angle and still completely cover the halftone original being photographed. Look at Fig. 16-46.

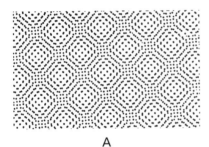

A

B

Fig. 16-44. A—A moiré pattern is an undesirable wavelike or checkerboard pattern visible across image. B—A moiré pattern is usually caused by a directional misalignment of dot pattern of original halftone print and dot pattern of halftone screen used for rescreening.

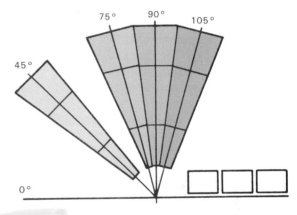

Fig. 16-45. Note Graphic Arts Technical Foundation screen angle guide. For purposes of rescreening, guide can be used to find angle of original screen ruling in a halftone print. (Eastman Kodak Co.)

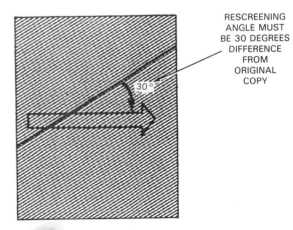

Fig. 16-46. For rescreening, contact screen must be pre-angled or large enough to be rotated to a different angle, generally 30 degrees from original copy. (Eastman Kodak Co.)

3. Clear acetate or fixed-out clear film, 8 x 10 inches in size, is recommended.
4. A pocket magnifier (loop).
5. A darkroom timer, Fig. 16-47.

Rescreening steps

The following steps should be followed when rescreening an already printed halftone.

1. Following the directions provided with the screen-angle indicator, determine the screen angle of the original halftone.
2. Place a sheet of halftone film on the camera back and cover it with a contact screen. If you are using a contact screen with an angle 30 degrees different than the original screen angle, align the screen and film together. See Fig. 16-48.

   If you are using a large enough screen, rotate the screen 30 degrees clockwise. Refer to Fig. 16-49. Refer to the instruction sheet packaged with the halftone screen to find the angle of the screen you are using.
3. Place the screened original in the camera copyboard.
4. Turn on the camera lamps, and open the shutter.
5. Set the camera lensboard and copyboard controls slightly out of focus. For same-size (100 percent) reproduction, set the lensboard at 101 percent.
6. Keep the lens f-stop setting as close to the chart listing as possible. For same-size reproduction, set the copyboard tape at 85 percent.
7. Set the timer for the main exposure time plus 20 or 25 percent. Start the timer.
8. During the main exposure, move the sheet of clear film in front of the lens in a circular motion, Fig. 16-50.
9. Make a normal flash exposure.

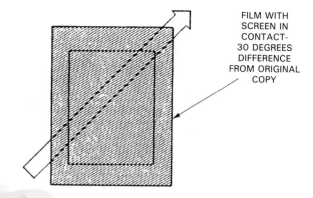

Fig. 16-48. Rescreening angle must be 30 degrees from that of original screen angle of halftone print. This special screen has a built-in angle of 30 degrees. (Eastman Kodak Co.)

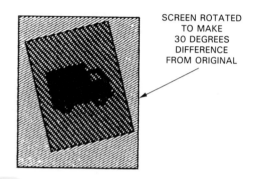

Fig. 16-49. If a regular gray or magenta halftone contact screen is used for rescreening, contact screen must be rotated 30 degrees clockwise to avoid moiré pattern. (Eastman Kodak Co.)

10. Develop the film. The film should be developed in the recommended developer and replenisher. Fix and wash the film. The result is a rescreened halftone.

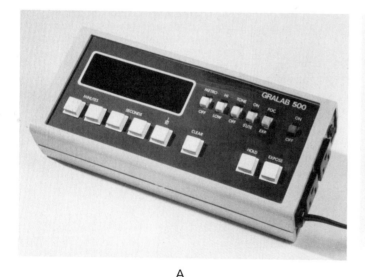

A

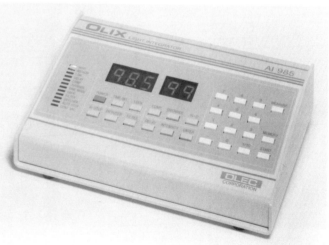

B

Fig. 16-47. A—This is an electronic digital darkroom timer. B—Computerized light integrator/darkroom timer is used for halftone reproduction work. (Bob Bright, Cal State L.A.; Olec Corp.)

Fig. 16-50. A sheet of clear acetate film is moved in a circular motion in front of lens during a rescreening exposure. (Eastman Kodak Co.)

POINTS TO REMEMBER

1. Halftone photography is a special photographic procedure used almost exclusively by the printing and publishing industry.
2. Continuous tone photographs have variations of tones of gray that blend together to form a picture.
3. To print a reproduction of a continuous tone photograph, the printer must break the image into halftone dots that allow ink to be placed on the offset plate.
4. Although the halftone dots are different sizes, all the dots have the same density so that the printer can print them with an offset press.
5. When continuous tone copy is printed as a halftone, the dots and the white, or light, paper fuse visually and give the illusion of a continuous array of tones.
6. A contact screen is a precision pattern of vignetted dots on a flexible support, such as polyester.
7. Contact screens are used for making halftone negatives on a process camera or on a scanner.
8. Contact screens are available as elliptical dot or square dot patterns. Other special screens are manufactured that can be used for halftone photography.
9. Contact screens are categorized as magenta, gray, or PMT.
10. Contact screens are classified as being either negative or positive.
11. As the name implies, a negative contact screen is used to make halftone negatives, NOT positives.
12. A positive contact screen is used to make halftone positives for direct color separation.
13. Contact screens must be handled with care and cleaned with a dry chamois.
14. There are three characteristics to be considered when evaluating a halftone negative—highlight, middletone, and shadow areas.
15. The highlights are the darkest part of the halftone negative and consist of small, clear openings in an otherwise solid area.
16. The middletones consist of 30 to 70 percent dots and typically represent light to medium gray tones.
17. The images of least density in the halftone negative, but greatest density in the printed halftone, are the shadow areas.
18. Original copy for halftone reproduction represents a tonal range from the whitest highlights through varied grays to the black of the shadows.
19. A good black and white photograph has a range of densities from paper white to a 1.7 density black or slightly darker.
20. Solid printing press black ink yields a density of about 1.40 to 1.60 on coated paper stocks and usually less on uncoated paper stocks. This is not completely black 2.0.
21. Various tests must be performed prior to using a halftone contact screen. The basic density range or BDR of the contact screen must be determined. In practice, the basic density range is usually modified to equal the density range of the copy.
22. When making halftone negatives, a halftone calculator is sometimes used to determine the main and flash exposures. A reflection density guide is also used to measure or read densities in the original continuous tone photograph.
23. Rescreening is a process in which an already printed halftone is rephotographed using a contact screen angled at 30 degrees to the original halftone print.

KNOW THESE TERMS

Basic density range, Contact screen, Continuous tone copy, Copy density range, Elliptical dot screen, Gray screen, Halftone calculator, Highlight, Magenta screen, Middletone, Negative contact screen, Positive contact screen, Reflection density guide, Rescreening, Shadow, Square dot screen, Tonal range.

REVIEW QUESTIONS

1. _____ _____ _____ consists of intermediate densities or tones of gray.
2. Halftone _____ screens are used primarily in offset lithography.
3. The difference between the lightest and darkest portions of a continuous tone photograph are the intermediate tones. True or false?

4. The _____ contact screen is a purplish shade of red.
5. When is a PMT gray contact screen used?
6. Negative contact screens are most often used to make halftone _____.
7. The highlights on the halftone negative should be approximately:
   a. Ten percent.
   b. 30 to 70 percent.
   c. 95 percent.
   d. 100 percent.
8. The darkest areas of the original continuous tone copy are called _____.
9. Explain the tonal range capability of a halftone contact screen.
10. A no-screen exposure is referred to as a:
   a. Bump exposure.
   b. Flash exposure.
   c. Intermediate exposure.
   d. Dodging exposure.
11. The halftone calculator is used to determine film development time in seconds. True or false?
12. A reflection density guide can be used much like a _____ scale.
13. Describe the function of a flash exposure.
14. Explain the purpose of a no-screen exposure.
15. The purpose of adjusting the basic density range (BDR) is to alter the tone scale to compensate for the type of continuous tone copy being photographed. True or false?

16. _____ is a process in which an already printed halftone is rephotographed using a contact screen angled at _____ degrees from the original.
17. What causes moiré patterns?
18. List the steps for rescreening a halftone.

SUGGESTED ACTIVITIES

1. Prepare a list of equipment, tools, supplies, and materials needed for halftone photography.
2. Prepare a list of the essential steps for making a halftone negative using a main and flash exposure.
3. Prepare a display of printed halftone samples from magazines and newspapers. Label each sample with its correct screen ruling (85, 100, 120, 133, 150, etc.).
4. After a demonstration by your teacher, determine the correct main and flash exposure times for a halftone.
5. Prepare satisfactory halftone negatives, using main and flash exposures. Check each of the halftone negatives with a 10X or 12X magnifier over a light table.
6. Prepare a photographic paper contact print from a halftone negative.
7. Prepare a satisfactory halftone negative from an already printed coarse-screened halftone print. You must avoid producing a moiré pattern when angling the contact screen.

Chapter 17

Special Effects Photography

When you have completed the reading and assigned activities related to this chapter, you will be able to:

O Identify the various camera reproductive methods known as special effects photography.

O Describe how various special effects are produced as negatives and positives on film and paper.

O Make special effects negatives and positives using the equipment and materials available.

O Describe the basic principles of the diffusion transfer process.

O Use the diffusion transfer process to make line positive prints, line film prints, and halftone prints.

O Describe the basic principles of the color copy diffusion transfer process.

Special effects photography refers to the creative side of darkroom photography and copy preparation. There are a number of procedures and materials available for producing special effects photography, Fig. 17-1. These techniques go beyond the familiar line and halftone processes described in Chapters 15 and 16.

Special effects are produced to meet two general needs:

1. The reproduction of creative images from otherwise average copy.
2. Adherence to the limitations created by equipment and materials in order to produce effective and economical printing.

The equipment found in most offset lithography darkrooms can be used for special effects photography. The most important piece of equipment is a process camera. The camera can be either a horizontal or vertical model. An enlarger is used to produce some types of special effects, Fig. 17-2. In addition, a contact printing frame and a point source light are needed for special effects. A diffusion transfer processor is also necessary for some special effects materials. See Fig. 17-3.

LINE CONVERSIONS

As you have learned, a common technique of making a halftone negative is using a halftone contact screen. In addition, special effects halftone screens can be used to convert continuous tone copy to *line conversions*. These special screens are used to create a variety of unusual effects. Among these are the mezzotint, steel engraving, circular, wavyline, and vertical straightline. Figs. 17-4 and 17-5 show some examples.

The layout artist can create unusual designs with the aid of special effects screened copy. The original copy is usually prepared on photoprint paper and attached directly to the paste-up. Refer to Figs. 17-6 through 17-8.

SPECIAL EFFECTS CONTACT SCREENS

The layout artist can specify contact screens having concentric circle patterns, vertical line, horizontal or diagonal line, wavyline, and mezzotint. All special effects contact screens require skillful handling and creativity.

Special effects contact screens are used much like halftone contact screens on continuous tone copy. It

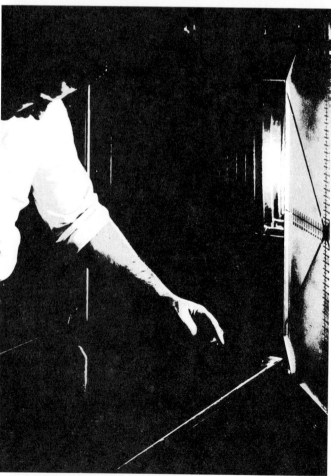 

Fig. 17-1. Special effects refers to creative side of darkroom photography. Many unusual special effects can be created photomechanically using continuous tone originals. (Eastman Kodak Co.)

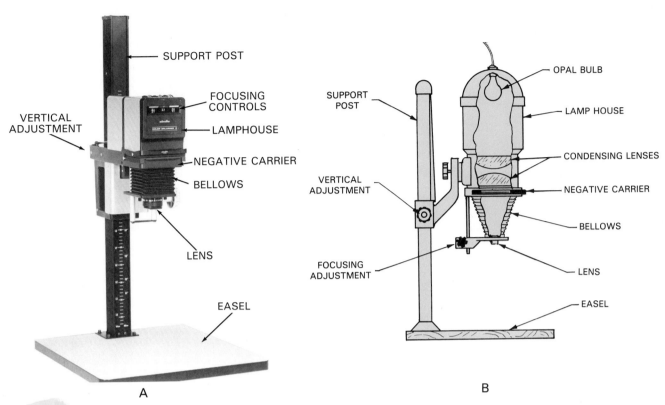

Fig. 17-2. A—Besides process camera, a standard darkroom enlarger can be used to produce some types of special effects materials. B—Cross sectional view of an enlarger shows basic components. (Minolta Corp.)

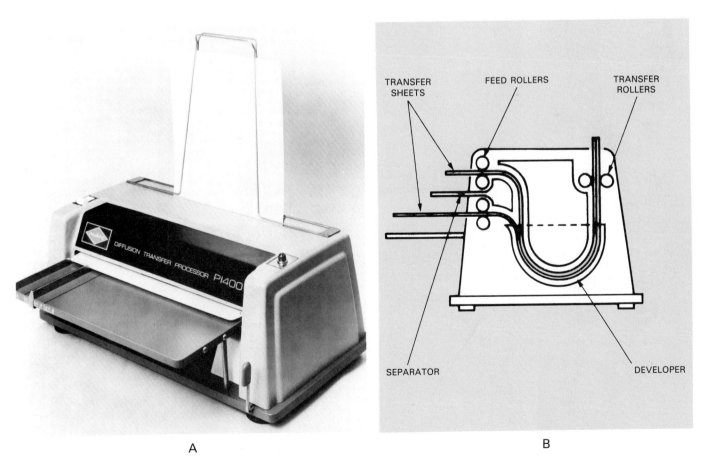

Fig. 17-3 A—A darkroom diffusion transfer processor is used to make some special effects materials. B—Cross sectional view of a diffusion transfer processor shows basic components. (nuArc Company, Inc.)

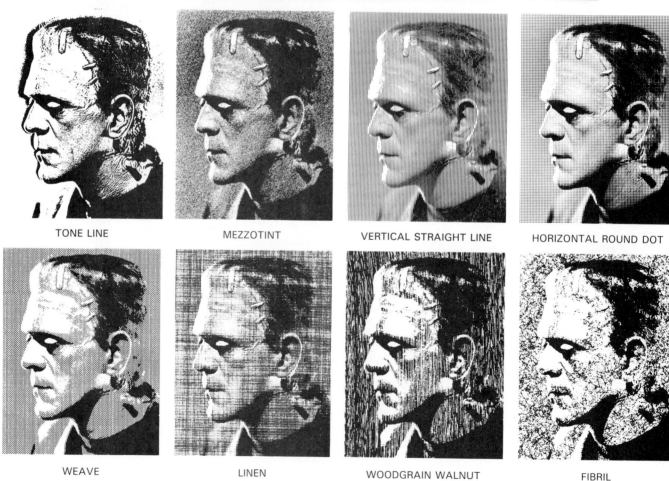

TONE LINE   MEZZOTINT   VERTICAL STRAIGHT LINE   HORIZONTAL ROUND DOT

WEAVE   LINEN   WOODGRAIN WALNUT   FIBRIL

(Continued)

Fig. 17-4. Special effects halftone contact screens are available in a variety of patterns. Unusual results can be achieved with the many special effects halftone contact screens available. (Courtesy of John N. Schaedler, Inc.)

CROSSHATCH

SQUARE HALFTONE

LINE RESOLUTION

HORIZONTAL WAVY LINE

CIRCLE LINE

STEEL ETCH

STEEL ENGRAVING

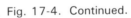

Fig. 17-4. Continued.

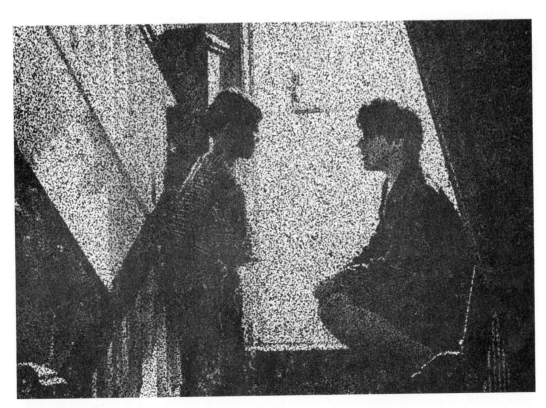

Fig. 17-5. Dramatic results can be achieved with conversion of continuous tone copy to line art.

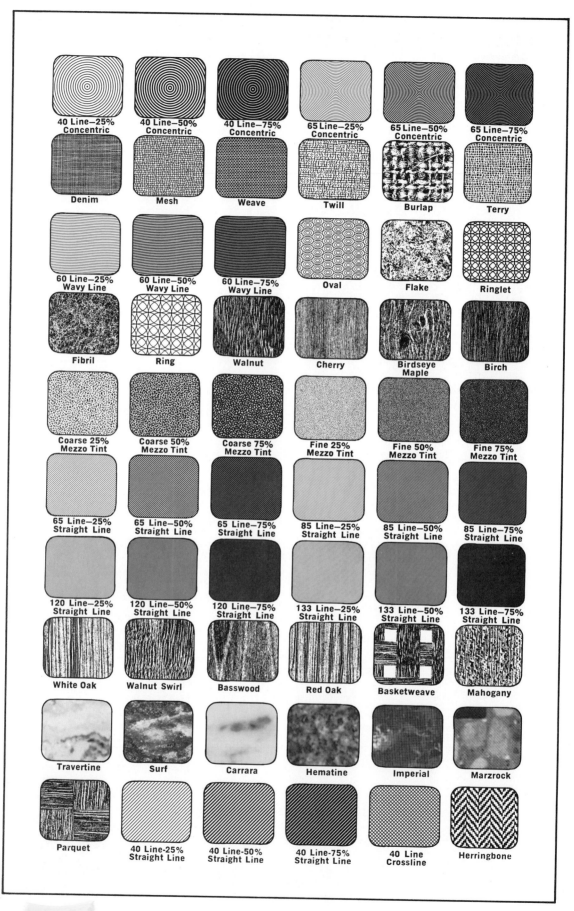

Fig. 17-6. Ordinary continuous tone copy can be converted to special effects by use of film tint screens or adhesive screen tint material directly over copy. (Formatt, Graphic Products Corp.)

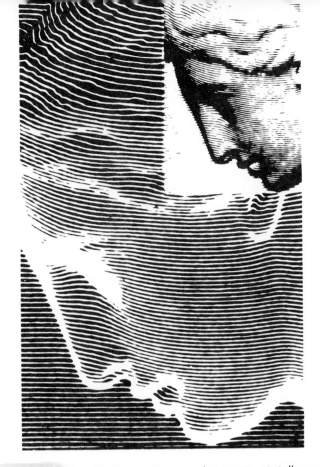

Fig. 17-7. Wavy line screen was used to create a totally new look to otherwise ordinary continuous tone copy. (Visual Graphics Corp.)

should be pointed out that these screens are NOT the same as the screen tints described later in this chapter.

Contact screens are generally used in a PROCESS CAMERA. In contrast, screen tints are used in a VACUUM FRAME at the platemaking stage. To demonstrate how special effects contact screens are used, the following example of the application of a concentric circle screen is given.

There are instances in which some continuous tone copy seems adaptable for use with a concentric circle treatment. It is untrue that any spherical object in a piece of continuous copy demands the application of this special screen effect. The most obvious quality of a concentric circle screen is that it draws the eye inwards to its center. This built-in focusing effect can be used to pinpoint one particular item from among similar ones.

Fig. 17-9A illustrates how one person has been singled out for attention by the zooming-in effect of the concentric circle screen.

Eyes make poor centers for these screens. Fig. 17-9B shows a woman, quite capable of generating adequate attention without a concentric circle focused in one of her eyes.

The image in Fig. 17-10 presented two problems before the special effects screening. First, there was no way of telling the time of day. Is it dawn, sunset, or

Fig. 17-8. Potential for special effects photography is unlimited. Layout artist can create unusual effects from average continuous tone copy. (Visual Graphics Corp.)

A

B

Fig. 17-9. A—Concentric circle special effects screen can be used to pinpoint one particular item from among similar ones. Example shows a person in crowd being singled out. B—In this example, a woman generates attention because concentric circle screen is focused in one of her eyes.
(Chemco Photoproducts Co.)

Fig. 17-10. Sometimes problems arise in special effects conversions. In this scene, is it dawn, sunset, or a moonlit night? Careful consideration to such elements of copy is essential when planning special effects.
(Chemco Photoproducts Co.)

a moonlit night? Second, there was a *halo spot,* caused by outside light (flare) striking the original camera lens. By centering the concentric circle screen in the middle of the halo, the second problem was solved, since the screen lines almost entirely absorb the halo. Closing the shadows made the scene darker and removed the question of time of day. Thus, the boat now sails on a moonlit sea with an artificial light-source in the sky. Also notice that the circular lines enhance the effect of the waves.

The techniques of using concentric circle screens involves two basic points: how short to make the tonal range, and how to center the screen. Tonal range depends on individual taste, nature of the copy, and screen range. Centering the screen is a simple procedure. The centering method described here has proven very accurate and fast.

1. Place the continuous tone copy in the copyboard of the process camera. Set the camera to the required percentage size.
2. Place the correct size sheet of lith ortho film on the camera back. Switch on the vacuum, and very carefully mark the exact outline with masking tape, as in Fig. 17-11(1).
3. Make an exposure without the concentric circle screen.
4. Develop the line negative. Fix, wash, and dry with minimal care and maximum speed. The piece of film must be thoroughly dry.
5. Lay the negative back on the vacuum back, lining it up with the pre-marked area. Fig. 17-11(2) shows how to match the image on the ground glass so that no mistake is made in orientation of the line negative. This should be upside down, emulsion-side up. Tape the negative down and slide a piece of white paper beneath it for good visibility. See Fig. 17-11(3).
6. Place the contact screen over the negative, and locate the center exactly where you want it. Make a hinge of masking tape for the screen, Fig. 17-11(4). Check to see that the center is not moved when the vacuum is switched on.
7. The copy and the precise screen position are now fixed. Retain the line negative in case you wish to experiment with different centers and the same copy.
8. Make the exposure and process the exposed film in the usual way.

POSTERIZATION

A *posterization* converts original tones of an ordinary, continuous tone black-and-white photograph into a high contrast or multi-color rendering, Fig. 17-12. The tones of the original are reproduced as a line negative.

Posterizations are classified according to the number of colors and the number of different tones reproduced on the final press sheet. Only the number of different ink colors is counted. One negative is required for each tonal color. In some applications, posterizations are reproduced in only one color.

A posterization is prepared by increasing the exposure approximately three times the normal of a line copy exposure. A second color may be added to the print by making a negative with the exposure reduced by approximately one-half. This records the middletones of the original copy and adds detail to the print. Midtones are between highlight and shadow tones. More colors may be added to record the highlight areas of the original. However, for each color, it is necessary to prepare a negative.

Special posterization techniques

Special effects contact screens are used when a full tonal range is important in the reproduction of the photograph and a pattern or texture is desired.

Fig. 17-13 shows how a special effects straight line contact screen can be used. It consists of a vignetted, special effects pattern. Results obtained in the final reproduction are determined by how light reflects from the various tones of the original photograph and penetrates into the vignetted portions of the contact screen.

Three-tone posterization

Fig. 17-14 shows a three-tone posterization with white, solid black shadow areas, and one middletone pattern. It was made by using a 25 percent straight line contact screen. The procedure for obtaining a posterization like this involves two camera exposures.

To obtain the midtones, the desired contact screen is placed in contact with the film on the camera vacuum

Fig. 17-11. Using a concentric circle special effects screen involves four steps. Notice that a concentric circle special effects screen was used in this illustration. (ByChrome Co.)

Fig. 17-12. Original tones of this continuous tone copy were converted into a high contrast rendering, called a posterization. (Dynamic Graphics)

Fig. 17-13. Example illustrates use of special effects straight line contact screen. (ByChrome Co.)

Fig. 17-14. This three-tone posterization includes white, solid black shadow areas, and one midtone pattern. It was converted from an original high contrast continuous tone photograph. (ByChrome Co.)

back. A base exposure (about same as that used for a normal halftone through a magenta contact screen) is made. This puts a pattern over the entire range of the photograph, except for the deep shadow areas. These shadow areas remain open in the negative and therefore print solid in the final reproduction. If a pat-

tern is desired further into the shadow areas, a greater amount of main exposure can be used.

The density of the midtones, on the other hand, is determined by the screen, NOT by the exposure. For example, a 25 percent screen produces a 75 percent value in the pattern areas of the posterization. Figs. 17-15 and 17-16 show examples of three-tone posterizations.

The pure whites (highlight areas) are produced by making a bump (no screen) exposure. This is done by exposing the original photograph without the screen over the film. The amount of exposure time is approximately 10 to 20 percent of the base exposure. The great amount of light reflected from the white areas causes the filling-in of the highlights on the negative. This results in a final reproduction with pure white areas, solid black shadows and a pattern or texture in the middletone.

Four-tone posterization

If a three-tone posterization does NOT contain enough detail for a particular photograph, a four-tone posterization can be made, Fig. 17-17. These posterizations are produced by using the same basic techniques as for the three-tone posterization. The difference is that the main exposure is divided into two parts. About two-thirds of the main exposure is made through the screen. The remaining one-third of the exposure is made with either a different screen or with the same

Fig. 17-15. This three-tone posterization was produced using a straight line special effects contact screen. (ByChrome Co.)

Fig. 17-16. This three-tone posterization was given a western-look by using a special effects mezzotint contact screen. (ByChrome Co.)

screen moved to a different position. The bump exposure will remain the same (10 to 20 percent of total main exposure).

Using conventional lith ortho film for posterization, the film negative can be stripped into position in the flat for exposure to the offset plate. An alternative method is to prepare a photographic print from the negative for paste-up and final line reproduction.

It should be noted that a posterized print can be produced quickly and easily using diffusion transfer materials. These prints are ideal for use as line copy on the paste-up board.

Multi-color posterization

Posterizations in more than one color can be very attractive. Good results can be obtained by dividing up a four-tone posterization onto two negatives, and running them in two different colors. For example, the vertical lines could run in one color and horizontal lines in another color by using a straight line screen.

When adding flat color to a posterization, the amount of color added can easily be varied by the exposure used for the line shot of the original

Fig. 17-17. Example of a four-tone posterization was produced using same basic technique as for three-tone posterization. (ByChrome Co.)

photograph. A normal line exposure will add color to the shadow and the midtone areas of the original. An underexposed line shot will result in color throughout the posterization except for the highlight areas. An overexposed line shot of the original will add color only in the shadow areas. See Fig. 17-18.

Posterization on a photographic enlarger

Three- or four-tone posterizations can also be produced on a darkroom enlarger. These are done directly from a continuous tone negative without first making a series of intermediate film tonal separation positives and negatives.

The darkroom enlarger should be equipped with a vacuum arrangement for good contact. The special effects screen is placed in contact with the photographic paper ready for exposure. It is done by placing the photographic paper on the vacuum surface emulsion-side up. The contact screen is placed over the photographic paper emulsion-side down. The paper must be a very high contrast material.

The only difference in this set-up, from a typical enlarger operation, is that a vacuum easel is used. Good contact is important and it is obtained by using an open-face vacuum easel. Neither a clear plastic overlay nor a glass cover is needed. Vacuum contact is achieved without adding extra surfaces which collect dust, contain abrasions, and slow production.

As in the case with the graphic arts camera, a three-tone posterization is achieved by making only TWO exposures—a base exposure and no-screen exposure, Fig. 17-19A. The base exposure is through the screen which is in contact with the photographic paper. If the print were to be developed after making just the base exposure, the result would be similar to that shown in Fig. 17-19B. The highlight areas remain pure white and the texture pattern covers all other areas of the print.

For this example, a 60 line, wavy line pattern was used in a 50 percent tone value. It should be noted that the density of the middletone area is controlled by the density of the texture screen, and NOT by exposure time.

In this example, if more white highlight areas are desired, the base exposure time should be reduced. If less whites are desired, exposure time should be increased. Test exposures are required to achieve the desired results.

For a pleasing print, the areas missing in this example are the solid areas for the deepest tones of the original. To obtain these solid shadows, an exposure of about 20 percent of the base exposure is made without the screen over the photographic paper. This no-screen exposure can be made before or after the base exposure since the results will be the same either way.

Fig. 17-19C shows the result obtained from just a no-screen exposure of five seconds. Increasing the exposure time produces more solid areas, and decreasing it produces less solids.

When both the base exposure and the no-screen exposure are made on the same paper, the result is as shown in Fig. 17-19D. This is a three-tone posterization. A four-tone posterization could also be produced as shown in Fig. 17-20.

In this example, an additional tone has been added to the deeper midtone areas of the original photograph. This added texture creates more detail and produces additional artistic effects. The additional midtone was obtained by using the same 60 line, wavy line 50 percent line pattern rotated so that the lines are in a vertical direction. This adds a crosshatch effect to the deeper midtone areas. To give you some idea of exposure times in this example, the primary base exposure time was twenty-five seconds. The secondary base exposure time was eight seconds. The no-screen exposure was five seconds.

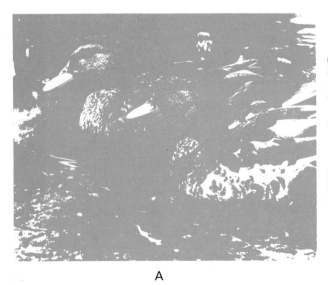

A

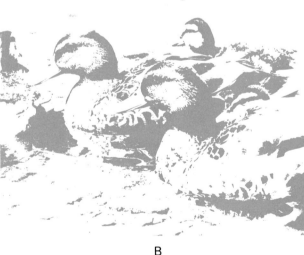

B

Fig. 17-18. A—An underexposed line posterization planned for flat color will result in color throughout posterization except for highlight areas. B—Overexposed line copy will add color only in shadow area. (ByChrome Co.)

Fig. 17-19. A—This example of a three-tone posterization required a base and a no-screen exposure onto photographic paper. B—Example was made with only a base exposure and results in little or no contrast. C—Example was made from just a no-screen exposure. D—When both base exposure and no-screen exposure are made on same paper, this effect results. (ByChrome Co.)

Fig. 17-20. Example of a four-tone posterization illustrates how additional tonal value deepens midtone areas.

Fig. 17-21. A duotone is prepared from a continuous tone photograph by making two negatives, called printers, one emphasizing highlights and other shadow detail. (ByChrome Co.)

Exposure times will vary with different negatives, depending on the density of the negative and the subject matter. A good guideline to follow is to first determine the primary base exposure to get the amount of pure whites desired. For the secondary base exposure, try one-third of the primary base exposure time. The no-screen exposure time generally runs about 20 percent of the primary base exposure. Processing can be done in lith ortho developer. Dektol developer produces equally pleasing results.

DUOTONES

The term *duotone* means the use of two layers of tones (colors) to produce one final image, Fig. 17-21. It is another type of special effects photographic process.

In this technique, a black-and-white continuous tone photograph containing good contrast is converted into a colored illustration of depth and beauty. This re-quires that two separate halftone negatives, called *printers,* be prepared for each color. The two negatives are made to different specifications. One negative emphasizes highlights and the other negative emphasizes shadow detail.

Duotone color combination selection is important. Most duotones are reproduced by using black ink and another color of ink. Sometimes a dark color (other than black) is used with a light color ink. This effect is sometimes more spectacular than using black ink and another color. A two-color duotone attracts more attention than a single-color halftone. The entire printed piece must reflect the mood that the duotone(s) attempts to achieve. See Fig. 17-22.

In preparing a duotone, the primary negative (black or dark color), is exposed and processed for nearly normal contrast. It should favor detail in the shadows and have only partial highlights. The second negative, printed in a lighter color, must have a contact screen angled at 30 degrees to the first. This is because a moiré effect will result from two negatives with tones printing directly over one another. The second negative should contain detail at the high end of the scale. This makes it look lighter when compared to the primary negative.

SCREEN TINTS

Adhesive acetate materials called *screen tints* are used by paste-up artists to provide emphasis and special effects to certain areas of the copy on the paste-up. Film

Fig. 17-22. Sparkle and emphasis have been placed on this flower by fake duotone method.   (ByChrome Co.)

work and type are screened, the printed result is a pattern of dots, Fig. 17-24.

There are two basic methods used to provide screen tints on copy elements. These include: film screen tint positives prepared by the camera department, and adhesive screen tint shading materials applied directly to the paste-up copy elements.

Film positive method

Screen tints are prepared by inserting a screen between a negative or positive and the offset plate. A positive is the opposite of a negative. The image on a positive is opaque on a transparent background. On a negative, the image is transparent on an opaque background.

In making an offset printing plate requiring a screen tint, the screen is placed between the negative or positive and the plate before exposure. In most instances, a positive screen tint film is used in which the dots are black.

If the desired tint is to be 60 percent, a 40 percent positive tint screen film is used. If the desired tint is to be 40 percent, a 60 percent positive tint screen film is used. These film tints are often taped behind the designated copy areas of the master film negative or positive during plate exposure.

It is important that each area of the paste-up to receive a screen tint be carefully marked with the desired percentage of tint and color of ink. This is done in red pencil on a tissue of vellum overlay attached to the original copy. If necessary, it can be done on the paste-up itself in nonreproducing blue pencil or pen.

screen tints differ from regular halftone contact screens in that there is no tonal variation within an area of the screen tint. See Fig. 17-23.

Tints are designated in percentage, on a scale of 0 percent representing WHITE and 100 percent representing solid BLACK. The range from 0 to 100 percent is usually in multiples of 5 or 10. Additionally, conventional percentage-calibrated screen tints are designated by the number of screen rulings per linear inch. These range from 65-line to 150-line. When art-

PERCENTAGE-CALIBRATED SCREEN TINTS

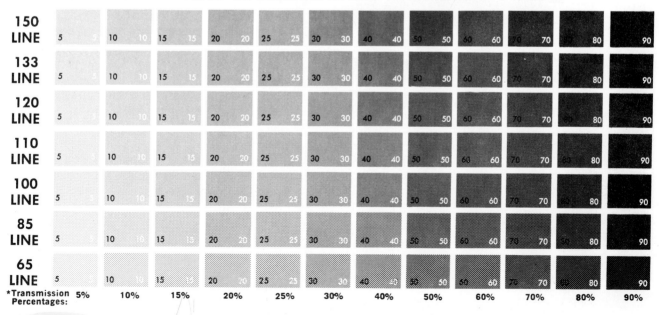

Fig. 17-23. Adhesive acetate materials, called screen tints, are used by paste-up artists to provide emphasis and special effects directly on paste-up copy.   (ByChrome Co.)

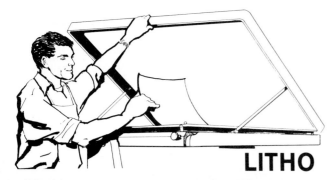

Fig. 17-24. When artwork and type are screened, printed material appears as a pattern of tiny dots. (ByChrome Co.)

Fig. 17-25 shows how a film positive tint method can be used to make normal line copy more interesting. This was done by providing an acetate (ruby film) overlay on the artwork in the shape of the tint areas. The line negative of the overlay was screened with a 150 line, 20 percent screen tint. It was double burned onto the offset plate. The shadow effect on the word litho was obtained by stripping a second line negative (screened with a 150 line, 30 percent tint) slightly out of register with the negative of the solid type.

Screen tints are ideal for business forms work. Printed materials such as office forms often become

heavy and hard to read when rules and type are only in black ink. Screen tints can be used to give write-in areas and headings importance and contrast.

When a photograph is to be used for a background effect, a simple method of subduing the photograph can be used. The screen tint is placed under the halftone negative to screen back the halftone, Fig. 17-26. A screen tint of the same line ruling as the halftone and angled at 30 degrees from the halftone is used. Selecting a screen tint of proper printing value will reduce the intensity of the halftone as much as desired.

Emphasis can be added to halftones by the addition of a color tint. The product or important subject matter of the photograph can be emphasized by using a screen tint over the product itself. Color can be added as the background around a product. To do this, an acetate overlay is prepared over the original photograph. The overlay is cut to provide the shape for the color tint area.

Screen tint texture patterns are available for a variety of printed products. These patterns provide a new dimension and modern appeal to artwork. Included among these special effect screen tints are several geometric patterns such as ring, ringlet, oval, wavy line, and concentric circle. These are suitable for checks, certificates, decorative panels, and borders.

Fabric design screen tints are eye-catching. Included in this group are the twill, denim, terry, mesh, burlap, and fibril patterns. These patterns can be used to create interesting and unusual effects for advertisements, booklet covers, menus, etc. Fig. 17-27 shows a brochure with a denim pattern printed on the cover to give it the richer look of a book.

The following information should be kept in mind when using screen tints prepared by the photo department:

1. When using positive copy, background tints should be light enough to keep the copy legible.
2. When using reverse copy, the background tint should be dark enough to carry the copy.
3. Screen tints tend to print 5 to 10 percent darker than specified.
4. Screen tints are difficult to butt together because the variation in angle produces a ragged edge. A black line can be used to separate adjacent screen patterns.
5. It is NOT advisable to superimpose one screen on another. The result is usually a moiré pattern. This is an objectionable wavering pattern that lacks tonal uniformity.
6. Tint screens should NOT be used behind type that is less than 8 points in size. The dots tend to destroy the letter structures of small type.

Adhesive shading method

Shading special effects tints can also be applied directly on the paste-up, as described in earlier chapters. These designs are printed on adhesive-backed

STOCK REQUEST		
No. of Sheets _____	SIZE: _____	
KIND _____		
WT. _____ COLOR _____		
CUT_____ OUT, TO PRESS SIZE_____		
NEEDED: DAY_____HOUR _____		
SPECIAL STOCK: FROM _____		
SPECIAL INK REQUEST		

Fig. 17-25. A film positive tint screen was used to make this line copy more interesting. This was done by preparing an acetate overlay on artwork in shape of tint areas. (ByChrome Co.)

USE A NORMAL
HALFTONE...

FOR BACKGROUND
OF CHARTS

NORMAL HALFTONE

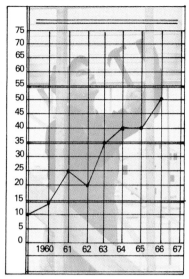

SAME HALFTONE NEGATIVE
SUBDUED WITH 150-30% SCREEN TINT

Fig. 17-26. This illustrates how a normal halftone can be subdued to create an effective contrast for type material.   (ByChrome Co.)

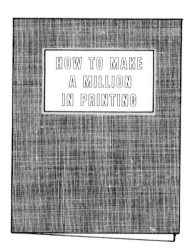

Fig. 17-27. Use of fabric design screen tints provides contrast and interest. Twill, denim, terry, mesh, burlap, and fibril are among fabric design tint screens available.   (ByChrome Co.)

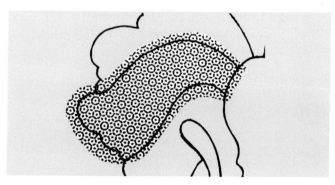

1. Place sheet over area to be shaded and cut piece of film slightly larger than needed. Use a sharp knife, but be careful not to cut through backing.

2. Lift cut area from backing sheet, position on art, and smooth down with heel of your hand.

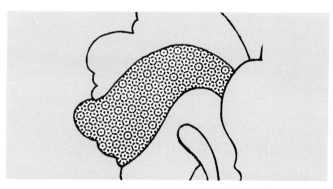

3. Trim away the excess pattern, cover Zipatone® with paper, and burnish firmly with a smooth broad-end burnishing tool.

4. Unused portion of sheet may be kept for future use without fear of deterioration.

Fig. 17-28. Adhesive screen tints are applied directly to artwork on paste-up board.   (Zipatone, Inc.)

transparent material mounted on a paper or acetate base. They are available in a variety of patterns and screen densities. A light tint might be 10 percent and a dark tint (almost solid) 90 percent.

In offset lithography, screens ranging from 65 to 100 lines per linear inch produce good reproductions without loss of dot detail. Screen rulings over 100 lines are difficult to photograph because of the smaller dots.

In using adhesive screen tints directly on the paste-up, the screen tint material is placed over the artwork or type to be treated. It is then cut to the desired shape, Fig. 17-28. It is secured by burnishing with a small hand-held burnisher.

Screen tint materials are usually applied to line drawings and outline type. By using adhesive shading materials, the paste-up can be presented for approval in the exact form in which it will appear when printed.

A sharp knife is used to cut around the desired image area. The knife should NOT penetrate the protective backing sheet. When cutting is complete, the knife is used to lift and peel off the screen tint cutout from the backing. The screen tint is positioned directly on the paste-up and burnished lightly. Any unwanted tint is trimmed off and removed. The image material is then burnished firmly.

The following procedures are recommended when attaching screen tint materials to the paste-up:

1. Carefully inspect all screen tint sheets for imperfections prior to use.
2. Make sure there are no pencil lines or dirt under the screen tint. The adhesive backing attracts dirt!
3. Do NOT attach screen tint materials over cut edges of artwork or type. When burnished, the edges distort the pattern and can be picked up by the camera.
4. With the exception of single-line patterns, screen tint materials cannot be effectively applied on top of each other. The result is a moiré pattern.
5. If a camera reduction in the paste-up size is planned, make sure that the screen tint dot sizes will accommodate the reduction without filling in and causing a loss of dots.

Surprinting

The printing of a solid image, such as type, over a lighter background is called *surprinting*, Fig. 17-29. Surprints are frequently combined with photographs to achieve special effects. A surprint is effective when combined with a light background.

A surprint usually consists of one or more lines of type matter. The type is composed and pasted up in the desired position over the base elements on an acetate overlay. The overlay is photographed to produce a line negative. A film positive is made from the negative.

The *base element* (either a photograph or screen tint) is then photographed. The positive containing the type matter and the negative containing the halftone or screen tint are combined and used to prepare an offset plate. A single exposure is made when preparing a reverse. The plate is then processed in the usual manner.

DIFFUSION-TRANSFER PROCESS

Photographic diffusion-transfer materials are intended to simplify production methods during paste-up and for making proofs. These materials are also suitable for preparing special effects. Diffusion-transfer materials are designed to produce photographic positive copies with a process camera. The positive copies are then used on paste-ups and other types of artwork.

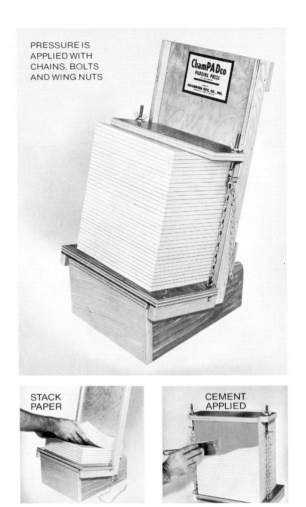

Fig. 17-29. Printing a solid image, such as type, over a lighter background area is called surprinting. (Cham Pad Co.)

The *diffusion-transfer* or *photomechanical transfer process* is a method to photograph copy and transfer the image directly to another carrier, such as a paste-up. It is unlike the typical offset procedure, such as: exposing the copy to make a film negative, developing the negative, then exposing the negative to photographic paper, film, or printing plate. With this process, copy is exposed on a negative paper and transferred directly to a receiver paper, film, or printing plate. The photomechanical transfer process permits the operator to begin with a positive image, make one exposure, and finish with a positive image.

The diffusion-transfer process can handle a wide variety of copy preparation operations. These include paste-ups, camera proofs, reflex proofs, line negatives, resized type, screened paper prints, halftone positives, in-position proofs, special effects, and reverses.

There are many advantages to preparing copy using the diffusion-transfer process:

1. You do not need a complex lith film system.
2. You can combine halftone copy with line copy to produce a single unit on the paste-up, permitting the copy-dot technique for preparing negatives

ready for platemaking. This eliminates the *stripping operation* (halftone taped onto film with line copy).

3. It can be used to make prints of many types of original copy. These include: yellow pages, newspapers, computer printouts, typewritten copy, pencil drawings, and phototypeset materials.
4. It allows the operator to drop out background colors and photographically clean up stains and dirt on the original copy.
5. It is less costly to retouch a reflection print than a film negative.
6. Changing the size of original artwork is easily accomplished.
7. The procedure permits the operator to make proofs.

The diffusion-transfer process originates from a donor photographic material, called the negative, which is exposed to the original copy in a process camera. The exposed negative material is then placed in contact with a receiver sheet and processed, Fig. 17-30. The image exposed on the negative material transfers to and appears on the receiver material. In this manner, a positive is made from a positive and a negative is made from a negative. With the new reversal materials, a positive can be made from a negative or a negative can be made from a positive. The process works because the donor material is light sensitive to accept exposures. The receiver material is chemically sensitive to accept the transfer. Because the donor material is light sensitive, the processing must be done under red safelight. See Fig. 17-31.

Paste-up artists realize the benefits of pasting up entire pages consisting of both line work and pre-screened halftone prints made through the diffusion-transfer process. This allows the entire paste-up to be treated as a line shot. It can be stripped up as one piece of film. This is often done in a fraction of the time it would take to assemble many line and halftone films on one or more flats.

The most reliable method of including halftones with line work is to first photograph individual halftone negatives. Then, photograph the line copy with white space left for the photographs. These are generally keylined with red or black lines indicating size and position. The final step involves assembly of the negatives into a comprehensive flat.

This method is time-consuming and has the potential for error unless the photographs are clearly marked for location. Additionally, assembling type close to halftones can, if the stripper is NOT careful, cause a build-up of tape which leads to halftone dot sizes spreading during plate exposure.

The problem of assembling type close to halftones can be simplified by stripping *complementary flats*. See Chapters 20 and 21. One flat is stripped consisting entirely of line copy and a second flat is stripped including all of the halftones. The plate is exposed twice. Often, a pin register system is used with this method to ensure speed and accuracy.

Another method that is popular for stripping halftones is to include black or ruby film windows on the paste-up. The windows indicate the size and location of each halftone. When the line shot is completed,

Fig. 17-30. In diffusion-transfer process, exposed negative material is placed in direct contact with a receiver sheet and inserted in a processor.   (Eastman Kodak Co.)

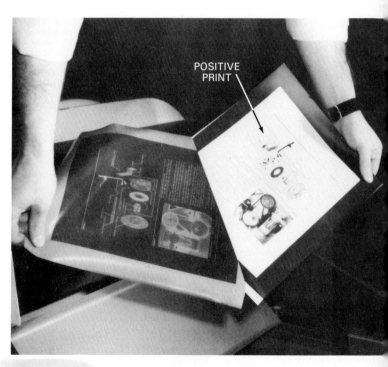

POSITIVE PRINT

Fig. 17-31. Diffusion-transfer processing results in a positive print of original copy.   (Eastman Kodak Co.)

the windows will be clear and the stripper positions each halftone into the appropriate window area. See Chapters 11, 20, and 21. One problem with this method is that the windows are not always clear. Lint and dust seem to collect on that part of the paste-up. A slight overexposure of the line shot can cause spotting in this area. Either method leads to imperfections in the halftone reproduction.

Using diffusion-transfer pre-screened prints to simplify the stripping procedure is also known as the *copy dot method*. Since the dots are already in the paste-up, they merely need to be copied with a line shot. Many printers hesitate to use this technique because they believe it produces poor results. There are time-saving advantages of being able to paste a positive print into position on a paste-up rather than cutting and taping films to a goldenrod flat. While the quality of the copy dot method is not high enough for some jobs, it can be useful for certain types of work.

Making a diffusion-transfer print

The *chemistry* (developing solution) used in the diffusion-transfer process is a monobath. *Monobath* means that the developer and fixer are combined in ONE solution. The solution is obtained from the manufacturer ready to use. A paper negative is exposed to light, and then run through a small diffusion-transfer processor along with a receiver sheet. Both sheets are quickly immersed in the monobath solution, and then pressed between rollers.

Areas on the negative sheet exposed to light harden. Those areas on the negative that were NOT exposed to light (corresponding to dark areas of copy) diffuse and transfer to the receiver sheet where they are reduced to metallic silver. This corresponds to a positive reproduction of the original, Fig. 17-32.

The procedure for preparing a diffusion-transfer print is described below:

1. Position the original line copy on the copyboard, Fig. 17-33A.
2. Place a sheet of photomechanical (PMT) negative paper on the camera back, Fig. 17-33B.
3. Set the camera timer for an exposure that is about 30 percent less than a normal line exposure. Make the exposure.
4. Remove the exposed negative paper from the camera back. Hold the negative material emulsion up, then place a sheet of PMT receiver paper above the negative material, emulsion to emulsion. See Fig. 17-33C.
5. Holding both sheets in an emulsion-to-emulsion orientation but separated by a fingertip, feed both sheets into a processor filled with PMT developer. The materials should enter the processor slightly apart, with the receiver above and the negative below the divider on the machine. The sheets will

Fig. 17-32. Diffusion-transfer process is extemely versatile since it can be used to prepare linework, reversals, and screened halftones. (Gilbert, Whitney, and Johns, Inc.)

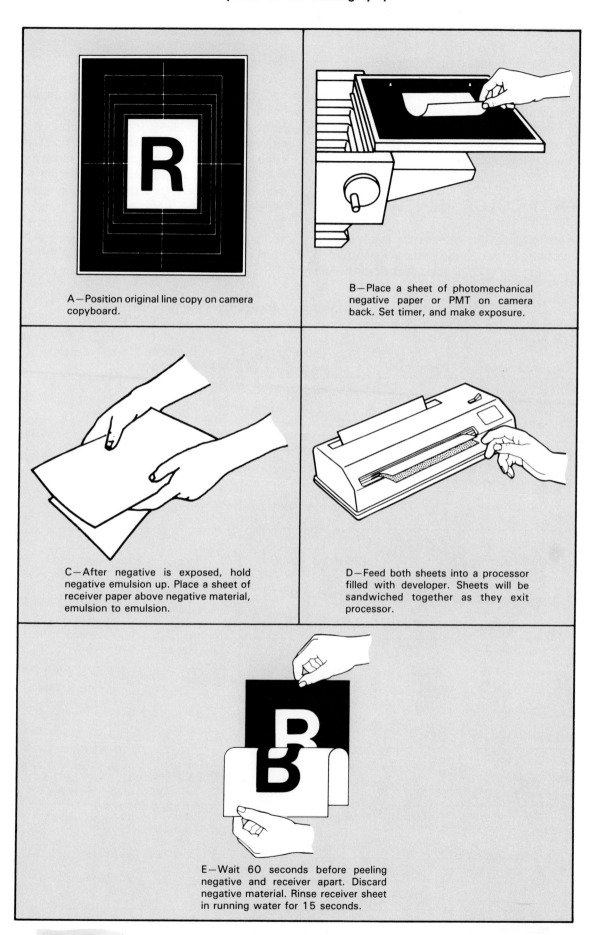

A—Position original line copy on camera copyboard.

B—Place a sheet of photomechanical negative paper or PMT on camera back. Set timer, and make exposure.

C—After negative is exposed, hold negative emulsion up. Place a sheet of receiver paper above negative material, emulsion to emulsion.

D—Feed both sheets into a processor filled with developer. Sheets will be sandwiched together as they exit processor.

E—Wait 60 seconds before peeling negative and receiver apart. Discard negative material. Rinse receiver sheet in running water for 15 seconds.

Fig. 17-33. Study procedure for preparing a diffusion-transfer print.   (Eastman-Kodak Co.)

be sandwiched together as they exit the processor, Fig. 17-33D.

6. After the sandwich has completely exited the processor, wait 60 seconds under safelights. Peel apart the sandwich and immediately rinse the receiver in running water for 15 seconds, then dry it. Look at Fig. 17-33E.

Fig. 17-34 gives a diffusion-transfer reference table. Study it. Fig. 17-35 summarizes diffusion-transfer operations.

PEEL-APART COLOR PRINTS

Color print reproduction systems are available for layout and paste-up artists. These systems are patterned after the diffusion-transfer process just described.

Using peel-apart diffusion-transfer color materials, artists can produce color prints to use for paste-ups and comprehensive layouts. By using the film material, overlays, transparent visuals for overhead projection, and other autiovisual aids, can also be produced.

The systems make use of equipment found in most printing firms. These include a process camera or enlarger and a diffusion-transfer processor. To make copies from transparencies, the camera should be equipped with an illuminator. Besides these main devices, materials, and the chemistry (depending on which system is being used), a few minor accessories are required. These include a color filter set, sodium vapor safelight, special contact screen, and an easel to hold the materials. See Fig. 17-36.

Some systems work almost the same as the familiar

DIFFUSION-TRANSFER REFERENCE TABLE

Your Need	Solution in Processor	Expose Donor Material	Transfer to Receiver Material	Exposing Device
1. Resized type 2. Resized screened halftone prints	A	PMT Negative Paper AD	PMT Receiver Paper A or PMT Receiver Paper, Type 2 AD	camera or enlarger
1. Resized type 2. Resized screened halftone prints	D	PMT Negative Paper AD	PMT Receiver Paper, Type 2 AD	camera or enlarger
1. Reversed type	D	PMT Reversal Paper D	PMT Receiver Paper, Type 2 AD	camera or enlarger
Resized 1. Overhead cells 2. Transparencies 3. Intermediates for proofs or positive working plates	A	PMT Negative Paper AD	PMT Transparent Receiver Sheet A	camera or enlarger
Same size 1. Overhead cells 2. Transparencies 3. Intermediates for proofs or positive working plates	A	PMT Reflex Paper A	PMT Transparent Receiver Sheet A	contact frame
Same size 1. Linework	A	PMT Reflex Paper A	PMT Receiver Paper A	contact frame
Short-run printing plate	A	PMT Reflex Paper A or PMT Litho Negative Paper A	PMT Paper Litho Plate A	contact frame or camera
Medium-run printing plate	A	PMT Litho Negative Paper A	PMT Metal Litho Plate A	contact frame or camera

Fig. 17-34. Study this diffusion-transfer reference table for various kinds of copy requirements. (Eastman Kodak Co.)

DIFFUSION-TRANSFER OPERATIONS

Donor Sheet	Activator/ Developer Temperature Range	Time in Diffusion-Transfer Processor	Waiting Time Before Stripping Donor Sheet From Receiver Paper
Reversal Paper D	65-80°F (18.5-26.5°C)	4½ to 6 seconds	60 seconds*
Negative Paper AD	65-80°F (18.5-26.5°C)	4½ to 6 seconds	30 seconds
Litho Negative Paper A	65-80°F (18.5-26.5°C)	4½ to 6 seconds	30 seconds
Reflex Paper A	65-80°F (18.5-26.5°C)	4½ to 6 seconds	30 seconds

IMPORTANT: Maintain the developer solution level at the recommended level. A high or low solution level will vary the development time.

NOTE: When KODAK PMT Negative Paper AD is processed with Transparent Receiver Sheet A, a higher maximum density can be obtained by leaving the sheets in contact for up to 90 seconds before stripping them apart.

If a uniform brown stain should occur on reverse prints, longer development time or a higher developer temperature will alleviate the condition.

*Reducing the waiting time before stripping the donor sheet from receiver material from 60 to 30 seconds can produce successful results in fresh developer. However, as the developer ages, the recommended strip time of 60 seconds is needed to produce optimum results and achieve recommended tray life. Extending stripping delay time from 60 seconds to 75 or 90 seconds will provide a means of extending the usable life of the developer.

Fig. 17-35. Study this reference table for various diffusion-transfer operating conditions. (Eastman Kodak Co.)

Fig. 17-36. Peel-apart, diffusion-transfer color materials are used for comprehensive layouts and some paste-up applications. Other uses include overhead transparencies, overlays, and audiovisual aids. (Agfa-Gevaert Inc.)

diffusion-transfer process. However, the diffusion-transfer processor is filled with color chemistry. Under a safelight, the camera operator places a negative in the camera back and exposes it at the proper size with the original in the copyboard. A typical exposure time for a continuous tone print is about 20 seconds at f/22. The operator places the exposed negative in contact with a receiver sheet and feeds them into the processor. After 60 seconds, the negative and receiver are peeled apart producing a color copy that can be used like any photographic print, Figs. 17-37 through 17-39.

Fig. 17-37. Exposed diffusion-transfer color negative and receiver materials are processed in a special one-step processor. (Agfa-Gevaert Inc.)

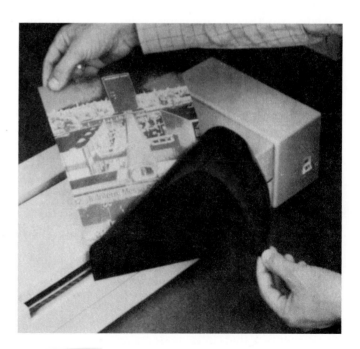

Fig. 17-38. Diffusion-transfer color negative and receiver materials are peeled apart, resulting in a color copy that can be used like photographic print. (Agfa-Gevaert Inc.)

Fig. 17-39. High quality photographic prints are produced by diffusion-transfer color process. This process is especially useful for advertising, comprehensive layouts, television, packaging, and architectural renderings. (Agfa-Gevaert Inc.)

Paper receiver sheets can be glossy (one-side imaging only) or matte (one or both sides). To make a color overhead transparency, all that is needed is the special film receiver sheet. The film receiver requires a two-minute transfer time.

These two systems have many applications in the printing and publishing industry. These include:

1. A process by which special effects can be prepared and presented for client approval.
2. A working medium in advertising research.
3. A final product in the photo-finishing industry.
4. In the presentation of comprehensive layouts to a client.
5. Packaging dummies for filming television commercials.
6. Documentation of artwork.
7. Colored cells for packaging design.
8. In photographic portfolios for illustrators, artists, and models.
9. Color copies from transparencies or drawings in the medical and engineering fields, among others.
10. Renderings in the architectural industry.
11. Color overhead transparencies for audiovisual presentations.

The following is a list of diffusion-transfer rules:

1. Remember that an increase in exposure time will lighten the print. For example, a shadow that has a 100 percent dot will need a longer flash exposure.
2. Be sure to allow the negative and positive sheets to transfer for the full time recommended by the manufacturer. The time is usually 30 seconds. While this may not be critical on line prints, it can make a difference with halftone prints.
3. Be sure to keep the chemistry in the processor fresh. This is essential when doing halftone work. Chemistry should not be replenished, but discarded when old. Oxidation ages diffusion-transfer chemistry rapidly. Keep it in a tightly sealed container, with all the air pressed out.
4. Special contact screens are manufactured for diffusion-transfer halftones. These are usually a positive type screen with a long screen range, designed to maintain detail and tonal separation in the middletones and shadows. Screen rulings range from about 65 to 133 lines per inch. For better grades of printing paper, use a positive screen with a fine screen ruling.
5. When determining exposure settings, remember that every three steps on the 20-step gray scale equals approximately one f-stop.
6. Keep in mind that the paper negative used in this process can be saved and sold for its silver content. This amounts to about 0.1 troy ounce per pound of scrap.

POINTS TO REMEMBER

1. Special effects photography goes beyond the familiar line and halftone processes.
2. Special effects refers to the creative side of darkroom photography and copy preparation.
3. Special effects are used for the reproduction of creative images from otherwise average copy.
4. Special effects contact screens are available in many different patterns including concentric circle, vertical line, horizontal or diagonal line, wavyline, and mezzotint.
5. Special effects contact screens are used much like halftone contact screens on continuous tone copy.
6. Posterization is another special effects technique. The different tones of an ordinary black-and-white photograph can be converted to a posterization. The tones of the original are reproduced as a line negative.
7. Posterizations are classified according to the number of colors and the number of different tones reproduced on the final press sheet.
8. The duotone is another type of special effects photography. The term duotone means the use of two layers of tones (colors) to produce a final image.
9. In the duotone technique, a black-and-white continuous tone photograph containing good contrast is converted into a colored illustra-

tion of depth and beauty. This requires that two separate halftone negatives, called printers, be prepared for each color.

10. Materials called screen tints are used by paste-up artists to provide emphasis and special effects to certain parts of the printed piece.
11. Screen tints are designated in percentages usually in multiples of 5 percent or 10 percent.
12. There are two basic methods used to provide screen tints on copy elements. These include film screen tint positives prepared by the camera department, and adhesive screen tint shading materials applied directly to the paste-up copy elements.
13. Surprinting is the printing of a solid image, such as type, over a lighter background. This requires a positive of the type matter and a negative of the halftone or screen tint.
14. In surprinting, both images are prepared on a single press plate and processed in the usual manner.
15. The diffusion-transfer process is suitable for preparing special effects materials for paste-up.
16. The diffusion-transfer process originates from a donor photographic material called the negative which is exposed to the original copy in a process camera. The exposed negative material is then placed in contact with a receiver sheet and processed. The image exposed on the negative material transfers to and appears on the receiver material.
17. Color print reproduction systems, similar to diffusion-transfer, are available for layout and paste-up artists.
18. By using the peel-apart color print reproduction systems, graphic artists can produce color prints to use for paste-ups, layouts, and comprehensive layouts.

KNOW THESE TERMS

Copy dot, Diffusion-transfer, Dot gain, Duotone, Four-tone posterization, Line conversion, Posterization, Screen tint, Special effects, Surprint, Three-tone posterization.

REVIEW QUESTIONS

1. Special effects refers to the creative side of darkroom photography and copy preparation. True or false?
2. One of the following is *not* a special effects screen:
   a. Mezzotint.
   b. Wavy line.
   c. Circular.
   d. Backslant.
3. In posterization, _____ negative is required for each tonal color.
4. _____ camera exposures are required to produce a one-color, three-tone posterization.
5. Three and four-tone posterizations can be produced on a process camera or darkroom _____.
6. A duotone consists of two halftones called:
   a. Stats.
   b. Printers.
   c. Mezzotints.
   d. Surprints.
7. Adhesive shading materials, called _____ tints, are used by paste-up artists to provide emphasis to certain parts of the paste-up.
8. Film positive screen tints are positioned and taped behind the designated copy areas of the stripped flat. True or false?
9. The printing of a solid image over a lighter background is called _____.
10. Diffusion-transfer materials are designed to produce photographic positive copies with the use of a _____ camera.

SUGGESTED ACTIVITIES

1. Obtain a black-and-white photograph suitable for making a background dropout. Make a dropout halftone negative by the regular high contrast film method. Make a dropout halftone by the diffusion-transfer method.
2. Obtain a black-and-white photograph suitable for making a tone-line posterization. Make a tone-line film positive. Make a negative from the positive.
3. Obtain a black-and-white photograph suitable for a 3-tone and a 4-tone posterization. Prepare the tones with care and note whether the correct detail was attained.
4. Obtain a black-and-white photograph with good contrast. Prepare a set of duotone negatives. Prepare color proofs of the negatives and note whether the correct detail was attained.
5. Obtain a black-and-white photograph. Use a suitable special effects contact screen to prepare a negative. Prepare a paper contact proof of the special effects halftone.

Four-color photography can be used to produce a dramatic effect.   (Eastman Kodak Co.)

Chapter 18

Full-Color Reproduction Photography

When you have completed the reading and assigned activities related to this chapter, you will be able to:

O Describe why full-color images are converted through color process photography.

O Describe the basic scientific facts of color as related to process photography.

O Describe how color is seen by the human eye.

O List and compare additive and subtractive color.

O Identify the three important dimensions of color.

O Select examples of the two basic kinds of copy that are used to make color separations.

O Name the color correction methods that are used to improve color separation.

O Describe the use of colored filters and different halftone screen angles in the preparation of process color separations.

Color adds visual impact and interest to the printed page. It is often used to influence consumers since it lends a lifelike quality to photographic reproductions, Fig. 18-1. As this chapter will explain, color can be used on the printed page to set an overall mood. This means that paper has an influence on color printing too. The uses of color vary, depending on the kind of product being printed.

UNDERSTANDING COLOR

Color is light. Light is a form of energy and, according to theory, travels in waves. Light waves start from a source, such as the sun, a light bulb, or a white can-

Fig. 18-1. Color lends a lifelike quality to photographic reproductions. Color is used to influence consumers and stimulate sales of a product. (W.A. Krueger Co.)

dle. In any of these sources of light, there are many different wavelengths (distance between waves).

Wavelength is measured from crest to crest in *nanometers* (billionths of a meter) or in *millimicrons* (millionths of a millimeter). The visible spectrum range is generally considered to be between 380 millimicrons and 770 millimicrons. This spectrum is part of the much larger electromagnetic spectrum.

In the seventeenth century, Sir Isaac Newton established an important principle of light. He discovered that a beam of sunlight passing through a prism is refracted or bent into separate bands of colors, Fig. 18-2. These are called the colors of the visible spectrum. Each color has its own wavelength.

When all the wavelengths are combined in suitable proportions, they produce *white light*. All individual and combinations of colors inherent (belong) in white light. When you see a rainbow, this is a way nature has of displaying color.

How we see color

All visible colors are contained in white or colorless light. Light energy travels in waves with each perceived color having a dominant wavelength. Each color's wavelength differs from the wavelength of any other perceived color.

The human eye processes color using color receptors. One set of receptors is sensitive to red wavelengths of light. A second set is receptive to green wavelengths. The third set of receptors is sensitive to blue-violet wavelengths. When the red and green receptors are stimulated, for example, you see yellow.

The way we perceive the various colors of the spectrum depends upon the strength and mixture of wavelengths that strike our visual receptors. Look at Fig. 18-3.

Colorless or white light is seen when our receptors are stimulated equally by all wavelengths. Color then is a sensation resulting from light energy striking the

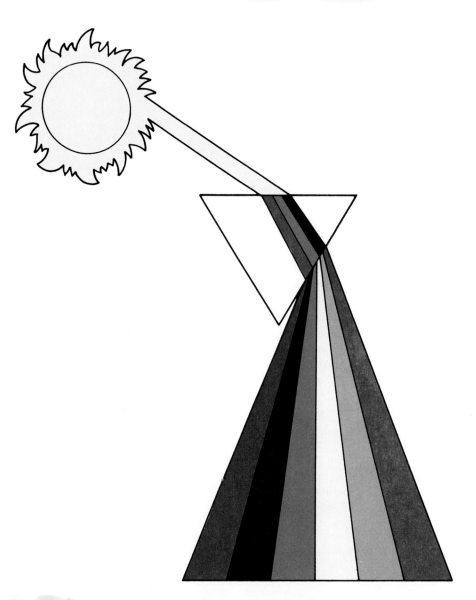

Fig. 18-2. Colorless light passing through a prism is reflected or bent into separate bands of colors. These are called colors of visible spectrum. (S.D. Warren Co.)

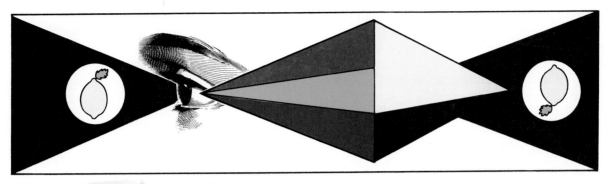

Fig. 18-3. When the red and green receptors of our eyes are stimulated, we see yellow, as in this example of a lemon. (S.D. Warren Co.)

receptors in our eyes. Our brain interprets these as being various combinations of red, blue, and green.

It can then be seen that perception of color is totally dependent upon light. Color perception depends upon the QUANTITY of light and the QUALITY of light available to the receptor in our eyes. An example of this is when we see the brilliant colors of sunset fade to a neutral gray and then disappear as light diminishes from daylight, to dusk, to darkness. Variation in the quality of light influences our perception of color, too. For example, have you ever purchased a sweater that appeared to be one shade of color under store lighting but different out in the sunlight?

Lightwaves reach our eyes in a number of ways: directly when we look at the sun or a light bulb, or indirectly when lightwaves pass through a transparent object held between the source of light and our eyes. This is called *transmission*.

Another way lightwaves reach our eyes is when light bounces from an object and into our eyes. This is called *reflection*.

The red of an apple is an illustration of color perceived by light reflection. Why do we see it as red? The apple absorbs all wavelengths but red, that is reflected from the surface of the apple to our eyes. The receptor, which is subject to stimulation by this particular wavelength, sends a signal to the brain. The brain upon receiving the signal, thinks: "Red." See Fig. 18-4.

In other words, an opaque substance like an apple appears to be a particular color. This is because it reflects the wavelengths corresponding to that color and absorbs those that do not. If an apple reflected all wavelengths without any one wavelength dominating the other, the apple would be perceived as white.

The principle remains the same for a transparent substance, such as colored glass. The transparent substance absorbs some wavelengths and transmits others. A transparent object, such as a green bottle, is seen as green by the same principle that the apple is seen as red. All wavelengths but green are absorbed. The green wavelength is transmitted rather than reflected to the eye. This is how you see color, Fig. 18-5.

Fig. 18-4. The red of an apple is a good example of color perceived by light reflection. (S.D. Warren Co.)

ADDITIVE AND SUBTRACTIVE COLOR

There are other elements of light and color. To understand them, we must return to Sir Isaac Newton's spectrum. Not all of the colors of the spectrum have equal status.

There is a basic difference between the color behavior of light and the color behavior of the red apple. It is basically a difference between something that adds or subtracts. Color that results from adding light energy is called *additive color*. Color that results when an object subtracts light energy is called *subtractive color*.

Additive color

A light source, such as the sun or a light bulb, is additive energy. Two separate light sources contribute more energy than one light source. Yellow is perceived when a green light source adds its energy to the energy of a red light source. The result is obtained by the addition of energy of differing light wavelengths.

It can be demonstrated that three colors such as red, green, and blue, when suitably combined, will reproduce white or colorless light, Fig. 18-6. Also, by combining red, green, and blue in different proportions, any of the other colors of the spectrum can be produced. As a result, red, green, and blue are called the *primary colors* of the visible spectrum. A combination of any two primaries produces an intermediate color which is called a *secondary color*. Red and blue

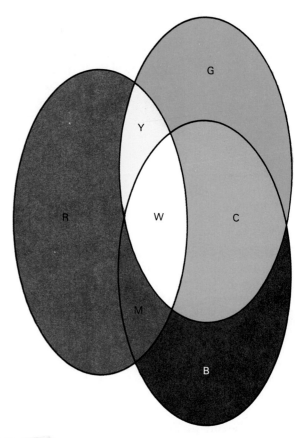

Fig. 18-6. When red, green, and blue are combined, they produce white or colorless light. Red, green, and blue are called the primary colors of visible spectrum. (Southwestern Publishing Co.)

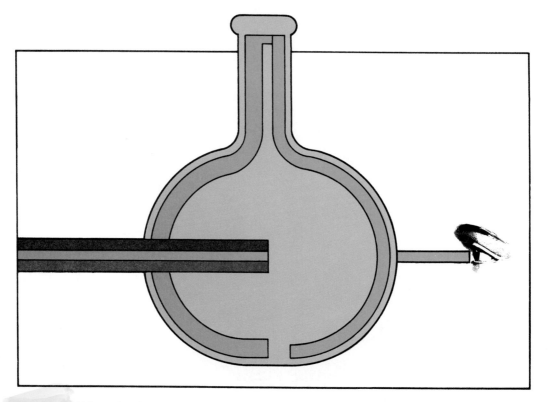

Fig. 18-5. When viewing a transparent object, such as this bottle, green wavelength is transmitted rather than reflected to eye. (S.D. Warren Co.)

382

produce magenta; green and blue produce cyan; green and red produce yellow. Refer to Fig. 18-7.

The red apple in Fig. 18-4 subtracts energy. It has no light energy of its own. However, it modifies light contributed to it from a radiating source, such as the sun or a light bulb. Inks, pigments, apples, and almost anything else are perceived as a certain color because you see the wavelengths they reflect or transmit. You do NOT see the wavelengths that are absorbed.

Subtractive color

It is also important that you understand the subtractive method of obtaining color. This is because in offset lithography you are concerned with such light absorbing materials as pigments, films, and paper surfaces. All materials that are NOT sources of light energy operate by the subtractive method.

There are three additive primary colors of light: red, green, and blue. These colors add their energies to produce all the other colors of the visible spectrum. Therefore, three subtractive colors can be selected which, in proper combination, duplicate the colors of the visible spectrum. This is shown in Figs. 18-1 and 18-7.

These subtractive secondary colors are commonly known as cyan, magenta, and yellow. Cyan contains blue and green; magenta contains red and blue. Each process printing ink (cyan, magenta, and yellow) ideally transmits two-thirds of the spectrum and absorbs one-third of the spectrum. This means that cyan

(blue/green) absorbs red; magenta (red/blue) absorbs green; and yellow (red/green) absorbs blue. Refer to Fig. 18-8.

Cyan has the property of transmitting blue and green and absorbing red. Yellow has the property of transmitting red and green and absorbing blue. The only color that both have in common or that both can transmit is green. Therefore, the visual result of overlapping these two subtractive secondaries is green.

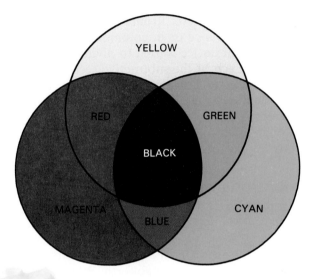

Fig. 18-8. Subtractive secondary colors are known as cyan, magenta, and yellow. As shown, cyan contains blue and green; magenta contains red and blue.

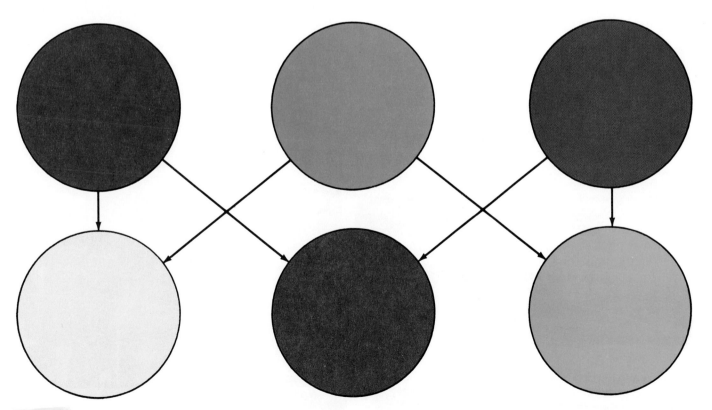

Fig. 18-7. This illustrates how a combination of any two primary colors produces an intermediate color, called a secondary color. Red and blue produce magenta; green and blue produce cyan; green and red produce yellow.   (S.D. Warren Co.)

When all three subtractive secondaries are over-lapped, there is no one color common to all three. As a result, they in effect cancel each other out and the result is black.

Subtractive secondary colors are needed to make the reproduction of other colors possible by the subtractive process. For example, consider the result when blue and red are overlapped. Each is a single color representing and transmitting only one-third of the spectrum, absorbing two-thirds. Therefore, they cancel each other out, for they in effect have nothing in common. A third color will result, however, from the overlapping of two colors that have a color in common.

A convenient device is used to divide the spectrum into thirds. Each of the subtractive secondary colors of light comprises, for practical purposes, one-third of the visible spectrum. Each of the subtractive primary colors represents two-thirds of the spectrum, which they must do in order to make the subtractive process work effectively. See Fig. 18-8.

Dimensions of color

Before going further, you should understand that color has three important dimensions: hue, value, and chroma.

1. *Hue* is the characteristic of color that you call red, green, yellow, blue, and so forth. By measurement, hue may be assigned a dominant wavelength.
2. *Value* is the lightness or darkness of a color as measured against a scale running from white to black. This means that there are light blues and dark blues, or light red, or dark greens, and so forth.
3. *Chroma* is descriptive of hue purity or cleanliness, which is another way of describing intensity or saturation. Therefore, you might call a color pure blue as compared to grayish blue.

PRINCIPLES OF PROCESS COLOR

The use of two or more colors to print line copy is usually called *mechanical color*. The preparation of the artwork for printing in different colors is done mechanically (by hand methods) usually by an artist. Multicolor photographs, paintings, transparencies, and other continuous tone copy cannot be separated by mechanical hand methods. Photographic methods are used to prepare film separations. It is referred to as *process color*. This method involves special filters, contact screens, and panchromatic film. In some cases, masking film is also used.

Three-color process

The *three-color process* printing method uses yellow, magenta, and cyan process inks. Black ink is NOT used as a printing color. When properly used, the three

process inks form a near-black color in the shadow areas. Fig. 18-9 is an example of three-color process printing.

Four-color process

In the *four-color process* printing method, yellow, magenta, cyan, and black process inks are used. The black increases the density range, improves shadow detail, and makes control of the other three colors less critical as to ink balance. Fig. 18-10 is an example of four-color process printing.

Purpose of color separation

The purpose of *color separation* is to separate the hues of a color original into four negatives. The negatives must be made to represent the quantities of density that can be used to prepare cyan, magenta, yellow, and black offset printing plates. Each plate is then used to print its color of ink. The four colors of ink combine to produce a full-color reproduction.

In color separation, the primary additive colors of red, blue, and green are used as filters to prepare cyan,

Fig. 18-9. Three-color process printing method uses yellow, magenta, and cyan process color inks. In this method, black is not used as a printing color. (The Lehigh Press, Inc.)

reflected light. Transmission copy is color separated by transmitted light.

Color separation filters

Colored filters must be used to prepare color separations on process cameras and on electronic scanners. The filters are the same colors as the three primary light colors (blue, green, and red), Fig. 18-11. Filters are constructed to absorb and to transmit light.

The blue filter absorbs green and red light. It transmits blue light. This creates the yellow printer. The green filter absorbs blue and red light. It transmits green light. This creates the magenta printer. The red filter absorbs blue and green light and transmits red light. This creates the cyan printer. The black printer is made by using all three filters.

Yellow, magenta, cyan, and black are the colors of the process inks used to create color reproductions.

Color separation halftone screens

Each process color separation must be converted into a halftone negative or positive before printing. This conversion requires that the separations be screened

Fig. 18-10. Four-color process printing method uses yellow, magenta, cyan, and black process color inks. Black increases density range and improves shadow detail in halftones. (The Lehigh Press, Inc.)

magenta, and yellow separations. Each of the required negatives are produced through a different color filter. A red filter produces the cyan negative, called *printer*. A blue filter produces the yellow printer. A green filter produces the magenta printer. All three filters are used to produce the black printer.

After the negatives are exposed to offset printing plates, the clear areas on the film become areas of density on the plate. Printing of all four plates on paper results in a reproduction of the range of hues of the color original.

Types of color originals

There are two basic types of continuous tone, color originals used to make color separations. These are reflection and transmission copy.

Reflection copy reflects light and includes such items as paintings and color photos. *Transmission copy* allows light to pass through the image and is usually 35 mm slides. Reflection copy is color separated by

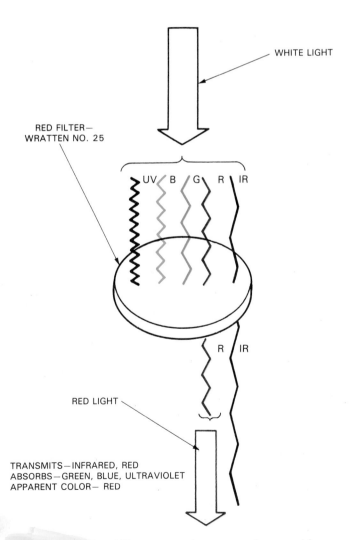

Fig. 18-11. Colored filters are used to prepare three- and four-color separations on process cameras and electronic scanners.

in the same manner as for a black and white halftone. Gray contact halftone screens are used to make process color separations. This is because gray is a neutral color. This means that the three colors of light pass through the screens without being affected or filtered. Magenta contact halftone screens would act as filters.

The screen for each separation must be positioned at a different angle, Fig. 18-12. The usual angles consist of the black printer at 45 degrees, the magenta printer at 75 degrees, the yellow printer at 90 degrees, and the cyan printer at 105 degrees. Pre-angled contact screens are available for the four angles.

Laser fiber optics are used on many color separation scanners instead of contact screens. Color separation scanner methods are discussed in Chapter 19.

COLOR SEPARATION METHODS

There are several devices used to prepare color separations for four-color process printing. Color separations can be made using a process camera, contact printing frame, or a photo enlarger. The most popular method of preparing color separations is by electronic color separation scanners. This is because of their great speed and versatility in the use of color originals. See Fig. 18-13.

Fig. 18-13. Most popular method of preparing color separations is by electronic color separation scanners. Scanners are fast and can handle a wide variety of color originals. (Crosfield Electronics, Inc.)

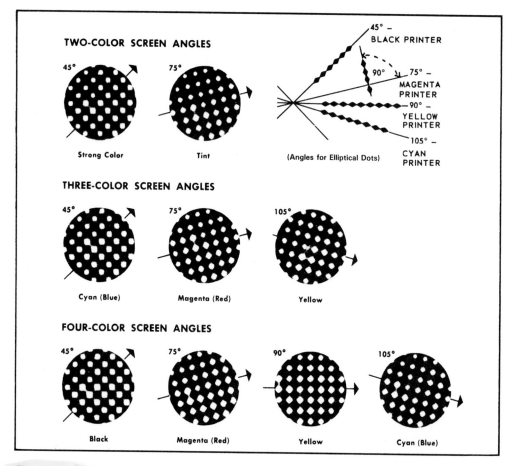

Fig. 18-12. Halftone contact screen for each separation must be positioned at a different angle. Shown here are screen angles for two, three, and four colors. (Eastman Kodak Co.)

Instruction books are available which cover the subject of color separation techniques in great detail. Refer to the Graphic Arts Technical Foundation (GATF) or the Eastman Kodak Company for more specific information.

There are two basic methods of making color separations. These include the direct and indirect methods.

The direct screen method produces each screen separation in one step. It is the simpler of the two methods. The indirect screen method requires three steps. First, continuous tone separation negatives are made. Then, halftone (screened) positives are prepared from the continuous tone negatives. Finally, the screened positives are contact printed to produce the final separation negatives. The additional steps in the indirect method can be used for color correction and proofing.

The direct color separation method is illustrated in Fig. 18-14. The two applications described here include the process camera and contact methods.

Process camera direct screen method

A process camera equipped with proper lights and filters is required for this method. The original color copy is set up on the process camera copyboard. Four separate exposures are made. Each exposure is made on a separate piece of panchromatic film. This is done with a different filter in the camera lens for each exposure and with the screen correctly angled for each exposure. The colors of the original color copy, recorded on the separation negatives, appear as black and white densities.

The first exposure will record the cyan in the original copy. It is made with a red filter in the lens and with the screen angled at 105 degrees over the film. This becomes the cyan printer. It is used to expose the offset plate for printing the cyan ink on the press.

The second exposure for the magenta printer is made with a green filter. The screen is angled at 75 degrees over the film.

The third exposure for the yellow printer is made with a blue filter. The screen is angled at 90 degrees over the film.

The fourth exposure for the black printer can be made as three partial exposures on the one piece of film. The first exposure uses the red filter; the second exposure the green filter; and the third exposure the blue filter. The screen is angled at 45 degrees over the film for these three partial exposures.

Contact direct screen method

The contact direct screen color separation method has a major advantage. It requires only simple, inexpensive equipment. There is no need for an enlarger or process camera. All that is needed is an open-face contact vacuum frame and a point source light.

The contact method, like the camera direct screen method, consists of two steps. These include making the mask and making the halftone separation negatives. This method is ideally suited to tri-mask film. This is because it contains three color-correcting masks in a single film. Each layer is sensitive to one of the primary light colors—red, blue, or green. The developed dye images form the desired mask for each of the separation negatives.

Fig. 18-15 illustrates how the mask is produced by exposure in contact with the transparency.

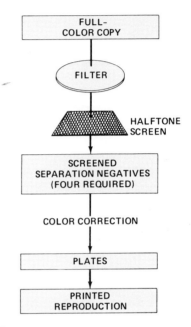

Fig. 18-14. Direct color separation method is one of several ways in which color separations can be produced. Procedures are repeated for each of the four colors. (Eastman Kodak Co.)

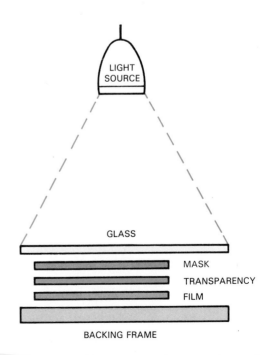

Fig. 18-15. In contact direct screen color separation method, a mask is prepared by exposure in contact with transparency. (Chemco Photoproducts Co.)

Once the mask is made, each of the four-color separations can be prepared. Each halftone negative is produced on halftone film by exposure through the masked transparency and contact halftone screen, Fig. 18-16. Exposure times, filters, and results are described by the manufacturer of the film.

Indirect screen separations

The indirect color separation method is illustrated in Fig. 18-17. The indirect color separation method requires more steps and uses more film than the direct method. This method is preferred when colors are to be corrected by hand. It also allows for color correcting on the continuous tone film positives. The same set of color corrected positives can be used to make enlargements or reductions before screening.

It is important to remember that the separation negatives are exposed without the contact screen in the camera. This results in continuous tone separation negatives. Continuous tone film positives are then made from the continuous tone separation negatives. Color correction is done before these continuous tone film positives are contact printed through the halftone screen. The end result is a set of color corrected halftone negatives from which the offset plates are prepared.

Color correction

The three functions of *color correction* are:
1. To compress the density range of the color original.
2. To compensate for color inadequacies in process inks.

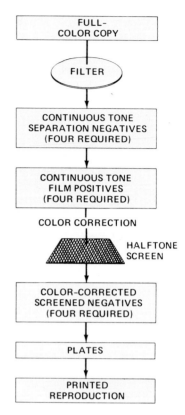

Fig. 18-17. Indirect color separation method requires more steps and uses more film than direct color separation method. (Eastman Kodak Co.)

3. To enhance the detail of the final reproduction.

Color correction can be done photographically, manually, or electronically. When color correction is done photographically, the operation is called *masking*. This method uses a special masking film. The mask is made directly from the original reflection or transparency copy. It is made by using special colored filters. The main purpose is to reduce the amount of light passing through some portions of the image when separations are made.

Manual color correction involves retouching by hand. Corrections are made in halftone positives by reducing the size of the dots with chemical reducers. This process is also called *dot etching*.

Electronic color correction can be done while the separations are being scanned. This process is very popular for making color separations. It has the versatility of color correcting by either adding or deleting color in any given area of the color originals. A color separation scanner is shown in Fig. 18-18.

Color separation proofs

After the color separation negatives or positives have been stripped but before the job is put into production on the press, proofs are prepared. *Proofs* are used to check for *register* (alignment), *imposition* (page location), accuracy, quality, and to obtain the customer's approval. The proofs also give a visual check of how the job will appear when it is printed.

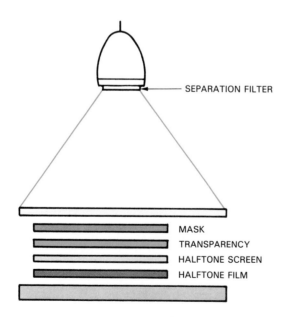

Fig. 18-16. Using mask, each of four-color separations can be prepared. Each halftone negative is produced on film by exposure through masked transparency and contact halftone screen. (Chemco Photoproducts Co.)

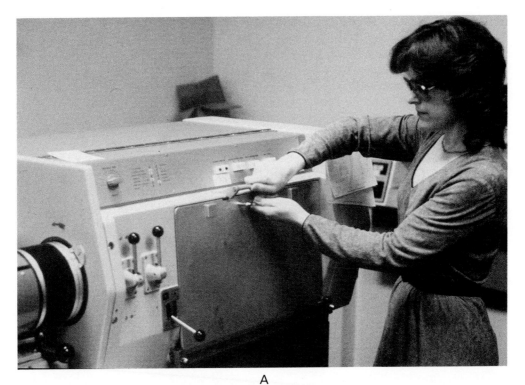

THE MAIN COMPONENTS OF THE CHROMAGRAPH DC 300

1. LAMP COMPARTMENT
2. HALOGEN LAMP HOUSING
3. FEED MOTORS
4. TRANSPARENCY ARM
5. SCANNING DRUM
   (INTERCHANGEABLE)
6. SCANNING HEAD
7. MASK SCANNING HEAD

8. MASK DRUM
9. RECORDING SPACE
10. DAYLIGHT CASSETTE
11. COLOR COMPUTER
    WITH CONTROL UNIT
    AND EXTENDED SELEC-
    TIVE CORRECTION
12. BASE FRAME

B

Fig. 18-18. A—Operator is using an electronic color separation scanner. B—Diagram of an electronic color scanner. (HCM Corp.)

There are basically several systems used to proof color separation negatives or positives. Some of these include: color proof press, color key, transfer key, cromaline, and cromacheck.

Two types of color proof press systems are in general use. These include press and off-press.

For some jobs, offset plates are made and run on four-color *proof presses,* Fig. 18-19. These presses hold all four printing plates. They print all four colors on a single sheet of paper in one pass through the press.

The press contains two large cylinders called the plate and blanket cylinders. The plate cylinder holds all four

A

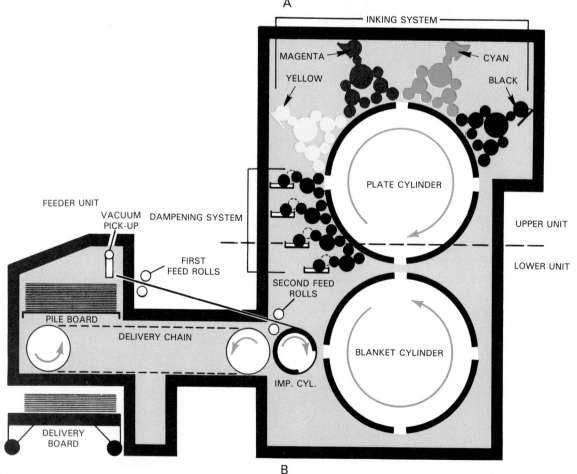

B

Fig. 18-19. A—Approval for printing of a color job may require proofing on a four-color proof press. Same plates, paper, and inks to be used for actual press run are used for proofing. B—Study schematic diagram of four-color proofing press. (Bobst Group, Inc.; Vandercook, Division of Illinois Tool Works)

offset plates. The blanket cylinder transfers each ink color to the sheet of paper. The impression cylinder (a third cylinder) makes four revolutions per sheet. This is required in order to print all four colors in register.

Off-press proofs are made directly from the corrected films. There are two types of off-press proofs called overlay and integral.

The *color key proofing system,* available as negative or positive working, consists of a light-sensitive polyester-base overlay film. It is available in any one of a number of colors. The process colors are used for making a proof of each color separation negative or positive. Each individual proof made for the separation is a color transparency. The four transparencies are taped in register and in printing sequence on a piece of white paper.

A standard platemaking exposure unit can be used to expose the film material. Yellow overhead lights in the exposing area are recommended. A 12-step platemaker's sensitivity guide with a density of about .97 at step 7 can be used to check correct exposure.

After exposure, the film material is processed with a one-step developer. The film material is washed in cold water and squeegeed. After the material is thoroughly dry, it is ready for viewing. This processing procedure is illustrated in Fig. 18-20.

The *transfer key (integral) proofing system* results in a single-sheet proof. This system involves the use of color dyes or pigments. Each of the four process colors is placed on a single base sheet, which may be the actual printing paper to be used for the job. The result is a proof on a single sheet of paper. These proofs appeal to customers since they appear realistic and precise.

The *cromaline (integral) proofing system* uses a patented laminator to apply a special photo-polymer to the proof paper. The laminated sheet is then exposed through a film positive using a conventional platemaking system. After exposure, the top mylar protective layer is removed and the entire sheet is dusted with a color toner, which only the exposed areas accept.

The dry toners are available in a wide variety of colors and can be mixed to match almost any press ink color. After all surplus powder has been removed, the proof sheet can be relaminated and exposed to additional flats to produce other color images.

The *chromacheck (overlay) proofing system* consists of four different films keyed to the four ink colors. Each of these films is composed of a color-coded, ultraviolet-sensitive photopolymer film sandwiched between two substrates of mylar polyester film. When, for example, the blue chromacheck film is exposed to the blue separation negative, a positive-reading image of the picture's blue components is produced instantly.

Chromacheck needs no processing. Once exposed to ultraviolet light, each color film simply is placed on a vacuum easel and the film sandwich peeled apart. The image adheres to the clear overlay film. When all four colors are peeled and assembled, the complete color page can be previewed. From start to finish, proofing requires only five minutes.

It should be noted that several integral systems come closer to matching the press print in appearance. However, most of these processes use special bases or substrates which do not look or feel like the printing paper. The color proofs are generally thought of as final proofs for customer approval.

All of the color proofs just described should be viewed under a common light source. Any variation in color temperature, light intensity, amount of reflected room light, evenness of illumination, or surrounding color environment will change human judgment with respect to color values. Many problems result when the printer and customer use different light sources for viewing the proofs. Special lighted viewing systems have been developed that meet the industry's specifications. Look at Fig. 18-21.

Process color charts are available that show designers and printers how process colors will appear when produced with various screen percentages, Fig. 18-22. A full size reproduction of a partial page from a process color chart is shown in Fig. 18-23.

POINTS TO REMEMBER

1. As with continuous tone black-and-white copy, continuous tone color copy must also be converted to a dot or halftone pattern.
2. In full-color reproductions, four sets of halftone negatives must be prepared.
3. The light around us is made up of the additive primary colors of red, blue, and green. When all of these colors are blended together, we see white.
4. The purpose of printing ink is to absorb light. Green ink, for example, absorbs red and blue light while reflecting green light to our eyes.
5. Ink works on the subtractive principle with the primary colors being yellow, magenta, and cyan.
6. Hue, tone, and saturation are other color principles that affect the way we see color.
7. Original color copy, such as original paintings or photographs, can be separated into four color printers using filters. Each printer represents a color absorbed in the original.
8. The four printers include: cyan, magenta, yellow, and black. As an example, the yellow printer represents the blue absorbed in the original.
9. Since process color inks are NOT perfect in their ability to absorb light, color correction must be used.
10. The direct method of color separation produces halftone negatives directly from the original color copy.

391

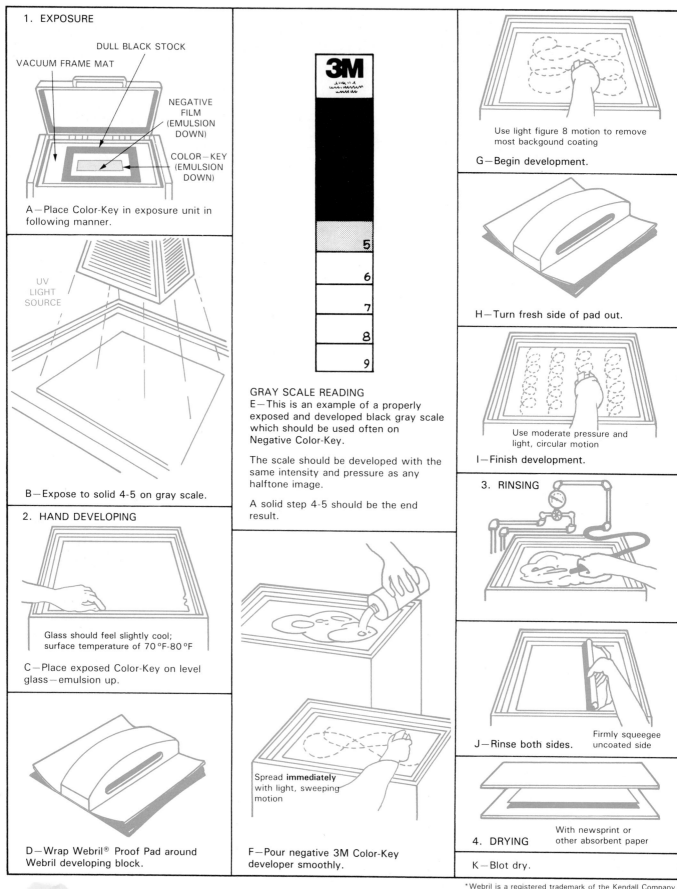

1. EXPOSURE

DULL BLACK STOCK

VACUUM FRAME MAT

NEGATIVE FILM (EMULSION DOWN)

COLOR—KEY (EMULSION DOWN)

A—Place Color-Key in exposure unit in following manner.

UV LIGHT SOURCE

B—Expose to solid 4-5 on gray scale.

2. HAND DEVELOPING

Glass should feel slightly cool; surface temperature of 70°F-80°F

C—Place exposed Color-Key on level glass—emulsion up.

D—Wrap Webril® Proof Pad around Webril developing block.

3M

5

6

7

8

9

GRAY SCALE READING
E—This is an example of a properly exposed and developed black gray scale which should be used often on Negative Color-Key.

The scale should be developed with the same intensity and pressure as any halftone image.

A solid step 4-5 should be the end result.

Spread **immediately** with light, sweeping motion

F—Pour negative 3M Color-Key developer smoothly.

Use light figure 8 motion to remove most backgound coating

G—Begin development.

H—Turn fresh side of pad out.

Use moderate pressure and light, circular motion

I—Finish development.

3. RINSING

J—Rinse both sides. Firmly squeegee uncoated side

With newsprint or other absorbent paper

4. DRYING

K—Blot dry.

*Webril is a registered trademark of the Kendall Company.

Fig. 18-20. Off-press, full-color proofs are made directly from corrected separation films. One method involves preparing polyester-base overlay transparency films for each color. Four transparencies are taped in register and in printing sequence on a piece of white paper.   (3M Company)

Chapter 19

Color Scanners

When you have completed the reading and assigned activities related to this chapter, you will be able to:
O Describe how early electronic scanners evolved.
O Describe the operating principle of a modern electronic color scanner.
O List and describe the main components of a modern electronic color scanner.
O List and describe the kinds of copy handled on an electronic color scanner.

A *color scanner* is a device that changes an image into electronic data that represents the image. The electronic data or image can then be modified, enhanced, or separated using a computer. The color scanner has drastically changed how many printers handle color work.

This chapter will describe the construction and operation of a color scanner. It will also explain the many new terms that are used when working with a scanner. Since scanners might someday replace the process camera, it is very important that you fully understand scanner theory and operation.

SCANNERS

As you have learned, the process camera is an exposing unit used to record pictures in both line and continuous tone form. Similarly, the color scanner is also an exposing unit used to record pictures. Refer to Fig. 19-1.

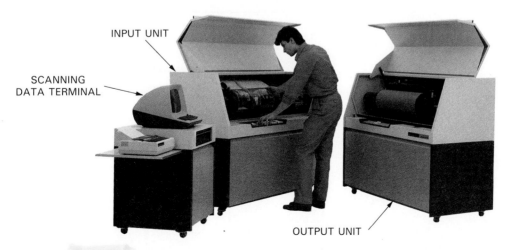

Fig. 19-1. A color scanner is an electronic exposing device used to prepare color separations for process color printing. Scanner's main components are identified. (Crosfield Electronics, Inc.)

A camera uses a lens, an exposing device, and film to record the tone scale and produce the halftone picture. The difference between it and a scanner is that the camera records the entire picture at one time. The color scanner records only tiny segments of the picture at one time. Because the color scanner records the picture bit by bit, it has greater flexibility when making desired changes in the original picture.

A color scanner performs the same basic operations of a process camera, but it works electronically. See Fig. 19-2. In the scanning process, a picture is examined by an optical device. This information is electronically modified and reproduced as pictures. Like the camera, the color scanner separates the color copy into negative or positive film of the standard four images. These images include cyan, magenta, yellow, and black. Just as photographic masks permit the correction of separations for the deficiencies of the printing inks, the scanner's light signal can be adjusted for these same deficiencies.

The basic components of a modern electronic color scanner are shown in Fig. 19-3.

Color scanner terms

The following is a list of terms that will help you understand color scanners. Study them closely.

1. *Argon-ion laser light* is a light source in which the argon-ion gas is stimulated to produce a monochromatic blue-green light.
2. *Condenser* is an optical element used to concentrate light on a small area.
3. *Detail enhancement* is the same as unsharp masking and peaking.
4. *Electro-optical modulator* is a crystal or device that

Fig. 19-2. A color scanner performs same basic operations as a process camera.

transmits light in proportion to an electrical impulse.

5. *Fiber optics* is an array of glass or plastic fibers that transfer light from one point to another.
6. *Full-page pagination* is the display of one full page of text and illustrations on a video display terminal for the purpose of making copy, text, or layout changes.
7. *Helical* means having the form of a spiral.
8. *Helium-neon laser light* is a light source in which the helium-neon gas is stimulated to produce a monochromatic red light.

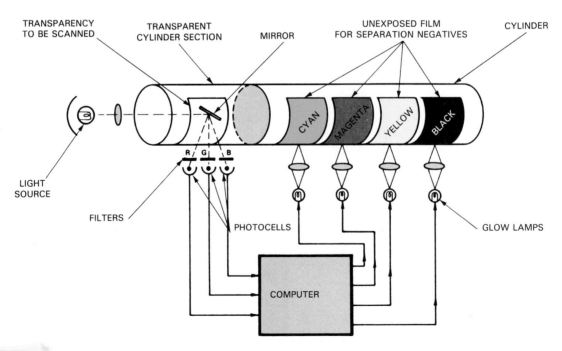

Fig. 19-3. Study basic components of modern electronic color scanner. (Printing Developments, Inc.)

9. *Input size* is the size of the copy used on the scanner. It is usually given as the largest size that the scanner can accommodate.

10. *Interactive color-correction feedback* is the reproduction of the original on a video display terminal to show the effects of color changes as they are being programmed.

11. *Laser* is an acronym for light amplification by stimulated emission of radiation. A device which produces an intense, single wavelength, unidirectional beam of light.

12. *Magnification range* is the amount of enlargement or reduction that the scanner is capable of providing. A 20 percent reduction and a 2000 percent enlargement are usually the extremes.

13. *Optics* is a branch of physics that deals with the properties of light and its interaction with devices used to redirect and alter its composition.

14. *Output size* is the size of film that is used for output with the particular scanner. Frequently, this will be referred to as 1-up, 2-up, or 4-up with a different size listed after each. This refers to the number of colors from the same transparency that can be separated. Therefore, 1-up, 16x20 means that is it possible to separate one color on a 16x20 inch size film, and it is necessary to rescan for the additional colors.

    In another example, 4-up, 8x10 means that all four colors can be scanned at the same time at the 8x10 inch size. Therefore, 1-up, 16x20 and 4-up, 8x10 may be equivalent if the scanner is capable of separating four colors simultaneously.

15. *Overlap* refers to how some scanners are built so that, when exposing the film, the exposing drum revolves in such a way as to have the scan lines overlap. This overlap may be very slight or as much as 50 percent.

16. *Scan pitch* is the number of sampling or exposing lines per inch or per centimeter of copy.

17. *Scan rate* is usually listed with two numbers, such as 70/500. The first number refers to the number of seconds that it takes to scan one inch. The second number is the scan pitch. Therefore, 70/500 means that it will take 70 seconds to travel one inch or 500 lines. With electronic dot-generating scanners, the scan rate depends upon the speed of the output and the halftone ruling being used. An 85 line-per-inch ruling scan is faster than a 150 line-per-inch ruling scan.

18. *Selenium cell* is an electronic element that varies its electrical voltage output in proportion to its intensity of illumination.

19. *Spectral sensitivity* is a response to the energy of light waves. The human eye responds only to the visible spectrum. Some photographic film and electronic components respond to infrared and ultraviolet energy as well as to the visible spectrum.

20. *Spectrum* refers to the broad range of electromagnetic radiation. The visible spectrum of light is defined from blue (400 nm) to red (700 nm).

21. *Specular highlight* refers to a portion of a picture containing intensely bright highlight, such as a spot of sunlight reflected from a strip of chrome.

22. *Unsharp masking* accentuates the contrast between adjacent tones and increases the edge effects where tones change. This accentuation gives the appearance of a sharper picture with more detail.

23. *Video display terminal* is a television-type screen (CRT or cathode ray tube) used for displaying text or graphic information.

Scanner development

The basic principles of modern photo-transmission were patented by Alexander Bain of England in 1843. This was more than 100 years before the first commercial use of the scanner.

The first practical transmission of pictures was developed in 1902 by Arthur Korn, a German physicist. Korn's development was based on the use of a selenium photoelectric cell. This development led to wire photos. American Telephone and Telegraph Company introduced wire photo service in the United States in 1925.

The first patent for a color separation scanning machine was issued to Alexander Murray. He was hired by the Eastman Kodak Company to do research in their engraving department. Murray later worked on the problem of improving masking methods in color separation. He invented the first color separation scanner in 1937.

About the same time, Arthur Hardy and F.L. Wursburg, of the Interchemical Corporation, independently developed a machine for photomechanical scanning. The Interchemical machine was taken over by RCA. Although later abandoned, the scanner's computer section has been used for analyzing color separation errors. See Fig. 19-4.

Early electronic scanners

The early electronic scanners scanned 35mm, 5x7 inch and 8x10 inch transparent originals. The light that scanned the original signaled the light that exposed the film. Scanning and exposing were both done at the same time. This type of scanner produced a continuous tone film of the same size as the original. It was necessary to scan the original once for each color.

The original, usually a transparency, was mounted on a transparent revolving drum. The exposing light, mounted inside the transparent drum, directed the light through the transparency to the scanning head. The scanning head was mounted in a stationary position. Each revolution of the drum moved it past the scanning head in a *helical* (spiral) pattern about 1/500 of an inch while a line was scanned.

On early scanners, the original and the final film were both mounted on drums of identical size. These drums revolved on a common axis at the same speed. These

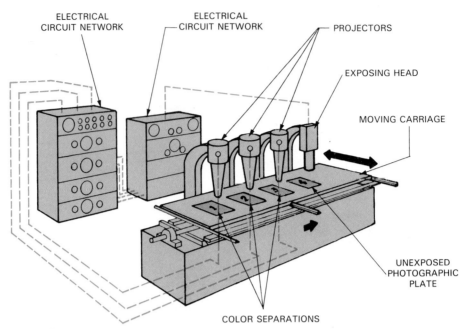

Fig. 19-4. This is an early model of a photomechanical scanner produced by Arthur Hardy and F.L. Wursburg of Interchemical Corporation. (Interchemical Corp.)

scanners produced continuous tone negatives. The negatives were then mounted on a process camera and screened into halftones. Size changes were made during the screening step.

Electronic scanners today

Most electronic scanners today use a principle similar to early models. The original transparency and the final film are mounted on drums. However, the drums do NOT need to be the same size. The transparency is mounted on an interchangeable input drum that is the appropriate size for the transparency. The final film is mounted on the output drum which is not changed and remains the same size. However, the scanning rate and the output rate are varied to change the size of the output image. Refer to Fig. 19-5.

To change size in the horizontal direction, the rate of travel of the scanning head is changed. To change

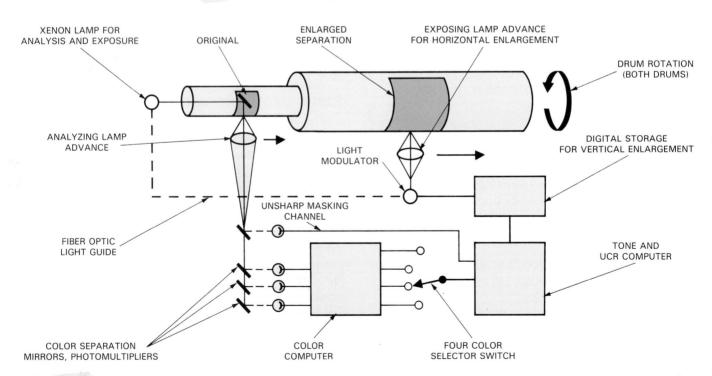

Fig. 19-5. On a modern electronic color scanner, original transparency is mounted on input drum and output image is located on output drum. (Crosfield Electronics, Inc.)

size in the vertical direction, the readout rate of the computer storage is varied. Therefore, on some scanners, it is possible only to change size in one direction. Since scanners can now change sizes in the scanning step, it is practical to scan directly to screened halftones. This eliminates one step from the process.

With the improvement in techniques for making size changes, the 35 mm and the 2 1/4 x 2 1/4 inch transparencies are most popular for making separations. The small format originals are sometimes enlarged up to 1000 times or more. They can be scanned at 1000 or 2000 lines per inch, but 500 lines per inch is more common. Today, many scanners can produce both continuous tone and screened films on positive or negative working films.

A scanner can be useful when the original copy is of poor quality. The scanner is equipped with numerous controls that allow for color correction and modification, Fig. 19-6. The scanner operator must have extensive experience in color theory.

Along with the scanner, a scanner language has evolved. For example, the word scanning may have different meanings to different people. The term *"scanning"* will be used here to describe the complete operation from analyzing the original copy to exposing the film output. The term *"analyzing"* will be used to describe the pickup of the information on the input function. The term *"exposing"* will be used to describe the recording of the image on film.

Scanner components

There are many types of scanners, each designed to perform a particular function or fit a particular need within the printing and publishing industry. Fig. 19-7 illustrates one type of scanner with components that are common to most scanners. It is not a particular

brand of scanner, but a representation of the basics of all scanners.

Reference has been made to the similarities of the scanner and the camera. They are both used to perform the same function—to expose film. Beyond that, they perform other necessary functions for color reproduction. Both camera and scanner can perform color correction and provide the tone required for the halftone system.

The major differences between the camera and the scanner are in how they perform these functions and the physical components that perform them. The major scanner components are described here.

Fig. 19-6. When original color transparency is poor quality, electronic color scanner can be used to color correct and modify images. (Crosfield Electronics, Inc.)

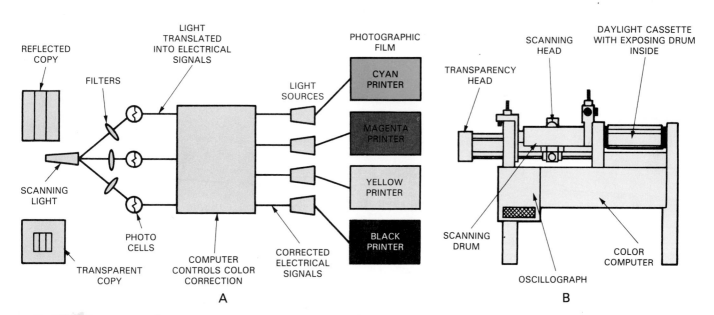

Fig. 19-7. Electronic color scanner components illustrated are basic to all scanners. A—Interior components. B—Exterior components.

Photomultiplier tube (PMT)

The basic electronic component that allows the scanner to interpret a color picture is the *photomultiplier tube* (PMT). It is the "eye" of the scanner. The photomultiplier tube changes light into an electrical signal. This signal, after being manipulated in a computer, controls the mechanism that exposes the film. In the computer, the signal may be changed by the operator to make color corrections or to modify the tone range.

The PMT has a spectral or color response just like film has a spectral sensitivity. Color separation and color correction are accomplished by the combined effects of color filter selection and spectral sensitivity of the film on a camera. This is done by color filter selection and spectral response of the PMT on a scanner.

The PMT can send a variable signal to the computer. The film in the process camera can record variable colors. In the case of film in the camera operation, that is the end of the color correction because the correction has been done by the mask. With the scanner, the computer signal can be modified by many controls for even more correction.

Lamphouse

Some scanners use a *lamphouse* to hold the analyzing light source and the main condenser lens. The lamphouse also contains the exposing light in scanners that use the same light for analyzing and exposing. Lamps are generally held in a sealed compartment to shield the other parts of the scanner from the heat and to eliminate flare. Look at Fig. 19-8.

There are four types of light sources used in scanners. These include: glow lamp, xenon lamp, tungsten-halogen lamp, and laser.

The *glow lamp* is one type of gas-discharge electric lamp used for exposing. The intensity of the light can be varied rapidly as the voltage from the PMT changes. A higher speed film is usually preferred with scanners that use glow lamps to permit operation of the lamp at a lower intensity.

The high-pressure *xenon lamp,* which can be used for both analyzing the original and exposing the film, is another type of gas-discharge lamp. The xenon lamp emits a constant high-intensity light and remains lit from the time the scanner is first turned on. The exposure variations are made by an electro-optical crystal modulator placed between the lamp and the film. The modulator effectively controls an electrical field. It varies the amount of light that reaches the film in response to the voltage applied to the crystal modulator. The electronic signals from the scanner do NOT affect the light but change the transmission characteristics of the crystal. The crystal functions like a rapidly changing neutral density filter. The xenon lamp gives a high concentration of light in a small area.

Fig. 19-8. A transparent scanning drum on color scanner holds original copy during analyzing or scanning operation. (Crosfield Electronics, Inc.)

The *tungsten-halogen lamp* is a low voltage, high current, tungsten light. It contains a halogen gas to prevent vaporized filament materials from depositing on the walls of the lamp and dimming its output. It offers a continuous spectrum of colors and is good for use as an analyzing light. One scanner uses it for the exposing light. It is a relatively inexpensive, easy-to-use, low-output lamp. However, it requires a higher speed film than most lamps.

A *laser* produces consistent, parallel rays of light that can be easily controlled electronically. It offers a high concentration of light in a controlled direction. It is more efficient and can use a slower speed film than other light sources. The laser is much cooler than other light sources for the equivalent amount of light. Laser light sources are used to produce both contact-screened halftones and electronic dot-generated halftones.

Cross-feed drive

The scanner's *cross-feed drive* controls the movement of the scanning head. This drive is reversible in most scanners to provide either right-reading or wrong-reading separations. The variable speed of the analyzing drive provides the lateral enlargement or reduction of the picture size. On many scanners, the adjustment for size regulates the scanner for a proportional change in the vertical and horizontal directions.

Therefore, setting the controls for a 100 percent increase in horizontal size would also increase the vertical direction 100 percent. However, on scanners that can distort the picture by changing only one dimension, it is necessary to make both a horizontal and vertical size adjustment.

Scanning drums

The transparent *scanning drum* holds the copy during the analyzing or scanning operation. It can be removed easily for mounting transparencies. Some scanners have scanning drums of different sizes for use with different size copy. They can hold small 35 mm slides or large 20x24 inch images, and can hold single or multiple pieces of copy. The copy is held on the drum by means of a vacuum or with tape, and can be mounted on a transparent carrier sheet.

Some scanning drums revolve in a fixed position while the analyzing head is being driven past the copy. The timing of the horizontally moving analyzing head and the revolving scanning drum is *synchronized*. This means that the analyzing head proceeds one scan line width for each revolution of the scanning drum.

The analyzing head on the outside of the scanning drum is connected to the analyzing light within the drum by a U-shaped arrangement. Therefore, the analyzing light and the analyzing head travel at the same speed, Fig. 19-9.

Fig. 19-10. Lamphouse on a color scanner holds exposing light. Lamps are generally held in a sealed compartment to shield other parts of scanner from heat and flare. (Crosfield Electronics, Inc.)

Fig. 19-9. Analyzing head on color scanner contains microscope optics through which picture is perceived. (Crosfield Electronics, Inc.)

Analyzing head

The *analyzing head,* sometimes referred to as the *pickup,* contains the microscope optics through which the picture is perceived. See Fig. 19-10. There are two types of analyzing heads.

One type of analyzing head travels alongside the scanning drum and moves horizontally, the width of the scan line, with each revolution of the scanning drum. This type of analyzing head requires four PMTs. These include one each for the three colors and one for the unsharp masking.

The second type of analyzing head is driven in such a manner that it travels the width of the scan line once for every three revolutions of the scanning drum. This type of analyzing head records each revolution with a different colored light. This means one revolution each with red light, green light, and blue light. Then the analyzing head travels to the next line and repeats the sequence. This type of analyzing head requires only two PMTs—one for the three colors and one for unsharp masking.

Scanning illumination arms

The *scanning illumination arms,* on scanners that use revolving drums to hold the copy, contain optical components that transmit the light inside the scanning drum and illuminate the transparency. On some scanners, the arm is NOT removable or interchangeable with a different size. Note that the arm is NOT used when scanning reflection copy.

Exposing compartment

The *exposing compartment* contains an exposing drum complete with register pins and, in some equipment, vacuum channels, Fig. 19-11. Some scanners require darkroom loading of the film and must be in a room that can be darkened. Other scanners use film holders. The film is loaded into the film holder in a darkroom. The compartment is then coupled to the scanner in room light, and the film transferred to the scanner drum.

A

B

Fig. 19-11. A—Electronic control and power supply may be an integral part of color scanner. B—It can also be separate from scanner. (Crosfield Electronics, Inc.; Chemco Photoproducts Co.)

Color computer

The *color computer* performs the color correction and editorial or selective color change of each of the four color separation signals. The signals relayed to the computer represent the proportions of each of the three colors in the original as perceived by the photomultipliers. The signals leaving the computer to expose the film represent the amounts of the four inks to be used on the press. Under the operator's control, the long tone-range of an original transparency is compressed to a usable tone-range for the press.

Special color changes to accommodate customer preferences, types of paper and press considerations, or different types of copy must be programmed by the operator.

Control console

Some scanners include a separate electronic control console. This may include a power supply unit separate from the scanner. There are other scanners that have the power supply and electronics as an integral part of the scanner. See Fig. 19-12.

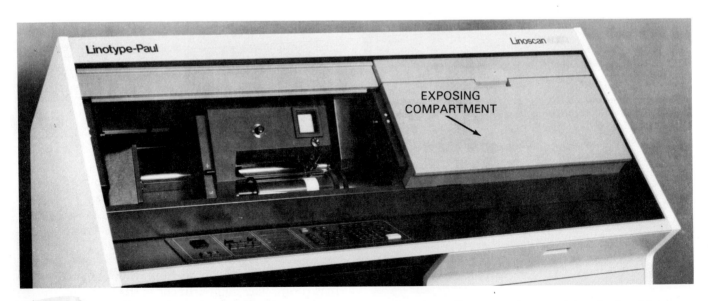

Fig. 19-12. Exposing compartment on color scanner encloses exposing drum. Drum may have register pins, and in some applications, vacuum elements. (Linotype)

DOT AREA AND DENSITY METER

Scanners do NOT have densitometers. However, they do provide metric readings which are the operator's primary source of information on how a particular job will be color separated. This is the *dot area* and *density meter* component. The meter will read either the values proportional to the light received or values that represent what will be exposed onto the output film in either the dot area or density.

The meter is a *comparative color photometer*. This means that the relative amounts of cyan, magenta, yellow, and black ink to be printed are predicted by viewing the original copy.

On most scanners, the meter is calibrated by exposing and processing a piece of test film. On some scanners, the meter reading is then adjusted to agree with the actual density or dot size on the test film. In this way, all variables of the film exposing and processing are accounted for in the meter calibration. Any change in film, contact screen, processing activity, etc., requires a new meter calibration.

Some scanners, having electric dot-generating capabilities, do NOT provide for changes in meter calibration. This is because the dot size is determined electronically, instead of by the film and processing. In practice, it is possible to experience significant deviations from the meter-predicted dot sizes with different films and with different exposing or processing conditions. The operator should be aware of these deviations, which, although small, are critical with high quality work.

UNSHARP MASKING SIGNAL

Unsharp masking is an increase of tonal contrast at the areas where light and dark tones come together. This can be referred to as an *edge effect*. This increased contrast gives the appearance of a SHARPER PICTURE with more detail. The unsharp masking signal regulates the color separation signal at the edges of tonal areas so that the lighter tone is slightly reduced and the darker tone is slightly increased.

The main signal is created from light striking on the PMT through a small aperture. Unsharp masking is accomplished by adding a second aperture and taking a second reading of the same area at the same time. The second aperture is larger, and its signal is used only to modify the first signal from the smaller aperture.

Where there is no tonal change, the second aperture is not significant because both apertures read the same and there is no modification of the main signal.

Where there is tonal change, the larger second aperture perceives more area. Because the pickup averages the density reading, this aperture has a different reading than the smaller main signal. The second signal from the larger aperture is subtracted from the first signal. The modified signal produces a reduction in density

in the lighter area and an increase in density in the darker area. This occurs at the point of change in tonal values. This creates the illusion of a sharper image.

ELECTRONIC GRAY SCALE

The *electronic gray scale* is an internally generated step tablet similar to the type used on a process camera. The gray scale value of each step is stored in the scanner and generated with a switch.

FOCUSING

The focusing on a scanner differs from focusing on a process camera. This is because a scanner typically has both an input focus and an output focus.

The *input focus* on some scanners is done visually, similar to focusing on the ground glass of a camera. However, on the scanner, the image is reflected from a mirror and the focus is on the mirror. On some scanners that have a prefocused setting, it is necessary to change the setting. This depends on whether the transparency is taped in place emulsion-up or emulsion-down.

The *output focus* is NOT an image focus but a focus of the method of image formation. A poor focus on the output will cause lines, streaks, or gaps in the halftone or continuous tone image. On the continuous tone scanner, individual scan lines must be blended to give the appearance of a smooth picture. Output focus adjustment may be needed because of the film thickness. The film is placed on the exposing drum emulsion-out, and a change in film thickness will change the focus.

COLOR SCANNER OPERATING PRINCIPLES

The color scanner reads copy densities. It then changes them to electronic signals. These signals are transmitted to the other end of the scanner which exposes the film.

Fig. 19-13 illustrates how the scanning drum spins so the scanning head can read the image. Both transparency and reflection copy can be used. The copy can be reduced or enlarged. This can be done while the separations are being made. Reductions may be 20 percent or one-fifth size and enlargements may be 2000 percent or 20 times.

Scanning transparency copy

In most cases, transparency copy consists of 35 mm (2 1/4 x 2 1/4 inch) slides. The transparency is positioned on a transparent revolving drum. The analyzing light is directed down a center optical path within the drum. The light passes through condenser lenses and is deflected by a mirror angled at 90 degrees. The light then illuminates the transparency on the drum.

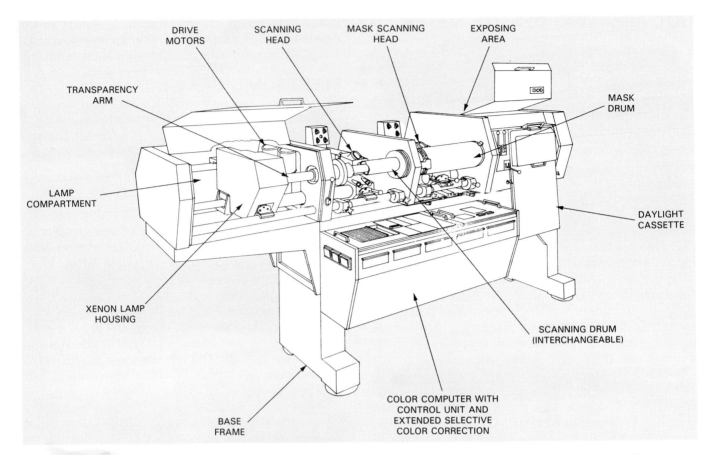

Fig. 19-13. Schematic drawing shows electronic color scanner operating components. (Hell-Color-Metal Corp.)

The illuminated transparency image is analyzed with microscope optics and is split into four light paths with the use of interference filters. See Fig. 19-14.

Each light path enters a photomultiplier tube (PMT). Three of the PMTs are covered with red, green, and blue separation filters to create the cyan, magenta, and yellow separation signals. The fourth light path activates a PMT used for unsharp masking. The black separation signal is produced by electronically combining information from the cyan, magenta, and yellow signals in a subsequent step.

Scanning reflection copy

Any reflection copy that is flexible enough to be mounted on the outside of the drum is scanned similar to transparency copy. The light beam is NOT directed through the drum but is projected onto the copy from the side. The analyzing light is then reflected to the scanning pickup.

Color computer

With both transparent and reflection copy, the microscope optics pick up the signal and relay it to a PMT. The PMT converts the beam of light into electronic signals, in proportion to the light it has received. The electronic signals next enter the *color computer* which enables the scanner operator to introduce color correction information at this point. The information

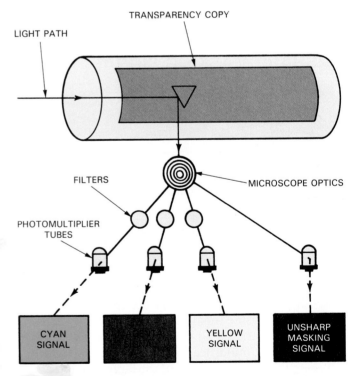

Fig. 19-14. Transparency copy is positioned on a transparent revolving drum. Analyzing light is directed down a center optical path within drum. Transparency is illuminated and then analyzed with microscope optics. Image is split into four light paths using interference filters which expose film. (Hell-Color-Metal Corp.)

may be preprogrammed instructions or value judgments of the operator. These instructions will modify the four signals.

The modified signals from the color computer are used to control the exposing mechanism. The film is exposed to a density (or dot size) proportional to the printing inks needed to reproduce the spot scanned in the copy. The black signal is then created by the computer from information taken from the three color signals.

Most scanners store approximately one line of information at a time when making size changes. The analyzed data is stored in order on the first revolution of the analyzing drum. This data is then stored to make sequential exposures on the second revolution of the drum, and the second line on the analyzing drum is stored. Only scanners that make use of a video display terminal store the entire picture.

The output signal from the computer is used to control the exposure. When the output is to be a continuous tone or a screened halftone, and the exposing light is a glow lamp, the output signal controls the intensity of the light. The light is focused on a sheet of photographic film mounted on a revolving drum. In most other systems, using halogen or xenon lamps, an electro-optical modulator acts as a rapidly changing neutral density filter. This blocks part of the light in accordance with the picture information.

For every scanned spot on the original, the light source gives the proper exposure. This produces a density on the film relative to the scanned spot modified by the color correction in the color computer. After the film is completely exposed, it is processed in a film processor. Refer to Fig. 19-15.

Many scanners can produce all four separations on one sheet of film or one large separation per piece of film. The number of separations depends upon the input copy size and the desired enlargement.

One important exception to the scanner just described is a drum that revolves 360 degrees three times successively to scan one line three times. This means it is scanned once each with red, green, and blue light before moving laterally one line width to the next position. This type of scanner requires only two PMTs, one for the three colors and one for the unsharp masking.

CLASSIFYING SCANNERS BY OUTPUT

Scanners may differ in the type of copy they accept. These may include transparencies, reflection, or both. The biggest difference between scanners is in their method of output. Most scanners in use today operate in one or more of the following output modes: continuous tone, contact screen halftone, and electronic dot-generating.

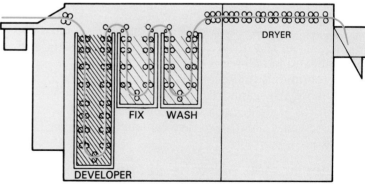

Fig. 19-15. Once film has been exposed on a color scanner, it is processed in an automatic film processing unit. Schematic drawing illustrates operation of a film processor. (Agfa-Gevaert Inc.)

Continuous tone scanner

The early scanners were all *continuous tone scanners*. This means that they produced continuous tone separations. It was then necessary to use a process camera to make screened halftones. Most continuous tone scanners used today have the option of producing continuous tones or screened halftones. However, it is possible to purchase a relatively inexpensive scanner that produces only continuous tone separations. Finer screening is required when producing continuous tones. For some purposes, the continuous tone scanners have been replaced by direct screen scanners. This is because they can go directly to screened halftones, eliminating the continuous tone step.

The continuous tone scanner is most practical when it is necessary to produce the same picture in many different sizes. One set of scanned negatives can be exposed on the camera to make several sets of screened positives at various sizes. The continuous tone scanner is also used to make positives for the gravure printing process.

Most continuous tone scanners use separate analyzing and exposing light sources. One model of a continuous tone scanner uses the same light source for both analyzing and exposing. The xenon lamp emits a path of light that is optically split into two paths. One path goes to the analyzing drum, after which the separation optics and PMTs produce signals which are sent to the color computer. The other light path goes to the exposing drum. The computer scale produces electronic signals that are used to control the exposing light to conform with the information scanned from the original copy.

Many of the continuous tone scanners also have the ability to make screened halftones. Several such scanners also contain a masking drum to permit photocomposing during the color separation step. Some scanners also feature a daylight-loading cassette, that permits the scanner to be operated in a light area. It should be pointed out that these features are not limited to continuous tone scanners.

The continuous tone scanners are among the least expensive of the scanners. Those that also have the capability to generate halftones electronically and/or contact-screened halftones are among the most expensive.

Contact screen halftone scanner

The *contact screen scanner* produces screened halftones in either negative or positive form. The same screen can be used for either positives or negatives. However, the tone reproduction may change and it may be necessary to adjust for different aim points in density. Special screens that are less dense are made for use with the scanner.

Many contact screen scanners are also capable of producing continuous tone output. This is a relatively simple and inexpensive adaptation. The only change in operation is to load the scanner with a continuous tone film instead of a lith film and to remove the contact screen. Of course, it is not quite that simple. The light intensity and the aperture must be changed because a different type of film is used. Scanners that offer both continuous tone and contact screen separations may have a different exposing head and/or different apertures for each use. This added capability is an advantage when the scanner may be used to produce separations for both offset lithography and gravure.

Two manufacturers of contact screen scanners use a laser light source for exposing films. One uses an argon-ion laser that emits blue-green light and requires a blue-sensitive film. The other uses a helium-neon laser that emits red light and requires a red-sensitive film. The contact screen scanners are usually in the medium to upper price ranges.

Electronic dot-generating scanner

The *electronic dot-generating scanner* is often referred to as a laser scanner. However, the term is misleading. Not all electronic dot-generating scanners use a laser light source. Of those that do, the laser is used as an exposing light, not as an analyzing light.

The analyzing light is converted by the PMT into an electronic signal. The color computer modifies the signal for tone compression and color correction. The signal then controls an electro-optical crystal modulator. This, in turn, controls the light that produces halftone dots of the proper size and shape for the desired screen ruling.

In one method, six fiber-optic cables aligned in a row carry the light to the film to produce micro dots which form the halftone dot. There may be as many as 12 rows of micro dots, both horizontally and vertically to form one row of halftone dots. These micro dots may have no overlap or as much as 50 percent overlap. Only a partial halftone dot is made on each revolution of the exposing drum. Depending on the screen angle used, it may take two or three revolutions to complete a halftone dot. The finished halftone dots may be square, rectangular, elliptical, or round.

Fig. 19-16 shows one type of electronically generated scanner dot and how it is formed by the laser light. The electronic dot is more like a hard contact dot than a soft dot exposed through a contact screen. The electronically formed dots can be chemically etched but not as well as dots made through a halftone screen. When over-etched, the structure of the dots will become apparent and reveal the individual scan lines where they form the dot.

There are variations of the method for forming electronically generated dots. One manufacturer uses programmed paper tapes. These contain information relating to dot shape, screen angle, and dot size to control the electronic screening computer. The customer purchases tapes to produce screen characteristics rather

CONVENTIONAL DOT   LASER SCANNER DOT

Fig. 19-16. Comparison of conventional halftone dots and laser scanned dots. Scanner dots are elliptical in shape, whereas conventional dots appear square. (Color Service, Monterey Park, CA)

than screens. Another manufacturer has this information permanently programmed into the computer and does not require a tape.

There is another type of electronic dot-generating scanner that uses a tungsten-halogen light source for exposing. This method requires the highest speed film. However, the scan rate is slow. The scanner using this method has been designed to produce all four separations at the same time, thereby saving considerable time.

Flatbed color input unit

To simplify color scanner setup, scaling, positioning, and make the best use of operator time and skills, a *flatbed color input unit* has been developed. The scanner input unit can be used as an interface to a regular drum scanner. It accepts transparencies and reflection copy, positives and negatives, in various sizes. Typically, images can be reduced to 20 percent and enlarged to 2400 percent in increments of less than one percent over the entire range.

Using an electronic cursor, the operator can determine highlight and shadow points and adjust for variations in color tone. A built-in digital density meter automatically provides density or dot percentage readings. The system uses a fluorescent white light source.

To operate the scanner, original art is placed in a special cassette (coded to indicate film type), which is then fed into the scanner. The input unit takes two minutes to scan the copy. The image data is transferred to a console for additional color work and image assembly. The final films for each of the process colors are produced on a high-resolution laser plotter.

A *flatbed scanner* mounts the image and film on a table-like surface, instead of a round drum. This offers the advantage of mounted stiff or rigid copy on the scanner without bending and creasing it around the scanner drum. This prevents damage to the original.

Large color flatbed scanners operate like the optics of conventional rotary scanners and the mechanics of small scanners discussed in the chapter on desktop publishing. Refer to the index for more information on flatbed scanners.

SCANNER ENVIRONMENT

A *scanner environment* refers to the air quality, temperature, and humidity conditions in the space for the scanner. It is very important, Fig. 19-17. Although it is not much different from the environment for a process camera, certain precautions should be taken. The scanner should be in a clean, dust-free area. Smoking may affect the fiber optics and/or the optical mirrors.

The electronic equipment requires control of temperature and relative humidity. The safety precautions for handling film are about equal to those required for a process camera.

Freedom from static discharge is important. If carpeting is to be used, a static-control carpet is required. Electronic equipment generates heat as do camera lights. Therefore, a need for cool air is necessary.

Fig. 19-17. Electronic color scanners must be maintained in a clean, dust-free environment for optimum output quality. Smoke may affect fiber optics and/or optical mirrors. (Color Service, Monterey Park, CA)

An appropriate climate for most scanners is a temperature of 70 degrees F, plus or minus five degrees. A 50 percent relative humidity, plus or minus ten percent is desirable.

Some of the larger scanners are made up of modules that contain the various components. The modules can be moved around to fit the location of the scanning room. Some scanners are built in one unit.

SCANNER QUALITY CONTROL DEVICES

The scanner electronically produces color separated films to exact specifications. There are variables such as input copy, ink, printing press, and paper that present judgment factors for the scanner operator. Special devices have been developed for maintaining quality control with scanners. Two such devices include: color reproduction guides and color previewers.

Color reproduction guides

Different types of original transparency copy are made with different dyes. Therefore, these dyes will color separate differently even though they may appear the same.

It is absolutely essential that guesswork be eliminated when adjusting the scanner for certain types of copy. To do this, test separations are made with different types of copy. A record is kept of the results of comparisons between the original and the reproduction. It is important to make these color comparisons in flesh tones, greens, and blues. This is because these are the areas that consumers are most apt to notice.

Grouping or *"ganging"* different types of transparencies for production is NOT recommended on the scanner. However, it is a good idea to gang different types for testing and compare the different results. This will give a reference for future production.

One type of *color reproduction guide* is a test kit that contains sample transparencies and a sample print on photographic paper. It shows different reproductions of different products. The reproduction guides contain these standard test objects, plus a booklet to identify specific films.

Color previewers

The development of *color previewers* has made it possible for the operator to simulate the planned reproduction on a video screen. The color simulation can be adjusted to reflect the variables of both the press and the prepress operations to give a close approximation of the finished reproduction.

Built into the previewers are circuits that modify the signal to compensate for the individual characteristics of paper, ink, and press. In operation, the original copy is scanned and entered into the previewer's memory. The original is then placed in position on a viewer next to the cathode ray tube (CRT) display unit for easy comparison. A control panel permits the operator to manipulate the image on the CRT to match the original. Color changes can be made overall or in a localized area of the copy.

Once the image on the CRT matches the original plus any changes requested by the customer, the settings used to obtain the correct color are recorded on a magnetic tape. The tape can then be used to modify the data in the scanner memory so that the exposing device on the scanner produces films that reproduce the original as seen on the CRT display unit.

POINTS TO REMEMBER

1. Like the process camera, the color scanner is also an exposing unit used to record pictures.
2. The difference between a process camera and a color scanner is that the camera records the entire picture at one time while the color scanner records only tiny segments of the picture at one time.
3. A color scanner does the same basic operations of a process camera but it works electronically.
4. The basic principles of modern color scanners started with Alexander Murray's invention in 1937.
5. Early color scanners produced a continuous tone film of the same size as the original. It was necessary to scan the original once for each color.
6. On early scanners, the original and the final film were both mounted on drums of identical size. These scanners produced continuous tone negatives. The negatives were then mounted on a process camera and screened into halftones.
7. Most electronic scanners today scan directly to screened halftones. In most instances, 35 mm transparencies are used as original copy.
8. The small format originals are sometimes enlarged up to 1000 times or more. These scanners can produce both continuous tone and screened films on positive or negative working films.
9. Scanners can be useful when the original copy is of poor quality. This is because the scanner is equipped with numerous controls which allow for color correction and modification.
10. Scanners can distort copy and make it smaller or larger on any dimension.
11. Scanner color previewers make it possible for the operator to simulate the planned reproduction on a video screen.
12. A color simulator can be adjusted to reflect the variables of both the press and the prepress operations to give a close approximation of the finished reproduction.

KNOW THESE TERMS

Analyzing, Analyzing head, Color computer, Color previewer, Color separation scanner, Contact screen scanner, Continuous tone scanner, Control console, Cross-feed drive, Dot area density meter, Dot-generating scanner, Exposing compartment, Glow lamp, Input drum, Lamphouse, Laser, Laser scanner, Output drum, Photomultiplier tube (PMT), Scanning, Scanning drum, Scanning illumination arm, Tungsten-halogen lamp, Unsharp masking, Xenon lamp.

REVIEW QUESTIONS

1. What is a color separation scanner?
2. What are the most popular kinds of originals for making separations on a color scanner?
3. The basic electronic component that allows a color scanner to interpret a color picture is the _____ tube.
4. On a color scanner, the _____ drum holds the copy during the analyzing operation.
5. The color computer on a color scanner performs the color correction and selective color change of each of the four separation negatives. True or false?
6. _____ _____ refers to an increase of tonal contrast at the areas where light and dark tones join.
7. The electronic gray scale in a color scanner is similar to the type used on a process camera. True or false?
8. The maximum copy enlargement size on a color scanner is generally:
   a. 300 percent.   c. 1000 percent.
   b. 500 percent.   d. 200 percent.
9. Color originals used to make color scanner separations are divided into _____ copy and _____ copy.
10. Color scanners are divided into three types of output modes. Explain them.

SUGGESTED ACTIVITIES

1. Plan to visit a printing firm and/or color separation house which produces 4-color process negatives and positives on a color scanner. Prepare a short report on your findings.
2. Assist your teacher in planning for a color scanner technical representative to visit the school and give a presentation before your graphic arts class.
3. Prepare a short research paper that gives some indication of how much color separation work is done on process cameras and how much is done on color scanners in the United States.
4. Photocells, digital signals, and a built-in computer are the primary components of the color scanner. Prepare a drawing (schematic) which shows the operational features of the color scanner.
5. Prepare an outline which can be used to compare the operations of making color separations on a process camera versus those on a color scanner.

Section V

Issues for Class Discussion

1. Darkroom films and chemistry are potential hazards and must be handled properly by all personnel. Discuss the precautions that should be followed with darkroom safety management, toxic waste disposal, and silver recovery systems. Include in your discussion information about the availability of silverless films and nontoxic chemicals developed for safe darkroom operations.
2. In setting up a darkroom for line, halftone, and special effects photography, specific quality control standards must be established to obtain consistent results. Discuss some of the procedures, equipment, and tools used for the purpose of establishing darkroom quality control standards.
3. Automated camera-processors are sophisticated machines now used in large and small darkroom operations throughout the industry. Identify and discuss the major applications to which these systems are being applied. Include in your discussion the implications these systems pose to skilled personnel in the prepress areas.
4. Color separation scanners are widely used throughout the printing and publishing industry. These machines have many advantages over the process camera in the preparation of color separation film negatives and positives. Participate in a class discussion aimed at identifying the processing features of both the scanner and process camera. Try to come up with a plus and minus list for both methods.

Section VI

Stripping and Platemaking

Two essential steps in the offset lithography production process include the assembly of a flat (stripping) and the processing of an offset plate (platemaking). This section will help you understand the fundamental procedures of stripping a flat and preparing both direct image and pre-sensitized plates.

Chapter 20 explains the techniques of positioning film images before platemaking, called stripping. In these procedures, image position must be determined from original designs while applying a knowledge of recommended production practices. Single and multiple-page stripping is common. Complementary flats are used for exposing a single plate, while multi-color flats are used for exposing separate plates.

There is an increasing use of computerized equipment in the stripping area. These machines generally perform imposition, pagination, and step-and-repeat operations without human assistance.

The various types of offset plates and platemaking procedures are discussed in Chapter 21. Plates are manufactured from various materials, including: paper, polyester, plastic-impregnated paper, aluminun, copper, chromium, and stainless steel. Most plates are imaged photographically using diazo or polymer resin compounds. Other plates use electrostatic imaging processes or silver halide emulsions. Both negative-acting and positive-acting plates, as well as subtractive and additive emulsion plates, are manufactured.

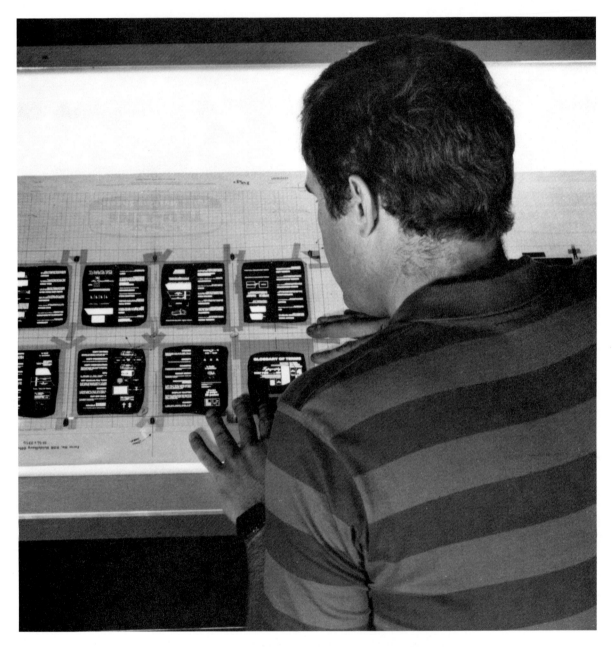

Stripping involves attaching films to a base sheet.

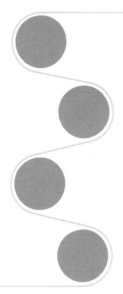

Chapter 20

Stripping

When you have completed the reading and assigned activities related to this chapter, you will be able to:

O List the common tools and equipment used in the stripping process.
O Explain the purpose of drawing the lead edge, gripper margin, and reference lines on a masking sheet.
O Strip a single negative flat using the emulsion-up method.
O Strip a halftone negative with a line negative on a single flat.
O Strip a halftone negative and a line negative on separate complementary flats.
O Strip a step-and-repeat flat with a single negative using the emulsion-up method.
O Strip a multiple-page flat using the emulsion-up method.
O List and describe the three single color proofing systems used prior to platemaking.
O Expose and process proofs of flats for single color work.
O Describe the need for imposition during the stripping process.
O Prepare flats for multiple page signatures.
O Describe the importance of a reliable registration system for stripping.
O Describe the equipment needed for a registration system.
O Use a pin register system in preparing single and multiple flats.
O Define and describe the automated stripping process.

Stripping, or *image assembly,* is the process of accurately positioning and fastening film negatives, or film positives, onto a masking sheet. The person who handles the task of stripping is called a *stripper* or *image assembler.*

The *masking sheet* is a thick sheet of paper that holds the negatives or positives. It is the same size as the offset plate for the job. The complete masking sheet with its taped-on film negatives, or film positives, is called a *flat.* The film images on the flat are exposed onto the offset plate in the required positions.

STRIPPING PROCESS

The basic steps in stripping a flat and making an offset plate are illustrated in Fig. 20-1. The stripping process is generally handled as described in the following paragraph.

Film negatives and film positives are prepared by the camera department. The stripper tapes them with the emulsion side up on a ruled masking sheet in the positions indicated on the layout. This procedure takes place on a light table. Refer to Fig. 20-2.

The stripped-up flat is then turned over on the light table so that it is *right-reading* (words and images can be read normally). This places the base side of the film up and the emulsion side down. The stripper then cuts out sections of the masking sheet, exposing the desired image elements on the film below. See Fig. 20-3.

The assembled flat is then placed over a sensitized offset plate in the platemaker. High-intensity light is used to expose the plate, Fig. 20-4. The plate is then developed with processing chemicals that bring out its image. This action causes the image to be ink-receptive on the offset press, Fig. 20-5.

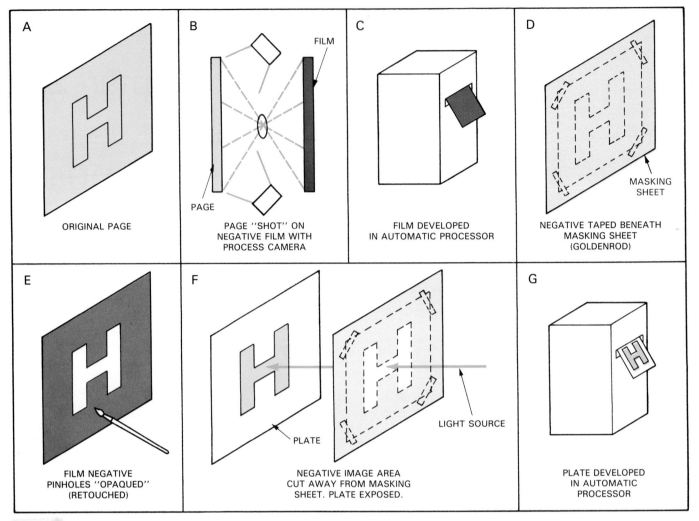

A	B	C	D
ORIGINAL PAGE	PAGE "SHOT" ON NEGATIVE FILM WITH PROCESS CAMERA	FILM DEVELOPED IN AUTOMATIC PROCESSOR	NEGATIVE TAPED BENEATH MASKING SHEET (GOLDENROD)

E	F	G
FILM NEGATIVE PINHOLES "OPAQUED" (RETOUCHED)	NEGATIVE IMAGE AREA CUT AWAY FROM MASKING SHEET. PLATE EXPOSED.	PLATE DEVELOPED IN AUTOMATIC PROCESSOR

Fig. 20-1. Stripping is process of accurately positioning and fastening film negatives, or film positives, onto a masking sheet. Basic steps in stripping a flat and making an offset plate are illustrated.

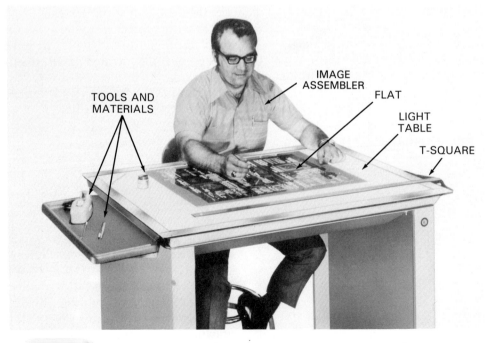

Fig. 20-2. Stripping is done by a stripper at a light table. (nuArc Company, Inc.)

Fig. 20-3. Unwanted sections of masking sheet are removed with a knife, exposing desired image areas.

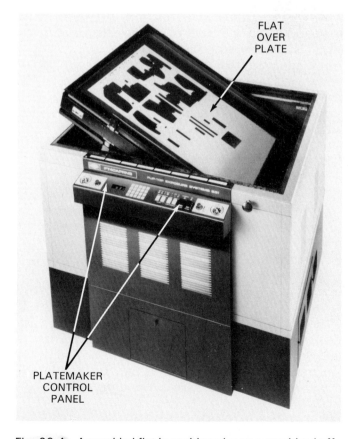

Fig. 20-4. Assembled flat is positioned over a sensitized off-set plate in platemaker. High-intensity light is used to expose plate. (nuArc Company, Inc.)

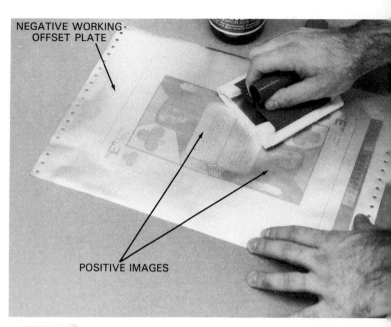

Fig. 20-5. A negative-stripped flat is used to expose a negative-working offset plate. Plate consists of positive images.

Negative and positive stripping

The term *negative stripping* refers to the preparation of flats that consist of film negatives. A negative-stripped flat is used to expose a negative-working offset plate. The plate consists of positive images, Fig. 20-5. When a film positive is stripped into a negative-stripped flat, a reverse image is produced on the negative-working offset plate.

The term *positive stripping* refers to the preparation of flats that consist of film positives. A positive-stripped flat is used to expose a positive-working offset plate. The plate consists of positive images exactly like the negative-working plate. When a film negative is stripped into a positive-stripped flat, a reverse image is produced on the positive-working offset plate.

Negative stripping is the most popular method in the industry. Detailed procedures for negative stripping are included in this chapter. Negative-working and positive-working plates are described in Chapter 21.

Stripping equipment and tools

It is necessary that the stripper organize the special equipment and tools so that they are easily located when needed. The following list of equipment, tools, and materials is by no means complete. It represents those items commonly found in the stripping areas of most offset lithography departments.

Most of the stripper's work is done at a *light table*. The working surface is translucent and lighted from below. The light shines through the film so it can be positioned on the flat more accurately. Light tables are equipped with working edges that allow for precise drawing of lines while using a T-square and triangle. Refer to Fig. 20-2.

In stripping the masking sheet, a *T-square* and *triangle* are needed to make layout lines, position negatives or positives, and cut film materials and masking paper. Most of the T-squares and triangles used for stripping are made of stainless steel.

A *magnifying glass* is used to examine film negatives and positives, Fig. 20-6. Damaged film materials can be detected more easily with the aid of a magnifying glass. Register marks on color flats can be accurately positioned and checked with a magnifying glass.

A *sharp knife* or *razor blade* is one of the most important tools used in stripping. Cutting is very critical when preparing the flat. Film materials are usually trimmed to size with a sharp pair of *scissors*.

Art brushes, varying in size from fine to medium, are used to opaque or cover up pinholes and scratches on film materials, Fig. 20-6. Since these brushes are quite small and delicate, they must be kept clean. Safe storage is important so that the bristles remain straight and not curled or crushed.

Goldenrod paper is used as masking paper when preparing a flat. Its name comes from its color, which is yellow-orange. Goldenrod paper allows *non-actinic light rays* (rays from fluorescent and incandescent bulbs) to penetrate the paper so that the images on the film can be seen during stripping. The paper does NOT allow *actinic rays,* those which cause photochemical changes, to affect the plate when the exposure is made.

Plastic masking sheet material is used where extreme accuracy in color printing is needed. Masking sheets are usually cut the same size as the offset printing plate being used.

Litho red and clear *pressure-sensitive tapes* are used for attaching negatives and positives to the masking material. Rolls of tape are kept in a dispenser close to the working area, Fig. 20-6.

Liquid opaquing solution is applied to the film materials to cover pinholes, scratches, and other unwanted areas. Opaque is available in water- or alcohol-base and in black or rust colors. The opaque solution must be kept thin because a thick solution prevents good contact between the film materials and the plate during the platemaking operation. Thinning can be done with water or solvent depending upon the composition of the opaque base.

NEGATIVE STRIPPING

As described earlier, *negative stripping* refers to the preparation of a flat that consists of film negatives. The location of the negatives on the masking sheet determines the accuracy of the image on the offset printing plate.

Before the stripper starts the job, certain preparations are necessary. At this point, most stripping errors occur because of stripping carelessness.

It is important that the stripper read the instructions that are included with the job. Specifications, including the original layout which accompanies the negatives, should be checked. By doing this, the stripper will learn all the details of the job.

Inspection and trimming

The negatives to be stripped should be examined for size and quality. This can be done by using a magnifying glass. This inspection is very important in the case of halftone negatives. The negatives are then trimmed with scissors, leaving at least 1/2 inch (12 mm) margin around the image or work area of the negatives. Crop and trim marks should NOT be cut off. If the negative image is to *bleed* (print in margin), do not trim the negative until it is taped in place on the masking sheet.

Select the T-square you will use, and be sure to use this same T-square for all work on the same flat. If another T-square is used, it may vary slightly in squareness. Clean the top of the light table and wash the glass top with glass cleaner and dry thoroughly.

Ruling the masking sheet

The stripper makes *reference lines* on the goldenrod paper to locate the lead edge, gripper margin, and other lines. This is done with a pen. Place a sheet of

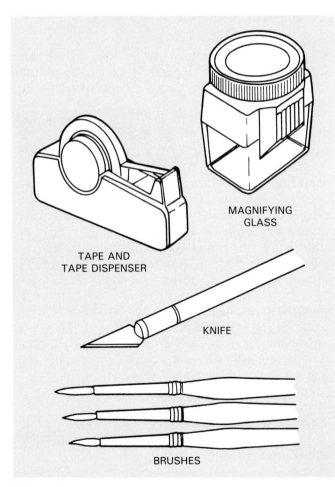

MAGNIFYING GLASS

TAPE AND TAPE DISPENSER

KNIFE

BRUSHES

Fig. 20-6. A few tools and some materials used by a stripper include magnifying glass, knife, tape, tape dispenser, and brushes. (Tobias Associates, Inc.; M. Grumbacher, Inc.)

goldenrod paper, cut to press-plate size, on the light table. One long edge of the sheet, referred to as the *gripper edge,* should be parallel to the edge of the table nearest the stripper. This edge corresponds to the lead edge of the plate and of the paper as it is fed into the press. For duplicator-size presses, the lead edge is the short dimension, and can be placed to the left of the stripper.

Line up the lower edge of the goldenrod paper with the upper edge of the T-square. Fasten it to the glass top with pieces of masking tape, working from diagonal corners. Smooth the paper before taping it down to avoid wrinkles and air traps. Mark the flat in a corner with the number or name of the job. This will identify the job for platemaking purposes.

Make sure you locate the film negatives accurately so that the images will be in the exact position on the printed sheet. First, you must determine from the instructions and layout which edge of the masking sheet will match the lead edge of the job, Fig. 20-7.

The *lead edge* (or gripper edge) is the edge of the press sheet that enters the press FIRST. It is held by a set of metal fingers in the press called *grippers.* The fingers hold the paper during its cycle on the press. The lead edge is also the point at which the offset plate is bent and attached to the plate cylinder on the press. Allowance must be made for the bend and lead edge dimension requirements. These are different for each press size. For most duplicator-size presses, the lead edge is approximately 1 inch (25 mm) from the top edge of the plate.

In addition to identifying the position of the lead edge, locate the gripper margin, Fig. 20-8. The *gripper margin* is the distance between the top edge of the

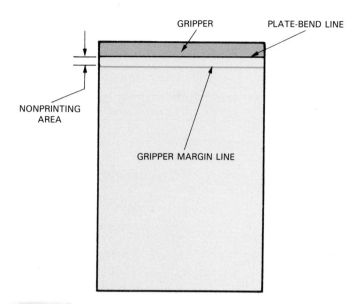

Fig. 20-8. In addition to identifying position of lead edge on flat, a gripper margin is drawn. Gripper margin is distance between top edge of press sheet and point where printed image begins.

press sheet and where the printed image begins. During the printing operation, this part of the sheet is held by the press grippers. The printed image cannot extend above this line.

The stripper locates and marks the lead edge and gripper margin on the masking sheet with a ballpoint pen. Since the gripper margin includes a portion of the top edge of the paper, an oversize press sheet is required for jobs with narrow margins or images that bleed.

After printing is completed, an oversize press sheet is trimmed to the final required dimensions. Gripper margins range from 5/16 inch (8 mm) for duplicator-size presses to 3/8 inch (10 mm) for larger presses. The stripper must know the gripper margin dimensions for the job and for each press in the press department.

In addition to the lead edge and gripper margin lines, the stripper draws other reference lines. A vertical line is drawn indicating the center of the printed sheet, Fig. 20-9. This is called the *centerline.*

Lines are also drawn to show the outline of the press sheet, Fig. 20-10. If the job is to be trimmed after it is run, lines representing *trim* (outside areas cut off) are drawn. The stripper may also draw lines to aid in positioning the negatives on the masking sheet. Some masking sheets are available cut to popular press sizes and preprinted with reference lines. These are commonly used for duplicator-size presses.

If the job is to be trimmed, folded, or cut after it is printed, the stripper cuts out a thin opening in the flat so that it will print on the press sheet. A portion of the flat may also be cut out and a piece of opaque negative can be taped onto the flat. Thin lines are scribed on the negative emulsion so they will print on the press sheet. These lines indicate where to trim, fold or cut the finished printed press sheets.

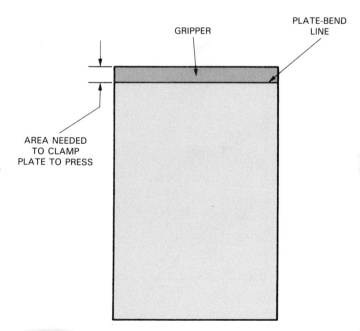

Fig. 20-7. Lead edge on flat represents edge of press sheet that enters press first. Press sheet is held by a set of metal fingers in press. These press fingers are called grippers.

417

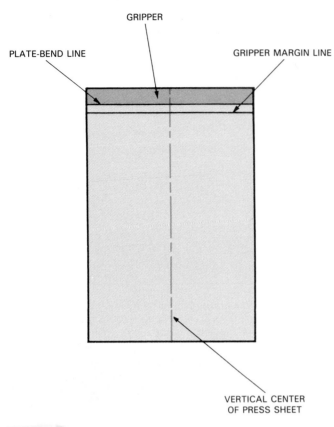

Fig. 20-9. Vertical line drawn on masking sheet identifies center of press sheet.

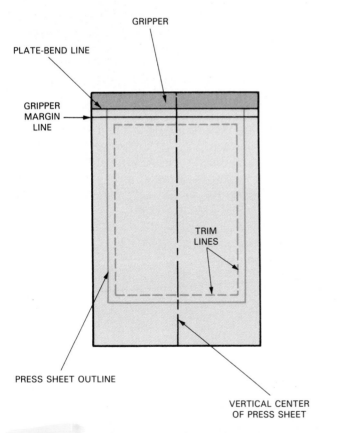

Fig. 20-10. Lines are drawn on masking sheet to show outline of press sheet. If job is to be trimmed after it is printed, lines for trim are drawn.

Emulsion-up stripping

Stripping is done by working with the negative either emulsion side up or emulsion side down. The emulsion-up method of stripping a masking sheet and positioning the film negative is outlined below. For this example, assume that the job is to be printed on 8 1/2 x 11 inch (210 x 297 mm) paper. A single film negative is used for this purpose.

1. Remove all tape and dirt from the surface of the glass on the light table. Clean the glass surface with a cloth and a recommended glass cleaner. Avoid using any cleaner that contains ammonia. Ammonia tends to leave the glass surface slippery so that tape will not stick properly. Gather all necessary tools and supplies. Use soap and water to clean the T-square, triangles, and other measuring tools.

2. Select a piece of goldenrod masking paper (unruled) the same size as the plate to be used on the press.

3. Line up the masking paper with the upper edge of the T-square. The lead edge should be to your left. This should be done in the lower left-hand corner of the light table for convenience in handling. Fasten the masking sheet to the glass surface with small pieces of masking tape. Work from diagonal corners when taping, smoothing the masking sheet before taping each corner.

4. Identify the lead edge of the masking sheet by drawing "Xs" along the lead edge (left side) of the masking sheet. See Fig. 20-11.

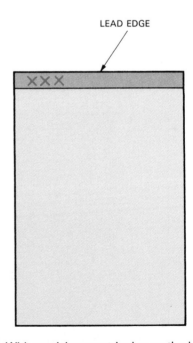

Fig. 20-11. With emulsion-up stripping method, lead edge of masking sheet is identified by drawing "Xs" along lead edge (left side) of masking sheet. A line, often 1 1/16 inch from edge of masking sheet, is drawn. This is referred to as lead edge. Note that different models and sizes of presses may require different lead edge allowances.

5. Draw a line 11/16 inch (18 mm) from the lead edge of the sheet. This line sets off the lead edge.

6. Draw another line 5/16 inch (8 mm) away and parallel to the first. The space between the lead edge line and the second line represents the gripper margin, Fig. 20-12. Remember that none of the image area of a negative can be positioned in the gripper margin.

7. Measure the width of the masking sheet and draw a centerline through the middle, perpendicular to the gripper margin line. Label it "₡" at the top.

8. Draw a line representing the paper length. Be sure to measure from the lead edge line on the masking sheet.

9. Draw the outer reference lines representing the paper width. Half the width will be on either side of the centerline.

10. Draw lines showing the exact size and position of the image area. Do NOT include the entire negative. Extend the lines approximately one inch (25 mm) beyond the corners of the rectangle to assist in positioning the negative.

11. When all the reference lines have been drawn, position the negative with the emulsion side up, Fig. 20-13. The negative will usually contain crop marks at four corners. These marks assist in locating the precise position within the paper outline area.

12. Using a T-square and triangle, carefully square the negative in the position required. Hold the negative firmly in position with one hand while using the other hand to reach for the red litho tape.

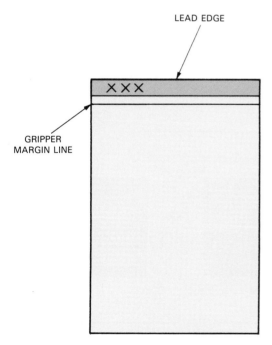

Fig. 20-12. A line, typically 5/16 inch away and parallel to the lead edge line, is also drawn. Space between lead edge line and second line is gripper margin. Note that different models and sizes of presses may require different gripper margin allowances.

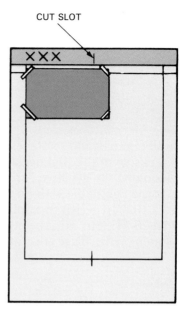

Fig. 20-13. After all reference lines are drawn, negatives are positioned with emulsion side up.

13. Place a one inch (25 mm) piece of red litho tape at each corner of the film negative and one piece along the center of each side of the negative. Use only enough tape to hold the negative in place. Excess tape prevents good contact between the film and offset plate during platemaking. The tape should be at least 1/8 inch (3 mm) from any image.

14. After the negative has been taped in position, make a wedge-shaped cut along the centerline at the gripper edge of the masking sheet. This cutout identifies the lead edge and helps the person making the plate to position the flat with the lead edge of the plate. It also assists the press operator to identify the lead edge and guide edge when attaching the plate to the press cylinder.

15. Once the negative has been taped in place, remove the tape holding the masking sheet to the table and turn the sheet over on the light table. With a sharp knife, cut away the goldenrod paper covering the image areas of the negative. Do not cut into the film negative since cut lines will reproduce on the plate. A small sheet of clear acetate inserted between the negative and goldenrod paper can be used to protect the negative during cutting.

SAFETY NOTE! Always use a sharp X-acto® blade. Do not force the cut. Handle the knife carefully; the blade is extremely sharp and can cause serious injury.

Cut the window at least 1/8 inch (3 mm) larger than the image areas. Allow the masking sheet to cover as much of the non-image areas of the negative as possible. This provides additional protection to the plate during exposure and reduces the time spent opaquing unwanted pinholes and scratches.

16. Small unwanted transparent openings, called *pinholes,* and scratches are often found on film negatives. These must be covered or they will show up on the offset plate as images. Also, some negatives may contain image areas you do not wish to print. They must be opaqued before the flat is sent to the platemaking department. Opaquing is usually done AFTER the windows have been cut.

Opaque all pinholes and scratches on the right-reading (base) side of the film negative whenever possible. Examine the negative for pinholes and scratches on the light table with its emulsion side down. Use a fine art brush to apply the thinned opaque, Fig. 20-14. Care must be taken not to get opaque solution on the image areas of the negative.

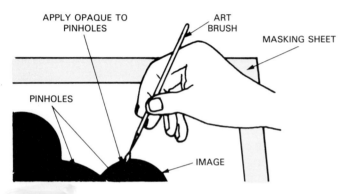

Fig. 20-14. To eliminate pinholes and scratches, a fine art brush is used to apply opaque to the base side of the negative after windows have been cut.

Emulsion-down stripping

The emulsion-down negative stripping method is used by some strippers who prepare single negative flats for duplicator-size presses. During the stripping operation, the negative is kept right-reading side up on the light table. This method is NOT practical when stripping more than one negative on a single masking sheet especially for large presses. Steps in the emulsion-down negative stripping method are given below.

1. Place the film negative on the light table surface with the emulsion side of the negative against the glass surface, Fig. 20-15. The negative will be right-reading. The side of the negative near the lead edge of the masking sheet should be placed away from you.
2. Tape the negative to the light table with small pieces of masking tape. See Fig. 20-16.
3. Place a pre-ruled masking sheet over the negative in the correct position, Fig. 20-17. The printed lines on the masking sheet should face up. Since the masking sheet is translucent, you can align it with the film images.

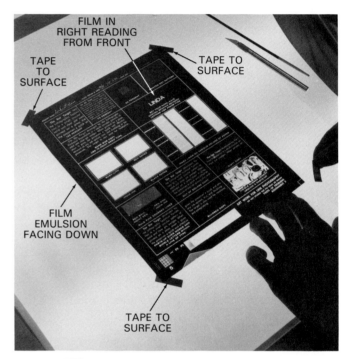

Fig. 20-16. With emulsion-down stripping, negative is taped to light table surface with small pieces of masking tape.

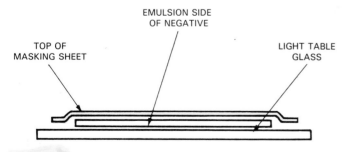

Fig. 20-15. With emulsion-down stripping, film negative is placed on light table with emulsion side of negative against glass surface. Negative will appear right-reading.

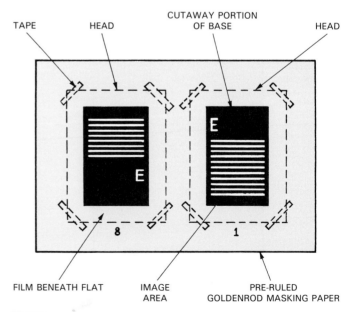

Fig. 20-17. This is a flat for an emulsion-down form of an 8-page signature.

4. When the masking sheet is in position over the negative, hold it firmly with one hand. Cut two small oblong windows in the masking sheet at opposite sides, top and bottom. Cut over the film negative but outside the image area. Be careful NOT to cut through the negative.

5. While still holding the masking sheet down firmly against the negative, place a piece of red litho tape over each window and press firmly.

6. Cut two more windows at the other sides of the negative and place tape over them. Continue to hold the masking sheet down firmly against the negative.

7. Remove the masking tape used to hold the film negative to the light table surface.

8. Turn the masking sheet over.

9. Place a one inch (25 mm) piece of red litho tape at each corner of the film negative and one piece along the center of each side of the negative.

10. With a sharp knife, cut away the goldenrod paper covering the image areas of the negative. Cut the windows at least 1/8 inch (3 mm) larger than the image areas.

11. Opaque all pinholes and scratches on the right-reading (base) side of the negative.

POSITIVE STRIPPING

The procedure for positive stripping, or stripping positive, varies slightly from negative stripping. The positive flat may consist of one or more individual film positives. These are taped to a transparent (film) *plastic base* the size of the press plate. It is also common practice to prepare a single film positive consisting of several different positive images. This is done in the camera department. The composite film positive is the same size as the press plate.

The basic steps for positive stripping are:

1. Tape a sheet of layout material, larger than press plate, to the light table. The layout material usually consists of ink-receptive matte acetate or white paper. This provides the base for a master layout. Refer to Fig. 20-18.

   Note! Another method of preparing a master layout is to make a film positive of the layout. Care should be taken to make sure that accuracy is maintained during film processing.

2. Draw all reference lines on the master layout sheet. These include press sheet and image limits, gripper, center, trim, fold, and cut marks. All the reference lines normally used in negative stripping are included for positive stripping. See Fig. 20-19.

3. Tape a sheet of transparent plastic (film base) in register over the layout. The transparent plastic must be the same size as the press plate. The transparent plastic takes the place of a masking sheet in the case of negative stripping.

4. Use cellulose tape to adhere the film positives,

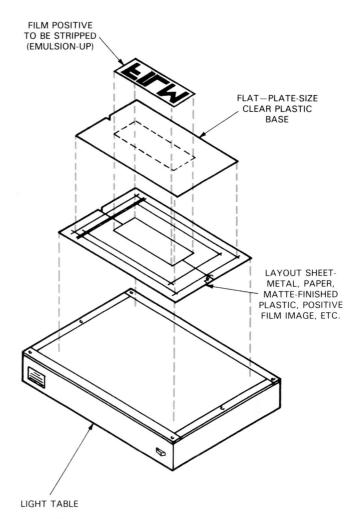

FILM POSITIVE TO BE STRIPPED (EMULSION-UP)

FLAT—PLATE-SIZE CLEAR PLASTIC BASE

LAYOUT SHEET—METAL, PAPER, MATTE-FINISHED PLASTIC, POSITIVE FILM IMAGE, ETC.

LIGHT TABLE

Fig. 20-18. Schematic illustrates basic elements and steps necessary for positive stripping.

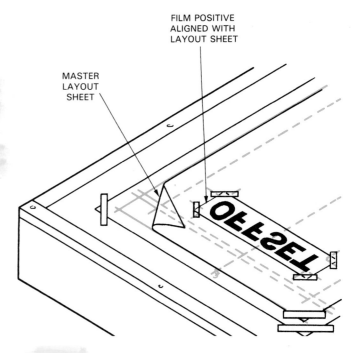

FILM POSITIVE ALIGNED WITH LAYOUT SHEET

MASTER LAYOUT SHEET

Fig. 20-19. Reference lines used in negative stripping are included in positive stripping method.

emulsion side up, to the transparent plastic. This is done by accurately aligning the positives with the layout underneath.

It is important that all film positives be of the SAME THICKNESS so film-to-plate contact is uniform during exposure in the platemaker. All adjoining film positives should be butted, rather than overlapped at the joints. A minimum of thin transparent cellulose tape should be used to fasten the film positives to the plastic base.

5. Remove the complete flat from the layout sheet. Carefully inspect the flat over a lighted table. Opaque pinholes and defects or scratches in the usual manner. Remove dust and dirt with film cleaner. A soft brush is used for this purpose.

MULTIPLE-PAGE STRIPPING

When stripping a job requiring two or more film negatives, the emulsion-up method is preferred. Since multiple-page stripping is almost the same as that for single-page stripping, only the major differences are detailed here.

1. Cut a piece of goldenrod masking paper to press plate size. Place its long edge parallel to the edge of the table nearest you. Align it with a T-square. This will be the lead edge.

2. Fasten the masking sheet to the glass surface with masking tape, working from diagonal corners.
3. Draw two "Xs" along the lead edge.
4. Draw a line setting off the lead edge using the recommended allowance for the press size being used.
5. Draw another line parallel to the first to show the gripper margin. This, too, must be the correct allowance for the job's press size. Remember the gripper margin area cannot contain an image.
6. Measure the width of the masking sheet and draw a centerline through the middle, perpendicular to the gripper margin. Label it "℄" at the top.
7. Draw the two reference lines which represent side margins of the paper size to be printed. Measure from the centerline out to both side margins.
8. Draw a line representing the paper length. Measure starting from the lead edge line on the masking sheet.
9. Inside the area, draw rectangles to the exact size and position of each of the image areas. Extend the lines approximately one inch (25 mm) beyond the corners of the rectangle to assist in positioning the negatives.
10. Position the film negatives emulsion side up. If the negatives contain crop marks, use them to locate the precise position within the paper areas. Look at Fig. 20-20.

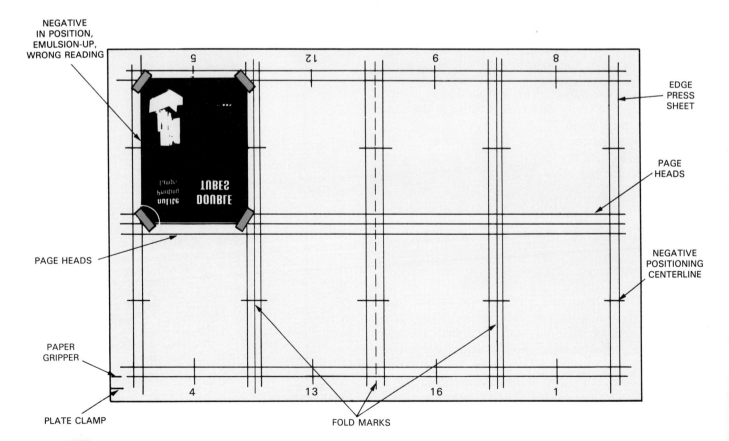

Fig. 20-20. With multiple-page stripping, film negatives are positioned emulsion side up over ruled masking sheet. If negatives contain crop marks, they are used to locate precise position within paper areas. (nuArc Company, Inc.)

11. Use the T-square to position and align the negatives one at a time.
12. Hold the negative with one hand while placing litho red tape at each corner. Also place one piece of tape along the center of each side.

When several negatives are being stripped close together, the film edges should NOT overlap. Overlapping usually creates contact problems between the negatives and plate when photographically exposing the plate. Edges of the film should be butted together rather than overlapped. See Fig. 20-21.

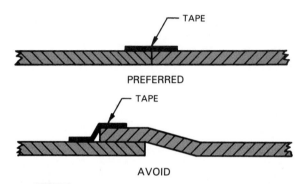

TAPE

PREFERRED

TAPE

AVOID

Fig. 20-21. When several negatives or positives are being stripped close together, film edges should not overlap. Edges of film should be butted together.

13. After the negatives have been taped in place, make a wedge-shaped cut along the centerline in the lead edge of the masking sheet. This is a reference for the platemaking and press departments.
14. Turn the combined masking sheet and negatives over on the light table and cut windows to expose the image areas.
15. Opaque any pinholes and scratches on the right-reading (base) side of the film negatives.

COMBINATION STRIPPING

Two basic stripping methods are used when combining line and halftone film materials. These are referred to as complementary flats and halftone windows.

Complementary flats

In stripping *complementary flats,* the line negatives and halftone negatives are stripped separately. This is often necessary when several negatives must be pieced too closely together to be handled in a single flat.

The line negatives are stripped up on one flat, called the *master flat.* A second piece of goldenrod is registered and taped over the master flat. The halftone negatives are stripped in register on the *second flat.* A set of complementary flats is shown in Fig. 20-22.

Two exposures are made in register on one plate. The platemaker makes the main exposure using the master flat containing the line negatives. This "burns" the line

copy into the printing plate. Without moving the exposed plate, the master flat is removed. The flat containing the halftone negatives is placed in register over the plate. A second exposure is made to expose the halftone images onto the plate. The plate is then processed in the usual manner.

Halftone windows

A line negative containing one or more clear (transparent) openings can be used to insert separate halftone negatives. These are commonly referred to as *halftone windows,* Fig. 20-23. The stripper generally

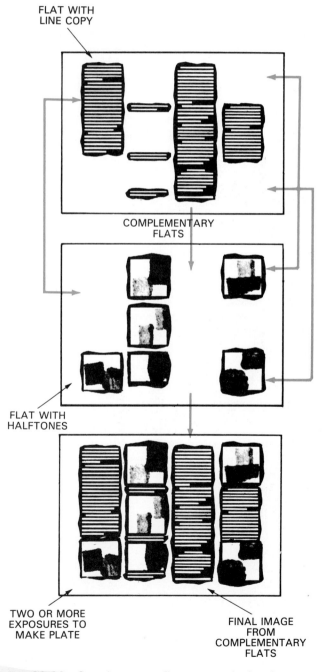

FLAT WITH LINE COPY

COMPLEMENTARY FLATS

FLAT WITH HALFTONES

TWO OR MORE EXPOSURES TO MAKE PLATE

FINAL IMAGE FROM COMPLEMENTARY FLATS

Fig. 20-22. Complementary flats are used when images are extremely close together. Each flat is exposed onto same plate separately. (nuArc Company, Inc.)

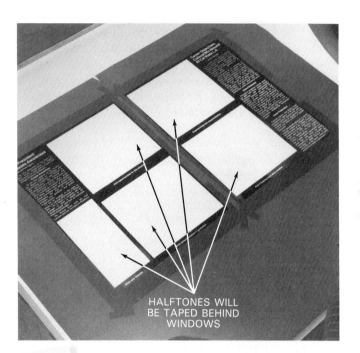

Fig. 20-23. This flat, containing a line negative, has five transparent window openings. During stripping process, halftone negatives are taped behind windows.

attaches the halftone negatives BEHIND the windows. It is important that each halftone be inserted in the window to which it was keyed on the layout.

The following procedure is used when attaching halftone negatives to window openings on a line negative that has been stripped and the image areas cut open.

1. Check to see that the masking paper has been accurately cut away from the transparent area where the halftone will be positioned.
2. Cut the halftone negative to a size that is slightly larger than the image that will be printed. From 3/16 to 1/4 inch (5 to 7 mm) is usually sufficient overlap.
3. Turn the flat over on the light table so the emulsion side is up. Place the halftone into the window opening, also emulsion side up. Attach the halftone negative to the line negative with short narrow strips of red litho tape. Make sure that the halftone is fully contained within the window opening and that it is square.

The camera department may provide the stripping department with two identical negatives containing line work and halftone windows. This procedure is especially useful when halftones and captions are close-fitting.

In this method, the stripper prepares a master flat with the line matter along with a goldenrod sheet cut to form a mask. The mask is cut to cover the halftone windows during platemaking. The second flat contains all halftone negatives taped behind the halftone windows. A goldenrod mask is then cut to cover all type matter on this flat. Platemaking is handled in the same way described for complementary flats.

STEP-AND-REPEAT STRIPPING

Because press time is shortened, it is economical to print two or more copies of the same job on a single sheet. This requires step-and-repeat stripping. With *step-and-repeat stripping,* one negative is exposed to a plate in two or more positions. The procedure is basically the same as regular stripping.

Step-and-repeat machines that automatically reposition the image are commonly used for this purpose. One is shown in Fig. 20-24. In this procedure, one negative is used to step-and-repeat a series of images on a single offset plate.

For most purposes, the following procedure can be used to step-and-repeat a job when a machine is not available for the purpose. Strip the negative on a masking sheet in the usual way, Fig. 20-25. Cut register marks, called *butterflies,* in the masking sheet. Lay out another masking sheet, large enough to cover the en-

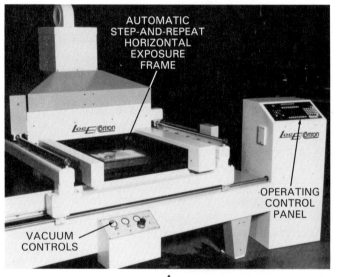

A

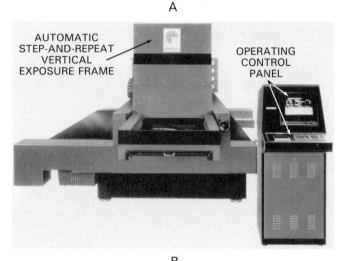

B

Fig. 20-24. A—This step-and-repeat machine has a computer control panel. B—In step-and-repeat process, one negative is used to expose a series of images on a single offset plate. (Log Etronics, Inc.; Krause)

tire plate, for as many exposures of the image as desired. Cut windows in the sheet wherever the negative is to be exposed. Cut corresponding register marks in both sheets so the negative will be positioned accurately.

During exposure, the masking sheet containing the negative is placed over one of the windows in the other masking sheet. The extra windows are covered. This process is repeated until the negative has been exposed over each of the windows.

Two other variations for step-and-repeat layouts are illustrated in Fig. 20-26.

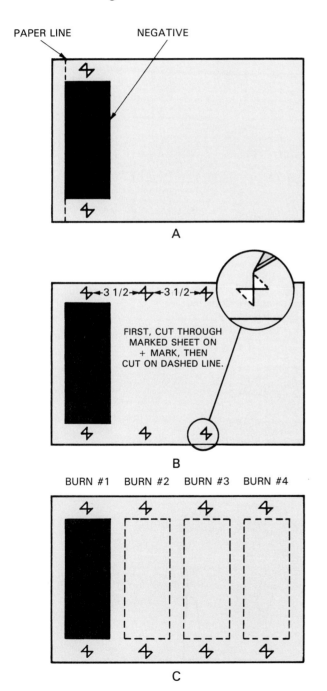

Fig. 20-25. This step-and-repeat method involves cutting register marks, called butterflies, in masking sheet. Butterflies are used to position flat for each exposure to plate.

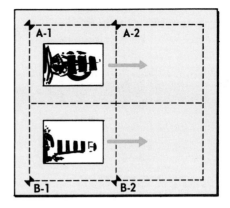

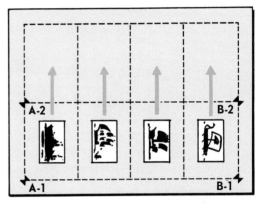

Fig. 20-26. Note two variations of butterfly step-and-repeat method. Platemaker moves A-1 and B-1 to positions A-2 and B-2 for second exposure. Uncovered portions of plate are masked off as each exposure is made.

MULTIPLE COLOR STRIPPING

If the printed job is to include two or more colors, all the negatives for the job can be stripped on one flat. For example, the stripper will indicate, in circled notations, the illustrations and type that are to print in black, and those that are to print in another color. Register crosses are marked in diagonal corners of this flat, called the main flat. The flat is then turned over on the light table.

A second sheet of goldenrod is placed over the main flat. The stripper traces register marks on this sheet, and cuts out small register windows at the center of the register marks to expose the register cross. Each part of the flat that is to print in black is cut open to form window openings in the goldenrod. The stripper marks the flat—BLACK FLAT.

The stripper then places another sheet of goldenrod over the main flat. Each part of the flat that is to print in blue is cut open to form window openings. Register marks are included as for the preceding flat. The stripper marks this flat—BLUE FLAT.

During platemaking, the main flat is taped to the plate. The goldenrod sheet with the window openings for the black flat is registered over the main flat. The plate, when exposed and developed, will contain all the images to print in black ink.

A second unexposed plate is placed in the exposure unit. The operator tapes the main flat to this plate. The goldenrod sheet with the window openings for the blue flat is registered over the main flat. The plate, when exposed and developed, will contain all the images to print in blue ink.

If more than two colors are required, this method can be adapted to include separate flats for each color. This requires a precise system of registration. A pin register system is used for this purpose.

PIN REGISTER SYSTEM

The term *register* describes the placement of one image in relation to another image. A *pin register system* uses punched holes and mechanical devices to produce an accurate register. This is especially true if the flats and plates are both punched identically. To complete the system, the offset press is fitted with a matching set of pins to receive the plate. See Fig. 20-27.

In using a pin register system, the acetate or plastic masking sheets are first punched on a two- or three-hole pin register punch, Fig. 20-28. Some punches have self-centering devices which automatically ensure accurate alignment. See Fig. 20-29.

Register pins and tabs are available in many sizes and styles, Fig. 20-30. These are taped to the surface of the light table. The first piece of masking material is then placed in position and fitted to the register pins, Fig. 20-31. After the first flat is stripped, the other flats are handled in the same manner as the first flat.

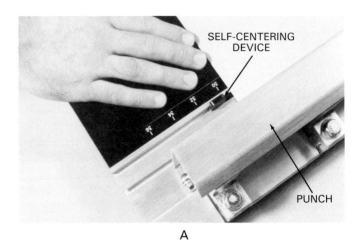

A

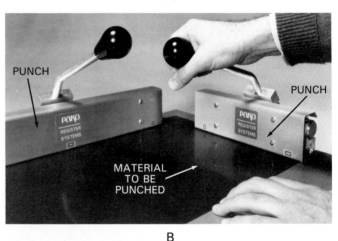

B

Fig. 20-28. A—Pin register board can accommodate several sizes of film and masking materials. B—Self-centering device, on this pin register board, automatically ensures accurate centering and alignment. (Pako Corp.)

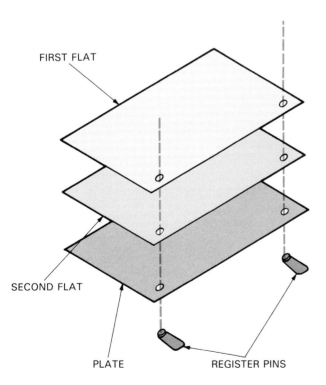

Fig. 20-27. With pin register system, flats and plate are punched. This keeps them in alignment when exposing plate.

Fig. 20-29. Film punch register system provides extremely accurate registration when preparing chokes, spreads, and graphic enhancements. (Byers Corp.)

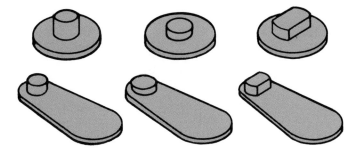

Fig. 20-30. These are a few types of register pins used for pin register stripping.

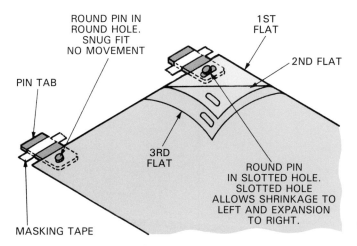

Fig. 20-31. Note use of register pins and punched flats to maintain register between flats.

Total pin register systems

Pin register systems are being used in all pre-press phases of offset lithography. This means that pin register begins with the paste-up operation. The paste-up artist assembles image elements onto a pin-registered base sheet. The base sheet material must be absolutely stable. *Base sheet stability* means that the material must be of such quality and construction that it will not expand or contract due to climatic conditions. Any change in the size of the base material will lead to *misregister* (images not in alignment with each other).

The paste-up is pin-registered on the process camera copyboard. The film is pin-registered to the camera back of the process camera. The processed film is pin-registered to the masking sheet. The flat is then placed over a set of pins during the plate exposure process. Finally, the plate is pin-registered onto the plate cylinder of the offset press.

A total pin register system is illustrated in Fig. 20-32.

COMPUTER-AIDED STRIPPING SYSTEMS

Computer-aided stripping systems are based on computer-aided design/computer-aided manufacturing (CAD/CAM) technology, Fig. 20-33. Drafting departments and design engineers have been using these devices to draw lines and geometric shapes with much more speed and precision than with manual methods.

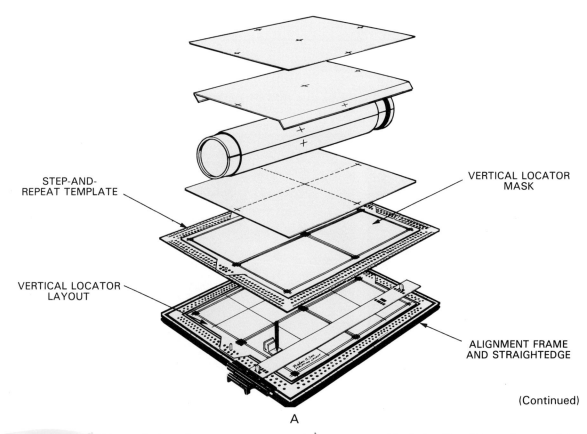

(Continued)

A

Fig. 20-32. A—This total pin register system starts with paste-up and includes film, flat, and press plate.

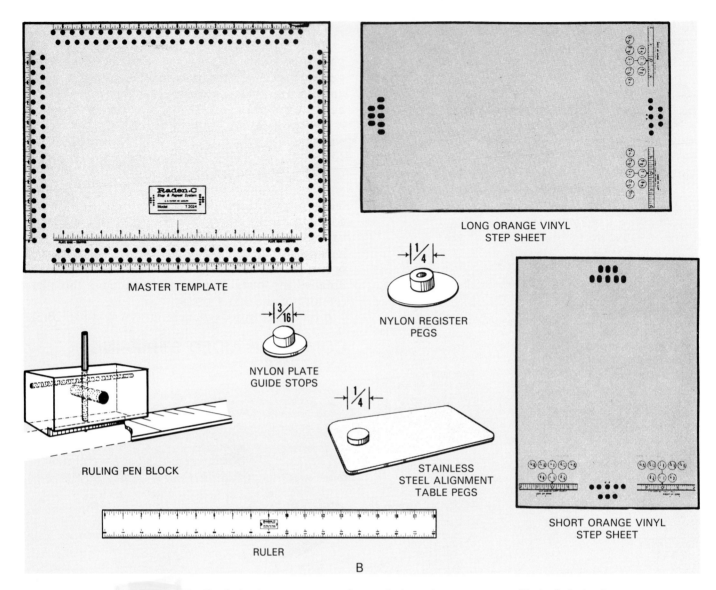

MASTER TEMPLATE

LONG ORANGE VINYL
STEP SHEET

NYLON REGISTER
PEGS

NYLON PLATE
GUIDE STOPS

RULING PEN BLOCK

STAINLESS
STEEL ALIGNMENT
TABLE PEGS

SHORT ORANGE VINYL
STEP SHEET

RULER

B

Fig. 20-32. B—Study basic components of a total pin register system. (Raden® C, Inc.)

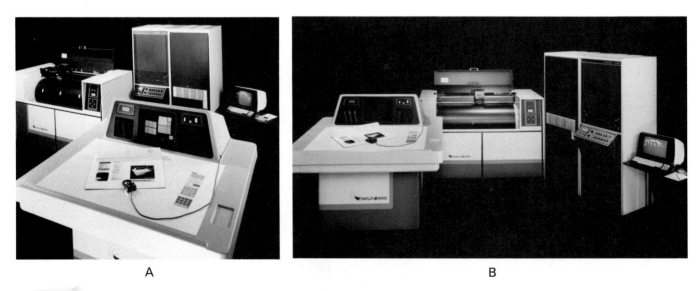

A

B

Fig. 20-33. A—Computer-aided stripping systems are adaptations of computer-aided design/computer-aided manufacturing. B—Computer-aided systems are extremely fast and can produce masks, dropouts, and undercuts for halftones and screen tint images.

The printing and publishing industry is using computer-aided stripping to produce masks, dropouts, and under-cuts for halftone and screen tint images. Computerized stripping can save over 50% of the total stripping time. Once the necessary mask films are made, all the stripper has to do is strip the halftones and screen tints on the films, using the masks to crop and position the images for platemaking.

Common features of computer stripping systems include:

1. Input station with an interactive image digitizer table, Fig. 20-34.
2. Output unit with high speed plotter to produce photographic, scribed, or masking film intermediates, Fig. 20-35.
3. Accuracy and resolution of ± 0.004 inch (± 0.102 mm) over a 48 inch (1220 mm) plotting area.

SCRIBING FILM NEGATIVES

The transparent areas on negatives will transmit light to produce the printing image. Wherever the blackened emulsion is removed from a film, light can pass through to expose the offset plate. To produce rules, grids, or borders on a negative, strips of emulsion can be shaved away with a scribing tool to form transparent lines. This process is called *scribing*. See Fig. 20-36.

Good scribing tools are ground to the correct widths for specified line thicknesses. When drawn across the emulsion side of a negative, they will shave away a strip of the emulsion and produce a sharply defined rule. A needle point or a blunt tool will furrow the emulsion and produce a ragged line. The emulsion for ruling must be clean and dry. It should not be excessively hardened in fixing or excessively dried by forced

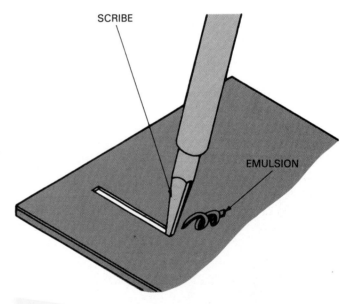

Fig. 20-34. This computer-aided stripping system includes input station with interactive image digitizer table. (Gerber-Scientific)

Fig. 20-36. Tool is used to scribe rules in film negative emulsion.

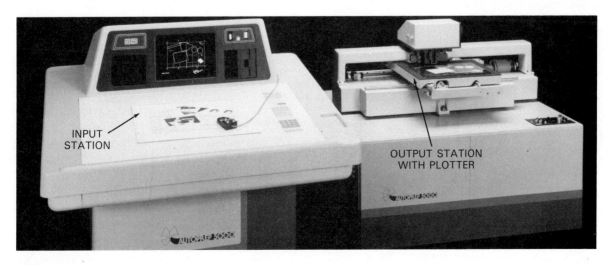

Fig. 20-35. This computer-aided stripping system incorporates output unit with high speed plotter to produce photographic, scribed, or masking film intermediates. (Gerber-Scientific)

heating. Before being used in the image area, the scribing tool should always be tested along the waste margins of a film. Check its ability to shave away a clean ruled line. Some emulsions are scribed more easily when slightly damp.

There are several operations which involve scribing. These include: adding rules on a negative, ruling uniformly spaced rules, duplicating a previously printed form, scribing horizontal and vertical rules, opaquing the ruled negative, bordering an illustration, bordering halftone inserts, and adding rules to positives.

Adding rules on a negative

To add rules on a negative, square up the negative on the light table. Select a light table area that is clean and free from scratches. Tape the film down securely with strips of red litho tape. Mark the location of the lines to be scribed along the margins of the negative, outside of the trim area of the job. Set the T-square to the location for the rule. Hold the scribing tool perpendicular to the edge and slightly inclined in the direction of ruling. Scribe the rule as needed.

Ruling uniformly spaced rules

If the rules for a job are to be uniformly spaced, tape down a thin steel rule alongside the form, Fig. 20-37. Position the T-square or triangle to the scale graduations for each line to be ruled. If the uniform spacing is not a full fractional value of an inch, such as when 212 rules are to be spaced uniformly in 5 1/4 inches, the steel rule can be tilted at an angle and taped down. The angle is selected so that the space is divided into the desired number of full graduations.

REMOVING THE EMULSION

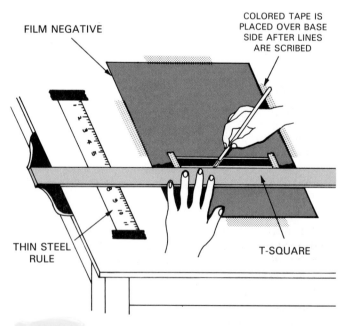

FILM NEGATIVE

COLORED TAPE IS PLACED OVER BASE SIDE AFTER LINES ARE SCRIBED

THIN STEEL RULE

T-SQUARE

Fig. 20-37. Rules are scribed in film negative emulsion using a steel T-square as a guide.

Duplicating a previously printed form

If a previously printed form is to be duplicated, the form can be squared up and taped down alongside the negative. Position the T-square at the printed rules. Before scribing the rules, pencil in an outline on the negative to show the limits for the widths of rules. Then spot out any large pinholes or other defects in the area to be ruled. Set the T-square to the location for the first rule. Hold the scribing tool perpendicular to the edge and slightly inclined in the direction of ruling.

Start slightly beyond the penciled limit and scribe with a steady uniform stroke across the negative to slightly past the other limit line. If the stroke was too light and not all of the line was cleanly removed, repeat the stroke without disturbing the T-square or the angle of the scriber. Continue with all the remaining lines to be scribed.

Scribing horizontal and vertical rules

If both horizontal and vertical rules are to be scribed, make sure the two edges of the light table are square with each other so that the rules will be at right angles. Otherwise, it is best to scribe the horizontal lines on the negative. Peel off the hold-down tape and rotate it 90 degrees. Square the negative with the T-square and steel triangle, and tape it down again. Then, scribe the other set of vertical rules while in the horizontal position.

Opaquing the ruled negative

After all the rules are scribed in, release the film and flop it over, emulsion side down. Square it with the layout table and tape it down. Use a ruling pen and opaque to draw the marginal borders that will mask out the ends of the successive ruled lines. Red litho tape can also be used if it is carefully applied and guided straight along the edge of the T-square or triangle.

Bordering an illustration

The same procedure can be used as in scribing rules on a negative. First, lightly pencil in the border on the negative. This will locate the lines and prevent scribing short or long lines. Use a brush to apply a thin covering of a colloidal-graphite opaque to any pinholes or unwanted transparent areas in the intended border lines.

When the opaque has dried, carefully scribe in lines of the required thickness. Follow the procedure previously detailed for adding rules. Opaque out the excess widths of the border rules at the corners. This is best done on the back side of the film. If opaque is applied to limit the width of a scribed rule on the scribed side of the film, the opaque may creep in along the line.

Bordering halftone inserts

When bordering halftone inserts, the scribed border may be applied either to the large negative or to the insert. This will depend on its required location and providing adequate clearance for the tape used to secure the insert. If the fit is so close that it will not permit scribing the border on the negative or insert, it may be possible to double print the border in location. This can be done by using a flap section on the flat or by preparing a complementary flat.

Adding rules to positives

Rules can be added to positive films by drawing them in with opaque and a ruling pen. First, lay out the rules on a sheet of paper. Then, position the positive over the layout and tape it down.

Always test the ruling pen for its ruling properties immediately before drawing on the positive. The positive should be clean and free from greasy fingerprints. If difficulty is experienced with ruling on positives, it is advisable to use acetate inks for ruling.

If only a few short lines are to be added to a positive, such as register or trim marks, the scriber can be used to engrave these marks. The scribed lines are filled in by rubbing them with a black crayon or litho crayon. The excess crayon is then wiped away with a soft clean tissue.

Cutting in film corrections

The stripper is often called upon to cut film corrections into the job. To do this, the stripper cuts out the incorrect section and replaces it with a corrected section. Care must be taken so that the corrected film section does NOT overlap the original film. This is because the overlap area can cause distortion of the images during platemaking.

A recommended procedure for cutting in corrections follows.

1. Securely tape the film to be corrected onto a light table, emulsion down. The film should be accurately aligned with the T-square before taping. Use red litho tape for this purpose. Place a sturdy, transparent cutting sheet beneath the film to protect the glass on the light table from becoming scratched by the knife.
2. Scribe lines in the emulsion surrounding the incorrect image area. These lines are used to help align the corrected film segment. They are not needed if line-for-line corrections are made.
3. Slide the film correction beneath the area to be removed. Align it so that the spacing and borders, if any, are correct.
4. Temporarily, tape the correction in place under the section to be removed, with transparent tape.
5. Using the knife, cut through both pieces of film.
6. Remove the incorrect section.
7. Tape the correction in place, using thin strips of red litho tape. Tape all nonimage areas that will transmit light, but do NOT tape over any image areas.

PROOFING FLATS

When a job has been stripped up and checked, it is ready for platemaking. Before the plates are prepared, however, proofs are made of each flat. The purpose of *proofing* is to check for quality, imposition, and typographic errors. The proof is also used to give the customer an opportunity to check and approve the job.

The proofing operations described here apply only to single-color flats. Pre-press color proofing was covered in Chapter 18.

Three common methods used to proof single-color flats are silverprint, blueline, and diazo.

Silverprint

Commonly called *brownlines, silverprint proofs* are formed on paper coated with a silver salt which is similar to that used on photographic films.

The brownline paper is exposed through the negative flat on a platemaker. The emulsion is developed in water, fixed in hypo, washed in water, and dried. Brownline prints have a brown image on white paper. The image is permanent, but the paper is not dimensionally stable. This means that it can easily change size due to climatic conditions. To make a brownline proof, follow these steps.

1. Cut a piece of brownline paper a little larger than the negative image area to be proofed.
2. Lay the piece of brownline paper on the bed of the platemaker, emulsion side up.
3. Place the negative flat on top of the brownline paper, emulsion side down. Turn on the vacuum pump and expose the image for the recommended time. Test exposures must be made to determine the correct exposure time.
4. Remove the brownline paper and develop it in a pan of water or under running water.
5. Place the developed brownline paper in a pan of regular photographic fixer for approximately three minutes.
6. Wash the brownline paper for ten minutes and then squeegee and allow it to dry thoroughly.

Blueline

Blueline is similar to brownline proofs but, the image is blue since blueprint paper is used for making the proof. The paper is coated with an organic iron compound that changes structure when light strikes it.

The blueline paper is exposed through the negative flat on a platemaker. The emulsion is developed in water until cleared. It is then fixed with a diluted hydrogen peroxide solution and dried. Blueprint paper is NOT dimensionally stable and the image tends to lighten with age.

To make a blueline proof, follow these steps working under subdued or yellow light conditions.

1. Cut a piece of blueline paper a little larger than the negative image area to be proofed.

2. Lay the piece of blueline paper on the bed of the platemaker, emulsion side up.
3. Place the negative flat on top of the blueline paper, emulsion side down. Turn on the vacuum pump and expose the image for the recommended time. Test exposures must be made to determine the correct exposure time.
4. Remove the exposed blueline paper. Place it in a plate processing sink.
5. Develop the blueline paper with water, allowing the image to become dark.
6. Rinse with water and squeegee.
7. Fix the blueline proof with a mild solution of hydrogen peroxide.
8. Rinse with plain water and squeegee.
9. Blot the blueline proof between sheets of newsprint paper and hang to dry.

Diazo

With *diazo proofing* material, a positive image is produced when exposed to a flat composed of transparent film positives. Similarly, a negative image is produced from film negatives. The exposure is made on a platemaker and the exposed emulsion is developed when placed in contact with ammonia fumes.

Diazo paper is relatively dimensionally stable. This is because the paper is not moistened during processing. Although economical, diazo paper is not made specifically for proofing flats.

POINTS TO REMEMBER

1. Stripping and imposition are two important processes in offset lithography.
2. The term stripping refers to the practice of positioning film negatives and/or positives on a masking sheet.
3. Imposition refers to the practice of positioning pages for proper sequence in a signature. The stripper uses the imposition layout to position film negatives in the correct sequence on a masking sheet.
4. Single and multiple page stripping is common. Frequently, different jobs are *ganged* (photographed together) to fill a single press sheet.
5. In stripping, masking sheets are laid out for the intended paper size and press to be used in the reproduction operation. Image area can be added or deleted during this stage of production.
6. Strippers use either the emulsion-up or emulsion-down methods of positioning film negatives and/or positives on the masking sheet.
7. Complementary flats are used for exposing a single plate, while multicolor flats are used for exposing separate plates.

8. Automation equipment, such as CAD/CAM, is now used in the stripping operation. These computer-aided design systems automate the stripping and masking function with fast, efficient design/drafting/masking capabilities.
9. An automated stripping system unit is operated from a single workstation. A stripper no longer must bend over a light table to cut masks and position negatives. Instead, the stripper sits in front of a digitizer table and inputs all graphic data with a hand-held cursor.
10. Strippers use several methods to scribe negatives and positives for purposes of inserting rules.
11. The stripper is often called upon to cut film corrections into the job.
12. When a job has been stripped and checked it is ready for platemaking. Before the plate is prepared, a proof is made of the flat.
13. The purpose of proofing is to check for quality, imposition, and typographic errors. The proof is also used to give the customer an opportunity to check and approve the job.

KNOW THESE TERMS

Blueline, Centerline, Complementary flat, Cut line, Diazo proof, Emulsion-down stripping, Emulsion-up stripping, Flat, Fold line, Gripper edge, Gripper margin, Lead edge, Negative stripping, Pin register system, Positive stripping, Reference line, Register, Scribing, Silverprint, Step-and-repeat stripping, Trim line.

REVIEW QUESTIONS

1. Positioning and fastening film negatives or positives to a masking sheet is called _____.
2. The preparation of flats that consist of film negatives is referred to as _____ stripping.
3. A positive-stripped flat is used to expose a negative-working offset plate. True or false?
4. Most of the stripper's work is done at a _____ _____.
5. Explain a few of the tools used by a stripper.
6. The edge of the sheet that enters the press first is called the _____ _____.
7. The distance between the top edge of the sheet of paper and where the printed image begins is called the _____ _____.
8. Lines drawn by the stripper to indicate the center of the sheet, the outline of the press sheet, and trim size are called _____ _____.

9. No part of the image area of a negative can be positioned in the gripper margin when stripping a flat. True or false?

10. The _____ stripping method is preferred when stripping more than one negative on a masking sheet.

11. How is positive stripping done?

12. When line and halftone images must appear together on a page, _____ _____ can be prepared by the stripper.

13. Printing two or more copies of the same job on a single press sheet requires:
    a. Emulsion-up stripping.
    b. Combination stripping.
    c. Step-and-repeat stripping.
    d. Emulsion-down stripping.

14. _____ is a procedure the stripper uses to place lines in a film negative or positive.

15. Describe silverprint, blueline, and diazo proofs.

SUGGESTED ACTIVITIES

1. List the equipment, tools, and materials commonly used for stripping flats.

2. Prepare a flat for the duplicator using the emulsion-down stripping method. Attach the negative(s). Open the masking sheet around the image area(s). Opaque all pin holes. Then, check the flat for accuracy by making a brownline or blueline proof.

3. Prepare a flat for the duplicator using the emulsion-up stripping method. Attach the negative(s). Open the masking sheet around the image area(s). Opaque all pin holes. Finally, check the flat for accuracy by making a brownline or blueline proof.

4. Prepare a set of complementary flats in which one flat contains the line work and the second flat contains at least one halftone. Check the flats for register by making a brownline or blueline proof.

5. Prepare a flat for an 8-page signature to be printed work-and-turn.

6. Prepare flats for a job that must be printed on both sides and then folded, gathered, stitched, and trimmed. Make a dummy layout for the job prior to the stripping operation.

7. Prepare a set of flats for a three-color or four-color process printing job. If possible, use a punch register system for this project.

8. Prepare a job that requires some hand film scribing. Practice scribing on a piece of discarded film before attempting the actual job.

9. Ask your teacher to provide you with a four-page folder so you can prepare a layout, make negatives, and strip the flat.

10. Visit a printing firm that uses the film positive method of stripping flats. Prepare a short report on your findings.

11. In consultation with your teacher, prepare layouts and carry out various stripping tasks for jobs in your shop.

Various pieces of equipment are required to efficiently strip flats in a production setting.
(nu Arc Company, Inc.)

Chapter 21

Platemaking

When you have completed the reading and assigned activities related to this chapter, you will be able to:

O Name and describe the equipment needed to expose lithographic plates.

O Describe the exposure conditions for surface plates.

O Expose and process subtractive and additive plates.

O Make plate deletions by the dry and deletion fluid methods.

O Describe the process of making lithographic plates by the photo-direct method.

O Describe the process of making lithographic plates by the electrostatic method.

O Describe the process of making lithographic plates by the diffusion transfer process.

O Describe the process of making direct-image masters.

O Describe automated systems used to make lithographic plates.

O Analyze a plate problem and describe how to prevent it from happening in the future.

O Repair a lithographic plate on the press.

O Store and handle plates correctly during processing.

O Organize a system to properly handle and store platemaking chemicals.

Offset or lithographic *plates* are thin sheets of metal, plastic, or paper used to duplicate an image on a printing press. The surface of the plate contains an image area and a non-image area. The *image area* is ink-receptive and the *non-image area* is ink-resistant. The inked image area is transferred to the offset press rubber blanket during the printing operation. The inked image on the rubber blanket is subsequently transferred (offset) to the sheet of paper passing through the press.

OFFSET PLATE CONSTRUCTION

Offset plates come in various sizes and surface finishes. Generally, the surfaces of most metal plates are grained to form a slight texture which makes them more water-receptive. See Fig. 21-1.

Grained plates are manufactured by a process that forms small peaks and valleys on the surface of the plate. These peaks and valleys are more capable of carrying water than smooth surface plates having no grain. The grain in the plate assists in maintaining consistent ink and water balance on the press.

It should be pointed out that some metal plates and all plastic and paper plates do NOT have a grained surface. These plates are referred to as *grainless plates*. Grainless plates are chemically treated during manufacture so that the surface retains a film of moisture during the printing operation. Fig. 21-2 illustrates the difference between grainless and grained plates.

Graining methods

There are several methods used to grain the surface of plates. These include: chemical, ball, brush, and sandblast methods.

The *chemical graining* method involves the application of an etching acid or anodizing solution to the surface of a plate. The solution is applied during the plate manufacturing process.

Fig. 21-1. Metal plates for offset printing are generally grained to form a slight texture to make them more water receptive.

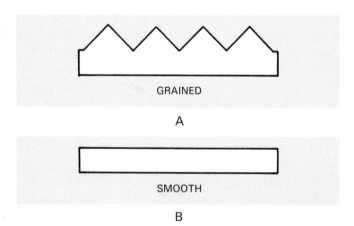

GRAINED

A

SMOOTH

B

Fig. 21-2. A—Grained offset plate has a slight texture for better water receptivity. B—Grainless offset plate has a smooth surface with no grain.

The *ball graining* process involves two steps. First, the plate is placed in the plate-graining machine where it is covered with an abrasive powder. Then steel or glass marbles are made to roll and vibrate over the plate's surface. The high-speed oscillation of the marbles produces the graining of the plate.

The *brush graining* method uses steel or soft brushes to abrade or roughen the surface of the plate. This creates finer, sharper, and more numerous peaks and valleys for a grained effect. This process provides more surface areas for improved ink and water balance on the press.

The *sandblast graining* process makes use of fine sand particles under high compression. This action creates a surface grain of fine peaks and valleys similar to ball graining.

Plate terminology

Offset plates are generally manufactured to fit standard press sizes. Smaller (duplicator) size plates are available in four different types of *plate ends*. These include straight, pinbar, slotted, and serrated, Fig. 21-3. The type of plate end required must correspond to the kind of plate clamps on the press.

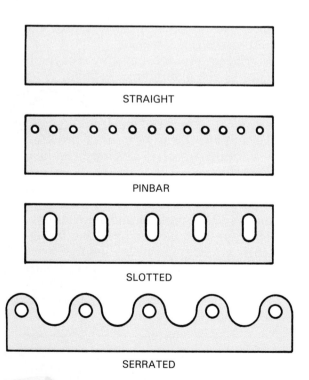

STRAIGHT

PINBAR

SLOTTED

SERRATED

Fig. 21-3. Offset plates are available in four types of plate ends. These include straight, pinbar, slotted, and serrated. (3M Company)

The term *master* is often used interchangeably with the term plate. Masters are made of metal, plastic, or paper. In general, masters refer to PLATE SIZES used on duplicator-size presses.

Plate composition

Offset plates are manufactured from many different types of materials. These include:

1. Paper, consisting of a cellulose base.
2. Plastic-coated paper.
3. Acetate.
4. Aluminum.
5. Paper base with a laminated aluminum surface.
6. Steel with a plastic surface.
7. Copper on chromium, stainless steel, or aluminum.

Multimetal plates consist of two or three layers of metal. These may be *bi-metal* (two metals) or *tri-metal* (three metals). Plates in this category are generally used for extremely long press runs.

HANDLING PLATES

Offset plates are manufactured under rigid quality control conditions. The manufacturer guarantees consistent printing qualities so long as the plates are handled and processed according to directions. The following points should be observed when working with offset plates.

Handle presensitized offset plates in subdued yellow light. Never subject plates to room white light or sunlight! Plates should be kept in original containers and stored in a dry, cool drawer near the platemaking equipment.

Plates should be removed from their container by lifting, not sliding. This avoids scratching one plate with the surface of another.

Handle plates by the ends, being careful to keep fingers away from the printing surface. Acid and moisture from fingers will cause marks on the plate's surface.

Plate storage

When offset plates are retained for future printing they must be stored to avoid damage and oxidation. Following the press run, the plate should be *gummed* to preserve it. This means that it is given a thin coating of gum arabic solution or similar gum preservative.

The gum coating prevents oxidation of the non-image areas of the plate. A gummed plate may be used again, since the gum coating acts as a preservative. Plates should be stored in a dry area in a hanging position. Plate cabinets are provided for this purpose. Refer to Figs. 21-4 and 21-5.

If the plates are NOT hung in this manner they should be placed in a large envelope and stored in a cabinet. Plates must be kept dry, since moisture attacks the protective gum coating and causes oxidation.

Fig. 21-4. When metal offset plates are to be reused, they should be stored in a hanging position or placed in a plate cabinet. (Foster Manufacturing Co.)

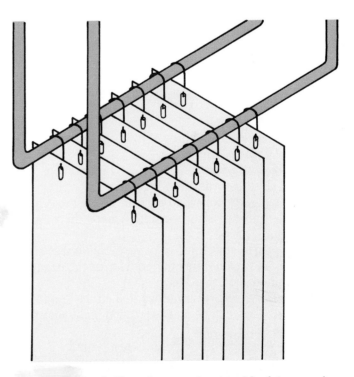

Fig. 21-5. Metal offset plates can be stored for future use by hanging them in a dust-free cabinet or locker.

Oxidized plates will attract ink over all the non-image areas affected. This usually requires a new plate.

Plate precautions on press

The offset plate is attached to the plate cylinder of the press. The operator should remove the gum coating from the surface of the plate before inking it. This is done with a damp sponge or cotton wipe.

If the press run is interrupted for more than a few seconds, the plate can dry. This is especially true when using alcohol in the fountain solution. The operator must dampen the plate again before proceeding. Such dampening is essential with ungrained plates such as a photo-direct paper plate. If the press operator must leave the press for an extended period of time the plate should be gummed immediately.

When the press operator is ready to resume the run, the gum coating must be removed with a sponge or cotton wipe. This procedure prevents deterioration of the clear non-image areas of the plate. If this procedure is not followed, the ability of the non-image areas of the plate to repel ink is destroyed!

Plate problems

Since the printing plate carries the image to be placed on the paper or substrate, it is one of the most critical components governing the quality of the final printed product. Various problems can result if the plates are processed or used improperly. The following is a list of definitions relating to plate problems.

1. *Black spots* can occur when plates do NOT print well on press due to poor storage, handling, age, or a combination of any of these factors.
2. *Plugging* results when images on plate fill in where they should be open and clean. Serifs and halftone shadow dots can suffer plugging, for example.
3. *Bridging* can happen when developer solution reacts to connect two or more image areas on the plate.
4. *Piling* is a problem due to excessive amount of developer solution allowed to build up on certain image areas of the plate.
5. *Faulty developer* is a result of outdated or improperly mixed developer solution.
6. *White spots* occur when parts of image areas have small, pinhead size openings where there is no image.
7. *Hotspots* are unwanted or overexposed image areas that result from improper contact of the negative and photo-offset plate.
8. *Halation* is soft, fuzzy unwanted circles similar to a halo. They can occur in screened highlight areas of the plate.
9. *Shadow plugging* results when small, 5-10% shadow dot openings in a halftone image combine to print as one solid. Poor vacuum contact between the negative and plate is generally the cause.
10. *Dot gain* is a term referring to halftone and other screened dots that enlarge during platemaking due to poor vacuum between the negative and plate.
11. *Gum blinding* results from an excessive amount of gum applied to the plate's image, causing images not to print properly.
12. *Streaking* of a press image can happen because of an excessive amount of gum on the plate's image, causing the ink to streak.
13. *Weak image* is a term indicating an image that is faint or lacking normal intensity on the plate.
14. *Streaky* generally refers to the discoloration of an image.
15. *Spotty* means lacking image uniformity.
16. *Contamination* is a loss of purity and intrusion of dirt or unwanted material in developer solution or other materials.
17. *Diazo* is a photo-sensitive emulsion usually coated on presensitized photo-offset plates.
18. *Decomposed* means the plate developer undergoes a chemical breakdown, rendering it useless.
19. *Deactivated* means not serving a useful purpose.
20. *Lacquer developer* is a colored, chemical solution placed on presensitized photo-offset plates so the image can be seen and will withstand long press runs.
21. *Developer acidic* is a solution having high acid content and can cause the plate image areas to not pick up the ink properly.
22. *Developer alkaline* means the solution is more alkaline than acid, thereby allowing non-image areas of the plate to pick up ink.
23. *Underdevelopment* occurs when exposure time and/or developer solution is inadequate.
24. *Overdevelopment* occurs when the exposure time and/or developer solution is used to excess.
25. A *pinhole* is an unwanted, pinhead-size opening in the film that forms a small dot-size image on the plate.

PLATE CLASSIFICATIONS

Common offset plate classifications in general use are described here. It is essential that you become familiar with the terms and end uses of these plate materials.

Negative-working plates

Negative-working plates are exposed with film negatives. Light passing through the clear open portions of the film hardens the coating on the surface of the plate. This action makes these areas insoluble in the developer used on the surface of the plate. The areas not exposed to light remain soluble in the developer. The exposed or hardened coating is ink receptive and becomes the image area. Areas of unexposed coating, NOT hardened by light, are removed during processing and become moisture receptive.

Positive-working plates

Positive-working plates require exposure with film positives. Light passing through the clear portions of the film begins decomposing the coating and these areas become soluble in the developer used on the surface of the plate. These areas become moisture receptive. Areas not exposed to light remain insoluble in the developer and become the ink-receptive image areas.

More specific classifications of offset plates require that three additional variables be considered. These include:

1. Type of carrier or base used.
2. Type of light sensitive coating.
3. Chemistry used to process the plate.

Using these variables as criteria, five groups of offset plates are discussed. These include: surface, deep-etch, photo-direct, dry-offset or relief, and electrostatic.

SURFACE PLATES

The surface plates are classified as either negative-working or positive-working. The surface plate has its light sensitive coating resting on the carrier or base, Fig. 21-6. After processing, the remaining coating is ink-receptive and rests on the moisture-receptive base material. Plates included in this grouping are additive or subtractive. In addition, additive and subtractive plates may be either *presensitized* (coating applied by manufacturer), or *wipe-on* (coating applied by user).

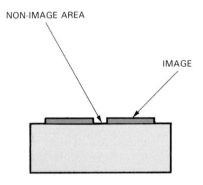

Fig. 21-6. Light sensitive coating of a surface offset plate rests on the carrier or base metal.

Additive plate

An *additive plate* requires that a lacquer be applied to the exposed image areas. Most *lacquers* for additive plates perform two functions. One is the removal of the unexposed coating to make these areas water-receptive. The second function is coating the exposed areas with an oleophilic lacquer to make them ink-receptive. *Oleophilic* means having an affinity for grease-based offset ink.

Press runs of over 20,000 impressions are possible with additive plates. Fig. 21-7 provides information which will be helpful when processing additive plates.

Subtractive plate

A *subtractive plate* requires a developer to make the unexposed coating soluble. This provides easy removal from the base material. Areas with the coating removed become moisture-receptive. The light sensitive coating exposed to light is hardened and becomes insoluble in the developer and is therefore ink-receptive.

Press runs of 15,000 to 200,000 impressions or more are possible with subtractive plates. Subtractive plates are now the most popular because of the ease in processing and avoidance of lacquer-base chemicals.

Fig. 21-8 provides information that is helpful when processing subtractive plates.

Surface plates can also be identified by the type of light sensitive chemicals used to coat them. The two main types of coating used are diazo and photopolymer.

Diazo coatings

Diazo sensitizers are used for presensitized and wipe-on plate coatings. Most of the diazos in use are condensation products of formaldehyde and diazo diphenylamine stabilized with a compound like zinc chloride. Some diazo oxides are also used. Diazos differ from collides such as albumin, casein, and gum arabic in that they are themselves light sensitive. Exposure to light converts them directly to insoluble resins that have good ink receptivity and reasonable wear characteristics for printing.

Photopolymer coatings

A popular *photopolymer coating* used for surface plates consists of a cinnamic acid ester of an epoxy resin sensitized with a suitable organic compound. This coating is insoluble in water but soluble in organic solvents. It can be used for presensitized or wipe-on plates. On exposure to light, the exposed parts of the coating become insoluble in the organic solvents which dissolve the unexposed portions. The resultant images are very strong, and the plates generally withstand long press runs.

Photopolymer coated plates are not affected by temperatures and relative humidity. The plates can be pre-coated and stored for long periods of time prior to use. High cost of materials and the necessity of using organic solvents in processing are disadvantages of photopolymer coatings.

Wipe-on sensitized coatings

Although this chapter is concerned with presensitized plates, it should be pointed out that plates without any sensitized coating can be purchased. The diazo and lacquer coatings can be applied in the shop.

Water-developed plates

Water-based or *aqueous plate* technologies have evolved as the result of environmental regulations.

Offset Lithographic Technology

PLATE PROBLEMS (ADDITIVE WORKING)

PROBLEM	PROBABLE CAUSE	REMEDY
Weak or faint images (usually due to lacquer developer not applying evenly).	1. Diazo coating on wipe-on plate too thin for grain. 2. Insufficient exposure. 3. Diazo coating not completely dissolved. 4. Contamination. 5. Diazo decomposed. 6. Diazo deactivated. 7. Lacquer developer has settled out in container. 8. Lacquer developer damaged in shipment or storage. 9. Lacquer developer out of date.	1. Try using higher dye load (diazo concentration) coating or relieve tension on roller coater. 2. Increase exposure time (use gray scale and shoot for a step 5-6). 3. Reshake bottle of coating mix; let sit 30 min. at room temp. (65-85 °F) until completely dissolved. 4. Use clean wipes when applying diazo; make sure rollers are clean on roller coaters; film should be clean when exposing. 5. Diazo too old, use fresh bottle; plates out-of-date or allowed to get too warm and/or too wet; use new plates and improve storage conditions. 6. Diazo came in contact with ammonia fumes (glass cleaner, film cleaner, floor cleaner); use new coating/plates; avoid ammonia. 7. Shake or stir lacquer developer until uniform color is achieved. If problem continues, use new developer. 8. Developer may have been frozen or overheated during shipment or storage; discard. 9. Developer too old, check shelf life recommendation of manufacturers; use new batch.
Black spots on background area (black marks on plate).	1. Plates stored too long and/or under hot, humid conditions. 2. Plates stored with slip sheet paper. 3. Diazo decomposed or contaminated. 4. Developer imbalance. 5. Old developer redepositing from processor. 6. Pinholes in negative.	1. Check shelf life and batch number of presensitized plates; if out-of-date, use new plates (counter etching or use of desensitizer before lacquering may help). 2. Paper chemicals plus humidity react with diazo and/or aluminum surfaces; avoid storage with slip sheet paper, especially during hot, humid weather. 3. Check age of diazo coating solution; diazo solutions should be stored cool (40 °F—70 °F) for only a few days; used and recycled coating decomposes faster, avoid use of old coating; check cleanliness of roller coater. Old decomposed diazo (staining on rollers) can catalyze the decomposition of fresh coating; clean rollers frequently with approved cleaners; use fresh wipes for hand application; if you must use a sponge, make sure it's clean. 4. If condition worsens with age of developer, discard and use fresh developer. 5. Clean brushes, belts, pans, lines, pump, etc., with approved cleaner. 6. Check negative for holes and opaque.
Plugging, bridging (images fill with developer), piling (lacquer build-up on images).	1. Too much developer. 2. Too much water in sponge or brush. 3. Developer too concentrated. 4. Developer has lost too much solvent and/or is too old.	1. Reduce quantity used by hand or reduce flow rate in processor. 2. Some developers are very sensitive to the quantity of water present while developing; monitor carefully. 3. Make sure developer is shaken/stirred uniformly; using the bottom portion of developer in a bottle that has settled will cause problems. 4. Check shelf life recommendations of manufacturer; solvents evaporate right through the plastic bottle; they will evaporate faster through an open bottle or in a processor and in hotter temperatures; avoid conditions that lead to evaporation; use fresh developer.
Development too fast or too slow (causes symptoms similar to piling and plugging).	1. Same causes as plugging and piling.	1. See cures recommended for piling and plugging.

(Continued)

Fig. 21-7. Additive offset plates require that a lacquer be applied to exposed image areas. Information in this chart provides helpful processing data.

440

Platemaking

PROBLEM	PROBABLE CAUSE	REMEDY
Developer scums, deposits on background (lacquer adheres to non-image areas).	1. Diazo coating not uniform, blank spots on plate. 2. Plate, diazo too old. 3. Plate contamination. 4. Faulty plates.	1. Make sure diazo completely coats plate; check to see if a wetting agent is included in the diazo coating; adjust rollers. 2. Use fresh plates or diazo coating. 3. Oil spots, fingerprints, defoamers, other chemicals can cause scumming; coat, expose, and develop plates in clean areas. 4. Improper anodization or silication can result in developer scumming.
White spots in image (light specks on type and solid images).	1. Dewetting of diazo on plate. 2. Water spots on diazo. 3. Dust and dirt on negative, plate or exposure frame glass.	1. Oil or chemicals can cause diazo to be repelled from plate leaving a void; clean rollers, use fresh wipes, new diazo. 2. Water spraying on plates or sneezing can cause white spots. 3. Carefully clean away any debris that can block light.
Halation (halftone dots plug or gain size, dark, hotspots in tone areas.	1. Dirt or debris between plate surface and negative. 2. Kinks in plate or film. 3. Insufficient vacuum. 4. Film flopped. 5. Overexposure.	1. Carefully clean all surfaces. 2. Carefully handle plates and film to avoid kinking. 3. Check vacuum gauge (25-27); allow longer times for pump-down (the larger the plate, the longer the time); check for vacuum leaks. 4. The emulsion side of the film should be in contact with the coated side of the plate. 5. Use a gray scale and shoot for a step 5-6; a contact scale or dot gain scale may be useful.
Blinding on press (some images do not take ink fully).	1. Too much gum on image.	1. Gum arabic must be buffed very thin on plate; use a lower Baumé concentration; e.g. cut a 14 Bé to 10 Bé or 7 Bé (caution—too thin a gum will cause scumming or tinting); use a gum-asphaltum (must be shaken well and buffed properly); use a non-blinding, non-gum finisher.
Streaking on press (marks across image). Scumming on press (non-image areas of plate pick up ink and transfer it to sheet).	1. Same as blinding, but in streaks. 1. Insufficient gumming. 2. Insufficient desensitizing.	1. Same as blinding. 1. Use a heavier concentration of gum on the plate. 2. Anodized plates are often harder to desensitize; use a more acidic gum; use a non-gum desensitizer.
Difficulty in achieving ink/H_2O balance on press (poor halftones, washed out solids). Heavy linting (uniform light colored tint over non-image areas).	1. Image not ink receptive enough (partial blinding). 2. Background not water receptive enough. 1. Background requires too much water to stay clean due to insufficient desensitizing.	1. Use less gum; buff down thinner; use gum-asphaltum; use a non-gum finisher. 2. Use a more powerful desensitizing finisher. 1. Use a more powerful desensitizing finisher.
Short plate life (image failure, short length of run, plate not properly processed and images break down to produce loss of detail).	1. Insufficient exposure. 2. Coating insufficient to fill grain. 3. Lacquer coating from developer too thin or too thick. 4. Plate faulty (poor graining, poor anodizing, poor silication). 5. Weak diazo coating (see weak image). 6. Harsh solvents, plate cleaners, roller cleaners, etc.	1. Use a gray scale, shoot for a step 5-6. 2. Use a heavier dye-load (concentration) diazo coating. 3. Shake developer thoroughly; use proper amounts while developing; use proper speed and brush press in processor to give uniform color in solids without piling. 4. Check another batch of plates. 5. Use fresh diazo or freshly coated plates. 6. Avoid using or splashing harsh solvents and cleaners on plates; some plate cleaners will etch and undercut dots.
Poor tonal reproduction (loss of halftone dots and image detail).	1. Faulty exposures. 2. Dots on film too soft, too much veil or fringe. 3. Images too thick (piling). 4. Improper desensitizing.	1. Check gray scale; check integrator, replace bulbs. 2. Use harder dots; change exposure level on film; change film or film developer; dupe the negative. 3. See remedies for piling. 4. Use a stronger desensitizing finisher.

Fig. 21-7. Continued.

441

PLATE PROBLEMS (SUBTRACTIVE WORKING)

PROBLEM	PROBABLE CAUSE	REMEDY
Slow development (excess time in development during plate preparation).	1. Developer temperature too low.	1. Most developers must be between 65-90 °F to operate properly, but even at 65 °F, development is much slower than at 90 °F; cold sinks, cold plates, or developer just in from warehouse need to be warmer; check heater and thermostat on processor.
	2. Developer activity too low.	2. In a processor, developer may have already processed too many plates (square feet of area subtracted); either replace part of developer, or flush and start a new batch; increased temperature may help but will lead to alcohol evaporation with some developers; slowing down throughput speed may be sufficient; in some cases, increasing replenisher concentration may help. (Caution: too much replenisher will undercut highlights.)
	3. Developer too old.	3. Out-of-date developer may have lost too much solvent (alcohol); increasing replenisher concentration may help (caution: too much replenisher will undercut highlights); change to newer batch.
	4. Overexposure (negative plates).	4. Use gray scale; shoot for step 5-6; overexposure will fill in shadows.
	5. Film flopped, out of contact.	5. Make sure film emulsion side is in contact with plate coating.
	6. Poor platemaker vacuum.	6. Check vacuum gauge (25-27); check for leaks; poor vacuum causes light striking (fogging, halation) with negative working plates.
	7. Plate exposed to light (negative plates).	7. Check safe light (yellow light) condition for source of stray ultraviolet light (doors, fluorescent fixtures, exposure frame, skylight, etc.); partially exposed plates are hard to develop; store plates in a dark area.
	8. Plate is too old.	8. Check shelf life date from manufacturer; use new lot.
	9. Developer diluted by rinse water.	9. Check for leaks in processor, adjust roller pressures to prevent feedback.
Dot loss on plate (loss of halftone dot structure).	1. Underexposure (negative plates).	1. Use a gray scale; shoot for step 5-6.
	2. Too much replenisher (alcohol).	2. Reduce replenisher concentration.
	3. Developer too hot.	3. Reduce temperature control on processor; temperature should read 80-90 °F.
	4. Development time too long.	4. Reduce development time to manufacturer's recommendation; make sure shadows are clean; speed-up processor throughput.
	5. Too much development pressure.	5. Reduce brush pressure; use a less abrasive developing pad.
	6. Overexposure (positive plate).	6. Use gray scale; shoot for a clear 2-3.
	7. Light striking or undercutting (positive plate).	7. Work in good yellow light; check for sources of light leaks; check for good contact between film and plate.
	8. Developer too acidic (negative plates).	8. Check pH; it should be between 4.5-9.5.
	9. Developer too alkaline (positive plates).	9. Check pH; it should be between 9.5-12.5; check plate manufacturer's recommendation.
Dot gain on plate (enlargement of halftone dots).	1. Overexposure (negative).	1. Use gray scale; shoot for step 5-6; use a dot gain scale.
	2. Under development (see causes of slow development).	2. Same as cure for slow development.
	3. Improper contact, vacuum problems.	3. Check vacuum gauge (25-27); look for leaks; clean glass and gaskets; increase draw-down time.
	4. Film flopped.	4. Emulsion side of film should be in direct contact with coated side of plate.
	5. Dots in film too soft, too much veil or fringe.	5. Use contact/dupe films; change film or film developer.
	6. Combination of too much replenisher and too long of an exposure.	6. Cut back on both exposure time and alcohol amount.

(Continued)

Fig. 21-8. Subtractive offset plates have become preferred offset plate in industry because of ease in processing. This chart provides helpful processing data.

PROBLEM	PROBABLE CAUSE	REMEDY
Spots on background (dark areas on plate).	1. Pinholes in film. 2. Image polymer depositing on background. 3. Weak developer.	1. Check film; opaque all pinholes. 2. Check developer or replenisher levels (see causes/cure of slow development). 3. Same as above.
Redepositing (developer deposit on image solids and/or screens).	1. Weak developer. 2. Developer solution too cold. 3. Developing pad, brushes, or rollers dirty. 4. Filter clogged.	1. Check developer age; add fresh developer or replenisher. 2. Check developer temperature; should be between 80-90 °F in processor. 3. Use new pad; clean brushes and rollers. 4. Change filter.
Halation (halftone dots plug, hot spots in tone areas).	1. Dirt on film or plate. 2. Kinks on plate or film. 3. Insufficient vacuum, uneven background. 4. Air trapped between plate or film.	1. Clean with appropriate cleaners. 2. Handle plates, film more carefully; use two hands, grasping opposite two corners. 3. Check vacuum gauge; check for leaks; adjust backboard. 4. Change filter.
White spots in image (light specks on type and solid images).	1. Dirt on film or plate or glass of vacuum frame.	1. Clean thoroughly with appropriate cleansers.
Heavy linting (uniform light tint over non-image areas).	1. Background requires too much water to keep clean.	1. Increase development time; use stronger developer or finisher with stronger desensitizer.
Blinding (image not taking ink).	1. Too much gum on image. 2. Contamination from sponge or brushes.	1. To prevent, gum must be buffed down thoroughly; use a lower concentration of gum (caution: too little gum can cause scumming, toning); use a non-gum, non-bleeding finisher; to correct, wet wash plate with hot water or mild acid solution; use gum asphaltum or plate cleaner to restore ink receptivity; make new plate. 2. Use thoroughly rinsed clean sponge; clean brushes; avoid other chemicals in sink while developing or gumming.
Scumming on press (tinting or toning on restart).	1. Insufficient gumming or desensitizing. 2. Light struck plate (negative plates).	1. Ensure complete coverage of plate with gum or finisher; allow sufficient time for developer and/or finisher to desensitize plate. 2. Check for sources of ultraviolet light leaks; film density too low, masking is burning through; check all light blocking materials for adequate performance.
Short plate life (short length of run causes image on plate and image to deteriorate).	1. Insufficient exposure (negative plate). 2. Light-struck. 3. Overdevelopment, developer temperature excessive. 4. Developer too harsh, too much replenisher. 5. Water pH too high (positive plate). 6. Harsh solvents in contact with plate. 7. Insufficient baking (positive plate).	1. Use gray scale; shoot for 5-6; check center to edge difference; may require change of distance from light source; use light source with higher UV output. 2. Check masking for light blocking ability; check for light leaks. 3. Cut back on development time and/or brush pressure in processor; temperature in processor too high, lower to 80-90 °F. 4. Cut back on replenisher concentration. 5. Check manufacturer's recommendations (typically keep between 9.5-12.5). 6. Keep film cleaner, blanket, wash, etc., from contacting plate. 7. Check baking time and temperature; one or both may require increase.

Fig. 21-8. Continued.

Printers in the United States are being forced to consider alternatives to traditional organic solvent-based plate technologies. The printers are finding that water-developed plates are often better than the plates they are replacing.

Organic solvent-based plate technologies present disposal problems for the printer. In many states and municipalities, it is illegal to simply flush spent solvents down the drain, as printers have done for years. Such chemistry contains large amounts of alcohol, in addition to a variety of other toxic substances. In some areas of the United States, it must be extensively pretreated or diluted prior to being dispensed down the drain. In other areas, where drain disposal is entirely illegal, printers must find alternate disposal methods.

Placing used chemistry in a container and having it hauled off by an Environmental Protection Agency-approved waste hauler is one alternative. Conforming to regulations in this manner can be time consuming and cost prohibitive, not to mention dangerous.

Water-developed plates utilize coatings that can be processed by developing chemistries that are primarily, if not entirely, water-based. Some solutions, however, do in fact contain small traces of organic solvents.

There are a number of water-developed offset plates available to printers. For example, the 3M Company manufactures the Hydrolith® plate. Citiplate offers the Aqua + photopolymer plate. These plates can be hand-processed or machine-processed. Each of the manufacturers includes detailed processing and handling instructions in the packaged plates. It is important that these instructions be carefully followed.

For purposes of this discussion, procedures for processing the 3M Hydrolith plate and Citiplate Aqua + plate are described here.

Hydrolith plate

1. Hydrolith (3M Registered Trademark) plates must be handled under yellow or subdued fluorescent lighting conditions. Plate exposure is made with a high-intensity point source ultraviolet light in a vacuum frame. Plate exposure is made to a solid step six on the 21-step platemaker's gray scale.
2. To process the exposed plate, place it on a flat surface in the developing sink. The sink area should be illuminated by an overhead yellow safelight. A special developing pad is used to spread plain tap water over the surface of the plate. This action releases the emulsion and develops the plate.
3. The plate is then rinsed with water and squeegeed dry.
4. To finish the plate, place it on a clean flat surface and gum it with Hydrolith plate finisher.

Hydrolith offset plates are run on the press using a conventional fountain solution with a pH factor of about 4.5. If the press run must be interrupted for any reason, apply plate finisher. Buff the plate dry with a cotton pad or soft cheesecloth pad.

Citiplate Aqua + plate

To determine the correct exposure, use a 21-step platemaker's gray scale, along with a halftone screen of known value. A correct exposure is indicated by a solid step six on the gray scale after the plate has been exposed and inked in on the press. Exposure time will vary according to the type of light source and the available voltage.

Aqua + plates may be developed either by hand processing, tank, or by automatic processor. All processing times quoted are based on a development temperature between 68° and 78°F. Developing temperatures below 68° or above 78°F will cause severe development problems.

The following procedure should be followed to hand develop an Aqua + plate.

1. Developer temperature must be 68° to 78°F. Apply approximately one ounce (30 mL) of subtractive hand developer per square foot of plate area, using an applicator pad. Spread the developer evenly over the entire plate using only the weight of the applicator.
2. Allow the developer to remain on the surface of the plate until there is a break in the image area. Wait 15-20 seconds before developing.
3. Gently develop the plate with the applicator pad using a circular motion. Avoid heavy pressure or extensive rubbing. Begin processing the plate in a circular motion across the length of plate with medium pressure, each time overlapping the strokes until plate has been completely processed lengthwise.
4. Repeat the same circular motion across the width of the plate.
5. Repeat development in both directions until a solid step six is produced on the 21-step gray scale.
6. After the plate has been developed to the proper step, pour another pool of developer on the plate and work gently in the same circular motion.
7. Rinse the plate with water using a piece of cheesecloth, rubbing in a circular motion to clean out any residue in the background and image areas. Repeat same step in both directions until surface is free of any residue.
8. Squeegee off excess water. The plate should be damp when applying finisher/desensitizer.
9. Place the plate on a clean table. Pour an adequate amount of Aqua + Finisher/Desensitizer to the surface of the plate (while plate is damp) and work in the same circular motion.
10. Apply another application of finisher/desensitizer and buff down to form a thin, even film. Then buff dry. The plate is now ready for the press or for storage.

DEEP-ETCH PLATES

Deep-etch plates are exposed with film positives instead of negatives. The coating on the plate is generally a bichromate type. Most deep-etch plates use aluminum, zinc, or stainless steel as the base or carrier. After exposure and processing, the ink receptive or unexposed areas are etched and coppered chemically, Fig. 21-9. This makes the ink-bearing image area slightly etched below the non-printing area.

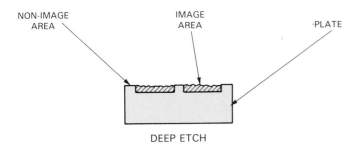

Fig. 21-9. Deep-etch offset plates are exposed with film positives rather than film negatives. Ink-bearing image area is etched below non-printing surface area.

The preparation of deep-etch plates requires considerable skill. These plates have a greater ink-carrying capacity than surface plates. Press runs over 500,000 impressions are common with deep-etch plates.

Multimetal plates

Multimetal plates consist of two or more metals in a sandwich-like combination, Fig. 21-10. The metals used in the plates are selected on the basis of their ink and water receptive characteristics. The combination of two metals each having outstanding ink and water receptivity is desired. This enables the press operator to work with a plate having optimum performance features.

Multimetal plates can be exposed with film negatives or positives. Both processes require an electro-chemical treatment of the plate to apply one or more of the metal surfaces to the base metal. The skills and cost required to prepare multimetal plates are considerable. However, the durability of the plate can exceed one million impressions.

The most common metals used for multimetal plates are copper for its ink receptivity, and chromium, aluminum, and stainless steel for their moisture receptivity qualities.

Metals for multimetal plates are often referred to as non-image metal and image metal. Multimetal plates have outstanding receptivity because of the electro-mechanical combination of two or more desirable metals. Single layer metal plates do not have this advantage and must rely on other means to achieve acceptable ink and water balance.

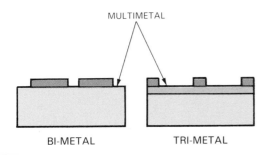

Fig. 21-10. Multimetal offset plates have outstanding ink and water receptivity, providing optimum performance on press.

PHOTO-DIRECT PLATES

Photo-direct presensitized plates are prepared in an automatic camera or projection-type platemaker, Fig. 21-11. These processors are designed to handle either ready-cut, press-size plates or plate material in roll form. A cross-sectional view of a photo-direct plate is shown in Fig. 21-12.

A

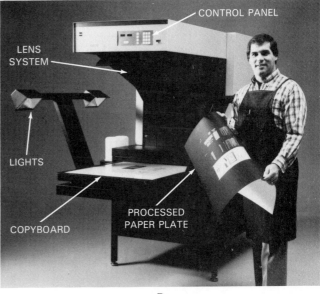

B

Fig. 21-11. A—This is a late model automatic, self-contained, daylight-operating camera platemaker. B—This photo-direct daylight-operating platemaker will reduce copy to 85% and enlarge to 105% with microprocessor-controlled automatic focus, sizing, and exposure. (Itek Graphix Corp.)

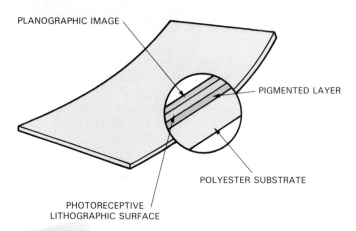

PLANOGRAPHIC IMAGE

PIGMENTED LAYER

POLYESTER SUBSTRATE

PHOTORECEPTIVE
LITHOGRAPHIC SURFACE

Fig. 21-12. Study this cross-sectional view of a photo-direct offset plate.   (3M Company)

The need for a negative is eliminated in the photo-direct process. A magazine holder is loaded with presensitized plates or a roll contained in a cassette. The original copy is placed on the copyboard, Fig. 21-13. The operator sets the lights, focus, size, and aperture and makes the exposure, Fig. 21-14. Light is reflected from the original copy through a prism lens to the plate inside the processing unit. The plate is then automatically processed and emerges ready for the press. A schematic diagram of a typical camera platemaker is shown in Fig. 21-15.

The photo-direct platemaking process is used primarily for fast copy duplicator operations. It is adaptable to this operation since no darkroom is required for all of the steps for plate processing. The processor is capable of making reductions and enlargements of original copy. Photo-direct plates can be used to print over 20,000 impressions. Some fully enclosed camera-platemaking units are equipped to produce press plates, stats, film negatives and positives, halftones, and phototypesetting material in cassette form, Fig. 21-16.

MAGAZINE

COPY LENGTH
CONTROL

1. SET PLATE LENGTH CONTROL

2. SET COPY LENGTH CONTROL

3. SET EXPOSURE TIMER

4. PRESS EXPOSURE BUTTON

Fig. 21-14. For most daylight-operating camera platemakers, operator must set lights, focus, size, aperture, and plate length before making exposure. Processor has a separate control console for operator.   (Itek Graphix Corp.)

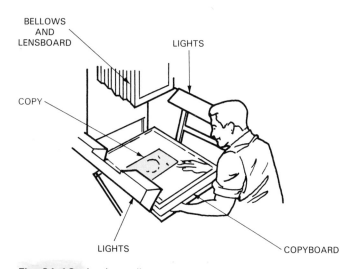

BELLOWS
AND
LENSBOARD

LIGHTS

COPY

LIGHTS

COPYBOARD

Fig. 21-13. In photo-direct process, original copy is placed on processor's copyboard.   (Itek Graphix Corp.)

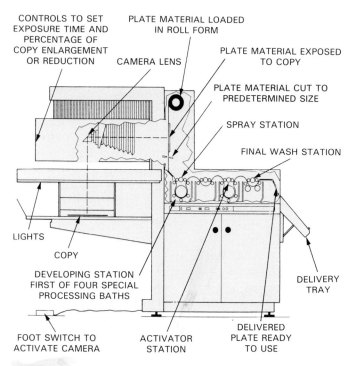

CONTROLS TO SET
EXPOSURE TIME AND
PERCENTAGE OF
COPY ENLARGEMENT
OR REDUCTION

PLATE MATERIAL LOADED
IN ROLL FORM

CAMERA LENS

PLATE MATERIAL EXPOSED
TO COPY

PLATE MATERIAL CUT TO
PREDETERMINED SIZE

SPRAY STATION

FINAL WASH STATION

LIGHTS

COPY

DEVELOPING STATION
FIRST OF FOUR SPECIAL
PROCESSING BATHS

FOOT SWITCH TO
ACTIVATE CAMERA

ACTIVATOR
STATION

DELIVERY
TRAY

DELIVERED
PLATE READY
TO USE

Fig. 21-15. Study this schematic diagram of a camera plate-maker.   (3M Company)

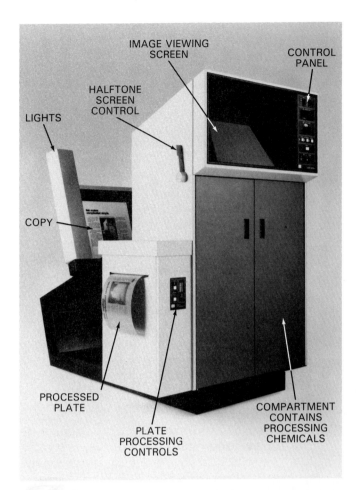

Fig. 21-16. Late model fully automatic camera platemaker. This model is equipped to produce offset plates, stats, film negatives and positives, halftones, and phototypesetting materials. (Itek Graphix Corp.)

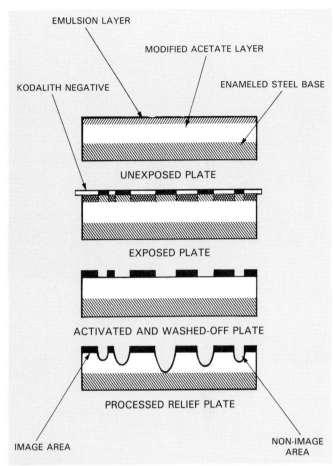

Fig. 21-17. Dry-offset plate is similar to letterpress plate, since images stand in relief. Plates, made of metal or plastic, are slightly thicker than conventional offset plate. (Eastman Kodak Co.)

DRY-OFFSET, RELIEF PLATES

The *dry-offset relief plate* does NOT use the press dampening system. It is completely different from any of the plate systems discussed so far. Dry-offset, (called *letterset*), involves a metal or plastic plate that is slightly thicker than the conventional offset plate. During processing, the non-image areas are etched away so that the image actually stands in relief. Look at Fig. 21-17.

The press dampening system does NOT operate when this type of plate is run on an offset press. Dry-offset plates are approximately 0.032 inch (0.8 mm) thick, with non-printing areas etched to a depth of about 0.015 inch (0.4 mm).

One of the more popular dry-offset photopolymer plates consists of a relatively thick layer of plastic bonded to a metal base. A negative is used to expose the plate. The plastic layer is hardened wherever it is struck by the light. The protected areas are not affected and are later dissolved or etched out by washing in a chemical solution.

Dry-offset plates are used on offset presses which have a plate cylinder that has been undercut to allow

for the extra thickness of the plate. Two offset presses adaptable to dry-offset are shown in Fig. 21-18.

ELECTROSTATIC PLATES

Electrostatic plates are prepared by a photoelectrostatic process that uses light and electrostatic properties. This process is sometimes referred to as *xerography*. Original copy is used to produce an image on the plate. The image may be same-size, enlarged, or reduced. Original copy may consist of a paste-up, typewritten copy, drawn images, printed images, or photographic prints.

Electrostatic plates are prepared on an automatic plate processor, Fig. 21-19. The plate processor's light-tight magazine holds a quantity of unexposed plates. These are either pre-cut to size or in roll form. The plate material has a light-sensitive, zinc-oxide resin coating designed to accept an electrostatic charge. A cross-sectional view of this plate shows three layers, Fig. 21-20.

The plate processor is prepared by first loading the copyboard with camera copy. The operator then sets

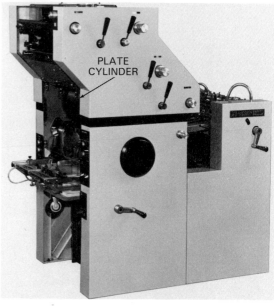

A   B

Fig. 21-18. A—This Heidelberg offset press can be used to print by dry-offset method. Regular offset plate cylinder is quickly replaced with special plate cylinder to accept dry-offset plate. B—This ATF Davidson offset press is adaptable to dry-offset printing process. (Heidelberg U.S.A.; ATF Davidson Co.)

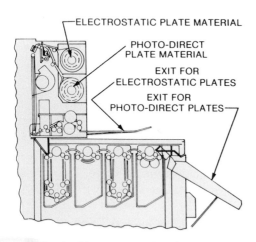

Fig. 21-19. Study this cross-sectional view of an automatic electrostatic offset plate processor. (3M Industrial Graphics Division)

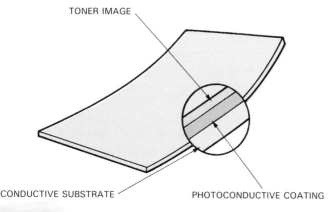

Fig. 21-20. Note cross-sectional view of an electrostatic offset plate. (3M Company)

the lens and copyholder distances for desired reproduction size. Lights, aperture (f-stop), timer, and plate length are set. After pressing a button, the processor automatically begins a series of steps that include charging, exposing, developing, and fusing. See Fig. 21-21.

Electrostatic plates are intended for uncoated papers. They should not be used when running clay coated enamel papers or non-porous materials such as plastics, foils, or pyroxylin coated cover stocks. This is because the special fountain solutions required for electrostatic plates contain chemicals that severely inhibit oxidation drying of the ink.

Since printing inks designed for use with uncoated papers dry primarily by absorption (ink is absorbed into paper like water into a blotter), electrostatic fountain chemistry poses no problem when running uncoated papers. However, when running coated papers, and the non-absorbent plastic and foil papers, the inks depend on oxidation to dry fully. Electrostatic fountain solution, which gets mixed into the ink during the printing process, will severely inhibit oxidation drying. This results in a soft, "cheesy" ink film that will remain wet for long period of time.

When printing a job on clay coated or non-porous papers, use a metal plate so that fast drying of the ink can be achieved. You can then use a fountain solution that will help speed drying, rather than slow it down.

PLATEMAKING EQUIPMENT

Most offset plates are light-sensitive and require the use of some type of exposure unit. This applies to both presensitized and surface plates coated in the shop.

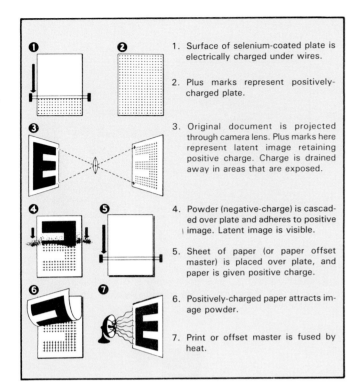

Fig. 21-21. Learn electrostatic offset plate processing steps by studying this illustration. (Xerox Corp.)

1. Surface of selenium-coated plate is electrically charged under wires.

2. Plus marks represent positively-charged plate.

3. Original document is projected through camera lens. Plus marks here represent latent image retaining positive charge. Charge is drained away in areas that are exposed.

4. Powder (negative-charge) is cascaded over plate and adheres to positive image. Latent image is visible.

5. Sheet of paper (or paper offset master) is placed over plate, and paper is given positive charge.

6. Positively-charged paper attracts image powder.

7. Print or offset master is fused by heat.

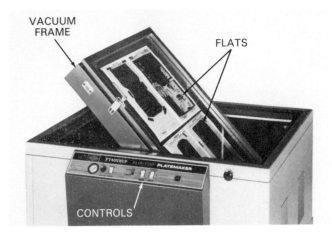

Fig. 21-22. This is a typical platemaker used to expose offset, deep-etch, multimetal, and dry-offset plates. (nuArc Company, Inc.)

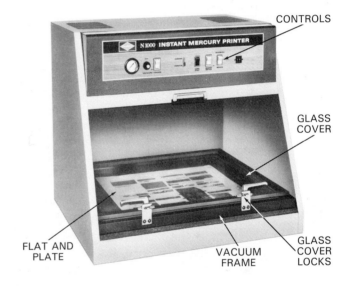

Fig. 21-23. Plate and flat are held together under vacuum. Vacuum blanket compresses plate and flat to produce an air-tight contact. (nuArc Company, Inc.)

Deep-etch, multimetal, and dry-offset plates also require some form of exposure unit. The exposure unit is a platemaker, Fig. 21-22. The processing of plates is done under subdued room light or in a room illuminated with yellow light.

The person who prepares plates has a vacuum frame with a rubber blanket and a glass cover. The blanket is bordered by a rubber sealing gasket. The blanket contains an outlet fitting connected to the vacuum pump by a rubber hose.

The offset plate is placed on the rubber blanket in the frame with the stripped-up flat on top of the plate. When the plate and flat are in position, the glass cover is locked in position, Fig. 21-23. The vacuum blanket compresses the plate and the flat to produce an airtight contact. The vacuum frame containing the plate and flat is then introduced to an ultraviolet light source, Fig. 21-24. The light source generally consists of pulsed xenon, mercury vapor, or carbon arc lamp.

Exposure control device

A platemaker's gray scale is used to determine the correct exposure time of an offset plate, Fig. 21-25. The *gray scale* is a measuring device used to obtain consistent exposure and development of offset plates.

The gray scale is a piece of film with 10 or 21 steps ranging in densities from clear to completely black. Each density is referred to as a *step*. The exposure time is influenced by the intensity of light, transparency of the negative, and thickness of the plate coating. The distance from light source to plate is kept constant.

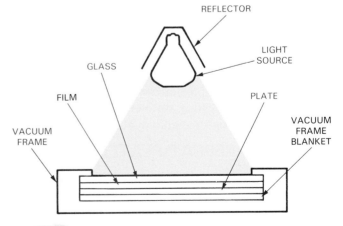

Fig. 21-24. Vacuum frame containing plate and flat is exposed to ultraviolet light for a specified number of seconds. Light sources include pulsed xenon, mercury vapor, and carbon arc.

Fig. 21-25. A platemaker's gray scale is used to determine correct exposure time of an offset plate. (Stouffer Graphic Arts Equipment Co.)

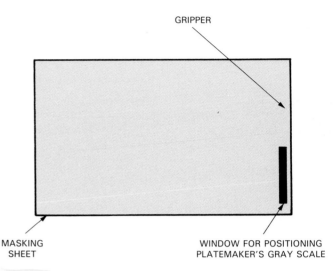

Fig. 21-26. Platemaker's gray scale is placed along gripper or tail edge of flat. Gray scale receives same amount of exposure as rest of image.

To determine correct exposure time for any given plate, the manufacturer's instructions should be followed.

The gray scale should be placed along the gripper or tail edge of the flat. To do this, a window opening is cut in the flat so that the gray scale is not covered. In this way, the gray scale will receive the same amount of exposure as the rest of the image. See Fig. 21-26.

After exposure, the plate image area, along with the gray scale, is processed with chemicals. If the plate has been correctly exposed, the correct numbered step will appear as shown in Fig. 21-27. In this example, step six should be as dark as step one.

If the plate is underexposed, the gray scale reading can be increased by increasing the exposure time. If the plate is overexposed, the gray scale reading can be reduced by reducing the exposure time. Charts are available to determine corrected exposure times, Fig. 21-28.

As a general rule, a gray scale reading of solid *step six* for a negative working plate indicates a properly exposed and developed plate. A higher gray scale reading will usually mean that the image has *spread* (become larger) enough to *plug up* (fill in) any halftones or screen images on the plate. A plate containing only coarse line work will generally NOT plug up even when taken to step seven or eight.

STEP-AND-REPEAT PLATEMAKING

It is frequently desirable to run a job two or more up on the press. This means that additional negatives must be used. An alternative method is to prepare one negative to be exposed in two or more different positions on the plate. This procedure is known as *step-and-repeat platemaking*.

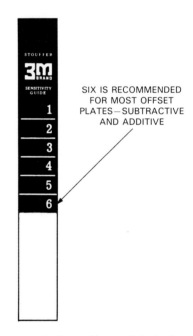

Fig. 21-27. Correct exposure of plate will reveal desired gray scale step. For most subtractive and additive working plates, solid step six is recommended. Always check manufacturer's recommendations for gray scale step. (3M Company)

To increase original gray scale reading by:	1 Step	2 Steps	3 Steps	4 Steps
Multiply your original exposure time by:	1.4	2.0	2.8	4.0

To decrease original gray scale reading by:	1 Step	2 Steps	3 Steps	4 Steps
Multiply your original exposure time by:	0.7	0.5	0.36	0.25

Fig. 21-28. This exposure modification chart can be used for correcting underexposed and overexposed offset plates. (3M Company, Printing Products Division)

With step-and-repeat platemaking, a single negative or positive is stripped and fitted into a special chase and frame on the platemaking machine, Fig. 21-29. The chase can be moved back and forth and up and down to any series of preset dimensions. An unexposed presensitized offset plate is loaded into the machine.

For exposure, the frame is closed and the vacuum is turned on. The frame is moved to a vertical position so that it faces the exposure light.

The step-and-repeat platemaking machine automatically moves to each required position on the plate and an exposure is made. During each exposure, all areas of the plate except the portion under the chase are covered and receive no exposure. After all exposures have been made onto the plate, it is removed and processed manually or in a plate processor.

The step-and-repeat platemaking machine is capable of performing these tasks since it is pre-programmed to perform each move via a built-in computer unit.

Fig. 21-30 illustrates how a step-and-repeat plate appears after multiple exposures and processing. This system greatly increases productivity in the platemaking department. The end result is maximum utilization of press time in printing items such as posters, cartons, letterheads, brochures, labels, shipping tags, business cards, and business forms.

USING DIRECT-IMAGE MASTERS

The usual methods of placing images on the direct-image master are by typewriter, drawing, lettering, and ruling. For correct positioning of copy on the master, space should be allowed from the top edge to the first line of copy. This is in addition to the space normally

Fig. 21-30. Step-and-repeat plate being developed and processed. Note multiple images on single plate. (Enco Products)

allowed when typing on regular paper. The extra space is used by the gripper fingers on the offset press. Guide lines on pre-printed masters allow for this additional space. Refer to Fig. 21-31.

For sharp images, a clean typewriter is necessary. Not only must the typeface be clean, but the platen, bail rollers, and the ribbon guides must also be free of ink and carbon grime. Any part of the typewriter that contacts the master should be clean.

Selection of the proper typewriter ribbon is most important. Carbon ribbons produce the best image on the surface of the master. Fabric ribbons are available in silk, cotton, and nylon.

The typewriter pressure setting should produce clean, smooth copy. Too heavy a pressure setting will indent the master, resulting in hollow type characters.

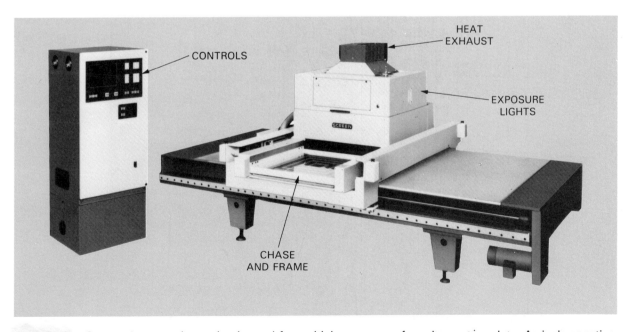

Fig. 21-29. Step-and-repeat platemaker is used for multiple exposure of one image to a plate. A single negative or positive is stripped and fitted into a special chase and frame on machine. Chase moves back and forth and up and down to expose a series of images on a single plate. (DS America)

Most direct-image masters can also be imaged by drawing, lettering, or ruling. Masters imaged this way generally produce fewer printed copies than masters imaged by typewriters. When a pen, pencil, or crayon is used, the master should be placed on a hard, smooth, flat surface. A medium firm pressure is applied. Excessive pressure may indent or damage the master's surface coating.

Neat, clean erasures are made with a clean, non-greasy abrasive eraser. The marks to be removed should be erased with a picking action to completely remove the visible image. Rough use of the eraser can damage the surface coating of the master. Any smudges will reproduce on the printed copies. The eraser should be kept clean at all times by rubbing it on a clean sheet of paper.

Running direct-image master

Before a direct-image master can be run on an offset duplicator, it must be chemically treated. This is done by moistening the entire surface of the master with an etch solution. This chemical prevents ink specks from appearing in the non-image areas of the master during printing. Etching is done on a smooth surface using a cotton pad to apply the solution.

Minimum amounts of moisture and ink are used on

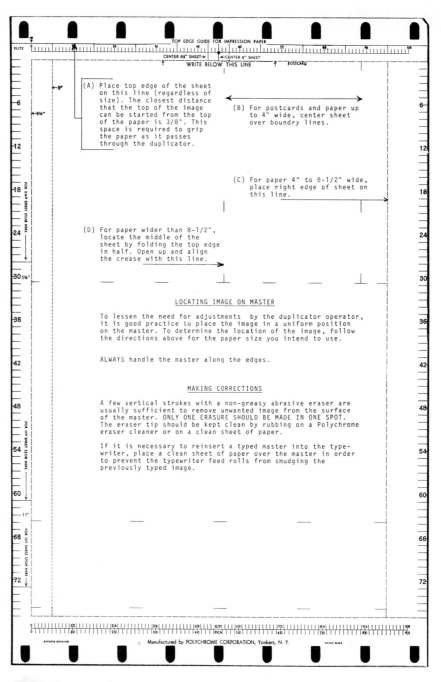

Fig. 21-31. Note example of a direct-image paper master. Best results are obtained by using a clean typewriter equipped with carbon ribbon to image master.

the press when running a master. This is because too much moisture may result in image breakdown and reduced run length.

Preserving direct-image master

Direct-image masters can be preserved after printing by allowing the ink on the surface of the master to run down. This is done by lifting the ink and water from rollers and running enough sheets through the duplicator to remove excess ink from the image areas of the master. The master is then removed from the duplicator and treated with a gum arabic solution. The solution should be applied with a cotton pad using horizontal and vertical strokes over the entire surface.

The master is then buffed dry and stored in a hanging position. A manila envelope may also be used for storing the master in a cabinet drawer.

USING PRESENSITIZED PLATES

Most *presensitized plates* consist of a base and a light-sensitive coating. As you have learned, *plate bases* are made of metal, plastic, or paper.

Aluminum-base plates are the most popular for offset printing. These are coated with an emulsion of light-sensitive compound, called *diazo*. Some plates are manufactured with a light-sensitive coating on both sides of the base. Presensitized plates are available for use with both negative and positive flats.

Presensitized plates are made to fit most duplicator and other press sizes. They are also designed for various lengths of press runs (impressions). Short-run plates are durable for about 5000-10,000 impressions. Other plates are available for printing runs of thousands of or even a million or more impressions.

Presensitized plates are exposed to ultraviolet light in a platemaker. This is done under subdued room light or in a room illuminated with yellow light.

Preparing presensitized plates

When removing a plate from the package, be careful not to scratch the presensitized coating. Keep your fingers off the plate surfaces. The plate should be handled by its outer edges and lifted rather than slid from the package. Plates must NOT be bent or creased in any way. Store unused plates in a cool, dry place.

Exposing presensitized plates

The plate should be placed emulsion side up on the rubber blanket of the platemaking unit, Fig. 21-32. The flat is positioned over the plate with the emulsion side of the negative(s) or positive(s) down. This places the flat in direct contact with the emulsion side of the plate. A platemaker's (film) gray scale should be placed in a cut-out portion of the flat along the gripper edge or tail edge of the plate, outside the printing area. Expose the plate to a solid step six, as shown in Fig. 21-33.

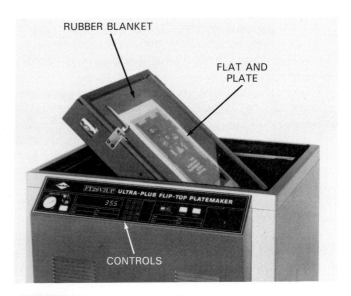

Fig. 21-32. Offset plate is positioned on rubber blanket of platemaker with emulsion side up. Flat is positioned over plate with emulsion side of negative down. (nuArc Company, Inc.)

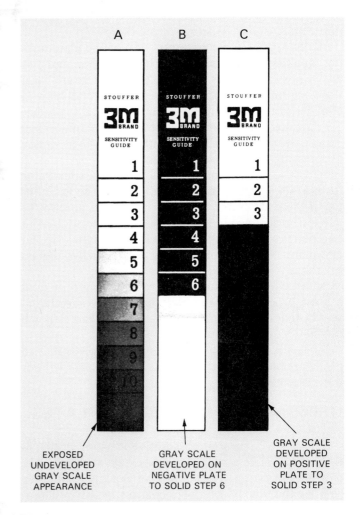

Fig. 21-33. Platemaker gray scale is used to verify correct exposure time. Step six (B) is recommended for negative working plates. Step three (C) is recommended for positive working plates. Example A illustrates how exposed undeveloped gray scale appears in flat. When developed, it looks like example B. (3M Company)

Step six is the recommended reading for most negative working plates, both additive and subtractive-working. Fig. 21-33 shows how a gray scale appears before developing.

NOTE! A solid step six means that step six is just as dense as step one on the developed image of the gray scale.

For most positive working plates, the gray scale reading should be an open step three or four, Fig. 21-33. This means that step three or four should be just as open as step one.

If a positive working place does NOT develop satisfactorily, it is probably underexposed. If the plate is underexposed, the gray scale should indicate this with an open step one or two.

If you expose through more than one layer of film, increase the exposure time. One extra layer of film usually requires one-half step more exposure.

Keep in mind that the lead edge of the flat should be aligned with the lead edge of the plate. In addition, the flat is generally aligned along the left-hand edge of the plate. When using a pin register system, be certain that the tabs fit snugly over the pins.

Always check the glass on the top of the platemaking machine for cleanliness. When the plate and flat are in position, lower and lock the glass cover. Turn the vacuum switch on and the vacuum blanket will compress the plate and flat to produce a tight contact. Set the timer for the recommended exposure time. Then expose the plate and flat to the ultraviolet light source by pressing the exposure button.

After exposure is complete, move the vacuum frame back to its normal position and turn off the vacuum. Open the glass frame carefully. A certain amount of vacuum resistance may be present and cause the flat and plate to stick to the glass. After removing the flat and plate, close and lock the glass frame to avoid glass breakage.

After exposure, process the plate by hand or with a plate processor. If the plate has been correctly exposed, the recommended numbered gray scale step will appear.

Since additive and subtractive working presensitized plates are handled differently, both processing methods are described here.

USING ADDITIVE WORKING PLATES

The diazo coating of an additive working presensitized offset plate hardens wherever light strikes it. The first step in processing an additive working plate is to remove the unexposed diazo coating. Apply a desensitizing gum to the entire surface of the plate, Fig. 21-34. Use a small cellulose sponge for this purpose. Move a sponge back and forth. This leaves the surface moist before starting the second step of processing.

Next, apply the developing lacquer to the plate. Pour the lacquer onto the surface of the plate in a small pool

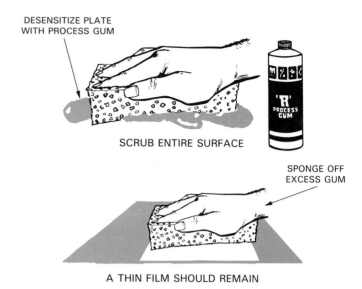

DESENSITIZE PLATE WITH PROCESS GUM

SCRUB ENTIRE SURFACE

SPONGE OFF EXCESS GUM

A THIN FILM SHOULD REMAIN

Fig. 21-34. Additive working offset plate is processed by first desensitizing entire surface with gum. (3M Company)

about 1 1/2 in. (38 mm) in diameter. Use a clean cellulose sponge in a circular motion to develop the image and gray scale to a ruby-red color. Refer to Fig. 21-35.

Rinse the plate with tap water in a plate sink, Fig. 21-36. This removes all the excess chemicals. If the plate is going directly on the press, no further preparation is necessary. If the plate is to be stored for a period of time, it must be preserved. This is done to protect it from oxidation, dirt, and handling.

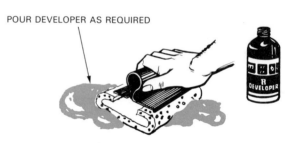

POUR DEVELOPER AS REQUIRED

SOAK UP DEVELOPER INTO PAD

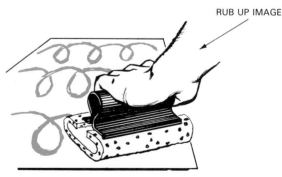

RUB UP IMAGE

USE FIRM CIRCULAR MOTION OVER ENTIRE SURFACE

Fig. 21-35. After desensitizing surface of additive working plate, developer is applied and a circular motion is used to develop image and gray scale to a ruby-red color. (3M Company)

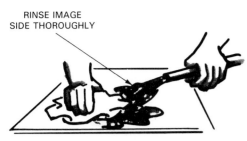

RINSE IMAGE
SIDE THOROUGHLY

USE DISPOSABLE PAPER WIPE
TO MOP SURFACE WHILE RINSING

Fig. 21-36. After development, additive working plate is rinsed with tap water in a plate sink. If plate is going directly to press, no further processing is necessary. If plate is to be stored, it must be preserved with gum.   (3M Company)

To preserve the plate, apply a thin coating of gum arabic or plate gum solution to the surface of the plate. Apply it with a lint-free cotton pad or soft disposable paper wipe. Buff the gum in the direction of the length of the plate. Continue the buffing action until the plate is completely dry.

NOTE! Never dry a plate under forced air heat since this action will penetrate the gum application and destroy the gum's usefulness.

The entire sequence of steps involved in processing an additive working presensitized plate is illustrated in Fig. 21-37. Note that both sides of the plate can be exposed and processed if desired. If a plate processor is used, follow instructions given by the manufacturer.

1. Apply a developing solution evenly over entire plate surface. Use a clean, soft cellulose sponge. Remove the excess gum, leaving only a thin film on plate. Gum removes unexposed diazo coating on plate or renders it water-receptive.

2. A special lacquer emulsion is applied before gum dries. Size of image area determines how much lacquer to use so that all image areas are covered. Continue rubbing lacquer until strong, uniform image appears.

3. Flush off excess lacquer with water and inspect plate. Image should not be streaked and should not rub off with the fingers.

4. Squeegee plate dry with a squeegee made expressly for this purpose.

5. Apply a small amount of special gum solution with a soft, clean cellulose sponge. Polish gum dry with clean, soft cheesecloth. Plate is ready for press or storage.

Fig. 21-37. Learn steps necessary to process an additive working presensitized offset plate.   (3M Company)

USING SUBTRACTIVE WORKING PLATES

A subtractive working presensitized offset plate is exposed in exactly the same way as an additive plate. However, a subtractive plate has a colored, light-sensitive coating over its entire surface. This coating must be removed from all the unexposed areas of the plate's surface.

Two solutions are required for subtractive working plate processing. These are the subtractive developer and the plate gum.

Place the exposed plate on a smooth, firm surface. Pour a small quantity of subtractive developer over the surface of the special pad. Then pour a liberal quantity of the developer on the surface of the plate. Spread the developer evenly over the entire surface with the pad. Use firm pressure on the pad with a tight circular motion to remove the coating from the non-image areas, Fig. 21-38.

The image areas, including the gray scale, must be completely developed. Fresh developer can be used to clean out halftone and tint areas that appear filled in. Do not allow the developer to dry on the non-image areas of the plate.

Then wipe the plate with a squeegee. Remove any remaining developer from the plate's surface with a soft disposable paper wipe. This will prevent image loss or

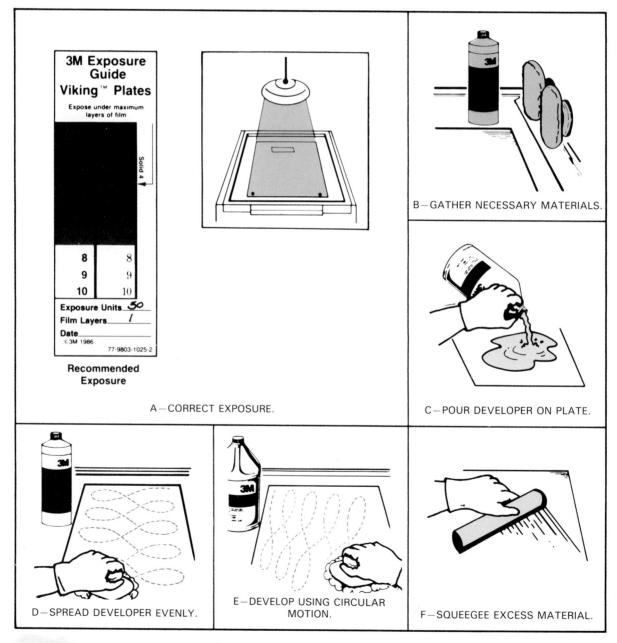

Fig. 21-38. Always follow manufacturer's recommendations when processing offset plates. For example, correct exposure for 3M Viking G-1 and G-2 subtractive working plates is a solid step four. Hand-processing of plate is done with a two-in-one developer gum.   (3M Company)

ink specks and streaks in non-image areas when the plate is run on the press. Rinse the plate in the plate sink with tap water. Also rinse the plate on the reverse side to remove any stray developer and dirt.

Use a soft disposable paper wipe to completely dry the surface of the plate. Pour a small pool of plate gum on the plate's surface and spread it with a soft disposable paper wipe. Buff the surface dry with a fresh paper wipe. Also wipe the reverse side of the plate dry.

Some subtractive working plates can be exposed and processed on both sides if desired. If a plate processor is used, follow the instructions given by the manufacturer.

PLATE ADDITIONS AND DELETIONS

After processing, a surface plate should be checked to determine the overall quality of its image and non-image areas. This must be done before the plate is released to the press department to avoid problems when on the press.

Plate corrections are categorized as either additions or deletions. Additions and deletions may be needed to repair minor problems on the plate.

Plate additions

Simple *additions* to lines, solids, and other image parts of an offset plate can be made either on or off the press. This is done by applying a tusche solution to the affected area of the plate. Pens are available for this purpose. *Tusche* is a form of greasy lithographic developing ink.

To make minor additions on the press, be sure the area to be treated is free of ink and completely dry. Apply tusche to the affected area with a tusche pen or cotton-tipped swab. The area tusched turns a brown color in 20 to 40 seconds.

Neutralize the tusched area immediately with water. Then dry the tusched area and apply offset ink to the area with your finger.

If the ink does NOT adhere to the affected area after resuming the press run, tusche the area again. This can be repeated as many times as necessary.

Plate deletions

After developing a surface plate, it may be necessary to make *deletions* to unwanted words, lines, spots, or solid image areas. Small areas can be deleted by using a clean, soft, rubber eraser with a lubricant, such as tap water.

First, dampen the area to be deleted. Moisten the eraser and rub the eraser on the unwanted area carefully. This is done by alternately wetting and then erasing the unwanted area. Keep the treated area wet with water throughout the deletion process. The deleted area is then desensitized and rinsed with water.

The plate should be preserved if not used on the press immediately. In most instances, deletions are needed after the plate is on the press and running.

In addition to the method just described, a solution called *deletion fluid* can be used to remove large areas of image on a plate. Apply the fluid using a cotton swab when the plate is clean and completely dry. After the fluid is allowed to work on the area for about one minute, clean the area with water.

Deletions on paper plates during a press run should be handled with care since the surface is easily damaged. Use only the recommended deletion procedure for the type of plate being used. In most instances, apply full-strength etch solution to the affected areas before resuming the press run.

POINTS TO REMEMBER

1. The image areas of the offset plate are grease receptive and water repellent. The non-image areas of the plate are water receptive and grease repellent.
2. Offset plates are made from several kinds of materials. These range from stainless steel to paper.
3. Some offset plates are multimetal, which means they are made from two or more kinds of metal.
4. Aluminum is the most popular kind of plate material, while other metals, such as chromium and copper, are also used.
5. Nonmetal plate materials include plastic, plastic-impregnated paper, and paper.
6. Direct-image masters are paper plates that are imaged by typing, drawing, lettering, and ruling.
7. Direct-image masters are used for short runs on offset duplicators and are generally used only once.
8. Photo-direct presensitized plates are prepared in an automatic camera projection-type platemaker. The need for a negative is eliminated in this process.
9. Photo-direct presensitized plate sizes are generally limited to duplicator size offset presses.
10. Electrostatic plates are prepared by a photoelectrostatic process. This is a Xerographic process.
11. Electrostatic plates are prepared on an automatic plate processor. These plates are generally run on duplicator size offset presses.
12. Most offset plates are manufactured ready to use.
13. Negative-acting and positive-acting plates as well as additive-working and subtractive-working emulsion plates are manufactured.
14. Offset plates are exposed using a strong actinic light in a platemaker machine.
15. A platemaker's gray scale is used to establish the correct exposure time.

16. Offset plates are processed either manually or in an automatic plate processor.
17. Once a plate has been processed, it is either placed on the press immediately or carefully stored for future use.

KNOW THESE TERMS

Additive plate, Ball graining, Brush graining, Chemical graining, Deep-etch plate, Deletion fluid, Desensitizing gum, Developing lacquer, Diazo coating, Dry offset relief plate, Electrostatic plate, Grainless plate, Gum arabic, Letterset, Master, Multimetal plate, Negative-working plate, Photo-direct presensitized plate, Photopolymer coating, Plate gum, Platemaker, Positive, Working plate, Sandblast graining, Step-and-repeat platemaker, Subtractive developer, Subtractive plate, Surface plate, Tusche, Xerography.

REVIEW QUESTIONS

1. One of the following is NOT a method used to grain metal plates:
   a. Chemical.
   b. Ball.
   c. Powder.
   d. Brush.
2. _____ offset plates consist of two or three layers of metal.
3. How should offset plates be handled?
4. How should offset plates be stored?
5. _____ working offset plates are exposed with film negatives.
6. Additive offset plates are developed by use of a _____.
7. Subtractive offset plates are the most popular since they avoid the use of lacquer-base chemicals. True or false?
8. The light-sensitive coating on presensitized aluminum-base plates is called _____ _____.
9. Deep-etch plates are exposed with film _____.
10. The image produced on both the negative-acting plate and the positive-acting plate is positive. True or false?
11. _____ presensitized offset plates do NOT require exposure to a negative flat.
12. A dry-offset plate consists of _____ images.
13. Explain electrostatic platemaking.
14. A platemaker's _____ _____ is used to determine the correct exposure time for a presensitized offset plate.
15. A step-and-repeat plate can be prepared with a single negative or positive. True or false?

16. The usual methods of placing an image on a direct-image master are by _____ and by _____.
17. Before a direct-image master can be run on an offset duplicator, the entire surface of the master is moistened with an _____ solution.
18. The first step in processing an additive-working presensitized offset plate is:
    a. Applying developing lacquer.
    b. Rinsing plate with water.
    c. Applying a gum arabic coating.
    d. Removing unexposed coating.
19. The two solutions required for processing a subtractive-working presensitized plate are subtractive _____ and plate _____.
20. A solution called _____ _____ can be used to remove unwanted image areas on an offset plate.

SUGGESTED ACTIVITIES

1. Prepare a direct-image master for printing on a duplicator. Use a typewriter with carbon ribbon, grease pencil, and reproducing ball-point pen to image the master. Prepare the images on the plate for an 8 1/2 x 11 inch press sheet. Observe all margins, including the allowance for the gripper.
2. Obtain a stripped flat. Cut in and tape a platemaker's 21-step gray scale along the bottom edge (tail) of the flat. Make the exposure to a presensitized plate of the type and size specified by your teacher. Process and preserve the plate for future use.
3. Prepare a plate using an electrostatic platemaker system.
4. Obtain a single negative flat for a step-and-repeat job. Make the step-and-repeat exposures. Process and preserve the plate for future use. Use the pin register system established for this flat during the stripping process.
5. Prepare offset plates by the subtractive and additive methods in your shop. Be sure you understand the terms subtractive and additive and negative-acting and positive-acting plates.
6. Gain experience in adding, deleting, and repairing images on presensitized metal plates. Use the solutions and tools discussed in this chapter.
7. If the equipment and materials are available, prepare a diffusion-transfer plate for the duplicator.
8. If the equipment and materials are available, prepare a photo-direct presensitized offset plate for the duplicator.

Paper and ink products are merged in the printing process.

Section VI

Issues for Class Discussion

1. The stripping process demands accurate, systematic, and efficient techniques. Stripping is often done by hand. However, computerization of the stripping process is becoming more common. The stripping of single and multicolor work can be done much faster and more accurately on computerized equipment. Discuss the implications of computerized layout and stripping equipment over hand-mechanical methods.

2. Certain single-color jobs require multiple flats as do jobs that require two or more colors. In both instances, some form of a pin register system is often used. Discuss the important elements of multiple-flat stripping and the need for a pin register system with single and multiple-color jobs.

3. Photo-direct offset plates are in common use throughout the printing and publishing industry, especially in the quick printing sector. Discuss how these plates are used, including the quality level that can be achieved.

4. Conventional plate processing generally requires the use of toxic chemicals, thereby creating many problems related to waste removal. Recent efforts by plate manufacturers have produced a number of water-developed plate materials. Research and discuss the applications and quality level of these plates. It may be necessary to contact a supplier of this kind of plate material for specific information.

Section VII
Paper and Ink

Together, paper and ink are usually the most expensive materials charged to a printing customer for a job. The huge paper and ink industries provide these critical supplies to every type of printer. The paper industry alone produces over 24 million metric tons of paper annually. Approximately 1200 million pounds of printing ink is produced and consumed annually.

Chapter 22 summarizes the manufacture of printing papers. This includes pulping, refining, forming, and finishing. The selection of paper for a printing job involves careful consideration of aesthetic and printing qualities. It is important for you to understand the important "role" that paper "plays" in the printing industry.

Ink is covered in Chapter 23. The manufacture of ink is carefully controlled and monitored to attain consistent quality control. Pigments, vehicles, and additives are the primary ingredients in ink. As you will learn, selecting the correct ink for the job is essential to quality reproduction.

This paper machine produces a roll of paper up to 238 inches wide at speeds to 3000 feet per minute. Combination Fourdrinier and twin-wire are features of this paper machine.   (Nekoosa Papers, Inc.)

Chapter 22

Paper

When you have completed the reading and assigned activities related to this chapter, you will be able to:

O Describe how paper pulp is prepared.
O Summarize the process of making printing paper.
O Identify and describe watermarked paper.
O Explain the several paper finishing operations.
O Describe how paper is packaged, stored, and conditioned.
O Distinguish between weights of basic papers.
O Describe what is meant by recycled paper.
O Compute weights of special sizes of paper.
O Explain why it is important to select the correct paper for the job.
O Determine the appropriate paper for various printing jobs.
O Write a paper order for a printing job.
O Calculate the cost of paper for a printing job.
O Describe the various sizes, shapes, weights, and styles of envelopes.
O Explain the terms and process used to determine metric paper sizes.

The manufacture of paper for printing is a basic industry that began 2000 years ago. Today, it is an industry experiencing technological change and growth. In the United States alone, there are nearly one million employees working in over 5000 paper manufacturing plants.

The word *paper* comes from the term "papyrus" which was a writing material made by ancient Egyptians. Papyrus reed was beaten and woven into a mat of thin, hard writing material.

Paper, as we know it today, was invented in 105 A.D. by Ts'ai Lun, a Chinese court official. He mixed mulberry bark, hemp, and rags with water. The ingredients were beaten, mixed, and pressed into flat sheets and hung to dry in the sun. See Fig. 22-1.

After nearly a thousand years, the Chinese technique of papermaking was brought into Europe by the Moors of North Africa. However, the method of making paper from wood was lost. Until the early 1800s, the Western world made paper from cloth and rags. Each sheet was individually made by dipping a screen into a vat of water-suspended fibers and then draining the water away.

Fig. 22-1. Ts'ai Lun invented paper in 105 A.D. The illustration on right depicts Chinese women boiling mulberry bark for paper manufacturing process used centuries ago.

The first papermaking mill in America was started by William Rottenhouse in about 1690. A typical papermaking machine of that period is shown in Fig. 22-2.

Papermaking machines were soon developed and the number of mills increased steadily. North America leads the world in papermaking because of its large forests and abundance of water.

Paper is the most important raw material for the printing process. It usually represents a substantial portion of the final cost of the printed product. Paper and paperboards are manufactured in numerous types and varying qualities. Paper is selected to meet the aesthetic needs and end use demands of the printed product.

As the offset printing process has evolved and as improvements in printing technology have occurred, papers have been developed with specific printing properties. With the increased types of paper and their varied characteristics, it is important for anyone employed in the printing industry to have a broad understanding of paper and its printing qualities.

HOW PAPER IS MADE

The manufacture of paper for printing involves two processes: pulp production and paper manufacture.

Pulp production includes debarking, chipping, pulping, bleaching, beating, refining, adding nonfibrous ingredients, and cleaning the pulp.

Paper manufacture includes forming the paper, producing watermarks, removing water, drying, and finishing the paper.

Wood is the basic raw material of all printing papers. To maintain a continuous supply of paper, the paper companies operate *tree farms*. Paper companies plant seedlings and care for them. When the trees are large enough (usually 15 years old), they are cut and transported to the paper mill.

Sulfur, magnesium, quick lime, salt cake, caustic soda, chlorine, and starch are also used in the papermaking process. Fresh water is also an essential element in papermaking.

PULPING

The manufacture of paper depends upon an adequate supply of pulp. *Pulp fibers* are obtained from wood, woody fibrous materials, and reclaimed products made of these materials.

The *nonwoody materials* used include cotton, flax, hemp, jute, sugar cane, esparto (long, coarse grass), and various straws. The wood pulp sources are divided between softwoods (coniferous) and hardwoods (deciduous).

The *hardwood trees* include poplar, gum, beech, maple, birch, and chestnut. *Softwood trees* include pine, fir, spruce, and hemlock. Reclaimed pulp fibers come from rags and recycled paper.

All paper fibers are made of a compound called *cellulose,* Fig. 22-3. This compound makes up the general structural content of plants and wood. Cellulose is found in varying amounts in different materials such as cotton, wood, and straw. Wood is the principle source of paper fibers. Besides cellulose, wood contains lignin and small amounts of minerals, resins, and oils.

Lignin serves as a binder and support for the cellulose fibers of woody plants. Wood usually contains 50 percent cellulose and 30 percent lignin.

Printing papers are made primarily from groundwood fibers called pulp. There are three popular types

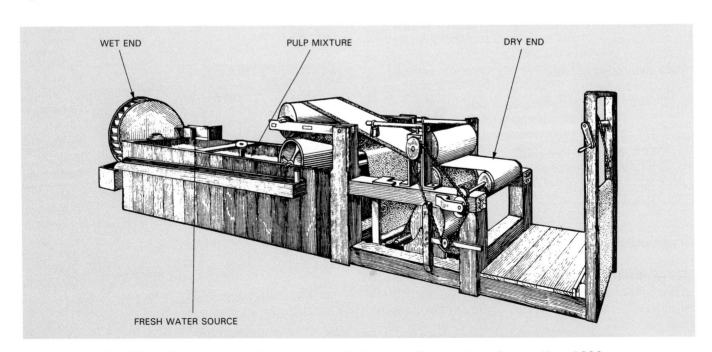

WET END   PULP MIXTURE   DRY END

FRESH WATER SOURCE

Fig. 22-2. Note crude wooden design of typical papermaking machine of around late 1690s.

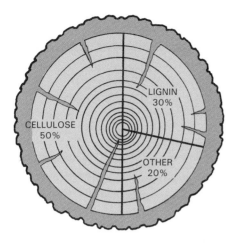

Fig. 22-3. Paper consists mainly of cellulose fibers. In their natural state, these fibers are tubular structures much finer than a human hair and shorter than a grain of rice. (S.D. Warren Co.)

of wood pulp: groundwood pulp, chemical pulp, and chemical-mechanical pulp. The process used to reclaim paper pulp is also described here.

Groundwood pulp

The *groundwood pulp process* uses almost all of the substance in wood to make paper. It starts at the paper-making mill from trees cut in surrounding forests. The tree logs are reduced to short lengths, Fig. 22-4. The bark is removed and the logs are cleaned and fed into machines that grind them into small chips, Fig. 22-5.

Groundwood pulp is often referred to as *mechanical pulp* because it has been reduced to fibers by a mechanical grinding process.

When the logs are further reduced to a mass of short, weak fibers, water is added to the pulp. This mass is similar to oatmeal in texture and color. This treatment helps eliminate slivers and large fragments of wood.

More water is added and the pulp flows onto the screen of a machine where it is drained. Through the use of giant presses, it is then squeezed into bundles resembling cotton bales. These are shipped to paper mills for final processing into paper.

The groundwood procedure has certain geographic, consumer, and economic advantages. The kinds and grades of paper manufactured in this manner are usually limited to popular items. Because lignin from the wood remains in the paper, the paper tends to darken when exposed to light.

Chemical pulp

The manufacture of *chemical pulp* uses chemical action to dissolve some of the wood's substances to make higher quality paper. It results from a more complex process. First the bark is removed from the logs and the wood is reduced to chips, as in the groundwood pulp process. Each chip is approximately three-quarters of an inch in length.

Fig. 22-4. Groundwood pulp process begins at papermaking mill from trees cut into four-foot lengths. (Hammermill Paper Group)

Fig. 22-5. After bark has been removed, logs are fed into machines that grind them into small chips size of a quarter. (Nekoosa Papers, Inc.)

Chemicals are added to the wood chips and the mixture is cooked under controlled temperature and pressure in a large cylindrical tower, called a *digester,* Fig. 22-6. Sulphite pulp is made by cooking the wood with lime and sulphuric acid. Sulphate pulp is wood cooked with caustic soda, Fig. 22-7. The chemicals remove most of the lignin and resin from the wood. As a result the wood fibers are longer and cleaner to

Fig. 22-6. Chemicals are added to wood chips and mixture is cooked under controlled temperature and pressure in a digester. (Nekoosa Papers, Inc.)

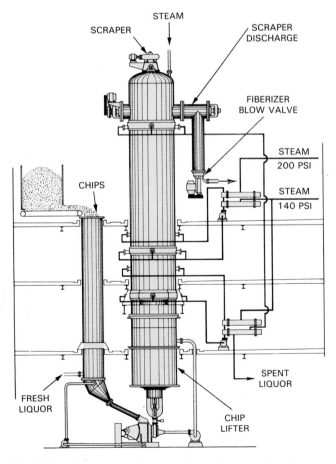

Fig. 22-7. Schematic diagram shows how wood chips are cooked with caustic soda in digester. (Hammermill Paper Group)

make a stronger, more stable paper. The wood pulp is then washed and screened thoroughly.

The sulphite process is the most popular in the United States. The chemical process delivers only about half as much pulp as the groundwood method. This and the cost of chemicals and length of treatment are the reasons for the higher price of chemical pulp.

Chemical-mechanical pulp

The *chemical-mechanical pulp* method combines mechanical and cooking separation processes. The chemical treatment varies according to the desired result. The fibers produced by this process are shorter than those from the chemical method and not as strong. Paper is rarely made from this process alone. Because of their shortness, chemical-mechanical pulp fibers are used to fill up the spaces between long fibers. This helps produce uniform sheets of paper and also improves its opacity or show-through quality.

Opacity refers to the amount of light that can be seen through a sheet of paper—the more opaque, the less light that passes through the paper. This process produces an intermediate pulp with both a relatively high strength and a high yield.

Reclaimed paper pulp

Reclaimed paper pulp is made from paper that has already been used in a product. The used paper is defibered by heating and then cooking in chemicals (lye) which dissolve the printing ink. The ink as well as other impurities are separated from the fiber by washing with huge quantities of fresh water and screening. This material is then reduced to pulp. The resultant pulp may also be bleached or whitened.

There are two classes of reclaimed paper pulps. Some of the reclaimed paper pulping processes deliver an inferior pulp with a poor color, and some processes yield excellent pulp.

Paper pulps of the inferior group are used in coarse papers and boxboard. Paper pulps of the better grade can be made into papers of excellent appearance with good formation and printing qualities.

TREATING THE PULP

Converting the pulp into paper requires several steps in treatment. These include: washing, bleaching, adding fillers, adding sizing, and adding dyes and pigments. A modern papermaking operation is illustrated in Fig. 22-8.

Washing

In all three pulping methods, *washing* is used repeatedly to remove all traces of chemicals and dirt from the wood fibers. The pulp is then flowed over a series of screens to remove small pieces of uncooked wood and grit.

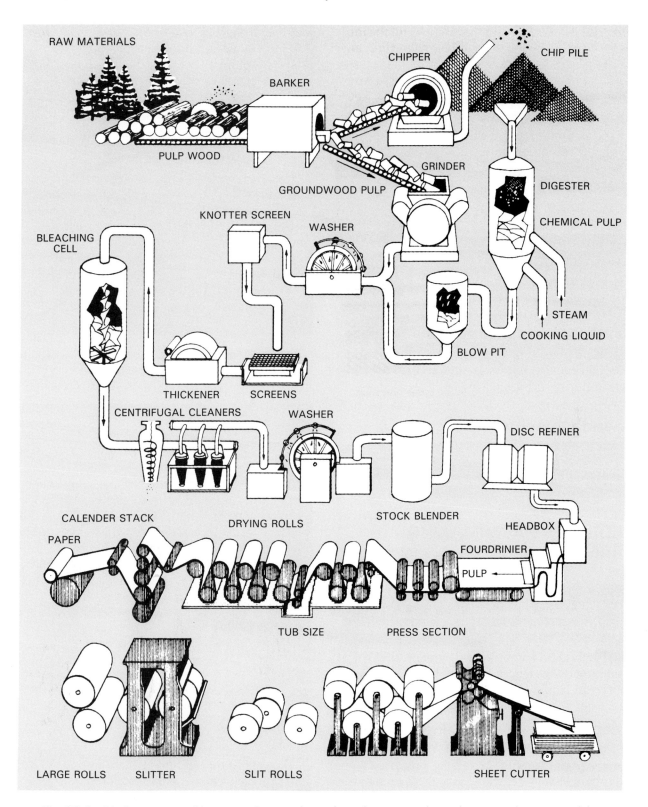

RAW MATERIALS

CHIPPER

CHIP PILE

BARKER

PULP WOOD

GROUNDWOOD PULP

GRINDER

DIGESTER

CHEMICAL PULP

KNOTTER SCREEN

WASHER

BLEACHING CELL

STEAM

COOKING LIQUID

BLOW PIT

THICKENER

SCREENS

CENTRIFUGAL CLEANERS

WASHER

DISC REFINER

CALENDER STACK

DRYING ROLLS

STOCK BLENDER

HEADBOX

PAPER

FOURDRINIER

PULP

TUB SIZE

PRESS SECTION

LARGE ROLLS

SLITTER

SLIT ROLLS

SHEET CUTTER

Fig. 22-8. Modern papermaking operation requires a large investment in equipment and raw materials.

At this point, chemical pulp may go directly to large driers. It is then cut into large sheets and pressed into bales weighing 400 to 500 pounds (180 to 225 kg). The bales are shipped to various paper mills for final processing. If a white paper is desired, the pulp is sent from the washers to the bleaching vats.

Bleaching

In a series of carefully controlled operations, the pulp selected for *bleaching* is treated with chlorine and similar chemical solutions to whiten the paper fibers. This process is continued until the desired whiteness is reached. Then the pulp is given a final series of washings and is sent to storage towers. In the storage towers, it awaits the addition of fillers, sizing, and dyes.

Fillers

The primary purpose of adding *fillers* to the pulp is to strengthen its properties, Fig. 22-9. Fillers are added to make an opaque paper with good ink absorbing qualities and a smooth, even surface. Typical additives or fillers consist of clay, talc, calcium carbonate, zinc sulfide and silicates.

Sizing

Sizing is used to help make paper less absorbent and more water repellent so it will not allow ink to spread. Rosin is the most widely used sizing and is applied to the pulp during beating. This is referred to as *internal sizing*. Sized papers are widely used in the offset printing process because of their water-repellent qualities.

External or *surface sizing* takes place on a large press and is applied to give greater stiffness. The addition of surface sizing also allows greater control over ink absorption during printing.

Dyes and pigments

Paper is made in a variety of colors from light tints to dark shades. *Dyes* and *pigments* are added to the pulp during washing to give the paper color. Almost 98% of all paper is dyed some color.

FOURDRINIER PAPERMAKING MACHINE

A Frenchman, named Nicolas Louis Robert, invented the papermaking machine in 1798. Robert and Sealy Fourdrinier improved the original design and later it became known as the Fourdrinier papermaking machine. A modern technological wonder, the typical machine extends the length of a city block and stands two stories in height, Fig. 22-10.

Forming the paper (Wet end)

The forming of paper starts at the "wet end" of the papermaking machine, Fig. 22-11. The processed pulp, which is about 99% water, flows rapidly over an apron the width of the machine onto a moving mesh wire screen. The water begins to drain from the pulp through the wire screen as it moves along. Table rollers and hydrofoils under the wire screen support the screen and carry off additional water.

Where the paper touches the screen, it shows a mesh pattern. This is called the *wire side* of the paper. The wire side develops a grain and greater openness as a result of losing short fibers, sizing, and fillers through the screen.

Fig. 22-9. Fillers are added to pulp to strengthen its properties. (Hammermill Paper Group)

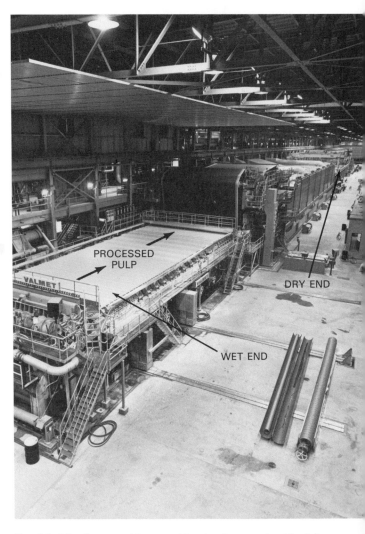

Fig. 22-10. Papermaking machine is about a city block long and two stories high. (The Mead Corp.)

468

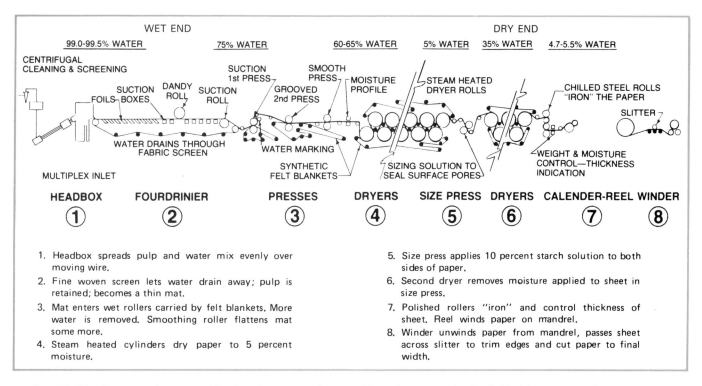

Fig. 22-11. Process of papermaking involves several steps. Note that operation is divided into wet end and dry end. (Hammermill Paper Group)

Dandy roll

As removal of the water continues, the paper begins to form. At this point, the pulp passes under a *dandy roll* to produce a surface finish on the paper, Fig. 22-12. If the paper is to be identical on both sides, the dandy roll is covered with fine wire. This produces a woven finish paper, much like woven cloth. To imitate a laid finish, or hand made paper, the dandy roll is covered with a different series of wires which produce parallel lines on the paper.

Designs called watermarks can also be produced by the dandy roll. A *watermark* is a symbol or logo which identifies the brand of paper or trademark of a company. An example of how watermarks are formed by the dandy roll is illustrated in Fig. 22-13.

The symbol design is made from brass or wire and attached to the dandy roll. The watermark leaves its impression on the top of the paper as it begins to form. The design can be formed because the paper still contains a high percentage of water as it makes contact with the dandy roll.

Press rolls and felt blankets

The paper begins to form into a *web* or long ribbon once it exits the dandy roll, Fig. 22-14. The web continues through *press rolls* and *felt blankets* which remove more water. The paper is then transferred onto a canvas-type dryer blanket. The side of the paper which touches the felt blanket is called the *felt side*. It is smoother and contains longer fibers and fillers. This brings it into contact with steam dryers.

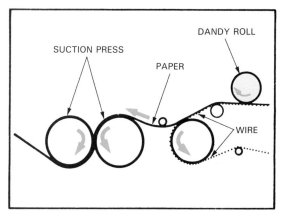

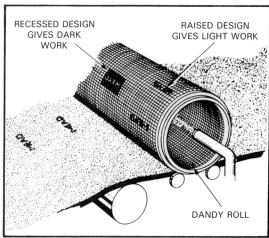

Fig. 22-12. Surface finish of paper is formed by a dandy roll just as paper begins to form on papermaking machine. (The Mead Corp.)

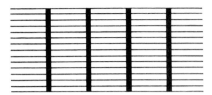

Fig. 22-13. Watermark is a symbol or logo formed by dandy rolls on papermaking machine. (Hammermill Paper Group)

PAPER FORMING INTO WEB

Fig. 22-14. After leaving dandy rolls, paper begins to form into a web or long ribbon. Web continues through steel press rolls and felt blankets which remove more water. (Hammermill Paper Group)

If the paper is to be surface sized, rollers apply the sizing at this point. The felt side is the side of paper preferred for printing when using only one side of the paper.

Finishing (Dry end)

As the web of paper comes off the dry end of the papermaking machine, it is forced between heavy rolls to smooth the paper surface and is then wound into large rolls.

If the paper manufacture is considered completed at this point, the paper is called *machine finished*.

If *calendered paper* is required, the paper is run through another series of rolls, called calender rolls. *Calender rolls* are made of smooth iron to polish the paper and give it a very smooth finish. This polishing is usually done on both sides at the same time. When *supercalendering* is required, it is generally done as a separate operation, one side at a time.

When leaving the papermaking machine, the finished web of paper is wound on huge steel reels. It is then hoisted onto a *rewinder-slitter machine* where it is cut into desired widths, Figs. 22-15, 22-16, and 22-17. When rolls of paper are required for printing (in the case of web press operations), the paper is cut into desired widths and rewound into separate rolls, Fig. 22-18.

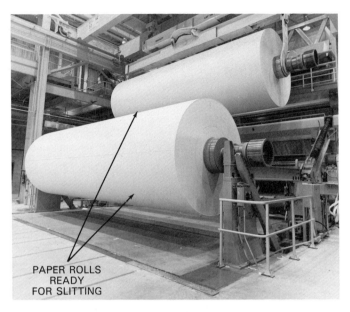

PAPER ROLLS READY FOR SLITTING

Fig. 22-15. When leaving papermaking machine, finished web of paper is wound on huge steel reels ready for rewinder-slitter machine. (The Mead Corp.)

SINGLE SHEETS CUT FROM ROLLS

Fig. 22-16. This rewinder-slitter machine cuts flat sheets from rolls of paper. Single sheets are sold for use on sheet-fed offset presses. (Hammermill Paper Group)

Fig. 22-17. Large single sheets of paper are taken from rewinder-slitter machine and further cut to standard sheet sizes. Sheets are then packaged and delivered to paper merchants. (Hammermill Paper Group)

Fig. 22-18. Large quantities of paper are sold in rolls for web press operations. Paper from papermaking machine is cut into desired web widths and rewound into separate rolls. (Hammermill Paper Group)

PAPER REQUIREMENTS

As discussed, various techniques can be used in the manufacture of paper. It is important for you to understand paper requirements for both sheet-fed and web presses. The paper used must have characteristics that match the printing method.

Sheet-fed paper requirements

Paper used for sheet-fed offset lithography must have higher surface and internal bonding STRENGTH than that used for other printing processes. This is because it must be able to withstand the tackier ink films used in sheet-fed offset lithography.

Water resistance is also needed for two reasons. One reason is the prevention of a softening and weakening of the paper surface, which can cause *picking* (lifting of paper fibers) and a transfer of fibers or coating to the blanket. A second reason is to avoid excessive moisture pickup from the press dampening system. This can cause curl and intolerable changes in paper dimension. However, excessive water resistance, such as in plastic-coated papers, presents other unusual problems for the offset press operator.

The complete contact of rubber blanket and paper, along with the inherent tendency of the blanket to lift any loosely bound materials from the paper, demands an exceptionally clean and strongly bonded surface.

Since offset lithography operates on chemical principles, paper must NOT release any active materials that will react unfavorably with the chemistry of the plate, ink, and dampening system.

To maintain register, paper must remain flat and not change its dimensions during printing. Its moisture content at the time of printing should be reasonably close to the pressroom relative humidity. The balanced moisture content of paper for any specific relative humidity will depend upon the type of paper and its moisture history. Consequently, it is important that the relative humidity of the paper be in balance with the relative humidity of the pressroom.

Web paper requirements

The requirements for web offset lithography are basically the same as those for sheet-fed offset lithography, except for the different aspects of sheet and web printing. The uniform tension maintained on web presses permits feeding and printing of lower basis weights or thinner paper than commonly used on sheet-fed presses. Basis weights of paper are described later in this chapter.

The basis weights of paper generally used on web offset range from about 20 to 80 pound. Heavier weights can be run, but the press delivers the finished product in sheets since they do not handle well in folders. Coated papers must be designed to resist blistering and to fold without cracking during heatset drying.

The moisture resistance for web offset papers need not be as high as that for sheet-fed offset. Papers having a lower pick and moisture resistance than those used for sheet-fed offset can be used. This is because web offset inks generally have lower tack.

In addition, less moisture is used in the web offset printing system, and less time exists for the paper to pick up moisture. Internal strength is required to resist the delamination forces of blanket-to-blanket presses.

For satisfactory runability, paper webs must be flat enough to pass through the squeeze impressions of the printing units without wrinkling or becoming distorted.

Paper rolls that unwind with even tension and flatness across the web, and without localized distortion, are required for good register and to prevent wrinkling. Paper rolls free of defects, with proper *splicing* (joining of two ends of rolls) are essential for good runability and for minimizing web breaks.

PROPERTIES OF PAPER

The various properties or ingredients of paper needed for offset lithography are those common to most papers. Paper is manufactured in many different forms and for many uses—the more specialized the paper, the more unique its properties.

The various *paper properties* include:
1. Grain.
2. Finish.
3. Pick resistance.
4. Basis weight.
5. Flatness.
6. Ink setting and drying.
7. Moisture absorbency and resistance.
8. Opacity.

Grain

As the paper moves from the wet end of the paper-making machine toward the dry end, the fibers in the paper tend to align themselves parallel to the direction they are traveling. This alignment of fibers is referred to as the *grain* of the paper. Paper has greater strength in the direction of the grain and folds more easily with the grain than against it. Grain is an important factor in all types of printing where folding is involved. Examples include magazines, brochures, and booklets.

Fiber length is important to paper tear strength. *Tear strength* in papers increases with increased fiber length. This is because longer fibers result in more frictional drag and thereby requires more effort to tear the paper.

Finish

When the paper leaves the papermaking machine, it will be wound into a huge roll weighing nearly 10,000 pounds (4 500 kg). Several operations must take place before the raw paper is ready for the user.

The term *finish* refers to any action performed to the surface of the sheet of printing paper that affects its surface.

For example, book papers are made in six finishes. These include:

1. Antique.
2. Eggshell.
3. Vellum.
4. Machine.
5. English.
6. Supercalendered.

Paper can also be run through metal rollers that are designed to produce artistic patterns in the surface. *Coated papers* are exceptionally smooth since they are subjected to an additional smoothing process.

Pick resistance

During printing, *picking* can result when fibers in the paper are lifted up and pulled from the paper's surface by the press action. These fibers enter the ink and end up on the plate as hickies. *Hickies* are doughnut-shaped specks which appear in the inked image areas of the press sheet. Fig. 22-19 shows magnified views of hickies.

Offset inks are usually stickier than other kinds of inks. They require paper with tighter, more pick resistant fibers. A paper with low pick resistance can also cause clogging in the inking system, leading to poor printing quality.

Brightness

A black cover paper is not really black in itself. It is black because it absorbs all of the colored rays in light and reflects none of the light rays. A blue cover paper appears blue only because it reflects the blue rays in light while absorbing the others.

Another example, grass absorbs all of the colored

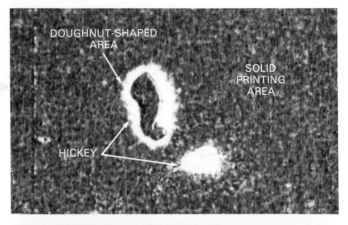

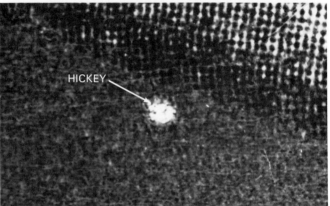

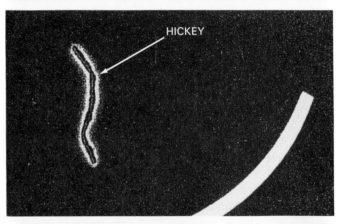

Fig. 22-19. Hickies are doughnut-shaped specks that appear in inked image areas of a press sheet. These magnified examples show hickies within solid and screened printing areas. (Graphic Arts Technical Foundation)

472

light rays except the green rays. The green rays are reflected to the eye and cause the grass to appear green.

A paper that would reflect the complete combination of colors as they are contained in sunlight would be a white paper. Therefore, to make paper pure white the paper manufacturer must find materials that will reflect all of the colors that are contained in light. To date, no paper manufacturer has been able to do this.

Any combination of known materials will either reflect a surplus of blue rays or orange rays. Excess blue reflection will cause the paper to appear blue-white. If the paper reflects a surplus of orange rays, the paper will appear cream-white in color. However, some progress has been made in the use of colorings. As a result, the color of white papers has been made to appear almost pure white.

The degree of *brightness* that can be captured in paper is proportionate to whiteness and directness of reflection.

Pure white light reflections are brightest. As the paper manufacturer approaches pure white, the brightness of paper improves.

Directness of reflection preserves brightness. If a pigment bounces the white light rays directly back to the eye, brightness is retained. If a pigment is constituted so that the light rays scatter along the surface or reflect to one side, there is a loss of brightness.

Scientists measure the reflecting qualities of different materials and give each a rating, called the *refractive index*. A material that bounces the reflection back to the eye and thus preserves brightness is given a *high index*. A material that bends the reflection to one side, causing a loss of brightness, is given a *lower index*.

Fig. 22-20 illustrates how refraction of light affects brightness of paper.

Paper weights and dimensions

Most printing papers are manufactured and identified according to their basis weight. *Basis weight* refers to the weight of a ream of paper in pounds. A *ream* is 500 sheets cut to the common size for that particular paper. See Fig. 22-21.

For example, a ream (500 sheets) of 70-pound book

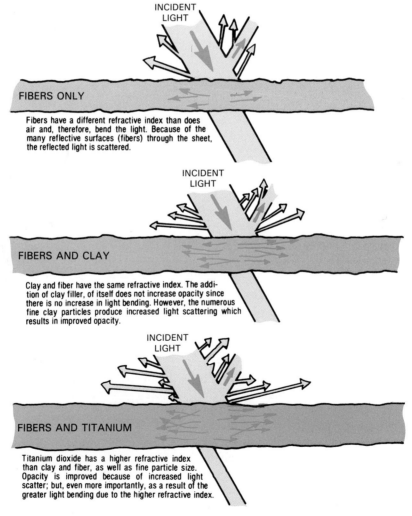

INCIDENT LIGHT

FIBERS ONLY

Fibers have a different refractive index than does air and, therefore, bend the light. Because of the many reflective surfaces (fibers) through the sheet, the reflected light is scattered.

INCIDENT LIGHT

FIBERS AND CLAY

Clay and fiber have the same refractive index. The addition of clay filler, of itself does not increase opacity since there is no increase in light bending. However, the numerous fine clay particles produce increased light scattering which results in improved opacity.

INCIDENT LIGHT

FIBERS AND TITANIUM

Titanium dioxide has a higher refractive index than clay and fiber, as well as fine particle size. Opacity is improved because of increased light scatter; but, even more importantly, as a result of the greater light bending due to the higher refractive index.

Fig. 22-20. Note how refraction of light affects brightness of paper.
(Graphic Arts Technical Foundation)

	Book 25 x 38	Bond and Ledger 17 x 22	Cover 20 x 26	Printing Bristol 22½ x 28½	Index 25½ x 30½	Tag 24 x 36
BOOK (Basis Weights in Bold)	**30**	12	16	20	25	27
	40	16	22	27	33	36
	45	18	25	30	37	41
	50	20	27	34	41	45
	60	24	33	40	49	55
	70	28	38	47	57	64
	80	31	44	54	65	73
	90	35	49	60	74	82
	100	39	55	67	82	91
	120	47	66	80	98	109
BOND and **LEDGER** (Basis Weights in Bold)	33	**13**	18	22	27	30
	41	**16**	22	27	33	37
	51	**20**	28	34	42	46
	61	**24**	33	41	50	56
	71	**28**	39	48	58	64
	81	**32**	45	55	67	74
	91	**36**	50	62	75	83
	102	**40**	56	69	83	93
COVER (Basis Weights in Bold)	91	36	**50**	62	75	82
	110	43	**60**	74	90	100
	119	47	**65**	80	97	108
	146	58	**80**	99	120	134
	164	65	**90**	111	135	149
	183	72	**100**	124	150	166
	201	79	**110**	136	165	183
	219	86	**120**	148	179	199
PRINTING **BRISTOL** (Basis Weights in Bold)	100	39	54	**67**	81	91
	120	47	65	**80**	98	109
	148	58	81	**100**	121	135
	176	70	97	**120**	146	162
	207	82	114	**140**	170	189
	237	93	130	**160**	194	216
INDEX (Basis Weights in Bold)	110	43	60	74	**90**	100
	135	53	74	91	**110**	122
	170	67	93	115	**140**	156
	208	82	114	140	**170**	189
TAG (Basis Weights in Bold)	110	43	60	74	90	**100**
	137	54	75	93	113	**125**
	165	65	90	111	135	**150**
	192	76	105	130	158	**175**
	220	87	120	148	180	**200**
	275	109	151	186	225	**250**

Fig. 22-21. Printing papers are classified by kind, basis weight, and basic sizes. (S.D. Warren Co.)

paper containing sheets 25 x 38 inches in size, weighs 70 pounds. The basis weight is indicated after the dimensions, such as 25 x 38 — 70. The figures that precede the dash (−) indicate the size of the sheets. The last figure indicates the weight in pounds of 500 sheets of paper. The underlined figures (38) indicates the paper grain direction.

The common size of paper is not the same for all kinds of paper. For example, bond paper is commonly 17 x 22 inches. Cover paper is 20 x 26 inches. Index bristol is 25 1/2 x 30 1/2 inches. Newsprint is 24 x 36 inches. These common sizes are referred to as the *basic size*. Look at Fig. 22-22. In addition, these papers are available in various thicknesses, called *calipers,* according to their weights. See Fig. 22-23.

Paper is manufactured in other than common sizes. These are listed in various tables available from paper dealers, Fig. 22-24. The tables usually list prices per thousand sheets. For example, a 20-pound bond paper would be listed as 17 x 22 — 40 M. The M represents a quantity of 1000.

To find the approximate ream weight of any irregular

Paper

TYPE	GENERAL USES	BASIC SIZE (Inches)	BASIS WEIGHTS (Pounds)
Bond	Business forms, letterheads, envelopes, direct-mail advertisements	17 × 22	9, 13, 16, 20, 24
Book/Offset	Catalogs, books, brochures, pamphlets, direct-mail advertisements	25 × 38	Coated: 60, 70, 80, 90, 100 Uncoated: 33, 40, 50, 60
Bristol (Card)	Tickets, postcards, covers, menus, novelty items	22½ × 28½	67, 80, 100, 120, 140, 160
Bristol (Index)	Index cards, folders, records	25½ × 30½	90, 110, 140, 170
Carbonless	Business forms, stationary, statements		
Cover	Covers, binders, menus, posters, tags	20 × 26	Coated: 60, 65, 80, 100, 125 Uncoated: 50, 65, 80, 130
Duplicator	Spirit duplicating	17 × 22	16, 20
Label	Labels	25 × 38	50, 60, 70
Ledger	Graphs, maps, bookkeeping, ruled forms	17 × 22	24, 28, 32, 36
Mimeograph	Mimeograph duplicating	17 × 22	16, 20, 24
Newsprint	Newspapers, advertisements	24 × 36	30, 32, 34
Onionskin	Carbon copies, airmail letter-heads	17 × 22	7, 8, 9, 10
Safety	Checks, securities	17 × 22	24

Fig. 22-22. Common sizes of paper are referred to as basic size. For example, basic size of cover paper is 20 x 26 inches.

The calipers listed are approximate averages. Variations will be found from one mill run to another, either to the light or heavy side of the basis weight, within trade custom tolerances. One point equals 1/1000 of an inch.

BOND, MIMEO, DUPLICATOR

bs. 17 x 22	13#	16#	20#	24#
Sulphite Bond	.003	.0035	.004	.0045
Cotton Fiber Bond				
Cockle Finish	.003	.0035	.004	.0045
Smooth Finish	.0025	.003	.0035	.004
Mimeo		.004	.005	.0055
Duplicator		.0025	.003	.0035

LEDGER

bs. 17 x 22	24#	28#	32#	36#
Smooth Finish	.0045	.005	.0055	.006
Posting Finish	.005	.0055	.006	.0065

BOOK PAPERS

bs. 25 x 38	45#	50#	60#	70#
Offset				
Regular	.0035	.004	.0045	.005
Antique	.004	.0045	.005	.006
Bulking		.0055	.0066	.0077
English Finish	.0032	.0035	.004	.0045
Supercalendered	.0022	.0025	.003	.0035
Gloss Coated		.0025	.003	.0035
Dull Coated		.003	.0035	.004
Coated 1 Side			.0032	.0037

bs. 25 x 38	80#	100#	120#	150#
Offset				
Regular	.006	.0075	.009	.011
Antique	.007	.009	.011	.013
Bulking	.0088	.011	.0135	
Gloss Coated	.004	.0055		
Dull Coated	.0045	.006		
Coated 1 Side	.004			

COVER PAPERS

bs. 20 x 26	50#	60#	65#	80#	90#	100#	130#
Uncoated							
Smooth			.0065		.011		.013
Antique	.007		.010				.020
Coated	.005	.0055	.006	.008	.009	.010	

bs. 20 x 26	50#		65#	80#	94#	110#	
Lusterkote	.0055		.0065	.008	.010	.012	

INDEX BRISTOL

bs. 25½ x 30½	90#	110#	140#	170#
Smooth Finish	.007 .0075	.008 .009	.0105 .0115	.013 .014

Fig. 22-23. Papers are manufactured in various thicknesses, called calipers, according to their kind and weight. For example, 17 x 22 inch 20-pound sulphite bond paper has a thickness of .004 inch. (Leslie Paper Co.)

475

Hammermill Bond Hammermill Paper
No. 1 Sulphite, Watermarked
Packaged: 500 Sheets

	Pkg	Price Per CWT		
		1 Ctn	4 Ctn	16 Ctn
White Bond, 13#	88.90	77.75	68.00	62.65
White Bond, 16#	72.20	63.20	55.25	50.90
White Bond, 20 & 24#	69.20	60.55	52.90	48.75
Colors Bond, 16#	76.25	66.75	58.35	53.75
Colors Bond, 20#	73.20	64.05	56.00	51.60
White Rippletone, 20#	78.30	68.50	59.90	55.20
Ivory Rippletone, 20#	82.30	72.05	62.95	58.05

17 × 22

Basis	Size	M Wt	Ctn	Price Per 1000 Sheets			
White							
13	17 × 22	26	5000	23.11	20.22	17.68	16.29
	17 × 28	33	4000	29.34	25.66	22.44	20.67
	22 × 34	52	2500	46.23	40.43	35.36	32.58
16	17 × 22	32	4000	23.10	20.22	17.68	16.29
	17 × 28	41	3000	29.60	25.91	22.65	20.87
	22 × 34	64	2000	46.21	40.45	35.36	32.58
	24 × 38	78	1500	56.32	49.30	43.10	39.70
	28 × 34	82	1500	59.20	51.82	45.31	41.74
20	17 × 22	40	3000	27.68	24.22	21.16	19.50
	17 × 28	51	3000	35.29	30.88	26.98	24.86
	17½ × 22½	42	3000	29.06	25.43	22.22	20.48
	19 × 24	49	3000	33.91	29.67	25.92	23.89
	22 × 34	80	1500	55.36	48.44	42.32	39.00
	22½ × 35	84	1500	58.13	50.86	44.44	40.95
	24 × 38	98	1500	67.82	59.34	51.84	47.78
	28 × 34	102	1500	70.58	61.76	53.96	49.73
24	17 × 22	48	3000	33.22	29.06	25.39	23.40
	17 × 28	61	2000	42.21	36.94	32.27	29.74
	22 × 34	96	1500	66.43	58.13	50.78	46.80
	24 × 38	118	1000	81.66	71.45	62.42	57.53
White Rippletone							
20	17 × 22	40	3000	31.32	27.40	23.96	22.08
	22 × 34	80	1500	62.64	54.80	47.92	44.16
Ivory Rippletone							
20	22 × 34	80	1500	65.84	57.64	50.36	46.44
Colors:							
Blue, Canary, Green, Pink							
16	17 × 22	32	4000	24.40	21.36	18.67	17.20
Buff, Gray, Ivory							
16	17 × 22	32	4000	24.40	21.36	18.67	17.20
Colors:							
Blue, Buff, Canary, Ivory, Cherry, Gray							
20	17 × 22	40	3000	29.28	25.62	22.40	20.64
Green, Pink, Russet, Cafe							
20	17 × 22	40	3000	29.28	25.62	22.40	20.64
Goldenrod, Greentint, Orchid, Melon, Salmon							
20	17 × 22	40	3000	29.28	25.62	22.40	20.64

Fig. 22-24. Paper is available in other than basic sizes. This catalog example shows a page with Hammermill No. 1 sulphite bond. (Hammermill Paper Group)

size, multiply the sheet length by the width. This will determine the square inch area of the irregular size sheet. Multiply this figure by the *weight factor,* shown in Fig. 22-25, to find the ream weight. To find the *M weight* (1000 sheets) multiply by two.

Flatness

Paper flatness is a measurement of how much the paper curls, waves, or bends. Since the entire printing surface of the offset plate is flat, an uneven or wavy sheet of paper can wrinkle during printing. Therefore, flatness is desirable in printing papers. Either *wavy edges* from moisture gain or *tight edges* and *curl* from moisture loss cause feeding problems on the offset press. Paper must be made, stored, and processed under ideal and uniform moisture conditions.

Pressrooms in large plants have controlled moisture and humidity systems. The paper is conditioned before going through an offset press. *Conditioned* means the paper is moved to the pressroom area before printing

---

TO FIND REAM WEIGHT (500 SHEETS)

To find the approximate ream weight of any irregular size, multiply the sheet length by the width. This will determine the square inch area of the irregular size sheet. Multiply this figure by the weight factor shown in the tables below, to find the *ream* weight. To find the M weight (1000 sheets), multiply by two.

BOOK PAPER
(basis 25 × 38) Ream of 500 sheets

30	Multiply the square inch area of the irregular size by	.03518
35	"	.03684
40	"	.04210
45	"	.04737
50	"	.05263
60	"	.06316
70	"	.07367
80	"	.08421
100	"	.10526
120	"	.12631

COVER PAPER
(basis 20 × 26) Ream of 500 sheets

25	Multiply the square inch area of the irregular size by	.0480
35	"	.0673
40	"	.0769
50	"	.0961
60	"	.1154
65	"	.1250
80	"	.1539
90	"	.1730
94	"	.1808

WRITING
(basis 17 × 22) Ream of 500 sheets

13	Multiply the square inch area of the irregular size by	.0348
16	"	.0428
20	"	.0535
24	"	.0642
28	"	.0749
32	"	.0856
36	"	.0962
40	"	.1070
44	"	.1177

BRISTOL
(basis 22½ × 28½) Ream of 500 sheets

110	Multiply the square inch area of the irregular size by	.1137
125	"	.1292
150	"	.1550
175	"	.1809
200	"	.2067
225	"	.2326

INDEX
(basis 25½ × 30½) Ream of 500 sheets

90	Multiply the square inch area of the irregular size by	.1157
110	"	.1414
140	"	.1800
170	"	.2185

TO FIND M WEIGHT (1000 SHEETS)

EXAMPLE:
Find the weight of 1000 sheets, 32" x 51", basis 60 lb., book paper.

```
     32                    1632   square inches
   × 51                  × .06316 Factor from table above for book paper
    ‾‾‾                   ‾‾‾‾‾‾‾‾
     32                  103.0771 = approximate ream weight (500 sheets)
   160                        × 2
  ‾‾‾‾‾                   ‾‾‾‾‾‾‾‾
  1632 square inches      206.1542 = approximate weight per 1000 sheets, 32 x 51
```

ANSWER:
1000 sheets of 32 x 51 — basis 60 lb. book paper will weigh approximately 206 lbs.

Fig. 22-25. These tables can be used to find the ream (500 sheets) weight and M (1000 sheets) weight of any irregular size paper. (S.D. Warren Co.)

takes place to balance moisture content of each. The asphalt-lined paper wrapping on the box or skid is opened prior to use on the press. This is done to avoid moisture loss or gain.

INK SETTING AND DRYING

Offset inks are manufactured to print and dry on a wide range of papers. The paper must react the same for a given set of printing conditions. The uniformity of paper is the responsibility of the paper mill, Fig. 22-26.

Uniformity must be maintained each time the same kind of paper is made. Chemists are responsible for quality control in papermaking mills. Ink setting and drying can be a serious problem when paper quality is poor or in cases where paper and ink are not compatible.

A

COMPUTERIZED QUALITY CONTROL SYSTEMS AND OPERATING CONSOLES

B

Fig. 22-26. A—Quality control of paper manufacturing is responsibility of chemists and technicians in paper mill. B—Computerized monitoring systems are used to maintain quality control on papermaking machine.
(Hammermill Paper Group; The Mead Corp.)

Moisture absorbency and resistance

Offset paper must be capable of absorbing small amounts of moisture from the inked image of the offset plate. *Moisture absorbency* is necessary so that the surface of the paper can accept and bond with the ink on the image. Paper *moisture resistance* is needed to prevent fiber separation from the sheet when the fibers come in contact with moisture during printing. A balance of moisture absorbency and resistance is necessary.

Papers with coatings too sensitive to moisture tend to stick to the offset press rubber blanket. This requires frequent press stops for blanket washings. Poor quality printing impressions are the result.

Paper opacity

The term *show-through* is commonly used in referring to paper's opacity. This means the less printing that shows through the reverse side of a piece of paper, the GREATER its *opacity*. Closely woven fibers in paper make it more opaque because less light can pass through it.

Paper bulk is a term used in connection with opacity. The combination of fibers and fillers provide the qualities to produce a given bulk of paper. In simple terms, bulk refers to the degree of thickness of paper.

PAPER GRADES

There are many kinds or *grades* of paper. The identifying features include use, appearance, quality, processing history, raw materials used, or a combination of these features.

The classification of all printing paper is determined by the general use of the particular grade. For example, this means bond papers will be thin, lightweight, made to print on one side, and with one-sided finish or texture.

Fig. 22-27 shows a chart which is broken down into eight sections. Each section represents the distinct characteristics and the various applications for the most commonly used printing papers.

Bond paper

Bond paper is used primarily for business letters and forms. Bond paper got its name because the "Old Lady of Treadneedle Street" (The Bank of England) was the first one to approve this new type of paper for the printing of bonds.

Bond paper has a medium finish surface that accepts ink easily from a typewriter or pen. Three grades of bond paper are available, including layout, medium, and heavy.

Most bond paper is used by forms printers, in-plant printers, and stationers. The basic size of bond paper is 17 x 22 inches.

Note! Do not confuse 8 1/2 x 11 inch business stationery with the basic size dimensions of bond paper.

Uncoated book paper

Magazines, journals, and hardbound books are printed on *book paper*. This type of paper is available in a variety of colors and weights. Book paper is manufactured in either antique, smooth, vellum, or woven finishes. The basic size of book paper is 25 x 38 inches.

Coated book paper

Offset lithography requires the use of water. Therefore, water-resistant paper is needed. Among these is *coated book paper*, which is stable during changes in humidity and temperature.

Sizing is added to coated book paper for moisture resistance. The surface of coated book paper is also

	Bonds			Duplicator		Mimeograph		Offsets					Covers	
	Hammermill Bond	Hammermill Scriptmark	Hammermill Fore Bond	Hammermill Duplicator	Hammermill Fore Duplicator	Hammermill Mimeo-Bond	Hammermill Fore Bond	Hammermill Offset Opaque Wove	Hammermill Offset Opaque Vellum	Hammermill Offset Opaque Lustre	Hammermill Offset Opaque Pearl	South Shore Offset	Hammermill Cover	Hammermill Dura-Glo
Announcements	(S)	(S)						(P)	(P)	(P)	(P)	(E)	(P)	(P)
Annual Reports (T)	(S)	(S)						(P)	(P)	(P)	(P)			
Annual Reports (C)													(P)	(P)
Assembly instruc.	(P)		(P)	(P)	(E)	(P)	(E)	(P)	(P)	(P)	(P)	(E)		
Bills (Consumer)	(S)		(P)									(E)	(S)	
Bill Stuffers	(P)		(E)	(P)	(E)	(P)	(E)	(P)	(P)	(P)	(P)	(E)		
Bonds	(S)	(S)												
Booklets (Text)	(S)	(S)	(E)	(P)	(E)	(P)	(E)	(P)	(P)	(P)	(P)	(E)		
Booklets (Covers)													(P)	(P)
Briefs	(P)							(P)	(P)	(P)	(P)			
Broadsides	(P)		(E)			(P)	(E)	(P)	(P)	(P)	(P)	(E)	(S)	
Brochures	(S)	(S)	(E)	(P)	(E)	(P)	(E)					(E)	(S)	
Business Forms	(P)		(E)	(P)	(E)	(P)	(E)	(S)	(S)	(S)	(S)	(S)	(P)	
Calendars	(P)		(E)	(P)	(E)	(P)	(E)	(P)	(P)	(P)	(P)	(E)	(P)	(P)
Calling Cards													(P)	(P)
Catalogs (Text)	(S)		(P)	(P)	(E)	(P)	(E)	(P)	(P)	(P)	(P)	(E)		
Catalogs (Covers)													(P)	(P)
Certificates	(P)	(P)	(S)					(P)	(P)	(P)	(P)	(S)		
Checks														
Circulars	(P)	(S)	(E)					(P)	(P)	(P)	(P)	(E)		
Co. Reports (C)													(P)	(P)
Co. Reports (T)	(S)			(P)	(E)	(P)	(E)	(P)	(P)	(P)	(P)	(S)		
Dealer Notices	(P)		(S)	(P)	(E)	(P)	(E)	(P)	(P)	(P)	(P)	(S)		
Direct Mail Pcs.	(P)	(P)	(S)					(S)	(S)	(S)	(S)	(S)		
Display Cards													(P)	(P)
Document Papers	(P)	(P)												
Deposit Slips	(S)		(S)									(S)		
Enclosures	(P)	(S)				(P)	(E)	(P)	(P)	(P)	(P)	(E)		
Envelope Stuffers	(P)		(E)	(P)	(E)	(P)	(E)	(P)	(P)	(P)	(P)	(E)		
Envelopes	(P)	(P)							(P)					
File Cards														
File Folders													(P)	(P)

(Continued)

(P)—Primary best looking results
(E)—Economy, good looking results at economy price
(S)—Secondary paper not specially designed for this purpose but will give satisfactory results

Fig. 22-27. This chart shows characteristics and various end-use applications for most commonly used printing papers.
(Hammermill Paper Group)

treated to resist picking. *Picking* refers to the lifting of the paper surface during printing. It occurs when the pulling force (tack) of the ink is greater than surface strength of the paper.

Coated book paper is coated with three types of *finishes,* including dull, semi-gloss, and high-gloss. Coated book paper comes in white and assorted colors. The basic size of coated book paper is 25 x 38 inches.

For consumers who are budget conscious, there has been an increasing interest in *ultra lightweight* coated book papers with a basis weight of under 34 pounds.

These lighter weight papers require less postage and the paper itself can be less expensive per page, resulting

	Lightweight Opaque			Translucents	P.P. Copier				Dual Purpose	Safety		Ledgers	
	Indexes												
	Hammermill Index	Hammermill Post Card	Hammermill Hy*O*Lite	Hammermill Translucent	Hammermill Xerocopy	Hammermill Fore Xerocopy	Hammermill Electrocopy	Hammermill Fore Electrocopy	Hammermill Savings Xeroprint	Hammermill Sentry Safety	Hammermill Scan-Mate	Hammermill Ledger	Hammermill Fore Ledger
Announcements	S		P										
Annual Reports (T)			P							P			
Annual Reports (C)													
Assembly Instruc.			P						E				
Bills (Consumer)		P	S									P	
Bill Stuffers			P						E				
Bonds										P			
Booklets (Text)			P						E				
Booklets (Covers)	S												
Briefs			P										
Broadsides			P						E				
Brochures			P						E	S			
Business Forms									E	P	P		
Calendars	P	S	P						E				
Calling Cards	S												
Catalogs (Text)			P						E				
Catalogs (Covers)	S												
Certificates									S	P			
Checks										P	S		
Circulars			P						E				
Co. Reports (C)	S												
Co. Reports (T)			P										
Dealer Notices			P						S	P			
Direct Mail Pcs.									S				
Display Cards	P	S											
Document Papers										P			
Deposit Slips									S		P		
Enclosures													
Envelope Stuffers			P						E				
Envelopes													
File Cards	P	S											
File Folders	P	S											

P —Primary best looking results
E —Economy, good looking results at economy price
S —Secondary paper not specially designed for this purpose but will give satisfactory results

Fig. 22-27. Continued.

in substantial savings. The Europeans, who have high postage, have developed and produced very low weight (as low as 22 pounds), high printability paper. They are not exporting this paper to the United States and Canada. In the United States, a few papermaking mills can also manufacture this type of paper.

Even though ultra light paper is more expensive per pound, it normally takes less paper (in pounds) to print a job. This too can reduce printing costs.

Text paper

As the name suggests, *text paper* is popular for its surface texture and attractive colors. Text paper is similar to book paper. It is often watermarked and deckle-edged. A *deckle edge* is a feathered edge left untrimmed when the paper is manufactured.

Text paper is used for books, booklets, and brochures. The basic size of text paper is 25 x 38 inches.

Cover paper

Cover paper is used primarily to cover and protect other printed material. It is heavy, strong, easy to fold, and available in a wide variety of colors. Cover paper is able to withstand rough handling. It is used for booklets, catalogs, pamphlets, covers, programs, and mailing pieces. Many different surface finishes are available in cover papers. The basic size of cover paper is 20 x 26 inches.

Index bristol is another type of cover paper. It is used for business forms, menus, booklet covers, postcards, and mailing pieces. It is manufactured in both smooth and antique finishes in many different colors. This kind of paper is used when rigidity, ruggedness, and erasability are important. The basic size of index bristol is 25 1/2 x 30 1/2 inches.

Tagboard paper is classified as a cover paper and is used primarily for jobs requiring sturdiness. This paper has long fibers which make it very strong. It is 100 percent sulphate pulp. Tagboard is used for file folders, shipping tags, and protective envelopes. It has excellent folding qualities, is water resistant, and has a surface adaptable to writing, stamping, or printing. Manila and white are the most common colors. The basic size of tagboard paper is 24 x 36 inches.

Carbonless paper

Carbonless papers are used to transfer and copy written, printed, or typed images between sheets. They have replaced many carbon copy interleaved forms. They simplify paper work, produce clean, smudge-free copies, and eliminate the messiness associated with the use and disposal of carbon papers.

Carbonless papers are technically different from regular papers. They incorporate a chemical transfer system and use a reaction between two different chemical coatings to transfer images.

The back side of the top sheet in the carbonless set is coated with encapsulated chemicals. This sheet is designated *CB* (coated back). The front and back sides of intermediate sheets have a receptor coating and an encapsulated coating respectively. This sheet(s) is designated *CFB* (coated front-and-back). The last sheet of the set has a receptor coating only. This sheet is designated *CF* (coated front).

When pressure is applied to the top sheet of the set by typing, writing, or letterpress crash printing, an image is formed. This occurs because of the reaction between the chemicals liberated from the collapsed capsules and the contacting receptor coating. Special precaution must be taken not to damage the encapsulated chemicals by excessive pressure and friction during handling, printing, cutting, and trimming.

The proper side of each paper used in the set must be printed so that the sheets are collated in their required sequence. Manufacturers of carbonless papers provide specific instructions for the printing and processing of their carbonless papers.

The basic size of carbonless paper is 17 1/2 x 22 1/2 inches. The most common cut size is 8 1/2 x 11 inches, which is used ideally by most quick printers for the printing of forms.

Newsprint paper

Newsprint paper is used primarily for the printing of newspapers. Newsprint is made almost entirely of groundwood pulp. Groundwood pulp has a low tearing strength so the paper manufacturer often adds some stronger pulp to hold the sheet together. Other uses for newsprint include direct-mail advertising, rough sketching, layouts, and all types of directories.

Newsprint is prepared in large rolls for rotary web-offset printing operations. Very little newsprint is manufactured for use in flat sheets. The basic size of newsprint is 24 x 36 inches. Roll sizes vary according to the type of press equipment.

Paperboard

A thick cardboard, *paperboard* is popular for many kinds of printed pieces. It is suitable for die cutting, creasing, and folding. Paperboard is usually formed into various shapes for its many end uses. It is manufactured in a wide range of densities. This makes paperboard very strong and durable.

The thinner paperboards are usually printable by offset. The thicker paperboards are printed on letterpress or screen process, flat-bed presses.

Some of the very heavy sheets of paperboard are called *blanks*. These are used in package printing, outdoor and transit advertising, point-of-purchase displays, and many commercial advertising programs.

The basic size of paperboard is 22 x 38 inches. Paperboard is manufactured in thicknesses up to 0.056 inch (1.4 mm). The various grades and thicknesses of paperboard are shown in Fig. 22-28.

Number of Plies	Caliper	Approximate Wt. (1000 Sheets)
	Plain Blanks	
3	.015	280
4	.018	330
5	.021	360
6	.024	420
8	.030	520
10	.036	600
	Coated Blanks	
3	.015	340
4	.018	420
5	.021	460
6	.024	530
8	.030	650
10	.036	760
	Railroad Board	
4	.018	400
6	.024	530
8	.030	650
	Tough Check	
3	.012	310
4	.018	430
6	.024	550
8	.030	680
	Thick China	
	.011	300

Fig. 22-28. Chart shows various grades and thicknesses of paperboard. (Boise Cascade Paper Group)

PAPER SWATCHBOOKS

Paper swatchbooks are valuable paper specification tools for printers because they provide samples of every color and each weight of paper available in a particular grade. However, not every combination of available weight and color will always be shown as a sample. This means that the caption next to a sample only identifies one available weight for the color shown. For this reason, always refer to the standards information for a complete listing of available combinations of weight and color.

While the charts that list standards may be structured differently by each manufacturer, they all provide the same basic details. The chart shown in Fig. 22-29 represents one way that standards might appear in a swatchbook. The standards are described as follows.

Swatchbook weight and grade

This information states the basis weights available, and defines the paper by grade. This is often followed by a standard dimension used throughout the industry for calculating basis weight in that particular grade of paper. These weights are always based on how much a *ream* (500 sheets) of the paper weighs when cut to the standard basis size. For example, the standard basis size for all text papers is 25 x 38 inches. This means that a ream of basis 80 text sheets in that size weighs 80 pounds.

Swatchbook sheet size and grain direction

In the example shown in Fig. 22-29, this is one of the two sheet sizes offered in a basis 80 text. The "35" has been underlined to indicate that the grain direction of this sheet runs along the longest dimension. This means that the paper has its grain running in the long direction, or *grain long*. Grain direction may also be indicated by setting the dimension in a **boldface type**, or by the letters *L* or *S* for long and short.

If the underline of boldface appears in the shortest dimension, the paper is *grain short*. When both dimensions are keyed to indicate grain, the paper is available in a choice of either grain direction.

Following the sheet dimensions, an *M weight* is often provided. This figure states the weight of 1000 sheets of this paper in the size listed. In the example in Fig. 22-29, it is 136 pounds. The M weight is helpful in estimating the cost since paper is priced by the pound.

Weight and Size	Sheets per Carton (or Package)	Weight per Carton	White Ivory Natural	Gray Blue Brown	Yellow Red Black
Basis 80 Text (25″ x 38″)					
23 x <u>35</u> (136M)	1000	136	X		
25 x <u>38</u> (160M)	750	120	X	X	X
Basis 80 Cover (20″ x 26″)					
<u>23</u> x 35 (201M)	750	150	X	X	
26 x <u>40</u> (320M)	400	128	X	X	X

Fig. 22-29. Note example of how important information about printing papers is displayed in paper merchant catalogs. Information like this is essential for printers and designers. (*Step-by-Step Graphics* Magazine)

Swatchbook sheet and weight per carton

Printers use the sheets per carton and weight information to prepare an accurate order for paper and to estimate costs. In the example in Fig. 22-29, the printer or estimator would know that a job requiring 10,000 sheets of 23 x 35 inches basis 80 text is an order for 10 cartons with a combined weight of 1360 pounds.

Swatchbook paper color

The printer must specify paper color on every printing order. It is important to check on the availability of colors since highly unusual papers are run as mill items and are probably not inventoried at the paper merchant's store. This means that the printer may be required to place a very large minimum order for the paper, and may have to wait ten to twelve weeks to get the paper.

UNIVERSAL BASIC SIZE (MM SYSTEM)

For a number of years, printers and paper manufacturers have proposed to substitute a *universal basic size (MM system)* for the awkward basis weight practice. A universal size of 25 x 40 inches (1000 square inches) would become the standard for weighing 1000 sheets. The weight of 1M (1000) sheets in the universal size, 1M square inches, would be used as an identifying number for each type of paper. It would also be a simple method of calculating any amount and size of paper of that number. It could be a means of comparing different kinds of paper by weight. Refer to Fig. 22-30.

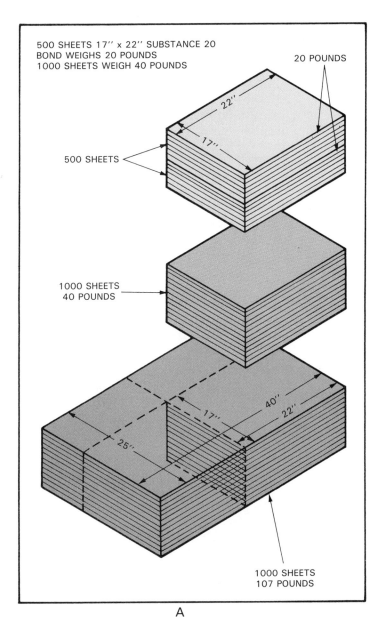

EQUIVALENT WEIGHTS (Pocket Pal)

In reams of 500 sheets, basis weights in bold type

Grade of Paper	BOOK 25 x 38	BOND 17 x 22	COVER 20 x 26	BRISTOL 22½ x 28½	INDEX 25½ x 30½	TAG 24 x 36	GRAMMAGE (gsm)
BOOK	**30**	12	16	20	25	27	44
	40	16	22	27	33	36	59
	45	18	25	30	37	41	67
	50	20	27	34	41	45	74
	60	24	33	40	49	55	89
	70	28	38	47	57	64	104
	80	31	44	54	65	73	118
	90	35	49	60	74	82	133
	100	39	55	67	82	91	148
	120	47	66	80	98	109	178
BOND	33	**13**	18	22	27	30	49
	41	**16**	22	27	33	37	61
	51	**20**	28	34	42	46	75
	61	**24**	33	41	50	56	90
	71	**28**	39	48	58	64	105
	81	**32**	45	55	67	74	120
	91	**36**	50	62	75	83	135
	102	**40**	56	69	83	93	151
COVER	91	36	**50**	62	75	82	135
	110	43	**60**	74	90	100	163
	119	47	**65**	80	97	108	176
	146	58	**80**	99	120	134	216
	164	65	**90**	111	135	149	243
	183	72	**100**	124	150	166	271
BRISTOL	100	39	54	**67**	81	91	148
	120	47	65	**80**	98	109	178
	148	58	81	**100**	121	135	219
	176	70	97	**120**	146	162	261
	207	82	114	**140**	170	189	306
	237	93	130	**160**	194	216	351
INDEX	110	43	60	74	**90**	100	163
	135	53	74	91	**110**	122	203
	170	67	93	115	**140**	156	252
	208	82	114	140	**170**	189	328
TAG	110	43	60	74	90	**100**	163
	137	54	75	93	113	**125**	203
	165	65	90	111	135	**150**	244
	192	76	105	130	158	**175**	284
	220	87	120	148	180	**200**	326
	275	109	151	186	225	**250**	407

500 SHEETS 17'' x 22'' SUBSTANCE 20 BOND WEIGHS 20 POUNDS
1000 SHEETS WEIGH 40 POUNDS

20 POUNDS

22''

17''

500 SHEETS

1000 SHEETS 40 POUNDS

40''

17'' 22''

25''

1000 SHEETS 107 POUNDS

A

B

Fig. 22-30. A—Universal basic size of 25 x 40 inches (1000 square inches). B—Equivalent weights of paper. (International Paper Co.)

ISO METRIC SIZES OF PAPER

Standard United States paper sizes have evolved over many years. Some sizes exist to satisfy a particular end-use requirement. For example, bond paper is often sold in size 17 x 22 inches because it will cut down to 8 1/2 x 11 inches for letterheads after printing. The most popular size for multi-color printing on coated paper is 25 x 38 inches because it will accommodate a 16-page, 8 1/2 x 11 inch signature with space left over for bleed and trim.

There are other influencing factors in paper sizes. Today, nearly all printing presses are manufactured outside the United States, and their sizes are expressed in metric terms. For instance, printers will commonly call a multi-color Heidelberg offset press a 40-inch press, but it is actually 102 centimeters or about 40 1/4 inches wide. A press of this size prints 25 x 38 inch paper with no difficulty, but it is capable of printing a sheet as large as 70 x 100 centimeters (approximately 27 1/2 x 39 3/8 inches). This is the preferred sheet size in Europe and other parts of the world.

In an effort to standardize printing paper sizes, the International Standards Organization (ISO) has developed a system being adopted in many countries using the metric standard. Look at Fig. 22-31.

The system is based upon the principle that a rectangle's side is in the ratio of 1:2. In the A Series, for example, A3 is half the area of A2 and double the area of A4. It has the unique advantage that the proportion of height to width remains the same regardless of size. This means that an illustration or photograph can be reduced or enlarged to fit the smaller or larger sizes in the same series.

While the United States still thinks of 8 1/2 x 11 inch as a standard letterhead and brochure size, ISO promoters have adopted the A4 size which is 8 1/4 x 11 1/2 inch.

Metric sheet dimensions are stated in millimeters (mm). To convert inches to millimeters, multiply the number of inches by 25.4 and round off to the nearest whole millimeter. For example, to convert 8 1/2 x 11 inches to millimeters, follow these steps:

1. 8 1/2 inches x 25.4 = 215.9 mm = 216 mm (rounded off).
2. 11 inches x 25.4 = 279.4 mm = 279 mm (rounded off).

Therefore, the familiar 8 1/2 x 11 inch size becomes 216 x 279 mm.

Fig. 22-32 shows the ISO sizes for envelopes. They are designed to hold the A Series sheets unfolded or folded. For example, an A1 sheet will fit unfolded into

Inches	=	mm	Inches	=	mm	Inches	=	mm	Inches	=	mm
6		152	14 1/4		362	22 1/2		572	30 3/4		781
6 1/4		159	14 1/2		368	22 3/4		578	31		787
61/2		165	14 3/4		375	23		584	31 1/4		794
6 3/4		171	15		381	23 1/4		591	31 1/2		800
7		178	15 1/4		387	23 1/2		597	31 3/4		806
7 1/4		184	15 1/2		394	23 3/4		603	32		813
7 1/2		191	15 3/4		400	24		610	32 1/4		819
7 3/4		197	16		406	24 1/4		616	32 1/2		826
8		203	16 1/4		413	24 1/2		622	32 3/4		832
8 1/4		210	16 1/2		419	24 3/4		629	33		838
8 1/2		216	16 3/4		425	25		635	33 1/4		845
8 3/4		222	17		432	25 1/4		641	33 1/2		851
9		229	17 1/4		438	25 1/2		648	33 3/4		857
9 1/4		235	17 1/2		445	25 3/4		654	34		864
9 1/2		241	17 3/4		451	26		660	34 1/4		870
9 3/4		248	18		457	26 1/4		667	34 1/2		876
10		254	18 1/4		464	26 1/2		673	34 3/4		883
10 1/4		260	18 1/2		470	26 3/4		679	35		889
10 1/2		267	18 3/4		476	27		686	35 1/4		895
10 3/4		273	19		483	27 1/4		692	35 1/2		902
11		279	19 1/4		489	27 1/2		699	35 3/4		908
11 1/4		286	19 1/2		495	27 3/4		705	36		914
11 1/2		292	19 3/4		502	28		711	36 1/4		921
11 3/4		298	20		508	28 1/4		718	36 1/2		927
12		305	20 1/4		514	28 1/2		724	36 3/4		933
12 1/4		311	20 1/2		521	28 3/4		730	37		940
12 1/2		318	20 3/4		527	29		737	37 1/4		946
12 3/4		324	21		533	29 1/4		743	37 1/2		953
13		330	21 1/4		540	29 1/2		749	37 3/4		959
13 1/4		337	21 1/2		546	29 3/4		756	38		965
13 1/2		343	21 3/4		552	30		762	*		
13 3/4		349	22		559	30 1/4		768			
14		356	22 1/4		565	30 1/2		775			

A

*To obtain other sizes — multiply inches x 25.4, and round off to nearest whole millimeter.

(Continued)

Fig. 22-31. A—International Standards Organization (ISO) has developed a universal system of metrics. B—Metric system used for paper is based upon the principle that a rectangle's side is a ratio of 1:2. This example shows A Series ISO sizes for paper.

484

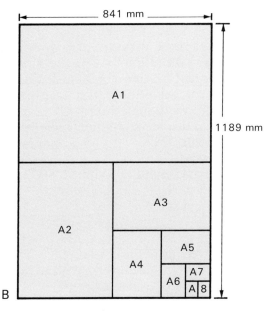

841 mm

A1

1189 mm

A2

A3

A4

A5

A6 A7

A8

B

Fig. 22-31. Continued.

a C1 envelope. If it is folded in half, it will fit in a C2 envelope.

ISO BASIC WEIGHT SYSTEM

Paper using the United States customary system of measurement is presently sold by the pound-weight per ream. The international paper sizes relate all basic weights to the A0 sheet size.

The ISO *metric basic weight* system for paper is expressed in grams per square meter (g/m^2). A rough conversion of United States customary weights to ISO metric for basic book weights can be made by multiplying the weight in pounds by 1.5. For example, a 50-pound book paper is about the same as 75 g/m^2.

PAPER REQUIREMENT CHECKLIST

Reproduction with small or large offset presses can be done effectively on almost any kind of paper. However, for quality work, good offset papers should meet certain basic requirements. These include the following:

1. **Precision trimming** — Inaccurate trimming of paper is perhaps one of the most common causes of trouble on the press. Misregister, jamming of paper as it feeds, and improper stacking are often the result of bowed or off-square paper.
2. **Flatness** — Absolute flatness is a requirement for offset work. Paper with wavy edges cannot pass through the cylinders without distortion that can result in misregister and wrinkles.
3. **Moisture** — Papers are packaged at five to six percent moisture content for best results under average conditions. If stored properly, careful control of moisture during manufacturer and packaging can help keep papers at the right moisture content until ready for use.
4. **Curl** — Offset papers should be received perfectly flat. However, exposure to moisture either in the shop or when the paper comes in contact with the moisture from the offset blanket can cause curl. A change in moisture content causes paper to expand or shrink more on the wire side than on the felt size. MORE MOISTURE makes paper curl toward the felt side and with the grain. LESS MOISTURE causes curl toward the wire side, also with the grain.
5. **Grain** — Paper is specified grain long or grain short. Therefore, if a sheet is grain long, it will widen

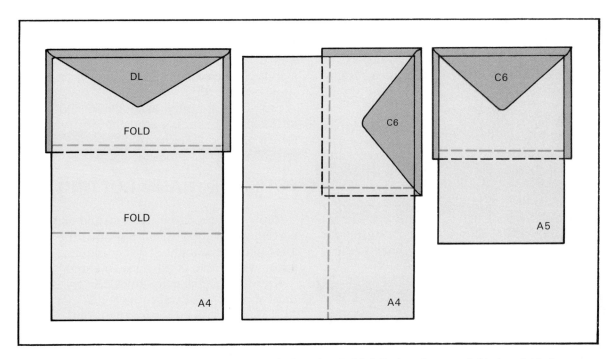

Fig. 22-32. ISO sizes for envelopes are designed to hold A Series sheets unfolded or folded.

more than it will lengthen under added moisture. Where grain is to be considered for good folding, where register is not a problem, or where economy of layout requires it, grain short papers can be used.

6. **Pick resistance** — Good pick resistance is required of offset papers because offset inks are usually tacky. Thinner ink films cause a greater pull on the surface of the paper. Also, resilient offset blankets conform well and cause more pull of the ink. Picking always means that the surface of the paper is ruptured either by lifting large areas of surface fibers or by blistering which causes roughness.

7. **Lint and dust** — Linting often occurs on uncoated papers. This is a result of loosely bonded fibers which break off and are carried by the offset blanket to the ink rollers. Ink becomes so contaminated with fiber particles that it will not flow evenly. This problem requires a complete press wash up. Paper dust usually comes from loose fibers or bits of paper lying between sheets. This is caused by improper cutting or trimming and by dull places. The blade tears the paper rather than cutting it cleanly.

8. **Piling** — Piling occurs in two ways. In uncoated papers, mineral fibers are picked up from the paper surface and form an abrasive layer on the blanket. This eventually wears the desensitized film off the printing plate and causes scumming. In the case of coated papers, the coating adhesive is too water soluble. Sufficient moisture resistance is a prerequisite of good offset papers to prevent separation of the coating even in extensive color work.

9. **Controlled ink drying** — In some cases, paper may contribute to drying difficulties if it contains an excessive amount of moisture. In addition, it may not possess uniform ink receptivity from sheet to sheet. Gloss inks require higher ink resistance than standard inks. This is also true in press varnishing operations.

ENVELOPES

Envelopes are manufactured in many styles and sizes, Fig. 22-33. Each style has a special use and is usually ordered by number. For example, a No. 10 envelope is for business use and measures 4 1/8 x 9 1/2 inches. A No. 6 3/4 commercial envelope measures 3 5/8 x 6 1/2 inches. Envelopes having a transparent window are used for business purposes, such as mailing statements and invoices.

Envelopes used for social invitations and wedding announcements are called *baronial envelopes*. A No. 5 baronial envelope measures 4 1/8 x 5 1/8 inches. Cards and blank sheets are available for printing and inserting in the envelopes.

Heavy Kraft or manila envelopes are manufactured

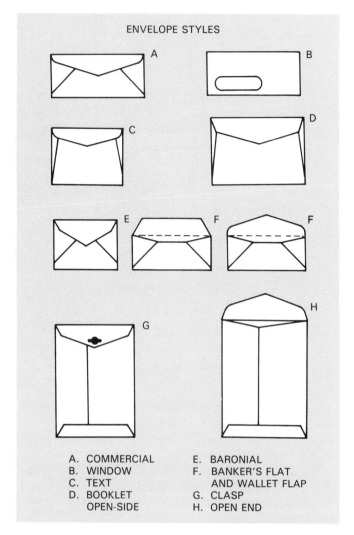

ENVELOPE STYLES

A. COMMERCIAL
B. WINDOW
C. TEXT
D. BOOKLET
   OPEN-SIDE
E. BARONIAL
F. BANKER'S FLAT
   AND WALLET FLAP
G. CLASP
H. OPEN END

Fig. 22-33. Each style and size of envelope fits a specific purpose. Envelopes are available in white and a few selected colors. (Leslie Paper Co.)

for mailing magazines, pamphlets, reports, books, and similar material. One of the most common types is the manila clasp envelope which comes in several standard sizes.

Envelopes are often available in colors. They vary in thickness according to their construction and the weight of paper. Envelopes are usually sold 250 or 500 to a box depending on style and size.

ESTIMATING AND CUTTING PAPER

Paper is an expensive material and makes up a large percentage of the cost of a printed job. The printer must plan the amount of paper required and how the paper will be cut in advance to avoid waste.

Keep in mind that many quick printers use precut standard sizes of paper, such as 8 1/2 x 11, 8 1/2 x 14, 11 x 17, and 5 x 8 inches for their jobs. These familiar cut sizes, and many more, are available for the quick printers at many paper supply outlets. In many instances, however, the printer must first determine the dimensions of the sheets needed for the printed job. Then, the size of the paper from which these will be cut is determined.

The printer then calculates how many pieces of paper can be cut from one full sheet of basic size paper. To do this, the dimensions of the sheet needed are written under the dimensions of the basic size sheet. The *cancellation method* is used to determine the greatest number of pieces of paper that can be cut from a single, full-size sheet.

In Fig. 22-34, the basic paper size is 17 x 22 inches and the size needed for the job is 8 1/2 x 11 inches. The paper can be cut in two ways, as shown in A and B. In A, the 11 inch lengths will be cut from the 17 inch side of the basic size sheet. Eleven goes into 17 once, so one is written below. Eight and one-half goes into 22 twice, so two is written below. The answers are multiplied to give the total number of press sheets (1 × 2 = 2).

17 x 22	17 x 22
11 x 8½	8½ x 11
1x2=2	2x2=4
A	B

Fig. 22-34. Printer uses simple mathematics to determine amount of paper required for a job and how paper will be cut in advance to avoid unnecessary waste.

Fig. 22-35 shows the paper cut this way. Notice the waste. However, if the 8 1/2 inch length is cut from the 17 inch side, as shown in Figs. 22-34B and 22-35B, two additional pieces can be obtained from the same basic size sheet. It is important that you check both dimensions to determine the best cut.

Estimating spoilage allowance

In cutting paper for a printed job, it is also necessary to add a percentage of extra sheets to the total pieces needed. This is called *spoilage allowance* and is included for replacing soiled or misprinted sheets and for use during *make ready* or when setting up the press.

In addition, allowance is made if more than one run is planned for the job through the press for a second color or if finishing operations, such as trimming, folding, drilling, and collating, are required. A table of allowances for spoilage is shown in Fig. 22-36.

One example, assume that 2000 finished copies of a single color invitation are required. Spoilage allowance is six percent. Multiplying six percent by 2000 yields a spoilage allowance of 120, and 2000 + 120 = 2120 sheets. You would need 2120 sheets to produce a run of 2000 sheets for the customer. Spoilage allowances are sometimes lower or higher, depending

primarily on the type of equipment and the competency of the press operators.

Estimating combination cuts

Some sheet sizes for various types of jobs do not divide evenly into the basic size paper from which they are to be cut. This requires a *combination cut* which provides the greatest number of press-size sheets while reducing waste to a minimum.

As an example, if a 32 x 44 inch sheet must be cut into the greatest number of 8 1/2 x 11 inch sheets with the least amount of waste, a combination cut is used. A and B in Fig. 22-37, shows the amount of waste involved. By drawing a cutting diagram, you can see that the waste is of such a size as to permit cutting four more pieces of 8 1/2 x 11 inch paper. These four pieces must be cut in opposite directions, as shown in Fig. 22-37C.

Drawing a cutting diagram before cutting the press sheets avoids unnecessary waste. In this example of a combination cut, it is assumed that the grain direction is not a factor in the job. This may not be the case for a job where grain direction is important to folding or other bindery operations.

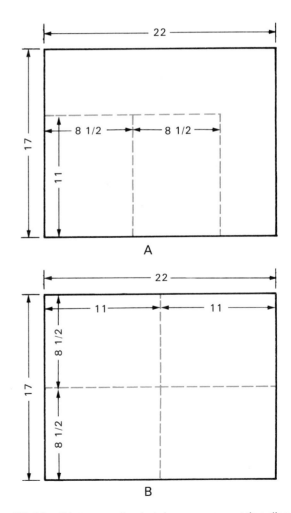

Fig. 22-35. Printer usually sketches a paper cutting diagram to verify mathematical calculations. In this illustration, method B is the best choice since there is no waste as in method A.

SHEET-FED OFFSET*	1000	2500	5000	10,000	25,000 AND OVER
Single Color Equipment					
One color, one side	8%	6%	5%	4%	3%
One color, work and turn or work and tumble	13%	10%	8%	6%	5%
Each additional color (per side)	5%	4%	3%	2%	2%
Two Color Equipment					
Two colors, one side	—	—	5%	4%	3%
Two colors, work and turn or work and tumble	—	—	8%	6%	5%
Each additional two colors (per side)	—	—	3%	2%	2%
Four Color Equipment					
Four colors, one side only	—	—	—	6%	5%
Four colors, work and turn or work and tumble	—	—	—	8%	7%
Bindery Spoilage					
Folding, stitching, trimming	4%	3%	3%	2%	2%
Cutting, punching or drilling	2%	2%	2%	2%	2%
Varnishing and gumming	7%	5%	4%	3%	3%

*Percentage represents press size sheets, not impressions. Figures do not include waste sheets used to run up color as it is assumed that waste stock is used for this purpose.

WEB OFFSET	WASTE % OF TOTAL IMPRESSIONS*
Press Run	
Up to 25M	18
Over 25M to 50M	15
Over 50M to 100M	13
Over 100M to 200M	11
Over 200M	9
Penalties to be added:	
For each additional web over 1	1
For using 2 folders	1
For 3, 4 or 5 colors	1
For Coated Paper	5
For Light Papers under 40#	2
For Heavy Papers over 60#	2

*Includes waste for core, wrappers, and damaged paper, which is estimated at 2½%.
The chart is for blanket-to-blanket presses running two colors on two sides of the web, on uncoated paper 40 to 60 lbs. and using one folder. The chart includes make ready spoilage.

Fig. 22-36. Table gives spoilage allowances for sheet-fed offset and web-fed offset.
(S.D. Warren Paper Co.)

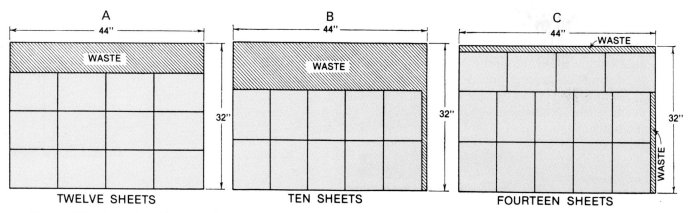

Fig. 22-37. A cutting diagram will reveal best cut with least amount of spoilage when calculating combination cuts.

Ordering paper

Paper is generally ordered after careful consideration of samples of catalog sheets supplied by the paper merchant. A price catalog is also supplied for costing purposes. Paper is generally ordered by direct telephone conversation with the paper merchant.

Sample price tables are shown in Fig. 22-38. The following information is required when ordering most types of paper.

1. The *stock number* appears on the catalog page and can be used to reference the kind of paper needed.
2. The *quantity* or number of sheets, packages, cartons, bundles, skids, or carload lots are specified when ordering paper.
3. The *sheet size* indicates the basic size required, or one of the standard sizes available. Size is important in relation to press sheet size and waste involved.
4. *Basis weight* is the weight of 500 sheets of paper cut to the basic size. The *M weight* is the weight of 1000 sheets of paper cut to the basic size. Always indicate text or cover weight for the type of paper requested.

5. Paper is manufactured in many different *colors.* Some colors are more expensive than others. Indicate the exact name of the color as shown on the sample in the catalog.
6. *Brand name* indicates the specific paper mill or manufacturer's name. An alternate brand may sometimes be necessary.
7. *Finish* indicates a dull, gloss, smooth, vellum, antique, or eggshell paper surface.
8. The *grain direction* or orientation of paper fibers should be specified, especially for jobs where folding is involved. The grain direction is indicated in the price catalog by a line under one of the two dimensions (25" x 38") or a notation stating grain long (L) or grain short (S).
9. *Packing* identifies the paper being supplied in standard size cartons or junior size (containing 10 reams of cut sizes 8 1/2 x 11 or 8 1/2 x 14 inches), in cases, or on skids. Whatever the packing, be sure to specify whether the paper is to be sealed in packages or unsealed (marked).
10. Be specific about *delivery date* or when the paper will be at your location. Specify the most desirable

Hammermill Wove Envelopes
Wove Finish
Boxed: 500

17×22 BASIS	SIZE	DIMENSION	CTN	BOX	PRICE PER 1000 1 CTN	5 CTN	10 CTN
White, Regular							
20	6-¾	3-⅝ × 6-½	5000	15.10	8.74	7.58	6.73
	10	4-⅛ × 9-½	5000	20.06	11.62	10.07	8.94
24	6-¼	3-½ × 6	5000	18.33	10.61	9.19	8.17
	6-¾	3-⅝ × 6-½	5000	16.27	9.42	8.17	7.25
	7-¾	3-⅞ × 7-½	5000	21.94	12.70	11.01	9.78
	9	3-⅞ × 8-⅞	5000	21.94	12.70	11.01	9.78
	10	4-⅛ × 9-½	2500	21.09	12.21	10.58	9.40
	11	4-½ × 10-⅜	2500	38.53	22.31	19.33	17.17
White, Glassine Window							
24	6-¾	3-⅝ × 6-½	5000	20.21	11.70	10.14	9.00
	7-¾	3-⅞ × 7-½	5000	26.31	15.23	13.20	11.72
	9	3-⅞ × 8-⅞	5000	26.31	15.23	13.20	11.72
	10	4-⅛ × 9-½	2500	26.31	15.23	13.20	11.72
	Check	3-⅝ × 8-⅝	2500	26.31	15.23	13.20	11.72
White, See-Clear Window							
24	9	3-⅞ × 8-⅞	5000	28.33	16.40	14.21	12.62
White, Remittance							
24	6-¼	3-½ × 6	5000	29.07	16.83	14.59	12.95
	6-¾	3-⅝ × 6-½	5000	29.21	16.91	14.66	13.02

Available in even Boxes

Fig. 22-38. Study example of page taken from paper price catalog showing envelope sizes and prices. Envelopes are generally sold 500 per box.

delivery time, such as early morning, midmorning, early afternoon, or late afternoon. When you know the receiving department is closed during the usual 12 to 1 lunch period, include that information.

11. When the printing facility has a railroad siding, include that information. Identify the railroad. When the printing facility has a receiving platform, be sure to indicate whether or not over-the-road trailers can be accommodated. When the paper is being shipped into a warehouse, identify it.

Figs. 22-39 and 22-40 illustrate two checklists for ordering paper in flat sheets and in rolls. By using these forms, the paper merchant is supplied with all the information needed to deliver satisfactory paper at competitive prices, to the assigned destination, and at the proper time.

Cutting and trimming paper

Paper can be cut to size either on a hand cutter or power paper cutter. The smaller *hand-lever cutter* is used in small shops where limited quantities of paper are cut. See Fig. 22-41.

Power paper cutters are used in larger printing firms, with large press runs, Fig. 22-42. Many of these cutters are equipped with programmable devices to accommodate a series of cuts and trims. Refer to Fig. 22-43.

Paper must be *cut* to the required size designated for the job. With books, booklets, and jobs that run more than one to a press sheet, a trim out is also made. The term *trim* refers to the actual outside dimensions of a sheet of the finished printed piece. This operation is done after printing is completed. A job run more than one up on a press sheet requires cutting and trimming.

CHECKLIST FOR ORDERING PAPER IN FLAT SHEETS

GRADE _____

QUANTITY _____ (Sheets) _____ (Pounds)

SIZE & WEIGHT _____ M; _____ basis weight

COLOR _____ FINISH _____ DESIGN _____

CALIPER THICKNESS _____ 1-sheet _____ 4-sheets _____ pages-to-inch

☐ WATERMARKED   ☐ UNWATERMARKED

GRAIN DIRECTION   ☐ long   ☐ short   ☐ optional (one dimension)

DESIGN OR PATTERN NUMBER   ☐ long   ☐ short   ☐ optional (one dimension)

TRIMMING   ☐ machine   ☐ trimmed-2-sides

TO ONE SIZE   ☐ trimmed-2-ends   ☐ trimmed-4-sides

PRESS SHEET   ☐ Sheetwise   ☐ Work-and-Turn   ☐ Work-and-Tumble

REPRODUCTION METHOD (indicate which)

  ☐ Letterpess   ☐ 1-color   ☐ 2-colors   ☐ 3-colors   ☐ 4-colors
                         ☐ wet   ☐ dry

  ☐ Lithography   ☐ 1-color   ☐ 2-colors   ☐ 3-colors   ☐ 4-colors
                         ☐ wet   ☐ dry

  ☐ Gravure   ☐ High-gloss inks   ☐ Metallic inks
  ☐ Photogelatine   ☐ Finishing   ☐ Varnishing   ☐ Lacquering   ☐ Embossing
  ☐ Acetate laminating   ☐ Liquid laminating   ☐ Polyethylene laminating

PLANT HUMIDITY REQUIREMENTS _____

PURPOSE FOR WHICH THIS PAPER IS INTENDED _____

PACKING   ☐ Ream-marked in cartons   ☐ Ream-marked in bundles
               ☐ Ream-sealed in cartons   ☐ Ream-marked on single-tier skids
               ☐ Ream-marked on double-tier skids   ☐ Ream-marked on 4-tier skids
               ☐ Felt-side-UP ☐ Felt-side-DOWN ☐ Cast-side-UP ☐ Cast-side-DOWN

SKID SPECIFICATIONS   ☐ 4-way-entry   ☐ 2-way-entry

runners   ☐ short-way   ☐ long-way

minimum distance between runners . . inches maximum height . . inches maximum weight . . pounds

     Note — MARK SKID NUMBER AND ORDER NUMBER ON RUNNERS

SHIPPING   siding on _____ RR

           plant can accommodate trailers up to _____ feet long

           ☐ sidewalk delivery by winch-truck

Most satisfactory delivery hours _____ AM to _____ AM   _____ PM to _____ PM

Receiving platform closed _____ to _____

SPECIAL MARKINGS ON SKID-WRAPPERS _____

SPECIAL INSTRUCTIONS NOT IDENTIFIED ABOVE _____

SAMPLE FOR MATCHING ACCOMPANYING

Fig. 22-39. Checklist for ordering paper in flat sheets provides paper merchant with all information regarding an order. (Walden-Mott Corp.)

CHECKLIST FOR ORDERING PAPER IN ROLLS

GRADE _____

QUANTITY_____pounds
Maximum Basis Weight _____pounds
Maximum thickness _____ pages-to-inch _____
Maximum Roll Width _____inches
Maximum Roll Diameter _____inches
Core Inside Diameter _____inches

Type of Press _____

Speed of Press _____

Maximum ink drying temperature _____

Type of Delivery _____

Type of Core
    ☐ returnable   ☐ non-returnable
    ☐ slotted   ☐ not-slotted
    ☐ slot in juxtaposition   ☐ Dimensions of Keyway _____x_____

Roll-winding   ☐ felt-side OUT   ☐ felt-side IN
        Note — MARK DIRECTIONAL ARROWS ON WRAPPERS

Splicing   Maximum acceptable to roll_____
    flag ☐ one-side   ☐ two-sides
           ☐ diagonally across roll
           ☐ use 3-M splicing material for heat-set reproduction
Wrapping ☐ moisture-proof   ☐ non-moisture-proof
Delivery   ☐ on side   ☐ on end
PLANT HUMIDITY REQUIREMENTS
Special instructions
    ☐ Rolls must be wound uniformly firm   ☐ Indicate weight of each roll on wrapper
    ☐ Surface of paper must be free from lint and ☐ Indicate roll-number and winding direction on
       other extraneous materials   ends of rolls
    ☐ Provide roll cards in core of each roll   ☐ Provide packing slips complete with roll-number,
    ☐ Number each roll on wrapper   weight-per-roll and number-of-splices-per-roll
SHIPPING   ☐ siding on _____RR
            plant can accommodate trailers up to _____ feet long
         ☐ side walk delivery by winch truck
Most satisfactory delivery hours _____AM to _____AM   _____PM to _____PM
Receiving platform closed _____ to _____
SPECIAL MARKING ON ROLL WRAPPERS _____
_____
_____

SAMPLE FOR MATCHING ACCOMPANYING
    Note — Be sure to send out-turned samples in advance of shipment

---

Fig. 22-40. Checklist for ordering paper in rolls is used by web offset printers.
(Walden-Mott Corp.)

In placing the paper in the paper cutter, the sheets are first *jogged* (bent into a curve and then squared up) to be sure all sheets are aligned and will be cut square, Fig. 22-44. The stack of paper is then placed against the back gauge of the paper cutter and aligned with the left-hand housing, Fig. 22-45. It is desirable to place the best edge of the paper against the back gauge so that the poorest edge will become the waste. All four edges should be trimmed to assure smoothness.

The *clamp* that holds the paper in position during the cut is lowered. On hand-operated paper cutters, the safety device on the cutting blade handle is released. Then the arm is lowered with firm pressure.

A power-operated cutting blade can only be lowered by pressing two separate buttons. This assures that both hands are out from under the powerful cutting blade.

SAFETY TIP! Use extreme care when using a paper cutter! A paper cutter must be operated with both hands to lower the blade. This is a safety feature that prevents accidental lowering of the blade which could result in serious injury.

Today's modern paper cutters, with computer memories, can cut repeatedly with extreme accuracy. The end result is increased productivity. Fig. 22-46 illustrates the relative time savings between hand-lever cutters and power cutters.

Paper cutting is the start and finish of many printed jobs. Paper cutter operators must have complete information about each job, including paper size, color, weight, amount of spoilage for each operation (press and bindery), and grain direction requirements. Paper for any single press run should be cut at one time. This

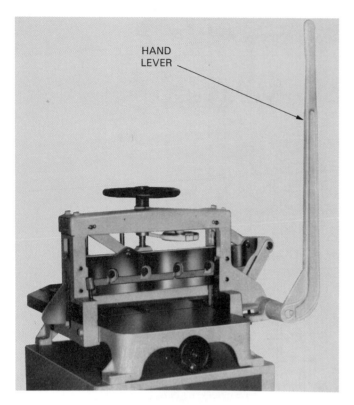

HAND LEVER

Fig. 22-41. A small printing firm may use a hand-lever type paper cutter to cut limited quantities of paper. (Challenge Machinery Co.)

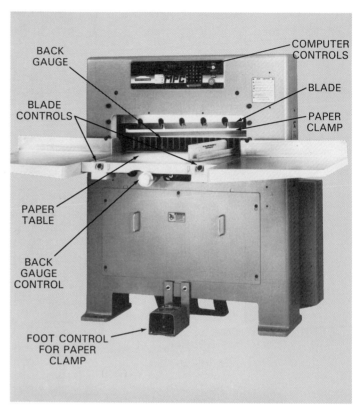

BACK GAUGE

COMPUTER CONTROLS

BLADE CONTROLS

BLADE

PAPER CLAMP

PAPER TABLE

BACK GAUGE CONTROL

FOOT CONTROL FOR PAPER CLAMP

Fig. 22-42. Power paper cutters are used in medium and large printing firms. (Challenge Machinery Co.)

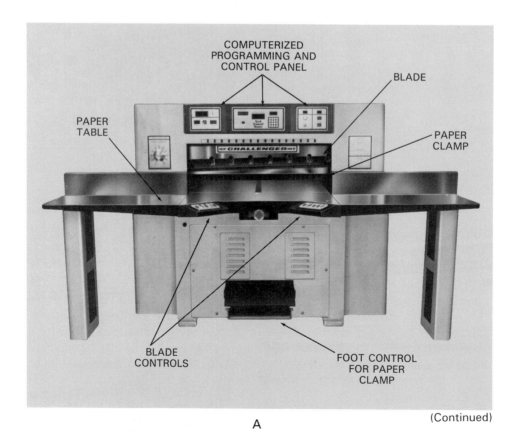

COMPUTERIZED PROGRAMMING AND CONTROL PANEL

BLADE

PAPER TABLE

PAPER CLAMP

BLADE CONTROLS

FOOT CONTROL FOR PAPER CLAMP

A

(Continued)

Fig. 22-43. A—Some large paper cutters are equipped with computerized programmable devices to accommodate a series of cuts and trims. B—High-volume programmable paper cutting setup utilizing two cutters; flow of work moves from left to right.

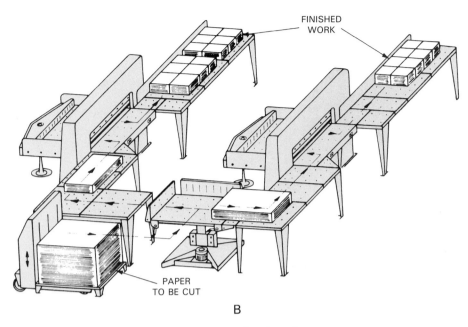

FINISHED
WORK

PAPER
TO BE CUT

B

Fig. 22-43. Continued.

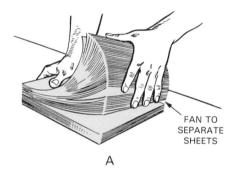

FAN TO
SEPARATE
SHEETS

A

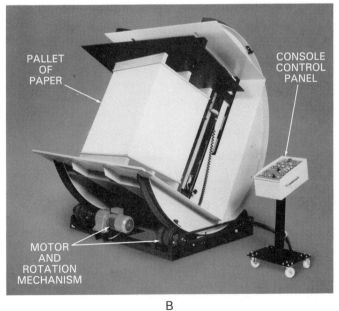

PALLET
OF
PAPER

CONSOLE
CONTROL
PANEL

MOTOR
AND
ROTATION
MECHANISM

B

Fig. 22-44. A—Before paper is placed in paper cutter, sheets are jogged to be sure all sheets are aligned. B—This machine allows an operator to jog, aerate, sort, inspect, dust, and square pallet-sized loads of paper. Pallet of paper is rotated to achieve various positions for jogging and alignment of paper. (Automation, Inc.)

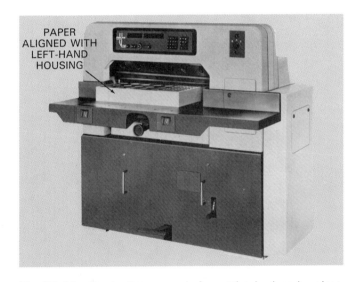

PAPER
ALIGNED WITH
LEFT-HAND
HOUSING

Fig. 22-45. Stack of paper ready for cutting is placed against back gauge of paper cutter and aligned with left-hand housing. (Heidelberg U.S.A.)

is because a paper cutter, except one with a memory computer, is difficult to reset to the exact same position.

After cutting, the operator generally stacks the paper on a utility truck or pallet. It is then ready for the press or for finishing if the sheets are already printed, Fig. 22-47.

Remember! When running color work, *ganged* (more than one up) jobs, or any close register work, paper size differences will cause serious problems.

Paper cutters must be properly maintained. Keeping the paper cutter lubricated and dust free is essential. Paper dust will transfer to the press and cause hickies to appear on the finished product.

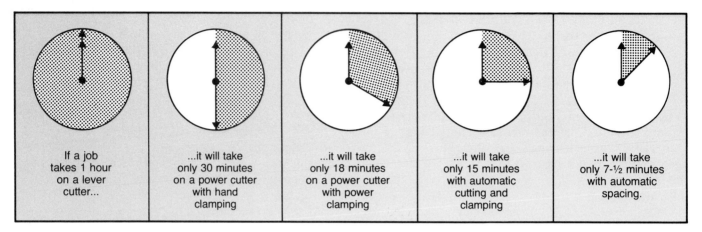

| If a job takes 1 hour on a lever cutter... | ...it will take only 30 minutes on a power cutter with hand clamping | ...it will take only 18 minutes on a power cutter with power clamping | ...it will take only 15 minutes with automatic cutting and clamping | ...it will take only 7-1/2 minutes with automatic spacing. |

Fig. 22-46. Note relative time savings between hand-lever paper cutters and power paper cutters. (Challenge Machinery Co.)

Fig. 22-47. After paper is cut, it is placed on a utility truck or pallet. It is then ready for press. (Kansa Corp.)

The blade must be sharp and free of nicks to keep the job dust free. Large plants sometimes change cutter blades at the end of each shift. This simply depends on how much the blade is used and the type of paper being cut. For example, cutting paperboard, chipboard, and similar heavy papers will dull the blade very quickly.

To check for cutter blade sharpness, use a piece of black velveteen cloth soaked with a little glycerine. Wipe the cut edges of the paper pile. If the cloth is FULL OF DUST, the blade is dull.

In addition, if the paper pulls or twists as the knife passes (under normal clamp pressure) the blade is dull. A dull blade makes a smacking noise when it strikes the pile of paper. A sharp blade makes very little sound as it cuts through the pile of paper.

The cutter stick should be kept in perfect condition. Wood or worn cutter sticks will ruin the last few sheets in every stack, or leave the bottom sheets uncut. Plastic cutting sticks are more expensive than wood. However, they stay fresh longer and provide an excellent return on investment.

To prevent paper damage while cutting, place a piece of chipboard under and on top of every stack. This ensures a clean bottom cut and eliminates clamp marks on the top sheet. If the cutting blade is sharp and clamp pressure is set correctly, the need for the chipboard is eliminated.

WASTE PAPER RECYCLING

Waste paper recycling involves saving and reprocessing unsalable paper materials. Recycling has been done since paper mills were first established in the United States. Annually, the paper industry is recycling or re-using, over 11 million tons of post-consumer waste. This is about 20 percent of the total consumption of paper and paperboard. In addition, the industry also recycles millions of tons of manufacturing waste, non-paper materials, wood residues, and fibers recovered from mill waste waters.

It has been estimated that the paper industry actually recycles about 50 percent of the fibrous waste materials that might otherwise enter the solid waste stream. Recycling is a vital part of the papermaking operation.

Waste paper recycling is a two-step process that involves:

1. The collection of waste paper at the point of generation or sorting, accumulation, and delivery to the point of conversion.
2. The conversion of collected waste paper into re-usable pulp by de-inking (if necessary), washing, and bleaching.

Where de-inking is used, the conversion process can be compared to a commercial or home laundry. This is because various chemicals, detergents, wetting agents, and bleaches are used to separate the dirt from the cellulose fiber. The term "dirt" refers to printing inks, dyes, pigments, and fillers.

Unlike metals and glass, which can be recycled indefinitely, all cellulose fibers suffer physical deterioration after each recycling. They lose some of their weight, length, physical strength, and papermaking quality. Even the strongest wood cellulose will lose most of its value as a papermaking material after several recyclings. It is important to keep in mind that recycled fiber is NOT a replacement for virgin pulp. Generally, recycled fiber should be thought of as a supplement for virgin pulps in the manufacture of new paper.

Fig. 22-48 summarizes a recycling system.

PAPER CHECKLIST

1. When planning your printing, take into account the standard dimensions of the paper to be used as well as the capacity of the press. This will help keep waste and cost to a minimum.
2. The finish of the paper selected can affect the mood of the completed job. Select a paper that prints well and harmonizes with the type and tone of the message.
3. Avoid selecting type designs with delicate lines for printing on rough-finish papers.
4. Remember that colored paper can add another dimension to the printed piece, but select an appropriate color.
5. Consider grain direction when figuring paper stock. If the finished piece is to be folded, the grain should be PARALLEL with the fold.
6. Process or full colors produce most accurately on neutral white paper.
7. *Runability,* or how efficiently or easily the paper can be printed, and print quality are important factors in selecting papers.
8. If colored ink is to be used on colored paper, check the compatibility of both the ink and paper colors.
9. Type is most easily read when printed on a soft, white paper.

POINTS TO REMEMBER

1. Paper is one of the most important materials used in printing.

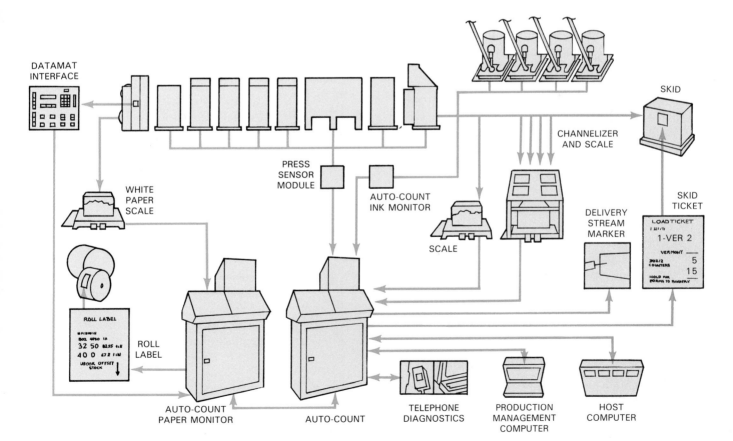

Fig. 22-48. Paper industry recycles about 50 percent of fibrous waste materials that might otherwise enter solid waste disposal system. (Auto Count)

2. There are over 5000 paper manufacturing plants in the United States.
3. The word "paper" comes from papyrus which was a writing material made by ancient Egyptians.
4. The first papermaking mill in America was started by William Rottenhouse in about 1690.
5. The manufacturing of paper involves two major processes: producing the pulp and manufacturing the paper from the pulp.
6. The pulp-producing process includes debarking logs, cutting logs into chips, pulping, bleaching, beating, refining, adding non-fibrous ingredients, and cleaning the pulp.
7. The manufacture of paper includes forming the paper, producing watermarks, removing water, drying, and finishing the paper. The manufacturing process takes place on a Fourdrinier papermaking machine.
8. Most printing papers are manufactured and identified according to their basis weight. This refers to the weight in pounds of a ream of paper.
9. As an example, a ream of 25 x 38 — 70 book paper weighs 70 pounds. The basis weight is indicated after the dimensions, such as 17 x 22 — 20. The figures that precede the dash (—) indicate the size of the sheets. The last figure indicates the weight in pounds of 500 sheets of the paper. The underlined figure (22) indicates the paper grain direction.
10. Papermakers and printers have proposed to substitute a universal basic size for the awkward basis weight practice. A universal size of 25 x 40 (1000 square inches) would become the standard for weighing 1000 sheets. The weight of 1M (1000) sheets in the universal size, 1M square inches, would be used as an identifying number for each type of paper.
11. Another method used to indicate paper sizes is mm (millimeter) specified by the International Organization for Standardization (ISO). ISO is a world body responsible for writing engineering and product standards.
12. Aesthetic considerations include such things as color, thickness, and texture.
13. Paper is graded according to use. Grades include offset, book, text, bond, cover, and index, among others.
14. Paper is packaged in standard sizes and quantities based on grade.
15. Extreme care should be taken when cutting and trimming paper, since a mistake in cutting dimensions is costly.
16. Usually, a paper cutter must be operated with both hands to activate the blade. This is a safety feature which prevents accidental

lowering of the blade resulting in serious injury.
17. Waste paper recycling is practiced in the United States.
18. Unlike metals and glass, that can be recycled indefinitely, all cellulose fibers suffer physical deterioration after each recycling process.
19. Recycled fiber is generally a supplement for virgin pulps in the manufacture of new paper.

KNOW THESE TERMS

Baronial, Basic size, Basis weight, Bleaching, Bulk, Calendering, Chemical pulp, Combination cut, Cover weight, Dandy roll, Deckle edge, Digester, Dye, Felt side, Filler, Finish, Fourdrinier papermaking machine, Grain, Groundwood pulp, Laid finish, Mechanical pulp, M weight, Opacity, Paper grade, Paper recycling, Papyrus, Pick resistant, Pigment, Ream weight, Semichemical pulp, Sizing, Spoilage allowance, Sulphate pulp, Sulphite pulp, Text weight, Watermark, Wove finish.

REVIEW QUESTIONS

1. Printing papers are made primarily from groundwood fibers, called _____.
2. Groundwood pulp is often referred to as _____ pulp because it has been reduced to fibers by grinding.
3. Chemical pulp, made from a mixture of wood chips and chemicals, is cooked in a tower, called a _____.
4. Pulp made by cooking the wood with sulphuric acid and lime is called _____ pulp.
5. Pulp made by cooking wood with caustic soda is called _____ pulp.
6. The amount of light that can be seen through a sheet of paper is referred to as its:
   a. Opacity.
   b. Deckle.
   c. Wove.
   d. Wire side.
7. The process of treating pulp with chlorine and similar chemical solutions to whiten the paper fibers is called _____.
8. Clay, talc, calcium, carbonate, zinc sulphide, and silicates are fibers that are added to pulp to strengthen its properties. True or false?
9. Paper can be made less absorbent and more repellent by adding:
   a. Filler.
   b. Dye.
   c. Sizing.
   d. Bleach.

10. _____ is the most widely used sizing.
11. Sizing added to the pulp during beating is referred to as internal sizing. True or false?
12. Surface sizing allows greater control over ink absorption during printing. True or false?
13. Approximately _____ percent of all paper is dyed in colors other than white.
14. The papermaking machine was invented in 1798 by:
    a. Ira Rubel.
    b. Ottmar Mergenthaler.
    c. Alois Senefelder.
    d. Nicolas Robert.
15. The side of the paper that touches the screen on the papermaking machine is called the _____ side.
16. As the paper begins to form on the papermaking machine, it passes under a _____ roll.
17. A _____ is a symbol or logo that identifies the brand of paper or trademark of a company.
18. Which side of the paper is preferred for printing when only one side of the sheet is used for printing?
19. Explain calendering.
20. The alignment of fibers is referred to as the _____ of the paper.
21. The degree of smoothness of a paper's surface is referred to as _____.
22. The weight in pounds of a ream of basic size paper is called its _____ weight.
23. The weight in pounds of 1000 sheets of a basic size paper is called its _____ weight.
24. Paper must be conditioned before it is run through an offset press. True or false?
25. A _____ edge is an untrimmed, feathery edge remaining on some kinds of paper.
26. Some of the very heavy sheets in the paperboard classification are called _____.
27. Good _____ resistance is required of offset papers since offset inks tend to be tacky.
28. Metric paper and envelope sizes are becoming more common in the printing industry. True or false?
29. Envelopes used for social invitations and wedding announcements are called _____.
30. The task of planning and cutting the required amount of paper for a job is called _____ and _____.
31. Adding a percentage of extra sheets to replace those ruined during a printing operation is called _____ allowance.
32. Drawing a _____ diagram, before cutting the paper for a job, avoids unnecessary waste.
33. To increase safety, a paper cutter must usually be operated with both _____ to activate the _____.
34. The paper industry recycles over 11 million tons of post-consumer paper waste each year. True or false?
35. In your own words, describe the various grades of paper.

SUGGESTED ACTIVITIES

1. Determine grain direction and felt and wire sides of several kinds of paper used in your shop. Use all the methods for testing grain direction described in this chapter.
2. Gather examples of letterheads from ten different business firms. Check to see which letterheads have watermarks. Write your findings on the back of each example.
3. Determine how many pieces can be cut from the sheet sizes listed below. Prepare a cutting diagram for each problem:

Piece Size (inches)	Basic Size (inches)
5 x 8	25 1/2 x 30 1/2
8 1/2 x 11	17 x 22
8 x 10	20 x 26
9 x 12	25 x 38
11 x 17	17 1/2 x 22 1/2

4. Assist your teacher in contacting a paper salesperson for purposes of giving a presentation to the class.
5. Determine the cost of paper for the following orders:
    a. For eight reams of 17 x 22 — 20 bond at 80¢ per pound.
    b. For 7000 sheets of 25 1/2 x 30 1/2 — 90 index at 92¢ per pound.
    c. For 40 reams of 17 x 22 — 24 bond at 83¢ per pound.
    d. For 2500 sheets of 25 x 38 — 70 book at $1.03 per pound.
    e. For 5000 sheets of 20 x 26 — 80 cover at $1.21 per pound.
6. Prepare a paper sample chart of the most common categories described in this chapter.
7. Give the millimeter sizes for the sheets listed below:
    a. 5 x 8 inch.
    b. 8 1/2 x 11 inch.
    c. 10 x 12 inch.
    d. 11 x 17 inch.
    e. 17 x 22 inch.
    f. 20 x 26 inch.
    g. 25 1/2 x 30 1/2 inch.
    h. 25 x 38 inch.

8. Visit a commercial printer and a quick printer. Determine how each orders paper and stores the paper. What differences can you find? Prepare a short report on your findings.

9. Assist your teacher in obtaining a motion picture, video cassette, or slide presentation that depicts the manufacture of printing paper.

HINTS FOR SELECTING PAPER

Paper may be purchased through the printer or ordered from a paper house. Paper houses do not manufacture paper. They stock papers made by several different manufacturers. Paper samples and prices are available to all clients. When selecting paper for a particular printed piece, the following factors should be considered:

- Select paper that will be compatible with the printing process and ink to be used. Think about durability, permanence, foldability, and exposure to various weather conditions.
- Price is extremely important. Expensive paper is not always necessary. Intended use is usually the key to determining quality and price.
- Paper that is heavier than necessary will lead to high mailing costs. A number of lightweight papers are available which have good opacity and exceptional printability.
- The paper surface and finish should be selected on the basis of the printing process to be used and the aesthetic qualities desired. Rough-textured paper gives a different feel and appearance than a smooth-textured paper.
- Paper is available in many different sizes. The printed piece size and the maximum press sheet size should be considered for maximum efficiency and minimum waste.
- Colored paper can add to the aesthetics of a printed piece. It may also add cost. Paper and ink color combinations should be studied carefully. Color is a psychological factor that every designer must understand thoroughly.
- Paper grain direction is most important when planning folded pieces. Paper tears and folds most easily with the grain. The grain direction may affect color registration, folding, and binding.

PAPER STORAGE

Proper storage is required to preserve paper and maintain its printing qualities. The "dos and don'ts" of proper paper storage are listed here:

- Store paper in an atmosphere that duplicates that of the pressroom.
- Use the older stock first.
- Reseal opened reams and opened cartons.
- Leave space between paper stacks for air movement.
- Keep the paper storage area clean.
- Do NOT store paper in direct sunlight.
- Do NOT store paper on concrete floors. Use pallets or a cabinet instead.
- Do NOT store cartons on end. Keep them flat.
- Do NOT stack paper so high as to compress the bottom sheets.
- Do NOT drop cartons.
- Do NOT store paper in a damp or unheated location.
- Do NOT destroy labels on opened paper packages.

Chapter 23

Lithographic Inks

When you have completed the reading and assigned activities related to this chapter, you will be able to:
O Describe the three basic ingredients of lithographic printing inks.
O Describe how ink is manufactured.
O List and describe the several types of lithographic printing inks.
O Mix and color-match ink.
O Identify the factors that affect quantity when estimating the amount of ink needed for a given printing job.
O List the data required by the ink formulator when ordering lithographic ink.
O Explain how ink is packaged, handled, and stored.

The United States Bureau of Census statistics indicate that there are over 230 ink companies in the United States. These companies produce inks in some 600 plants throughout the country. Industry sales of printing inks are over $700,000,000 annually and growing at an average rate of about five percent a year. Refer to Fig. 23-1.

Although the printing ink industry is relatively small, it is one of the most complex and advanced in the nation. The quality of its products and services is maintained through the industry's high level of technical competence. Many ink companies have complete ink research laboratories which are staffed by chemists, ink technicians, and color specialists. These laboratories are equipped with a variety of testing equipment. See Fig. 23-2.

The *National Association of Printing Ink Manufacturers* (NAPIM) has been the only national trade association for the printing ink industry since its founding in 1914. It is made up of commercial printing ink manufacturers engaged in the production and sale of printing inks in the United States.

Fig. 23-1. Annual ink sales are over $700,000,000 and growing at an average rate of about five percent a year. Ink manufacturers are among the many firms that exhibit their products at printing industry trade shows.

Fig. 23-2. Ink chemists and technicians use a variety of test equipment to maintain product quality control and research.

PURPOSE OF PRINTING INKS

Lithographic *printing inks* are colored coatings graphically applied to the surface of a substrate. Discussed previously, a *substrate* is any kind of surface for printing, such as paper, plastic, cellophane, metal, and glass. Paper is the most familiar substrate.

Inks are basically chemical compounds, different types of which have specific physical properties. Inks are manufactured from a variety of natural and synthetic materials.

Lithographic inks are manufactured in many colors for use on all types of offset presses. In addition, there are many ink formulations to serve offset printers.

Fig. 23-3 illustrates various ink formulations for different presses and substrates. Study each formulation.

Lithographic inks are stronger in color value than inks used in other printing processes. This is because less ink is applied to the printed sheet, so its effect must be strong. The average lithographic plate-to-paper ink deposit is only half that of letterpress (relief) printing. This is because the offset press rubber blanket picks up only a small deposit from the offset plate. The transfer of ink from the offset plate to the blanket is called *split-film action*.

INGREDIENTS OF INK

Lithographic inks contain three important ingredients. These include: pigment, vehicle, and modifiers.

Lithographic Inks

Sheet-Fed Presses	Web-Fed Presses
Substrates Paper Foil Film Thin Metal	*Substrates* Mostly Paper
Ink Vehicle Class Oxidative—Natural or synthetic drying oils.	*Ink Vehicle Class* Oxidative—Drying oil varnish.
Penetrating—Soluble resins, hydrocarbon oils & solvents, drying and semidrying oils and varnishes.	Penetrating—Hydrocarbons, oils & solvents, soluble resins.
Quick Set—Hard soluble resin, hydrocarbon oils and solvents, minimal drying oils and plasticizers.	Heat Set—Hydrocarbon solvents, hard soluble resins, drying oil varnishes, and plasticizers.
UV Curing—Highly reactive, cross-linking proprietary systems that dry by UV radiation.	UV Curing—Highly reactive, cross-linking proprietary systems that dry by UV radiation.
Gloss—Drying oils, very hard resins, minimal hydrocarbon solvents.	Thermal Curing—Dry by application of heat and use of special cross-linking catalysts.

Fig. 23-3. Note ink formulations for different offset presses and substrates or printing surfaces and materials. Paper is most familiar substrate. There are significant differences between sheet-fed and web-fed operations.

Pigment

The solid coloring in lithographic ink is the *pigment*. When you see black, red, green or any other color on the printed page, you are usually seeing only the pigment. Because lithographic inks must be strong in color value or strength, the pigment in an ink determines to some degree whether or not the ink is suitable for the offset process. Pigments also determine if the ink is safe or nontoxic enough for such printed products as butter and meat wrappers. These inks must be more water-resistant so they will not bleed through the paper and affect the food.

Black pigments are made from organic ingredients such as lamp black. *Opaque white ink* is made from organic compounds such as zinc sulfide and zinc oxide. *Transparent white ink* is made from magnesium carbonate, calcium carbonate, and clays. The various other colors available are made from inorganic and organic compounds.

Vehicle

The purpose of the *vehicle* is to carry the pigment or ink color. It also works as a BINDER to hold the pigment on the printed surface. The most common vehicle is *linseed oil*. It is sometimes referred to as *litho varnish*. Raw linseed oil is NOT suitable as a printing ink vehicle. It must first be changed by boiling. During *boiling,* the flow qualities of the oil are changed.

Vehicles for lithographic inks must resist water and the slight acidity of fountain solutions. They must have high resistance to *emulsification* which happens when too much fountain solution is mixed with the ink. *Emulsified ink* must be removed from the press and replaced with fresh ink.

The popular quick-set inks dry rapidly on contact with the paper. The vehicle for this kind of ink is a carefully balanced combination of resin, oil, and solvent. The *solvent* is absorbed by the paper very rapidly and leaves a dry ink film.

The vehicle of an ink must remain in a liquid condition while it is on the printing press. However, it must also dry rapidly when it touches the paper.

Modifiers

Ingredients called *modifiers* include driers, waxes, greases and lubricants, reducing solvents, and antiskinning agents. Some are placed directly in the litho varnish while it is being prepared. Others are added while the ink is being manufactured. Many can be added to the finished ink to modify the ink for special conditions during a press run, Fig. 23-4.

To help speed up drying of the ink on the printed paper, *driers* are added to the ink. Usually the correct amount of drier is placed in the ink when it is manufactured. There are certain occasions when the press operator adds additional drier to the ink.

The use of too much drier can be just as harmful as not enough. *Too much drier* can cause the ink to

Fig. 23-4. Ingredients, called modifiers, are added to ink for special conditions during a printing run.

dry on the press, fill in halftones, or cause the printed sheets to stick and setoff in the delivery pile. *Setoff* is the transfer of ink from one sheet to the back of the next. Too much drier can also cause the ink and paper to become waterlogged and dry even slower.

To help prevent setoff and sheet sticking, *waxes* are added to the ink during manufacture. Some waxes will do one job while others will do both. The most commonly used waxes are paraffin wax, beeswax, carnauba wax, and polyethylene.

Grease, Vaseline, and *tallow* reduce the tack or stickiness of an ink and cause it to dry more quickly. These ingredients also help lubricate the ink so that it will distribute and transfer easily. Too much of this type modifier will cause an ink to become greasy and print poorly.

Reducing solvents and *thinners* are used in lithographic inks to reduce *tack* (ink stickiness). These ingredients also help the ink to dry faster on the paper. Only solvents that blend well with the vehicles used in lithographic inks are acceptable.

MANUFACTURE OF PRINTING INK

The manufacture of lithographic printing ink involves a number of different types of machines and may also involve from one to several steps. Ink is often prepared in small batches or the ingredients may be fed continuously into and through the manufacturing system. The large majority of all printing inks are made by the batch process. Certain large-volume standardized inks, such as news inks, use continuous manufacturing processes.

Mixing

The *ink mixing* step or introduction of the coloring agent is a major part of ink manufacture. This is because the material must be broken down and thoroughly combined with the vehicle. Some coloring materials are made in chip, pulp, or other forms. Mixing is the first large input of energy.

The nature of the vehicle and the pre-dispersion characteristics of the coloring are main factors. These factors largely decide whether the ink can be made by the mixing process. Many inks are made in a one- or two-step mixing process. If it cannot, then a milling or grinding process must be used to make the ink.

The mixing step is done in *batch containers* with large mixing blades, Fig. 23-5. Small containers may hold 5, 10, or 20 gallon (19, 38, or 76 liter) batches. Large containers may hold batches of 100 to 1 000 gallons (380 to 3 800 liters).

The mixing speed depends on the kind of ink. It also depends on the pigment being blended into the vehi-

cle. The speed of the blades ranges from a few revolutions per minute to several thousand revolutions per minute. Many advances have been made in engineering and equipment. This includes the range of mixing speeds, blade sizes, blade positions, and ratio of blade size to container diameter.

Milling

Many printing inks cannot be reduced to their final product specifications with simple mixing stages. A second stage, called *milling* or *grinding*, is needed to crush and further blend the pigment into the vehicle. See Fig. 23-6. There are many devices for further dispersion of the pigment. Some are three-roll mills, ball mills, colloid mills, sand mills, and shot mills. There are also turbine devices and other types of color dispersers.

Three-roll mills

Most *three-roll mills* have steel rollers that revolve in opposite directions to act upon the pigment. They revolve at three speeds, in a ratio of about 1:2:4. Look at Fig. 23-7.

The ink mixture, called *slurry,* is fed to the rear roll hopper. It may be done continuously from a large tub. It may also be done in small batches by hand. As the

Fig. 23-5. The first step in ink manufacture is mixing of pigment with vehicle. This is done in a batch container with large mixing blades.   (General Printing Ink Co.)

Fig. 23-6. After mixing step is complete, the ingredients are further blended in an operation called milling. Steel rollers in mill revolve in opposite directions to act upon pigment and further disperse it into vehicle.   (General Printing Ink Co.)

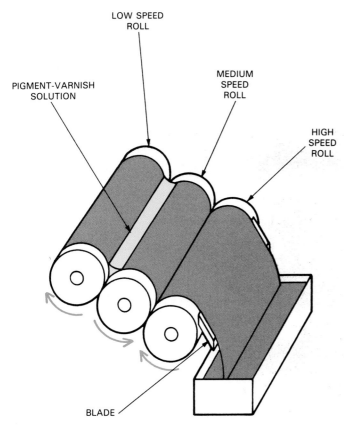

LOW SPEED ROLL

MEDIUM SPEED ROLL

PIGMENT-VARNISH SOLUTION

HIGH SPEED ROLL

BLADE

Fig. 23-7. This three-roll ink mill has steel rollers that revolve in opposite directions to act upon pigment. Note that rollers revolve at three different speeds.

ink moves through the rollers, the pigments are dispersed. This is done both by grinding and by a shearing force. Shearing force occurs between the steel rolls as they revolve in opposite directions. These rolls are kept in close contact. Positive pressure is applied to the front and rear rolls. The ink is removed at the front by a knife-edged apron. It then flows down the apron to a container or tub.

The speed of the mill, temperature of the rolls, pressure between the rolls, and general care and maintenance of the mill determine its grinding efficiency. Harder pigments sometimes need more than one pass through the mill.

Ball mills

Ball mills are large cylindrical structures with closed ends that contain steel or ceramic balls for acting on the pigment. Other shapes also are used for grinding. These mills grind materials containing volatile solvents which would rapidly escape in more open mixing containers. Ball mills are also used for large batches.

The ink ingredients are placed in the mill so the balls, or other grinding shapes are covered. The mill is then sealed and revolved. The balls cascade and tumble within the mill. This grinds the ink by the friction between their surfaces. The longer the mill rotates, the finer the ink is generally ground. Choice of mill con-

tainers and grinding media are determined by the nature of the ink and the need for color exactness.

LITHOGRAPHIC INK PERFORMANCE

There are many factors that influence the printing performance of a lithographic ink. The ability of ink to print well, however, depends primarily on how well its body and working properties are adjusted to the press, the printing speed, and the surface being printed. Generally these conditions require a careful selection of pigment, resin, varnish, drier, and the addition of modifiers.

Lithographic inks must work well with water. This is the principal reason why lithographic ink must differ from inks used in other printing processes. From a practical point of view, it is NOT easy to make a good lithographic ink. The formula for a good working ink may contain twelve to fifteen different ingredients, and is developed only after many experiments.

The ink manufacturer must select ingredients that are more readily moistened by greasy varnish than by water. The pigments in lithographic ink should NOT dissolve in water. This means that water should NOT be able to separate the pigment from the varnish vehicle. Likewise, the varnish should NOT emulsify or break up and mix into the water. Further, the combination of pigment and varnish must NOT emulsify or mix excessively with the water.

There are two types of ink emulsification in offset lithography. One is where there is water in the ink. A small amount of water mixes into the ink during printing. This type of emulsification does not usually create a problem. However, the other type of emulsification where ink mixes into the dampening water is more serious. This produces an *overall tint* on the printed surface.

If the ink is too greasy, the non-image areas of the printing plate may become ink-receptive and grease or scum. The greasiness of an ink is determined mainly by the greasiness of the ink vehicle. Some pigments, compounds, and driers also affect the greasiness of the inks. These materials must all be carefully chosen and controlled by the ink manufacturer.

There must be nothing in a lithographic ink that can react chemically with the printing plate. Undesirable chemical reactions between the ink ingredients and the plate may cause scum, discolor the plate, or even destroy the grain of the plate. There should be no grit in a lithographic ink. Grit can wear the printing plate, damage the image, and cause scumming.

The selection of a suitable ink for a printing job requires careful consideration. The printing process, kind of press, paper, and end use of the product must all be considered. Each of these factors is influenced by ink performance. Ink performance involves characteristics such as viscosity, body, tack, length, opacity, and permanency.

503

Viscosity

Ink viscosity is the degree to which it resists flow under applied force. This means that if ink flows easily (thin or runny), it has a *low viscosity*. If it resists flow (thick), it has a *high viscosity*.

Inks tend to change viscosity when used on a press. This is because friction in the ink rollers creates heat and the heat lowers the ink viscosity. The same principle applies to engine oil in an automobile. Its viscosity changes with cold and heat.

Body

Lithographic inks change their consistency under varied press operating conditions. The *consistency,* called *body,* of an ink changes according to how much it is used on the press. This is due to heat caused by friction in the ink rollers. When ink becomes thinned, its tack, length, and fluidity are automatically changed.

When running line and halftone work in a single press operation, it is necessary that a medium bodied ink be used. It should flow well once the distribution has begun on the rollers, and continues to do so until it has reached the sheet. A finely ground ink will do a better job of reproduction on either line or halftone work. The cost of ink increases when it is finely ground during the manufacturing process.

Medium and softer bodied inks are usually used for softer papers. Ink penetration is greater when printing on soft, absorbent papers. The ink must have the ability to be absorbed into the fibers of the paper. Under these conditions, inks dry more by absorption than by oxidation.

Tack

The amount of stickiness or resistance of a thin film of ink to undergo a split-film action indicates the amount of *tack* present. The split-film action occurs when the ink from the plate is transferred to the rubber blanket. Not all of the ink is transferred in the process, thus splitting it up. Too much ink tack in lithographic ink reduces its splitting ability. If the ink is NOT tacky enough, the print quality may NOT be clean and sharp.

Length

The ability of an ink to flow determines its *length.* The test for length is its capability to form a string when drawn (spread) out by an ink knife. Some length is necessary to allow an ink to feed properly to the fountain roller and then transfer. Too much length may cause an ink to *mist* (fly around) in the press area. Length of an ink changes when used on the press. This is due to heat and friction.

Opacity

Most ink films are not completely opaque or completely transparent. *Opacity* refers to the covering ability of an ink. What a person sees on a printed sheet is a combination of light reflected in different ways. An opaque ink will show more of its own color when printed over another color. A transparent ink allows previously-printed colors to show through or change by combining with the other ink.

Permanency

Some inks require that the printed image resist light. This is called *ink permanency.* Permanency of an ink increases when there is a large amount of pigment. Ink with a small amount of pigment tends to fade rapidly causing decreased permanency. One example, because of exposure, a high permanency ink should be used on jobs meant for outdoor displays.

FORMULATION OF INKS

Ink formulation refers to the amount and types of ingredients mixed together to make an ink. The formulation or mixing of various inks for lithography is a highly technical process. The subject is presented here only briefly and covers most of the common inks available to printers.

Sheet-fed inks

Sheet-fed lithographic inks dry primarily by a chemical or polymerization reaction initiated by oxygen and catalyzed (quickened) by cobalt and manganese ions from the drier. Sheet-fed lithographic inks have a higher tack than web inks. The relatively low speed of the press permits use of a high tack ink, which is required for sharp image definition. This also helps to avoid filling in of halftone dots, and avoids setoff in the printed pile of paper.

Web offset heatset inks

Web offset heatset inks must contain a litho varnish that must work satisfactorily with a fountain or dampening solution. The primary drying process is evaporation followed by chilling or coldsetting. Absorption and chemical reaction may also be involved. Control of evaporation temperature of the heatset oil used in the ink is critical. If the oil evaporates at too low a temperature, the ink will be unstable on the press. If it evaporates at too high a temperature, the ink will not dry properly. Some drying oil and drier may be added to these heatset inks to provide hard drying. Otherwise, the rosin and hydrocarbon resins remain soluble in the dried ink film and can be smeared by fingers, which have normal skin oils on them.

Web offset nonheatset inks

Web offset nonheatset inks include web offset news inks. Where carbon black is used for the pigment, a blue or violet toner is often added to overcome the brown shade of most carbon blacks. The vehicle contains a high-viscosity oil and a low-viscosity oil, blended to give the desired body.

Nonheatset web offset inks are usually used on highly absorbent, uncoated, groundwood pulp papers. Coated papers provide too much ink holdout and usually cannot be printed with nonheatset inks. *Holdout* refers to the amount of ink on the surface of the paper that does not dry by absorption.

Quickset inks

Either sheet-fed or web inks can be made into *quickset inks* by incorporating a quickset varnish into the formulation. A *quickset varnish* consists of a resin, usually combined with a drying oil, dissolved in a low-viscosity, high-boiling hydrocarbon oil. If the varnish and ink are properly formulated, the hydrocarbon oil will be absorbed by the paper coating, and the viscosity of the printed ink film will rise rapidly.

Quickset inks not only permit the printed sheets to be handled more quickly than they would if printed by a non-quickset ink, but they can set rapidly enough to increase the tack of the printed ink between units on the press. If they set sufficiently between units, it is possible to print process colors using four inks that all have the same tack rating. This means that the printer can use one set of process colors in any printing sequence.

Rubber-base inks

Rubber-base inks are used on both coated and uncoated papers when flexibility is a concern. These inks consist of a heavy formulation that gives quick setting and drying. Rubber-base inks can remain on the press (open) for several days without drying. These inks are compatible with conventional or aquamatic dampening systems.

When leaving rubber-base ink on the press several precautions are needed. After standing overnight, the ink on the rollers might appear to be drying. To overcome this problem, leave a heavy ink film on the rollers. This can be done by placing extra ink on the large oscillating roller and running the press. Run the press slowly when starting. If the press has an aquamatic system, remove the dampening system. Should the press be overinked, manually feed (power off) paper in and out of the ink train. The paper will collect the excess ink and bring it back to normal.

When using rubber-base inks on aquamatic systems, use just enough ink to cover the ink rollers and run low on the dampening solution. Then increase both ink and water to acquire the desired density. Avoid overinking and/or overdampening. Keep the pH (acid content) of the fountain solution between 4.5 and 5.5. The pH factor keeps the nonimage areas of the printing plate clean.

Nonporous inks

Nonporous ink is suited for metallic or plastic-coated papers. Ink such as Van Son's Tough Tex is an example of an ink with a nonporous formulated vehicle. It dries by oxidation rather than by absorption. It is important not to overdampen this ink. Because the substrate (such as paper) is nonporous, the fountain solution remains in the ink. Excessively dampened ink will not dry or set and will easily smear or set off to other sheets. Using an acid level of less than 4.5 will also retard drying. Ink additives are NOT recommended with this type of ink. Piles should be kept short and only a small amount of spray powder should be used.

Low-solvent, low-odor inks

In an attempt to reduce the amount of *effluent* (outflow) from the stack of a web offset press drier, ink manufacturers have studied ink and varnish formulations to reduce the amount of varnish used and to minimize its odor. By careful choice of resin and by careful formulation, it has proven possible to produce inks containing 30-35 percent less solvent than that used in conventional heatset inks. These types of inks are called *low-solvent, low-odor inks*.

With careful treatment of the solvent, it is possible to remove both the components that carry the odor and the ones that are known to react with sunlight to create smog or irritating air contaminants. However, complete destruction of ink solvent is usually the only way to comply fully with stringent environmental laws.

Metallic inks

Metallic inks actually contain small metal particles. Gold and silver inks are produced from bronze or aluminum powders. In comparison to the usual ink pigments, they are very coarse. Metallic inks perform poorly on the press. A typical problem is piling on the blanket, the plate, and the rollers. However, if the powders are ground finer, they lose their luster. Bronze, which has higher specific gravity than aluminum, causes more trouble. It is sometimes possible to apply two coats of a very dilute pigment. Actually, these powders are handled better by the gravure printing process, where the pigment level is much lower.

Magnetic inks

Magnetic inks are designed primarily for imprinting various products, like checks, which enables the items to be sorted electronically, Fig. 23-8. Magnetic inks are formulated with pigments that can be magnetized after printing. The printed characters can

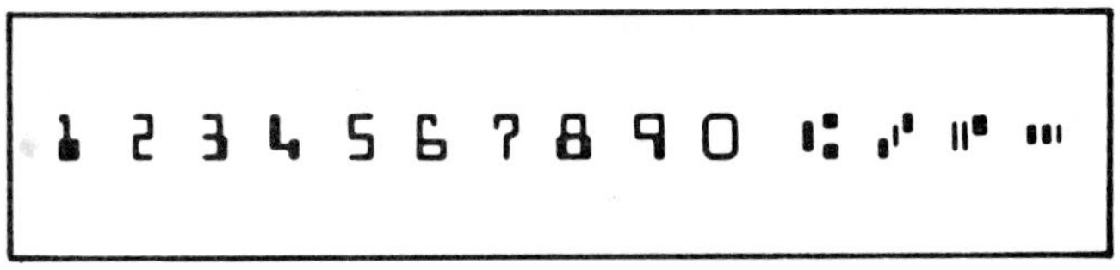

Fig. 23-8. Magnetic inks are designed primarily for imprinting various products like checks. Magnetic inks are formulated with pigments that can be magnetized after printing. (International Paper Co.)

then be recognized by electronic scanning equipment. The magnetic properties of the ink are derived from a special crystalline form of magnetic iron oxide. Proper performance requires careful and precise formulation of the ink. Magnetic inks can be formulated as conventional, heatset, or quickset lithographic or letterpress inks.

Soybean based ink

Soybean based ink is produced from soybean derived vegetable oil. This new type of offset ink is now under study by scientists in the printing and publishing industry. The new vegetable oil ink is biodegradable. *Biodegradable* means the ink is capable of being broken down and reused without causing further environmental problems.

Soybean ink is already being used experimentally as a web-fed offset ink for color newspaper and magazine publishing. The thin base ink is also adaptable to high-speed rotary offset presses.

Overprint varnishes

Overprint varnishes are inks without any pigment. The ideal overprint varnish is colorless and transparent, yielding good gloss and scuff resistance when dry. Overprint varnishes also must be stable on the press, dry rapidly, and adhere well to the print.

Water-based overprint varnishes may be applied on the fifth unit of a sheet-fed offset press. The inking system is disconnected and the varnish is applied to the blanket from a coater or sometimes from a specially designed dampening system. These varnishes consist of emulsions of acrylic resins. They form films very rapidly after they are applied to the print. Varnished pieces can be cut or folded a few minutes after printing.

INK-RELATED PROBLEMS

There are a number of common problems on the press related to ink. These occur largely because of the fountain solution necessary in the offset process. Problems also arise due to inaccurate setting of press roller pressures. Since ink performance is such a vital factor in the printing process, ink problems can affect the efficiency of the printing operation.

Some of the common ink problems encountered include: poor color, mottling, non-drying, sticking in the pile or rewind, picking, fill in, poor binding and rub, setoff, piling and caking, trapping, show-through and strike-through, ink not following the fountain, ink flying and misting, ghosting, ink drying on rollers, plate wear, and crystallization.

Poor color

Poor color occurs when the color of the printed image does not match the original image. The correct ink color is an important function of printing ink. The press department should supply the ink technician with samples of the paper for the job. This should include a print of the standard color desired and, if possible, a wet sample of the standard ink. When the ink is matched, quality control checks are made to be certain that the production batch matches color specifications for the established standard.

Quality control measures may ensure that the proper color of ink has been produced. However, other factors can cause the ink to be off color during the printing process. Some of these conditions include:
1. INK FILM THICKNESS—too much or not enough ink is being carried on the paper.
2. CONTAMINATION—press rollers and fountain are not properly cleaned.
3. PAPER—a variation in the paper that the ink is printed on can cause color problems.
4. COLOR DRIFT—some pigments used to obtain correct color have a tendency to change shade (called drift) upon aging or when used in tints.

To keep color problems to a minimum, the press operator should always check the printed results against a known standard at the start of a press run. If variations exist, the press operator should check the factors listed above.

Mottling

A condition known as *mottling* occurs when the solid portions of the dried print appear uneven and speckled. The ink does NOT produce a uniform film, Fig. 23-9. There are several causes for mottling:
1. Hard surfaced, non-absorbent papers.
2. Faulty distribution of the ink on the printing plate.
3. Poor wetting or dispersion of the ink.
4. Ink may be too soft.
5. Uneven ink absorption.

The mottling problem can be improved or eliminated by making the ink heavier by adding a body gum. Use a more opaque ink. In some cases, use of a flat ink, instead of glossy ink, will overcome the problem.

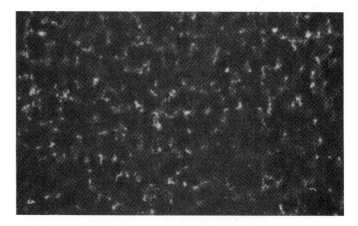

Fig. 23-9. Mottling occurs when solid portions of dried printed piece appear uneven and speckled.
(National Association of Printing Ink Manufacturers, Inc.)

Slow drying

Slow drying problems are evident when the ink smears or sets off onto the back of the next sheet. Depending on the type of ink and printing process, the method of ink drying occurs by one of the following methods:

1. ABSORPTION INTO THE PAPER — if slow drying occurs in inks that dry by absorption, the best solution is to apply a thinner film of ink.
2. SOLVENT EVAPORATION — slow drying, that occurs in solvent evaporation inks, can be improved by adding a more volatile solvent to the ink. The drying oven temperature can be increased on web-fed offset presses.
3. OXIDATION — slow drying problems in oxidizing inks require the addition of more drier.
4. PRECIPITATION — inks that dry by *precipitation* (evaporation) can be speeded up by use of steam or addition of a faster glycol solvent.

Paper can also contribute to slow drying especially if it contains excessive moisture or has high acidity. Temperature and high humidities can also have an influence on proper drying. See Figs. 23-10 and 23-11.

Sticking in pile or rewind

Ink sticking is associated with both setoff and poor drying. Prints become marred by transfer of ink or by picking of paper fibers when the sheets are separated. Transfer of ink to the under side in the delivery pile or rewind may result in sheets actually sticking firmly together. The ink acts as an adhesive.

With drying oil type inks, sticking may be due to poor penetration of the ink into the paper. The ink may be too heavy-bodied or contain too much non-penetrating vehicle. Lack of ink absorbency of the paper can also cause sticking. Excessive drier can contribute to this type of setoff and sticking since driers decrease penetration of the ink into the paper. Inks that dry by penetration may stick because of insufficient heat or insufficient cooling of the paper web.

All problems of sticking can be eliminated by using as thin an ink film as possible. Thin ink films tend to dry more rapidly or penetrate into the paper more completely.

Picking

Picking refers to small particles of the printing paper being picked off by the offset rubber blanket. These particles collect on the plate and print. White specks appear in the solid printed area. The result is a doughnut-shaped image imperfection, called a *hickey*. Refer to Fig. 23-12.

The most common causes of picking include:

1. Too much body and tack in the ink.
2. Ink drying too fast.
3. Coating of the paper is poor.
4. Rollers are too hard.
5. Temperature is low and press is cold.

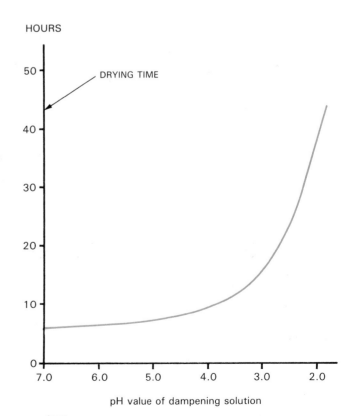

Fig. 23-10. Chart shows how temperature and relative humidity affect ink drying time. (Graphic Arts Technical Foundation)

Fig. 23-11. Chart shows how different pH values of a dampening solution can influence ink drying time.

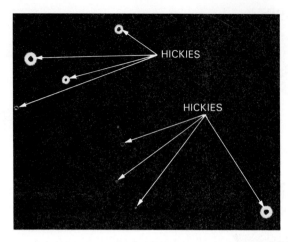

Fig. 23-12. When picking occurs on press, it can result in image imperfections called hickies. Hickies are doughnut-shaped areas that usually form in solid printed areas on press sheet.

To remedy picking, reduce the ink tack or use a paper with a higher rated pick resistance.

Fill-in

The problem of *fill-in,* or *specking,* produces a "muddy look" in the middletones and shadows. It is often detected in the printing of the small, nonprinting area between the halftone dots. It also occurs in the open area in fine type and line work. This gives the screen highlights a specky appearance and a loss of detail. This problem may be caused by the presence of foreign material in the ink. This could include lint, fiber, and paper coating picked up from the paper itself. Other foreign material such as dirt or spray dust may have also worked back into the ink.

Fill-in is more serious if a heavy ink film is carried. Problems arise from excessive impression, poor makeready, hard glazed rollers, and heavy roller settings. Sometimes the ink itself may be at fault. This is true if it is poorly ground, contains too high a pigment content, or is made from a low viscosity, tackless vehicle with poor pigment carrying properties. In this case, the ink must be either reground or reformulated.

Poor binding and rub

The problem of poor binding and rub should not be confused with ink drying slowly. *Poor binding* and *rub* results when the ink is dry on the printed paper but does not hold to the surface when rubbed under pressure. Some of the possible causes of rub include:
1. Not enough binder in the ink for the particular paper being used.
2. The vehicle portion of the ink penetrates into the paper too much, leaving only the pigment on the surface.
3. Ink is too soft.
4. Paper is very absorbent.

A possible remedy is to add a heavy drying varnish or overprint varnish to the ink. Drier can be added to

speed drying of the ink. Use a heavier ink with additional wax to produce better rub properties. If the job has been printed, the best solution is to overprint the printed portion with a good overprint varnish.

Setoff

The problem of *setoff* occurs when ink transfers to the under side of the next sheet of paper in the delivery pile or rewind roll. There are several causes for setoff:
1. Ink sets to slowly.
2. Static electricity in the paper.
3. Too much ink is being carried.
4. Too much impression.

Setoff can be decreased by adding offset compound to the ink and/or the use of anti-offset spray.

Piling and caking

A dry build-up or cake on the printing plate, both on solid and screen areas, is called *piling* or *caking.* Printed sheets show blotchy areas outlined by a thin, nonprinted line. Frequent press wash-ups are needed to eliminate fill-in of screens and reverse type.

Sometimes the pigment is too coarse or poorly wet. This prevents the vehicle from carrying the pigment properly. This results in a short, pasty ink.

The build-up on the plate, or rubber blanket in offset printing may be due to cutter dust or paper coating. A paper coating, with insufficient binder, or with a weak binder that attracts moisture, may cause trouble when humidity is high. The simplest remedy is to change the paper.

When the ink is at fault, the solution is reformulation of the ink. If time is critical, caking may be minimized by improving the pigment carrying capacity of the vehicle. This is done by adding body gum, heavy varnish, or lubricating compounds. Helpful press adjustments might include running the ink as spare as possible and adding another form roller for better distribution.

Trapping

To obtain proper *trapping* or register of colors in multicolor printing, the first color printed must have a higher tack than the second color printed. Besides the tack of the ink, factors affecting trapping are the absorbency of the paper and the ink film thickness. Absorbent papers assist in trapping. Hard, nonabsorbent papers make good trapping more difficult and may require more ink tack. Thin ink films will trap better than thick films.

If the tack sequence is not correct, the best remedy is to reformulate the inks. This will ensure that the first down color has more tack than the second down color.

Show-through and strike-through

The lack of opacity in a paper may cause *show-through* which allows the printed image to be seen or show through on the opposite side of the sheet. Show-through is generally a fault of the paper and not the ink. The best remedy is to use a more opaque paper. A strong heavy-bodied ink printed with a thin film will help reduce the show-through but will not eliminate it.

The problem of *strike-through* occurs when the ink or vehicle penetrates through the paper. The image becomes visible on the reverse side of the sheet. The excessive penetration is usually due to the ink being too soft or too slow when drying. The ink may also contain an excessive amount of non-drying oils. When the ink film is too heavy, it will cause strike-through. To improve the strike-through problem, a heavier ink should be used and a thinner ink film printed.

Ink not following fountain

When ink does not flow to the fountain roller, difficulty is encountered in holding proper color. The printing is streaky or uneven and the color is light. This problem is caused by improper ink consistency. The ink is too short or has a very *thixotropic* (thin) body.

To remedy this condition, the addition of a varnish to the ink should improve flow. In addition, the ink should be stirred while in the fountain.

Ink flying and misting

In severe cases of *ink flying* and *misting,* a spray, fog, or mist of ink is visible in the air. This problem usually occurs on high-speed presses. The ink mist settles and covers press frames, floors, and walls with a fine deposit of ink.

Flying is classified into two types. In one case, the ink particles are extremely small. They appear as a fog or mist floating over the press area. In the second case, the ink actually is thrown off in large particles by centrifugal force. This occurs because of faulty roller settings, excessive ink on the rollers, lack of affinity of ink for the rollers, or cuts or breaks in the rollers. Long, stringy inks, which tend to "cobweb" between the rollers, will usually fly.

One remedy for the problem is to shorten and soften the ink. Replacement of defective rollers and proper roller setting often completely eliminate flying. The use of a fountain cowling in critical areas prevents travel of the flying ink. Electrical grounding of the press to eliminate static often helps. Air conditioning in the pressroom can help remove the ink mist from the air.

Ghosting

A faint image design in unwanted areas of the printed images refers to *ghosting,* Fig. 23-13. This occurs when the metal rollers do not supply enough ink to the form rollers to stop the design before it contacts the form rollers a second time. Ghosting can appear when the ink has more affinity for metal than for the form rollers.

The addition of a long, sticky varnish may help to prevent this complete transfer. A tight setting of the

Fig. 23-13. Ghosting appears as a faint design in unwanted areas of image.

form rollers to cause a slight slippage of the roller will blur the design and make it less apparent.

Ghosting occurs most often on presses with relatively few form rollers. The addition of another form roller can help stop ghosting. Use of form rollers of different diameters also helps prevent ghosting. The positioning of the job on the plate may also be a problem. Solids in line with other solids, halftones or heavy type should be avoided. This is especially true on duplicator-size presses since only two ink form rollers are usually provided.

Ink drying on rollers

Ink drying on rollers may increase the ink tack to the point where the ink picks or tears the sheet. It may pull sheets from the grippers or even break the web on a web-fed press. Examination of the ink on the rollers, besides revealing the degree of tackiness, may show that the ink is skinning or drying, particularly on the ends. There it may have a dull, dry look rather than the shiny, wet appearance seen toward the center of the rollers.

With oxidizing inks, this difficulty may indicate the use of too much drier, or a vehicle that dries too fast. It may be corrected by small amounts of an anti-oxidant. Non-drying compounds can be added at the ends of the rollers. With heat-set inks or quick-setting type inks, drying on the rollers may be caused by solvents that are too volatile.

Rollers sometimes tend to absorb solvent from the ink. They must be replaced by a more suitable type roller. When using new rollers, pretreatment with the solvent in the ink is helpful. Rapid evaporation of an ink solvent causes drying at the end of the rollers. It may be corrected by dripping some solvent on the ends while the press is running. Also helpful is the use of a light solvent spray during start-up and shut-down.

High humidity can cause improper drying of moisture-set inks. This difficulty is sometimes helped by adding special compounds to increase the moisture tolerance of the ink. Otherwise the ink must be completely reformulated for greater moisture tolerance. With moisture-set inks, solvent absorption into the rollers is less often the cause of drying on the rollers. The remedies suggested for heat-set inks may be used.

Plate wear

In offset lithography, the problem of *plate wear* is often noticed when halftones lose contrast and fill-in occurs. In severe cases, the plate actually wears away. The ink may be at fault because it is poorly ground, is too highly pigmented, or lacks lubrication. Pigments or extenders may be too abrasive. The vehicle may also be too acidic.

Press faults contributing to wear are excessive or uneven impression, poor makeready, and form rollers set too tightly to the plate. Some papers are more abrasive than others. The coating or paper dust may work into the ink, and contribute to excessive wearing of the plate.

Crystallization

The problem of *crystallization* can occur in multicolor printing when the first down ink dries too hard and the second down ink does not adhere to the dried ink surface. Another problem occurs when the second down ink tends to mottle or rub off when dry. The basic cause for this problem is too much drier, especially if cobalt is used in the first down ink.

To remedy crystallization, retard the drying of the first down ink or any ink which is to be overprinted. This may be done by reducing the drier content and eliminating the cobalt drier.

PURCHASING OFFSET INK

The high cost of printing supplies makes it very important that they be purchased in the best possible way. This includes printing ink. The decision on purchasing ink is usually left to the pressroom. It will be helpful, therefore, to review some of the considerations that should be made when purchasing lithographic ink. These include the following:

1. Purchasing ink to do the job required with particular reference to quality definitions and specifications.
2. Purchasing ink in the most economical way, including consideration of price, quality, and coverage.
3. Establishing and maintaining inventory goals. This refers to good turnover of inventory in relation to realistic and flexible reorder levels.

Tell the ink manufacturer as much as possible about the proposed printing job, especially the type of work

Fig. 23-14. Printers purchase ink in various size containers. Tanker-truck quantities are used by newspaper and magazine publishers.   (General Printing Ink Co.)

being printed. This includes the type and make of press, running speeds, type of paper, and what method of drying is being used. It should also include such variables as product end use requirements, product resistance, etc.

Where possible, establish certain basic specifications. These include color, strength, color rotation, and tack preferred. Some optional specifications such as gloss, rub resistance, and fade resistance can also be requested. Wet samples should be retained for future reference. This is because there may be questions as to whether the ink manufacturer has maintained the correct shade, etc.

Buy ink in maximum quantities since this will result in the best price. The larger the batch, the less the cost of production for the ink manufacturer. Ink can be obtained in a wide range of quantities. These include by the pound in 1-, 3-, 5-, and 10-pound cans, or by the gallon in 2-, 3-, 5-, 15-, 30-, and 55-gallon containers. Containers are also available in 2500-pound reusable tote bins. Examples of containers are shown in Fig. 23-14.

Remember that ink is sold by the pound and paid for by the pound. However, you print with ink by the square inch. The true value of a pound of ink is how far it will go—not how much it costs!

Estimating ink coverage

Ink coverage is a measurement of how much area a specific amount of ink will print over. With this information, the amount of ink required for a printed job can be closely estimated. Allowance must be made for makeready, spoilage, and wash-ups. The ink manufacturer can assist you in determining the amount of ink needed for a job.

The ink estimating chart in Fig. 23-15 shows coverage for certain colors of ink on various types of paper. The estimates are for sheet-fed presses. Keep in mind that this method of estimating ink coverage is a rough indication. Five important variables should be considered when using the estimating chart. These include: color strength, specific gravity, paper finish, type of job, and press makeready.

Color strength

The *color strength* (opacity or transparency) of an ink depends on the pigments. The true color of ink results from light reflected by the paper and transmitted through the ink film.

Tinting strength is coloring power, or the amount that an ink can be reduced or diluted with a white pigment dispersion to produce a tint of a given strength. No printed ink is completely opaque or completely transparent. If an ink is weak, a heavier film must be used to achieve the desired color.

Specific gravity

The weight of a given volume of ink, compared to the weight of an equal volume of water refers to an *ink's specific gravity*. The volume of ink instead of weight is the critical factor. As specific gravity INCREASES, volume DECREASES. This requires a greater amount (pounds) of ink.

For example, the specific gravity of Poppy Orange ink is 2.30. The specific gravity of Marigold Orange ink is 1.10. Therefore, Poppy Orange will cover twice as much as Marigold Orange pound for pound. The assumption is that all other variables are equal.

Paper finish

The type of paper *finish* or surface characteristics influences the amount of ink required. A smooth, hard

Lithographic Ink Coverage Chart — Sheet-Fed

GRADE OF STOCK	Enamel	Litho coated	Dull Coated	Machine Finished	Antique
Black	425	380	375	400	275
Rubberbase Black	445	430	425	435	335
Purple	360	350	320	350	235
Process & Transparent Blue	355	340	335	340	220
Transparent Green	360	350	335	350	235
Process & Transparent Yellow	355	355	340	340	220
Chrome & Lemon Yellow	285	260	250	250	150
Persian Orange	345	325	310	325	225
Process & Transparent Red	350	345	340	340	225
Semi-Transparent Red	350	340	325	340	175
Brown	345	335	325	335	225
Silver	335	300	285	295	220
Gold	125	115	115	115	75
Opaque White	200	175	165	175	135
Tint Base	400	380	375	385	250
Flourescent	135	120	120	120	85
Over Print Varnish	450	425	415	425	—

Fig. 23-15. Ink estimating charts similar to this are used by printers to estimate amount of ink required for any given job. The figures above are the number of thousand square inches that one pound of ink will cover. Measure with a ruler the amount of square inches of solid coverage of copy. Multiply this figure by the number of impressions to be printed. Determine the amount of thousand square inches to be obtained from the color to be printed from one pound of ink as shown on the chart. Divide this number into impressions X square inches covered. (National Association of Printing Ink Manufacturers, Inc.)

surfaced paper, such as enamel, requires LESS ink film thickness than a rough textured paper such as antique. The absorbency rate is higher for a rough textured paper than for a smooth, hard finished paper.

Type of job

A job that consists mainly of text matter requires less ink than a job with solids and halftones. The amount of ink needed may double or triple for jobs requiring heavy coverage.

Press makeready

Every job requires some type of *press makeready* (preparing to run press). Therefore, any estimate of ink must also include a percentage for makekready. Web-fed presses usually require more paper for makeready than sheet-fed presses.

REMOVING INK FROM CONTAINERS

When taking ink from a container, it should be scraped smoothly from the top surface with an ink knife. If ink is gouged out from the middle, it will dry up and be wasted. When using a cartridge, the first quarter of an inch of ink that is pushed through the spout should be discarded. Dried or lumpy ink should NEVER be placed on an offset press.

Ink container lids should be securely replaced when not in use. A small amount of oil on the inside rim of the lid helps to seal the can. When using cartridges, always replace the cap.

MIXING COLORED INKS

A wide assortment of ready-mixed colored inks are available. Sometimes, it is necessary to prepare special colors and tints. This is done before placing the ink on the offset press.

Yellow, red, and blue are the *primary pigment colors*. This is because they can be mixed together to produce any other colors needed. Remember the following:
1. Red and yellow produce ORANGE.
2. Red and blue produce PURPLE.
3. Blue and yellow produce GREEN.

Black and white are NOT considered colors. They can be mixed with colors to lighten or darken them. When black ink is added, the color becomes a *shade*. If a color is added to white, a *tint* results. The color should always be added to the white, and NOT white to the color. If the lighter color is added last, it takes much more ink than is necessary for the job.

Mixing procedure

Ink mixing must be done on a clean table near the press area. The following equipment should be used for this purpose:

1. A piece of glass or plastic on which to mix the ink.
2. Ink knives to handle and mix the ink.
3. An ink scale to weigh the inks, Fig. 23-16.
4. A color chart or color sample of the ink required.

To mix a desired ink color, compare the color chart and the sample to be matched. Select an ink from the chart that most closely matches. Prepare a small test batch first. Make the first color comparison using an ink knife to form a swatch of ink. Do this on the paper that will be used for the printing.

After the text, add what is needed to more closely match the color desired. When the color has been mixed, reducers, body gum, or drier can be added as needed.

To mix a tint, start with mixing white ink in an amount that will complete the printing job. Then, add color in small amounts. Continue to add color until the correct tint is achieved.

Never start with a colored ink and add white ink! This will result in mixing far more ink than is required. Several small mixed batches may be needed to achieve the correct match.

You should now try mixing a small batch of ink. Mix less than is needed for the job. It is best to weigh all ingredients. Maintain a record of all additions until the color is correct. The ingredients must mix well and remain stable.

Ink should be mixed thoroughly before placing it in the ink fountain. Never mix an ink in the press ink fountain. This can cause a matching problem because ink cannot be thoroughly mixed and matched in the fountain. All ink going into the fountain must be well mixed to avoid streaking. Ink fountains on offset presses are NOT designed for ink mixing.

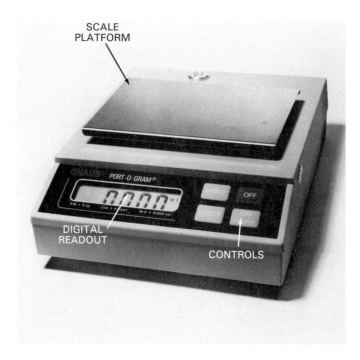

Fig. 23-16. This computerized ink scale is used to mix desired amount of ink for a job. (Ohaus Corp.)

After the ink has been mixed, it is stored in a small metal or plastic container. The need for cleanliness when mixing and storing the ink is important. If the ink becomes contaminated or dirty, the color will change. All inks should be stored on shelves at a stable temperature. They should also be marked by kind, color, and delivery date. All cans should be checked for leaks. New containers may be required if cans are found to be damaged.

Pantone Matching System®

The *Pantone Matching System®* is a method universally accepted for specifying and mixing ink colors. Using this technique, printers, artists, and customers can select any of the more than 900 shades from a swatch book. The printer can then mix the desired ink color by using the swatch number and referring to a formula guide. The guide gives the formula for making the color. The *ink formula guide* identifies the basic colors and indicates how much of each to mix together.

The *ten basic colors* in the Pantone Matching System® are black, green, process blue, rubine red, warm red, yellow, reflex blue, purple, rhodamine red, and transparent white.

All of the formulas in the formula guide are equally balanced and designed to assist the printer in mixing the desired color under controlled conditions. Colors can be matched using volume or weight by following the basic steps below.

1. Gather your equipment. Obtain a measurement device, mixing slab, ink knives, test papers, cleaning solvents, and utensils. Be sure all equipment is clean to avoid color contamination.
2. Choose the desired Pantone color that best meets the specifications required for the printing job.
3. Analyze the formula for that color and determine the proportion of each ink needed as indicated under parts or percentage, whichever is easier for you.
4. Gather the necessary basic color inks at your work station, and prepare them for mixing the required color. Be sure to review the basic ink colors for possible contamination, dried ink, and sufficient amount for the job.
5. Using a clean ink knife, measure the required amount of ink (volume or weight) as indicated in the formula and place it on your ink mixing slab or ink mixing equipment.
6. Record all ingredients on your personal formula card for this particular printing job and/or customer. File it for future use. Refer to Fig. 23-17.
7. Mix all ingredients until all ink colors are blended properly with none of the basic ink colors visible.
8. After the mix is blended properly, test against the standard (sample in Pantone Color Formula Guide). Either pull a pre-press proof, or complete a drawdown (swatch) test and compare it to the approved ink sample.

It should be noted that there will be a visual difference between wet ink compared to dry ink in the formula guide. In addition, the color of the paper used will affect the final color because the matching system is a transparent ink system. This means that there is always a slight degree of ink show-through on most colored papers.

The color formula guide contains simple formulas for colors on coated and uncoated papers. A color formula comprises a specific number followed by a letter "C" for coated, or "U" for uncoated paper. By simply mixing the formula given, the user can achieve the same color every time, in any quantity, regardless of geographic location.

Under each standard color is the mixing formula divided proportionally and identified in parts followed by the percentage of the total for each ingredient. Every Pantone color in the color formula guide can be matched using various combinations of eight basic colors plus black and transparent white.

The following color formulas are examples illustrating simplicity of mixing and matching a specified color:

Color	Color Combination	Ratio
Pantone 165C	8 parts yellow 8 parts warm red	50.0% 50.0% (Ratio 1:1)
Pantone 164C	4 parts yellow 4 parts warm red 8 parts white	25.0% 25.0% 50.0% (Ratio 1:1:2)
Pantone 168C	8 parts yellow 8 parts warm red 4 parts black	40.0% 40.0% 20.0% (Ratio 2:2:1)
Pantone 163C	2 parts yellow 2 parts warm red 12 parts white	12.5% 12.5% 75.0% (Ratio 1:1:6)

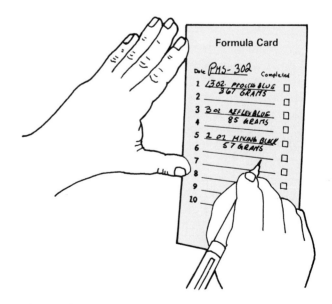

Fig. 23-17. All ingredients for a printed job requiring a special mixed ink color should be noted on a formula card. Card should be filed for future reference in case of reruns.

NEW INK TECHNOLOGY

There are several new methods of ink drying being used throughout the printing industry. Most of these methods are directly related to pollution and environmental control regulations. With the faster drying inks, there is less adverse effects on paper and the atmosphere. Some of these methods in use, or under study, include microwave, ultraviolet, overcoating, and chemical.

Microwave drying

The lower energy *microwave drying* generally generates heat in the ink vehicle but NOT the paper. The cost and health hazards associated with microwave drying have limited its use in the printing industry.

Microwave ink drying occurs when heat causes the resins or monomers to form into solids. This process requires expensive equipment and is used primarily on web offset presses.

Ultraviolet drying

With *ultraviolet drying,* a light source, supplying ultraviolet energy, is used to agitate molecules in the ink film thereby causing them to cross-link and dry. The molecules are monomers. Their atoms vibrate and heat up when exposed to ultraviolet light. This means that the heating causes the monomers to join together, thereby cross linking to cure or dry the ink.

Ultraviolet drying is generally used with web offset presses and requires special inks. These inks dry in milliseconds, without effluents.

The various energy areas of the electromagnetic spectrum are illustrated in Fig. 23-18. The illustration shows that the longer the wavelength and the lower the frequency, the less energy the wave produces.

Overcoating drying

With *overcoating drying,* drying takes place by solvent loss, which causes the resin to harden. The process known as overcoating and *heated drying* has been used for several years. A newer method involves coating wet ink with resins mixed in an emulsion with water and alcohol. This process is used on web offset presses.

The coating is referred to as an *alcohol-soluble propinate.* It is valuable because of the reduction in stock effluents and the speed of drying. The coating mechanism is generally positioned between the press units and the folder. The coating is 10 to 45 percent solids with a mixture of 75 percent isopropyl alcohol and 25 percent water. A high gloss results when a heavy coating is applied. Application of thin coatings appears to be uneconomical with respect to this process.

Chemical drying

Chemical drying refers to the deliberate process of causing a change through an activator or catalyst. The activator causes monomers to cross-link and form a

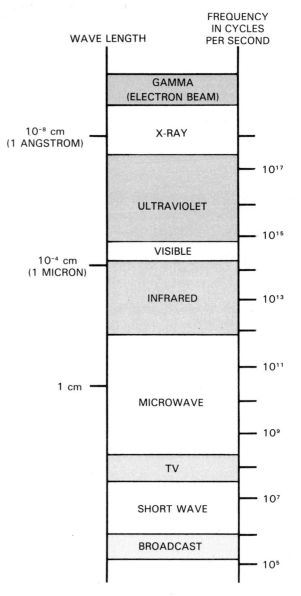

Fig. 23-18. Energy areas of electromagnetic spectrum. Longer wavelength and lower frequency produce less wave energy. Microwave ink drying uses ultraviolet band.

chain (a polymer), thereby becoming a solid. The activator (catalyst) itself is generally released to change monomers through the application of heat. The most common polymers are unsaturated polyesters and epoxy resins. These catalytic inks are commonly referred to as solid-state and solventless inks.

POINTS TO REMEMBER

1. Lithographic printing inks are colored coatings graphically applied to a surface or substrate.
2. Inks are manufactured from a variety of natural and synthetic materials.
3. The components or ingredients of lithographic ink include a pigment, vehicle, and modifiers.

4. Pigments provide color, opacity, and permanence to inks. Both organic and inorganic pigments are used for this purpose.

5. Vehicles serve to carry and bind the pigment to the substrate. They dry by absorption, oxidation, evaporation, and polymerization.

6. Modifiers or additives, such as dryers, waxes, oils, and varnishes, are used to adjust the basic ink characteristics.

7. The manufacture of lithographic printing ink involves a number of different types of machines and processes.

8. Ink is prepared in small batches or the ingredients are fed continuously through the manufacturing system.

9. In the ink manufacturing process, the ingredients are first mixed together. The second stage involves milling or grinding of the ingredients.

10. Chemists monitor the manufacture of every batch of ink.

11. Color matching is a crucial stage of the ink manufacturing process.

12. Printers encounter a number of common problems related to ink. These occur largely because of the fountain solution and the roller pressures on the offset printing press.

13. Some of the more common ink problems encountered on the press include poor color, mottling, non-drying, sticking in the pile, picking, fill-in, setoff, piling, and caking.

14. Special ink colors can be mixed by the printer using a color system such as the Pantone Matching System®. Over 900 color shades are possible with the system.

15. Future changes in ink technology are likely to center around new vehicles for inks in an attempt to reduce harmful pollutant emissions.

KNOW THESE TERMS

Antiskinning agent, Ball mill, Binding and rub, Body, Caking, Color strength, Crystallization, Driers, Fill-in, Ghosting, Grinding, Hickey, Ink absorption, Ink misting, Length, Litho varnish, Milling, Modifiers, Mottling, Opacity, Oxidation, Permanency, Picking, Pigment, Piling, Precipitation, Reducing solvent, Setoff, Shade, Show-through, Specific gravity, Split-film action, Solvent evaporation, Strike-through, Tack, Thinner, Three-roll mill, Tint, Trapping, Vehicle, Viscosity, Waxes.

REVIEW QUESTIONS

1. Lithographic inks are usually stronger in color value than most other inks. True or false?

2. The purpose of the _____ in lithographic ink is to carry the pigment (color).

3. What is the most common ink vehicle?

4. The solid coloring in lithographic ink is the:
   a. Varnish.
   b. Pigment.
   c. Resin.
   d. Modifier.

5. Driers, waxes, greases, lubricants, reducing solvents, and antiskinning agents are all called _____.

6. The unwanted transfer of ink from one printed sheet to the back of the next is called _____.

7. How can sheet sticking be prevented?

8. Explain the mixing of ink.

9. A three-roll mill is used to _____ the _____ _____.

10. The consistency of lithographic ink is referred to as:
    a. Viscosity.
    b. Strength.
    c. Tack.
    d. Body.

11. The ability of an ink to flow determines its _____.

12. Ink _____ is required for printed jobs meant for outdoor displays.

13. Uneven and speckled ink drying in solid areas of the printed sheet is called:
    a. Scumming.
    b. Mottling.
    c. Chalking.
    d. Piling.

14. A dry build-up or cake on the offset plate, both on solid and screen areas, is called:
    a. Piling.
    b. Scumming.
    c. Chalking.
    d. Mottling.

15. Accurate register of ink colors in multicolor printing is called _____.

16. An incorrect balance between the tack of the ink and the strength of the paper surface can cause:
    a. Chalking.
    b. Setoff.
    c. Ghosting.
    d. Picking.

17. Describe ink flying and misting.

18. The problem of ink crystallization occurs in _____ printing.

19. Ink _____ charts show the coverage for certain colors of ink on various types of paper.

20. If a color is added to white ink, the color becomes a _____.

SUGGESTED ACTIVITIES

1. Familiarize yourself with an ink swatch book. Learn the terms and procedure for ordering lithographic ink.
2. Mix and match three colors provided by your teacher. Be careful not to mix too much ink for the job.
3. Obtain a printed one-color flyer. Estimate how much ink would be used to print 10,000 flyers. Use the ink coverage chart in this chapter.
4. Prepare several swatches of transparent and opaque inks on white index paper. Note the difference in appearance of these inks.
5. Visit your local newspaper plant and ask to see the pressroom. Note how the ink is supplied to the newspaper plant and the method used to fill the ink fountains on the press. Write a short paper on your observations.

INK STORAGE AND HANDLING

Proper storage and handling is required to preserve ink and maintain its printing qualities. The "dos and don'ts" of proper ink storage and handling are listed here:

- Store ink in an enclosed, fire-resistant cabinet.
- Use older inks first.
- Keep inks covered to prevent oxidation.
- Group inks according to color and usage.
- Remove skin from ink before using.
- Keep the ink storage area clean and free from ink spills.
- Do NOT destroy labels on opened ink tubes and cans.
- Do NOT dig out ink from the ink can. Instead, scrape the required amount of ink from the surface layer.
- Do NOT damage the lid when opening an ink can.

A large web-fed offset press.   (Metroliner Press)

Section VII

Issues for Class Discussion

1. The process of papermaking involves four major steps: pulping, refining, forming, and finishing. Discuss the various factors that must be considered in a papermaking mill related to environmental impact and toxic waste management.
2. Recycling of paper has increased during recent years. Over one-half of all new products are produced from recycled paper. Discuss factors that encourage the recycling of paper by consumers and by the paper-producing industry.
3. Two major changes are forecast for ink production. Due to the increasing cost of energy, other ink types will be developed and substituted for heat-set inks. The second major change will come about as a result of increasing pollution controls. Discuss these issues and give recommendations and alternative strategies.
4. Discuss applications of the Pantone Matching System® with respect to its use and benefits to the designer and printer.

Section VIII

Image Transfer and Finishing Operations

In offset printing, the image areas of the plate are grease receptive, while the non-image areas are water receptive. The offset press consists of three cylinders: plate, blanket, and impression. Offset presses are grouped according to their size and function. The smallest of these are the duplicator presses which generally take sheets up to 11 x 17 inches or smaller. Offset presses generally handle sheet sizes 14 x 20 inches and above. Presses that print from rolls of paper are referred to as web presses.

Section VIII contains six chapters. Chapter 24 covers the six basic systems and fundamentals of offset duplicators and presses. General operating procedures and safety precautions related to duplicators and offset presses are described in Chapter 25. Chapter 26 describes several kinds of duplicators and offset presses and explains simple press adjustments and maintenance. Chapter 27 introduces web press design and operating principles. Greater emphasis is being placed on the mini-web press. Chapter 28 identifies numerous printing-related problems that can occur on a duplicator or offset press. After the job has been printed, it is usually necessary to perform finishing and/or binding operations. Chapter 29 explores the varied and generally specialized operations undertaken in the finishing and binding area.

Offset presses range in size and capability.　(A.B. Dick Co.; Rockwell International Corp., Graphic Systems Division)

Chapter 24

Offset Press Fundamentals

When you have completed the reading and assigned activities related to this chapter, you will be able to:

O Distinguish between an offset duplicator and an offset press.

O List examples of different kinds and sizes of offset lithography duplicators.

O Describe the features common to offset presses and duplicators.

O Describe the six major operating units of offset presses and duplicators.

O Describe special systems that can be added to offset duplicators.

O Identify the controls of an offset press and duplicator.

O Identify the roller and cylinder locations on an offset duplicator.

O Recognize the safety regulations that apply to offset duplicator operation.

Sheet-fed offset duplicators and offset presses have similar operating features. There are many brands of offset duplicators and offset presses. In addition, manufacturers of this equipment offer a number of models. The different brands and models vary in certain details of construction, operating controls, cylinder arrangements, feeding, and delivery. However, the principles of operation among the brands and models are basically the same.

To understand press operation, the press operator must be familiar with the major systems of the press. Before operating the press, the location and function of the press operating controls must be understood.

In addition, the press operator must understand the proper operation and adjustment of each part of each major system of the press.

This chapter describes the systems that make up an offset press. It also provides an understanding of the general procedures used in setting up these systems prior to operating the press. Operation of specific presses commonly found in commercial and quick printing is covered in Chapter 26.

PRESS CLASSIFICATIONS

There are many offset press classifications. Offset printing presses are constructed according to the maximum size sheet the press will handle. Some presses print only a single color at a time. Others print several colors and also print on one or both sides of the paper. Most offset presses operate on the principle of one printed copy for each revolution of the cylinders.

Sizes and types of presses

Small offset presses, up to a maximum sheet size of about 11 x 17 inches (28 x 43 cm), are generally referred to as *offset duplicators,* Fig. 24-1. Presses that print on sheets LARGER than this are referred to as *offset presses,* Fig. 24-2.

The operating principles are basically the same for both kinds of presses. Most offset presses and duplicators are classified as either sheet-fed, web-fed, multicolor, or perfecting.

Sheet-fed presses

Offset presses and duplicators designed to print a single sheet of paper at a time are referred to as *sheet-fed presses.* These presses are built to pick up individual

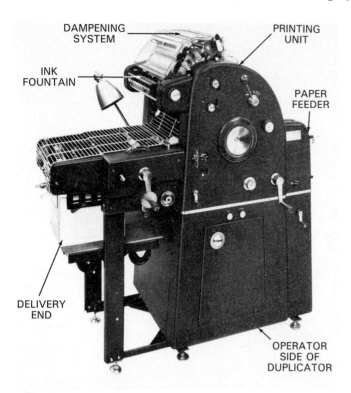

Fig. 24-1. Duplicators are small offset presses with a maximum sheet size of about 11 x 17 inches. (A.B. Dick Co.)

Fig. 24-2. Presses larger than duplicators are referred to as offset presses. Operating principles are basically same for both kinds. (Miehle Products, Graphic Systems Division, Rockwell International)

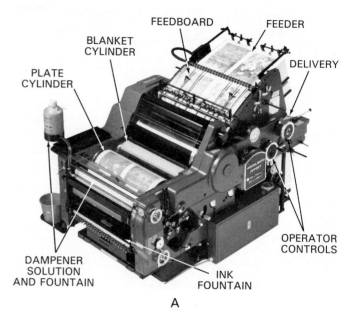

A

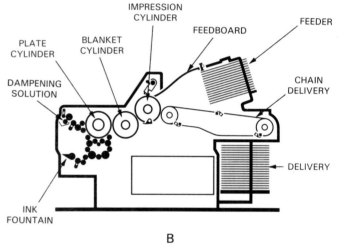

B

Fig. 24-3. A—Typical sheet-fed offset press, note that unlike most presses, feeder and delivery are at same end. B—Study schematic diagram of same offset press. (Heidelberg U.S.A.)

sheets of paper from a *feed table,* Fig. 24-3. Sheets are delivered, one at a time, down a *feedboard* to the printing unit. A small space separates the tail of one sheet and the head of the next sheet.

On other sheet-fed presses, the sheets of paper overlap one another as they travel down the feedboard, Fig. 24-4. This is referred to as *stream feeding.* By overlapping the sheets, it is possible to run the sheets of paper down the feedboard at a slower speed than that required for individually-fed sheets. This improves the registration of each sheet as it enters the printing unit. The sheets are less likely to bounce away when they strike the register guides at the end of the feedboard.

Web-fed presses

Presses that feed from a roll of paper are referred to as *web-fed presses,* Fig. 24-5. These presses are used for work requiring long runs, such as magazines, newspapers, and other similar publications. Although the method of feeding and the design of web presses are different from sheet-fed presses, the operating principles are the same.

Web-fed presses are extremely large, sometimes half the length of a football field and reaching one or two stories in height. A smaller version press, called the

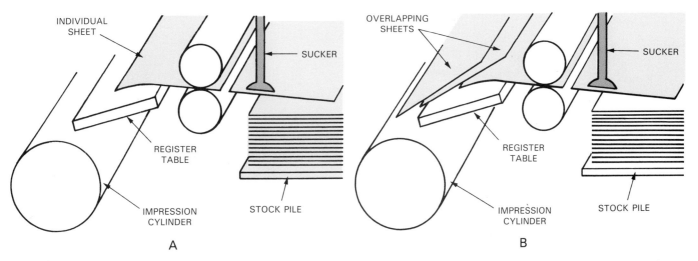

Fig. 24-4. A—Note press feeder that picks up single sheet at a time and delivers it down feedboard to printing unit. B—Note press stream feeder in which sheets of paper overlap one another as they travel down feedboard.

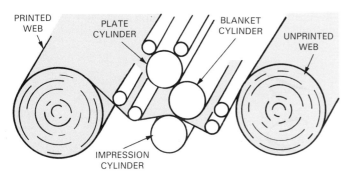

Fig. 24-5. Offset presses that feed from a roll of paper are referred to as web-fed presses.

mini-web, is rapidly gaining popularity especially with in-plant and small commercial printers. Chapter 27 contains a detailed description of web-fed presses.

Multicolor presses

Many offset presses and duplicators have only one printing unit and therefore can print only one color at a time. If a second or third color is required on a job, the sheets are allowed to dry after the first color is printed and then run through the press again to print the next color. This process is repeated until all the colors have been printed. This is a slow process but is acceptable for short press runs.

Presses that print more than one color at the same time are referred to as multicolor presses. These presses consist of a series of single printing units connected in a row (tandem) to form one press. Each unit can be set up with a different plate and a different color of ink. The paper can be fed in either sheets or rolls. See Fig. 24-6.

Townsend Color Head®

The Townsend Color Head® is a unique self-contained printing unit allowing second color printing on most offset duplicators, Fig. 24-7. The unit permits two

Fig. 24-6. Multicolor web presses like this can print several colors in one pass through press. Web presses consist of a series of single printing units connected in a row to form one press. (Miehle Products, Graphic Systems Division, Rockwell International)

colors to be printed in one pass through the press. The unit is capable of good registration and affords one- and two-color flexibility. These color heads are used largely by quick printers and in-plant printers.

Until recently, the color head was available only as a bolt-on unit specifically made to fit on several models of offset duplicators. With this version, the entire unit requires removal when not needed for two-color work. When the unit is again bolted on the press for color work, several adjustments must be made. The latest version is a swing-away model that eliminates the

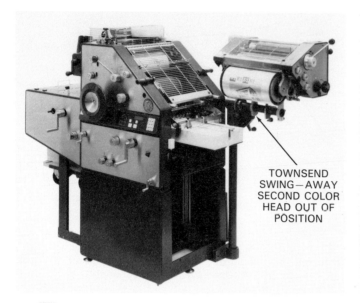
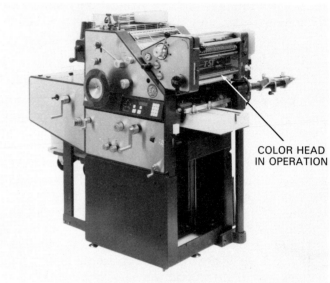

TOWNSEND
SWING—AWAY
SECOND COLOR
HEAD OUT OF
POSITION

COLOR HEAD
IN OPERATION

Fig. 24-7. Study self-contained second color printing unit on a duplicator. A—Unit has been swung aside out of position. B—Unit is in printing position ready for operation.   (Townsend Industries)

necessity of installing and removing the unit each time two colors are required, Fig. 24-8. Similar to the older version in operation, the newer swing-away model is a totally self-contained printing unit with its own plate cylinder and ink and dampening system. The image is transferred from its plate to the duplicator's blanket in register with the image from the duplicator's plate.

The swing-away color head has a single-lever control. It also features a variable control water dampening system. The swing-away capability along with the ability to remove and replace the color head without going through the old readjustment steps make the unit easier to operate.

Perfecting presses

Most offset presses print on only one side of the paper at a time. In order to print on the reverse side of the paper, the ink from the first side is allowed to dry and the paper turned over and run through a second time. Presses that print on both sides of the paper at the same time are known as perfectors. These are either sheet- or web-fed and may be single or multicolor. The operating principle of a perfecting press is illustrated in Fig. 24-9.

Main printing unit classification

The main printing unit of an offset press or duplicator includes three, but sometimes two, large cylinders. The plate cylinder carries the printing plate. It brings the inked plate image into contact with the blanket on the blanket cylinder. The inked image on the plate is transferred to the blanket. The inked image on the blanket is then transferred to the paper as the paper passes between the blanket and impression cylinders.

Fig. 24-8. Schematic diagram shows major components of second color head. Note segregated ink and dampening systems.   (Townsend Industries)

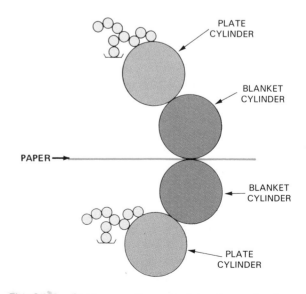

PLATE
CYLINDER

BLANKET
CYLINDER

PAPER

BLANKET
CYLINDER

PLATE
CYLINDER

Fig. 24-9. Study operating principle of a perfecting press.

The operation and arrangement of the cylinders in the main printing unit vary. Presses with three main cylinders are the most popular.

The *three main cylinder press* or duplicator has a plate cylinder, which holds the image (plate) to be printed. A blanket cylinder carries a rubber blanket onto which the image is transferred or offset from the plate. An impression cylinder presses the paper against the blanket. The image is then transferred from the rubber blanket to the paper. See Fig. 24-10.

On the plate, the image reads left-to-right, or *right reading*. When transferred to the blanket the image is backwards, or *wrong reading*. When the image is then transferred from the blanket to the paper, it is right reading once more.

The *two main cylinder offset press* or duplicator has a double-sized upper cylinder containing a plate section and impression section. The smaller lower cylinder, which is half the diameter of the upper cylinder, is called the *blanket cylinder*. Look at Fig. 24-11.

Since the blanket cylinder is half the diameter of the upper cylinder, the blanket cylinder revolves twice for each revolution of the upper cylinder. On the first revolution, the blanket contacts the impression section of the upper cylinder. The impression section of the cylinder forces the paper against the blanket cylinder. This causes the image to offset to the paper. At about the same time, the plate again receives more water and ink for another printing cycle.

OFFSET PRESS SYSTEMS

Offset presses and duplicators are divided into six operating units. These include:
1. Feeder unit.
2. Register board.
3. Printing unit.
4. Inking unit.
5. Dampening unit.
6. Delivery unit.

Feeder unit

There are basically two types of feeders on sheet-fed offset presses. These include pile feeders and continuous feeders.

With *pile feeders,* the paper to be run is piled on a platform in the press feeder, Fig. 24-12. This pile must be straight and neat. It must also be positioned in relationship with the settings of the register and insertion devices and for placement of the sheet.

The pile feeder raises the paper automatically to a constant level during printing. The pile feeder is the most versatile and flexible method for handling sheets.

The main disadvantage of the pile feeder lies in the fact that the press must be stopped for its loading. In very long runs, this drawback is a very important one. In short run work, where the total quantity of paper

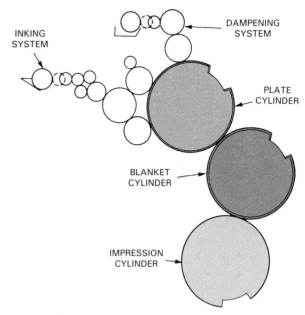

Fig. 24-10. Three main cylinder offset press or duplicator includes a plate cylinder, blanket cylinder, and impression cylinder.

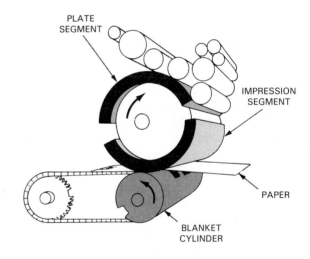

Fig. 24-11. Two main cylinder offset press or duplicator has a double-sized upper cylinder containing plate and impression segments. Lower cylinder, which is half diameter of upper cylinder, acts as blanket cylinder.
(A.B. Dick Co.)

to be printed does not exceed the capacity of the pile feeder, this limitation is NOT important.

A *continuous feeder* makes it possible to reload a pile feeder while the press is running. It brings the fresh load into feeding position before the preceding load has run out. The fresh load is placed into elevating position when the preceding load still has some time to run. By means of supplementary elevator bars, the weight of the small load is temporarily supported on rods inserted under or near the bottom. The pile elevator bars are lowered to the fresh load and the paper is elevated up to the bottom of the preceding load. When the rods are withdrawn, there is a single pile in the feeder.

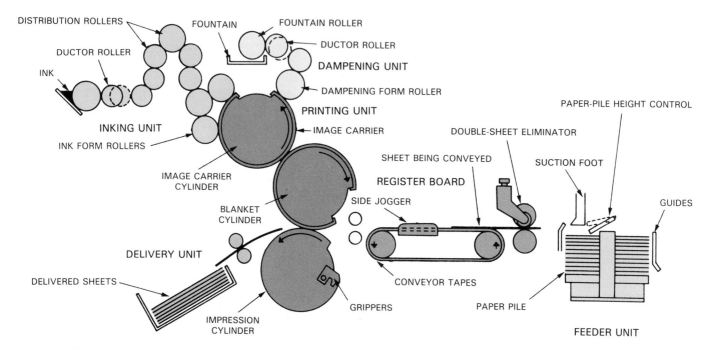

Fig. 24-12. Learn these six systems, typical of offset presses and duplicators. Note that paper to be run is piled on a platform in feeder. (A.B. Dick Co.)

If care is used in these operations, there is no interruption to the press run. Fig. 24-13 shows a continuous feeder.

There are several advantages to continuous feeding aside from increased productivity. There is also lower spoilage and consistent quality. It is well recognized that the more steadily an offset press runs, the easier it is to maintain consistent color. This is because the ink and water balance can be set for optimum productivity without the need for stopping to reload the press.

The *paper feeder unit* (on presses with pile or continuous feeders) consists of a feed table that holds the paper and can be raised and lowered by turning a handle on the side of the press. The feed table is lowered manually when loading the feeder with paper. It is manually raised again to operating height before starting a press run.

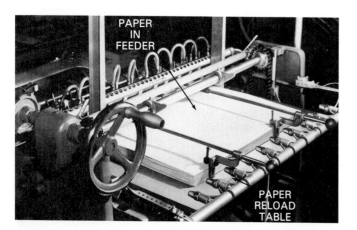

Fig. 24-13. This offset press continuous feeder makes it possible to reload feeder with paper while press is running.

Most presses are equipped with a *release lever,* usually located near the paper table. This lever permits the operator to lock the paper table so that it will not rise. Two metal bars, known as *piling bars,* are positioned along a scale at the front of the feed table. Their position corresponds to the width of the paper to be run. The *piling bars* steady the paper as it is fed into the press.

After the piling bars are adjusted, paper is placed on the paper table. The *paper table* rises automatically as the sheets of paper are fed into the press, keeping the height of the stack constant.

The amount of paper loaded on the paper table depends on the length of run and type of press. For example, most duplicators are equipped to handle a stack of 20-pound, bond paper about 20 inches deep.

The paper must be fanned and jogged prior to placing it onto the feed table. *Fanning* is the procedure which introduces a blanket of air between the sheets of paper, separating them and making feeding easier. *Jogging* refers to squaring and aligning all the sheets just before loading onto the feed table.

The *pile height regulator* controls the speed of elevation of the feed table. This allows the table to run higher or lower, depending on the thickness of paper being used. As the paper is fed into the press, the pile height regulator rides lower and lower. This action moves the feed table up one notch at a time, the thicker the paper—the faster the table rises.

A *vacuum blower pump* inside the press forces air through plastic tubes near the front of the paper stack. The air is regulated so that it will separate the top sheets of paper, providing better feeding. A set of *suction feet* drops down against the front edge of the top sheet of paper on the pile and lifts it to the edge of the feed-board. See Fig. 24-14.

Sometimes, the suction feet pick up two or more sheets of paper at a time. To avoid jamming the press or damaging the rubber blanket, a *double-sheet detector* is provided, Fig. 24-15. This safety device is set to the thickness of a single sheet of the kind of paper being used. When more than one sheet of paper is fed through the feeder, the double-sheet detector automatically rejects all but one sheet. The rejected sheet(s) fall into a metal tray beneath the feedboard.

On larger offset presses, the double-sheet detector is adjusted until a single thickness just goes through, but a double thickness of paper jams or trips the device. There is no metal tray beneath the feedboard on larger offset presses.

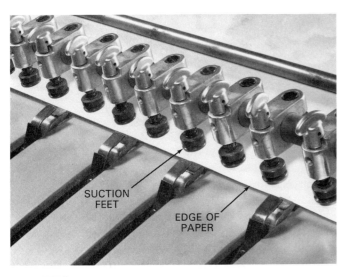

Fig. 24-14. Suction feet on offset press feeder unit drop down against front edge of top sheet of paper on pile and lift it to edge of feedboard.   (Heidelberg U.S.A.)

Fig. 24-15. Offset presses and duplicators are equipped with double-sheet detectors to avoid jamming press or damaging rubber blanket.   (Heidelberg U.S.A.)

The amount of forced air and suction is usually adjusted by controls on the side of the press. The blower is usually adjusted to separate 5-6 sheets of paper in the pile. Suction is set to regulate the suction feet. Light suction is used for thin paper and heavy suction is used on thick paper and cardboard. See Fig. 24-16.

Fig. 24-16. Operator of an offset press or duplicator can control force of feeder air blower and suction.
(Royal Zenith Corp.)

NOTE! Check the oil level in the vacuum blower pump jars weekly. Clean the filters and usually add nondetergent SAE No. 10 weight oil as needed.

Most offset duplicators and some offset presses are also equipped with sheet separators. These are spring steel fingers that extend over the lead edge of the pile of paper in the feeder, Fig. 24-17.

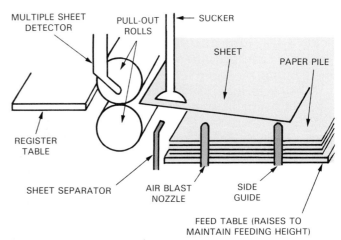

Fig. 24-17. Sheet separators are spring steel fingers that extend over lead edge of paper pile in feeder unit. These fingers hold top sheet down as it is separated by front blowers.

Sheet separators hold the edge of the top sheet down as it is separated by the front blowers. When the suction feet grip the top sheet on the pile, the sheet separators prevent the next sheet from being picked up at the same time. In order to have the separator devices function properly, the pile must be at the proper height in the feeder, and the contour of the top of the pile must be reasonably flat. This is especially important on offset presses with the capacity to handle large sheets.

The sheet separators must be properly adjusted to provide the correct amount of resistance. If the separators drag too heavily against the paper, they may interfere with the action of the suction feet. If the tension is too light, the separators may fail to separate the sheet properly. This can cause more than one sheet to enter the press.

At this point, the difference in feeders of various designs is most noticeable. The manufacturer's instructions must be followed in all instances. It is well, however, to keep in mind several principles that apply, generally speaking, to all feeders.

On many offset presses, separation and forwarding of sheets may be accomplished entirely by air. Air blast nozzles fluff the rear corners of several sheets on the top of the pile. Suckers drop down and pick the top sheet up. Another set of air nozzles moves under this top sheet and floats it on a cushion of air. Then the same suckers, or a set of separate forwarding suckers, guide the sheet into the forwarding devices at the top of the feed ramp.

On differently designed feeders, especially the larger ones, separation is generally started at the rear corners of the sheet usually by a combing device of some kind. See Fig. 24-18.

Telescopic suckers come down and pick up the corners of the top sheet while it is being combed. Blower tubes, shaped like a foot, descend and enter under the top sheet, resting firmly on the pile. Then, a blast of air from the blower feet separates the entire top sheet. This blast of air floats the sheet up to a second set of suckers while the first pair still have the sheet under control. The second set of suckers are the ones that generally advance the sheet to the forwarding mechanism.

While principles remain the same, there are considerable variations in the structure and operating details of the feeders supplied by the various manufacturers. Even within the line of presses of one manufacturer, there are wide differences, especially between the larger and smaller offset presses. The brief descriptions given of two basic types, apply generally to the larger presses. Most of the modern smaller presses and duplicators use other separation devices.

It is obvious that control of the amount of air and the direction of the blower tubes or feet are critical. The sheet must be positively floated without riffling, waving, or flapping. The sheet being forwarded must

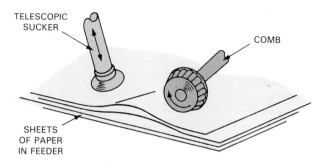

Fig. 24-18. On large sheet-fed offset presses, sheets in feeder are initially separated by a combing device. Telescopic suckers descend on sheet and lift it to a second set of forwarding sucker feet. (Graphic Arts Technical Foundation)

be completely separated from the pile so as not to drag the next sheet, or stumble on its leading edge, especially in feeders where no blower foot device is provided.

The sheet must be under such complete control that it reaches the forwarding devices at the right time and perfectly straight. If both of these conditions are NOT met, proper timing of the sheet is lost. Also, a sheet coming down the ramp or board in a cocked position will either jam at the side guide mechanism or reach the guides or insertion devices in such a way that it will jam and/or trip the press.

In addition to the pile and continuous types of feeders just described, many continuous feed presses are also equipped with streamfeeders. See Fig. 24-19. The slow travel of the sheet to the registering position on stream feeders simplifies some of the sheet control problems found on a successive-sheet feeding mechanism.

Register board

In the preceding discussion, you learned how each sheet of paper is separated from the pile and forwarded to the registering position. Once this has been accomplished, the feeder unit has fulfilled its function. At this point, a completely different set of devices takes over control of the sheet. Among these devices, there are differences between equipment manufacturers as well as between some models of the same manufacturer.

The *register board* accepts a sheet from the feeder unit and inserts it in register ready for printing.

Before going further it is necessary to clearly understand the significance of register. *Register* means the proper positioning of the sheet with respect to the image on the blanket. On all large offset presses and some duplicators, the press inserts and aligns the sheet so that when the impression cylinder grippers draw the sheet through the printing nip, the front edge of the paper is a set distance from the image. This distance is determined when the plate is clamped into proper position and the front guides are set.

One side edge is also a fixed distance from the image edge. This distance is also determined when the plate

Fig. 24-19. This two-color, sheet-fed offset press is equipped with a stream feeder. This simplifies some of the sheet control problems. (Miehle Products, Graphic Systems Division, Rockwell International)

is put on the press and the side guide is set in its proper position.

The purpose of all these operations is important. Each and every sheet must be aligned exactly alike. With the sheets under proper control, and the registering devices properly set and operating, good register will be accomplished.

Register and insertion methods

In general, there are three press register and insertion methods. These include: three-point guide system, feed-roll system, and swing-feed or transfer-gripper system.

With the *three-point register system,* the sheet is forwarded to the front guides (two to four). While it is held against the front guides, a side guiding mechanism pushes or pulls the sheet into proper side alignment. In the three-point register system, the sheet is held in this position until the impression cylinder grippers take hold. At this time, the front guides or stops are lifted out of the sheet's path of travel. As soon as the entire sheet clears the feedboard, the guides drop back into position to align the next sheet.

In the simple three-point register system, the guides are mounted on a shaft above the feedboard, Fig. 24-20. The guides can be moved sideways along this shaft to properly balance the particular size of the sheet being run. The guide itself is essentially a flat face plate. In the registering (lowered) position of this guide, its bottom edge straddles a metal tongue protruding under the feedboard, or a register plate.

It should be pointed out that the three-point guide system is not an insertion device. However, the system does fix the sheet in printing position just as is done for any insertion device. However, on presses equipped with insertion devices, the positioning of the sheet BEFORE the insertion device starts its cycle of operation is generally referred to as *pre-registering.*

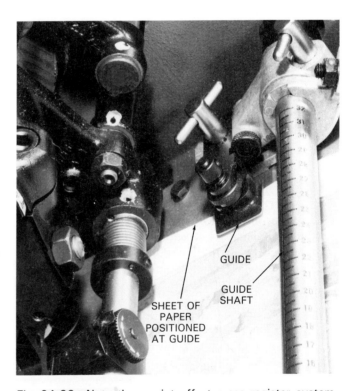

Fig. 24-20. Note three-point offset press register system. Guides are mounted on shaft above feedboard. (Heidelberg U.S.A.)

Where the three-point register system is supplemented by an insertion device, the general principle and construction are similar. However, the front guides are generally constructed so as to drop down out of the forward path of the sheet after the grippers on the insertion device take hold of the sheet. See Fig. 24-21.

With a *feed-roll system,* the impression cylinder grippers do NOT take the sheet from the feedboard. The device actually inserts the sheet into the grippers of the impression cylinder or an intermediate device. After the sheet is forwarded down the feedboard, it is pre-

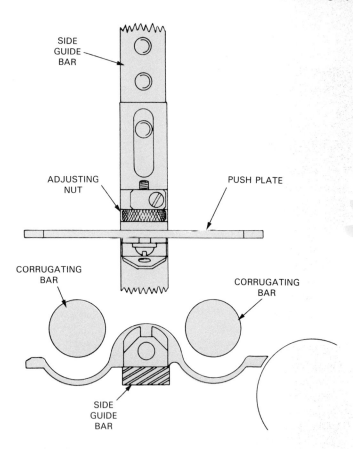

SIDE GUIDE BAR

ADJUSTING NUT

PUSH PLATE

CORRUGATING BAR

CORRUGATING BAR

SIDE GUIDE BAR

Fig. 24-21. Cross-sectional view shows parts of a push-type offset press side guide. (Graphic Arts Technical Foundation)

registered against a number of stops (front guides) spaced across the entire width of the feedboard. While the sheet is held against these stops, it is side-guided.

While the sheet is at rest, it is firmly gripped in its pre-registered position by being pinched between upper and lower feed rolls or cams. The front guides then drop down or lift up. As soon as they are clear, one set of the feed rolls (upper or lower depending on press design) starts rotating and drives the sheet against *gauge pins* or *stops*. Depending upon press design, the gauge pins or stops are located on either an intermediate cylinder (sometimes called a feed or transfer cylinder), or on the impression cylinder. These stops may be nothing more than slight lips protruding above the gripper pad. When the sheet is inserted into the grippers, it is held against these cylinder stops until the grippers have closed firmly on the sheet. At this instant, the low points of the upper feed-rolls have come around, and the sheet is completely under the control of the grippers.

Where an intermediate feed or transfer cylinder is used between the feedboard and the impression cylinder, complete control of the sheet is maintained by having the grippers of the intermediate device positively grip the sheet while the impression cylinder grippers close on it. For a short distance of travel, BOTH sets of grippers are actually holding the sheet.

This point is worthy of special mention, since on all presses, where a sheet is being transferred from one

set of grippers to another, transfer is never accomplished "on-the-fly." This means that at some point during the transfer both sets of grippers are actually holding the sheet.

The design of a *swing-feed* or *transfer-gripper system* is completely different from that of feed-roll insertion devices. In this system, the sheet is forwarded down to the feedboard and brought to rest against front stops. Then it is side-guided. After being properly guided, the sheet is picked up by a set of grippers, usually mounted over the feedboard. The front guides move out of the forward path of the sheet. This pick-up mechanism is called a *swing-feed, transfer cylinder,* or *transfer-gripper assembly*.

After the sheet has been picked up, the mechanism forwards the sheet and inserts it into the impression cylinder grippers. In this system, positive control over the sheet is maintained during the transfer of the sheet from the insertion device to the impression cylinder.

Front and side guides

Guides are divided into front and side guides. Front guides are classified as multiple-stop, or two-point drop guides or stops. Front guides may form part of the insertion devices. Side guides operate independently of the front guides or guiding device.

Neither the design of the side guide or the style of its operation are affected by the type of front guides, insertion device, or gripper action on the press. Side guides can be described according to their motion or by the manner in which they operate.

As far as their motion is concerned, side guides can be either push-guides or pull-guides. This is true disregarding the side edge (left or right) of the sheet to be guided. Some presses are so equipped that when one side guide is pulling, the guide on the opposite side may be set to push at the same time.

All *push-guides* are essentially the same. The sheet is forwarded into the front guides or insertion device. Then the side guide acts by pushing the sheet into its predetermined position. Whether the side guiding is done at the near side, or at the far side of the press, does not alter the actual guiding operation.

Push-guides are generally found on smaller presses and those running heavy, rigid material such as metal and cardboard. They do NOT function very well where large size, light-weight paper must be handled. Along with the push-guide type of mechanism on paper presses, there is usually some type of corrugating or stiffening device. This device serves to put a slight buckle in the sheet to stiffen it against the action of the push-guide.

Pull-guides are of three general types. These include: finger-type pull-guide, rotary or roller guide, and a combination of these. In all types, the side-guide plate is fixed in the desired position, and the sheet pulled up against it after the sheet has been positioned at the front guides or stops.

The *finger-type pull-guide* advances over the sheet, closes on it by pinching the sheet against a lower plate and pulls the sheet to the side-guide plate. The tension on this finger is adjustable. The finger must be adjusted so that it slips over the surface of the sheet without buckling it once the sheet is stopped by the side-guide plate. This adjustment of the finger pull-guide is possible for any weight paper. Even though the mechanism is rather simple, the adjustment is quite critical.

The *rotary* or *roller guide* is common on most presses. Built into this mechanism is a lower roller which constantly rotates when the press is running. The sheet is forwarded to the front stops so that the side edge rides over this lower guiding roller. After front register has been accomplished, a second spring-loaded roller, mounted above the lower one, drops down and pinches the sheet against the rotating lower roll. Friction causes the sheet to be pulled against the side-guide plate. As with the finger-type of side guide, the tension is adjustable to accommodate a wide range of paper thicknesses.

With a *combination guide,* a finger slides under the front-guided sheet. When it has reached the end of its forward movement, a spring-loaded upper roller pinches the sheet against the lower finger. Then the finger starts to move back and pulls the sheet with it until the side edge of the sheet is stopped by the face plate of the side guide.

Grippers

Grippers are classified either as tumbler grippers or as low-lift grippers. Presses using a three-point register system, without an insertion device, require tumbler grippers. Presses which have either a feed-roll or swing-feed (rotary gripper) insertion device use a low-lift gripper.

A *tumbler gripper* rotates through a rather large arc when it opens and actually drops back into the impression cylinder gap. It must drop below the cylinder surface to clear the gripper edge of the sheet as the impression cylinder comes around to its sheet-taking position. At this point, the gripper rotates into the closed position, pinching the sheet against gripper pads, and pulling the sheet into the impression nip.

The term *"tumbler"* is derived from the mechanism that imparts the opening and closing motions to the gripper shaft.

The term *"low-lift gripper"* refers to the grippers on presses with insertion devices that open just enough to allow the sheet to be inserted. Instead of a mechanism that imparts a complete tumbling action, low-lift grippers are mounted on a shaft rotated slightly through a cam and cam-roller device at the end of the gripper shaft.

Grippers are generally of one-piece or two-piece construction. A *one-piece gripper* has its finger constructed as an integral part of, and an extension of the base or gripper clamp. The base is often a solid piece of metal

drilled to fit snugly on the gripper shaft. A set screw is installed into the base. When the set screw is tightened, it clamps the gripper firmly on the shaft. In other cases, the base may be a split collar around the shaft. The set screw, instead of being turned down on the gripper shaft, or gripper-shaft shoe, is threaded through both ends of the collar. When the set screw is tightened, it closes the collar around the shaft thereby clamping the gripper firmly to the gripper shaft. The finger itself may be an extension of one side of the collar, or the collar is formed on the opposite side of the gripper shaft from the gripper finger.

In two-piece construction, the finger and the collar are actually two separate pieces. They design so that while the finger opens and closes as an integral part of the assembly, provision is made for spring loading the gripper finger itself. This simplifies attaining a more perfect, uniform gripper "bite pressure" on the sheet. This same objective is accomplished in the case of some one-piece grippers by spring loading the gripper pad set in the top of the gripper post. Three common types of gripper construction are shown in Fig. 24-22.

An important detail in the construction and function of grippers is the character of the gripping face on the top of the gripper pad. It is obvious that there is considerable pull on the sheet as it passes through the impression nip. Not only must the grippers guide the sheet through the nip but they must also hold the sheet against the pull of the nip. In some instances, the pressure of the gripper against the pad may not be enough. Too much bite pressure may mark the sheet and cause difficulty in later passes of the sheet through the press.

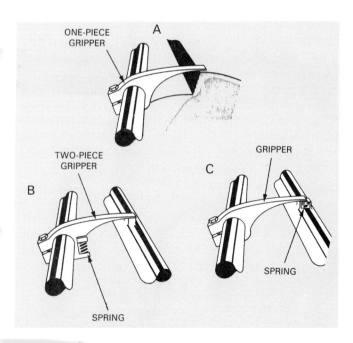

Fig. 24-22. Three common types of gripper construction. A—One-piece gripper. B—Two-piece spring-loaded gripper. C—One-piece gripper with a spring-loaded gripper pad. (Graphic Arts Technical Foundation)

Several techniques have been used to accomplish correct gripper pressure. Some of the newer offset presses have grit-faced grippers to prevent slippage of the sheet during the critical printing cycle, Fig. 24-23. On older presses, the grippers or gripper pads may be faced with light-weight emery cloth if proper bite pressure cannot be developed through the spring-loaded grippers or gripper pads.

Duplicator register system

Having covered the topic of register and insertion devices on larger offset presses, the following discussion concerns register as it relates to duplicators and small offset presses.

During the feeding cycle on many offset duplicators, the suction stops briefly, releasing the top sheet to a pair of rubber forwarding rollers. These rollers deliver the sheet of paper onto a set of moving *conveyor tapes* that feed each sheet to the grippers.

NOTE! A few press designs do NOT have conveyor tapes. Instead, the sheet of paper is delivered directly to the grippers.

The narrow belts or conveyor tapes are located on the feedboard. The fabric tapes are threaded over a roller at each end of the feedboard. When the press is operating, the rollers turn and the tapes travel down the feedboard and back under it in one continuous motion. Individual pulleys located under the feedboard keep the tapes tight. The tapes can be adjusted individually inward or outward to handle any width paper. The tapes are moved by sliding the pulleys from side to side or by rotating a knob to move the tapes. Fig. 24-24 shows a set of conveyor tapes on a duplicator.

To keep the sheet of paper under control during its trip down the conveyor tapes, other devices are required. Most duplicators are equipped with *metal strips* that extend the full length of the feedboard and are usually positioned over each of the conveyor tapes. These are set so that there is a slight space between the strips and conveyor tapes to allow for passage of the paper. The metal strips can also be adjusted to handle any width of paper.

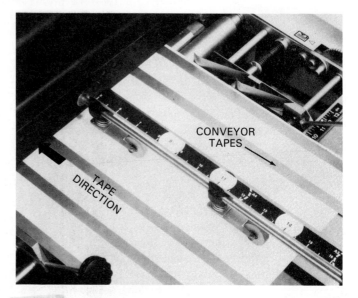

Fig. 24-24. Most duplicators are equipped with conveyor tapes that transport each sheet from feeder to front guides. (A.M. International, Multigraphics)

Some duplicators are equipped with rods instead of metal strips. The *rod* holds small glass or plastic balls similar to marbles. The balls press on the sheet of paper as it travels down the feedboard and keep it under control.

The *stop fingers* or *front guides* are two metal posts or pins that stop the sheet of paper and hold it on the feedboard while the side guide moves it into position. Look at Fig. 24-25. When the sheet is in position at the sides, the front guides either rise above or drop below the surface of the feedboard and allow the sheet to pass. The paper is then caught by the impression cylinder and carried into the printing unit.

On some duplicators and most larger offset presses, the front guides can be adjusted forward or backward to regulate the gripper on the paper. If the image is slightly crooked on the plate, the press operator can move the front guide in or out on one end. This changes the position of the image on the sheet. Some presses have other types of devices for the same purpose. The sheet of paper is made to skew slightly as it feeds into the printing unit.

The *side guides* push the sheet of paper sideways to its proper position. Some duplicators have only one side guide while others have one on each side of the feedboard. Only one guide is used at a time. The guide not being used is locked out of operation or moved to the edge of the feedboard.

In setting the side guide, the duplicator is operated manually by turning a flywheel until the guide to be used is at the end of its inward travel. The guide is then moved to a position along the *feedboard scale* that is the same width as the width of the paper to be run.

When the active side guide has been set to its proper position, a sheet of paper is run through the press. This is done to check that the active guide is pushing the

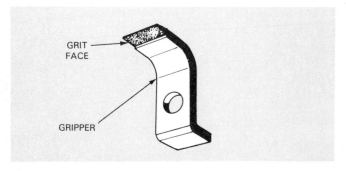

Fig. 24-23. This grit-faced gripper prevents slippage of sheet during printing cycle. (Graphic Arts Technical Foundation)

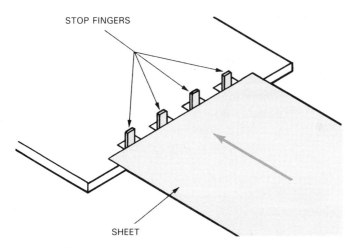

Fig. 24-25. Stop fingers are metal posts or pins that stop sheet of paper and hold it on feedboard while side guide moves it into final position.

sheet of paper about 1/8 to 1/4 inch (3 to 6 mm) sideways. If the duplicator is equipped with a second side guide, it is adjusted to rest against the paper with a slight pressure.

A metal *sheet detector* finger senses when a sheet of paper fails to feed through the duplicator. If no sheet of paper is present, the main cylinders pull away from one another. This prevents the blanket cylinder from printing on the impression cylinder. On some duplicators, the sheet detector finger is located near the center of the feedboard while on other presses it is over to one side.

The *impression cylinder grippers* are small metal fingers attached to the impression cylinder and operated by a cam, Fig. 24-26. The gripper fingers open to receive a sheet of paper and close again along the paper's lead edge. Duplicators are equipped with up to six gripper fingers. The grippers pull the paper into the printing unit between the blanket and impression cylinders, called the *nip*.

PRINTING UNIT

The printing unit is the heart of the offset press. Everything about the actual printing cycle centers around the three cylinders in the *printing unit*. The three cylinders are: the plate cylinder, the blanket cylinder, and the impression cylinder. The surfaces of these three cylinders must travel at the same speed (same distance during each cylinder revolution). They must be adjusted for minimum cylinder-to-cylinder pressure capable of producing accurate printed reproduction and long plate life.

Plate cylinder

The *plate cylinder* holds the printing plate and its image. Most duplicators can take offset plates ranging in thickness from about 0.0025 to 0.006 inch (0.06 to 0.15 mm). Large offset presses take thicker offset plates. The attaching device on duplicators is usually a pinbar clamp. The clamp can be changed for use with other types of plate-end attaching methods.

Several kinds of plate-end attachments are illustrated in Fig. 24-27.

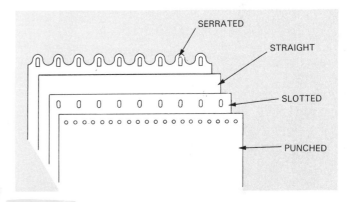

Fig. 24-27. Press and duplicator plate-end attaching methods include serrated, straight, slotted, and punched.

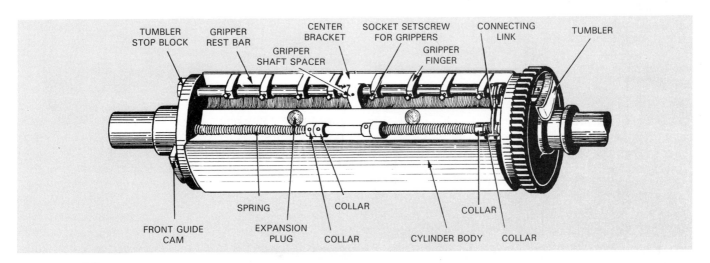

Fig. 24-26. Impression cylinder grippers are metal fingers attached to offset press or duplicator impression cylinder and operated by a cam. (U.S. Government Printing Office)

533

The plate clamps are set so that one clamps the leading edge of the plate while the other holds the tail edge of the plate firmly. Both clamps are designed for a slight sideways cocking movement of the plate. This is sometimes necessary when a plate must be shifted to obtain proper register. On large presses, there is also some latitude for forward and backward movement.

The procedure of attaching an offset plate to the plate cylinder of a duplicator is illustrated in Fig. 24-28. The lead edge of the plate is attached to the lead edge clamp of the plate cylinder. The cylinder is then rotated to draw the plate around the cylinder. The tail clamp is raised by pushing up on a lever and locking the tail edge of the plate over the pins. When the lever is released, spring tension on the tail clamp is enough to hold most paper and foil plates tightly. For metal plates, the tail clamp is tightened slightly by turning the tightening dial clockwise. The process is reversed when removing a plate from the plate cylinder.

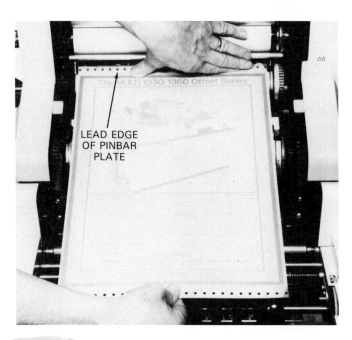

Fig. 24-28. First step in attaching a duplicator plate to plate cylinder is fitting it to lead edge clamp of plate cylinder. Tail edge of plate is then attached to tail clamp.

On some duplicators and larger presses, the plate can be twisted slightly on the cylinder. This is done to compensate for an image that is NOT exactly square on the plate. Other presses are equipped with a paper stop bar. The bar can be adjusted so that one end is set farther in than the other. The paper registers against the fingers on the bar when it reaches the end of the feedboard. Setting the bar at an angle causes the sheets of paper to feed into the printing unit at a slight angle or skew. This aligns the paper with the image on the plate and produces a square image on the paper. The paper stop bar can also be slightly adjusted forward or backward to raise or lower the printing image on the paper.

Adjusting the image more than about 3/16 inch (5 mm) means the plate cylinder must be moved by turning it manually. Turning the cylinder moves a scale containing an indicator pin, Fig. 24-29. When the cylinder has been turned the proper distance along the scale, the cylinder is locked in place.

After adjusting the plate cylinder, the rubber blanket on the blanket cylinder must be thoroughly cleaned before making another impression.

Undercut cylinders

On large offset presses, the bases of the plate cylinder and the blanket cylinder are each *undercut*. This means that they are made smaller in diameter than the narrow band or metal ring at each end, called the *bearers*. When properly adjusted, the bearers maintain the two cylinders parallel to each other at the correct level, Fig. 24-30. The amount that the cylinders are undercut below that of the bearers is stamped in the cylinder gutters. It indicates, in thousandths of an inch, how much the cylinder body is undercut below the bearers.

The undercut allows for adding packing sheets beneath the plate and blanket. This helps compensate for variations in the thickness of plates and blankets. It also maintains the two cylinders at virtually equal diameters. If one were larger than the other, the image size could be changed during image transfer. The total thickness of the packing sheets, plus the thickness of the blanket and the plate should equal 0.003 inch (0.08 mm) more than the combined undercut of the two cylinders. A packing gauge is used for this purpose, Fig. 24-31. This excess is referred to as *squeeze*.

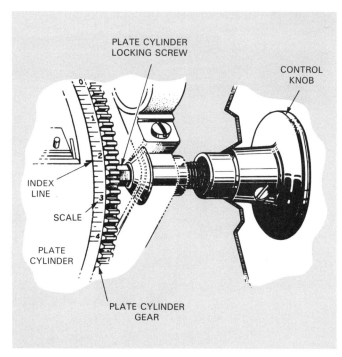

Fig. 24-29. Offset duplicators are equipped with control device for raising or lowering image on paper. (A.M. International, Inc.)

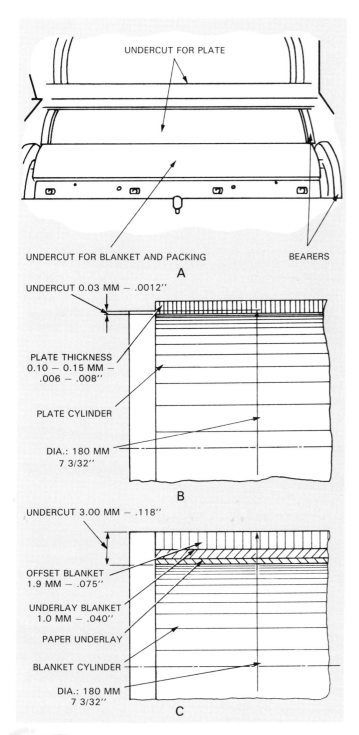

UNDERCUT FOR PLATE

UNDERCUT FOR BLANKET AND PACKING   BEARERS

A

UNDERCUT 0.03 MM — .0012''

PLATE THICKNESS
0.10 — 0.15 MM —
.006 — .008''

PLATE CYLINDER

DIA.: 180 MM
7 3/32''

B

UNDERCUT 3.00 MM — .118''

OFFSET BLANKET
1.9 MM — .075''

UNDERLAY BLANKET
1.0 MM — .040''

PAPER UNDERLAY

BLANKET CYLINDER

DIA.: 180 MM
7 3/32''

C

Fig. 24-30. Offset press bearers (A) maintain the plate cylinder (B) and blanket cylinder (C) parallel to each other at the correct level. (Heidelberg U.S.A.)

Blanket cylinder

The *blanket cylinder* holds a rubber blanket that accepts an image from the plate. The *blanket* attaches to a set of pins at each edge of the cylinder, Fig. 24-32. The rubber blanket is attached to the upper, or lead edge, pins first. The cylinder is turned manually to draw the blanket around. The blanket is then attached to the tail edge of the cylinder. The blanket is drawn tight by turning the clamp tightening dial counter-clockwise.

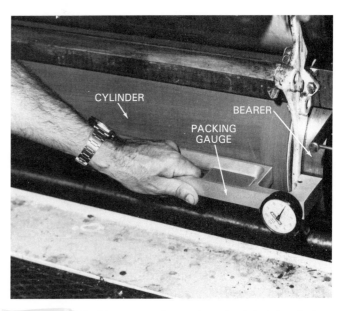

CYLINDER   BEARER

PACKING
GAUGE

Fig. 24-31. Packing gauge is used to check height of plate and blanket cylinder packings. (Baldwin-Gegenheimer Corp.)

NOTE! The blanket may stretch when it is first used on the press. After several hundred impressions, the press should be stopped and the blanket re-tightened.

It is not necessary to change a blanket unless it becomes damaged or worn. Some press operators keep several blankets on hand and rotate them. This gives each blanket a chance to rest. This procedure helps to prolong blanket life and gives better printing quality. Most press operators use a separate blanket for perforating and running envelopes.

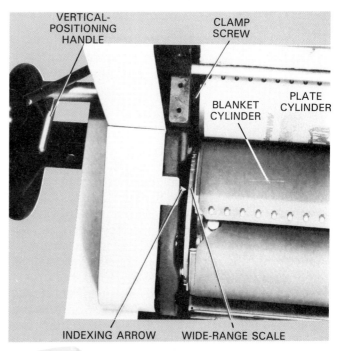

VERTICAL-
POSITIONING
HANDLE   CLAMP
SCREW

BLANKET
CYLINDER   PLATE
CYLINDER

INDEXING ARROW   WIDE-RANGE SCALE

Fig. 24-32. Rubber blanket on this offset duplicator attaches to a set of pins at each end of blanket cylinder.

Blanket care

The surface of the rubber-covered fabric on an offset blanket receives the image of ink from the plate. This image is offset to the paper as it is pressed against the blanket by the impression cylinder. The impression cylinder is a sturdy steel surface. Therefore, the quality of the impressions on the paper depends primarily on the quality of the blanket surface.

Blankets are made of a number of layers. The top layer is natural or synthetic rubber. The second layer is generally a polyester film. This film keeps the bottom two layers from becoming solvent-soaked. The third layer is a fabric. Cotton is often used for this purpose since it is flexible and strong. The fourth layer is a waffle design, Fig. 24-33. This layer is either rubber or a synthetic material.

Blankets with waffle backing can withstand minor smashes. A *smash* is a low spot on the blanket. When a wrinkled sheet goes through the press, it will have areas of two or three thicknesses. This can make a low spot in the blanket. In some instances, a *blanket fix solution* can be used to repair the low spot. This is a rubber-based solution which adheres to the blanket, filling in the low spot.

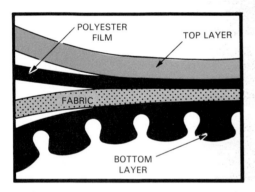

Fig. 24-33. Cross-sectional view of offset blanket shows four primary layers of material. (A.B. Dick Co.)

Installing a new blanket

The blanket to be used should be recommended by the manufacturer of the press or duplicator. It is also important that the packing beneath it, if required, should conform to the manufacturer's recommendations. Always inspect a new blanket carefully. Remove the powder usually found on new blankets by washing it with press wash. If there are imperfections on uneven surfaces, it may be necessary to scour the surface with pumice powder or return the blanket to the manufacturer.

The blanket must be attached to blanket bars at each end. Bars are made for each kind and size of press and duplicator. They tightly grip the blanket at all points of the lead and trailing edges. Some bars use bolts to apply pressure while others use special teeth and gripping devices.

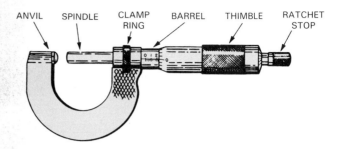

Fig. 24-34. A micrometer is used to measure thickness of offset blankets and to check packing thickness.

Blanket thickness must be measured, Fig. 24-34. This is done at a number of key points. The press operator must be precise when using the micrometer. If it is needed, the packing is chosen based upon the blanket thickness. The packing sheets are placed under the blanket.

NOTE! Most duplicators do NOT require packing under the blanket or plate.

Tightening the trailing blanket bar is the final step. The blanket bars must be secure since they could release during the press run. This could cause severe damage to the duplicator or press.

The blanket must be tight, but not too tight. When a blanket is over-tightened, it can stretch the fabric base and rubber surface. This will make the blanket thickness uneven. This causes poor ink transfer from the plate to the blanket. Poor ink transfer will also occur from blanket to paper.

Always adjust the blanket and cylinder pressure or remove some of the packing when running heavy paper. On heavy paper, adjust the feed mechanism properly to avoid feeding two or more sheets through the press at once. This can smash a blanket easily.

Types of offset blankets

There are two basic types of offset blankets. The conventional offset blanket bulges at the impact point. A compressible offset blanket compresses (squeezes) and instantly rebounds, returning to its original shape, Fig. 24-35. Because of this instant rebound factor, better dot for dot image transfer without distortion can be achieved. This produces a higher quality printing job.

Keep in mind that several types of blankets are available, Fig. 24-36. For example, some work better with coated papers than others. Specially formulated blankets must be used with some types of ink. For example, on large presses, using heatset inks, special blankets are required. Always select the blanket best suited to the paper and ink being used on the press. The blanket manufacturer is the best source for questions relating to which blanket is best for the job.

Preserving blanket life

If the duplicator or press will not be used for several days, release the tension on the blanket to keep it

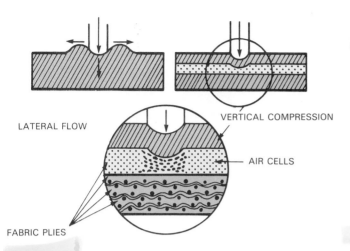

LATERAL FLOW

VERTICAL COMPRESSION

AIR CELLS

FABRIC PLIES

Fig. 24-35. Note cross-sectional view of a compressible off-set blanket. This type of blanket compresses and instantly rebounds, returning to its original shape. (A.B. Dick Co.)

resilient. All ink should be removed after each press run. To keep the blanket clean and ready for the next job, do NOT allow fluids to dry on the blanket. Wipe the blanket clean with a clean dry cloth.

To lengthen the life of the blanket, use a high quality

blanket wash, Fig. 24-37. Inexpensive blanket washes tend to dry out a blanket and make it less resilient. See Fig. 24-38.

Danger! Never use gasoline as a blanket wash! Make sure that the blanket wash you are using is NOT a flammable material.

Warning! Exercise extreme care when using solvents and inks. Place solvent-soaked rags in a metal safety can with a tight metal lid. Spontaneous combustion is always a threat!

Impression cylinder

The *impression cylinder* forces the sheet of paper against the blanket to transfer the image to the paper, Fig. 24-39. It is also called the *back cylinder* because it is usually located behind the blanket cylinder.

A set of grippers on the impression cylinder grip the sheet of paper when it reaches the end of the feedboard. This draws the paper around forcing it against the blanket to make the impression. The grippers then open to release the sheet for delivery.

The pressure between the blanket and impression cylinders must be adjusted each time a change in paper thickness is made. When going from a thin to a thick

BLANKET GAUGE AND COLOR CHART

Blanket Name	Ply	Gauge	Color Availability			
Diamond	2	.050″	Red			
Conventional[1]	3	.066″	Red	Blue	Black	Green
	4	.076″	Red	Blue	Black	Green
Double Diamond	3	.066″	Indigo			
Compressible[2]	4	.076″	Indigo			
Diamond Underblanket	—	.039″	Black			
Mercury Duplicator	2	.050″	Red			Green
	3	.066″	Red			Green
	4	.076″	Red			Green
Fujikura Conventional	3	.065″	Sapphire Blue			
	4	.075″	Sapphire Blue			
Fujikura Compressible	3	.065″	Jade Green			
	4	.075″	Jade Green			
Dunlop Consul	3	.067″	Purple			
	4	.077″	Purple			
Dunlop XXX	3	.066″	Blue			
	4	.076″	Blue			
Dunlop GP	3	.066″	Grey			
	4	.076″	Grey			
Dunlop Underlay	—	.039″	Black			
	—	.066″	Black			
	—	.076″	Black			

[1]DIAMOND CONVENTIONAL BLANKETS are also recommended for use with heat set and oil inks and for carton printing, metal, decorating and Dycril" dry offset.

[2]For use with heat set and oil inks or carton printing and Dycril" dry offset.

" Registered trademark of DuPont

Fig. 24-36. Chart gives information regarding several types of offset blankets, including name, plies (layers of material), gauge (thickness), and color.

A

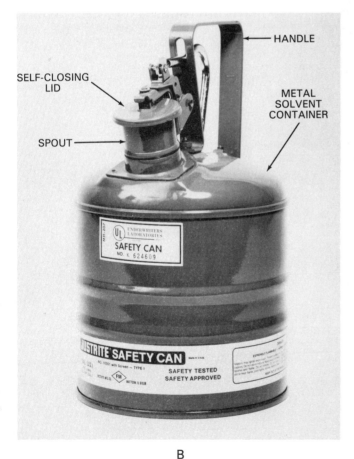

B

Fig. 24-37. A—Offset blankets should be cleaned with a high quality solvent to preserve life and resiliency of blanket. B—Solvents should be kept in metal safety dispensers. This container has a self-closing lid on spout.
(Harry H. Rogers Company, Inc.; Justrite Co.)

paper, the pressure between the blanket and impression cylinders is decreased. This is necessary so that the rubber blanket is not flattened from the increased thickness of paper.

Setting the press *on impression* means to bring the impression cylinder into contact with the paper and blanket. This causes the image to print on the paper.

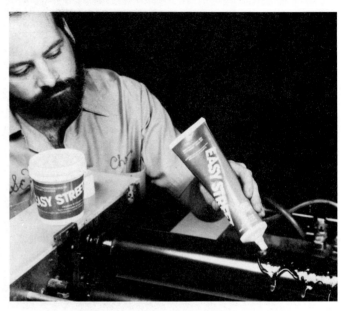

Fig. 24-38. This cleaner is used to wash up offset press and duplicators. Liquid is much thicker and more concentrated than regular blanket and roller solvents.

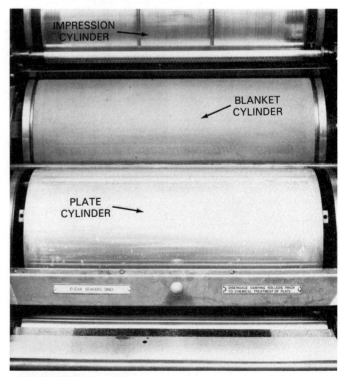

Fig. 24-39. On an offset press or duplicator, impression cylinder forces sheet of paper against blanket to transfer image to paper.

Cylinder and roller pressures

There are several pressure checks that must be made to cylinders and rollers each day. These checks should also be made when changing rollers, roller coverings, blankets, plate thicknesses, and paper thicknesses. Check the specific recommendations given by the manufacturer for the equipment.

General instructions for the sequence of pressure checks are given below and in Fig. 24-40:

1. Check dampener form roller-to-plate pressure.
2. Check ink form roller-to-plate pressure.
3. Check plate cylinder-to-blanket cylinder pressure.
4. Check impression adjustment, called *squeeze.*

Make sure that you have properly installed and gummed the plate on the duplicator or press. Place the dampener form rollers in the OFF position. Place the ink rollers in the ON position, and allow the gummed plate to ink up over its entire surface. Stop the press and place the ink rollers in the OFF position. Allow the plate cylinder to come into contact with the blanket cylinder. Take the plate cylinder out of contact. Do this in several different locations on the blanket. You should have a uniform ink band line 1/8 to 3/16 inch (3.2 to 4.8 mm) wide. If an adjustment is required, correct the parallel first, and then the overall pressure adjustments. See Fig. 24-41.

Some duplicators and presses have a self-compensating adjustment which is a spring-loaded impression. It provides for variations in paper thicknesses. Consult the press manufacturer's instruction book. The following method can be used to test impression.

With the press stopped, insert two 2 x 8 inch (50 x 200 mm) test strips between the blanket and the impression cyinders. Insert one strip near each end. The test strips should be made of the kind of paper to be used for the printing job. Adjust for parallel impression until the pull on each strip indicates the strips are held firmly and equally between the cylinders. Add approximately 0.003 inch (0.08 mm) more squeeze either by adjustment or by the addition of packing sheets.

INKING UNIT

The *inking unit* consists of an ink fountain, which holds the ink supply and several rollers for feeding the ink, Fig. 24-42. These include the fountain roller, ductor roller, three or four metal oscillating rollers, four or more rubber intermediate rollers, and two to four rubber form rollers.

The supply of ink to the plate is regulated at the ink fountain by a set of screws, called *keys.* By adjusting

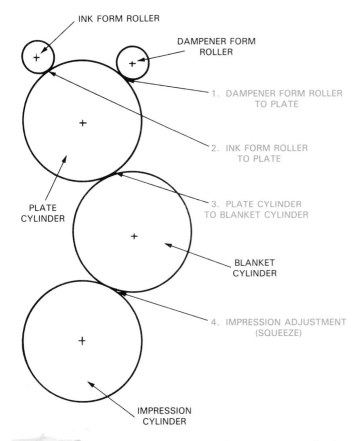

INK FORM ROLLER

DAMPENER FORM ROLLER

1. DAMPENER FORM ROLLER TO PLATE

2. INK FORM ROLLER TO PLATE

3. PLATE CYLINDER TO BLANKET CYLINDER

PLATE CYLINDER

BLANKET CYLINDER

4. IMPRESSION ADJUSTMENT (SQUEEZE)

IMPRESSION CYLINDER

Fig. 24-40. Learn offset press and duplicator pressure checks. These pressure checks should be performed each day prior to starting presswork.
(A.M. International, Inc.)

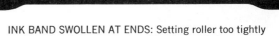

INK BAND SWOLLEN AT ENDS: Setting roller too tightly against the vibrator causes rubber to break loose from the roller shaft. This allows solvent to swell rubber at roller ends. Grind rollers or replace.

TOO HEAVY AT ONE END, TOO LIGHT AT OTHER END: Uneven setting. Reset rollers to obtain uniform happy medium.

HEAVY IN CENTER, LIGHT AT ENDS: Roller bowed or worn at ends. Caused by form roller being set too tightly against vibrator roller. Correct by resetting rollers. Regrind or replace rollers if it's no longer possible to obtain the desired setting.

LIGHT AREAS IN INK BAND: Indicates improper grinding or manufacture of roller. Regrind or replace.

IDEAL SETTING: Uniform parallel bands.

Fig. 24-41. Parallel adjustment of ink form rollers is essential for uniform ink distribution. (A.B. Dick Co.)

the keys, the press operator moves a blade against or away from the fountain roller. Look at Fig. 24-43.

The *blade* controls the flow of ink to the roller. The farther away the blade is moved, the more ink will be carried on the fountain roller. The keys are adjusted beginning at the center and working outward, alternating left to right.

The adjustment of the ink keys allows the press operator to control the amount of ink delivered to the plate. The keys also allow control of ink distribution. Ink can be distributed evenly across the entire plate or made heavier or lighter in any area on the plate.

Keyless inking is a relatively new method of ink control which is unlike the conventional ink fountain just described. *Keyless inking* uses a metering roller to deliver a uniform coating of ink to the plate. Once this is done, a ductor roller scrapes off excess solution. Ink is continuously circulated, and little evaporation takes place. The process is much simpler than conventional ink fountains and produces acceptable press sheets, with fewer makeready sheets.

Ink feed control

The supply of ink can be further regulated by a gear which controls the speed of the fountain roller. Turning the control dial clockwise allows a maximum amount of ink. Turning the control counterclockwise reduces the feed. The usual procedure is to operate with

a minimum of ink on the fountain roller. The ink feed control dial is then regulated as needed. The ink supply can be completely stopped by turning a lever.

When the press is in operation, the metal fountain roller revolves in the fountain. This mixes the ink and picks up a measured amount of ink. Large offset presses have an *ink agitator* which continuously mixes the ink in the fountain, Fig. 24-44.

The ink is then transferred to the ductor roller. The *ductor roller* moves back and forth contacting the fountain roller and a distributing roller. The distributing roller is in continuous contact with the other rollers as it receives ink from the ductor roller.

The *distributing* and *idler rollers* break down the ink before it reaches the oscillating rollers. The *oscillating rollers* move from side to side as they rotate. This spreads the ink evenly over the form rollers. The *form rollers* then ink the image on the plate.

Controlling ink form rollers

Duplicators generally have two ink form rollers, while larger offset presses have three or more. A handle, called the *night latch,* is provided for lowering the rollers against the plate and for lifting them out of contact with the plate. This is usually done manually by the press operator.

The night latch is usually used to raise the rollers at the end of the working day. On some duplicators,

Fig. 24-42. Inking unit on offset press or duplicator consists of a fountain, fountain roller, ductor roller, metal oscillating rollers, rubber intermediate rollers, and form rollers. (A.M. International, Multigraphics)

Fig. 24-43. Ink fountain is equipped with a set of keys used to adjust flow of ink to fountain roller. (A.M. International, Multigraphics)

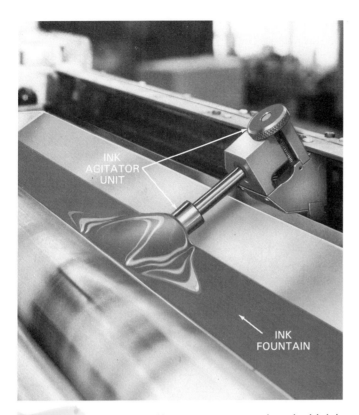

Fig. 24-44. Many large offset presses are equipped with ink agitators. These devices continuously mix ink in fountain, giving good distribution and dispersion.
(Baldwin-Gegenheimer Corp.)

the ink form rollers operate automatically. They rest against the plate when the duplicator begins to print and rise again when it is not printing.

On some duplicators, the ink fountain roller is turned off when sheets of paper are not being fed through the press. This prevents excess ink from accumulating on the rollers. An automatic control lever allows the operator to distribute ink over the rollers without turning on the paper feeder. When the lever is pushed down, the fountain roller will turn, even though no paper is feeding through. Raising the lever brings the automatic control back into use.

DAMPENING UNIT

The *dampening unit* is similar to the inking unit, except that it moistens the non-image areas of the plate so ink will not adhere to them. Refer to Fig. 24-45.

The *fountain roller* rotates in a pan of water mixed with a chemical, known as *fountain solution*. This chemical, when mixed with water keeps the non-image areas of the plate ink-free. The fountain roller transfers a thin film of fountain solution to the ductor roller. The *ductor roller* then transfers the solution to a distributing roller which transfers it onto the form roller.

Since the fountain is only a pan, it has no blade or keys for controlling the distribution and flow of the

Fig. 24-45. Dampening unit on offset press or duplicator consists of a fountain solution bottle, fountain pan and solution, fountain roller, ductor roller, distributing roller, and form roller. (Heidelberg U.S.A.)

solution. The fountain solution is controlled by an adjustment which regulates the rotating speed of the fountain roller in the solution.

Mixing fountain solution

The fountain solution on a duplicator or press is extremely important in achieving quality printing. Keep in mind that it is a delicately balanced chemical solution. Many press problems arise as a result of improperly prepared fountain solutions. Many times the preparation of the fountain solution is given minimal attention. This often results in printing or press problems.

Fountain solution is prepared so that a proper balance is made with the ink. Instructions provided with each package of offset plates tell how to mix the solution for that particular plate. For example, direct-image paper masters require a stronger fountain solution than presensitized metal plates.

Water is used as the ingredient to keep the non-printing areas of the plate ink repellent. Offset ink is greasy and does not stick to a surface having a film of water covering it. Enough water must be transferred from the dampening system to the plate to protect the non-printing areas from accepting ink.

When the duplicator is running, water becomes mixed with the ink to a slight degree. This is normal. Running an excess amount of water, however, reduces the quality of printing and weakens the ink's natural color. The definition between printing and non-printing becomes fuzzy. This can also lead to *emulsification*. This occurs when ink becomes saturated with water. The press operator must stop the press and change the ink. Too much ink or not enough water will cause the plate to ink in the non-printing areas. Too much ink may also cause characters to thicken and halftone dots to fill in. See Fig. 24-46.

To help the water keep non-printing areas free of ink, a small amount of concentrated acid is added to the solution. In preparing the fountain solution, the concentrated acid is added in a small amount to the total volume of distilled water required. This is usual-

Fig. 24-46. Adjustments of dampening and ink are even more critical on a large web press like this one. Improper adjustment can cause huge amount of spoilage in matter of minutes. (Metroliner Press)

542

ly in a ratio of fifteen parts of water to one or two parts of concentrate.

When the press or duplicator is equipped with variable control water stops, the press operator can regulate the amount of fountain solution being distributed to the plate. Some areas of the plate may require less dampening than others.

The importance of water

Ordinary tap water, depending upon the part of the country where it is found, sometimes contains a great variety of chemical compounds. The most common compounds are salts of calcium, magnesium, iron, and sodium which occurs in the form of chlorides, silicates, or carbonates. Some salts might be neutral and some might be acid or alkaline. The ideal water for use in offset lithography is *distilled water* because it is neither acid nor alkaline.

There is a numerical formula of expressing the hardness or softness of water. For example, if a water is classified as 70 ppm, it means there are 70 parts of a salt in one million parts of water. The abbreviation *ppm* stands for parts per millions.

The hardness or softness of water is classified as follows:

A. 0-50 ppm is *soft water.*
B. 50-100 ppm is *medium-soft water.*
C. 100-200 ppm is *medium-hard water.*
D. 200-500 ppm is *hard water.*

This grouping is not precisely defined to geographical locations. It refers more to the type of water generally found in any region. The degrees of softness and hardness of water in the United States is illustrated in Fig. 24-47.

The exact degree of acidity or alkalinity is expressed all over the world in terms of *pH units.* The term pH was proposed by a Danish chemist, Sorensen, in 1909. It represents the power of the hydrogen ion concentration.

The pH values are measured on a scale much as temperature is indicated by the degrees on a thermometer scale. On a thermometer, the larger number corresponds to a higher temperature. A lower number corresponds to a lower temperature. In this way, any given temperature can be compared with familiar temperatures.

In scientific terms, the pH scale actually measures acidity or hydrogen ions. This scale runs from 0.0 for *strong acids* to 14.0 for *strong alkali.* The halfway point (7.0) represents a *neutral solution.* Refer to Fig. 24-48.

The pH scale differs from the temperature scale in that the smallest number (0.0) is the strongest acid. The largest number (14.0) is the least acid. Another difference is that each pH unit represents a ten-fold dif-

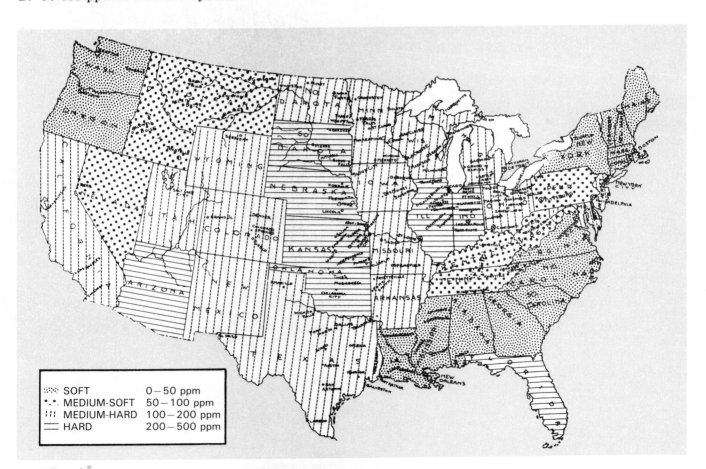

SOFT	0—50 ppm
MEDIUM-SOFT	50—100 ppm
MEDIUM-HARD	100—200 ppm
HARD	200—500 ppm

Fig. 24-47. Map of United States shows degree of water softness and hardness. Water quality is an important factor in offset lithography.

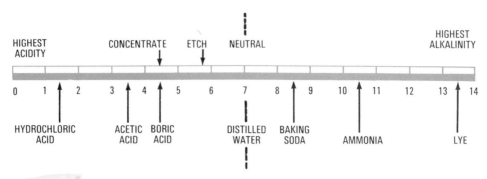

Fig. 24-48. The pH scale measures acidity and runs from 0.0 for strong acids to 14.0 for strong alkali. Half-point (7.0) represents a neutral solution. (A.B. Dick Co.)

ference in the number of free hydrogen ions. Keeping these two items in mind, it becomes apparent that a solution at pH 6.0 has ten times as many hydrogen ions (acidity) as a solution at pH 7.0. Similarly, a solution at pH 5.0 has 100 times as many hydrogen ions as a solution at pH 7.0

The weight of hydrogen ions in a liter of solution for each pH is illustrated in Fig. 24-49.

Measuring pH values

Most offset press operators use *pH test paper* as a simple, fast and economical method of pH measurement. For most solutions, it is only necessary to dip a short strip of test paper in the solution for a second or two, Fig. 24-50. The color of the wet test paper is matched with its color chart to read the pH value. See Fig. 24-51.

All pH test papers are made with special indicator dyes that change color at specific pH values. The useful limit of color change for most individual indicators is about 2 pH units. Beyond this limit, no further noticeable color change takes place, regardless of a further change in pH.

For example, with a pH 6.0 to 8.0 range test paper, a reading of a solution which falls within these points

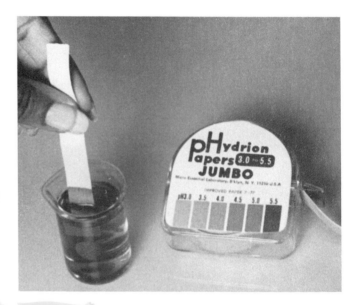

Fig. 24-50. Offset press operators generally use pH test paper to check fountain solution. This should be done at start of day's work and continued periodically throughout day. (Micro Essential Laboratory, Inc.)

Weight of Free Hydrogen Ions in a Liter of Solution in Grams		pH
1.0	(10^0)	0.0
0.1	(10^{-1})	1.0
0.01	(10^{-2})	2.0
0.001	(10^{-3})	3.0
0.0001	(10^{-4})	4.0
0.00001	(10^{-5})	5.0
0.000001	(10^{-6})	6.0
0.0000001	(10^{-7})	7.0
0.00000001	(10^{-8})	8.0
0.000000001	(10^{-9})	9.0
0.0000000001	(10^{-10})	10.0
0.00000000001	(10^{-11})	11.0
0.000000000001	(10^{-12})	12.0
0.0000000000001	(10^{-13})	13.0
0.00000000000001	(10^{-14})	14.0

Fig. 24-49. Chart shows weight of hydrogen ions in a liter of solution for each pH. (Micro Essential Laboratory, Inc.)

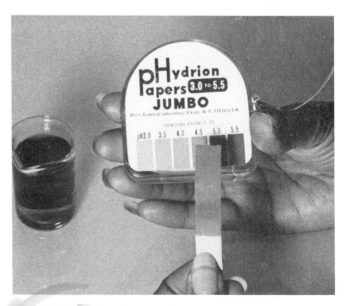

Fig. 24-51. Color of wet pH test paper is matched with its color chart to read pH value of fountain solution. (Micro Essential Laboratory, Inc.)

can be taken correctly. However, if the pH of the solution is lower than 6.0 or higher than 8.0 the test paper cannot indicate the correct pH value. It will, however, show the 6.0 or 8.0 color of the chart.

Some materials, such as proteins and alkaloids, interfere with the indicator color changes. Strong salt solutions (2 percent or more) may also have an adverse effect. In these cases, pH test papers may not give reliable results.

Alcohol fountain solutions

On some duplicators and most offset presses, alcohol is used in the fountain solution. The two systems commonly found in industry are Dahlgren and Varn. Both dampeners are continuous dampeners in that they do not have a ducting moisture control.

The Dahlgren and Varn alcohol systems re-circulate the fountain solution. This makes it possible to maintain consistency in pH of the fountain solution. It also assists the press operator in maintaining the proper fountain solution level in the fountain pan. The entire fountain system generally runs cleaner than a conventional dampening unit.

The duplicator version of the *Dahlgren unit* is a scaled down and much more simplified version of the units made for large offset presses. See Fig. 24-52. It consists of a large diameter (20 to 25 durometer form roller) that is gear-driven from the plate cylinder gear. A rubber fountain roller spins in the fountain solution and is mated against a chromed metering roller.

The fountain and metering roller are independently powered by a variable speed motor. This variable speed capability allows the press operator to vary the surface speed of the rollers of the moistening system. This means that proper dampening can be provided to the plate at all press speeds, and with all ink tacks.

In addition to the speed control, the operator can increase or decrease the pressures between the chrome metering roller and the fountain roller so as to vary

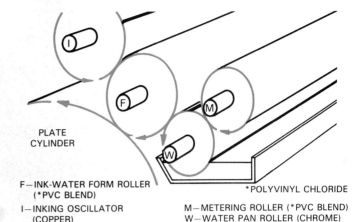

PLATE
CYLINDER

F—INK-WATER FORM ROLLER
(*PVC BLEND)
I—INKING OSCILLATOR
(COPPER)

*POLYVINYL CHLORIDE

M—METERING ROLLER (*PVC BLEND)
W—WATER PAN ROLLER (CHROME)

Fig. 24-52. A Dahlgren system makes it possible for a press operator to maintain consistency in pH of fountain solution. Entire fountain system generally runs cleaner than a conventional dampening unit. (Graphic Arts Technical Foundation)

the film of dampening solution metered to the form roller. This adjustment provides very precise control that contributes to excellent print quality.

The Dahlgren dampener offers several positive benefits over the standard dampeners on most duplicators. It allows for fine line printability up to 200 lines. It reduces ghosting, increases ink lay down with large solids, and allows the use of inks that set faster than the standard duplicator ink formulations.

The *Varn Kompac dampening unit* is designed to give automatic ink-water balance to offset duplicators and presses. The unit is installed after the original dampening system is removed. The unit can be used with duplicators or presses equipped with conventional ink-water systems and those with integrated systems. The Varn unit is shown in Fig. 24-53.

The Varn dampening unit is a continuous-feed, two-roller system that dampens the plate directly from a plate form roller. The plate form roller is gear-driven by the plate cylinder gears to operate at the same sur-

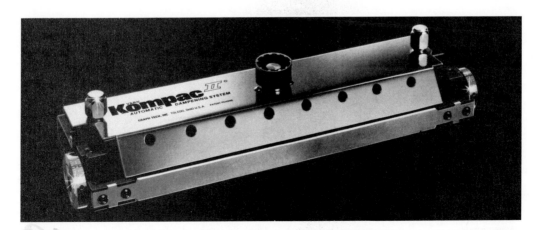

Fig. 24-53. Varn Kompac dampening unit is designed to give automatic ink-water balance to offset duplicators and presses. In this system, fountain solution is milled (combined) into ink under pressure, thus delivering an extremely thin and uniform film of both fountain solution and ink. (Varn Products Company, Inc.)

face speed as the plate. Once the Varn Kompac unit is installed, no other adjustments are necessary from job to job and paper to paper.

The unit consists of a metering roller and a rubber plate form roller. The nip between these rollers carries a constant level of fountain solution. The solution is adjusted by the tapered metal wick built into the dampening cover itself. Both the plate form roller and the metering roller split and share a film of ink with the plate, and both carry ink in the image area.

The fountain solution is milled into the ink under pressure at the inward nip, forming microscopic droplets of fountain solution in the ink. The emulsified fountain solution in the ink is rolled across the plate by the plate form roller. This functions as a giant squeegee, delivering an extremely thin and uniform film of both fountain solution and ink.

Since the emulsification takes place away from the ink form and distributing rollers, there is virtually no chance of emulsification taking place in the ink train (rollers). The plate form roller builds up a small surplus of fountain solution on its trailing edge. This is automatically returned to the nip reservoir when the plate gap passes under the plate form roller. This automatic-return feature eliminates a build-up of water in the system, which would require a rebalancing of ink and water.

By having this reserve of fountain solution available at every plate revolution, the system offers almost instant start-up from one job to another. The system is unaffected by changes in room temperature, press stoppages, and drafts of air. By splitting the ink film with the image on the plate, the Varn unit adds to the inking capabilities of the duplicator or press. This, along with the finer ink film and the lack of emulsification in the inking system, permits the unit to run larger solids, finer screens and halftones, and smaller reverses.

Since a thinner film of fountain solution and ink is run without sacrificing ink density, a more fluid ink can be used. Difficult jobs such as onionskin, carbonless paper, and tissue can be run at normal press speeds because of the thinner ink film and the reduction in dampening solution.

Advantages of alcohol

The use of alcohol in the fountain solution of a duplicator or large offset press has several advantages. It eliminates the problem of too-warm a solution during hot weather operations in a shop that is not air conditioned. It lowers the surface tension of the water, allowing a finer film of water to be fed to the plate. Alcohol causes faster evaporation of moisture from the offset plate and ink rollers. This results in less ink emulsification and faster drying of printed sheets.

Adding alcohol makes the dampening film more uniform and allows a thinner application. This means that less dampening solution is required during a printing run. Alcohol evaporates faster than water. Due to

this property and the more sparing feed of dampening solution, less dampening fluid reaches the printing paper via the rubber blanket. At the same time, the cooling effect of evaporation counteracts any unwanted heating up of the respective machine parts.

Fig. 24-54 shows that beyond a certain percentage, further addition of alcohol has no significant effect on the surface tension. Drastic reductions of surface tension are achieved by alcohol addition up to about 20 percent. Beyond this point, the effect is minimal. It is therefore useless and unnecessarily costly to add more than 20 to 25 percent alcohol to the dampening solution.

The alcohol recommended is 20 to 25 percent (anhydrous) isopropyl (Isopropanol) and is 99 percent pure. The press operator should follow the manufacturer's instructions when mixing a fountain solution containing alcohol. Cleanliness of the entire fountain system is essential.

Dampener rollers

The dampener rollers are either solid metal, solid rubber, or cloth-covered rubber, Fig. 24-55. Fountain rollers are usually solid metal, while ductor and dampener form rollers are usually cloth covered. This is because cloth increases the water-carrying capacity of the roller surface.

The dampener form and ductor rollers on most duplicators are covered with a cloth, called *molleton*. Molleton covers are available ready-cut in tubular or wraparound strip form. This material is also available in continuous tubular rolls. The molleton in Fig. 24-56 has a seam. The molleton in Fig. 24-57 is seamless.

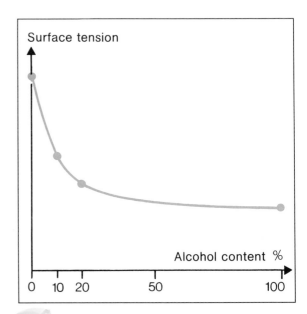

Fig. 24-54. Adding alcohol to dampening system makes dampening film more uniform and allows a thinner application. Chart illustrates that, beyond a certain percentage, further addition of alcohol has no significant effect on plate's surface tension. (Heidelberg U.S.A.)

A   B

Fig. 24-55. A—Dampener rollers are either solid metal, solid rubber, or cloth-covered rubber. Rollers in this example are cloth-covered. B—Various kinds of cloth used to cover rubber rollers are illustrated here.

Molleton covered dampener rollers should be clean, the molleton evenly applied, and the rollers properly *set* (pressure adjusted). Dirty rollers should be removed from the duplicator or press and washed first with solvent, then detergent and water. If they are excessively greasy or dirty, the covering should be replaced.

Molleton dampening covers must be installed with care. An uneven cover will not spread the fountain solution smoothly over the plate. The steps listed below can be used to install a new molleton cover.

1. Remove the old cover. Do this carefully, so the roller is not damaged.
2. Clean the roller. Use soap and water. Solvent can be used to remove ink build-up.
3. Slide the new cover over the roller. An acetate sheet can be placed around the roller to make it easier to pull the molleton sleeve over the roller.
4. Tie both ends of the sleeve. Do this with care so the molleton is smooth. Cut off the extra string, Fig. 24-58.

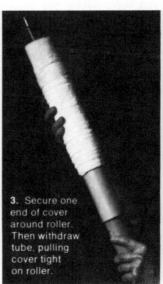

1. Cut material and pull over tube.

2. Slide tube and cover over roller.

3. Secure one end of cover around roller. Then withdraw tube, pulling cover tight on roller.

4. Trim off excess and secure other end of cover around journal.

Fig. 24-56. This is correct procedure for installing a molleton cover with seam. (Jomac, Inc.)

547

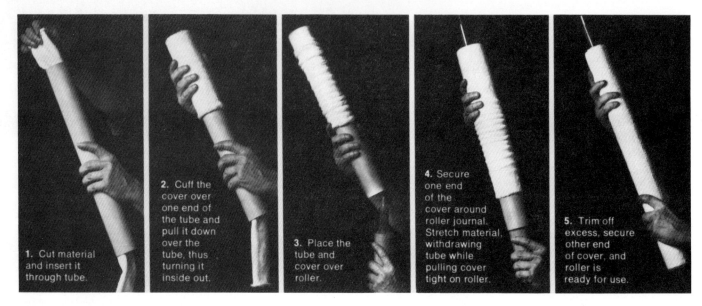

Fig. 24-57. Note correct procedure for installing a molleton cover with no seam. (Jomac, Inc.)

5. Wet the entire molleton in a sink. Smooth down the nap with a cellulose sponge.
6. Install the roller in the press. Run the press to let the roller run in and smooth the surface.

There is another type of dampener roller sleeve made of preformed synthetic fiber. It is frequently used instead of molleton covers. The seamless fiber design eliminates lint and the bulge of seams, Fig. 24-59. The sleeves are designed to be used with a matching dampener roller to ensure maximum performance.

Dampener roller pressure settings

New dampener covers must be set for proper pressure after they are installed. A pressure check must also be made after the new dampener covers have been run in for a while. The press manufacturer's instructions should be followed for plate packing (if used), order of procedure, rollers to be checked, and pressure required.

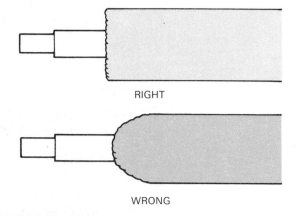

RIGHT

WRONG

Fig. 24-58. When installing molleton dampener sleeves, ends of sleeve must be square. Extra string is cut off. (Graphic Arts Technical Foundation)

The following procedure can be used to check dampener roller pressure settings:
1. Prepare the dampening system for operation. Check for dampness of the form roller. Let the dampener form roller run against the plate on the plate cylinder for a few minutes. Lift the form roller and stop the duplicator.
2. With the duplicator stopped, place two, 1 x 8 inch (25 x 200 mm) strips of 0.005 inch (0.13 mm) thick acetate between the form roller and the plate. Place one strip near each end, as illustrated in Fig. 24-60.
3. Drop the form roller to the ON position.
4. With one hand holding each test strip, pull toward yourself with a slow, uniform tension. There should be a slight, even drag on both strips. If one strip pulls easier than the other or if there is too much, or too little, drag on both, adjust the roller on one end, or both, as indicated by the test procedure.
5. After pressures seem satisfactory, drop the form roller to the plate and lift it up again. You should see a faint damp line across the entire width of the plate.

It should be noted that the basic dampening system described so far has been the conventional or separate type. This means that the ink and dampening solution are applied to the plate separately from individual fountains and rollers. However, the dampening systems on some duplicators are different from the conventional dampening system. When these duplicators are operating, the ink is distributed over all the rollers, including those in the dampening system. This is referred to as a *combined* or *integrated dampening system.*

Water and ink are fed to the plate from the same set of form rollers, Fig. 24-61. The fountain solution is distributed, not only to the rollers in the dampening system, but also to some of the ink rollers.

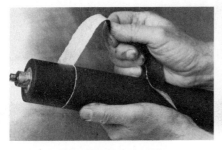

1. Remove old sleeve from rolls.

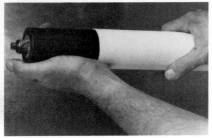

2. Slide seamless sleeve onto roller.

3. Install dampening sleeve and dampen completely.

Fig. 24-59. Study correct procedure for removing old sleeve and installing new seamless fiber sleeve.   (3M Company)

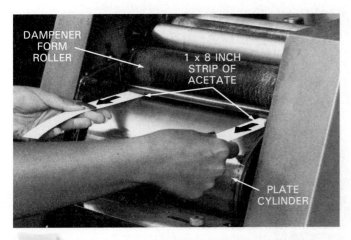

Fig. 24-60. New dampener covers must be set for proper pressure after they are installed on press or duplicator. Thin strips of acetate can be used to check pressure between dampener form roller and plate cylinder.

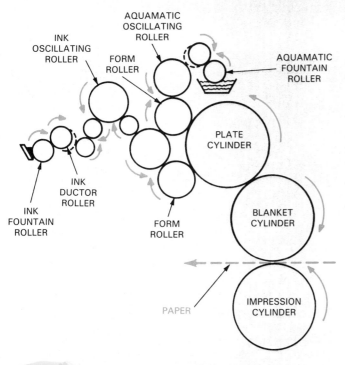

Fig. 24-61. In a combined or integrated dampening system, water and ink are fed to plate from same set of form rollers.   (A.B. Dick Co.)

DELIVERY UNIT

The *delivery unit* on most duplicators and presses consists of a chain delivery, receding delivery table, joggers, and some type of anti-setoff spray attachment. On some duplicators, the delivery unit consists of a simple ejector mechanism in which the printed sheets of paper are forced into a paper receiver tray.

Chain delivery

Most duplicators and all large presses are equipped with *delivery grippers* consisting of a series of small metal fingers. The fingers are attached to a bar extended between two continuous chains, Fig. 24-62. The chains are guided and driven by a pair of sprockets. The chains contain segments and special devices for preventing the sheet from rubbing, whipping, or waving during the delivery. There may be two or more sets of these grippers depending on the speed of the duplicator or press and the distance between the delivery platform and the impression cylinder. They are spaced at regular intervals along the chains so that one set of grippers is receiving a sheet while another set is delivering the preceding sheet.

The grippers are usually held closed by spring pressure. However, as the chains carry the grippers toward the impression cylinder, they pass over a cam. The cam forces the grippers open long enough to receive the sheet. The grippers then carry the sheet to the end of the duplicator or press where a trip cam forces the grippers to open again to release the sheet to the delivery platform. The sheet comes to the delivery pile face-up, with the gripper edge toward the front of the duplicator or press.

On most duplicators and larger presses, the *delivery platform* is constructed to lower automatically as the sheets are delivered. This is part of the receding delivery table, called the *receding stacker*. The press operator can set the platform to operate at any suitable speed. The receding stacker mechanism must be set so the sheets are jogged as perfectly as possible. In addition, the printed sheets must be floated onto the pile in such a way as to prevent smudging, smearing, or set-off. The adjustable joggers, usually on three sides of the

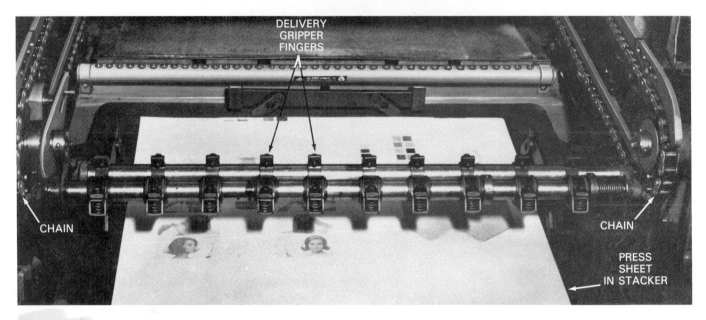

Fig. 24-62. This press is equipped with chain delivery and receding stacker. Small metal gripper fingers are attached to a bar extended between two continuous chains.

sheet, are set so that the sheet falls exactly between them when it drops downward.

Static eliminators, that reduce static electricity on the sheets, are generally provided on most offset presses and duplicators. These may be located near the delivery area or plate cylinder, as illustrated in Fig. 24-63.

Ejector delivery

On many duplicators, a simple *ejector mechanism* is used to deliver the printed sheets to a tray. After impression takes place, a cam on the impression cylinder causes the cylinder grippers to open and release the sheet. Ejector fingers then rise to lift the edge of the sheet away from the cylinder and force it out over a set of stripper fingers. The *stripper fingers* separate the sheet from the impression cylinder and direct it under two ejector rollers which force it into the paper receiver tray. See Fig. 24-64.

Paper guides and a *paper retainer* in the receiving tray help direct the sheets into the tray. On some duplicators, the left paper guide does NOT move. The right guide moves back and forth with a jogging action while the duplicator is operating. The front paper guide can be adjusted for any depth of paper being run.

POINTS TO REMEMBER

1. In offset printing, the image areas are grease receptive and the non-image areas are water receptive.
2. To understand press operation, the press operator must become familiar with the six major operating systems of the press.
3. Offset presses are classified by size and function.

Fig. 24-63. Most offset presses and some duplicators are equipped with static eliminators (arrow). These devices make it possible to overcome paper-feeding and delivery problems caused by static electricity buildup within press. (Baldwin-Gegenheimer Corp.)

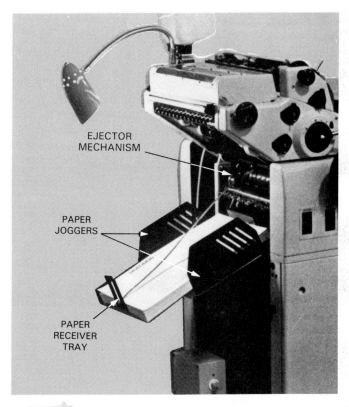

EJECTOR
MECHANISM

PAPER
JOGGERS

PAPER
RECEIVER
TRAY

Fig. 24-64. On offset duplicators equipped with ejector paper receiver trays, stripper fingers separate printed sheets from impression cylinder. Stripper fingers direct sheet under two ejector rollers which force it into paper receiver tray.
(A.M. International, Multigraphics)

4. Duplicator-size presses typically print sheets 11 x 17 inches or smaller.
5. Presses that print sheets larger than 11 x 17 inches are referred to as offset presses, not duplicators. These presses are capable of producing a full range of large printed materials in one or more colors.
6. Most offset presses and duplicators are classified as either sheet-fed, web-fed, multicolor, or perfecting.
7. Offset presses and duplicators are designed with either three main cylinders or two main cylinders.
8. Offset presses and duplicators are divided into six units. These include: feeder unit, register board, printing unit, inking unit, dampening unit, and delivery unit.
9. Offset press and duplicator roller systems deliver ink and dampening solution to the plate in the proper metered quantity.
10. There are two major variations in offset press and duplicator dampening systems.
11. A conventional dampening system wets the plate independently of the inking rollers.
12. A combined or integrated dampening system combines the dampening solution and ink to wet the plate.

13. The Dahlgren system uses alcohol and water in the dampening solution.
14. The Varn Kompac dampening system is designed to give automatic ink-water balance to offset duplicators and presses.

KNOW THESE TERMS

Back cylinder, Bearers, Blanket cylinder, Chain delivery, Combined dampening system, Compressible offset blanket, Conventional offset blanket, Conventional dampening system, Dampening unit, Delivery unit, Distributing roller, Ductor roller, Ejector delivery, Emulsification, Fan, Feeder unit, Form roller, Fountain roller, Printing unit, Register board, Idler roller, Impression cylinder, Ink agitator, Inking unit, Jog, Molleton, Multicolor press, Offset duplicator, Offset press, Oscillating roller, Perfector, pH test paper, pH value, Plate cylinder, Printing unit, Sheet-fed press, Stream feeding, Web-fed press.

REVIEW QUESTIONS

1. Offset presses operate on the principle of one printed copy for each revolution of the cylinders. True or false?
2. Small offset presses up to 11 x 17 inches are usually referred to as _____.
3. When the sheets of paper overlap as they travel down the feedboard of a sheet-fed press, it is called _____ _____.
4. Web-fed offset presses feed from a _____ of _____.
5. Offset presses that can print more than one color at the same time are called _____ presses.
6. Offset presses that can print on both sides of the sheet of paper at the same time are called _____ presses.
7. A three main cylinder offset press has a _____ cylinder, a _____ cylinder, and an _____ cylinder.
8. Describe a two main cylinder offset press.
9. The printing unit on an offset press consists of all but one of the following:
   a. Plate cylinder.
   b. Chain cylinder.
   c. Blanket cylinder.
   d. Impression cylinder.
10. Why are some press cylinders undercut?
11. A _____ offset blanket has the ability to squeeze and return to its original shape.
12. The impression cylinder is also called the _____ cylinder.
13. The inking unit on an offset press contains two to four form rollers. True or false?

14. Explain the function of the ink fountain keys.
15. A _____ test is taken to determine the alkalinity or acidity of the fountain solution.
16. How is water hardness or softness measured or expressed?
17. The kind of alcohol used in some fountain solutions is called isopropyl. True or false?
18. Describe the operation of a duplicator dampening system.
19. Squaring and aligning the sheets of paper before they are loaded into the feeder unit of the press is called _____.
20. The safety device that allows only one sheet of paper at a time to enter the press is called the _____ _____.
21. _____ _____ hold edge of the top sheet in the stack down as it is separated by the front blowers.
22. Rollers in a conventional dampening system are covered with:
    a. Felts.
    b. Molletons.
    c. Shrink wrap.
    d. Blankets.
23. On most offset duplicators, the _____ unit consists of a simple ejector mechanism that delivers printed sheets to a tray.
24. Small metal _____ _____ attached to the impression cylinder, hold the sheet of paper as it is printed.
25. On large offset presses, the delivery gripper fingers are part of the _____ delivery mechanism.
26. On large offset presses, the receding stacker is part of the delivery platform. True or false?
27. In a combined dampening system, water and ink are fed to the plate from the same set of _____ rollers.

SUGGESTED ACTIVITIES

1. Ask your teacher for a press operating manual for each press and/or duplicator in the shop. Familiarize yourself with the basic operating controls and systems of each.
2. Draw a schematic diagram of the six operating systems of each press and/or duplicator in the shop. Include the following:
   a. Feeder unit.
   b. Register board.
   c. Printing unit.
   d. Inking unit.
   e. Dampening unit.
   f. Delivery unit.
3. With your teacher's assistance, perform the following:
   a. Install dampener covers on an offset press and a duplicator.
   b. Make the necessary dampening system pressure checks and adjustments.
   c. Mix fountain solution to specifications given by your teacher. Test the solution for correct pH.
   d. Install and remove a blanket on one of the offset presses or duplicators in the shop.
   e. Install and remove a plate on one of the offset presses or duplicators in the shop.
   f. Perform all the necessary pressure checks and adjustments on an offset press or duplicator in the shop.
   g. Install the proper thickness packing on the plate and blanket cylinders for a given job. Use the micrometer for determining proper packing thicknesses.
   h. Clean and check the vacuum blower pump jars. Refill with nondetergent SAE 10 weight oil.

Chapter 25

Offset Press Operation

When you have completed the reading and assigned activities related to this chapter, you will be able to:

O Operate the offset duplicator and/or small offset press while following safe work procedures.

O Prepare the feeding unit of an offset duplicator in preparation for running a job.

O Prepare the register board of an offset duplicator.

O Prepare the printing unit of an offset duplicator.

O Prepare the inking unit of an offset duplicator.

O Prepare the dampening unit of an offset duplicator.

O Prepare the delivery unit of an offset duplicator.

O Operate the offset duplicator and/or small offset press to successfully complete several jobs.

O Clean and secure an offset duplicator and/or small offset press at the end of a job or at the end of the working day.

The previous chapter explained the six main operating systems of offset duplicators and small to medium offset presses. Press operating procedures vary from one press to another. However, they are similar enough so that when the fundamental principles are learned, you should be able to undertake the operation of almost any type of duplicator or small offset press. The following material will give you an understanding of the sequence of operations involved in setting up and running a job on a duplicator.

Keep in mind that your teacher is the expert in this area. There is no substitute for the teacher's qualified professional guidance. This applies to procedures and safety precautions.

Pre-starting procedure

Press operator's instruction manuals for the particular presses in your shop should be available. Familiarize yourself with the manual for the duplicator or press you will be using.

> Danger! Never operate any press until you have been thoroughly instructed in the safe operation of that press! It is vital that you be able to demonstrate a complete understanding of press operation. In addition, always obtain the teacher's permission before operating a press. Unauthorized personnel should not be allowed around a running press!

All press adjustments should be supervised by your teacher. It is better to ask questions first rather than to make wrong adjustments that result in injury to you and/or damage to the press.

PRESS PRELIMINARY INSPECTION

To complete a *press preliminary inspection,* examine the press to make sure that all safety devices and operating components are in place. Set all controls in the OFF position. Remove any loose articles from the press.

Install any rollers and fountains that may have been removed earlier. If necessary, remove any protective gum coating from cylinders or plate. If the blanket has been loosened to allow it to rest, adjust and retighten it.

Wipe away any lint or paper dust from the feeder and register board. In addition, wipe off any dust and dirt from the rollers, cylinders, and blanket. You can use a cloth dampened in solvent for this purpose.

Turn the handwheel on the press two or three revolutions by hand. Check to see that nothing interferes with the operation of the press. If all is clear, turn on the power. Set the speed selector at a slow speed. Check to see that all parts are functioning properly.

NOTE! Never adjust the speed selector while the press is stopped!

Pressure checks must be made prior to the press run. The sequence of checks are found in the operator's manual. General instructions for doing these checks are found in Chapter 24. You should use either the gummed plate for the job or a *test plate* of the same thickness to do the pressure checks.

Once all press checks have been made, determine the specifications for the job you will run using the information on the *job docket*. These include:

1. Correct ink and color.
2. Adequate fountain solution.
3. Correct paper stock and size plus makeready and waste sheets.
4. Plates on hand.
5. Adequate press wash-up solvent and blanket wash.
6. Special instructions.

If you are in doubt about any aspect of the job, consult your teacher before proceeding. Do not attempt to out-guess instructions found on the job docket. Clarification of instructions takes only a brief moment, whereas running the entire job over to correct mistakes is costly.

PRESS LUBRICATION

Correct *press lubrication* will reduce wear on all moving parts and lengthen press life. The productive life of the press will be shortened by lack of lubrication, use of inferior lubricants, or by carelessness in the use of lubricants. Study the manufacturer's lubrication chart to locate each oil hole, grease fitting, and point of lubrication. These locations on the press should be marked with red paint to distinguish them easily.

Always begin your lubrication at a certain point on the press and work your way back to the same point. Use a good non-detergent lightweight oil, SAE No. 20, for example. Wipe off excess oil with a clean cloth.

Lubricate the feeder and delivery drive chains once a week. Use gear grease compound on all gears. Use penetrating oil where applicable. Motors should be oiled sparingly during the weekly lubrication. Motors with grease fittings require grease at least twice yearly. Follow the manufacturer's recommendations when lubricating motors.

Check the oil level in the vacuum-blower pump jars weekly. For most pumps, SAE No. 10 weight oil is recommended. Clean the air inlet holes and pump filters.

Check the paper required for the job. Make sure the grade, size, and quantity of the paper are correct. It is assumed the paper conditioning, so far as moisture content is concerned, has been taken care of within the limitations of the shop. The paper should have been in the press operating area long enough in advance of the press run to reach a temperature balance. Twenty-four hours is considered a minimum length of time for this purpose.

The operator should obtain the required ink for the job. This may be a PMS numbered ink or a special color that must be mixed. If color mixing is necessary, mix the proportions on a piece of glass or ink palette. An ink knife is used for this purpose. If ink driers or other ink modifiers are required, add them at this time. Never mix inks in the press ink fountain. See Chapter 23 covering lithographic inks.

PRESS STARTING PROCEDURE

A standard procedure of starting a press should be followed. In this way, you will find that a minimum of time will be consumed in setting up the press. This will result in a uniform quality of work. The following are the general sequence of steps in setting up and running a job on an offset duplicator.

Prepare inking unit

The ink rollers should be clean of all lint and dust before filling the ink fountain. If the duplicator has been standing idle for several days, look over the rollers carefully for accumulated dust. Clean the rollers with blanket and roller wash.

Remove the fountain and clean it thoroughly with solvent. Clean the fountain roller with blanket and roller wash. Replace the fountain and turn the duplicator by its handwheel until the ink ductor roller is out of contact with the ink fountain roller.

WARNING! Most solvents are flammable and must NOT be used near an open flame! Make sure solvents are used in well-ventilated areas. Do NOT splash solvent in your face or breathe fumes for an extended period of time. Dispose of solvent-covered rags or paper in a metal fire safety container.

Fill the ink fountain by applying small amounts of ink with the ink knife against the fountain roller. Ink should be removed from the ink can with an ink knife. Skim any hardened crust from the top before removing ink, Fig. 25-1. The ink fountain should be at least half full.

The ink fountain is equipped with a number of adjusting *keys* or screws, Fig. 25-2. Turning these screws inward (clockwise) will DECREASE the flow of ink. Turning the screws outward (counterclockwise) will INCREASE the flow of ink.

While turning the ink fountain roller control knob, adjust the fountain screws. Start at the center of the

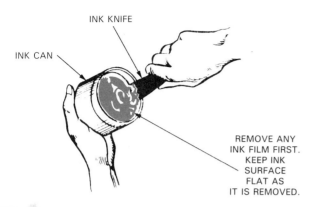

INK KNIFE

INK CAN

REMOVE ANY INK FILM FIRST. KEEP INK SURFACE FLAT AS IT IS REMOVED.

Fig. 25-1. Ink should be removed from ink can with an ink knife. Skim any hardened crust from top before removing ink.

fountain and adjust the fountain screws. Establish an even flow or pickup line between the ductor roller and the fountain roller across the width of the fountain.

Turn the duplicator on and set the ink feed volume control to the third or fourth notch. Hold the ink fountain roller automatic control lever in the down position. Let the duplicator run until all ink rollers are properly coated with a thin film of ink. Make final adjustments of the fountain screws to establish the desired flow of ink.

NOTE! If the revolving ink rollers sound like "eggs being fried," there is too much ink on the rollers. Turn the press off and remove some ink from the rollers with a clean cloth and a slight amount of solvent. Restart the press and recheck the ink. Push up on the automatic control lever to release and turn the press off. The ink unit is now ready for use.

Prepare dampening unit

The dampener fountain and fountain roller should be clean and free of dirt. Fill the fountain bottle with the proper mixture of fountain solution. Hold the bottle, with the spout down, over a sink to be sure that it does not leak. Bring it over the side of the press, but not over the ink rollers, and insert it into the holder.

If there is a night latch for the dampener roller, turn it off. This drops the rollers into operating position. The dampener form roller should remain off and should NOT contact the plate.

Start the duplicator, allowing the rollers to pick up moisture. Operate the fountain roller knob by hand to help the moisture along. You can also use a sponge to drip a little fountain solution on the oscillating roller.

You may also pre-dampen the dampener ductor roller. With the press stopped, turn the handwheel to bring the water ductor roller into contact with the fountain roller. Turn the fountain roller knob by hand. This operation transfers fountain solution from the water fountain roller to the ductor roller. Continue until the ductor roller is sufficiently dampened.

When the water form roller(s) is sufficiently damp, stop the press. You may touch the form roller with your knuckles to determine proper form roller dampness. This should be done only when the press is stopped. Your teacher can demonstrate the preferred method of determining proper form roller dampness.

Now you should set the ratchet control for the fountain roller at its normal setting. Be sure the ductor control lever is on. The dampening unit is now ready for use.

Attaching the plate

Always be sure that the plate cylinder is thoroughly cleaned before attaching a plate. A small particle of dirt under a plate can ruin a halftone or solid in only a few impressions. The plate cylinder should be dry.

NOTE! Moisture under a paper master will cause wrinkles and deterioration of the image.

Turn the handwheel until the lead plate clamp is in position. Attach the lead edge (gripper edge) of the plate to the plate clamp, Fig. 25-3. Start at the right

INK KEYS OR SCREWS

Fig. 25-2. Offset press ink fountain is equipped with a number of adjusting keys or screws. Flow of ink to fountain rollers is controlled by turning keys or screws. (Heidelberg U.S.A.)

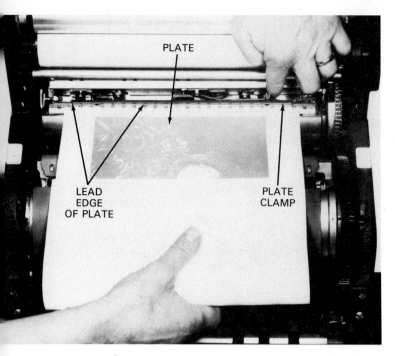

Fig. 25-3. On this duplicator, offset plate is attached to plate clamp.

side of the clamp and work over to the left side. Hold the tail of the plate square, tight, and down near the feedboard to prevent the plate from becoming detached.

Holding the tail end of the plate with the right hand, turn the handwheel clockwise. Do this until the tail clamp is in position for attaching the plate. Hold the plate with your left hand and bring the tail clamp up to position with the right hand.

Attach the plate to the clamp. Tighten the plate clamp by turning the plate clamp tightening dial clockwise.

NOTE! If a paper master is being used, the spring tension of the clamp alone will hold it in place.

Prepare the feeder unit

With your right hand, depress the paper platform lock release. With your left hand, turn the paper platform crank counterclockwise to lower the platform. Set the inside of the left paper stack side guide to the position indicated on the scale for the size sheet to be run.

Lay a sheet of paper on the platform close to the front and left side guide. Set the right paper stack guide so that it will be about 1/16 inch (1.6 mm) away from the edge of the paper.

For a quick, accurate way of positioning the front guide and center blower, fold a sheet of paper for the job in half (longways). Fold it in half again. Unfold the paper and you will have creases as represented by lines A and B in Fig. 25-4. Lay the open sheet on the platform and center the front guides on line B with the center blower on line A.

Set the front paper-pile guides. The front paper-pile guides are about on a line representing one-quarter of the front edge of the sheet. Suction feet should be centered on the front guides directly above the sheet separators, called *cat's whiskers*. The front air blowers should be in the center of the sheet.

Fig. 25-5 illustrates the relative positioning of the vacuum feeder and paper platform.

Load the paper feeder. *Fan and jog* the sheets of paper to separate any that might be sticking together, Fig. 25-6. Place the paper in the center of the feed table using the scale as a guide for centering. This scale corresponds with a similar scale on the head clamp of the plate cylinder. Move the paper stack forward so that it touches, but does NOT press too heavily against the separator blocks.

Bring the paper stack up to within 1/4 inch (6.4 mm) of the sheet separators. Set the paper stack tail back guide so that the front set of wings are on top of the paper stack. The back set of wings should be behind and up against the stack.

Set the pile side guides. The pile side guides are set so that they are about one inch (25 mm) from the back

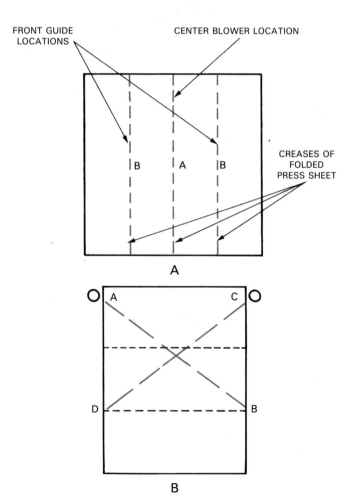

Fig. 25-4. A—Examples of a press sheet folded to locate front guides on line B, and center blower on line A. B—Location and positioning of front side blowers should be A to B and C to D. (A.B. Dick Co.)

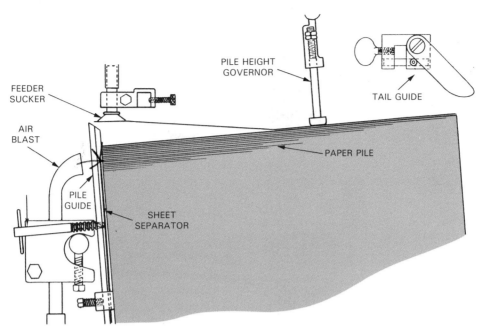

Fig. 25-5. Study relative positioning of vacuum feeder and paper platform.

of the sheet. Set the side air blower paper separators so that they blow about two-thirds of the way back and across the paper, Fig. 25-4A. The second hole in the air blower tube should be in line with the top of the paper when the paper is 1/4 inch (6.4 mm) below the sheet separators. See Fig. 25-7.

Position the paper platform. The paper platform should be lowered about one inch (25 mm). Turn the duplicator ON to permit the paper platform to raise automatically. Adjust the elevator control knob so that the paper stack stops about 1/4 inch (6.4 mm) below the sheet separators. Turning this knob clockwise increases the height and turning it counterclockwise will decrease the height.

Position the pullout roller and double-sheet detector. The paper pullout roller and double-sheet detector (eliminator) are attached to the same unit. Whenever possible, the unit should be positioned so that the pullout roller is close to the center of the paper.

The pressure of the paper pullout roller is controlled by a thumb screw. This should be set so that there is just enough pressure to move the sheet of paper through to the feedboard.

Too much paper pullout roller pressure will distort the sheet. If already printed sheets are run for a second color, the pullout roller may mark the sheet if it is too tight. Insufficient pressure will cause sheets to delay or arrive late at the paper stops. A pullout roller release lever is provided should paper jam under the pullout roller.

The double-sheet detector is also controlled by a thumb screw. It should be set by starting up the duplicator and vacuum motors and feeding sheets through the duplicator. The screw should be turned counterclockwise until single sheets are ejected. The

screw is then turned slightly clockwise until the ejector does NOT act on single sheets.

Fig. 25-8 illustrates the various parts of the feeder unit.

After passing through the double-sheet detector, the sheet is then conveyed down the register board until it reaches the front stop fingers. Stopped at this point, the sheet is guided across the board to a predetermined register position.

Set the register board and sheet controls. Paper may be jogged from either the left or right side on the register board. The right-hand jogger (stationary guide) should only be used when running jobs to be printed on both sides of the paper. In this case, it is desirable to use the left-hand jogger the first time through, and the right-hand jogger the second time through. In this

Fig. 25-6. All paper to be loaded in press feeder should be fanned and jogged to separate any that might be sticking together.   (United States Government Printing Office)

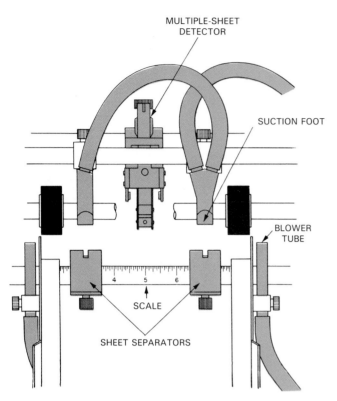

Fig. 25-7. This is correct positioning for multiple-sheet detector, suction feet, sheet separators, and side blower tubes.

one time. These should be adjusted only when the duplicator is running or the tapes will be distorted. This can cause inaccurate positioning (register) of the paper.

Set the feedboard joggers. To set the left jogger into action, start the press. Hold the right jogger with your left hand and pull it toward you while moving the jogger selector back toward the feeder. This will set the left jogger into action.

To set the right jogger into action, start the press. Hold the left jogger with your left hand and push it away from yourself while moving the jogger selector into its forward position. This will set the right jogger into action.

Both joggers can be set into action by moving the jogger selector into a position midway between the left-hand and right-hand positions. This is often used for small cards and envelopes.

Set the leaf springs. When the jogger leaf springs are used, the spring on the jogger being used should be raised. This is done by pressing down on the BACK END of the spring retainer. The spring on the stationary jogger should be lowered. This is done by pressing down on the FRONT END of the spring retainer. Fig. 25-9 shows the jogger and leaf spring.

When in use, the stationary jogger should be set so that, when the paper is moved against the spring, it does not move the spring more than 1/16 inch (1.6 mm). On most duplicators, the angle of either jogger can be regulated by turning the jogger angle control screw. Feed a few sheets of paper through the duplicator to check jogger action. The paper must be square with the edge of the jogger.

Position the image on the paper. The image must be positioned square on the paper by adjusting the stop bar control knob. This bar can be tilted in either direction. This eliminates any need for adjusting the angle of the plate on the plate cylinder.

way, the backup run utilizes the same paper edge as was used the first time through.

For jobs to be printed on one side only, the left-hand jogger is used. The stationary jogger should be set to act as a guide. It may have a leaf spring. The spring should be set to compress slightly as the sheet is jogged against it.

On most duplicators, the jogger control knob moves the side jogger, ball race, and paper feed tapes all at

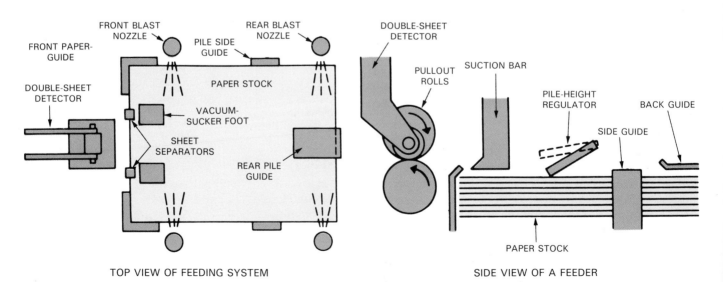

TOP VIEW OF FEEDING SYSTEM

SIDE VIEW OF A FEEDER

Fig. 25-8. Study component parts of duplicator feeder unit. Proper positioning of feeder components is essential to troublefree feeding. (A.B. Dick Co.)

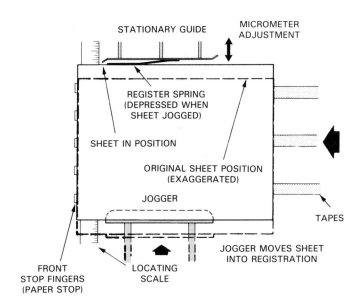

Fig. 25-9. Study parts of jogger and leaf spring mechanisms. For most printing applications, one of two joggers is locked in stationary position with leaf spring lowered. (A.M. International, Multigraphics)

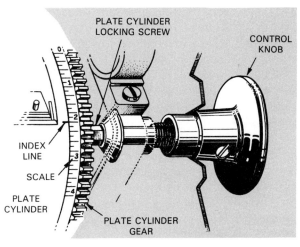

Fig. 25-10. Vertical movements of image on press sheet are made by moving plate cylinder. A plate cylinder lock control is provided for this purpose. (A.M. International, Multigraphics)

NOTE! Some duplicators are not equipped with an adjusting stop bar. This would require an adjustment of the plate on the plate cylinder to change the position of the printed image.

Set gripper bite control. Small vertical movements of the image can be made by moving the paper stop bar forward or backward. This allows a larger or smaller gripper bite. This movement is limited to about 3/16 inch (4.8 mm), or 3/32 inch (2.4 mm) in each direction off center (zero).

Larger vertical movements of the image on the paper are made by moving the plate cylinder. This is done by turning the handwheel until the zero position on the plate cylinder scale is opposite the mark on the plate cylinder guard, Fig. 25-10. The press plate cylinder lock control is pressed until it sits on the plate cylinder lock nut. The lock control is then turned to the left to unlock the nut while holding the handwheel. Holding the lock control in position, the handwheel is turned until the indicator line shows that the cylinder has been moved to the desired position on the scale. The lock control is then turned to the right to lock the cylinder in position.

After the sheet has registered properly against the side guide and stop fingers, the front stop fingers drop down. The feed rolls propel the sheet forward into the impression cylinder grippers. The timing is adjusted so that the leading end of the sheet usually buckles slightly as it meets the gripper stops. This buckling ensures positive positioning at the gripper, Fig. 25-11.

Then the impression cylinder grippers close on the sheet. They carry it between the blanket and impression cylinders under pressure. At this point, the sheet receives the inked image from the blanket.

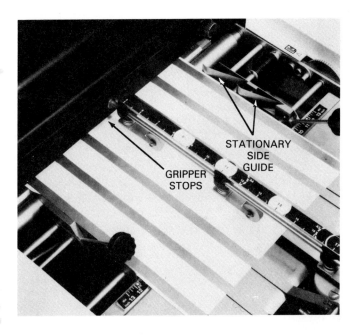

Fig. 25-11. Once sheet of paper has registered against side guide and stop fingers, lead end of sheet buckles slightly as it meets gripper stops. (A.M. International, Multigraphics)

Setting the delivery unit

Set the delivery tray, jogger and/or stacker for sheet width and length required, Fig. 25-12. Feed a sheet of paper into the duplicator. Stop the sheet short of the delivery tray, just as the jogger reaches the closed position. Set the jogger, tray, or stacker to the sheet position.

The lower ejector wheels on a duplicator equipped with delivery tray must be moved to at least one inch (25 mm) inward from the sides of the sheet, Fig. 25-13.

The upper ejector wheels should be moved outward 1/2 inch (13 mm) from the lower wheels. Positioning the ejector wheels directly over each other causes the sheets to become wrinkled.

A chain delivery system has gripper bars mounted across two endless chains. The chains revolve between a delivery cylinder and the delivery end of the duplicator, Fig. 25-14. Chain-delivered printed sheets are released into the stacker and jogged automatically. The stacker descends automatically to accommodate a large quantity of printed sheets. The chain delivery system may also be equipped with an anti-offset spray attachment. When adjusted properly, the *anti-offset spray attachment* prevents smudging and transfer of wet ink to the back side of the next sheet.

If the duplicator is equipped with a chain delivery system, make the necessary adjustments to the unit. Use extreme care when working around the chain

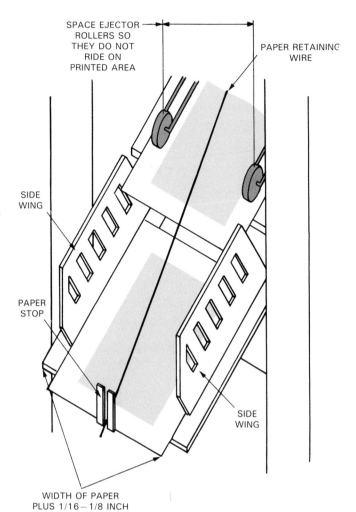

Fig. 25-13. Lower ejector wheels of delivery tray-equipped duplicator should be positioned, if possible, at least one inch inward from sides of press sheet.

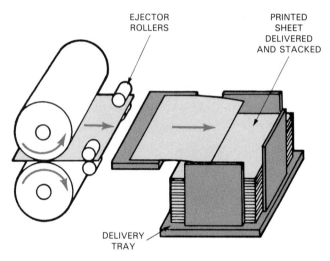

DELIVERY TRAY

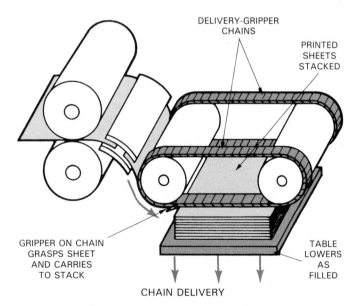

CHAIN DELIVERY

Fig. 25-12. Two kinds of paper delivery systems include delivery tray and chain delivery with receding stacker. Delivery tray system is used exclusively on some models of duplicators.

delivery gripper fingers, since they are somewhat pointed and sharp. The chain delivery system should be protected with safety covers.

FEEDING TEST SHEETS

Assuming that the duplicator has been prepared as described, several test sheets should be run through the press to determine if any adjustments are required.

With the duplicator set up according to the previous directions, feed several test sheets as follows:

1. Using a cotton pad, moisten the plate with fountain solution.
2. Start the duplicator under power.
3. Lower the dampener form roller to the plate cylinder.
4. Allow the duplicator to run for several more revolutions while checking the plate for dampness.
5. Lower the ink form rollers to the plate cylinder.
6. Start the vacuum and blower motor.
7. Set the duplicator on impression.
8. Run a sheet of paper through the duplicator.
9. Take the duplicator off impression.

Fig. 25-14. Chain delivery system has gripper bars mounted across two endless chains. Chains revolve between delivery cylinder and delivery end of duplicator. (ATF Davidson Co.)

10. Raise the ink form rollers.
11. Turn off the vacuum and blower motor.
12. Stop the duplicator.
13. Check the position of the image on the sheet.
14. Make all necessary adjustments to achieve position, ink and water balance, and ink coverage desired. This may take additional trial runs to achieve the desired results.

CHECKING TEST SHEETS

Inspect the test sheets carefully. Here are some of the important points to examine on the test sheets.

The image on the paper should be clear, well-inked and free from background tone.

If the plate tends to pick up ink in the non-image areas, it indicates an excessive amount of ink or insufficient amount of dampening solution. The ideal ink/water balance occurs when each printed sheet removes exactly the same amount of ink and water as is being metered to the plate. Too little ink or water will show up immediately on the copy and is easily remedied. However, too much ink or water is harder to detect, and the situation can cause chronic quality problems.

In the case of tone, turn the dampener fountain roller knob one or two turns clockwise, at the instant the ductor roller contacts the fountain roller. This will momentarily supply more moisture to the plate. If this causes tone to disappear, advance the fountain feed control lever one or two notches.

If the image appears over-inked and tone remains, retard the ink feed control. An excessive amount of ink will cause toning of the non-image area and filling-in of the image.

An excessive amount of moisture, or lack of ink tends to produce a gray, washed-out image. The ideal point of balance is to run the plate as dry as possible. This means to use the minimum amount of moisture with only enough ink to produce a dark image.

The initial symptoms might appear very much like ink starvation: light, washed out copies. In attempting to correct this, the press operator can trigger a vicious cycle. The print looks too light so the operator adds more ink, which causes the background to start toning, so the operator adds more water, which causes the ink to print light again. Eventually, the excess ink and water becomes whipped together by the action of the rollers and the ink that prints the copy encapsulates excess water and air bubbles. This causes a mottled and washed out print.

The water in the fountain picks up bits of emulsified ink causing dirty copy, and in severe cases, a buildup of ink on the blanket outside the paper area. On duplicators equipped with integrated dampening systems, in which the same form rollers carry both the ink and water to the plate, the emulsified ink builds up and piles on the top water roller. It does this instead of going to the plate. This causes the copy to print light again.

As previously mentioned, the press operator should feed only the amount of ink to the plate as can be picked up by the paper with each impression. This is accomplished by cutting back slightly on the water setting and watching the copy closely. Keep cutting back on water until the background begins to tone. Once this happens, bring the water setting up slightly so that it is just barely enough to keep the background clean. Cutting back on the water will result in denser, smoother copy even though no ink has been added.

Check the position of the image. The test sheets should be checked for image position using a line gauge. These checks should include lateral (left to right) and vertical (up or down) positions. When a position adjustment is necessary to the plate or plate cylinder, wash the image from the blanket. If this is not done, the next print will show a double impression.

Check for proper impression. The impression check should have been done before running test sheets. Impression that is too heavy results in image spread. This is especially true in the case of halftone dots which spread in size due to improper impression. Too little impression results in a light image, faint halftone dots or uneven solids.

Refer to the duplicator instruction manual for specific impression requirements.

HANDLING THE DELIVERY

Observing the run on a duplicator involves more than maintaining specific settings and adjustments. The problems concerning specific components of the duplicator have been discussed. The press operator must also be concerned with maintaining the appearance and quality of the job produced on the duplicator.

It is assumed that the mechanical aspects of running the duplicator are correct. The feeder is functioning properly; the sheet is registering; the color is okay; and the printed sheets are delivering and jogging in a satisfactory manner. Observing the run then involves seeing to it that no hickies or spots appear on the printed sheets, and that the proper amount of ink is being deposited on the sheets.

DUPLICATOR WASHUP AND CARE

At the end of the period, day, or when a color change is required, an offset duplicator must be washed up. To wash the duplicator, follow these steps:

1. Remove the fountain solution bottle.
2. Remove the fountain solution from the fountain with a sponge or syringe.
3. Use an ink knife to remove and discard the ink from the ink fountain.

4. Remove the ink fountain from the duplicator by lifting it up to a vertical position. Clean the ink fountain with a cloth moistened with an appropriate solvent.

WARNING! Make sure solvents are used in well-ventilated areas. Do NOT splash solvent in your face or breathe fumes for an extended period of time. Wear eye protection. Dispose of solvent-covered rags or paper in a metal fire safety container.

5. Clean the ink fountain roller with blanket and roller wash.
6. Remove the excess ink from the rollers by manually feeding a sheet of paper into the rollers. Hold on to the end of the sheet while manually turning the handwheel. Back the sheet out and discard it. Repeat this process several times to make the washup easier and faster.
7. Attach a cleanup mat to the plate cylinder and start the duplicator. Turn the speed down to slow.

NOTE! Some duplicators are equipped with an automatic washup device. Refer to Fig. 25-15.

8. Squirt a small amount of blanket and roller wash over the ink oscillating roller, allowing it to work into the ink system.
9. Move the two form rollers to the ON position. Continue to add small amounts of wash until the rollers appear clean. Do not use too much wash at one time. Work one side of the rollers at a time so the rollers do not slip and slide.
10. Turn the ink form rollers (and aquamatic control if applicable) to the OFF position. Stop the duplicator, remove the cleanup mat and discard it. Repeat the operation with another cleanup mat, and a third if necessary.
11. After all ink is removed, moisten a soft, lintless cloth with blanket and roller wash. Wipe all the rollers clean of any ink residue or lint. Wipe the ends of all ink rollers. This will prevent rollers from cracking and hardening on the ends.
12. Use a cloth to clean the duplicator of any ink that may have accumulated on interior and exterior parts. Clean off spots caused by fountain solution. Wipe off all chrome-plated parts including cylinders and trim. Wipe the ejector rollers and collars and lubricate them so that they move freely along the shaft.
13. Clean the floor around the duplicator of all ink, solvents, and loose papers. Check beneath the duplicator by opening the motor access doors or covers. Visually check the vacuum pump fluid level and fill it if necessary.
14. Check the ink and fountain rollers for *glaze* or a shiny, glossy accumulation on the surface of the rollers. It is caused by ink and fountain solution combining during a press run. The ink rollers should be treated with a deglazing solution or mild pumice. The fountain roller should be removed

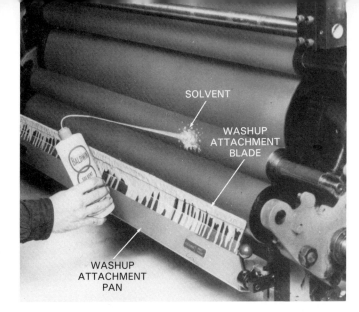

Fig. 25-15. Press washup attachment should only be used at a slow press speed. Blanket and roller wash is applied sparingly to rollers. (Baldwin-Gegenheimer Corp.)

and scrubbed with a deglazing agent made for this purpose.

15. At the end of the day, loosen the tension on the blanket to rest it. Retighten the blanket at the start of the next day.
16. Also, at the end of the day, set the night latch to release pressure on the form rollers.
17. Pick up all tools and articles around the press.

PRESS OPERATING SAFETY

Safe duplicator and press operation means that the duplicator and/or press is always operated in a sensible manner. Personal safety factors, as well as mechanical and general shop conditions, should be considered. Observe the following practices.

1. When operating any duplicator or press, always dress appropriately! Remove coats and sweaters. Roll shirt sleeves above the elbows. Remove or tuck in neckties. Remove all jewelry, such as chains, bracelets, and rings. Tie back any long hair. Do not carry wiping cloths in your pockets. Keep shirttails tucked into dresses, slacks, trousers, jeans, and shorts.
2. Jewelry, especially watches and rings, should NOT be worn when working around the duplicator or press! The desire to keep one's hands intact should take precedence over appearance, convenience, or sentiment.
3. Wear eye protection when needed.
4. Never lean against the duplicator or press or rest hands where there is any chance of getting them caught in moving parts! Use hand tools whenever they are provided, particularly when working at the ink fountain.
5. Place knives in safety sheaths when not using them for cutting, packing, or other press operations! If razor blades are used, they should have holders on them.

6. Never carry tools in your pockets. They could cause puncture wounds or be dropped into the duplicator, press, or other hazardous locations.
7. Use the proper size tools when making adjustments! Keep wrenches and other hand tools in good condition.
8. Operation of a duplicator or press that you have never run before is dangerous and should not be attempted!
9. Watch finger clearance when lifting rollers in and out of the duplicator or press and with similar operations. Inspect the edges of the plate before handling it, especially if it has been cut down from a larger size. Rough metal edges should be filed smooth so they do not cut your hands.
10. Oiling the duplicator or press should NOT be attempted unless the press is standing still. Wipe up oil from floors and platforms to reduce slipping hazards.
11. Never reach into the duplicator or press while it is running! Stop the machine to clear jams, lubricate, make adjustments, repairs, or clean the press. Never attempt to remove hickies while the duplicator or press is running. Always stop the machine before picking off the hickies.
12. All guards should be in place before the duplicator or press is started. See Fig. 25-16.

Guards

Offset duplicators and presses are equipped with adequate mechanical safeguards when they are shipped by the manufacturer. However, in the process of transferring equipment from one owner to another, the guards may become misplaced or lost. Also, through experience and use, the operator may find additional areas which may need mechanical protection.

Fig. 25-16. Offset duplicators and presses are equipped with adequate guards. All guards must be in place before starting any duplicator or press. (A.B. Dick Co.)

Sometimes, guards are unwisely removed because of their inconvenience.

A good operator learns what kind of mechanical protection the machine provides, noting especially such danger points as:

1. In-running cylinders and rollers.
2. Between the blanket cylinder and the plate cylinder.
3. Gears at the edge of the plate and blanket cylinder and near the bearers.
4. At the foot board and at the rear of the blanket and plate cylinder.
5. At feeder end over-insertion devices.
6. Chain delivery. See Fig. 25-17.
7. Cam that controls height of pile in the feeder.
   DANGER! Never operate a press unless all of its guards are in place!

Washing safety

Cleaning parts of the duplicator or press require the following precautions:

1. Do not smoke while using solvents!
2. Take only small amounts of solvent to the duplicator or press and then only in approved safety containers.
3. Avoid spilling flammable liquids! The vapors may flow long distances along the floor or ground before they are ignited and flash back.
4. Never pour used solvents down a drain! They should be transported in a closed container to an approved safe dump where they will not contaminate streams or drinking water supplies, or be accessible to children.
5. Observe health measures such as protection against inhalation of fumes! Do not use solvents on your body, and always use protective gloves when handling solvents.
6. Place rags and materials used with solvents into self-closing approved safety containers!
7. When removing paper stuck to the blanket, use water with a sponge and then the solvent with a rag. Make sure that the power is turned off before beginning this operation.
8. Operators using bichromate fountain solutions should report any skin irritations promptly. Non-bichromate fountain solutions for use with aluminum plates are helpful to those press operators who are sensitive to bichromate.

Housekeeping safety

Poor housekeeping in and around duplicators and presses contributes to many accidents, especially falls. For your own safety, and those who work around you, check the area for the following conditions:

1. Are the oil drip pans emptied periodically?
2. Do you have safety containers for solvents?
3. Are metal containers for cleaning rags available?
4. Are waste containers for paper provided?
5. Are there ink cans on the floor or platforms that could be spilled?

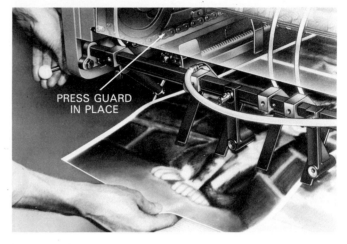

PRESS GUARD IN PLACE

Fig. 25-17. Removing a sheet from a chain delivery press with receding stacker. Note transparent safety guard covering gripper bar cylinder. (Heidelberg U.S.A.)

6. Are there tools left on walkways that could cause an accident?
7. Is there wearing apparel dangerously hanging on control boxes or on duplicator or press frames?
8. Are rollers properly racked so they cannot be accidentally knocked from their holders?
9. Are the floor, platform, and steps free of grease and oils?
10. Are there any empty skids standing on edge or leaning against equipment, walls, or columns?
11. Are the air hoses in reels or racks?
12. Are all aisles free of debris?

POINTS TO REMEMBER

1. Press operator's manuals are available for all offset duplicators and presses. Use them!
2. The operator of an offset duplicator or press should be thoroughly familiar with the procedures as found in the manual for the specific press.
3. Before starting a duplicator or press, it is important that the operator prepare the machine by checking for loose articles on or in the duplicator or press.
4. Offset duplicator and press pressure checks should be made prior to running the unit.
5. The duplicator or press should be thoroughly lubricated and wiped down at periodic intervals.
6. All paper required for a press run should be conditioned so far as moisture content is concerned.
7. Paper should be delivered to the pressroom area long enough in advance to reach the proper temperature balance.
8. The press operator must adequately prepare the duplicator or press for operation. This includes preparing the dampening and inking units, the plate, the feeder unit, double-sheet

detector, register board, and the delivery unit.

9. When the duplicator or press has been prepared for operation, several test sheets are run. Adjustments are made if required. The test sheets are inspected for overall image quality and image position. The press operator also checks for correct impression.

10. When the press run has been completed, the press operator must clean the duplicator or press of all ink, paper lint, and setoff spray powder residue. The machine should then be ready for the next job.

11. Safe duplicator and press operation means that the machine is always operated in a sensible manner.

12. All duplicator and press guards should be installed and in working condition.

13. Cleaning blankets, plates, rollers, and parts of the duplicator or press, such as the ink fountain, require special precautions.

14. Proper housekeeping in and around the duplicator and press area is essential to safety.

KNOW THESE TERMS

Night latch, Register board and sheet controls, Pullout roller, Conveyor tapes, Front stops, Movable side guide, Stationary side guide, Test sheet, Impression check, Run down, Roller glaze.

REVIEW QUESTIONS

1. Never operate any press or duplicator until you have been thoroughly _____ in the _____ operation of that machine.

2. What safety precautions regarding personal dress must be observed when operating an offset press or duplicator?

3. Describe the various operations that are part of the offset press or duplicator preliminary inspection.

4. Identify the parts of the duplicator shown.

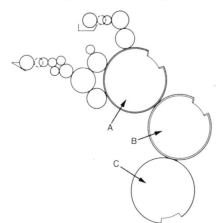

5. A press operator's instruction manual should always be available for every model of press or duplicator in the press area. True or false?

6. The press feeder and delivery drive chains should be lubricated:
   a. Once a week.
   b. Twice a week.
   c. Once every two weeks.
   d. Once a month.

7. Explain how ink should be placed in the offset press or duplicator ink fountain.

8. Most vacuum-blower pumps require SAE No. _____ weight oil.

9. Most solvents are flammable and must NOT be used near an open flame. True or false?

10. What is the dampener roller called that makes contact with the plate?
    a. Ink.
    b. Distributing.
    c. Form.
    d. Ductor.

11. The lead edge of the offset plate is attached to the _____ _____.

12. Press pile side guides should be set so that they are about _____ inch or _____ mm from the back of the sheet.

13. On most offset presses, press register board includes left- and right-hand paper joggers. True of false?

14. A stationary paper jogger should be set so that when the paper is moved against the spring, it moves the spring no more than:
    a. 1/32 inch (0.8 mm).
    b. 1/16 inch (1.6 mm).
    c. 1/8 inch (3.2 mm).
    d. 1/4 inch (6.4 mm)

15. List some important areas of the duplicator that are protected by guards.

16. Describe some of the important essentials of good press housekeeping.

SUGGESTED ACTIVITIES

1. Under supervision of your teacher, lubricate an offset duplicator and/or press to prepare for a print job. Inspect for loose tools, paper and other objects in and around the machine.

2. Obtain a copy of the duplicator operating manual. Review the essential steps in its operation. You may need to ask your teacher for help.

3. With your teacher's assistance, perform the necessary duplicator pressure checks.

4. Install an offset plate on the duplicator plate cylinder. Check to be sure the tail clamp is secure.

5. With teacher help, set up an offset duplicator for operation.
6. Feed test sheets through the duplicator and check for proper water and ink balance, and image position. Get a press sheet OK from your teacher. Complete the press run for the required number of impressions.
7. When the duplicator run has been completed, remove the plate and clean it thoroughly. Preserve the plate and store it for future use. Remove the fountain solution and save it.

Note! some press operators do not save fountain solution since it becomes somewhat contaminated during the day. Remove and discard the ink by wrapping it in paper towels.
8. Wash the duplicator thoroughly and have your teacher check it for cleanliness. Be sure roller ends are completely clean of ink.
9. Store extra ink rollers in a covered rack or small wooden box.
10. Carefully inspect for loose tools, paper and other objects in and around the press.

PRE-MAKEREADY TASKS CHECKLIST

The following things should be done before the press or duplicator is made ready for printing. *Makeready* refers to the procedures that are used to prepare the press to print a particular job. Makeready covers all the activities of the press operator between starting the job and running the job. The following activities are pre-makeready tasks. These should be completed before any job is attempted on a press or duplicator.

- Oil the press or duplicator according to the recommendations of its manufacturer. Properly lubricated mechanisms function efficiently and dependably. Mechanisms that are not lubricated correctly cause frequent problems and shortened service life.
- Clean the press of dirt, dried ink, grease, and excess oil. Cleaning the press in this manner helps to guard against paper spoilage and improves the machine's performance.
- Check pressure settings of all rollers and cylinders. Pressure requirements and checking procedures are described in your press or duplicator operating manual.
- Examine the blanket for defects. Repair or replace the blanket as required. Procedures for removing glaze and correcting smashes are in the operating manual.
- Examine the inking rollers for glaze. Glazed rollers will not transfer a uniform film of ink to the offset plate. If glaze has developed, remove it with one of the commercial deglazing compounds available for this purpose. Directions for use are provided with the deglazing compound. After deglazing, roller surfaces should exhibit the velvety, nonsmooth appearance that is required for efficient ink transfer.
- Examine dampener roller covers and change them if required. Replacements for molleton (towel-like) and fiber covers are available for most of the common size dampening rollers. Complete installation instructions are usually supplied by the manufacturer of the replacement covers.
- Check the pH of the fountain solution.
- Mix or otherwise prepare the ink that is to be used for the immediate job.
- Examine the image and non-image areas of the plate to be printed. Repair or replace the plate if defects are found.

CLEAN-UP TASK CHECKLIST

The following tasks are performed after the press run is completed. These are required to maintain the offset duplicator:

- Remove printed sheets from the delivery tray and place them in a rack to dry.
- Remove unprinted sheets from the feed table. Also, remove any deflected sheets from beneath the register table.
- Remove the fountain solution bottle from the duplicator; empty and wash it.
- Drain the fountain solution tray and wipe it dry.
- Remove excess ink from the ink fountain by wedging it between two cards. Discard the excess ink.
- Remove the ink fountain from the duplicator and clean it with blanket wash.
- Remove the ink ductor roller and clean it with a cloth moistened with blanket wash.
- Wipe the ink fountain roller clean with a cloth moistened with blanket wash.
- Attach a cleaner sheet (plate-size blotter) to the master cylinder. Turn on the machine and set it to run at its slowest speed. With the form rollers contacting the cleaner sheet, apply a small amount of blanket wash to the uppermost ink rollers. Use a plastic squeeze bottle to apply the wash to the rollers. Continue to add small amounts of wash to the rollers until they appear clean. *Note:* Repeat this step several times, using a fresh cleaner sheet each time, until all traces of ink have been removed from the rollers. Used cleaner sheets should be set aside to dry. When thoroughly dry, these sheets may be reused on the reverse side.
- With the duplicator stopped, inspect all rollers, especially their ends. Clean as required, by hand. Do not allow a film of blanket wash to remain on the rollers as this can cause glazing. Dry the rollers with a clean cloth.
- Clean all dampening rollers.
- Clean the blanket with blanket wash.
- Clean the impression and plate cylinders with blanket wash.
- Replace fountain solution bottle, ink fountain, and rollers. Make sure that the rollers are returned to their correct positions.
- Turn off all duplicator controls and set the form rollers to the NIGHT LATCH position.
- Wipe the outside of the machine clean with a cloth moistened with an appropriate solvent. Then polish dry.

Chapter 26

Sheet-Fed Offset Presses

When you have completed the reading and assigned activities related to this chapter, you will be able to:

O Describe the general procedures for running a duplicator and a sheet-fed offset press.

O Describe the common components among the various duplicators and sheet-fed offset presses.

O Adjust and test the dampener form rollers on a duplicator and an offset press.

O Adjust and test the ink form rollers on a duplicator and an offset press.

O Adjust and test the plate-to-blanket pressures on a duplicator and an offset press.

O Adjust the impression cylinder on a duplicator and an offset press for various weights of paper.

O Accurately pack the undercut cylinders of an offset press.

This chapter will continue your study of duplicators and small sheet-fed offset presses. Chapters 24 and 25 summarized the major parts and operation of duplicators and small offset presses. This chapter will explain how to run these types of machines in more detail. It will describe the adjustments and methods needed to operate a duplicator or a press properly.

Note! Press manufacturers' instruction manuals will be helpful for further study of this topic. The press manual will give information that is more specific, or possibly even unique, to the particular press.

There are many kinds of duplicators and small sheet-fed offset presses available. All are equipped with features that represent the latest technology in press design and operation. It would be impossible to include all the duplicator and press models in this book. This makes it difficult to include the operating features of any given duplicator and/or press. See Fig. 26-1.

For purposes of this book, very common duplicators and presses were selected for expanded explanation of operating procedures. This decision was based on data reflecting current use of these particular duplicators as the basic equipment of choice in commercial shops and schools throughout the United States. The chapter concludes with a brief description of several specific brands of duplicators and small offset presses.

Fig. 26-1. Many kinds of duplicators and small sheet-fed off-set presses are available. The best source of information for any particular offset press or duplicator is the operator's manual. (Ryobi, Inc.)

A.B. DICK 9800 SERIES

The A.B. Dick 9800 series duplicators are sheet-fed with vacuum feeders. These duplicators are derivatives of the earlier models. These are equipped with integrated dampening systems. In this system, the ink and the dampener solutions originate in separate fountains. They are then *integrated* (combined) and fed through the same form rollers to the plate, Fig. 26-2.

These duplicators have a single control lever for operating the dampener form roller, the ink form rollers, and for providing plate-to-blanket contact. The 9835 is equipped with chute delivery. The 9840 is equipped with chain delivery and receding stacker. The 9850 is basically the same as the 9840 except that it is equipped with additional rollers and adjustments for alcohol use in the fountain solution.

Specifications

1. Paper size: 3 x 5 inches to 13 1/2 x 17 3/4 inches (76 x 127 mm to 343 x 451 mm).
2. Paper weight: Minimum 12-pound bond to maximum 110-pound index (5.4 kg to 50 kg).
3. Printing area: 12 1/2 x 17 1/4 inches (318 x 438 mm).
4. Metal plate size: 13 x 19 3/8 inches (330 x 492 mm).
5. Gripper margin: 1/4 inch (6.4 mm).

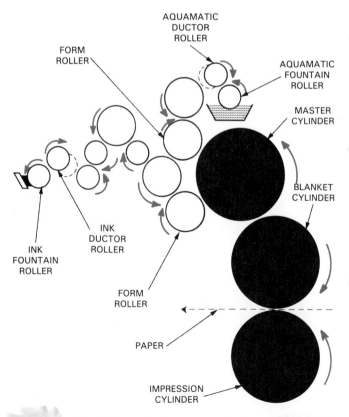

Fig. 26-2. The ink and dampener solutions originate in separate fountains with integrated dampening systems. The solutions are then combined and fed through the same form rollers to the plate. (A.B. Dick Co.)

6. Blanket size: 12 5/8 x 19 3/16 inches (321 x 487 mm).
7. Feed table capacity: 20 inches or 5,000 sheets of 20-pound bond paper.
8. Receiving tray capacity (9835 only): 3 inches or 500 sheets of 20-pound bond paper.
9. Chain delivery capacity (9840/9850 only): 21 inches or 5,000 sheets of 20-pound bond paper.
10. Speed: 4,500 to 9,000 impressions per hour (iph).

Operating procedure

The location of operating controls and major components for the A.B. Dick 9840/9850 is shown in Fig. 26-3. The control panel contains a series of panel switches and indicator lamps for copy count, machine speed, fault detection, and low paper conditions. Refer to Fig. 26-4. Note! For purposes of classroom instruction, it is recommended that an A.B. Dick 9810 (without keypad) be used.

A panel *clear switch* is used to clear the light-emitting diode (LED) count/speed display or to reset jam and doubles indicators. The jam and doubles indicators will remain illuminated, disabling the feed, until the paper path is clear and the indicators are reset by pressing the clear switch.

Note! Once the indicators are reset, pressing the switch a second time will clear the copy count in the LED display. Pressing the clear switch during machine operation will turn off the paper feed and reset the LED display to zero (0).

During machine operation, the initial copy count can be recalled for several seconds by pressing *recall switch*. The counting function will continue uninterrupted while the initial count is being displayed.

The *keyboard* has ten panel switches numbered "0" to "9." The number of copies selected will be entered into the count display (up to 99999). See Fig. 26-4.

The *display window* has a five digit LED display that indicates copy count and machine speed. When a count is entered, the duplicator will count down until it reaches zero. Then the paper feed and vacuum pump will turn off and the LED display will be reset to the initial copy count. If no count is entered, the duplicator will count up until paper feeding is manually stopped.

The *RPH* (rotations per hour) will be displayed (to the nearest hundred) in the count display for 8-10 seconds. To return to the count mode prior to this delay, press the clear button. Counting is not interrupted while the duplicator speed is displayed.

The *feed switch* turns the vacuum pump and paper feed ON and OFF. The feed switch will not override the pump switch. This means that if the vacuum pump is turned ON using the pump switch, it can only be turned OFF by re-engaging the pump switch.

The *On/Off switch* turns the drive motor ON and OFF.

The *jam indicator* shows that the paper feed is shut off and the indicator illuminates when a sheet does not

Sheet-Fed Offset Presses

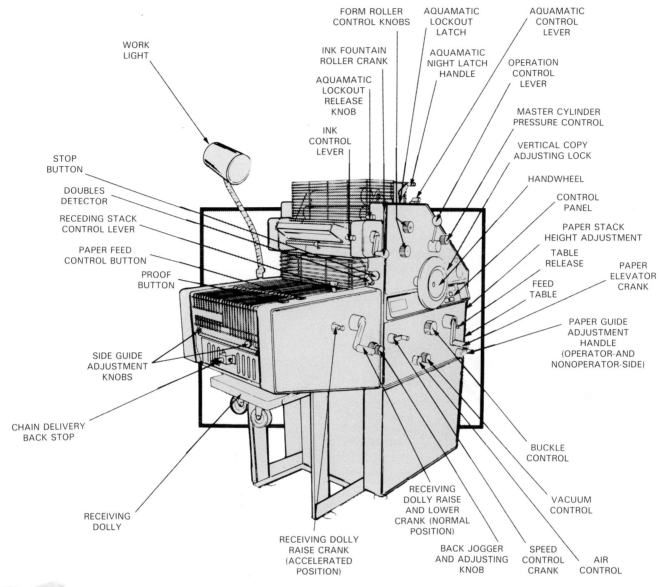

Fig. 26-3. The operating controls and major components of the A.B. Dick 9840/9850 series duplicators. (A.B. Dick Co.)

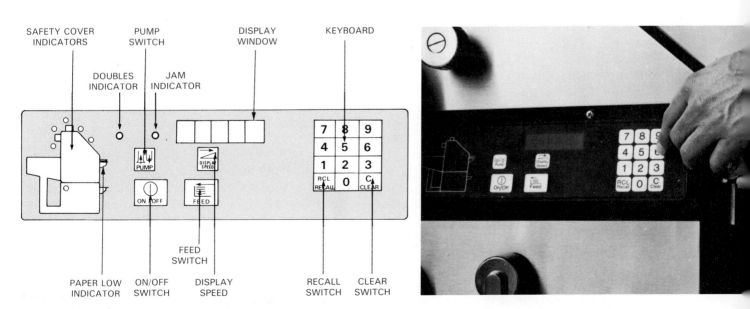

Fig. 26-4. The 9840/9850 series A.B. Dick duplicators are equipped with keypad control panels. (A.B. Dick Co.)

exit the duplicator within a preset time period. When a jam occurs in the duplicator, turn OFF the machine. Clear the paper path. Press the clear button to reset the jam indicator, and restart the duplicator.

The *doubles indicator* illuminates the panel when more than one sheet is sensed entering the duplicator and the paper feed shuts OFF. When a double is fed through the duplicator, turn the machine OFF (unless there is access to double sheet, as with duplicators equipped with receiving trays). Remove the extra sheet(s). Press the clear button to reset the doubles indicator, and restart the duplicator if stopped.

The *pump switch* turns the vacuum pump ON and OFF. If the pump is turned ON using this switch, it can only be turned OFF by this switch.

The *safety cover indicators* are five LEDs which illuminate to show the location of an open safety cover. Only one guard will be displayed at a time, even if more than one guard is open. After the first guard is secured, the next open guard (in a clockwise sequence) will be displayed.

SAFETY NOTE! The drive motor will not operate when any safety guard is open. If any guards are opened during machine operation, the drive motor will turn off automatically.

A *paper low indicator* flashes when there is less than 100 sheets (20-pound bond paper) remaining on the feed table. An audible "beep" is provided to indicate this condition. The machine will continue to print copy when a low paper condition is detected. The audible signal can be shut off by pressing the clear button once. However, the indicator lamp will continue to flash until the feed table is lowered.

Operation control lever

The *operation control lever* has five positions, as shown in Fig. 26-5.
1. *Night latch*—This position keeps the rollers from making full contact during idle periods.
2. *Neutral*—This lever position brings the rollers into full contact. This position is used for inking or cleaning the roller system and preparing the duplicator for operation.
3. *Ink*—This engages the form rollers on the plate cylinder to supply ink to the image areas and water to the non-image areas of the master or plate.
4. *Image*—This lever position engages the plate cylinder to the blanket cylinder to transfer the image to the blanket.
5. *Feed*—This turns on the paper feed and vacuum pump to engage paper feed. After the feed position is engaged, the operation control lever is spring-loaded and will normally return to ink after being released from feed for producing printed sheets. There is a small detent in the image position, however, so that if the operation control lever is returned to image from feed, the alternate sheet feed is actuated.

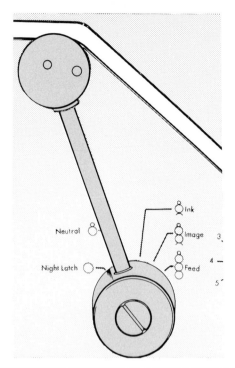

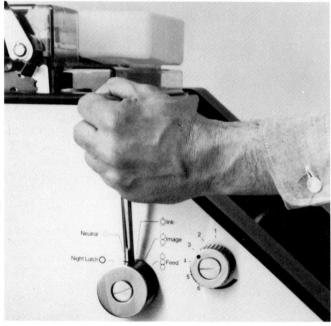

Fig. 26-5. The operation control lever has five positions including night latch, neutral, ink, image, and feed. (A.B. Dick Co.)

Duplicator set-up

1. Raise the ink roller safety guard. Refer to Fig. 26-6.
2. Place the Aquamatic oscillating roller (A) in position in the ink system. Make sure the hole in the roller shaft is placed over the drive pin on the nonoperator-side with the rounded-side of shaft up. The opposite end of the shaft seats in the channel on the operator-side of the inside frame.
3. Position the primary ink oscillating roller in contact with the two distributor rollers.

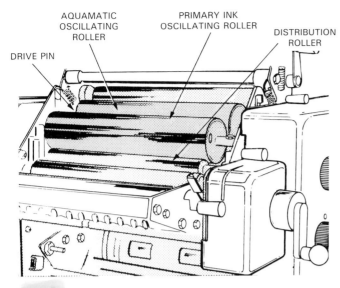

Fig. 26-6. Rinse the ink roller safety guard to begin duplicator set-up. Place the Aquamatic oscillating roller in position in the ink system. Position the primary ink oscillating roller in contact with the two distributor rollers. (A.B. Dick Co.)

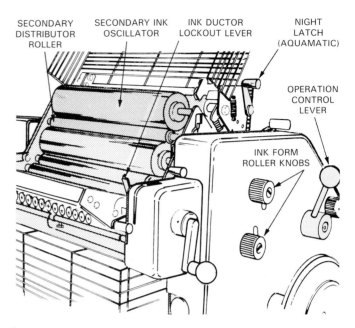

Fig. 26-8. Install the secondary distributor (or rider) rollers in the brackets over the ink and water oscillators. Install the secondary ink oscillator and move Aquamatic night latch handle. Move the operation control lever to neutral and move the ink ductor lockout lever to allow the ink ductor to come into contact with ink fountain roller. (A.B. Dick Co.)

4. Position the lock levers in the down position to lock the primary ink oscillating roller in place. See Fig. 26-7.
5. Install the secondary distributor (or rider) rollers in the brackets over the ink and water oscillators, Fig. 26-8.
6. Install the secondary ink oscillator and position it in the channel above both distributor rollers.
7. Move the Aquamatic Night Latch handle towards the receiving-end of the duplicator (9835 and 9840 only).
8. Move the operation control lever from Night Latch to Neutral and turn the ink form roller knobs to ON.

9. Move the ink ductor lockout lever toward the receiving-end of the duplicator to allow the ink ductor to come into contact with the ink fountain roller.
10. Lower the ink roller safety guard and turn ON the main power switch, Fig. 26-9.

Paper feed
1. Using the paper guide handle, on the nonoperator-side of the duplicator, move the paper guide to the

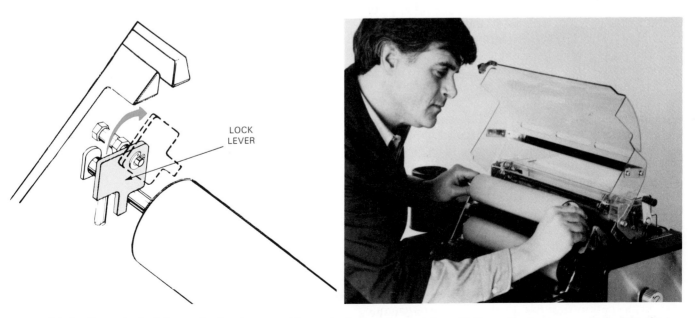

Fig. 26-7. Place the lock levers in the down position to lock the primary ink oscillating roller in place. (A.B. Dick Co.)

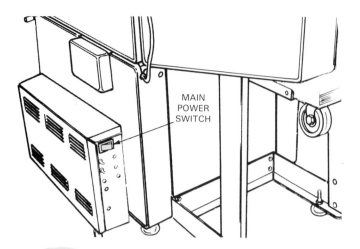

Fig. 26-9. Lower the ink roller safety guard and turn on the main power switch.　(A.B. Dick Co.)

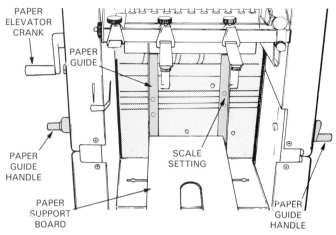

Fig. 26-10. Use paper guide handles on the nonoperator and operator side of the duplicator to move the paper guide to the correct scale setting for width of the paper being used.　(A.B. Dick Co.)

correct scale setting for the width of the paper being used. Refer to Fig. 26-10.

2. Using the paper guide handle on the operator-side of the duplicator, move the paper guide to the correct scale setting for the width of the paper.
3. The paper support rails slide from side-to-side. The paper support rails should be positioned to clear both backstops when the table is at its maximum height, Fig. 26-11.
4. Place the paper support board on the paper support rails, Fig. 26-10. The paper support board should be slightly smaller than the paper size.
5. Press down on the upper lever of the table release until it locks to disengage the automatic table raise mechanism, Fig. 26-11. Push in the paper elevator crank to engage it, then turn the crank counterclockwise to lower the feed table, Fig. 26-10.
6. Turn the handwheel until the paper height regulators are in the lowest position, Fig. 26-12.
7. Fan the paper and load it on the feed table, being careful to keep the reams neatly stacked. The assembly holding the backstops, spring guides, and micromatic may be lifted and latched out of the way to assist in paper loading. When loading is completed, release the latch and bring the assembly down to its operating position. See Fig. 26-13.
8. Squeeze the two levers of the table release and raise them to engage the automatic table raise mechanism. Push in the paper elelvator crank to engage it, then turn the crank clockwise to raise the feed table until the top sheets contact the paper height regulators. Pull out the paper elevator crank to disengage it. Look at Fig. 26-14.
9. Square the top few sheets of paper and adjust the micromatic paper guide so it just touches the stack of paper on the nonoperator-side, Fig. 26-14. The lead edge of the paper stack should be flush with the front plate.
10. Each suction foot has a valve to control vacuum, Fig. 26-14. The feet that contact the sheet should

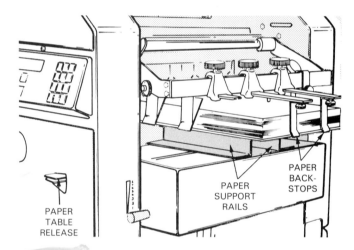

Fig. 26-11. The paper support rails slide from side-to-side. They should be positioned to clear both backstops when the table is at its maximum height. Place paper support board on the paper support rails. Press down on upper lever of the table release until it locks to disengage automatic table raise mechanism. Push in the paper elevator crank to engage it and lower feed table.　(A.B. Dick Co.)

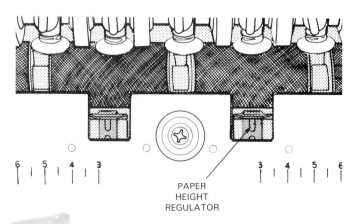

Fig. 26-12. Turn the handwheel until the paper height regulators are in the lowest position. Fan the paper and load it on the feed table.　(A.B. Dick Co.)

have their valves positioned vertically, while the feet NOT contacting the sheet should have the valves positioned horizontally. See Fig. 26-15.

11. Adjust the backstops so that they touch the tail end of the stack but do not bind the stack against the front plate, Fig. 26-16. On heavy card or index stock, there should be approximately 1/8 inch (3 mm) between the backstop and the paper stack. There are two positions for attaching the backstop to the retaining bar. The bar and stop may also be reversed. Use the best position for the length and type of paper stock being used.

12. Place the paperweight in a forward notch for light papers and farther back for heavier papers, Fig. 26-17.

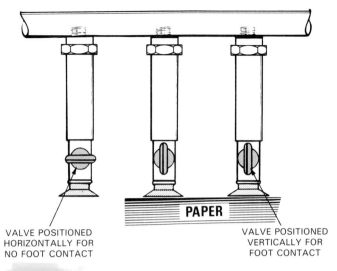

VALVE POSITIONED HORIZONTALLY FOR NO FOOT CONTACT

PAPER

VALVE POSITIONED VERTICALLY FOR FOOT CONTACT

Fig. 26-15. Each suction foot has a valve to control vacuum. The feet that contact the sheets should have the valves positioned vertically. The feet that do not contact the sheets should have the valves positioned horizontally. (A.B. Dick Co.)

PAPER FEEDER ASSEMBLY

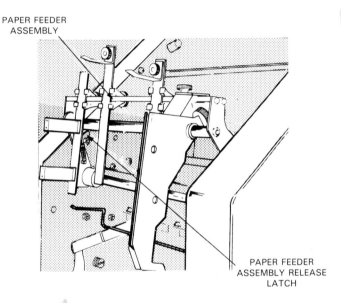

PAPER FEEDER ASSEMBLY RELEASE LATCH

Fig. 26-13. The feeder assembly holding the backstops, spring guides, and micromatic may be lifted and latched out of the way to assist in paper loading. When loading is completed, release the paper feeder latch and bring the assembly down to its operating position. (A.B. Dick Co.)

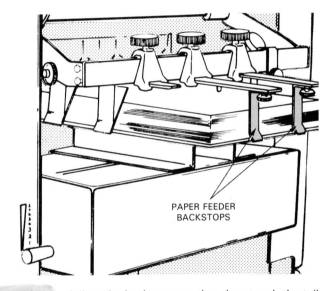

PAPER FEEDER BACKSTOPS

Fig. 26-16. Adjust the backstops so that they touch the tail end of the stack, but do not bind the stack against the front plate. (A.B. Dick Co.)

SUCTION FOOT

PAPER HEIGHT REGULATOR

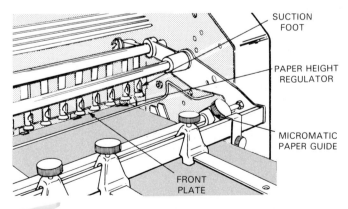

MICROMATIC PAPER GUIDE

FRONT PLATE

Fig. 26-14. Squeeze the two levers of the paper feeder table release and raise them to engage the automatic table raise mechanism. Push in the paper elevator crank to engage it, then turn the crank clockwise to raise the feed table until the top sheets contact the paper height regulators. Pull out the paper elevator crank to disengage it. (A.B. Dick Co.)

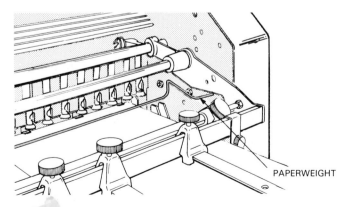

PAPERWEIGHT

Fig. 26-17. Place the paperweight in a forward notch for light paper and farther back for heavier papers. Refer to the paperweight settings table for proper alignment. (A.B. Dick Co.)

See the paperweight settings table below.

Paperweight Settings

12 lb. bond	1 (special light paperweight)
13 lb. bond	2
16-20 lb. bond	2-3
24 lb. bond	3
60 lb. book	3
65 lb. cover	4
110 lb. index	4 (or remove paperweight)

13. The feeding level of the paper stack should generally be LOW for lightweight papers and HIGH for heavier weight papers and card stocks. The automatic raising mechanism can be used to control the stack by raising or lowering the paper height control lever. This lever has eight gradations with position #8 representing MAXIMUM table raise, Fig. 26-18.

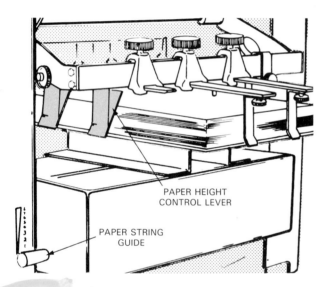

Fig. 26-18. The feeding level of the paper stack should generally be set low for lightweight papers and high for heavier papers and card stocks. The automatic raising mechanism can be used to control the stack by raising or lowering the paper height control lever. The spring guide, for most papers, should be depressed about 1/8 inch by the paper stack. (A.B. Dick Co.)

The paper height control lever settings table below indicates the approximate position for optimum feeding. With a particular paper or card stock slight deviations may be necessary.

Paper Height Control Lever Settings

12 lb. bond	1-3
13 lb. bond	2-4
16-20 lb. bond	4-6
24 lb. bond	5-7
60 lb. book	5-7
65 lb. cover	6-8
110 lb. index	7-8

14. Make sure all safety guards are closed.
15. Momentarily press the ON/OFF switch on the control panel and allow the paper stack to reach the proper height. Then press the ON/OFF switch again to stop the duplicator.
16. Place the spring guide equidistant between the lead and trail-edges of the sheet. The spring, for most stocks, should be depressed about 1/8 inch (3 mm) by the paper stack, Fig. 26-18. Make any necessary final adjustments to the rear guide and back stop.

Air and vacuum

1. Turn the air and vacuum control knobs clockwise as far as possible to the full position. Refer to Fig. 26-19.
2. Turn both knobs counterclockwise three half-turns.
3. To check the feed operation, close all safety guards, press the pump switch momentarily. Turn the handwheel until the paper height regulators are at their highest position at the top of the front apron. The top few sheets of the paper stack should fluff up and follow the paper height regulators. It may be necessary to increase the air setting by turning the air control knob clockwise until the sheets fluff up and follow the regulators. Turn the pump switch OFF.
4. To check the operation of the sucker feet, press the feed switch and turn the handwheel counterclockwise to manually feed one sheet of paper through the duplicator. If the suction feet do not pick up the paper properly, increase the vacuum setting by turning the vacuum control knob clockwise.

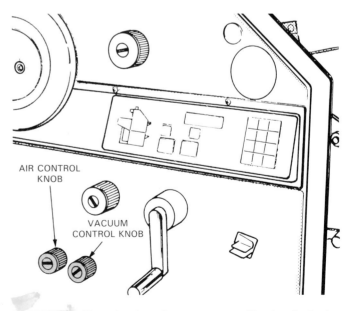

Fig. 26-19. Turn the air and vacuum control knobs clockwise as far as possible. Turn both knobs counterclockwise three half-turns. Close all safety guards and check feed and suction feet operations. (A.B. Dick Co.)

See the air and vacuum control knob settings table below.

Air and Vacuum Control Knob Settings
Air Adjustments

12 lb. bond	8 to 12 half-turns counterclockwise
13 lb. bond	2 to 3 half-turns counterclockwise
16-20 lb. bond	two half-turns counterclockwise
24 lb. bond	two half-turns counterclockwise
60 lb. book	two half-turns counterclockwise
65 lb. cover	two half-turns counterclockwise
110 lb. index	one-half turn counterclockwise

Air and Vacuum Control Knob Settings
Vacuum Adjustments

12 lb. bond	two half-turns counterclockwise
13 lb. bond	3-4 half-turns counterclockwise
16-20 lb. bond	3-4 half-turns counterclockwise
24 lb. bond	2-3 half-turns counterclockwise
60 lb. book	2-3 half-turns counterclockwise
65 lb. cover	2-3 half-turns counterclockwise
110 lb. index	one-half turn counterclockwise

Buckle control

Adequate buckle is required to deliver each sheet firmly against the paper stops in the paper grippers, Fig. 26-20. This assures accurate vertical registration.

Zero (0) on the buckle control produces minimum buckle and 15 produces maximum buckle. The buckle is set too high if the stock is being nicked at the lead edge, Fig. 26-21. Changing the buckle may slightly alter the registration. Therefore, the buckle control should NOT be changed during a press run when close registration is required. Normal buckle setting is 7-8.

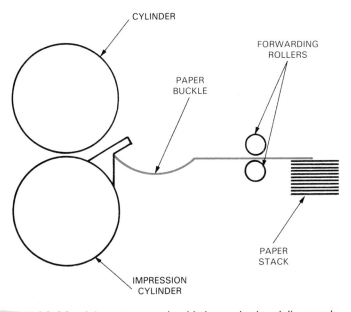

Fig. 26-20. Adequate paper buckle is required to deliver each sheet firmly against the paper stops in the paper grippers. Zero on the buckle control produces minimum buckle; 15 produces maximum buckle. (A.B. Dick Co.)

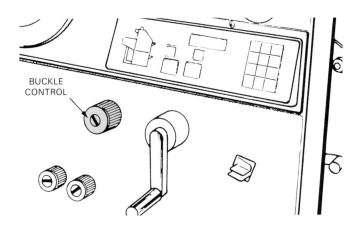

Fig. 26-21. Registration may be affected by changing the buckle setting. Do not change buckle control during a press run when close registration is required. Normal buckle setting is 7-8. (A.B. Dick Co.)

See the buckle settings table below.

Buckle Settings

12 lb. bond	15
13 lb. bond	12
16-20 lb. bond	7-8
24 lb. bond	5-6
60 lb. book	5-6
65 lb. cover	3-4
110 lb. index	0-2

Receiving tray (9835 only)

DANGER! The drive motor must be OFF whenever an adjustment is made to the receiving tray or the delivery wheels and rings.

1. Set the stationary guide on the nonoperator-side of the duplicator for the selected width of paper. Look at Fig. 26-22.
2. Place one sheet of paper in the receiving tray. Adjust the backstop for the length of paper being used, and position the sheet against the stationary guide.
3. Rotate the handwheel so that the jogging guide is in the inward position. Using the lock knob, adjust the guide so that it just touches the paper. See Fig. 26-22.

WARNING! Keep hands clear of receiving end of duplicator when turning handwheel.

4. Position the tray bail into the slot in the backstop and shape it to provide the best stacking for the type of paper being used, Fig. 26-22.
5. Open the safety guard over the delivery system; then turn the feed switch ON.
6. Turn the handwheel counterclockwise to manually feed one sheet of paper through the duplicator until it is just past the upper delivery wheels. Then, turn OFF the vacuum pump motor, Fig. 26-23.
7. The upper and lower delivery wheels control the paper as it leaves the cylinders and guides it into

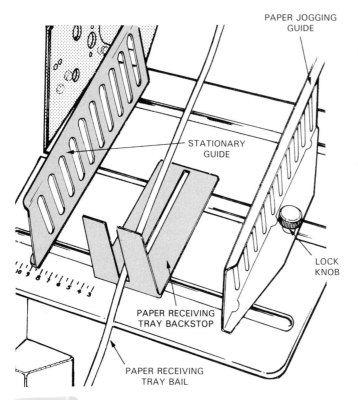

PAPER JOGGING GUIDE

STATIONARY GUIDE

LOCK KNOB

PAPER RECEIVING TRAY BACKSTOP

PAPER RECEIVING TRAY BAIL

Fig. 26-22. Set the stationary guide on the nonoperator side of the duplicator for the selected width of paper. Place one sheet of paper in the receiving tray. Adjust the backstop for the length of paper being used, and position the sheet against the stationary guide. Adjust the paper jogging guide so that it barely touches the paper. (A.B. Dick Co.)

the receiving tray. The delivery arms are movable so that the wheels can be positioned over the margins of the paper. They should NOT ride over the image area. In cases where it is necessary that the delivery wheels travel over portions of the image, an optional *wiper kit,* which keeps the delivery rollers free of ink, is available.

8. After the delivery wheels are set, turn the handwheel to forward the paper until it is just under the lower delivery wheels. See Fig. 26-23.

9. Position the delivery rings either to the inside or outside of the delivery wheels, and as close as possible to the wheels without crimping the paper. Refer to Fig. 26-23.

10. The purpose of the *delivery rings* is to control paper curl so that the paper will be delivered properly into the receiving tray. When using flat or curl DOWN paper, position the rings to the outside of the delivery wheels. For paper that curls up, position the rings to the inside of the delivery wheels.

Sheet feeding will be better if the paper is fed in the direction of the grain (long grain). When paper is fed against the direction of the grain (short grain), it has a tendency to tumble or flip when entering the receiving tray. It is not always possible to identify curl up or curl down paper until it is fed through the duplicator. Curl up and curl

down paper can most often be identified by the way the lead edge of the paper lifts *(curl up)* or drops *(curl down)* into the receiving tray.

11. Continue turning the handwheel until the sheet falls into the receiving tray.

12. Close the safety guard.

Chain delivery and receding stacker

SAFETY NOTE! Be sure the drive motor is OFF whenever either safety screen is open and exercise caution. At no time should anything be allowed to rest on the closed screen.

1. Place the dolly on the yoke assembly with the swivel wheels in toward the duplicator, Fig. 26-24.

2. Place the two-position handle on the raise/lower stud (for accelerated raise) and lift the dolly until it is just below the guides of the chain delivery. Be careful not to jam the dolly into the jogging arms, Fig. 26-24. Seat the handle on the normal raise/lower stud (as shown), place the table raise lever fully toward the delivery end of the duplicator, and raise the dolly until it is flush with the bottom of the chain delivery guides. Refer to Fig. 26-24.

3. Press the feed switch ON and hand feed one sheet into the duplicator, then press the feed switch OFF. Turn the handwheel counterclockwise until the paper is held by the chain grippers.

4. Open the delivery end safety screen, exercising caution while the screen is open, Fig. 26-24.

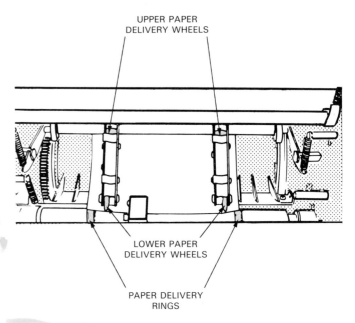

UPPER PAPER DELIVERY WHEELS

LOWER PAPER DELIVERY WHEELS

PAPER DELIVERY RINGS

Fig. 26-23. Manually turn handwheel counterclockwise to feed one sheet of paper through the duplicator until it is just past the upper paper delivery wheels. Turn off the vacuum pump motor. When delivery wheels are set, turn handwheel to forward the paper until it is just under the lower paper delivery wheels. Position delivery rings either to the inside or outside of delivery wheels, and as close as possible to the wheels without crimping the paper. (A.B. Dick Co.)

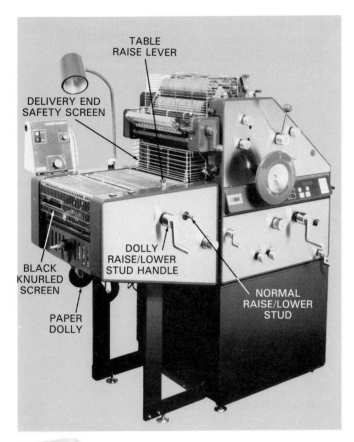

Fig. 26-24. When using a chain delivery operation, place the dolly on the yoke assembly with the swivel wheels in toward the duplicator. Place the two-position handle on the raise/lower stud (for accelerated raise) and lift the dolly until it is just below the guides of the chain delivery. Place the handle on the normal raise/lower stud, place the table raise lever fully toward the delivery end of the duplicator, and raise the dolly until it is flush with bottom of the chain delivery guides. (A.B. Dick Co.)

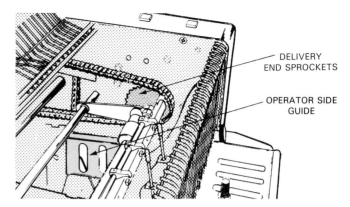

Fig. 26-25. Hand feed one sheet into the duplicator. Turn the handwheel counterclockwise until the paper is held by the chain grippers. Open the delivery end safety screen. Exercise caution while the screen is open. Turn handwheel until gripper bar reaches the delivery end sprockets. Adjust nonoperator-side guide and operator-side guide. (A.B. Dick Co.)

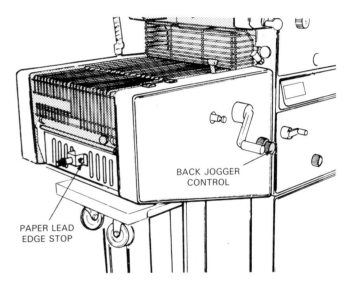

Fig. 26-26. Turn the duplicator handwheel until the jogger moves fully toward the paper. Loosen the center locking knob of the back jogger control. Move the back jogger until it is almost touching the trail edge of the paper. Lock the back jogging guide in place and check to see that the back jogger and side guides do not bind on the dolly. (A.B. Dick Co.)

5. Continue to turn the handwheel until the gripper bar reaches the delivery end sprockets, Fig. 26-25. Loosen and move the black knurled screw at the delivery end of the duplicator and adjust the nonoperator-side guide to just touch the edge of the paper. Look at Fig. 26-24.

6. Again, continue to turn the handwheel until the paper is released and drops on the receiving dolly. Adjust the operator-side guide so it just touches the paper, Fig. 26-25.

7. As the gripper releases the paper, the back jogger is still away from the paper. Turn the handwheel until the jogger moves fully toward the paper and loosen the center locking knob of the back jogger control, Fig. 26-26. Turning the outer adjusting knob of the back jogger control, position the back jogger until the jogger is almost touching the trail edge of the paper, Fig. 26-27. Lock the back jogging guide in place and make sure the back jogger and side guides do not bind on the dolly.

8. Loosen the knurled knob on the hinged lead edge stop and adjust the stop so that the back jogger

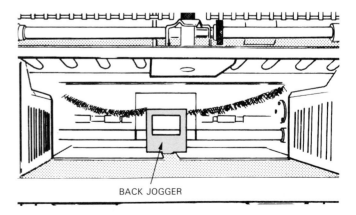

Fig. 26-27. Adjust back jogger so it stacks sheets without forcing the backstop to move. (A.B. Dick Co.)

577

can fully stack the sheet without forcing the back stop to move. See Fig. 26-26.

9. Pull the knurled knob toward the nonoperator-side of the sprocket shaft and position the cam of the variable paper release sprocket so that the pin is in the center hole (position 4), Fig. 26-28. When feeding lightweight paper, position the cam at a HIGHER NUMBER, and reseat the pin in another hole. This will allow the gripper to hold the paper longer and provide better stacking on the dolly, Fig. 26-28. When feeding heavier stock, rotate the cam to a LOWER NUMBER.

10. The paper guides are normally positioned one to two inches from the edges of the printed sheet. They can be adjusted in or out for various types of copy and stocks, Fig. 26-29.

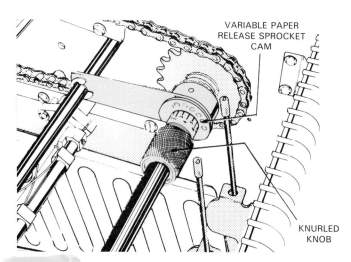

Fig. 26-28. Pull the knurled knob toward the nonoperator side of the sprocket shaft. Position the cam of the variable paper release sprocket so that the pin is in the center hole (position 4). Lightweight paper requires a higher number setting. (A.B. Dick Co.)

Warning! The delivery wheels must be installed correctly to avoid damage to the grippers. If the wheels are removed, reinstall them as follows:

a. Turn the handwheel until the groove of the delivery roller shaft is facing up (12:00 position). See Fig. 26-29.

b. Place the open delivery wheels onto the shaft so that the spring-loaded latches are on the nonoperator-side of the delivery wheels.

c. Lock the delivery wheels on the shaft and slowly rotate the handwheel to assure that the grippers will clear the shaft.

11. Adjust the receding control lever toward the feed end of the chain delivery for heavier stocks and toward the receiving end for lighter weight stocks. For normal lever position, center the lever in its slot, then bring the lever back toward the receiving end of the duplicator 5-6 notches, Fig. 26-30. Since the lever controls the rate of descent of the dolly, readjust the lever after running 400-500 sheets so that the top of the paper stack is centered between the top and bottom of the paper joggers when producing printed sheets.

Double sheet detector

1. Turn the double sheet detector knob clockwise three to four turns. See Fig. 26-31.
2. Press the ON/OFF and feed switches to power the drive and feed systems.
3. Turn the double sheet detector knob counterclockwise until the feed stops.
4. Turn the adjusting knob clockwise 1/8 turn.
5. Press the clear switch on the control panel to reset the detector and press the feed switch to resume feeding.
6. Repeat steps 4-5 until the feed is uninterrupted.

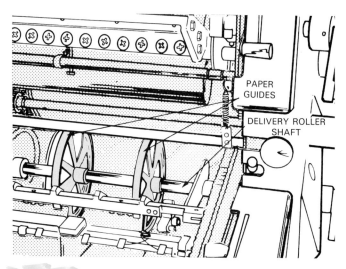

Fig. 26-29. The paper guides are normally positioned one or two inches from the edges of the printed sheet. They can be adjusted in or out for various types of copy and papers. (A.B. Dick Co.)

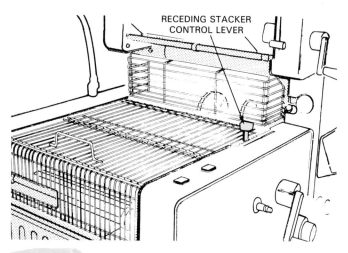

Fig. 26-30. Adjust the receding stacker control lever toward the feed end of the chain delivery for heavier stock, or toward the feed receiving end for lighter stock. For normal lever position, center lever in its slot and bring the lever back toward the receiving end of the duplicator five to six notches. (A.B. Dick Co.)

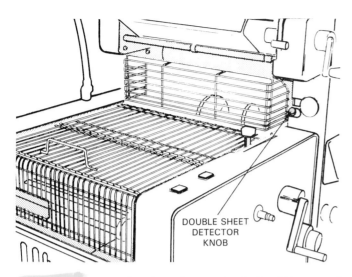

Fig. 26-31. The double sheet detector knob should be turned clockwise three to four turns. Press the ON/OFF and feed switches to power the drive and feed systems. Turn the double sheet detector knob counterclockwise until the feed stops and turn the adjusting knob clockwise one-eighth turn. Press the clear switch on the control panel to reset the detector and press feed switch to resume feeding. Repeat the procedure until sheets run uninterrupted. (A.B. Dick Co.)

Inking

Note! If setting up a 9850, always be sure to add the fountain solution before inking the rollers.

1. Turn the ink fountain screws all the way clockwise, shutting off the ink supply. See Fig. 26-32.
2. Turn all the screws one-half turn counterclockwise. These screws provide an even flow of ink to the rollers.
3. Using an ink knife and required can of ink, dispense ink evenly across the ink fountain. Keep a smooth, level surface in the ink can.
4. Spread the ink in the ink fountain. Using the ink fountain roller handle, rotate the ink fountain roller counterclockwise at least one revolution to carry the ink down into the ink fountain and coat the roller with ink. Refer to Fig. 26-32.
5. Rotate the handwheel counterclockwise until the ink ductor roller contacts the ink fountain roller, Fig. 26-32. While rotating the ink fountain roller counterclockwise, adjust the ink fountain screws until an even, finely *stippled effect* is obtained on the ink ductor roller (similar to an orange peel texture).
6. Raise the Aquamatic lockout lever toward the receiving end of the duplicator to allow full contact between the ductor and oscillating rollers (9835 and 9840 only). See Fig. 26-32.
7. Lift the ink fountain control up to the full (number 11) position and the Aquamatic control to number 45. Look at Fig. 26-32.
8. Make sure all safety covers are closed.
9. Press the ON/OFF button to turn the duplicator ON.

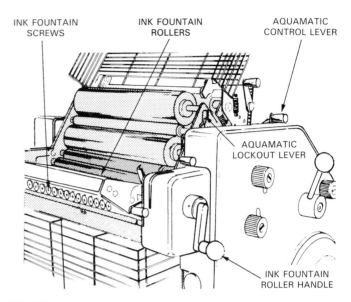

Fig. 26-32. Set up ink fountain and adjust ink fountain keys for even ink distribution. Raise the Aquamatic lockout lever toward the receiving end of the duplicator to allow full contact between ductor and oscillating rollers. Set the ink fountain control to the number 11 position and the Aquamatic control to 45. (A.B. Dick Co.)

10. The speed should be adjusted only when the duplicator is running. Adjust the speed control to the slowest speed (range is from 4,500 to 9,000 impressions-per-hour). See Fig. 26-33.
11. Allow the duplicator to run until all the rollers are evenly covered with a thin film of ink.
12. Move the ink fountain control to one or two notches above the OFF position.
13. Move the Aquamatic control to the operating position (usually number 20).

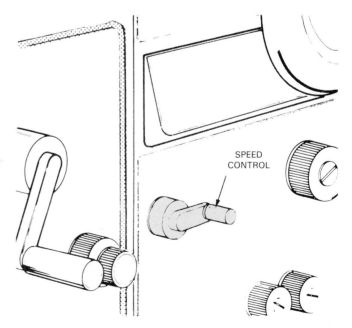

Fig. 26-33. Speed of the duplicator should be adjusted only when it is running. (A.B. Dick Co.)

14. Move the Aquamatic lockout latch toward the receiving end of the duplicator to allow automatic ducting while printing. See Fig. 26-34.

Fountain solution (9835 and 9840)

Note! Always be sure the ink rollers, including the two in the Aquamatic unit, are inked before fountain solution is added.

1. Fill the fountain solution bottle with properly mixed fountain solution.
2. Make sure the drain hose is properly secured, Fig. 26-35.
3. Close the safety cover and place the filled bottle in position in the Aquamatic unit, Fig. 26-35.

Attaching master or metal plate (pinbar)

1. Open the safety guard over the plate cylinder.
2. Attach the lead edge of the master or metal plate to the head clamp. While holding the master/plate

Fig. 26-34. Move the Aquamatic lockout latch toward the receiving end of the duplicator to allow automatic ducting while printing. (A.B. Dick Co.)

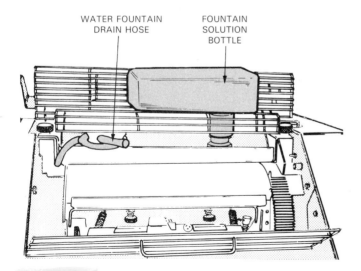

Fig. 26-35. Fill the fountain solution bottle and check to see that the drain hose is properly secured. Close the duplicator safety cover and place the filled bottle in position in the Aquamatic unit. (A.B. Dick Co.)

taut, use your finger or thumb to crease the master/plate at the head clamp. See Fig. 26-36.

3. While still holding the master/plate taut with the right hand, rotate the handwheel counterclockwise until the tail clamp is about three inches (75 mm) from the Aquamatic fountain. Hold the master/plate taut to avoid contact between the master/plate and the form rollers.

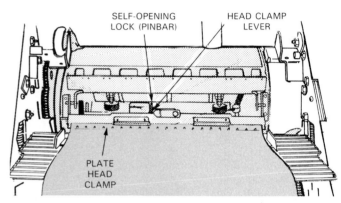

Fig. 26-36. Attach the lead edge of the metal plate or master to the head clamp. Use your finger or thumb to crease the plate at the head clamp while holding it taut. (A.B. Dick Co.)

4. Holding the master/plate against the cylinder surface, crease the tail end of the master/plate on the anvil of the cylinder. Lift the tail clamp and insert the pins of the clamp over the holes of the pinbar master/plate. Look at Fig. 26-37.

WARNING! Keep fingers clear of the tail clamp spikes.

5. Finger tighten the two knurled knobs to secure the master/plate on the plate cylinder, Fig. 26-37.
6. Apply the correct type of etch for the master; if using a metal plate, remove the gum coating with a cotton wipe and water. Close the safety guard.

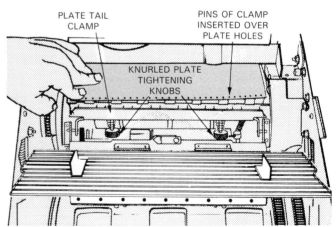

Fig. 26-37. Crease the tail end of the plate or master on the anvil of the cylinder while holding it against the cylinder surface. Lift the tail clamp and insert the pins of the clamp over the holes of the plate. (A.B. Dick Co.)

Note! When using the pinbar clamp, the self-opening lock must engage the lever, Fig. 26-36.

Self-opening clamp

1. Open the safety guard over the plate cylinder. The side guides on the inside of the open guard are used as a master/plate loading table.
2. Set the side guides to the width of the master/plate, using the scale on the loading table to center the master/plate. Actuate the self-opening clamp by disengaging the lock from the clamp latch. See Fig. 26-38.

   Note! The self-opening clamp may also be actuated by pushing in on the clamp lever, Fig. 26-38.
3. Position the straight edge master/plate on the loading table.
4. Turn the plate cylinder clockwise until the clamp opens, Fig. 26-38.
5. Slide the master/plate squarely into the clamp and continue turning the cylinder clockwise until the clamp closes.

   WARNING! Since the clamp snaps shut as the handwheel is rotated, keep fingers clear of the clamp area.
6. Apply the correct type of etch (if using a master) on a cotton pad and thoroughly wet out the master. If using a metal plate, use fountain solution to thoroughly dampen the surface.

Sequence of duplicator operation

1. Close all safety guards. Be sure that the Aquamatic control is at 20 and the ink fountain control is between 4-5.

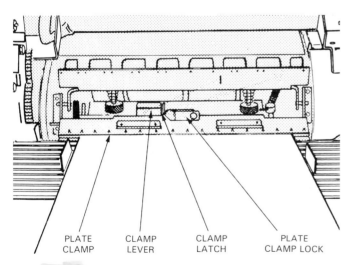

Fig. 26-38. Secure the plate or master to the cylinder by tightening the two knurled knobs. When using the pinbar clamp, the self-opening lock must engage the lever. To use the self-opening plate clamp, use the side guides on the inside of the open guard as a plate or master loading table. The side guides are set to the plate width and the self-opening clamp is actuated. The plate is inserted into the clamp until it closes. The clamp snaps closed as the handwheel is rotated. (A.B. Dick Co.)

PLATE CLAMP · CLAMP LEVER · CLAMP LATCH · PLATE CLAMP LOCK

2. Press the ON/OFF switch to start the duplicator. Depress the Aquamatic lockout release for 3-5 revolutions to apply fountain solution to the rollers. Refer to Fig. 26-39.
3. Move the operation control lever to the ink position, Fig. 26-39. Check the master/plate as the duplicator is idling in the ink position to be sure the master/plate is clean of ink in the non-image areas. If it tends to pick up ink in the non-image areas, there is too much ink or too little water in the ink system and adjustments should be made as required.

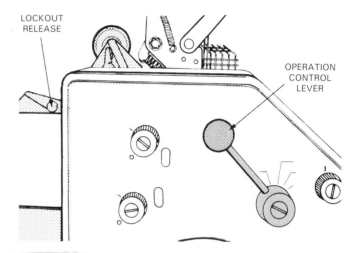

LOCKOUT RELEASE · OPERATION CONTROL LEVER

Fig. 26-39. Turn the duplicator on after closing all safety guards. Move the operation control lever to the ink position. Then, move the control lever to image position. When ready, move the control lever to feed and allow one sheet to be fed, and then press feed switch off. (A.B. Dick Co.)

4. Move the operation control lever to the image position and hold it there for several revolutions (generally two to four revolutions are required).
5. Move the operation control lever to feed. When released, the lever will automatically return to ink. Allow one sheet of paper to be fed, then press the feed switch OFF. If operating a chain delivery model, single-sheet feed can be accomplished by pressing the proof button after imaging the copy, Fig. 26-40.
6. Remove the copy from the receiving tray or, if a chain delivery model, place the control lever in neutral and press the ON/OFF switch to shut the duplicator down before removing the copy from the hinged lead edge stop door, Fig. 26-40.

   DANGER! Always exercise caution when removing printed sheets from the hinged lead edge stop door.
7. Inspect the copy for position, clarity, and general acceptability. If copy adjustments are necessary, make sure the operation control lever is in neutral and the duplicator is OFF, and see copy adjustments below.

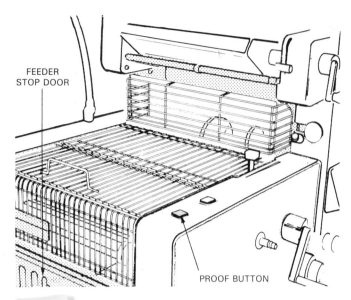

Fig. 26-40. When operating a chain delivery duplicator, single-sheet feed can be accomplished by pressing the proof button after imaging the copy. Remove the copy from the receiving tray or from the hinged lead edge stop door of the chain delivery model. (A.B. Dick Co.)

8. If the trial copy is acceptable, enter the count for the required number of impressions. Turn the duplicator ON. Place the operation control lever in ink. Then, press the feed switch ON. If necessary, lighter and/or darker copy in various areas of the master/plate can be compensated for by adjusting the ink control screws. Clockwise will DECREASE the ink supply and counterclockwise will INCREASE the flow of ink.

9. If a count has been entered, once the required number of copies has been printed, the feed will automatically shut off. Move the operation control lever to neutral.

10. If the master/plate is to be retained, remove the excess ink from the master/plate and preserve it.

11. Clean the blanket after the run using a cloth dampened with blanket wash.

DANGER! Blanket wash is usually very flammable! Do not use it near fire or open flame and avoid prolonged skin contact. Always use with adequate ventilation.

Stop button

The duplicator may be stopped during a printing cycle by pressing the feed switch OFF and engaging the red stop button which interrupts the safety circuit and cuts current to the drive motor. See Fig. 26-41.

Place the operation control lever in neutral in order to move the form rollers out of contact with the master/plate while the duplicator is idle.

In order to restart, press the ON/OFF switch, move the operation control lever to ink until the non-image areas of the master/plate are clean. Then press the feed switch ON.

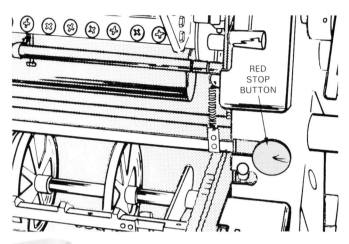

Fig. 26-41. The duplicator may be stopped during a printing cycle by pressing the feed switch OFF and engaging the red stop button. The stop button interrupts the safety circuit and cuts current to the drive motor. (A.B. Dick Co.)

Copy adjustments

WARNING! The duplicator must be STOPPED when making any copy adjustments!

1. Angular copy adjustments can be made by removing the tail end of the master/plate from the clamp and rotating the shaft and pin assembly right or left, Fig. 26-42.

If the copy is running downhill from left to right, turn the pin counterclockwise to raise the right side of the copy. If the copy is running uphill, turn the pin clockwise to lower the right side of the copy. Reinsert the tail end of the master/plate in the tail clamp and always clean the blanket after making this adjustment.

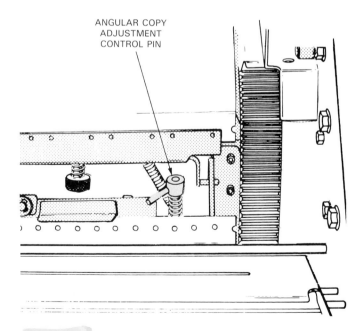

Fig. 26-42. Angular copy adjustments can be made by removing the tail end of the plate or master from the clamp and rotating the shaft and pin assembly. (A.B. Dick Co.)

2. Lateral copy adjustments of up to 1/4 inch (6.3 mm) can be made by removing the tail end of the master/plate from the clamp and turning the knurled knob so the head clamp moves in the desired direction. See Fig. 26-43. Reinsert the tail end of the master/plate in the tail clamp and always clean the blanket after making this adjustment. For a lateral adjustment greater than 1/4 inch (6.3 mm), move the paper stack.

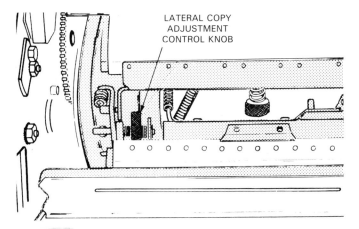

Fig. 26-43. Lateral copy adjustments of one-quarter inch maximum can be made by removing the tail end of the plate or master from the clamp and turning the knurled knob so the head clamp moves in the desired direction.   (A.B. Dick Co.)

3. To make a vertical copy adjustment, lift the safety cover over the blanket and rotate the handwheel until the locking gear lines up with the built-in printing adjusting tool. See Fig. 26-44. Push the print adjusting lock knob in and loosen the locking gear by turning it counterclockwise. Holding the knob in, move the scale on the opposite side of the blanket cylinder by turning the handwheel and following the arrows to raise or lower the copy. Refer to Fig. 26-44.

After the adjustment is made, tighten the locking gear. It is not necessary to clean the blanket after raising or lowering the copy image, as the relationship of the plate cylinder to the blanket cylinder remains the same. For normal operating procedures, set the scale at "0" as no change should be necessary, Fig. 26-45.

Adjusting micromatic paper control

1. To adjust the micromatic, loosen the outer guide knob and turn the inner knob toward the mounting bar to move the stack to the nonoperating-side. Look at Fig. 26-46.
2. To move the guide to the operating side of the duplicator, loosen the inner guide knob and turn the outer guide toward the mounting bar.
3. Hold the knob used to make the adjustment and tighten the opposite knob to secure the micromatic.

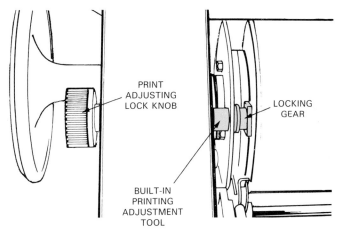

Fig. 26-44. To make a vertical copy adjustment, lift the safety cover over the blanket and turn the handwheel until the locking gear lines up with the built-in printing adjustment tool. The lock knob is depressed and the locking gear is loosened by turning it counterclockwise. The scale on the blanket cylinder is moved to raise or lower the copy while holding the knob in. (A.B. Dick Co.)

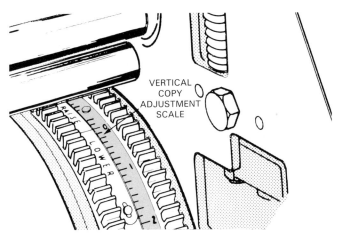

Fig. 26-45. Set the vertical copy adjustment scale at 0 for normal operating procedures.   (A.B. Dick Co.)

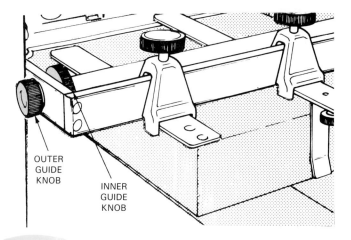

Fig. 26-46. To adjust the micromatic paper control, the outer guide knob is loosened and the inner knob turned toward the mounting bar. This moves the stack to the nonoperating side. (A.B. Dick Co.)

When adjusting the feed table side guides, be sure the guides clear the paper stack, support rails, and paper board before raising the feed table. For adjustments greater than 3/16 inch (5 mm), the guide plate assembly and the paper guide must be moved before the fine adjustment can be made.

Impression and plate cylinder adjustments

The impression and plate cylinders automatically adjust to the blanket cylinder to compensate for changes in master/plate or paper thickness. Some applications, however, may require an adjustment in the impression or plate cylinder control. For instance, the copy quality in rough textured papers, particularly copy which contains halftones and solids can be improved with increased pressure from the impression cylinder. Make these adjustments when the master or plate is on the plate cylinder.

SAFETY NOTE! Keep hands and the Allen wrench clear of rotating handwheel while making the impression cylinder adjustment.

Impression cylinder adjustment

1. Insert the Allen wrench into the control dial opening. Turn the dial to a lower number (clockwise) to increase pressure and to a higher number to decrease pressure. Refer to Fig. 26-47.
2. While running the required copy, turn the dial counterclockwise to decrease the cylinder pressure until the copy becomes *broken* (not printing fully). Then turn the dial clockwise until an optimum copy image is achieved.
3. Adjust only until optimum copy is achieved. Overadjusting will shorten the life of the blanket.

Plate cylinder adjustment

This adjustment is specifically designed for adjusting to different thicknesses of masters and plates. As the number is increased, the pressure is decreased.

Cleaning the duplicator

1. Remove the fountain solution bottle.
2. On the 9835 or 9840 only, raise the safety guard and unfasten the hose to drain fountain solution from the fountain. Discard any unused solution. Close the safety cover.
3. Remove the ink from the ink fountain by wedging the ink between two strips of card stock and removing. Discard the ink.
4. Move the ink ductor lockout lever up and toward the feed end to lock the ductor away from the ink fountain. See Fig. 26-48.
5. Remove the ink fountain by turning it up to a vertical position and then lifting it off the duplicator, Fig. 26-49. Clean the ink fountain and fountain roller with a cloth moistened with blanket wash.

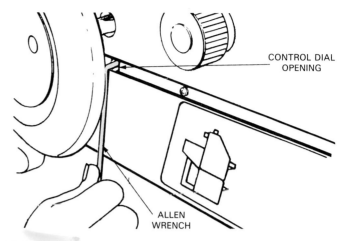

CONTROL DIAL OPENING

ALLEN WRENCH

Fig. 26-47. Impression cylinder adjustment is accomplished by inserting an Allen wrench into the control dial opening and turning the dial. A lower number increases pressure and a higher number decreases pressure. (A.B. Dick Co.)

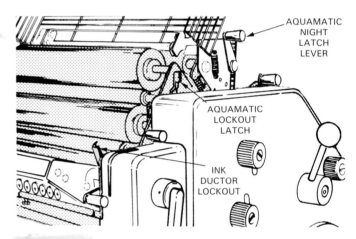

AQUAMATIC NIGHT LATCH LEVER

AQUAMATIC LOCKOUT LATCH

INK DUCTOR LOCKOUT

Fig. 26-48. To clean the duplicator, raise the safety guard and unfasten the hose to drain the fountain solution from the fountain. Remove ink from the ink fountain. Move the ink ductor lockout lever up and toward the feed end to lock the ductor away from the ink fountain. (A.B. Dick Co.)

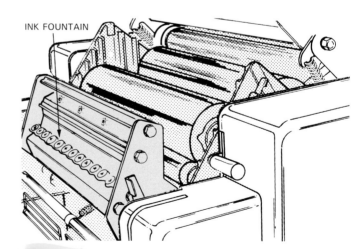

INK FOUNTAIN

Fig. 26-49. Remove the ink fountain by turning it up to a vertical position and lifting it off the duplicator. Ink fountain and fountain rollers are cleaned and the blanket is washed. (A.B. Dick Co.)

6. Move the Aquamatic lockout latch and night latch lever toward the receiving end of the duplicator (9835 and 9840 only). Look at Fig. 26-48. Position the Aquamatic control to number 45 (9835 and 9840 only).

Note! On 9850 duplicators, be sure that the night latch lever is positioned toward the feed end to keep the Aquamatic rollers out of contact with the rest of the ink system. Refer to Fig. 26-48.

7. Attach a clean-up mat to the plate cylinder just as you would a master or plate.

8. Be sure all safety covers are closed, press the ON/OFF switch to start the duplicator. Turn the speed down to minimum. From a container filled with blanket wash, apply a small amount to the ink oscillating roller through the opening in the guard. See Fig. 26-50.

9. Move the operation control lever to ink.

10. Continue to apply small amounts of blanket wash over the ink rollers until the clean-up mat is saturated with ink. This will occur almost immediately with the first clean-up mat.

11. Move the operation control lever to neutral. Stop the duplicator. Then remove the clean-up mat. When thoroughly dry, the mat may be reused on the reverse side.

12. Install another clean-up mat and repeat steps 8-11 until the last mat is clean and the rollers are dry.

Note! Do NOT allow a film of blanket wash to remain on the rollers, as the film will cause glazing or sensitizing of the rollers.

13. Move the Aquamatic night latch lever toward the feed end of the duplicator, Fig. 26-51.

14. Remove the secondary ink oscillator and distributor rollers. Place them in support cradle accessory, Fig. 26-51.

15. Remove the Aquamatic oscillating roller and place it on its support bracket, Fig. 26-52.

16. Unlatch the lock levers by moving them toward the feed end of the duplicator, Fig. 26-52.

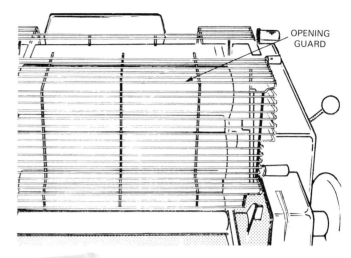

Fig. 26-50. Attach a clean-up mat to the plate cylinder. Put all safety covers in place. Adjust duplicator speed to low. Apply a small amount of blanket wash to the ink oscillating roller through an opening in the guard with a squeeze container. Continue to apply small amounts of blanket wash over the ink roller until clean-up mat is saturated. Repeat the process with new mats until the last mat is clean and the rollers are dry. (A.B. Dick Co.)

17. Lift and tilt the ink oscillating roller towards the ink fountain so it does not contact any other rollers.

18. Drain the fountain solution from the 9850 only.

19. Move the operation control lever to night latch and the upper form roller OFF, Fig. 26-52. Remove the two knurled knobs and lift the Aquamatic tray up and away from the ink system and clean thoroughly. Replace the tray.

20. Clean the impression and plate cylinders.

21. To keep the finish of the impression and plate cylinders clean, apply hand cleaner to the stained areas and let it stand for a few minutes. Rub the area with a soft cloth or cotton pad until the stain dissolves. Remove the residue with a damp cotton pad.

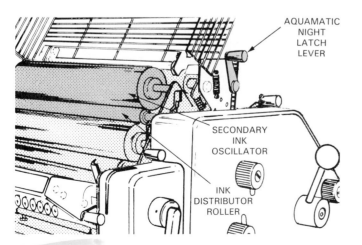

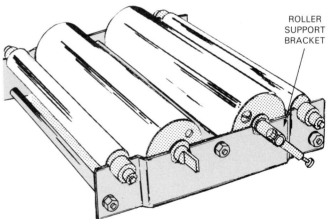

Fig. 26-51. Move the Aquamatic night latch lever toward the feed end of the duplicator. Remove the secondary ink oscillator and distributor rollers. Place them in the roller support cradle. (A.B. Dick Co.)

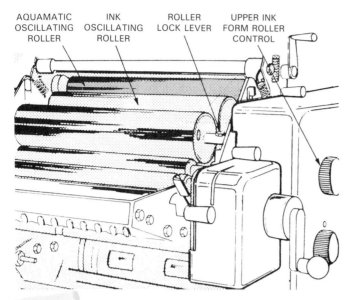

Fig. 26-52. Remove the Aquamatic oscillating roller and place it on its support bracket. Unlatch the lock levers by moving them toward the feed end of the duplicator. Move operation control lever to night latch and lift the upper form roller off. Remove Aquamatic tray and clean thoroughly. Clean impression and plate cylinders. Thoroughly clean and polish the plate and impression cylinders. Remove all ink from the blanket. (A.B. Dick Co.)

ATF CHIEF 115/117

The ATF Chief 115 and 117 are chain delivery duplicators manufactured in two sizes, Fig. 26-53. The Chief 115 has a maximum sheet size of 11 x 15 inches (27.9 x 38.1 cm). The Chief 117 has a maximum sheet size of 11 x 17 inches (27.9 x 43.2 cm). There are also two chute delivery models known as the Chief 15 and Chief 17 respectively. The Chief 115 and 117 models have separate inking and dampening systems which are controlled individually.

Specifications

1. Paper size: Model 115, 3 x 5 to 11 x 15 inches (7.6 x 12.7 cm to 27.9 x 38.1 cm); Model 117, 3 x 5 to 11 x 17 inches (7.6 x 12.7 cm to 27.9 x 43.2 cm).
2. Paper weight: (115) Minimum 11-pound bond to maximum 2-ply card, .012 inch (41 g/m² to .30 mm); (117) Minimum 13-pound bond to maximum 2-ply card, .012 inch (49 g/m² to .30 mm).
3. Printing area: (115) Up to 9 3/4 x 13 1/4 inches (24.8 x 33.7 cm); (117) Up to 10 1/2 x 16 1/2 inches (26.7 x 41.9 cm).
4. Metal plate size: (115) 10 x 15 inches (25.4 x 38.1 cm); (117) 11 x 18 inches (27.9 x 45.7 cm).
5. Gripper margin: (115) 3/16 to 5/16 inches (.5 to .8 cm); (117) 1/4 to 5/16 inches (.6 to .8 cm).
6. Blanket size: (115) 10 x 15 3/16 inches (25.4 x 38.6 cm); (117) 10 31/32 x 18 9/16 inches (27.9 x 47.1 cm).
7. Feed table capacity: (115) 21 3/4 inches (55.2 cm); (117) 21 3/4 inches (55.2 cm).

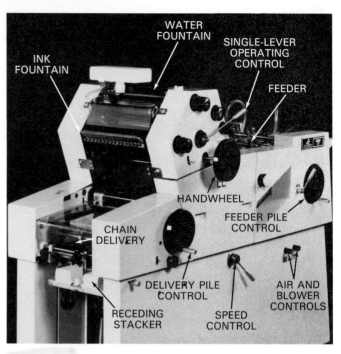

Fig. 26-53. The Chief 117 has a maximum sheet size of 11 x 17 inches. This duplicator is equipped with chain delivery and receding stacker, and operates from separate inking and dampening systems. (ATF Davidson Co.)

8. Chain delivery capacity: (115) 20 3/4 inches (52.7 cm); (117) 20 3/4 inches (52.7 cm).
9. Speed: (115) 4,500 to 9,000 impressions per hour; (117) 3,400 to 7,200 impressions per hour.

Operating procedure

The basic operating procedures are summarized below. These procedures should be conducted only after the operator has been thoroughly checked out on the duplicator.

Preparing feeder

The function of the *feeder* is to feed single sheets of paper in a continuous uninterrupted flow to the printing unit. Well regulated paper feeding is essential for good duplicator operation.

1. Lower the paper platform by pushing in on the paper platform handwheel. Then turn it counter-clockwise.
2. Set the inside of the left paper stack side guide on the scale marking for the size sheet to be run, Fig. 26-54.
3. Place a sheet of the proper size paper in position. Set the right-hand paper stack side guide so that it will be about 1/16 inch (1.6 mm) away from the edge of the paper.
4. Set the front guides so they are about on a line representing one-quarter in from the edge of paper, Fig. 26-54.
5. Center the suction feet over the front guides directly above the thin metal sheet separators.

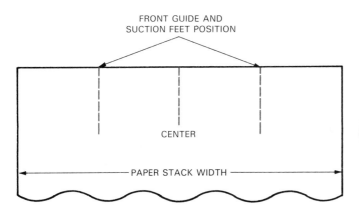

Fig. 26-54. Setting the inside of the left paper stack side guide on the scale marking for the size sheet to be run. (ATF Davidson Co.)

6. Position the front air blower in the center of the sheet edge, Fig. 26-54.
7. Fan the paper and load it on the paper platform and raise the paper stack to within 1/4 inch (6.4 mm) of the sheet separators.
8. Set the paper pile back guide so that the front wings are on top of the paper pile and the back wings are behind and up to the paper pile. See Fig. 26-55.
9. Set the backside guide extensions so that they are about 1/2 inch (13 mm) from the back of the sheet. They should be set at their extreme point when running larger size stock.
10. Set the side air blower paper separators so that they blow about two-thirds of the way back and across the paper. The second hole in the air blower tube should be in line with the top of the paper when the paper is 1/4 inch (6.4 mm) below the metal sheet separators.

For quick, accurate positioning of the air blow direction, fold the paper to be run in thirds, as shown in Fig. 26-56. Unfold and lay the sheet of paper on top of the paper stack and turn the blow direction indicator so that air is directed to the second crease as represented in lines A-B and C-D.

11. Lower the paper platform about one inch (25 mm). Turn the drive motor switch ON to allow the paper platform to raise automatically. Adjust the elevator control knob so that the paper stack stops about 1/4 inch (6 mm) below the sheet separators. Turning this knob clockwise lowers the height, and counterclockwise raises the height.
12. Set the feedboard joggers. The paper is picked up from the paper pile, moved to the paper pull-out roll by the sucker feet, and then delivered to the paper feedboard. It travels down the feedboard by way of the conveyor tapes to the paper stop bar and joggers. It is then jogged into correct feeding position.

Paper may be jogged from either the left or right-hand side. When running copy on both sides of the paper, use the left-hand jogger the first time through and the right-hand one for the second time through the backup. This procedure uses the same paper edge for jogging during each printing cycle. For work printed on one side only, the left-hand jogger is most often used. Moving the jogger selector to center position sets both joggers in motion while the press is running. The center position is frequently used when printing small-sized jobs.

To use the left-hand jogger, hold the right jogger with your left hand. Pull it toward you while you move the jogger selector back toward the feedboard. This sets the left jogger in motion.

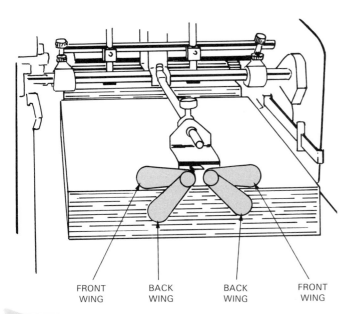

Fig. 26-55. Paper pile back guide is set so that the front wings are on top of the paper pile and the back wings are behind and up to the paper pile. (ATF Davidson Co.)

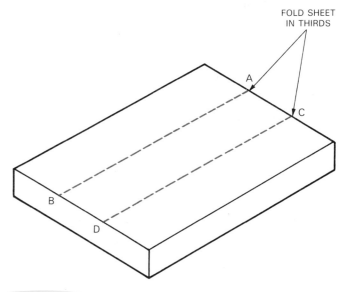

Fig. 26-56. To position air blow direction, fold a sheet to be run in thirds. Unfold and lay the sheet on top of the paper stack. Adjust the blow direction indicator so that air is directed to the creases as represented by lines A-B and C-D.

587

To use the right-hand jogger, hold the left jogger with your left hand and push it away from you while you move the jogger selector into its forward position, toward the front of the duplicator. This will set the right jogger into motion.

13. Set the feathers. Lower the feather on the idle edge jogger by pressing down on the front end of the feather retainer. The feather on the jogger being used should be raised by pressing down on the back end of the feather retainer. The feather on the idle jogger should be depressed no more than 1/16 inch (1.6 mm) when the active jogger moves a sheet for registration.

Delivery system

The paper is picked up from the paper stop bar by the tumbler grippers. It is carried through the impression phase and then transferred to the delivery grippers where it is delivered to the receding stacker.

1. Raise the delivery board to the uppermost position by depressing the delivery lock release. Then, turn the delivery pile handwheel clockwise.
2. Feed a sheet of paper through the duplicator, letting it drop into normal delivery position on the delivery board.
3. Loosen the delivery side guide control knobs. Bring the stationary side guide against the sheet of paper. Tighten the control knob. Turn the press handwheel until the automatic side-jogging guide is in the extreme forward position. Slide it to within 1/16 inch (1.6 mm) from the paper when the paper is against the stationary guide. Tighten the control knob.
4. With the paper against the front guide, turn the handwheel until the back jogging guide is in the extreme inward position. Loosen the guide setscrew. Slide the guide to the paper, and retighten the setscrew.
5. Run several sheets of paper through the duplicator to check all adjustments. Turn the drive motor switch ON. Slow the duplicator to its slowest speed. Turn the air-vacuum pump switch ON. Engage the paper feed control lever. Inspect the sheets as they pass through the duplicator and make necessary adjustments. The paper must feed through the duplicator efficiently and smoothly to give maximum print quality.

Inking system

1. Before filling the ink fountain, inspect the ink rollers to see that they are clean of all lint, dust, and glaze, especially if the duplicator has been idle for several days. See Fig. 26-57. Use blanket wash to clean the rollers if necessary.
2. Fill the ink fountain at least half full with the type of ink required for the job.
3. Turn the ink fountain roller control knob counterclockwise. Turn the ink fountain adjusting

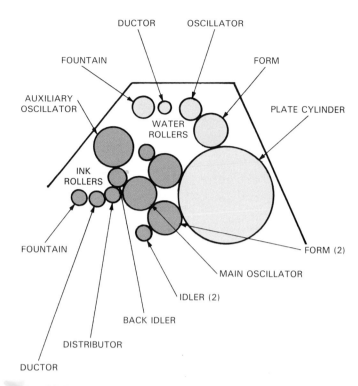

Fig. 26-57. Inspect all rollers to see that they are clean of lint, dust, and glaze before filling the ink and water fountains. Use a small amount of blanket wash to clean ink rollers, if necessary. Use a clean cloth with dampener solution to clean water rollers. (ATF Davidson Co.)

screws until a thin film of ink is spread evenly over the fountain roller.

4. Place the ink feed volume control on the fourth notch. Turn on the power switch. Depress the ink fountain manual control override. Run the duplicator until all ink rollers are covered with a thin film of ink.
5. Raise the manual control lever. Turn the duplicator OFF. The inking system is now ready to transfer ink to the plate.

Dampening system

1. The dampener fountain roller should be clean and free of dirt. Clean it if necessary.
2. Fill the dampener fountain bottle with the proper mixture of fountain solution and water for the plate to be run.
3. Turn the bottle upside down with the spout down over a sink or wastebasket. Be sure that the float valve does not leak. Bring the bottle over the side of the duplicator, being careful not to spill any solution on the ink rollers. Insert the bottle in its holder.
4. Moisten a cotton pad with fountain solution. Wipe the entire surface of the dampening rollers until each is thoroughly moistened but not wet.
5. Set the dampener volume control on the second or third notch. Run the duplicator for about 20 revolutions. The dampener system is now ready to transfer solution to the plate.

General operating procedure

1. Check all systems of the duplicator using the information previously covered.
2. Inspect the prepared plate for damage and/or dirt and correct if necessary.
3. Inspect and clean the plate cylinder. It must be thoroughly clean before the plate is attached.
4. Turn the duplicator with the handwheel until the plate cylinder lead clamp is in the correct position. Look at Fig. 26-58.
5. Attach the lead edge of the plate to the lead plate cylinder clamp. Attach the plate while holding the tail or bottom edge of the plate taut with your right hand.
6. Hold the tail of the plate with your right hand and turn the duplicator clockwise with the handwheel until the tail clamp is in view. See Fig. 26-59.
7. Bring the tail clamp up with spring tension and attach the trailing edge of the plate to the clamp. Be sure the plate is securely attached.
8. Tighten the plate clamp by turning the plate clamp vernier wheel clockwise. Secure the locknut to keep the plate from loosening during long press runs.
9. If a metal plate is used, moisten it thoroughly with a cotton pad dampened with fountain solution. This removes the protective gum coating.
10. Turn the duplicator power switch ON.
11. Turn the form roller control lever to the second position. This places the dampening form roller against the plate. Let the duplicator make several

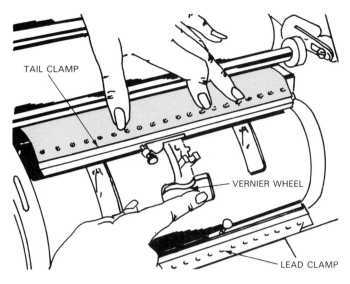

Fig. 26-59. Hold the tail of the plate and turn the duplicator handwheel clockwise until the tail clamp appears. Attach the trailing edge of the plate to the clamp. Tighten the plate clamp and secure locknut to keep plate from loosening during press run. (ATF Davidson Co.)

revolutions. This places an even layer of dampening solution over the non-image area of the plate.

12. Turn the form roller control lever to the third position. This places the ink form rollers against the plate along with the dampening form roller. Let the duplicator make several revolutions to put an even layer of ink on the image areas of the plate.
13. Turn the air-vacuum pump switch ON. Run three or four sheets through the duplicator.
14. Return the form roller control lever to the first position and turn off the duplicator.
15. Inspect a printed sheet to be sure that the image is in the correct position and that the image is satisfactory. Make any required adjustments for: vertical positioning, horizontal positioning, or crooked copy.

Vertical adjustment

a. Turn the duplicator with the handwheel until the vertical positioning control knob is aligned with the plate cylinder locknut located at the end of the plate cylinder. Refer to Fig. 26-60.
b. Engage or mesh both by pushing and holding in the vertical positioning control knob.
c. Turn the control knob to the left to unlock the cylinder. By turning the handwheel, the image can be either raised or lowered to the correct position.
d. Turn the control knob to the right. Relock the plate cylinder. Never attempt a vertical adjustment while the duplicator is in motion.

Horizontal adjustment

a. Move the paper to the left or right. This adjustment is made by moving the right and left

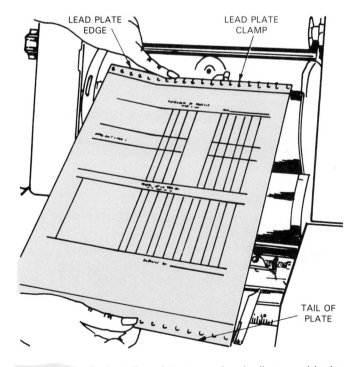

Fig. 26-58. To install a plate, turn the duplicator with the handwheel until the plate cylinder lead edge clamp is in the correct position. Attach the lead edge of the plate to the lead plate cylinder clamp. Hold the tail of the plate taut with your right hand. (ATF Davidson Co)

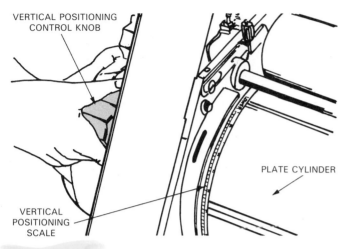

Fig. 26-60. For vertical positioning of the image on the sheet, use the vertical positioning control knob with the plate cylinder locknut located at the end of the plate cylinder on the operator's side. The image can be either raised or lowered by unlocking the cylinder and then turning the handwheel. (ATF Davidson Co.)

jogger control knobs simultaneously in the correct direction.

b. When the image has been centered correctly, adjust the paper stack accordingly.

Crooked copy adjustment

a. Square the crooked copy on the plate to the paper by using the paper stop bar straightening adjustment. See Fig. 26-61.

b. Loosen the binder screw that lets the bar tilt in either direction. This eliminates the need to adjust the angle of the plate on the plate cylinder.

16. Repeat steps 11 through 15. Continue this procedure until the image is printing in the correct

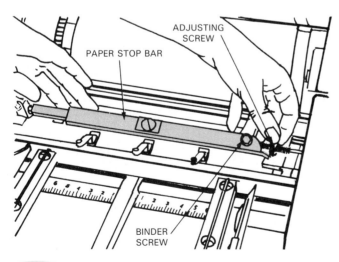

Fig. 26-61. For angular copy adjustments, square the copy on the plate to the paper by using the paper stop bar straightening adjustment. Loosen the binder screw that allows the bar to tilt in either direction and then turn the adjusting screw to correct copy. (ATF Davidson Co.)

position on the sheet. Check image quality again. If you are printing halftones or solids on rough-textured stock, it may be necessary to adjust the blanket-to-paper impression.

17. Check for sufficient ink coverage. Increase or decrease the supply of ink as needed.

18. Set the counter to zero and print the required number of sheets. Check the quality of the printed sheets as the duplicator runs. Use minimum amount of water with just the right amount of ink for complete coverage and density.

Note! Refer to Chapter 28 for press operating problems.

Cleaning duplicator

1. After the run is completed, stop the duplicator. Remove the plate and clean off the ink with plate cleaner and a soft cotton pad. Preserve the plate if it is to be saved.

2. Wash the blanket with blanket wash.

3. At the end of the day or for an ink color change, remove all ink from the ink fountain. Clean rollers, dampeners, and the blanket. Remove the ink fountain.

4. Remove fountain bottle and solution from the fountain.

5. Clean rollers using a wash-up attachment or cleaner mats. Check to see that roller ends are thoroughly clean of all ink.

6. Clean the plate and impression cylinders.

7. If the duplicator is to be secured for the day, set the night latch and turn off all electrical power.

MULTILITH 1360

The Multilith 1360 is a sheet-fed duplicator with a vacuum feeder. It has separate dampening and inking systems. The Multilith 1360 has the three-main-cylinder design, Fig. 26-62. Like most duplicators, it has a single control lever for operating the dampener form roller, the ink form rollers, and for providing plate-to-blanket contact. See Fig. 26-63.

Specifications

1. Paper size: 3 x 5 to 13 x 17 inches (76 x 127 mm to 330 x 431 mm).

2. Paper weight: Minimum 11-pound bond to maximum 110-pound index (40 g/m² to 200 g/m²).

3. Printing area: 12 1/2 x 16 1/2 inches (318 x 420 mm).

4. Metal plate size: 12 3/4 x 18 1/16 inches (324 x 432 mm).

5. Gripper margin: 5/16 inch (8 mm).

6. Feed table capacity: 5,000 sheets of 16-pound paper.

7. Chain delivery capacity: 5,000 sheets of 16-pound paper.

8. Speed: 5,000 to 10,000 impressions per hour.

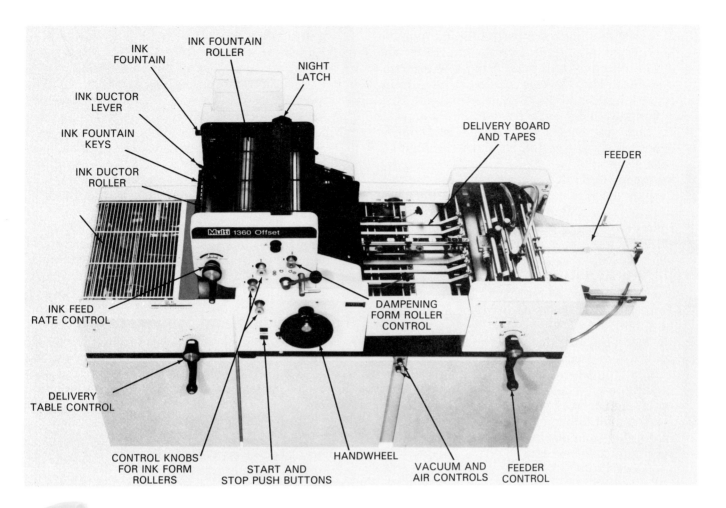

Fig. 26-62. The Multilith 1360 is a sheet-fed duplicator with a vacuum feeder and chain delivery. It has separate inking and dampening systems. Maximum sheet size is 13 x 17 inches. (AM International, Multigraphics)

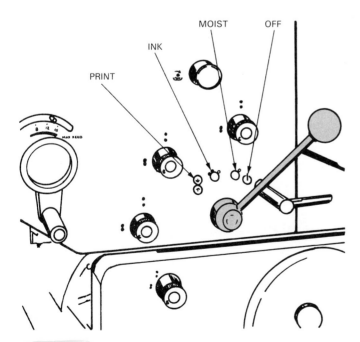

Fig. 26-63. The Multilith 1360 has a single control lever for operating the dampener form rollers, the ink form rollers, and for providing plate-to-blanket contact. (AM International, Multigraphics)

The location of controls and interlocked covers are shown in Fig. 26-64. The basic operating procedures are detailed below. These procedures should be conducted only after you have been thoroughly checked out on the duplicator.

Preparing ink unit

Refer to Fig. 26-64 as the steps for operations are summarized.

1. Move the single-lever control to the OFF position.
2. Place the night latch in the horizontal position.
3. Set the control knobs for the ink form rollers so that indexing dots are located at the ON position.
4. Set the control knob for the moisture form rollers so that the indexing dot is located at the OFF position.
5. Place the ink ductor shut-off lever into the up position.
6. Fill the ink fountain. Use an ink knife to deposit ink along the entire width of the ink fountain. Turn the ink fountain roller crank counterclockwise to work ink across the full width of the fountain.
7. Turn the handwheel until the ink ductor roller contacts the ink fountain roller.

8. To feed an even film of ink across the full width of the ink fountain roller, adjust the ink fountain keys, while turning the ink fountain roller. The ductor roller must be in firm contact with the fountain roller. Turn the keys counterclockwise to increase ink flow and clockwise to decrease the flow. Adjust the center keys first and then work outward on each side until a thin, even film of ink is deposited across the full width of the fountain roller.

9. Set the ink feed rate lever to number eight setting.

10. Push the machine start button and allow the rollers to revolve for a short period of time until all ink rollers are covered with a thin, even film of ink.

11. Push the machine stop button.

Preparing moisture unit

Refer to Fig. 26-65 as these steps are discussed.

1. Place the ink ductor shut-off lever in the down position.

2. Replace any rollers that were removed during the last cleanup procedure.

3. Lift all the wipers off the fountain roller (if duplicator is so equipped) by turning the wiper mounting bar as required.

4. Fill fountain solution bottle with correct mixture of distilled water and fountain concentrate for the type of plate being used.

5. Cap the fountain solution bottle firmly and place it in the bracket.

6. Adjust the moisture feed rate control knob to a position in the range of 4-6 on the scale.

7. Raise the moisture ductor lever to the ON position.

8. Turn the handwheel until the ductor roller contacts the fountain roller.

9. Spin the fountain roller knob until a water bead appears completely across the width of the two rollers.

10. Push the machine start button and allow about 75 revolutions to thoroughly moisten the rollers.

11. Push the machine stop button.

12. Check the moisture form roller to see if it is damp enough. If not, spin the fountain roller several more times to dampen the roller further.

13. Turn the ink feed rate control to the number eight position before printing any copy.

Paper loading and feeder setup

The procedures below are meant to illustrate a feeder setup for 11 x 17 inch (27.9 x 43.2 cm) paper feeding with the 11 inch dimension as the leading edge. Adjustments for leading edges of different widths are similar.

WARNING! Be sure duplicator and pump are OFF while the following procedures are performed.

The controls of the press are illustrated in Fig. 26-66.

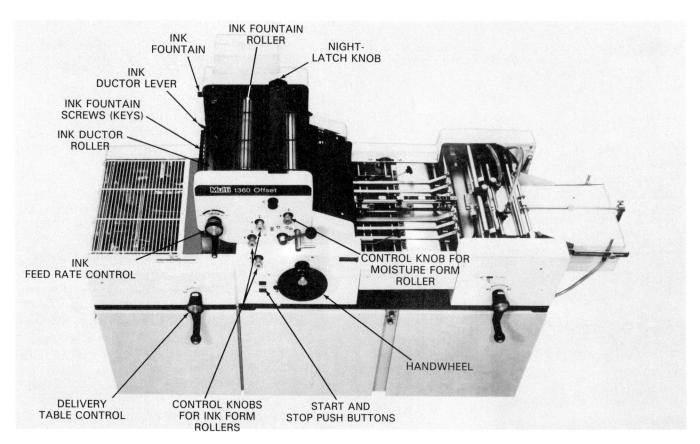

Fig. 26-64. Location of controls referred to in the basic operating sequence of the Multilith 1360. (AM International, Multigraphics)

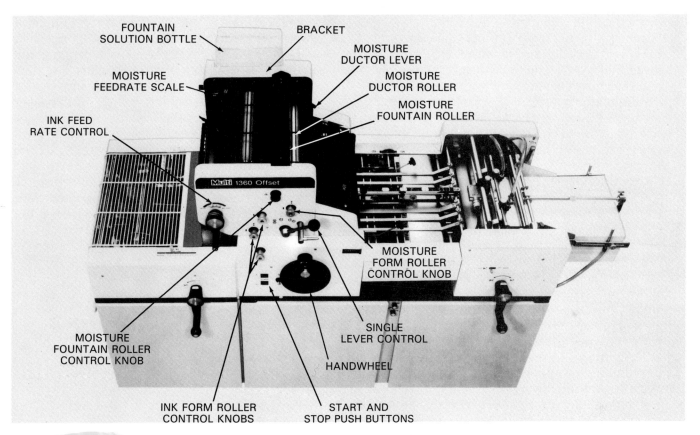

FOUNTAIN
SOLUTION BOTTLE

BRACKET

MOISTURE
DUCTOR LEVER

MOISTURE
FEEDRATE SCALE

MOISTURE
DUCTOR ROLLER

MOISTURE
FOUNTAIN ROLLER

INK FEED
RATE CONTROL

Multi 1360 Offset

MOISTURE
FORM ROLLER
CONTROL KNOB

MOISTURE
FOUNTAIN ROLLER
CONTROL KNOB

SINGLE
LEVER CONTROL

HANDWHEEL

INK FORM ROLLER
CONTROL KNOBS

START AND
STOP PUSH BUTTONS

Fig. 26-65. Location of the controls referred to in the preparation of the Multilith 1360 moisture system.
(AM International, Multigraphics)

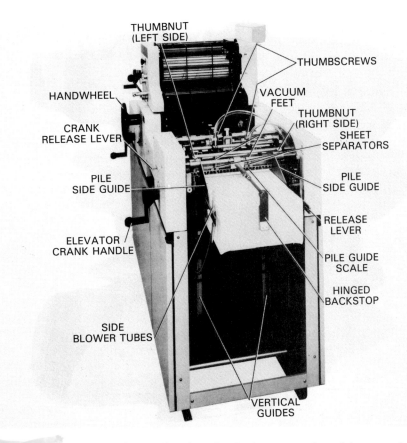

THUMBNUT
(LEFT SIDE)

THUMBSCREWS

HANDWHEEL

VACUUM
FEET

THUMBNUT
(RIGHT SIDE)

CRANK
RELEASE LEVER

SHEET
SEPARATORS

PILE
SIDE GUIDE

PILE
SIDE GUIDE

RELEASE
LEVER

ELEVATOR
CRANK HANDLE

PILE GUIDE
SCALE

HINGED
BACKSTOP

SIDE
BLOWER TUBES

VERTICAL
GUIDES

Fig. 26-66. Location of controls referred to in the preparation of the paper feeder.
(AM International, Multigraphics)

1. Raise the hinged backstop.
2. Set paper elevator to mid-height position. To raise the table, turn the elevator crank clockwise. To lower table, hold crank release lever to the right and turn crank counterclockwise.
3. Position the paper pile guides. Unlock the thumbnuts and the release lever. Reset pile side guides to a scale reading on the pile guide scale that indicates the width of paper to be run. Lock the thumbnuts and the release lever.

   Note! Side guides and vertical guides are linked together. Resetting the side guides automatically resets the vertical guides.
4. Position the sheet separators. The sheet separators are fastened to brackets which can be moved horizontally along their scale. Slide the brackets to a scale reading that indicates the width of sheet to be run. This will position the brackets to the proper location, which is inward from the sheet margin by a distance of one-fourth the sheet width. Examine the sheet separators for proper shape and deflection.
5. Position the vacuum feet directly in line with the sheet separators. When the vacuum feet are in the raised position, they should be directly above the separators. Unlock the thumbscrews and move the vacuum feet as required. Relock the thumbscrews.
6. Install the pile support board. Notice which end of the board is equal to or less than the width of the paper to be used. Then push that end snugly against the vertical guides, letting the pile support board rest on the paper supports.

   If a pile support board is not available, make one to the dimensions shown in Fig. 26-67. It may be cut from the paper supply box or a piece of plywood. A board made to these dimensions can accommodate the largest sheet size the paper path can handle.

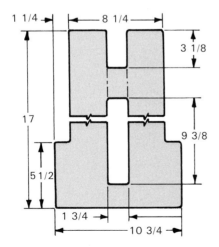

Fig. 26-67. A paper pile support board made to these dimensions can accommodate the largest sheet size the paper path can handle. The board and paper stack is supported on the feeder platform.

7. Load the paper. Examine the paper and remove any sheets with damaged or deformed edges. Fan the paper as shown in Fig. 26-68. Fanning will reduce the possibility of more than one sheet being fed at a time. Place the fanned paper correct side up on the pile support board. As the paper pile becomes higher, the elevator may have to be lowered so that sufficient clearance is maintained for the addition of more paper. Snug the pile all the way into the vertical edge guides and make sure that the leading edges lie squarely against the guides. Crank the elevator up until the top sheet on the pile is even with the bottom of the slot in the sheet separator bracket.

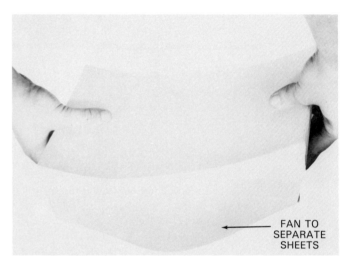

FAN TO SEPARATE SHEETS

Fig. 26-68. Paper should be fanned and jogged before loading it in the feeder.

8. Lower the hinged backstop and adjust it so that the pile end guide (spring) just deflects perceptibly.
9. Position the pullout wheels. Depress the release lever and slide the wheels to the position on the pullout wheel scale that indicates the width of the sheet to be run.
10. Check the positions of blower tubes and adjust if necessary. The blower tubes aid in separating the uppermost sheets of paper by floating them up toward the suction feet. Position the center blower tube on the centerline of the paper path. Adjust the side blower tubes in their brackets so that the third set of holes from the top is even with the bottom of the slot in the sheet separator bracket, and the rearward rows of holes face each other across the paper pile. The forward column of holes is then necessarily positioned to blow diagonally across and toward the front of the pile. This is shown in Fig. 26-69. If the duplicator is equipped with rear blower tubes, these should be positioned so that the forward holes face each other and the rear holes blow diagonally across and toward the rear of the pile. See Fig. 26-70.

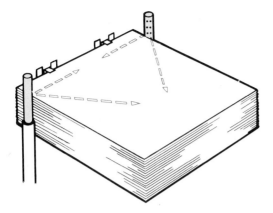

Fig. 26-69. Front blowers are positioned to blow air diagonally across and toward the rear of the paper pile. (AM International, Multigraphics)

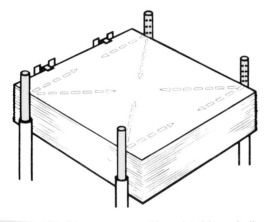

Fig. 26-70. Rear blowers are positioned to blow air diagonally across and toward the front of the paper pile. (AM International, Multigraphics)

11. Adjust the double-sheet detector. Loosen the thumbscrew and position the double-sheet detector in line with the center of the paper path. Tighten the thumbscrew. Turn the handwheel countercockwise until the suction feet are in their lowest position. Lower the paper pile far enough for clearance. Tear a strip of paper approximately one inch (25 mm) wide and six inches (150 mm) long. Fold the strip off center to create a single-thickness/double-thickness gauge. Turn the pump motor switch ON. Insert the leading edge of the single-thickness portion of the folded sheet under the double-sheet detector. If the eliminator kicks the deflector gate open, turn the raise-lower adjustment knob in the raise direction until the gate no longer opens. Insert the folded strip farther until the second thickness causes the gate to open. Then adjust the raise-lower knob until the deflector gate just ceases to open. Lower the raise-lower knob one click. The final check is that the gate opens when it senses a double thickness of the folded sheet but not a single thickness. Turn the pump motor switch OFF.

Register board setup

Refer to Fig. 26-71 when the following sequence of steps are described.

1. Remove the jogger and the spring side guide. First unlock the side guides by turning the locking knobs counterclockwise. Lift the guides out.

2. Space the tapes and the sheet retainers. Turn the duplicator ON. With the duplicator running, move the tape guides in or out to position them for the width of sheets to be run. This means that the tapes should lie within the graduation marks on the register board scale that correspond to the intended sheet width. Not all the tapes will be needed if you are going to run 8 1/2 x 11 inch (21.6 x 27.9 cm) paper or smaller with the narrow edge leading. The extra tapes can be positioned to the outside. Reposition the sheet retainers to lie directly over the tapes. Turn the duplicator OFF.

3. Set the duplicator to the setup phase of its rotational cycle. Turn the handwheel until the red dot on the edge of the blanket cylinder lines up with the arrow on the side casting, as in Fig. 26-72. Then, the register board jogger is at its full inward stroke.

4. Set the micrometer wheel to the neutral position, Fig. 26-71. Turn the micrometer wheel all the way counterclockwise. Then turn it clockwise exactly two full turns. This is the neutral position.

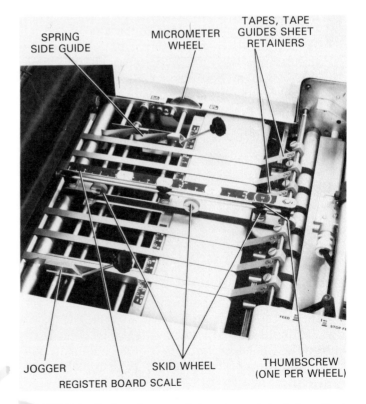

Fig. 26-71. Register board and sheet controls include pullout roll and wheels (skids), double-sheet detector, conveyor tapes, riders (steel balls and metal bands), front stops, movable side guide (jogger), stationary side guide, and forwarding rollers. (AM International, Multigraphics)

595

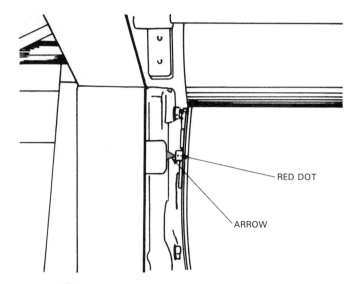

RED DOT

ARROW

Fig. 26-72. Micrometer wheel feeder adjustments are made by setting the duplicator to the setup phase of its cycle. Turn the handwheel until the red dot on the edge of the blanket cylinder lines up with the arrow on the side casting. (AM International, Multigraphics)

5. Reinstall the jogger and the spring side guide. Reinstall the jogger side guide and position it so that the locating arrow lines up with the graduation mark on the register board scale that corresponds to width of sheet to be run. Then lock the jogger in place.

6. Position the skid wheels, Fig. 26-71. The skid wheels should be positioned so that the rear wheel (nearest receiver end of duplicator) lies on the graduation mark of the register bar scale indicating sheet length. The other wheels should then be spaced out evenly. If the sheet is too long to permit this arrangement, the rear wheel should be positioned farther toward the receiver end of the duplicator, and the next wheel should be moved up into position. Then, that wheel and any others which lie toward the feeder (front of duplicator) should be lowered. Those which lie toward the printing head, should be raised. To position a wheel, take hold of the thumbscrew and slide the wheel forward or backward as desired. To raise or lower a wheel, turn the thumbscrew clockwise or counterclockwise as viewed from above. When lowering a wheel, start with the wheel in a raised position. Turn the duplicator ON, and then lower the wheel until it just begins to turn. Then turn the duplicator OFF again.

Note! If the paper wrinkles or jams upon entering the printing area during a test of the paper path setup, or if vertical register is poor after operation, the skid wheel may have to be repositioned very slightly. This adjustment is made with the duplicator running. Ideally, this skid wheel should lie off the trailing edge of the sheet by no more than a couple thousandths of an inch when the sheet lies against the stop fingers. Also,

the adjustment should be made only after the stop finger adjustment and the feed roller level and pressure are known to be satisfactory.

Paper receiver setup (chain delivery)

1. Set the duplicator to the setup phase of the rotational cycle. Turn the handwheel until the red dot on the edge of the blanket cylinder is lined up with the arrow on the side casting. Then, the jogging side guide and the end jogger in the chain delivery unit are at the correct phase of their respective strokes.

2. Position the chain delivery side guides. Unlock the stationary side guide and position it so that it lines up with the graduation mark on the cover scale that corresponds to the width of sheet to be run. Do the same for the jogging side guide.

3. Adjust the paper stop. It is advisable to position the stop bracket initially so that the knurled locking nut lies at the center of the adjusting slot. This allows ample adjustment range if further adjustment is needed after operation has begun. Loosen the knurled locking nut and slide the bracket to the desired position. Then relock knurled nut. When paper is delivered at high speed, it will carry farther. In this case, the stop bracket should be pulled to the extreme outward position.

4. Position the end jogger. The end jogger is positioned with respect to the scale on the paper table. Set the descending speed control to the release position. Push the crank handle IN to engage it with the shaft and turn it counterclockwise to raise the paper table. Raise the table to a level about one inch below the jogger. Swing the cover up and unlock the end jogger by means of the locking lever. Manually position the end jogger so that it lines up with the graduation mark on the paper table scale that indicates the sheet length. Lock the jogger and lower the cover.

5. Position the paper guide wheels. The paper guide wheels, sometimes called *skeleton wheels* or *star wheels*, should be positioned inward from the edges of a sheet as far as possible without tracking over an image area. Loosen the lockscrews on the wheels with the T-wrench. Then slide the wheels to their proper position and tighten the lockscrews. If it is necessary to feed a sheet into position to make the adjustment, use the following technique.
   a. Turn the duplicator ON and turn the speed down as far as possible.
   b. Turn ON the sheet feed control.
   c. With the paper feeder properly set, turn ON the pump motor. Let one sheet of paper feed onto the register board. Then turn the duplicator OFF before the sheet is transported into the nip.
   d. Flick the duplicator switch ON and OFF as necessary to nudge the sheet forward.

Note! The technique may require a little practice at first, but it is soon easily mastered.

Operating procedure

1. Check control settings. The roller controls should already have been set properly. Be sure the moisture form roller knob and the ink form roller knobs are turned ON. The indexing dots on all the knobs should be set at the nine o'clock position. The form rollers will now make contact with the plate cylinder whenever the single-lever control is in the moist, ink, or print position.

   Be sure the night latch knob is in the operating (horizontal) position. This knob is located on the gear side of the duplicator. Be sure the moisture ductor lever is set to the ON position, or UP. The ink ductor shut-off should be in the up position, Fig. 26-73.

   The pump motor switch should be ON. The sheet feed control should be OFF (pointing towards feeder).

   The ink feedrate control should be set to the number eight position as a nominal setting for startup. If it becomes evident during operation that a different setting would be better, the control can be readjusted at that time.

   The moisture feedrate control, Fig. 26-74, should be adjusted somewhere in the range of the number 4-6 setting as a nominal adjustment for startup. If it becomes evident during operation that a different setting would be better, the control can be readjusted at that time.

2. Install the plate on the plate cylinder. The standard plate clamp for the 1360 is a universal type, which accepts pinbar plates and straight-edge paper masters. To install the plate, turn the handwheel counterclockwise until the plate cylinder clamp

assembly is accessible. Turn the plate tension screw counterclockwise until it is fully extended. This will allow the trailing edge of the plate to be more easily attached later. Hold the plate at the trailing edge with the image side up.

If you have the pinbar-type plate, place the leading edge of the plate over the pins on the pinbar clamp so that the holes in the plate are engaged with the pins. Make certain all holes are engaged with the pins, Fig. 26-75.

If you have a straight-edge master, slide the leading edge clamp release to the left (toward center of cylinder) to open the clamp. Slide the leading edge of the master under the pinbar so that it inserts between the pinbar and the sliding bar, Fig. 26-76.

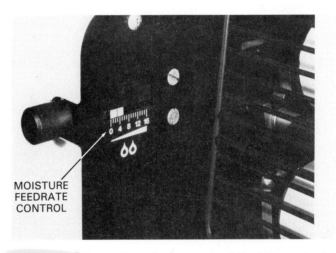

MOISTURE FEEDRATE CONTROL

Fig. 26-74. The moisture feedrate control should be set in the range of 4-6 for nominal startup adjustment position. (AM International, Multigraphics)

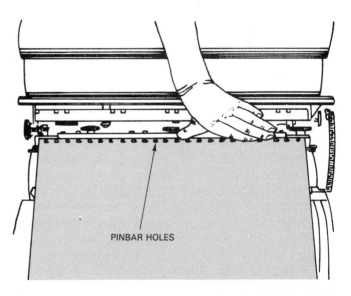

PINBAR HOLES

Fig. 26-75. The standard plate clamp for the Multilith 1360 is a universal type, which accepts pinbar plates and straight-edge paper masters. When using the pinbar-type plate, place the leading edge of the plate over the pins on the pinbar clamp, so that the holes in the plate are engaged with the pins. (AM International, Multigraphics)

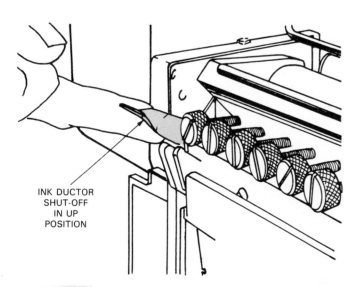

INK DUCTOR SHUT-OFF IN UP POSITION

Fig. 26-73. The ink ductor shut-off switch should be in the up position when setting up the duplicator for operation. (AM International, Multigraphics)

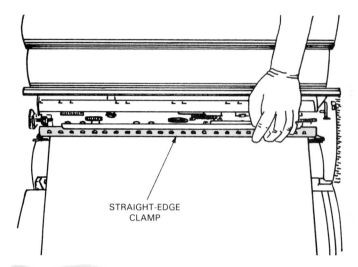

Fig. 26-76. When using a straight-edge metal plate or master, slide the leading edge clamp release to the left (toward the center of the cylinder) to open the clamp. Slide the leading edge of the plate under the pinbar so that it inserts between the pin bar and the sliding bar. (AM International, Multigraphics)

When the plate is definitely inserted as far as it will go, slide the clamp release to the right to lock it. Hold the plate taut with your right hand to prevent the leading edge from disengaging. With your left thumb and forefinger, form the plate over the leading edge of the plate cylinder, as shown in Fig. 26-75. While continuing to hold the plate taut, turn the handwheel counterclockwise one-half revolution.

3. Move the single lever control to the moist position. This will contact the moisture form roller to the plate and keep the plate on the cylinder so that holes at the leading edge remain engaged with the pins in the pinbar clamp. Stop turning when the trailing edge clamp is accessible.

Swing out the release lever for the trailing edge clamp and then pull it towards you to lift the upper trailing edge bar to its fully-raised position. While holding the bar in its fully-raised position, slide the trailing edge of the plate under the spikes in the bar. Then lower the bar so that the spikes either enter the holes of the plate (if plate is pinbar type), or pierce the surface of the plate (if plate is straight-edge type). Return the release lever to the locked position. Form the plate over the trailing edge of the cylinder. The spring tension on the pinbar is sufficient to hold a paper master taut.

If a paper master is too short to be clamped at the trailing edge, the trailing edge can be left unclamped. A metal plate, however, requires additional spring tension. The additional tension is applied according to the following two steps.

a. Turn the tension screw counterclockwise until you feel resistance.

b. Then turn the tension screw an additional turn clockwise, but no more.

4. Moisten the plate. When you are ready to make a trial copy, apply a thin even film of the proper solution over the entire surface of the plate with a cotton pad. See Fig. 26-77.

5. Run several sheets through the duplicator. Check for position, copy quality, and proper amount of impression. If the impression requires adjustment, follow the steps below.

a. Turn the power OFF.

b. Slide open the door that covers the impression cylinder pressure adjustments, located below the handwheel.

c. Loosen the clampscrew with the T-wrench.

d. Remove all pressure by turning the adjusting screw clockwise until it stops turning.

e. Using the handwheel, position a sheet of paper (size and thickness to be used) between the blanket and impression cylinders. Turn the handwheel counterclockwise until the cylinders latch, as shown in Fig. 26-78.

f. Turn the adjusting screw counterclockwise to increase pressure by pushing your middle finger across the bottom of the knurled edge until the knurling slips on your fingertips. At this point, an additional 1/4 turn of the screw by means of thumb and forefinger may be necessary.

g. Tighten the clampscrew and then close the cover.

Note! The quality of the copy will be the final guide in determining impression cylinder pressure. If a change in pressure still seems to be required after the cylinders have been adjusted, it is a good indication that the blanket is worn and should be replaced. A dirty impression cylinder also may produce poor copy quality but might incorrectly suggest that impression cylinder

Fig. 26-77. Prior to making a trial press copy, moisten the plate with a thin, even film of fountain solution (or plate etch for a paper master) using a cotton pad.
(AM International, Multigraphics)

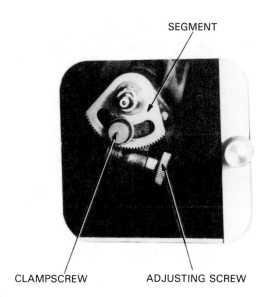

SEGMENT

CLAMPSCREW   ADJUSTING SCREW

Fig. 26-78. Pressure adjustments for a standard impression cylinder device. (AM International, Multigraphics)

pressure is wrong. Clean the impression cylinder and check copy quality again before making a final judgment about either blanket condition or impression cylinder pressure.

6. Set the sheet counter to zero.
7. Position all interlocks in the safeguard position. Make sure that the interlocked covers for the ink unit and moisture unit are down, the ejector guard is down, and the doors are closed (safeguard positions). An open interlock would prevent the duplicator from being started.
8. Start the duplicator. Turn the duplicator ON with the start button.
9. Apply moisture to the plate. Place the single lever control into the moist position and allow the plate cylinder to make one or two revolutions to moisten the plate thoroughly.
10. Apply ink to the plate. Place the single lever control into the ink position and allow the plate cylinder to make four to six revolutions to fully ink the image on the plate.
11. Transfer image to the blanket. Place the single lever control into the print position and allow the plate cylinder to make two to four revolutions to properly develop an inked image on the blanket.

Note! The exact number of blanket-inking revolutions required depends largely on the ink being used and the type of image being printed. Too many revolutions may cause the first sheet to stick to the blanket, especially if the image contains a large solid, reverse, or halftone. On the other hand, too few revolutions may result in the insufficient inking of the blanket to print the first sheets with adequate image density.

12. Run several test copies. Move the sheet feed controls towards the receiver. Paper will begin to feed immediately. If the first few sheets are not printed with adequate image density, you can start the

sheet feeding process before moving the single lever control to the print position. Let a few sheets feed through the duplicator and then move the sheet-feed control towards the feeder. Paper feeding then stops.

13. Set the single lever control back to ink position. Now you can inspect the printed sheets and evaluate the copy quality.

Note! The single lever control is moved back to the ink position only because the duplicator will probably be back into production within a few moments. Otherwise, this control would be put into the OFF position.

14. Inspect the copy. If the copy is acceptable, go to step 15 (on next page) and continue the press run. However, if the copy is not acceptable, make whatever adjustments are necessary to bring copy quality up to production standards. Adjustments may include the following.

Raising or lowering image

1. Make sure the release pin is in one of the holes in the interposer ring, and the centerline of the micro-adjust scale lines up with the scribe mark on the interposer ring.
2. Turn the handwheel until the clamp screw on cylinder is in line with the socket on vertical positioning handle. This should happen when the zero setting on the wide-range scale lines up with the indexing arrow on the side casting.
3. Push the vertical positioning handle inward to engage the socket with the clamp screw. Then turn the knob about 3/4 turn counterclockwise to loosen the screw.
4. With a screwdriver, turn the micrometer screw as required, counterclockwise to raise image on the sheet or vice versa. The movement can be read from the micrometer scale, with the scribe mark on the interposer ring used as a reference.
5. After making your adjustment, tighten the clamp screw by turning the vertical positioning handle clockwise.
6. After the adjustment has served its purpose, reset the cylinders so that the centerline of the micro-adjust scale again lines up with the scribe mark on the interposer ring.

Note! If the image must be relocated more than this adjustment allows, relocate the release pin to another hole in the interposer ring and repeat the adjustment.

Moving image left or right

1. Raise the adjustment lock for the jogger micrometer disc.
2. Turn the micrometer disc to move the jogger in the desired direction: IN to move image to the right; OUT to move image to the left.
3. Lower the adjustment lock for the jogger micrometer disc.

Rotating image

If an image is skewed slightly about the vertical axis on the plate, the best solution is to burn a new plate with the image square to the axis. If it is impractical to burn a new plate, or if the plate is to be repositioned ever so slightly for close register, the existing plate must be counter rotated in the clamp to compensate for the misaligned image. As shown in Fig. 26-79, this can be done by using the universal clamp.

1. Turn the handwheel as required to make the clamp accessible.
2. Loosen the tension screw on the trailing edge clamp.
3. Determine the direction the angular adjusting screw must be turned to bring about the desired image rotation. Turning the outward edge of the screw to the right, away from the center of the cylinders, raises the right leading edge of the plate so that the image is rotated to the left. Turning the screw the other way has the opposite effect.
4. Turn the angular adjustment screw as required and simultaneously turn the angular stabilizer screw in the same direction and to the same extent.

Note! Be careful that the final setting of the angular stabilizer screw allows the screw to just make contact with the cylinder. If it bears on the cylinder with too much force, the pinbar will bend and the leading edge holes in the plate will be distorted.

5. Turn the moisture form roller knob ON and set the single lever control to the moist position. This contacts the moisture form roller to the plate.
6. Turn the handwheel counterclockwise to rotate the cylinders one revolution while the moisture form roller keeps the plate lying flat on the cylinder.
7. When the plate cylinder has made one complete revolution, examine the trailing edge to see whether the plate is buckled.
8. Adjust the trailing edge lateral adjustment, as necessary to eliminate any buckling or warpage.

The plate must lie smooth and snug on the cylinder.
9. Tighten the tension screw.
10. Print trial copies and evaluate the adjustment. Readjust if necessary.

Note! When the printing run for this plate is done, reset the plate clamp to lie parallel to the leading edge of the plate cylinder.

15. After making the necessary adjustments, print the number of copies required for the job. Return the single lever control to the print position. Move the sheet-feed control towards the receiver again. Now the printing resumes. Print the number of copies required.
16. Turn the sheet-feed control OFF, towards the feeder.
17. Move the single lever control to OFF.
18. Turn the duplicator OFF. Press the stop button. Remove the printed sheets from the stacker.
19. Remove the plate and preserve it if desired. If the metal plate is to be preserved, follow the remainder of the sequence below. If not, simply remove the plate and proceed to clean the blanket as per step 20.

Set the ink feedrate control lever to zero or turn the ink ductor shut-off down. Turn OFF the moisture ductor lever. Turn OFF both ink form roller knobs, but leave the moisture form roller ON. This will supply moisture to the few sheets that are yet to be fed through the duplicator and prevent static buildup in the paper and help avoid paper jams.

Turn the duplicator ON and set the single lever control to the print position. Turn ON the pump motor, and the sheet-feed control. Allow 10-15 sheets to run through the duplicator. This will remove most of the ink from the image on the plate. Turn OFF the sheet feed control and the pump motor. Set the single lever control to the OFF position. Turn the duplicator OFF.

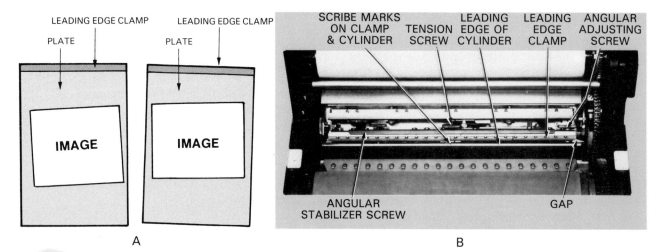

Fig. 26-79. A—Rotating plate in clamp to compensate for a skewed image. B—Angular adjustments on plate clamp. (AM International, Multigraphics)

Clean the metal plate surface with plate cleaner and buff dry. Turn the handwheel as required to make the trailing edge clamp accessible. Unclamp the trailing edge of the plate. Then, holding the trailing edge firmly and pulling on it to keep the plate under tension, turn the handwheel clockwise to unwind the plate off the cylinder.

20. Clean the blanket. Clean the image off the blanket every time a plate is removed or repositioned.
21. Clean the duplicator as follows.

Cleaning ink fountain

1. Cut two strips (about 3 1/2 x 12 inches [8.9 x 30.5 cm]) from card stock or a cleaner sheet.
2. Insert the strips into the ink fountain so that as much ink as possible is wedged between them.
3. Carefully remove both strips, with the ink wedged in the middle, and discard in an approved safety container. Repeat as necessary to remove the bulk of the ink.
4. Place the ink ductor lever in the down position to lock the ductor roller against the distributor roller.
5. Lift the ink fountain to disengage both end brackets from the end bearings and remove the fountain.
6. Remove any remaining ink from the fountain with an ink knife.
7. Triple-fold a clean cotton pad. Fold it over the blade of the ink knife and saturate the folded pad with blanket wash.
8. Finish the cleaning of the fountain with the padded knife blade. A thorough cleaning will probably require that you replace the pad with a fresh one when the cleaning is about half done.
9. After the ink fountain has been cleaned, dry it with a clean cloth wrapped around your hand. Be sure to remove any remaining ink from the very ends of the fountain.
10. Clean the fountain roller with a cotton pad or a cloth dampened with blanket wash.
11. Replace the ink fountain.

Cleaning ink roller system

There are two methods of cleaning the ink roller system. These include the use of a cleanup attachment or the cleaner sheet. For purposes of this book, the cleaner sheet method is detailed. Follow the operator's manual if using a cleanup attachment.

1. Remove the moisture oscillator roller and the moisture form roller.

   Note! Some press operators remove all the rollers in the moisture system at this point so that they can wipe the ductor frame clean. The rollers are then left out until the ink system cleanup is completed.
2. Turn the moisture form roller knob back to the OFF position.

3. Be sure the ink form roller knobs are ON.
4. Clamp a cleaner sheet (mat) onto the plate cylinder.
5. Shut OFF the ink feedrate control.
6. Set the ink ductor lever to the ON or down position so that the ink ductor locks against the distributor roller.
7. Pull out the lower cover for the ink system. The exit gate is part of the lower ink system cover. First, position the exit gate in the UP position. This unlatches the assembly. Then, pull the assembly straight out away from the ink system, Fig. 26-80.
8. Lay a cleaner sheet in the cover.
9. Replace the cover. Guide the interlock actuator into the opening slots and then push the assembly in until the cover snaps into place over the ink unit tie bar. Set the interlock actuator and pivot. When the pivot and actuator are seated, the exit gate must be swung down into the safeguard position before the duplicator will start. The cleaner sheet in the cover will protect everything underneath it from the blanket wash that will be used to clean the ink rollers.
10. Start the duplicator.
11. Reduce speed to minimum.
12. Apply a small amount of blanket wash to the ink rollers completely across their width through the lower slot in the covers.
13. Place the single lever control in the ink position.
14. Watch the cleaner sheet as it picks up ink. As soon as it becomes saturated, move the single lever control to OFF. Stop the duplicator and replace the dirty cleaner sheet with a fresh one.
15. Repeat the washing procedure until the last cleaner sheet no longer picks up ink. Then, move the single lever control to OFF and turn the duplicator OFF.

LOWER INK SYSTEM COVER   EXIT GATE

Fig. 26-80. Removing the lower ink system cover-and-exit gate assembly.

601

16. Remove the cleaner sheet from the lower ink unit cover and replace the cover.
17. Turn the form rollers OFF and set the night latch knob to the vertical position.
18. Push the spring posts for the lower oscillator to the night latch position (if duplicator is so equipped).
19. Check to see that the impression and plate cylinders are clean.
20. Replace all tools and clean up around the duplicator area.

SPECIFIC PRESS CONFIGURATIONS

The last section of this chapter will discuss and illustrate specific types of presses. This will help you be more familiar with brand names and their unique features.

ATF-DAVIDSON

The larger ATF-Davidson Chiefs include the 25, 215, and 217, Fig. 26-81. They are considered compact presses useful for a variety of job requirements. The

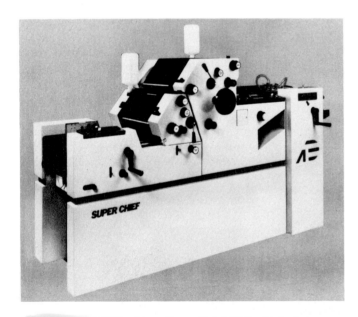

Fig. 26-81. ATF Davidson Super Chief 217 with built-in two-color offset duplicator will print maximum sheet size of 13 x 17 inches. (ATF Davidson Co.)

25 is a single-color press, with a sheet size of 19 x 25 1/4 inches (48.3 x 64.1 cm). The press is equipped with roller-type pull guides, Spiess sheet separator, one-piece frame and base, automatic oiling system, four front stops, quick-change plate clamps, helical gears, external gear train, Dahlgren dampening system, and powder spray unit. The 215 and 217 are two-color versions of the smaller Chiefs. These feature integral second color units which are not add-ons. These presses have independent second ink and water units.

HAMADA

Hamada introduced Japanese-manufactured presses into the United States in a duplicator-size format. There are several models offered in assorted sizes, Fig. 26-82.

The Hammadastar is an automatic sheet-fed perfector equipped with two inking systems. Its automatic features include two sets of plate loading and ejection devices, auxiliary etching rollers, automatic blanket cleaners, and a copy counter board. Other models feature double-sheet detectors and eliminators, ink roller washups, a spray device with a separate compressor, and static eliminators.

A

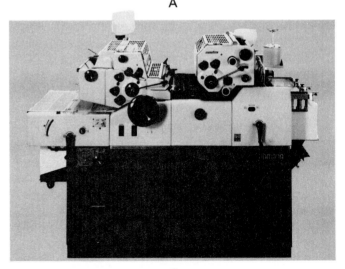

B

Fig. 26-82. A—Hamada Model 661 duplicator is an automatic sheet-fed perfector equipped with two inking systems. It will print on sheet sizes up to 14 3/8 x 20 1/4 inches. B—Hamada Model 885 is a large duplicator equipped with two color heads. (Hamada of America, Inc.)

HCM GRAPHIC SYSTEMS

Champion presses, offered by HCM Graphic Systems, are built in printing sizes ranging from 20 3/32 x 29 1/8 inches (50.8 x 74 cm) in the Perle series, to the larger range of 28 x 40 (71 x 101.6 cm), up to 35 x 52 inches (88.9 x 132.1 cm). The larger Champions come in models 29, 38, and 42. These are offered in four, five, and six colors with special equipment such as gravure units added to the sheet-fed offset systems.

Features of the Champion presses include a newly designed ink film system, continuous-flow dampening system, helical gearing throughout, large impression cylinders, and cam-forced gripper system. All presses are general purpose with capabilities up to 0.032 inch (.8 mm) board and heavy ink coverage.

HEIDELBERG

The T-Offset electronic duplicator from Heidelberg will accept stock sizes from 3 x 4 1/8 to 11 x 15 1/2 inches (7.6 x 10.5 to 27.9 x 39.4 cm). See Fig. 26-83. The single color, T-Offset features bearer-to-bearer pressure between plate and blanket cylinders and a combined dampening and inking system for improved start-up and ink coverage.

Heidelberg also offers the K models (KOR, KORA, KORD, KORS), Fig. 26-84. Presses larger than the K models are the MO series (19 x 25 1/2 inches [48.3 x 64.7 cm]); the S series (SORK, SORM, SORD, SORS), and Z models (two colors) in sizes 19 x 25 1/2 up to 28 x 40

Fig. 26-83. Heidelberg T-Offset is a single-color duplicator featuring bearer-to-bearer pressure between plate and blanket cylinders, and a combined dampening and inking system. It will print on sheet sizes up to 11 x 15 1/2 inches. (Heidelberg U.S.A.)

Fig. 26-84. Heidelberg offset or letter press, Model KORD. Removable segments on the plate cylinder can be interchanged for either process. Maximum sheet size is 15 3/4 x 22 1/2 inches. (Heidelberg U.S.A.)

inches (48.3 x 64.7 up to 71 x 101.6 cm). After these come the 72 models, which are 20 1/2 x 28 3/8 inch (52.1 x 72.1 cm) straight-line presses convertible to perfecting. The 102 series are two, four, five, and six colors in the larger 28 x 40 1/2 inch (71 x 102.9 cm) size.

IMPERIAL/KOMORI

Sheet-fed offset presses offered by Imperial/Komori come in several models. These include a single-color 20 x 26 inch (50.8 x 66 cm) press, named the Sprint L25C; a two-color Sprint L225B; a four-color Sprint L425BP; a larger 26 x 40 inch (66 x 101.6 cm) multicolor Kony Super; and a two-color Kony in the same size. Features of these presses include exceptional register and minimum maintenance due to centralized oil injection and bath. The presses are equipped with pin register control, centralized operation controls, quick-release split-plate clamps, and highly responsive feeders.

To aid in reducing spoilage and downtime, Imperial/Komori presses have crooked and misfed-sheet detectors, feeder pile ascent timers for compensation of variance in ratio between paper thickness and pile speed, and fully adjustable pull guides without tools. The presses have exceptional register capability resulting from a patented double-size transfer cylinder configuration.

ITEK

Itek's 975 series of small offset duplicators includes the XLE and XLD models. The 975 series offers a printing area of 12 x 17 1/4 inches (30.5 x 43.8 cm), a maximum sheet size of 13 3/4 x 17 3/4 inches (34.9 x 45.1 cm), and four-sided 11 x 17 inch (27.9 x 43.2 cm) bleed jobs. Both duplicators feature two-speed water fountain adjustment for control over quality at high speeds, flip-up feed guides for easy loading and a micrometer side guide. See Fig. 26-85.

KOENIG & BAUER

Koenig & Bauer offers the Rapida SRO 20 x 28 inch press line, and the Rapida SR III 28 x 41 inch press line. Both are available in two- and four-color models, and are rated at speeds up to 15,000 impressions per hour. The Rapida presses have console controls at both ends of the machine for all important operations. They feature inking roller adjustments located outside the press, a continuously rotating ink ductor roller, malfunction indication by signal lamps, a five-cylinder printing unit, a stop drum infeed system, an Alcomatic film dampening unit, electronic sheet-fed control, and a nonstop stream feeder.

MIEHLE

Miehle offers a range of Miehle-Roland presses. Sizes are from 20 1/2 x 28 3/8 up to 47 1/4 x 63 inches (52.1 x 72.1 up to 120 x 160 cm) in single-, two-, four-, five-, and six-color units, Fig. 26-86. The Miehle-Roland series 800 has been featured in packaging plants and has a reputation for combining speed with short makeready time, Fig. 26-87. These presses feature an instant-response remote dampener control and a remote running register control device. The 800 is available in three sizes: 50, 55, and 63 inch (127, 139.7, and 160 cm), in two-, four-, and six-color units. Miehle now offers the M.A.N.-Roland series of presses. These range from 35 x 49 5/8 to 47 1/2 x 62 7/8 inches (88.9 x 126 to 120.7 x 159.7 cm). All have press speeds of 10,000 impressions per hour. These presses are offered in two- to six-color configurations.

MILLER

The Miller Company offers a range of presses from 20 x 29 to 28 x 41 inches (50.8 x 73.7 to 71 x 104 cm), Fig. 26-88. All except the two single-color presses in the line are convertible to perfector. Standard features on most Miller presses include the Spiess feeder, feedroll register for consistent register, powerful inking, Miller-Meter dampening system, and fast makeready. The most popular of the Miller presses is the TP-38 American, a 25 x 38 inch (63.5 x 96.5 cm) convertible perfector. It is designed to a standard for more savings in purchasing plates, blankets, and in the use of ink.

Fig. 26-85. Itek Graphix duplicator is a single color offset press with a maximum sheet size of 13 3/4 x 17 3/4 inches. (Itek Graphix Corp., Graphic Systems Div.)

Fig. 26-86. Miehle 800 series presses are available in two-, four-, and six-color units in sheet widths of 50, 55, and 63 inches. (Rockwell International, Miehle Products)

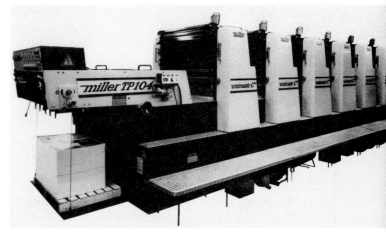

Fig. 26-88. Miller six-color sheet-fed offset press features special dampening system, powerful inking unit, accurate register, special feeder, and fast makeready system. (Miller Printing Equipment Corp.)

ROYAL ZENITH

Royal Zenith offets the Planeta Variant, a 28 x 40 inch (71 x 106 cm) sheet-fed press available in from two to six colors, Fig. 26-89. It can be equipped with or without convertible perfecting. In the larger sheet sizes, Royal Zenith offers the Planeta Super Variant in three sizes: 38 x 50, 41 x 55, and 44 x 64 inches (96.5 x 127, 104 x 139.7, 111.8 x 162.6 cm). All are available in two to seven or eight-color units. Speeds for the 50-inch and 55-inch presses are in the 10,000 impressions per hour range while the larger 64-inch Super Variant is rated at 9,000 impressions per hour.

RYOBI

Ryobi offers a number of duplicators which are equipped either as integrated or separate ink and water systems. It now offers the 500N and 500N-NP offset presses. The 500-NP features an impression cylinder for numbering, perforating, slitting, creasing, and imprinting in one pass. The only other difference is a standard sheet controller for the 500N-NP, which is

Fig. 26-89. Royal Zenith offers a variety of offset press sizes in single- and multi-color. The larger press sizes are capable of printing on sheets up to 44 x 64 inches at speeds of 10,000 impressions per hour. (Royal Zenith Corp.)

Fig. 26-87. Miehle-Roland four-color sheet-fed offset press is capable of speeds up to 10,000 impressions per hour. (Rockwell International, Miehle Products, Graphic Systems Division)

optional for the 500N. Standard features of both include an automatic centralized oiling system, a maximum sheet size of 19 11/16 x 14 3/16 inches (50 x 36 cm) and a single lever control for dampening, inking, printing, and paper feed functions. The inking and dampening system consists of ten inking rollers, with two form rollers of different diameters for higher performance. Refer back to Fig. 26-1.

POINTS TO REMEMBER

1. Sheet-fed offset presses are classified on the basis of press size, feeding arrangement, number of colors (units), and whether one sided or perfecting.
2. Small offset presses, up to a maximum sheet size of approximately 11 x 17 inches (27.9 x 43.2 cm) are classified as duplicators.
3. Offset presses that print on sheets larger than 11 x 17 inches (27.9 x 43.2 cm) are classified as offset presses.
4. The number of colors a sheet-fed press will print is derived from the number of printing units it contains.
5. Multicolor presses are designed as either add-on tandem systems or integrated permanent systems.
6. Sheet-fed offset presses deliver sheets to the printing unit either one at a time or by stream feeding.
7. The term perfecting refers to an offset press that prints on both sides of the sheet in one pass through the press.

KNOW THESE TERMS

Integrated, Clear switch, Recall switch, Keyboard, Display window, RPH (rotation per hour), Feed switch, Jam indicator, Doubles indicator, Pump switch, Safety cover, Operation control lever, Backstop, Wiper kit, Delivery rings.

REVIEW QUESTIONS

1. _____ _____ refers to the maximum size sheet that can pass through the press.
2. Each printing _____ on a sheet-fed multicolor press is capable of printing a different color.
3. Most _____ offset presses print from cut sheets of paper.
4. _____ feeding refers to the overlapping of sheets as they travel down the feedboard of a sheet-fed offset press.
5. A _____ press can print an image on both sides of the press sheet with one pass.
6. The form rollers on a duplicator or offset press make contact with the plate. True or false?
7. The parts of the ink fountain that can be adjusted to allow more or less ink flow are called _____ _____.
    a. Impression keys.
    b. Doctor blades.
    c. Fountain keys.
    d. Ductor rollers.
8. Duplicators are generally capable of printing sheets up to _____ x _____ inches.
9. Offset presses are generally capable of printing sheets _____ x _____ inches and larger.
10. Describe what is meant by the term *tandem* as it refers to multicolor presses.

SUGGESTED ACTIVITIES

1. Plan to visit a commercial printer to observe a press operator. Closely watch what is happening. Should problems develop, observe how the operator solves them. Do not disturb the operator during press operation. Write a short report on your findings.
2. Find examples of the following dampening systems: conventional, integrated, alcohol, no-molleton, and "bare back." Examine each system and familiarize yourself with their operating features.
3. Obtain several printing trade journals. Check for articles that feature new presses, press modifications, and accessories for presses. Learn to use the journals for research and information gathering.
4. Pick one of the duplicators or presses in the shop and list the features you like and dislike. What is it about this particular machine that impresses you more than the others?
5. Try to arrange a meeting with a press technician. Ask questions that will assist you in operating the equipment more proficiently. Ask the technician for tips on running more difficult jobs such as onion skin, carbonless, envelopes, etc. Make a chart of the technician's recommendations for future use.

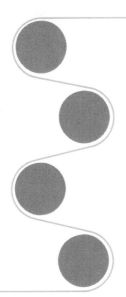

Chapter 27

Web-Fed Offset Presses

When you have completed the reading and assigned activities related to this chapter, you will be able to:
O Describe the basic components of a four-unit blanket-to-blanket web offset press.
O Describe the purpose and operational differences between a flying splicer and a zero-speed splicer on web offset presses.
O Describe common methods that register can be achieved on web offset presses.

Much of the expansion in the offset lithography industry over the past fifteen years is credited to the growth of web offset printing. Web offset printing produces single color and multicolor work for small and medium run newspapers, magazines, business forms, mail order catalogs, gift wrapping, books, inserts, and all types of commercial printing. See Fig. 27-1.

Offset presses that feed from a roll of paper are generally referred to as *web offset presses*. The term "web-fed" is commonly used to separate these presses from "sheet-fed" presses. Printing presses designed to feed paper from a roll of paper rather than single sheets are extremely FAST. Speeds of 1800 feet per minute (fpm) are common. Most web offset presses print on rolls 36 to 38 inches (914 to 965 mm) wide. Larger size rolls up to 76 inches (1 930 mm) are available, Fig. 27-2. Much of the work produced on web offset presses proceeds to a folder which is an integral part of the press. At this point, various combinations of folds convert

Fig. 27-1. A double-width web-fed offset newspaper press capable of printing 70,000 tabloid-size newspapers per hour. (Rockwell International, Graphic Systems Division)

Fig. 27-2. Most web-fed offset presses print on rolls 36 to 38 inches wide. Larger size rolls up to 76 inches in width are also used. (Rockwell International, Graphic Systems Division)

the web into folded signatures ready to be gathered, bound, and trimmed. Other operations that can be performed on web offset presses include paste binding, perforating, numbering, and rotary slitting and cutting. All of these operations can be handled while the press is running at high speed.

Web offset presses are two to four times faster than the fastest sheet-fed offset presses. Fig. 27-3 shows a schematic diagram of a typical four-unit web offset press.

WEB PRESS CONSTRUCTION

Most web offset presses are of *unitized construction*. This means that the presses are constructed of printing units (in tandem) selected to do the type of work intended. A web press can be ordered, designed, and constructed to feed from one or more rolls at the same time. The press may have up to eight or more printing units. It may be equipped to print on one or both sides (perfecting) of each *web* (roll) of paper. It may also be capable of printing in one to six or more colors in one pass through the press. The press can also be equipped to fold, cut, perforate, slit, imprint, varnish, glue, etc. These presses are capable of delivering printed work as signatures, single sheets, or rewound on rolls. Refer to Fig. 27-4.

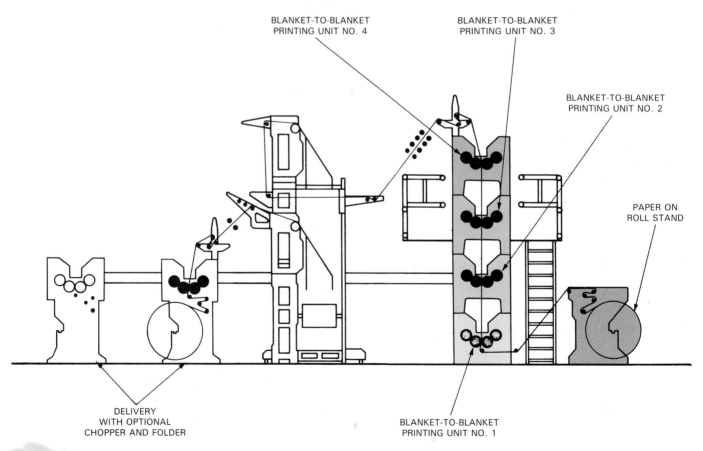

Fig. 27-3. This web-fed offset press has four stacked printing units at right, fed from a freestanding paper roll stand. The blanket-to-blanket design of this press produces four colors on both sides of the web in one pass.

Fig. 27-4. This high-speed four-unit commercial web-fed offset press delivers complete printed signatures. This is a perfecting press since it prints on both sides of the web simultaneously.
(Rockwell International, Graphic Systems Division)

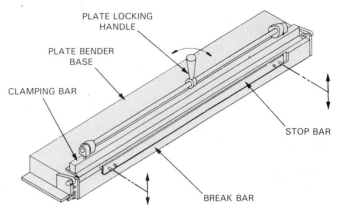

Fig. 27-5. A plate-bending fixture is used to accurately make the leading and trailing edge bends in the web-fed offset press plates. (Harris Intertype Corp.)

Because the paper is pulled through the press in a continuous web, a web offset press has no transfer or cylinder grippers like the sheet-fed presses. Not having sheet grippers allows for narrower cylinder gaps (as small as 3/8 inch or 9.5 mm) and faster running speeds.

On a web offset press, the circumference of the plate cylinder determines the length of the printed page. This is referred to as the *cut-off*. This is a fixed dimension on most web offset presses.

The plates for web offset presses are received from the plate room after being exposed, developed, and gummed. A plate-bending fixture is used to accurately make the leading and trailing edge bends in the plate. See Fig. 27-5.

Automated press controls are common on most web offset presses, Fig. 27-6. Controls for speed, temperatures, water, ink, register, and on-off are located at a console, Fig. 27-7. Some consoles are used to control more than one web offset press at the same time.

WEB PRESS DESIGNS

It is important to remember that web offset presses are often custom-built to the specific needs of the printer or publisher. Considerable thought and planning must be given to the size of the web offset press required. Each web offset press is designed for a specific purpose. There are three types of web offset presses. These include:
1. Perfecting blanket-to-blanket.
2. In-line open.
3. Drum common impression.

Perfecting blanket-to-blanket

The *perfecting blanket-to-blanket web* offset press does NOT have impression cylinders. The blanket cylinder of one unit acts as the impression cylinder for the other unit, and vice versa.

Fig. 27-6. This web-fed offset press is equipped with computerized controls. The press operator monitors all press functions from this control console. (Harris Intertype Corp.)

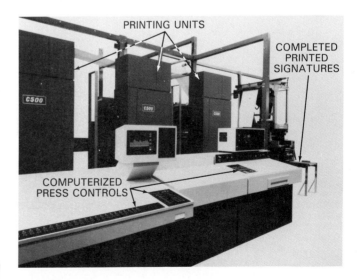

Fig. 27-7. Computerized press controls on this console regulate speed, temperatures, water, ink, register, and color.
(Rockwell International, Graphic Systems Division)

Illustrated in Fig. 27-8, each printing unit has two plate and two blanket cylinders. The paper is printed on both sides at the same time as it passes between the two blanket cylinders. The plate cylinders allow for quick plate changes. One, two, or four-page (tabloid) plates are locked into position by means of a quick set and release plate lockup. The rubber-faced blanket cylinders transfer plate images from the printing plates to the web. One blanket cylinder acts as an impression cylinder to the other, Fig. 27-9.

Each printing unit has an upper and lower ink fountain, Fig. 27-10. These ink fountains are adjustable through a series of fountain keys (25 keys typical) which allow the press operator to control ink flow across the plate. Newer web offset inking systems have *keyless fountains* which are controlled by an operator at the console.

Each ink fountain roller is driven by a separate gear motor to provide close control of inking. The inking system consists of one *ink fountain roller* that rides in the ink fountain, *one transfer roller* that conveys ink from the fountain roller to the rest of the system, three distributor rollers, one vibrator roller, and two form rollers.

Each printing unit also has an upper and lower *water system,* Fig. 27-10. This system applies the proper amount of dampener solution to the non-image areas of the plate surface. The dampener solution repels ink, thus enabling the non-image areas to run clear.

The *water fountain roller* is driven by a motor. It collects dampener solution on its surface as it rotates. As it turns, the fountain roller carries the solution to the cloth-covered transfer roller, which is driven by the vibrator roller. The solution is then transferred from

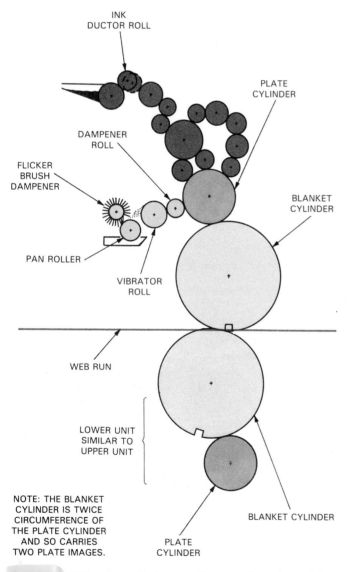

Fig. 27-8. This schematic diagram illustrates how the perfecting blanket-to-blanket web-fed offset press operates. The blanket cylinder is twice the circumference of the plate cylinder and thereby carries two plate images.
(Harris-Cottrell Company, Division of Harris Intertype Corp.)

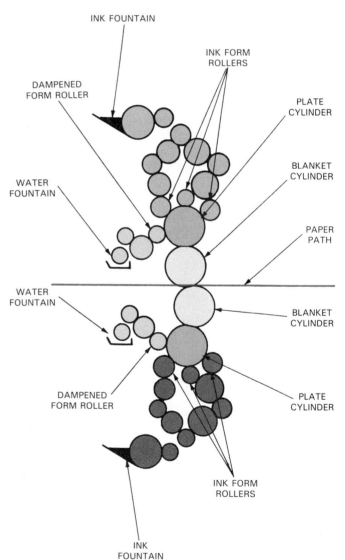

Fig. 27-9. This schematic diagram illustrates the principle of how the rubber-faced blanket cylinders on a perfecting blanket-to-blanket offset press transfer plate images from the printing plates to the web of paper.
(Rockwell International, Goss Division, MGD Graphics Systems)

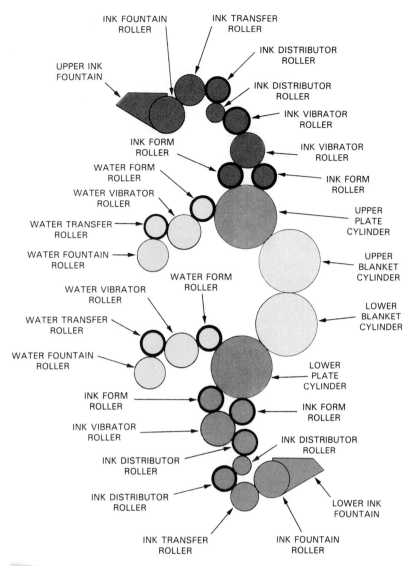

Fig. 27-10. A perfecting blanket-to-blanket web-fed offset press has upper and lower ink fountains. This schematic diagram illustrates how ink is transferred from the ink fountains to the plates.   (Harris Intertype Corp.)

the vibrator to the water form roller, which in turn applies it to the plate. The water system is designed so that there is continuous circulation of the filtered solution.

As described earlier, the *folder* is capable of producing either a standard, tabloid, or quarter-folded printed piece. The folder and quarter-folder are engaged and disengaged by separate clutches on the lower folder assembly. One clutch is for the folder assembly, the other for the quarter-folder. An electrical counter registers the quantity of folded signatures produced.

The standard folding operation begins when the web is routed over the web lead-in rollers. The web then moves under the gathering roll, over the roller top of the former, through the *trolleys* (where a slitter blade cuts web to tabloid width if desired), and down the former into the nipping rollers. See Fig. 27-11.

The former-folded web, or slit web for tabloids, is pulled around the cutting cylinder by six pins spaced across the width of the cutting cylinder. These *pins* are projected from the cylinder by a cam mechanism to pierce the web and keep the sheets correctly positioned until they pass through the cutting mechanism.

The *tucker blade* on the cutting cylinder tucks the web into the jaw blades of the jaw cylinder. Then the cutting knife slices the former-folded web at the proper point. The *jaw cylinder* takes the cut and folded signature from the cutting cylinder. It carries the paper around and through the hold-down roller and either into the *fan,* which lays it down on the conveyor assembly, or to the quarter-folder table.

In-line open

The *in-line open web* offset press is similar to a sheet-fed offset press. The main difference is that the cylinder gap is VERY NARROW. Each unit prints one color on one side. Additional printing units are required for more colors. To print the reverse side, the web of paper is turned over (180 degrees) between printing units by means of turning bars. This exposes the unprinted side of the web to the remaining units, Fig. 27-12. This press design is used for printing business forms.

Drum common impression

The *drum common impression web* offset press has all the blanket cylinders grouped around a large common impression cylinder. This type of press is also called a *satellite press*. One to five colors are printed at the same time on one side of the web, Fig. 27-13. The web is immediately dried, turned, and the reverse side is printed on the same printing unit by a process known as *double ending*. A web, one-half the width of the drum, is printed, dried, turned over, and brought back through the other half of the drum for printing the reverse side.

Mini-web offset press

A smaller version of the larger web offset press is called a *mini-web,* Fig. 27-14. It has become popular with in-plant and small commercial printers because of its smaller size and excellent speed. Mini-web roll widths are usually 14 inches (355 mm) with an 8 1/2 or 17 inch cut-off. These presses are capable of delivering 30,000 8 1/2 x 11 inch sheets per hour, or 15,000 11 x 17 inch sheets per hour.

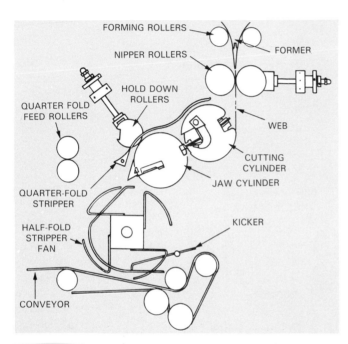

Fig. 27-11. The folder on a web-fed offset press is capable of producing either a standard, tabloid, or quarter-folded printed piece. This schematic diagram illustrates the operation of a folding unit. (Harris Intertype Corp.)

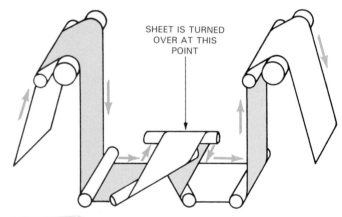

Fig. 27-12. The turning bars turn the web of paper over between printing units on a web-fed offset press.

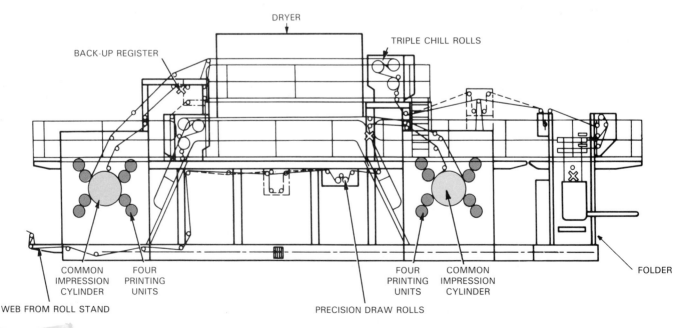

Fig. 27-13. A drum common impression web-fed offset press. Note that all the blanket cylinders are grouped around a large common impression cylinder. (Baker Perkins Limited)

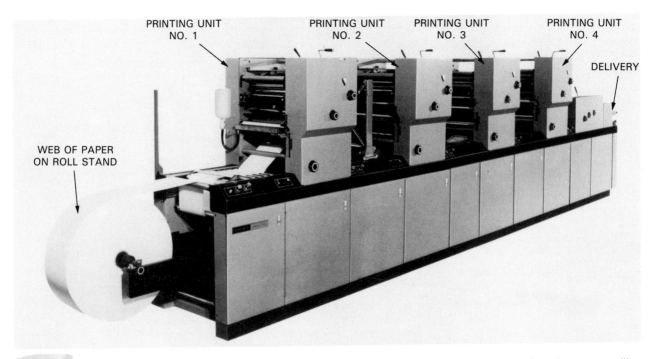

Fig. 27-14. This smaller version of a web-fed offset press is called a mini-web and is sometimes referred to as a satellite press. (Didde Graphic Systems Corp.)

Mini-web offset press prices are similar to sheet-fed offset presses, Fig. 27-15. The cost varies depending on accessories desired. For example, a fully equipped four-color, mini-web can print four colors on one side or two colors on each side of the web *(two-over-two)*. It can also print three colors on one side and one color on the other side *(three-over-one)*. This press would be equipped with a turn bar for printing both sides (perfecting). It might also be equipped with folding, perforating, punching, and numbering attachments.

A mini-web offset press requires about the same space as a multicolor sheet-fed press. Like most larger web offset presses, mini-webs are built in modular form. Color sections can be added as work demands, as can folders and other attachments.

A mini-web offset press can turn out the same amount of work as two to four duplicators. Besides speed, a mini-web offset press has other advantages.

Fig. 27-16 illustrates a common size commercial printing job and compares sheet-fed to web-fed operations for a two-color, two sided 11 x 17 inch sheet.

In addition to the savings shown in the illustration, other benefits can be realized from a mini-web press. For example, paper purchased in rolls is at least ten percent less than sheet-fed paper. Paper inventory for a mini-web takes up no more space than storing flat sheets for sheet-fed presses. Since most in-plant firms print 80 percent or more of their work on one type of paper, stored rolls are usually the same kind of paper.

WEB PRESS OPERATION

Using a typical mini-web press model, Fig. 27-17, the following is a summary of operation for an 8-page, mini-web offset press.

Paper is fed into the press from large rolls on a *roll stand,* replacing the pile feeder of a sheet-fed press.

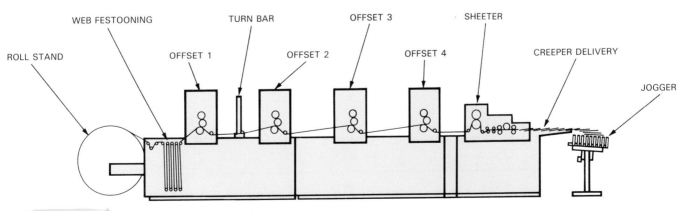

Fig. 27-15. A schematic diagram of the mini-web offset press illustrated in Fig. 27-14. (Didde Graphic Systems Corp.)

50,000 — 11 x 17 INCH PRINTED TWO-COLORS, TWO SIDES

	Single Color 25 inch	Two-Color 38 inch	Four-Color Mini-Web Press 11 x 17 1/2 inch Roll width
Press Makeready	19″ x 25″	25″ x 38″	11″ x 17″
Running Speed	.50	1.00	1.00
Number of Makereadies	7500	7500	30,000
Sheet Run	4	2	1
Press Run	25,000 — 17x22	12,500 — 22x34	50,000 — 11x17
	100,00	25,000	50,000
Total Makeready Time	2.00	2.00	1.00
Total Running Time	13.33	3.33	1.66
Total Press Time	15.33	5.33	2.66
Finishing	2 out 17x11	4 out 22x34	0
Cutting Time	1.06	1.30	0
Total Time	17.39	6.63	2.66
Total Operators	2	3	1
TOTAL LABOR HOURS	17.39	11.96	2.66

Fig. 27-16. Study this example of a common size commercial printing job comparing sheet-fed to web-fed operations for a two-color, two-sided 11 x 17 inch sheet. What benefits are derived from running the job on a mini-web?

When a roll begins to run out, a new roll can be spliced onto the old, using an *automatic splicer*. The web is threaded into the press while the press continues to run at full speed.

The large *festoon* is a paper storage device that unwinds paper to keep the press running at the same speed while the splice is made and the new roll accelerates to press speed.

One of the keys to high quality web printing is *tension control* or the amount of pull to keep the paper from having slack. To run a continuous web of paper at high speed between numerous rollers and cylinders, it is necessary to control the tension of the web precisely. This ensures accurate register of the sequential colors as they are laid down. On a web press, this exact control is accomplished electronically at the infeed unit. Sensitive monitoring equipment automatically controls tension of the web once the press operator has dialed the proper tension setting.

Web guides control side-to-side register of the web as it passes through the press, just as a side guide exercises this control on a sheet-fed press.

Printing units, that apply ink to the web, may number from one to as many as eight or ten on a given press. These are blanket-to-blanket perfecting units containing a pair of plate and blanket cylinders. Each has its own inking and dampening mechanisms. The blankets face each other and act as mutual impression cylinders to produce one color on each side of the web. Four units, therefore, provide process color, two-side capability.

Additional units are often used to add a special color or a layer of varnish or to perform some finishing operations. Inking and dampening systems on these presses are specially designed to assure sharp dot reproduction for a high-quality printed image.

The press operator can make adjustments to the various units manually or from an *electronic console* where many functions can be monitored automatically. Running adjustments allow a high level of operator control over the finished product.

From the printing units, the paper web flows into the *dryer* to evaporate the ink solvents, leaving a soft ink film. Automatic controls assure that the proper

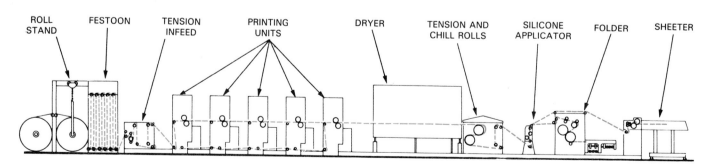

ROLL STAND · FESTOON · TENSION INFEED · PRINTING UNITS · DRYER · TENSION AND CHILL ROLLS · SILICONE APPLICATOR · FOLDER · SHEETER

Fig. 27-17. A schematic diagram illustrating the various operating components of a mini-web offset press. Note the roll stand on the left, allowing the web to travel through the printing units, dryer, chill rolls, folder, and finally into the sheeter at the right. (Harris Intertype Corp.)

temperature is maintained regardless of paper weight and speed of the web through the press.

The *chill rolls* then cool the web to room temperature, hardening the ink film. This results in a glossy ink finish, even on a dull paper, usually eliminating the need for *varnish* (clear overcoat). These chrome-plated cylinders are filled with a coolant solution, and they also adjust the web tension beyond the dryer for optimum in-line finishing control.

The printed web is then run either into a *sheeter,* which chops the roll into 8-page signature-sizes, or into a folder. Based on a 26 inch width and 17 3/4 inch cut-off, the most frequently ordered web roll, Fig. 27-18 shows sheet layouts for some of the most economical

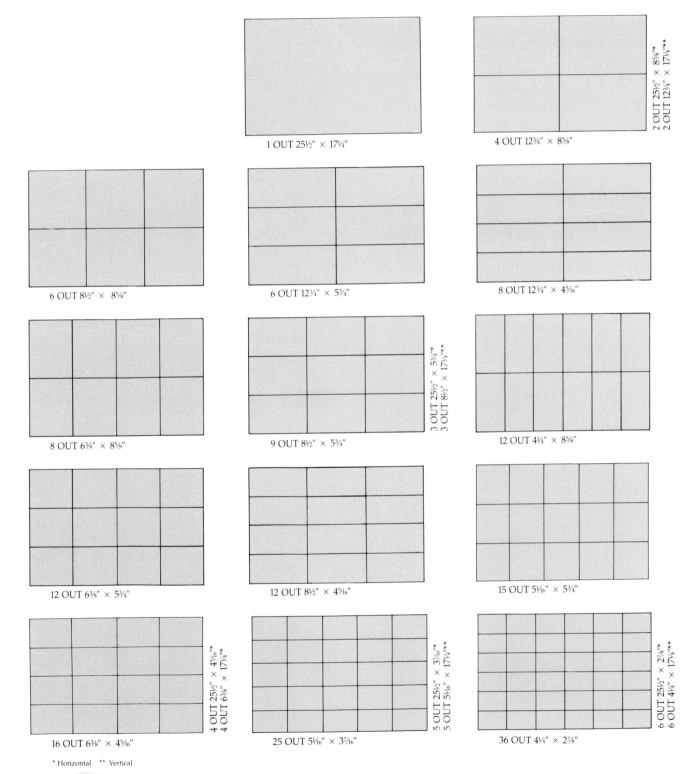

1 OUT 25½" × 17¼"

4 OUT 12¾" × 8⅝"
2 OUT 25½" × 8⅝"**
2 OUT 12¾" × 17¼"**

6 OUT 8½" × 8⅝"

6 OUT 12¾" × 5¾"

8 OUT 12¾" × 4⁵⁄₁₆"

8 OUT 6⅜" × 8⅝"

9 OUT 8½" × 5¾"
3 OUT 25½" × 5¾"**
3 OUT 8½" × 17¼"**

12 OUT 4¼" × 8⅝"

12 OUT 6⅜" × 5¾"

12 OUT 8½" × 4⁵⁄₁₆"

15 OUT 5¹⁄₁₆" × 5¾"

16 OUT 6⅜" × 4⁵⁄₁₆"
4 OUT 25½" × 4⁵⁄₁₆"**
4 OUT 6⅜" × 17¼"**

25 OUT 5¹⁄₁₆" × 3⁷⁄₁₆"
5 OUT 25½" × 3⁷⁄₁₆"**
5 OUT 5¹⁄₁₆" × 17¼"**

36 OUT 4¼" × 2⅞"
6 OUT 25½" × 2⅞"**
6 OUT 4¼" × 17¼"**

* Horizontal   ** Vertical

Fig. 27-18. Sheet layouts for some of the most economical unit sizes for work processed through a web-fed offset press in-line sheeter.   (Harris Intertype Corp.)

unit sizes for work processed through the in-line sheeter.

The folder is typically capable of delivering a variety of folded products at speeds up to 40,000 impressions per hour and sheet-fed products up to 35,000 impressions per hour. Changeover from folder to sheeter or from one fold configuration to another can be accomplished in minutes. In the past, these operations had to be done by other pieces of equipment.

With the addition of *special modules,* many other finishing operations can be accomplished on the press. These include the following: die cutting, punching, scoring, slitting, gluing, imprinting, plow folding, rotary cutting, and numbering.

The ability to perform these operations while on the press greatly magnifies the range of products available off the press. This makes these small web presses extremely versatile in a competitive market.

Web flying paster

A unique mechanism for bringing a new roll of paper into the web press feed cycle without stopping the press is called a *flying paster.* One is pictured in Fig. 27-19.

As the main feeding roll nears its end, the roll stand is rotated to bring the next full roll of paper into running position. This is done with the press running at full speed. Double-sided tape is applied to the leading edge of the new roll. Then, the full roll is moved into contact with the running roll of paper. The taped edge of the full roll is pressed against and immediately adheres to the running roll. The paper from the depleted roll is cut off and the roll brought to a stop. The new roll begins feeding the press. See Figs. 27-20 and 27-21.

On some newer web offset presses, *paste* is used instead of double-sided tape. Refer to Fig. 27-22. Fig. 27-23 illustrates an efficient method of handling webs in the pressroom.

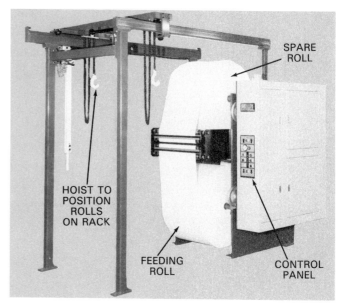

Fig. 27-20. Typical flying paster unit used for high-speed web-fed offset press operations. (Enkel Corp.)

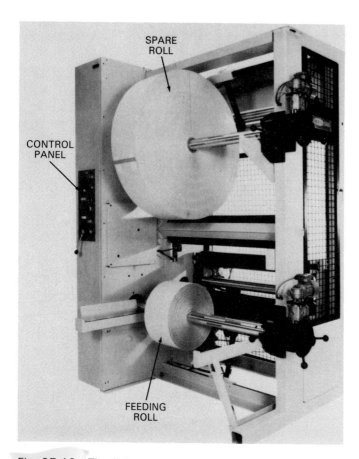

Fig. 27-19. The flying paster (splicer) operates at full press speed without stopping the paper rolls. (M.E.G. Corp.)

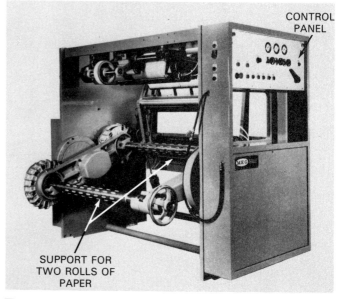

Fig. 27-21. A flying paster unit with a roll stand for two rolls of paper. As the feeding roll nears depletion, the roll stand is rotated, bringing the spare roll into position for splicing to the end of the feeding roll. (M.E.G. Corp.)

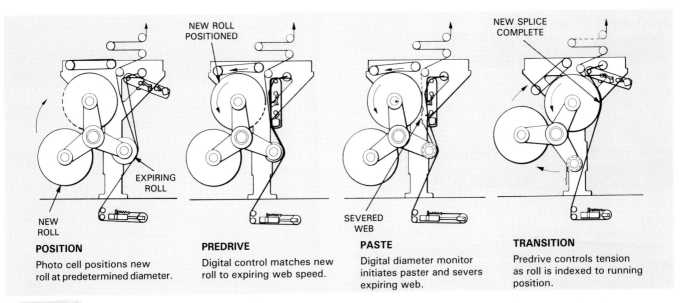

POSITION

Photo cell positions new roll at predetermined diameter.

PREDRIVE

Digital control matches new roll to expiring web speed.

PASTE

Digital diameter monitor initiates paster and severs expiring web.

TRANSITION

Predrive controls tension as roll is indexed to running position.

Fig. 27-22. A photocell positions a new roll of paper at predetermined diameter on this web offset press. Digital control then matches the new roll to expiring web roll speed. Automatic controller initiates paster unit and adjusts tension on roll for running speed. (Rockwell International)

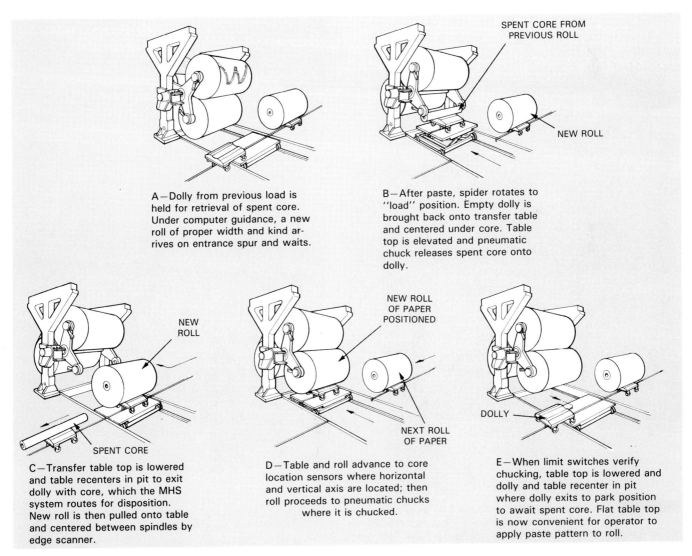

A—Dolly from previous load is held for retrieval of spent core. Under computer guidance, a new roll of proper width and kind arrives on entrance spur and waits.

B—After paste, spider rotates to "load" position. Empty dolly is brought back onto transfer table and centered under core. Table top is elevated and pneumatic chuck releases spent core onto dolly.

C—Transfer table top is lowered and table recenters in pit to exit dolly with core, which the MHS system routes for disposition. New roll is then pulled onto table and centered between spindles by edge scanner.

D—Table and roll advance to core location sensors where horizontal and vertical axis are located; then roll proceeds to pneumatic chucks where it is chucked.

E—When limit switches verify chucking, table top is lowered and dolly and table recenter in pit where dolly exits to park position to await spent core. Flat table top is now convenient for operator to apply paste pattern to roll.

Fig. 27-23. An automated computer—controlled paper web handling system. These systems are used for large newspaper, magazine, and commercial web-fed offset operations. (Rockwell International)

617

Web zero-speed paster

A *zero-speed paster* differs from a flying paster because the rolls are stopped momentarily, Fig. 27-24. A new roll of paper is brought to feed as follows. When the feeding roll nears its end, the paster unit rollers draw out considerable slack in the running web. The end of the new roll is taped. A splice is made while both rolls are STATIONARY. The paper is cut from the old roll. The new roll is brought up to press speed. These operations are performed before the slack in the splicer rollers is absorbed or taken up. The change in feed rolls occurs without stopping the press.

Web-break detector

Web offset presses are equipped with *web-break detectors,* Fig. 27-25. This is a device that activates a control to automatically stop the press in case the paper web breaks during operation. The web-break detector activates an electrical circuit to automatically bring the press to a FULL STOP. At the same instant, the web-break detector activates a mechanism that cuts the web just ahead of the printing unit. This propels the oncoming paper out of the printing unit. This prevents any free paper from whipping back and wrapping around the cylinders. This entire procedure is accomplished in a fraction of a second, Fig. 27-26.

There are five main causes for *web breaks.* These include:

1. Wet spots from fountain solution or washup are a principal cause of web breaks, particularly at startup. *Wet spot breaks* should diminish as the press crew gains experience in handling washups neatly.
2. Edge cuts from careless handling are a cause of web breaks. This calls for some training effort with paper handlers.
3. Closely related to edge cuts are felt-hair, calender, and fiber cuts. These are usually problems which originate at the paper mill and are a cause of complaint to the supplier.
4. Tension control malfunction can cause web breaks, but this is relatively infrequent as compared to the previous three causes of web breaks.
5. Humidity problems in dry winter climates cause a seasonal increase in web breaks. Some paper manufacturers have increased moisture content to counter the problem.

Web image alignment and register

Image alignment and register are important in sheet-fed and web-fed offset printing. The web press system is demanding because of its basic design. Unlike the sheet-fed press, rolls of paper must be run through the web press perpendicular (square) to the axes of the press cylinders. On a sheet-fed press, the sheet can be twisted considerably and still run through the press.

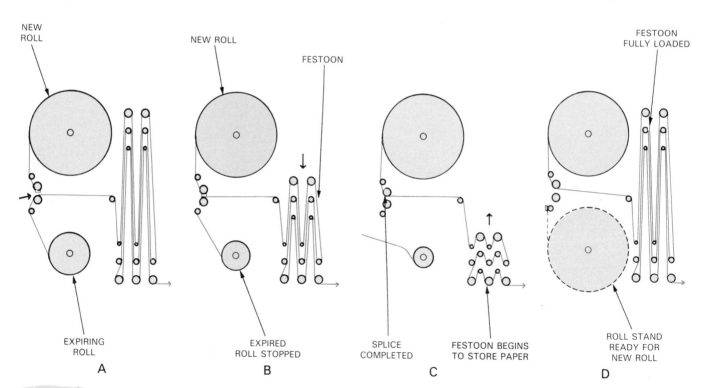

Fig. 27-24. Note sequence of operation of a "zero-speed" automatic splicer. A—Festoon (right) stores paper. New roll has been mounted with its lead edge ready for splicing (arrow). B—The expired roll has stopped and paper is being fed into the press from the festoon. Lead edge of the new roll has been moved under pressure against the expiring roll to make glue contact. C—Splice has been completed, the expired roll cut free and the newly-spliced roll accelerated to press speed. Festoon is again storing paper. D—Festoon is fully loaded. Roll stand is ready for mounting of a new roll and repetition of splicing sequence. Press has remained at constant speed during entire splicing operation.   (Harris Corp.)

Fig. 27-25. Web-fed offset presses are equipped with web-break detectors. These devices automatically stop the press in case the paper web breaks while running. (Baldwin Technology Corp.)

For purposes of illustrating web alignment, the press may be viewed as a series of precisely parallel printing units. The units are intersected by a web of paper traveling exactly perpendicular to the cylinder axes. An imaginary perpendicular intersection is created by the web of paper and the printing cylinders at each unit, Fig. 27-27. The folding and sheeting units fold and cut the web of paper parallel and perpendicular to the cylinder axes. Therefore, they form their own imaginary perpendiculars.

This means that there is little room for anything that is not going through the press in alignment with these imaginary perpendiculars. Everything must fit the perpendicular design which is fundamental in all web offset press arrangements. There is very little adjustment for regulating image alignment on most web offset presses. Any adjustment that is available is measured in thousandths of an inch or hundredths of a millimeter. This minor adjustment is primarily a control for paper, moisture, tension, or other small varia-

Fig. 27-26. When a break occurs in the web of paper, the detector automatically stops the press. (Baldwin Technology Corp.)

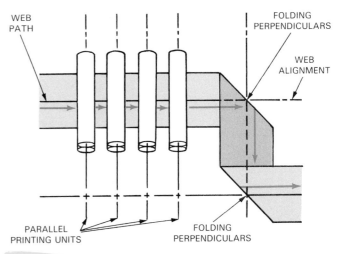

Fig. 27-27. To achieve perfect web alignment, the web press consists of a series of precisely parallel printing units which are intersected by a web of paper traveling exactly perpendicular to the cylinder axes. Folding and sheeting units fold and cut the web of paper parallel and perpendicular to the cylinder axes. (Raden C® Inc.)

tions that normally occur on the press. The adjustment cannot cure the problems of plates with improperly aligned images. There are also many web offset presses which have no adjustment for image squareness. Proper preparation procedures for web offset press plates are essential!

An ideal register control system for the web offset press is a unit that can consistently align images with the exact imaginary perpendiculars formed on the press. This exact alignment can be tested on any press.

A horizontal line is printed on the front and back of the sheet, parallel to the cylinder axis. Two plates are used, with one printing on the front of the sheet and the other printing the identical line on the back. In this test, as with production runs, the back-up plate is turned over when placing it on the bottom unit. Any alignment error is doubled by turning the back-up plate over. Alignment error is easily determined by the amount the lines vary from being parallel, Fig. 27-28.

Before any alignment system can produce work that is consistently aligned to the press perpendiculars, it must begin with precisely perpendicular stripping layouts. Exact alignment of a stripping layout can be tested by placing two identical layouts back to back. Any alignment variation can easily be detected by the mismatch of the layouts.

There are a number of register and alignment systems ideally suited to the web offset operation. These systems provide the web printer with consistently aligned stripping layouts and flats. The stripping department is further provided with a means of locating these stripped images, without deviation, to the plate and plate bender.

One such system, known commercially as the Raden C® Register and Alignment System, is widely used for web offset press operations, Fig. 27-29. It consists of an alignment frame used for layout preparation and stripping. In addition, there are standard format stripping layouts, prepunched masking material to accommodate various plate sizes, and a step-and-repeat template for accurately positioning flats to the plate and the bending location.

Web printing unit adjustments

There are a number of adjustments that can be made to web press printing units, Fig. 27-30. This is necessary to place printed images correctly on the paper web and achieve register in back-up and with images printed by other units on the press. As mentioned earlier, there is very little adjustment available for regulating image alignment on most web offset presses. These adjustments are usually made while the press is idle. Remote control master consoles are also used for this purpose allowing adjustments to be made while the press is in operation.

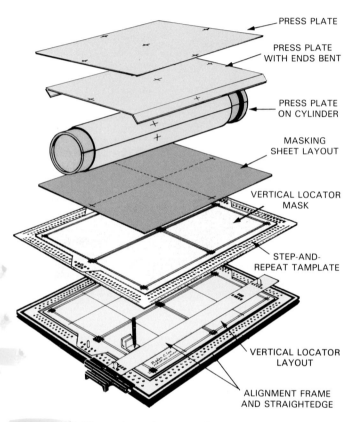

Fig. 27-29. The various elements of a register and alignment system. This system provides the web printer with consistently aligned stripping layouts and flats. (Raden C® Inc.)

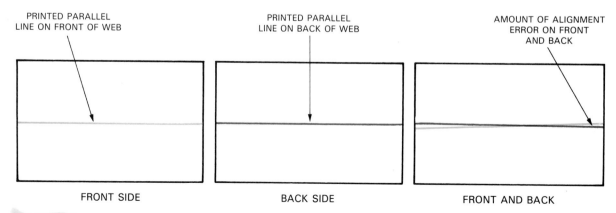

Fig. 27-28. Web press alignment can be checked by printing an identical line on both sides of the web. Alignment error is easily determined by the amount the lines vary from being parallel. (Raden C® Inc.)

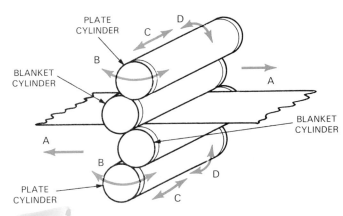

Fig. 27-30. The standard adjustments that can be made to web press printing units to achieve register. A—Forward or backward movement of entire printing unit to allow for register with other printing units. B—Angular (plate-cocking) adjustment of plate cylinder. C—Lateral (side-lay) adjustment of plate cylinder. D—Circumferential adjustment of plate cylinder. (Rockwell International, Goss Division, MGD Graphic Systems)

A press operator can check the printing image quality and register on one or both sides of the web while the press is running. This is done by fixed or portable high-intensity lights positioned above and below the running web of paper. See Fig. 27-31.

The high intensity *strobe light,* controlled by the printing cylinder, flashes a beam of light once for every revolution of the printing cylinder. This makes the lighted image area appear to be standing still. The light is synchronized to give a stop-motion effect of the web of paper. This allows the press operator to view and adjust the register if necessary. The automatic action of the light as it checks register and alignment acts as though the press were idle or off.

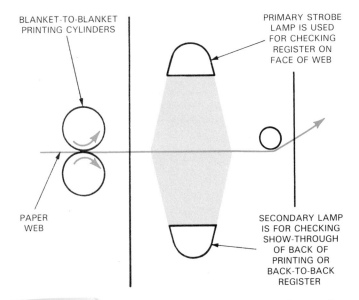

Fig. 27-31. Strobe lamps are used to check printing quality and register on both sides of the web. The primary strobe lamp is used for checking register on face of web. The secondary lamp is used for checking show-through of back of printing or back-to-back register.

The press operator, after checking the register and alignment, might be required to make any one of the following adjustments:

1. Unit-to-unit register.
2. Angular or skewing adjustment.
3. Lateral adjustment of plate cylinder.
4. Circumferential register (back-to-back register with opposite plate image).

It should be pointed out that web offset presses do NOT use plate clamps. Instead, the plate slips into a small slot in the plate cylinder. The plate is jogged into place while under blanket pressure and the tail edge is slipped into the reel rod slot. The reel rod is rotated to snug down the plate and then locked by a ratchet or clamp outside on the cylinder end.

Web ink and drying systems

There are a number of special considerations required when using inks on web offset presses. Even though web presses operate on the same principle as sheet-fed presses, their faster speeds create special problems. One example, the ink used for web offset printing is different from that in sheet-fed operations. Besides the common oxidative (evaporative) and penetrating inks, three other more sophisticated types of ink and drying systems are used in web offset. These include: heat-set, UV-curing, and thermal curing inks.

Heat-set ink

Most of the ink used in web offset printing is of the *heat-set* variety, which requires heating units for drying. This type of ink is formulated with high-boiling, slow-evaporating synthetic resins and petroleum oils. These ingredients are dissolved or dispersed in hydrocarbon solvents. The solvent has a narrow boiling range with a low volatility at room temperatures. The solvents are vaporized in the press heating units at tempertures of 400 to 500°F (204 to 260°C). This has the effect of leaving the resins of the ink on the paper.

Web offset presses using heat-set inks must be fitted with an *exhaust system* to eliminate solvent vapors. In addition, chill rolls are used to set the heated resins in the ink. This procedure is illustrated in Figs. 27-32 and 27-33. The exhaust system is constructed to comply with air pollution control regulations. Additional exhaust from the press area is provided in the form of roof exhaust fans or a thermal incinerator, Fig. 27-34.

UV-curing ink

Unlike heat-set inks, *UV-curing infrared inks* create no fumes which must be exhausted and dry when exposed to ultra-violet light. Special inks must be used for this drying process. The system uses ultra-violet (UV) light which dries the ink instantly. The infrared system allows for maintenance of web temperatures generally at between 158 to 176°F (70 to 80°C). Temperatures are never higher than 212°F (100°C). The

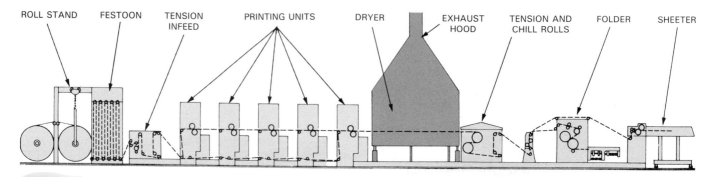

ROLL STAND   FESTOON   TENSION   PRINTING UNITS   DRYER   EXHAUST   TENSION AND   FOLDER   SHEETER
                          INFEED   HOOD   CHILL ROLLS

Fig. 27-32. Web-fed offset presses that use heat-set inks must be equipped with an exhaust system to eliminate ink solvent vapors. The exhaust hood system is constructed as a part of the press dryer unit.   (Harris Corp.)

drying follows immediately after printing, before the web enters the folder, sheeter, or rewinder.

With the infrared process, very little heat is absorbed by the atmosphere. Instead, it penetrates deeply into the web of paper. For perfecting (two-sided) printing, the press is usually fitted with a reflector plate on the other side of the web. This allows the paper to be dried in a single operation using a sandwich type dryer design.

Thermal curing ink

Web offset inks classified as *thermal curing* are similar to UV-curing inks but are formulated to dry instantly by the use of heat. Thermal curing inks are highly reactive formulas containing little or no solvent. They dry by application of heat and use of special interacting catalysts.

Filtration systems

Ventilation and filtration systems are commonly found in the web offset press area to overcome the problem of ink misting. This is common in newspaper pressrooms and occurs periodically in commercial operations.

Commercially produced ink mist suppressors are used when conditions require them. These are generally electrically charged devices that repel the charged ink back to the ink roller.

WEB PRESS MANAGEMENT SYSTEMS

Developments in the printing industry have accelerated and become more complicated. It is extremely important to obtain complete production information rapidly. In this way, management knows where it is at the present, and what steps should be taken to improve efficiency of operations in the future.

To fill this need in the web press area, several computerized *management* or *information systems* are used to help monitor and control the web press printing operation. These systems differ somewhat in the type of information collected, Fig. 27-35. In general, the information obtained includes items such as: running time, running speed, downtime, cause of downtime, number of good copies, number of poor copies, makeready time, and cause of web breaks. The computer software also allows for reporting of press delays by day, week, or month. Such items as ink and plate problems, press stops for customers and maintenance are all functions capable of being computerized and monitored.

Most press management systems are designed for a plant with several web presses. A press monitoring and analysis console is located at each press. This console accepts running data from sensors on the press, and job data from the press operator and production supervisors.

For example, the information on a simple job ticket is entered into the console by an office worker before

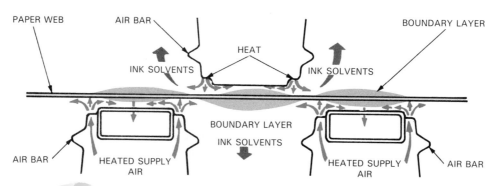

PAPER WEB   AIR BAR   BOUNDARY LAYER
HEAT
INK SOLVENTS   INK SOLVENTS
BOUNDARY LAYER
INK SOLVENTS
AIR BAR   HEATED SUPPLY   HEATED SUPPLY   AIR BAR
           AIR   AIR

Fig. 27-33. Web-fed offset press exhaust systems are constructed to comply with environmental air pollution control regulations. This schematic diagram of a web-fed press exhaust system illustrates how ink solvent vapors are exhausted.   (TEC Systems)

Web-Fed Offset Presses

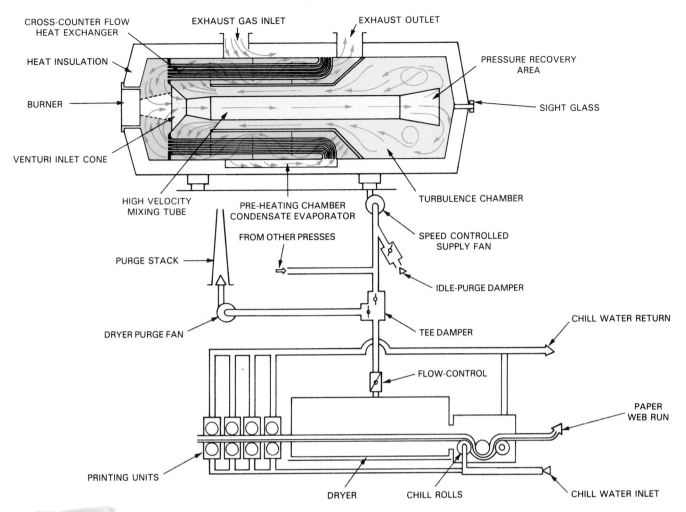

Fig. 27-34. Some web-fed offset press exhaust systems use combustion chambers to move ink vapor gases at high velocity. This velocity produces turbulence, thoroughly mixing the solvent-laden air. A more uniform temperature reduces the ultimate operating temperature. (TEC Systems)

	Make Ready Hours	Run Hours	Delay Hours	Total Labor Hours	% Delay
Goreville Plant					
Press #101	22.2	460.4	45.4	528.0	8.5
Press #105	37.2	433.9	56.0	528.0	10.7
Press #106	8.0	417.2	78.8	504.0	15.0
Vienna Plant					
Press #200	98.3	394.2	11.5	504.0	2.2
Press #205	31.3	448.1	25.6	504.0	5.0
Press #206	45.8	323.9	135.3	504.0	26.8
Simpson Plant					
Press #302	6.7	515.5	5.8	528.0	1.0
Press #305	53.4	414.9	49.7	528.0	9.4
Press #306	75.0	413.9	29.1	528.0	5.5
Cairo Plant					
Press #403	82.8	327.9	69.3	480.0	14.4
Press #404	69.4	273.7	136.9	480.0	28.5
Press #407	10.1	249.6	196.3	456.0	43.0

Fig. 27-35. Computerization of web press management is essential to maximum production and cost efficiency. This shows information relating to several presses at various locations.

the run is started. This includes the job number, kind of paper, number of webs, quantity to be run, press layout, how folded, quality level, and type of ink. Much of the information requested is transferred automatically from press sensors to the console. This includes items such as running speed, good and poor sheet count, web breaks (including location), production versus standard, waste, downtime, etc. Some details, such as the cause of a web break or the reason for a press stop, are entered manually into the console.

The information collected by the console is transferred on-line to a remote entry computer console. The console is generally located in the production manager's office. The console manipulates the data it receives, and prints out various reports on a line printer. See Fig. 27-36.

WEB OFFSET PRINTING PAPERS

Most web offset printing is done on good quality papers. Cleanliness and uniformity of moisture content, *caliper* (thickness), basis weight, and finish over the entire paper surface is critical. The paper ink receptivity should be highly compatible to the quick drying qualities of the inks. The paper rolls usually possess no weak areas incapable of withstanding the great stresses imposed by the press equipment. Paper grain direction should parallel the web, and the rolls should arrive well protected. The rolls should be wound to the proper tension on well constructed cores of suitable diameter.

Web offset papers should measure up to the standard requirements of sheet-fed papers. Because of the different mechanical characteristics of the two processes and certain demands placed on the offset papers for each, paper manufacturers make papers with special qualities for each printing process.

For example, less moisture is present on web offset plates and blanket cylinders during the printing cycle. This means that less moisture is picked up by the web of paper. This factor, combined with the high speeds at which webs move through the press, tends to lessen the danger of excessive paper softening. This means that the inherent water-resistance of web paper can be LOWER than might be considered satisfactory for sheet-fed papers.

All paper is generally printed with the grain running in the long dimension, as shown in Fig. 27-37. In sheet-fed offset, the mechanical stress undergone by the paper as it peels from the blanket cylinder is across the grain. The paper is parallel to the grain in web offset. In the latter case, little or no stretch takes place. In the case of sheet-fed papers, excessive mechanical stretch can be a real problem.

Web press paper undergoes a different kind of tension than sheet-fed paper because of the way it separates from the blanket cylinder. Web paper releases from the blanket without the severe bend and flex that sheet-fed papers must undergo. However, they get severe surface scuffing when passing through the blanket cylinders.

Handling rolls of paper

There are several recommended procedures associated with the handling and storage of rolls of paper for web presses. Recommended roll handling and storage procedures are illustrated in Fig. 27-38. In particular, observe the following precautions when handling and storing rolls of paper.
1. Stack rolls on their edges with the largest diameter at the bottom.
2. Use protective pallets, roll racks, or wooden floor strips instead of storing rolls directly on bare cement or shop/warehouse floors.

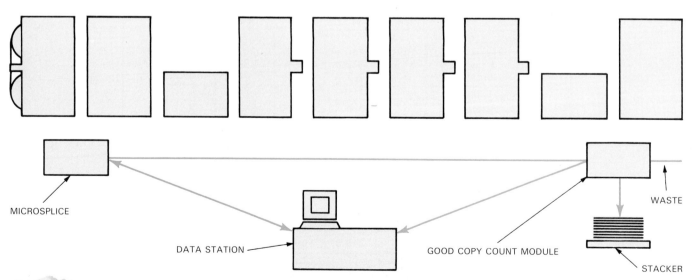

MICROSPLICE

DATA STATION

GOOD COPY COUNT MODULE

WASTE

STACKER

Fig. 27-36. Typical web-fed offset pressroom configuration. The "closed-loop" process control concept in which waste from a splice, blanket wash-up or other cause is automatically counted and diverted by an on-line module. This ensures that each skid contains only a known quantity of acceptable printed sheets.

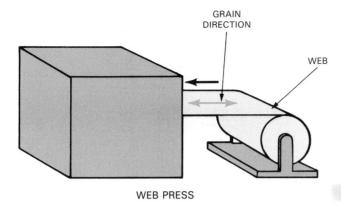

WEB PRESS

3. Use a roll clamp truck with minimum pressure to lift rolls. Narrow rolls can be crushed with clamp pressure.
4. Use floor markings along the storage area aisles to designate danger zones.
5. Use scrap paper as dividers when stacking unwrapped rolls.
6. Keep the storage area and pressroom floors clean at all times.

Practices which should be AVOIDED during roll handling and storage are illustrated in Fig. 27-39. They include the following.

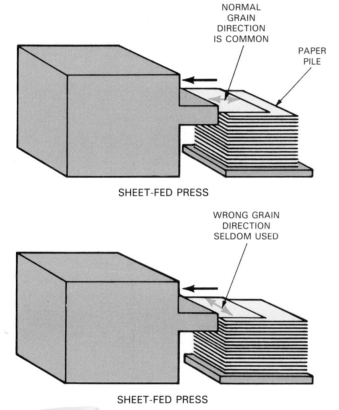

SHEET-FED PRESS

SHEET-FED PRESS

Fig. 27-37. Paper grain direction is an important factor when running web-fed or sheet-fed presses. In web-fed work, paper is generally printed with the grain running in the long direction. In sheet-fed work, paper is normally run across the grain.

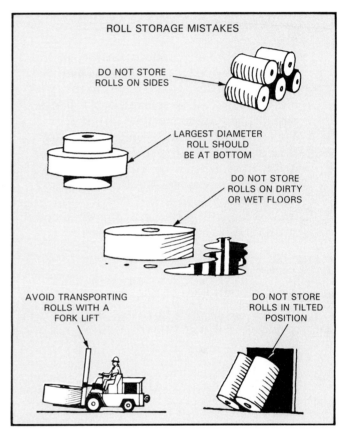

Fig. 27-39. These practices should be avoided when handling and storing rolls of paper. (Hammermill Paper Co.)

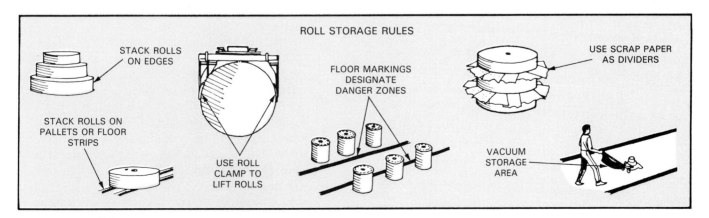

Fig. 27-38. Study these recommended procedures for handling and storing rolls of paper. (Hammermill Paper Co.)

1. Do NOT store rolls by placing them on their sides!
2. NEVER store small diameter rolls under rolls of larger diameter when stacking on end.
3. Do not store rolls on dirty floors or on floors which could become wet.
4. NEVER store rolls in a tilted position or tilt rolls on edge at any time as edge crimping will result.
5. Do not risk damage to roll edges through use of a fork lift truck.

Press operators and production personnel frequently need information relating to web roll weights and number of feet in rolls. Figs. 27-40 and 27-41 can be used as handy tables for determining such information.

POINTS TO REMEMBER

1. Web offset printing produces single color and multicolor work for small and medium run newspapers, magazines, business forms, mail order catalogs, gift wrappings, books, inserts, and all types of commercial printing.
2. Web offset presses feed from a roll of paper.
3. Besides printing at extremely high speeds, web offset presses can be equipped to fold signatures, paste bind, perforate, number, slit, and cut.
4. Most web offset presses are of unitized construction. This means that the presses are con-

APPROXIMATE ROLL WEIGHTS PER INCH OF WIDTH

Dia. In.	All Bond	Regular Ledger	Posting Ledger	Regular Offset	Regular Tagboard
10	1.86	2.34	1.98	2.10	2.55
12	2.83	3.37	2.94	3.12	3.74
14	3.92	4.58	4.07	4.32	5.07
16	5.13	5.99	5.38	5.71	6.86
18	6.60	7.58	6.86	7.27	8.01
20	8.23	9.35	8.51	9.04	10.21
22	10.00	11.16	10.28	10.90	12.47
24	11.90	13.34	12.30	13.05	14.56
25	13.10	14.62	13.42	14.22	15.96
26	14.10	15.80	14.53	15.41	17.24
27	15.30	17.05	15.68	16.63	18.60
28	16.40	18.33	16.88	17.90	20.00
29	17.70	19.66	18.12	19.22	21.46
30	18.90	21.05	19.40	20.58	22.96
31	20.02	22.45	20.73	21.99	24.62
32	21.60	23.93	22.10	23.45	26.10
33	22.90	26.52	23.52	24.95	27.78
34	24.20	27.89	24.98	26.49	29.49
35	25.70	29.86	26.48	28.09	31.25
36	27.00	31.60	28.03	29.73	33.06
37	28.70	33.39	29.61	31.41	34.93
38	30.20	35.23	31.25	33.15	36.83
39	31.90	37.13	32.93	34.92	38.81
40	33.70	39.06	34.65	36.75	40.82

Fig. 27-40. This chart can be used to determine approximate roll weights per inch of width for five kinds of paper. (International Paper Co.)

APPROXIMATE NUMBER OF LINEAR FEET IN ROLLS

To determine the approximate number of linear feet in a roll of paper, use the formula and factors shown below:

FORMULA:

$$\frac{\text{(Net Weight} \times 12 \times \text{(Factor)})}{\text{(Basis Weight)} \times \text{(Width)}} = \textbf{LINEAR FEET}$$

FACTORS:

Bond	1300
Cover	1805
Book or Offset	3300
Vellum Bristol (22½ x 28½)	2230
Index (25½ x 30½)	2700
Printing Bristol (22½ x 35)	2739
Wrapping, Tissue, Newsprint, Waxing (24 x 36)	3070
Tag	3000

EXAMPLE:

To find the number of linear feet in a roll of Form Bond—20″ width, sub. 16 lb., net weight 750 lbs.

$$\frac{750 \times 12 \times 1300}{16 \times 20} = \quad \begin{array}{c} 36,562.5 \\ \text{or} \\ 36,563 \end{array} \quad \textbf{LINEAR FEET}$$

To obtain a more exact approximation of linear feet, use the formula and average calipers given below:

$$\frac{[(\text{Roll Rad.})^2(3.1416)] \text{ less } [(\text{Core Rad.})^2(3.1416)]}{\text{Caliper}} = \text{Linear Inches}$$

$$\frac{\text{Linear Inches}}{12} = \textbf{LINEAR FEET}$$

Fig. 27-41. This chart can be used to determine the approximate number of linear feet in a roll of paper. (Zellerbach Paper Co.)

structed of printing units (in tandem). These presses may have up to eight or more printing units.
5. Web offset presses are equipped to print on either one or both sides of the paper.
6. There are three basic types of web offset presses. These include: perfecting blanket-to-blanket, in-line open, and drum common impression.
7. A smaller version of the larger web offset press is called a mini-web. It has become popular with in-plant and small commercial printers because of its smaller size and speed.
8. Mini-web offset presses can typically print four colors on one side of the sheet or two colors on each side of the sheet (two-over-two). The press can also print three colors on one side and one color on the other side (three-over-one).
9. Like the larger web offset presses, mini-webs can be equipped to fold, perforate, punch, number, cut, glue, imprint and die cut.

10. There are two common methods used to bring a new roll of paper into the web press feed cycle. These include: flying paster and zero-speed paster. Both methods allow the press to remain running while a new roll of paper is prepared.
11. In addition to pasters, most web offset presses are equipped with a web-break detector. This is a device that activates a control to automatically stop the press in case the paper web breaks during operation.
12. Web presses are equipped with image alignment and register mechanisms to keep the paper perpendicular or square to the axes of the press cylinders.
13. There are a number of adjustments that can be made to the press printing units of a web offset press. This is necessary in order to place printed images correctly on the paper and ensure register.
14. There are at least three kinds of ink drying methods used on web offset presses. These include: heat-set, UV-curing, and thermal curing.
15. Fast-drying inks are required on web offset presses because of the faster running speeds. Ventilation and filtration systems are also part of the inking system on web offset presses.
16. Web offset press computer management systems are used to significantly improve overall efficiency of press operations. These systems provide day-to-day, weekly, and monthly operational reports.
17. Web offset printing is generally done on good quality papers. Cleanliness and uniformity of moisture content, caliper, basis weight, and finish over the entire paper surface is critical.
18. Proper handling and storing of paper rolls is essential in web offset press printing operations.

KNOW THESE TERMS

Blanket-to-blanket perfector web press, Cut-off, Drum common impression web press, Flying paster, Heat-set ink, In-line open web press, Mini-web press, Thermal curing ink, UV-curing infrared ink, Web-break detector, Web offset press.

REVIEW QUESTIONS

1. Explain other operations, besides printing, done on a web press system.
2. Web offset presses do NOT have sheet grippers. True or false?

3. The sheet _____ is the length of the printed page which is determined by the circumference of the plate cylinder on a web offset press.
4. In most cases, web offset presses are of _____ construction.
5. What is a perfecting blanket-to-blanket web offset press?
6. To print on the reverse side of the paper on an in-line open web offset press, the web is turned _____ degrees between the printing units.
7. The drum common impression web offset press is also known as a/an:
   a. In-line open press.
   b. Converter press.
   c. Hemisphere press.
   d. Satellite press.
8. Describe a common impression cylinder web press.
9. The _____ offset press is popular with in-plant printers because of its smaller size and speed.
10. A _____ paster brings a new roll of paper into the web press feed cycle without stopping the press.
11. A _____ detector activates an electrical circuit which automatically brings the press to a full stop.
12. How is paper fed through a web press?
13. Why are exact alignment of stripping and platemaking extremely important factors in web offset press operation?
14. High-intensity lights positioned above and below the running web of paper check for image _____ and _____.
15. Web offset presses do NOT use plate clamps. True or false?
16. Web offset presses that use heat-set inks must be fitted with a system to eliminate _____.
17. Ultra-violet-curing infrared inks do NOT create fumes and operate at low temperatures. True or false?
18. _____ curing inks dry by application of heat and use interacting catalysts.
19. Explain the purpose of web offset press management systems.
20. The _____ direction of paper rolls should parallel to the web direction on the press.

SUGGESTED ACTIVITIES

1. Gather examples of web press printing. Try to get a broad selection of items, such as: throwaway publications, magazines, and expensive four-color work. Prepare a bulletin

board that attractively illustrates the products of web offset press work.

2. With the assistance of your teacher, arrange a field trip to a printing firm or local newspaper having a web press. Record your visit in a one or two page written report.

3. Prepare schematic diagrams of various web presses showing rollers, cylinders, and sheet flow arrangements of each.

4. Prepare a list of the types of jobs a web offset press is generally best suited to handle. Make a report to the class.

5. Prepare a short report on the application of mini-web presses for in-plant and commercial printing operations. You will need to examine printing journals and other print-related media to obtain information on this subject.

Chapter 28

Offset Press Operating Problems

When you have completed the reading and assigned activities related to this chapter, you will be able to:
O Install and properly maintain dampening roller covers on an offset duplicator or press.
O Clean and care for the inking system on an offset duplicator or press.
O Mix dampening solution and measure its pH value.
O Install a rubber blanket on an offset duplicator or press.
O Care for offset lithography blankets.
O Describe the lubricants and solvents used to help maintain and clean offset duplicators and presses.
O List and describe general safety rules to be followed while lubricating, cleaning, and working around offset duplicators and presses.
O Use one or more of the available quality control devices to achieve optimum results on an offset duplicator or press.
O Recognize and control problems related to static electricity on an offset duplicator or press.
O Identify common printing problems and their solutions while operating an offset duplicator or press.

Presswork involves turning all of the preparatory work into finished printed sheets. The printed sheets must be exact reproductions of the originals. There are many factors which enter into the quality of the finished printed product. Most important of these is the presswork itself. Just as in the preparatory work, strict attention to presswork fundamentals is required for quality printing.

Note! Several other chapters discuss printing problems. Refer to the index for more information on this topic.

PROBLEM AREAS

Presswork and its related operating problems can be divided into three areas. These include: paper, ink, and press. Exact press procedures in these instances is complicated by the number of makes and models of presses on the market. Some of the most common printing problems encountered on duplicators and presses are illustrated in Figs. 28-1 through 28-9. There are certain basic procedures that apply to most duplicators and presses.

Paper problem considerations

Most printing papers will run satisfactorily on a duplicator or larger offset press. The press operator should recognize that there is a wide variety of papers made just for the offset process. There are considerable differences between papers made for letterpress printing and those made for offset printing. These differences are due primarily to the use of water in offset lithography. In addition, the offset process utilizes the close contact of a rubber blanket to the paper.

Offset paper starts with a strong, tough base pulp blended with fillers for good running properties and opacity. Offset lithography papers are also *surface-sized* to prevent the tacky offset inks and close blanket contact from pulling or picking the fibers from the surface. Dust from paper cutters and slitters and stray fibers from other sources must be carefully eliminated

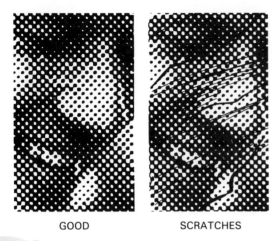

GOOD   SCRATCHES

Fig. 28-1. When scratches occur in the offset plate, it is necessary to prepare a new plate. Care should be taken to avoid scratching the plate with your fingernails and dry particles of dirt or developer in the plate processing pad or sponge. (3M Company)

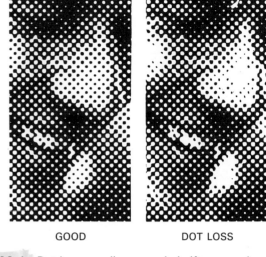

GOOD   DOT LOSS

Fig. 28-4. Dot loss usually occurs in halftones and screened image areas. The dots disappear in the highlight (light) areas of the offset plate image. (3M Company)

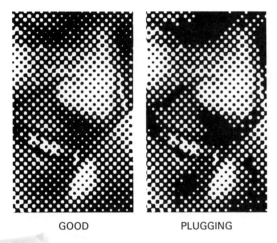

GOOD   PLUGGING

Fig. 28-2. Plugging is a condition in which the image areas are filling in with ink on the printed sheets and plate. (3M Company)

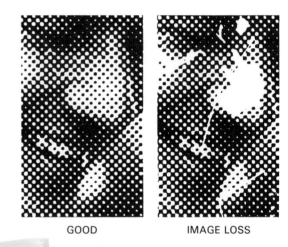

GOOD   IMAGE LOSS

Fig. 28-5. Image loss is the gradual disappearance of some lines and some halftone dots from the plate image. (3M Company)

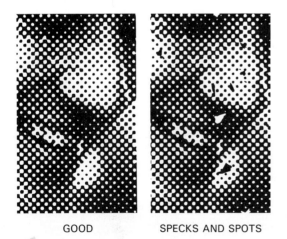

GOOD   SPECKS AND SPOTS

Fig. 28-3. Specks and spots are usually referred to as hickies which are particles of foreign matter that attach themselves to the offset plate and/or blanket. They cause an undesirable black or white spot on the printed sheet or the printed image. (3M Company)

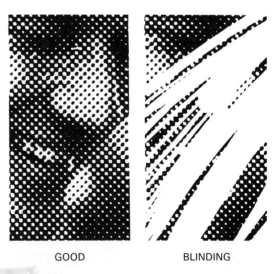

GOOD   BLINDING

Fig. 28-6. Blinding appears as weak spots in the image areas. It is generally caused by dried plate gum, strong fountain solution, or a glazed blanket. (3M Company)

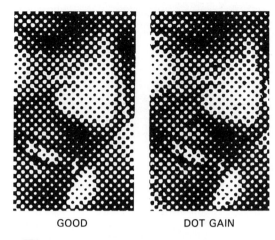

GOOD   DOT GAIN

Fig. 28-7. Dot gain usually occurs in halftone and screened image areas of the printed sheet. It is due to either too much impression pressure, too much ink, or excessive pressure causing slippage of the blanket against the paper just before the impression stroke. (3M Company)

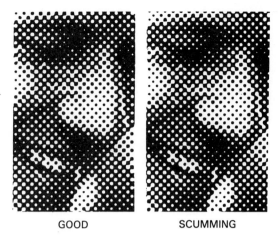

GOOD   SCUMMING

Fig. 28-8. Scumming occurs when the plate picks up ink in the non-image areas and transfers ink to the non-image areas of the sheet. This may be caused by glazed ink rollers, glazed blanket, too much ink-form-roller pressure, or too much dampener-form-roller pressure. (3M Company)

TINTING	MISTING OR FLYING	SCUMMING
Ink has emulsified into the water fountain. Pigment is being put on the plate from the water dampeners. Tint can easily be washed off plate.	Fine droplets or filaments of ink are formed on the roller train during film splitting. Ink may form a mist, or it may actually be sprayed or thrown off the press rollers.	The inability of water to keep the nonimage area of the plate clean. Scum cannot be easily washed from the plate.
BLANKET EMBOSSING & ROLLER SWELLING	**POOR ROLLER TRANSFER AND/OR GLAZING**	**PREMATURE PLATE BLINDING**
The blanket develops a relief image of the image on the plate, and/or the rollers swell so that they no longer stay within their normal settings.	The ink appears to dry on the rollers or the inking system seems to be unable to adequately transport the proper ink/water emulsion down the roller train onto the plate.	A strong image area of the plate is progressively losing its receptivity to the ink.
INADEQUATE DRYING	**INK PILING**	**POOR TRAPPING**
Printed ink film is wet or tacky for an unreasonable length of time.	Ink builds up on areas of the rollers, blankets, and/or plate, creating a dry accumulation known as ''caking'' or ''piling.''	Superimposed inks are being improperly laid down on the previously printed colors, causing poor color balance and poor overall appearance.
DOT SPREAD	**POOR RUN & SCRATCH RESISTANCE**	**LINTING/PICKING OR PAPER ''PILING''**
The halftone dots increase in size, causing the printed signature to lack sharpness.	The printed ink film appears dry but exhibits poor rub and/or scratch resistance nonetheless.	Paper surface is roughed or torn. ''Linting'' refers to pulling of the fibers on uncoated stocks, ''picking'' to lifting of the coating on coated stock onto blankets and plates.

Fig. 28-9. Refer to these illustrations for clues to problems that occur while operating offset duplicators and presses.

from offset papers. This is necessary to prevent pickup of such particles by the rubber blanket as sheets run through the press.

Hickies — A *hickey* (unwanted white or black spot) usually results from a small piece of foreign matter that attaches itself to the offset blanket or plate. The piece of material causes an unwanted black or white spot on the printed sheet or the image itself. The small particles of foreign matter may be fibers from a molleton cover, dried ink, paper coating, or dirt.

As discussed in earlier chapters, hickies appear as either white spots or doughnut-shaped black spots with a white spot directly around the hickey. The press operator can prevent hickies by practicing the following methods of operation.

1. Avoid excessive ink tackiness.
2. Pay careful attention to the condition of molleton rollers.
3. Replace damaged and cracked ink rollers.
4. Do NOT use setoff spray powder or liquid excessively.
5. Avoid ink skinning in cans and in the duplicator or press fountain.
6. Keep all rollers clean, especially at the ends.
7. Use cleaner mats with care. They can leave paper particles on ink rollers.

The Rodel Hickey Picker® is a unique ink form roller used as a substitute for conventional rubber rollers on an offset press, Fig. 28-10. It removes dirt, ink skin, paper dust, and other foreign material from the offset plate thereby preventing hickies. The hickey picker has a special surface layer made from synthetic fibers, plastics, and elastomers permanently vulcanized to the base roller. Dirt, dried ink, and other foreign matter are removed from the plate and either stay with the roller until washup or travel back up the ink train to the fountain. Plate life is unaffected due to the thick nap of the roller and the light setting required. Due to the good inking qualities of the roller, there is no problem with pattern from the roller. See Fig. 28-11.

The careful elimination of paper dust and stray fibers also helps ensure quality impressions. Sheets used in offset lithography must be stiffer than those for letterpress. This is because of the faster operating speeds of offset duplicators and presses.

Fig. 28-10. The Rodel Hickey Picker® is a unique ink form roller used as a substitute for conventional rubber rollers on an offset press. (Rodel, Inc.)

Fig. 28-11. The Rodel Hickey Picker® roller removes dirt, dried ink, and other foreign matter from the offset plate. These materials either stay with the roller until washup or travel back up the ink train to the fountain. (Rodel, Inc.)

Ink problem considerations

Inks used for offset lithography have a high concentration of pigmentation. In addition, they possess adequate water repellency and tack to produce a uniform print. Much of the lack of uniform ink coverage in duplicator and small offset presses can be traced to improper ink in either of these two areas. This is equally true on large offset presses.

Quality offset inks are expensive. It is false economy to use cheap inks. There are a number of problems usually blamed on ink. However, these problems are generally caused by other sources.

Three of the most common misconceptions are described here. These include: setoff, drying, and scumming.

Setoff — *Setoff* (inked image prints on back of another sheet) usually occurs as a result of running too much ink on the duplicator or press. For example, in order to achieve the desired ink color, a press operator runs the ink heavy (thick). However, in most instances, the problem is too much water. Instead of increasing the ink, the press operator should REDUCE the amount of water. It may be necessary to use a deeper ink color in order to permit the running of a thinner ink film to get the correct color results.

Setoff can also be caused by improper handling of the freshly printed sheets. Stacking freshly printed sheets too high also causes setoff. This is especially true where a setoff spray attachment is NOT used. While setoff sprays are desirable in some cases, they are not a necessity in running all jobs. Such sprays can sometimes be a source of trouble in themselves. For example, the spray may build up on the blanket and cause hickies.

Drying — Most offset lithography inks are manufactured to be press ready. This means that they have drier added to them. There are times when it is desirable to add a small amount of a cobalt drier or combination cobalt and manganese paste drier to the ink. In most cases, the only time it is necessary to add driers is when HIGH HUMIDITY situations are encountered. Moisture, whether in the form of humidity or dampener solution, slows drying. Care must be taken because adding too much drier can actually retard the drying of the ink.

Many of the ink drying problems are the result of running too heavy an ink film. In addition, carrying an excessive amount of water on the dampener rollers can cause emulsification and also retard the drying process. Some duplicators and presses that feed the water directly to the ink do not use a dampening roller.

Scumming — A problem related to ink and water balance is called *scumming* (ink prints lightly in non-image areas). As noted earlier, running an excessive amount of water is a common cause. This sequence usually starts out as the press operator gets a light print on the sheets. As the amount of ink is increased, a filling or scumming problem begins. The problem is made worse by adding more water which again produces a light print. The recommended procedure is to carefully REDUCE the amount of water until a sharp full density print is obtained. If this produces filling of fine letters, halftones or other fine detail, the water should be carefully increased until the condition disappears. The same procedure should be followed if the background begins to pick up ink.

Press operators use the term *catchup* to describe this condition of scumming. Some press operators fight a light print condition for an entire press run. Later they discover that the problem was NOT enough back cylinder (impression cylinder) pressure that actually caused the light print problem.

Note! The chapter on lithographic ink contains a complete description of how inks are manufactured along with many of the problems associated with ink on an offset duplicator and press.

Printing on carbonless papers

Printing on carbonless papers can sometimes create problems on the duplicator or press. For best results, follow these typical procedures:

1. Fan the paper from the corners (not the ends) before loading. Check (using CFB test sheet in carton) to assure printing side is up.
2. Paper height should be set to about 1/8 inch to 1/2 inch below sheet separators. Blower tubes should be adjusted to evenly float sheets just below the separator fingers. Suction feet should be positioned about 2 inches from the edges of the paper.
3. Pull-out roller should be centered on the paper stack in order to forward the sheet without cocking.
   a. Place the paperweight bar in one of the two forward notches which will help control the amount of air separating the sheets. Use the lightweight paper kit for best results.
   b. Air should be adjusted to float the top 3 or 4 sheets. Air and vacuum may have to be adjusted as duplicator speed increases.
   c. Set the buckle control at number 7 and move up as far as number 12 until the paper runs smoothly.

4. The side register guide should barely tap the side of the sheet. Pushing the sheet excessively may cause it to buckle and wrinkle as it passes through the duplicator.
5. Reduced impression pressure and quick release compressible type blankets are recommended to minimize capsule damage of sheets. This will also reduce any tendency for the paper to stick to the blanket. Results can be checked by spraying a CB surface with instant replay damage indicator.
6. Use quality inks that dry by oxidation and have lower tack (in 10-12 range on a FATF Inkometer) to reduce the possibility of setoff and coating buildup on the blanket.
7. Run with a minimum amount of fountain solution necessary for quality printing.

Press problem considerations

Offset duplicators and presses rely on accurate pressure settings of the various rollers and cylinders. The setting of pressures should be a daily routine performed by the press operator. Each press manufacturer has recommended settings listed in the press operating manual. However, the symptoms caused by inaccurate pressure settings are common. Here are the important checkpoints that require attention on most offset duplicators and small offset presses.

Ink form rollers and plate — Too little pressure between the ink form rollers and the plate can cause incomplete transfer of ink to the plate. Too heavy a setting can result in image spread and slur. This can also cause excessive plate wear.

Plate to blanket — Too little pressure between the plate and blanket cylinders can cause improper transfer of the image to the blanket. This will result in a light print. Excessive pressure can cause image spread, horizontal streaks from blanket slippage, and slurred dots.

Back cylinder (blanket-to-impression cylinder) — Too little impression between the blanket and impression cylinders can result in a broken or light print. Excessive pressure can cause spread of the image.

Dampener form roller-to-plate — If the pressure between the dampener form roller/s and plate is too light, dampening of the plate will not be uniform. The tendency will be to run too much water due to the poor transfer of moisture to the plate. In addition, any variation in the diameter of the dampener roller/s will show up very quickly. Excessive pressure can produce slurring of the image. This condition can also cause a bounce of the dampener form roller/s as it goes through the gap of the plate cylinder.

CYLINDER PRESSURE QUALITY CONTROL DEVICES

There are quality control devices made especially for checking proper press cylinder pressures. They are available as film negatives or positives. This allows them to be stripped into the flat and made part of the page image. The printed image is checked by the press operator to verify correct cylinder pressures.

Star target

The *star target* is used to check print quality of black and white as well as color printing. It consists of a circle about 3/8 inch (9.5 mm) in diameter, Fig. 28-12. This device is placed at the center or at the tail end of the printed sheet.

The star target contains thirty-six segments within a circle. When these segments are filled with ink, NOT giving a clear print, the press operator is alerted to print quality failure. The imperfections of this symbol indicates that the printed material is doubling and/or slurring, and/or dot gaining.

The star target can also be used in platemaking to check contact of film and plate. It is also used by the camera department to check for copy resolution. See Fig. 28-13.

Dot gain scale and slur gauge

The *dot gain scale and slur gauge device* is used to control dot area reproduction of halftones and process color, Fig. 28-14. The dot gain scale and slur gauge gives a visual indication of dot changes by the displacement of an invisible number to a higher or lower graduated value. For a satisfactory condition, the scale should give an apparent sense of movement of the invisible number.

Fig. 28-12. A star target is used to check print quality of black and white, as well as color printing. The target is used to check such areas as doubling, slur, and dot gain. (Graphic Arts Technical Foundation)

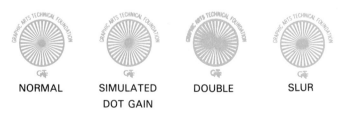

| NORMAL | SIMULATED DOT GAIN | DOUBLE | SLUR |

Fig. 28-13. The star target can be used by the camera department to check for copy resolution. This illustration compares how the printed target appears as normal, with dot gain, doubling, and slur. (Graphic Arts Technical Foundation)

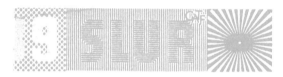

Fig. 28-14. The dot gain scale and slur gauge is used to control dot area reproduction of halftones and process color. (Graphic Arts Technical Foundation)

The slur gauge segment of the scale indicates dot gain by ink spreading in one direction. When slur occurs, the word "SLUR" appears to the right of the value scale, Fig. 28-15. *Slur* is a change in sharpness caused by a slippage of the cylinders usually in the around-the-cylinder direction. This causes the dots to become football shaped and to sometimes have a comet shape with a tail. If this happens in only one color, it can cause a color change. If it happens in all colors, it can darken and muddy the colors.

Fig. 28-15. When slur occurs on the printed sheets, the word "SLUR" appears to the right of the dot gain scale and slur gauge. This identifies a change in sharpness caused by slippage of the cylinders usually in the around-the-cylinder direction. (Graphic Arts Technical Foundation)

Scanning densitometer

On all three and four-color process printing, a strip of colors, called *color bars,* are included along one edge of the sheet. This is an ink-measuring density system developed by the Graphic Arts Technical Foundation. It has become the industry-accepted standard.

When a color job is being printed, the press operator uses a unit called a scanning densitometer, Fig. 28-16. The *scanning densitometer* measures the density of the ink on the printed piece so a comparison can be made to the original approved color proof. The bar and densitometer also show whether there is dot gain in halftones and if the ink is trapping properly.

IMPORTANT PRESS CHECKS

Offset duplicators and presses are precision machines. They can only be as precise as you keep them. Do not take duplicator and press settings for granted just because everything ran well on the last job. This is especially true if the previous job was run using a different paper, ink, or plate.

WARNING! Check that all safety guards are in place and operating properly! Keep fingers clear of all pressure (nip) points, clamps, and cylinders! Wipe up all excess oil, grease, and solvents!

Recheck the duplicator or press settings BEFORE and DURING the run. The following items should

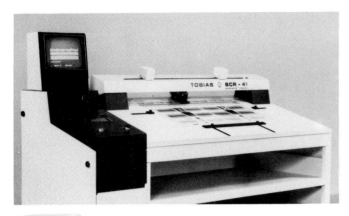

Fig. 28-16. This scanning densitometer measures the density of the ink on the printed sheets so a comparison can be made to the original approved color proof. Color bars are included along the edges of the press sheets for this purpose. (Tobias Associates, Inc.)

receive careful attention when preparing to run a job on the duplicator or press.

1. Oil press periodically! Refer to the press operator's instruction manual for lubrication requirements.
2. Inspect duplicator/press cleanliness! Check fountains, rollers, and all surfaces for dried ink, gum, dirt, or mechanical contamination.
3. Check the printing plate! Inspect for correct exposure, development, completeness of image, and damage.
4. Adjust plate clamps! Adjust plate clamps back to same position each time.
5. Mount plate carefully! Pack cylinder to manufacturer's specifications if required. Check tightness of the plate at the edge of cylinder gap. Register all plates to the same cylinder gutter.

6. Discard dry ink skin in the can! Never save any of this problem ink.
7. Set the ink fountain! Adjust the ink fountain keys to average image ink flow requirements using just over half the catches on the ratchet. Check evenness of ink film as the fountain roller turns while in contact with the ductor roller.
8. Test the water fountain! Check the pH with litmus paper. If using alcohol, also check specific gravity. Clean the fountain completely every day.
9. Measure dampener pressures! When testing dampener form roller-to-plate pressure, a uniform, firm pull should be attained on plastic gauges or paper strips at two points along the rollers.
10. Inspect ink form roller-to-plate pressure! Gently drop the inked form rollers to the plate surface. Check printed bands for uniformity and thickness.
11. Check plate-to-blanket pressure! Transfer the ink band from the plate to the blanket. They should match perfectly.
12. Adjust impression cylinder setting! Set for the thickness of paper being run.

Note! These topics are discussed in detail elsewhere in this textbook. Refer to the index as needed.

TROUBLESHOOTING

The troubleshooting reference guide included here is intended to help you eliminate many common problems associated with duplicator and press operations. It is divided into two columns: probable cause and cure. When printing problems occur, refer to this data. The guide was prepared by the Minnesota Mining and Manufacturing Company (3M), Printing Products Division.

TROUBLESHOOTING REFERENCE GUIDE	
BACKING AWAY—Ink does not follow the fountain roller. Printed areas become light in color.	
Probable Cause	Cure
1. Ink rolls up in fountain and will NOT follow ink roller.	1. Work ink with ink knife more often.
2. Ink is too short and heavy bodied.	2. Add reducer or varnish to increase flow of ink body.
CHALKING—Ink rubs completely off printed sheets when dry.	
Probable Cause	Cure
1. Ink vehicle is absorbed by paper before ink has time to set properly on surface of sheet.	1. Add a body gum or binding base to control absorption by paper.
2. NOT enough drier in ink.	2. Add 3-way drier to the ink.
3. Job printed with wrong ink.	3. Overprint with varnish.
COLOR FADES—In the pile, color bleaches out.	
Probable Cause	Cure
1. Not enough oxygen between sheets to dry ink properly. Heat is produced and bleaches color.	1. Fan stock often during first few hours after printing. Use anti-setoff spray.
DRYING—Ink dries on the rollers.	
Probable Cause	Cure
1. Too much drier.	1. Reduce amount of drier.
2. NOT enough printed copy to use all ink from ink rollers.	2. Allow ink fountain to carry fresh ink to rollers, even a small amount will soften the old ink.
3. Wrong ink formula for the job.	3. Call the ink manufacturer for recommendations.

(Continued)

FILLING IN—Screens and reverses appear "muddy."

Probable Cause	Cure
1. Too much ink being used on press.	1. Cut back on ink at ink fountain.
2. Ink too soft.	2. Add body gum or a heavy varnish.
3. Too much drier.	3. Use an ink made for the paper.
4. Press pressures incorrect.	4. Check and reset all press pressures.
5. Paper coating not good quality.	5. Change to another type of paper.

GHOSTING—Indistinct image patterns appearing generally in solid image areas.

Probable Cause	Cure
1. Too much water being applied.	1. Recheck roller dwell and pressure settings.
2. Glazed ink rollers.	2. Clean rollers thoroughly.
3. Engraved blanket.	3. Replace blanket.
4. First printed color is sensitizing plate on next unit.	4. Desensitize plate. Change ink.
5. Excess ink from a reverse in preceding image (called a positive ghost).	5. Correct imposition of job layout.
6. Lack of ink in heavy tone caused by heavy preceding image (called negative ghost).	6. Run with minimum water. Use opaque ink if possible. Correct job imposition layout.

HICKIES—Doughnut-shaped spots appear in solid image areas.

Probable Cause	Cure
1. Specks of dried ink or skin from can.	1. Remove all hard or skinned ink from can before putting ink in fountain.
2. Cutting or slitter dust.	2. Dust paper pile edges before putting on press.
3. Dust from overhead objects, such as lights or ventilation units.	3. Vacuum overhead objects in work area.

INK ROLLER STRIPPING—Rollers in ink unit do NOT accept ink.

Probable Cause	Cure
1. Fountain solution too acid.	1. Use a pH tester. If it does not read between 4.0 and 5.6, adjust solution.
2. Rollers glazed.	2. Deglaze rollers with a glaze remover.
3. Too much water being run.	3. Cut back water at fountain.
4. Rollers have become desensitized.	4. Pumice and etch ink rollers as recommended by roller manufacturer.

MOTTLING—Occurs as a muddy image on the printed sheets. Ink does not cover evenly.

Probable Cause	Cure
1. Too much reducer in ink.	1. Add heavy varnish, liquid glass, or magnesia to ink.

PICKING—Paper fibers are pulled up by ink.

Probable Cause	Cure
1. Ink is too tacky.	1. Use an ink reducer.
2. Impression cylinder pressure is too great.	2. Readjust cylinder pressure.
3. Paper of poor grade or NOT suited to type of ink being used.	3. Replace with a different grade of paper.

PILING—Occurs when ink builds up, or piles, on rollers, plate, or blanket.

Probable Cause	Cure
1. Pigment is NOT carried by vehicle.	1. Add a #3 or #4 varnish to ink.
2. Ink has not been sufficiently ground in manufacturing process.	2. Do the same as above.
3. Ink is over-pigmented.	3. Do the same as above.

PLATE WEAR—Plate image areas deteriorate.

Probable Cause	Cure
1. Improper ink form roller pressures.	1. Readjust ink form roller pressures.
2. Too much drier in ink.	2. Use less or different type of drier.
3. Fountain solution too acid.	3. Use a pH tester. If it does not read between 4.0 and 5.6, adjust the solution.
4. Too little ink.	4. Carry more ink on plate.
5. Improperly developed plate.	5. Make a new plate.

POOR SPREAD—Ink distribution on rollers is uneven.

Probable Cause	Cure
1. Ink it too stiff or tacky.	1. Add a reducing compound.
2. Ink rollers are out-of-round.	2. Replace bad rollers.
3. Ink rollers are glazed.	3. Deglaze ink rollers with a glaze remover.
4. Uneven spread of ink from fountain.	4. Adjust ink fountain keys and clean small ink rollers of excess ink.

RUNNING or BLEED—Paper shows image on back of printed sheets.

Probable Cause	Cure
1. Poor grade of paper, one which is too absorbent.	1. Replace paper with paper of a better grade.
2. Ink has been thinned with solvent that is absorbed too quickly into paper.	2. If solvent has been added to ink by press operator, it is best to start with a fresh ink.

SCUMMING—This condition is evidenced by plate picking up ink in non-image areas and transferring ink to non-image areas of sheet.

Probable Cause	Cure
1. Non-image areas of plate NOT etched properly. This results in non-image areas not being able to repel ink.	1. Re-etch plate or make a new plate.
2. Too little fountain solution.	2. Increase flow of fountain solution.
3. Dampener form rollers NOT parallel, or set too far from plate cylinder.	3. Set rollers according to instruction manual.
4. Dirty dampener rollers.	4. Clean or replace dampener rollers. Check pH of fountain solution. Feed less ink.
5. Unevenly applied molleton cover on dampener roller.	5. Even out dampener cover, or re-cover if necessary.
6. Dampener roller NOT parallel with plate.	6. Re-set dampener roller according to manual. Make a new plate if it is damaged.
7. Dampener roller dragging or bouncing on plate.	7. Re-set dampener roller according to manufacturer's manual. Make a new plate if it is damaged.
8. Fountain solution too strong or too weak.	8. Prepare new solution at recommended pH.
9. Plate and blanket cylinders NOT properly packed.	9. Set packings to specifications.
10. Flattened form rollers.	10. Replace rollers.
11. Plate NOT perfectly concentric with cylinder.	11. Adjust plate slack. Make a new plate if image is damaged.
12. Skidding form roller.	12. Maintain proper contact between rollers and their riders.
13. Too much ink during makeready.	13. Reduce amount of ink being used and re-etch plate.
14. Loose blanket.	14. Adjust plate slack.
15. Oxidation of non-image areas on plate.	15. Make a new plate or re-etch plate being used.
16. Overdeveloped plate.	16. Make a new plate.
17. Plate contains too much gum arabic or coating solution.	17. Make a new plate.

SETOFF—Ink transfers to back of sheets in delivery pile or stacker.

Probable Cause	Cure
1. Too much ink on paper.	1. Use less ink by resetting fountain keys.
2. Fountain solution too acid.	2. Use a pH tester. If it does not read between 4.0 and 5.6, adjust fountain solution.
3. Wrong ink for kind of paper being used.	3. Contact ink company to find proper ink.
4. NOT enough drier.	4. Add a 3-way drier to speed up drying of ink.
5. Paper coating of poor quality.	5. Use paper of different quality.
6. Ink too heavy, does not absorb into paper.	6. Add some varnish to ink.
7. Delivery pile too high.	7. Reduce height and remove smaller piles from press.
8. Press operator squeezes paper pile when removing from press.	8. Use a paper delivery board under pile and lift paper on board.
9. Static in delivery pile is attracting sheets to each other.	9. Use a static eliminator in delivery area.

SLOW DRYING—Ink does not dry on printed sheets.

Probable Cause	Cure
1. Fountain solution too acid.	1. Use a pH tester. If it does not read between 4.0 and 5.6, adjust fountain solution.
2. Wrong ink for kind of paper being used.	2. Contact ink company to find proper ink.
3. NOT enough drier added.	3. Use more drier in ink.
4. Paper is too acid.	4. Replace with a different grade of paper.

SLUR—Image on sheet appears double or smeared.

Probable Cause	Cure
1. Blanket is loose.	1. Tighten blanket clamps or replace with a new blanket.
2. Gears worn on press.	2. Check gears for wear and replace if necessary.

STICKING—Sheets may stick together in delivery pile or stacker, causing sheets to tear when pulled apart.

Probable Cause	Cure
1. Ink being run too heavy.	1. Decrease ink flow. Use a setoff device.
2. Too much drier in ink.	2. Use less drier. Use a setoff device.
3. Paper will NOT absorb ink.	3. Use a different ink. Run with less ink. Remove sheets from delivery end in small piles. Change grade of paper being used. Use a setoff device.
4. Relative humidity is too high in press area.	4. Adjust and maintain correct relative humidity. Add drier to ink.
5. Moisture content of paper is too high.	5. Condition paper before using. Adding drier may help.
6. Paper surface is too acid.	6. Add drier to ink.

TINTING—Overall background of ink color appears in nonprinting area.

Probable Cause	Cure
1. Ink is too soft.	1. Add varnish to ink.
2. Plate NOT properly processed.	2. Prepare a new plate.
3. Fountain solution is too acid.	3. Use a pH tester. If it does not read between 4.0 and 5.6, adjust fountain solution.
4. Ink and water are out of balance.	4. Clean up press and readjust ink and water.
5. Paper coating is NOT of the right grade.	5. Change grade of paper.

VANISHING—Gradual disappearance of some lines and halftone dots from plate.

Probable Cause	Cure
1. Friction of form rollers.	1. Readjust form rollers.
2. Chemical action of certain inks.	2. Change ink and/or grade of paper being used.

POINTS TO REMEMBER

1. There are a number of critical factors that enter into the quality of the finished printed product. Most important of these is the presswork itself.
2. Presswork and its related operating problems can be divided into three areas: paper, ink, and press.
3. Exact duplicator and press operating procedures are complicated by the number of makes and models of presses in use.
4. There are certain basic operating procedures that apply to most offset duplicators and presses.
5. To assist press operators, troubleshooting reference guides are available.

KNOW THESE TERMS

Blinding, Backing away, Catchup, Chalking, Dot gain, Dot gain scale, Dot slur, Doubling, Ghosting, Hickies, Misting, Mottling, Picking, Piling, Plugging, Roller stripping, Running, Scumming, Setoff, Slur, Star target, Sticking, Tinting, Trapping, Vanishing.

REVIEW QUESTIONS

1. A gray, washed out image is generally caused by too much moisture and not enough ink. True or false?
2. Too much ink and NOT enough water on the plate will cause _____.
3. What problems can a loose blanket cause?
4. _____ appears as a uniform, light image color over the entire sheet of paper.
5. Ink build up on the rollers is called:
   a. Piling.
   b. Filling up.
   c. Tinting.
   d. Sticking.
6. A problem called _____ occurs when a printed sheets picks up an image on its reverse side from the sheet below it in the delivery pile.
7. Loss of image on the plate is called:
   a. Running.
   b. Picking.
   c. Vanishing.
   d. Stripping.
8. Tacky ink can cause picking. True or false?
9. Mottling generally causes a _____ image and is caused by too much _____ in the ink.
10. What can cause an ink not to follow the fountain roller?
11. A condition known as _____ occurs when completely dry ink rubs off of the press sheets.
12. Doughnut-shaped spots appearing in solid areas of the printed image are called _____.
13. Define the term "filled in."
14. A smeared or double image is called _____.
15. Explain causes and corrections for sticking.

SUGGESTED ACTIVITIES

1. Check the ink rollers on all duplicators and/or presses for glaze. Thoroughly clean and deglaze all suspect rollers.
2. With your teacher's assistance, perform preventative maintenance on the duplicators and/or presses in the shop. Keep records of all work performed on each unit.
3. Check the impression (back) cylinder on all duplicators and/or presses. Any ink accumulation, glaze, etc., should be removed according to the manufacturer's directions.
4. Gather samples of printed jobs that show out-of-register, hickies, smudges, poor ink coverage, ghosting, and other problems. Label each sample and make a display that can be used for the entire class.
5. Prepare a set of 3 x 5 inch reference cards that give the symptoms, probable cause, remedy for most of the problems that commonly occur during a press run. Place one problem on each card and fasten the set together with one screw post binding device in the upper left corner. The pack of reference cards will then be accessible and handy to use.

YOU'RE OKAY
Crisp, dark lines and solids • A clean background • Clean halftones, screens and reverses • Good registration • Each sheet dried completely.

GRAY, WASHED OUT?
Not enough ink • Too much moisture • Wrong color of ink • Incorrect dampener form roller pressure • Incorrect plate-to-blanket pressure • Incorrect impression-to-blanket pressure.

PAPER CURLING IN RECEIVER
Too much moisture • Curl in paper.

PAPER MISSING GRIPPERS
Stop fingers incorrectly set • Feed rollers out of adjustment.

PAPER NICKING ON EDGE
Paper stop fingers too high • Feed rollers not set properly • Paper hitting back stop in receiver too hard.

PAPER WRINKLING
Too much moisture • Paper damp • Too much pressure between blanket and impression cylinder • Register board not set properly.

(Continued)

Study these illustrations before reading the individual captions and try to determine what caused the printing problems.
(3M Company)

BACKGROUND DIRTY—SCUMMING
Too much ink • Not enough moisture • Dirty dampener roll covers • Dampener covers tied too tightly on ends.

GRAY, WASHED OUT PLUS DIRTY BACKGROUND
Glazed ink rollers • Glazed blanket • Too much ink form roller pressure • Too much dampener form roller pressure.

IMAGE BREAKS DOWN WHILE PLATE IS RUNNING
Too much dampener form roller pressure • Too much ink form roller pressure • Too much plate-to-blanket pressure • Fountain solution too strong • End play in form rollers.

IMPROPER REGISTER
Loose blanket • Side guides not set properly • Paper not cut straight • Cam band not set.

STREAKING
Incorrect ink form roller pressure • Incorrect dampener form roller pressure • Incorrect plate-to-blanket pressure • Incorrect impression-to-blanket pressure • Improper ink • Loose blanket.

NO IMAGE AT ALL
Not enough ink roller form pressure • Not enough plate-to-blanket pressure • Not enough impression-to-blanket pressure • Too much moisture • Glazed blanket • Glazed ink rolls.

(Continued)

COPY TOO DARK
Too much ink • Too much impression-to-blanket pressure • Not enough plate-to-blanket pressure • Too many revolutions on blanket without paper going through (build up on blanket).

DOUBLE IMAGE (BLURRED COPY)
Loose Blanket • Too much ink and fountain solution • Not enough plate-to-blanket pressure • Loose plate • Incorrect impression-to-blanket pressure.

UNEVEN PRINTING
Incorrect ink distribution • Glazed rollers • Incorrect dampener form roller parallel pressure • Poor paper (surface of paper) • Incorrect ink form roller parallel pressure • Incorrect plate-to-blanket parallel pressure • Incorrect impression-to-blanket parallel pressure • Dirty impression cylinder.

WEAK SPOTS (SPOTTY COPY)
Incorrect plate-to-blanket pressure • Incorrect impression-to-blanket pressure • Low spots in blanket • Tacky ink • Tacky blanket • Dirty impression cylinder • "Blind" image on plate caused by dried gum or too strong fountain solution.

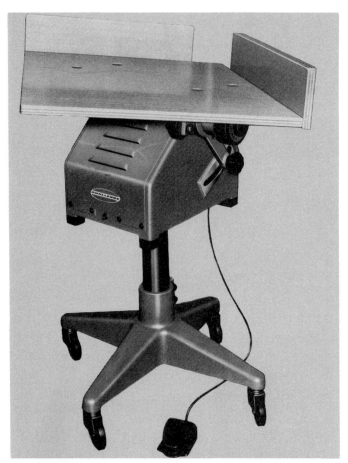
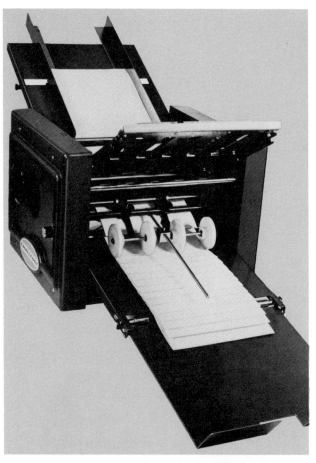
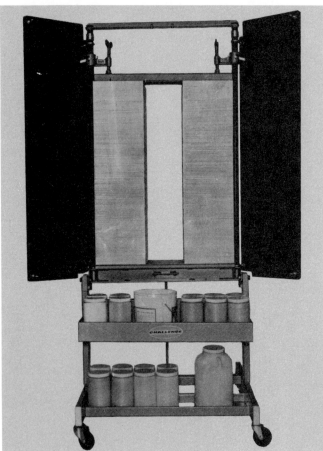
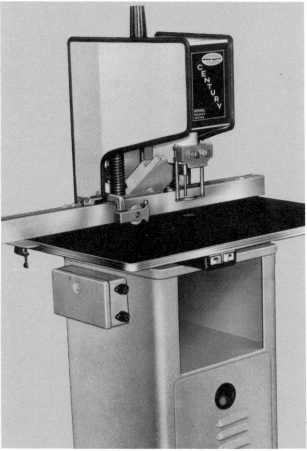

A variety of finishing and binding equipment is used in industry.   (Challenge Machinery Co.)

Chapter 29

Finishing and Binding Operations

When you have completed the reading and assigned activities related to this chapter, you will be able to:
O Describe paper cutting and trimming operations.
O Use a hydraulic paper cutter to cut and trim paper.
O Describe the die cutting process.
O Determine the number of sheets of paper and cutting sequence required for a given printing job.
O List, describe, and perform basic operations included within the finishing category.
O List, describe, and perform basic operations included within the binding category.
O Describe the hot stamping process.
O Describe the several methods used to package printed products ready for delivery.

Transferring the image to the paper is sometimes thought of as the end of the printing process. However, other production steps are often required before the customer receives the product. The final processing of a printed product generally includes some form of finishing and binding operations. The equipment used to perform these operations is either automatic or semi-automatic. Since the equipment is varied and usually specialized, these operations tend to be labor intensive. Finishing and binding operations are areas in which strict attention to detail is required. A mistake in this production phase can lead to a costly rerun of the job.

FINISHING OPERATIONS

Finishing is the term used to describe all operations performed on printed materials beginning at the delivery end of the printing press until the final shipment to the customer. Most printed jobs have as a final step in production one or more finishing operations. This might only be the packaging of the product for delivery.

Explained in this chapter, finishing operations include:
1. Cutting and trimming.
2. Folding.
3. Punching.
4. Drilling.
5. Gathering.
6. Collating.
7. Jogging.
8. Scoring.
9. Perforating.
10. Slitting.
11. Die cutting.
12. Hot stamping.
13. Embossing.
14. Varnishing.
15. Thermography.
16. Packaging of the product.

CUTTING AND TRIMMING

After pages for magazines, books, and pamphlets are printed, they are cut and trimmed. The printed sheets are *cut* so that they can be folded into signatures. A *signature* is a large press sheet that is generally printed on two sides and folded to make 4, 8, 16, 32, or 64 pages, Fig. 29-1. When signatures are *bound* together they form a book, booklet, or magazine.

Trimming is done after the binding on a three-knife trimmer. Paper cutters range from the single-knife hydraulic type to the fully automatic three-knife book trimmer. See Fig. 29-2.

DANGER! Keep hands away from the pressure clamp and cutting blade on the paper cutter. Most accidents on a paper cutter occur as a result of misusing the paper clamp.

For most printing, it is common practice to print or repeat the same image on a large sheet of paper. This method saves time on the press. For example, if a job requires 100,000 pieces, it can be printed eight to each large press sheet or *form* to reduce the press run to 12,500 impressions. Large press sheets or forms printed in this manner usually end up being cut apart and trimmed to final size. This is sometimes done on a single-knife paper cutter, Fig. 29-3.

Some paper cutters are programmable. This means that they can be adjusted to automatically change settings for each of the various cuts required in the sequence. Refer to Fig. 29-4.

FOLDING

Signature sheets, booklets, flyers, and programs require some type of *folding* which bends the sheet over onto itself. This operation is accomplished on automatic folding equipment. Some machines are of the table-top variety for light- to-medium-duty folding by in-plant and quick printers. See Figs. 29-5 and 29-6. Heavy-duty folders can handle extremely large sheets of paper, Fig. 29-7. Folders usually produce the fold by forcing the paper between metal rollers.

Most machine folding involves sheets printed as signatures. The signature sheets are fed through the folder and come out folded with the pages *(folios)* in the correct numbered sequence. Some of these machines are capable of stitching and trimming all in one operation. See Fig. 29-8.

Kinds of folds

A complete listing of all kinds and uses of folds would fill several chapters in a book. However, this subject can be simplified if it is understood that all possible folds are modifications of TWO basic folds. These *two basic folds* are known as parallel and right-angle folds.

Parallel folds are characterized by all folds being made parallel to each other. In *right-angle folds,* each fold is made at a right angle to each preceding fold. Fig. 29-9 illustrates typical applications of parallel and right-angle folds.

Machine folding

Machine folding is performed on modern high-speed equipment. To fully understand how the folds are ac-

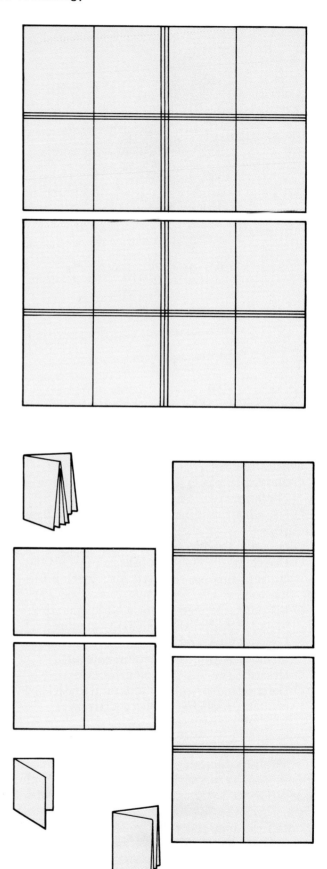

Fig. 29-1. A signature is a large press sheet that is generally printed on two sides and folded to make 4, 8, 16, 32, or 64 pages. These examples illustrate 4-, 8-, and 16-page signatures.

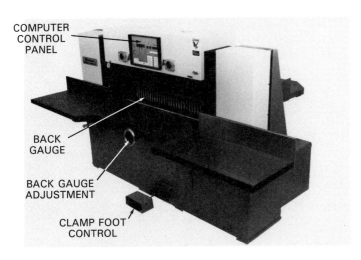

Fig. 29-2. A hydraulic paper cutter with a computer control panel that can be programmed for several successive cuts or trims.

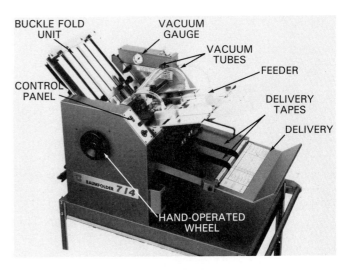

Fig. 29-5. Table-top folders are used for light- to medium-duty folding by in-plant and quick printers. (Baumfolder Corp.)

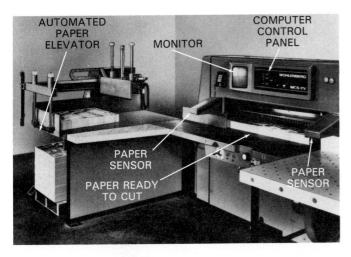

Fig. 29-3. An automated paper cutter with a monitor to evaluate data relating to various cuts and trims on a job. The automated paper lift elevates and delivers a lift of paper for trimming. Low-pressure air is forced upward through holes on paper work surfaces to assist operator in moving heavy lifts of paper. (Bauman, Inc.)

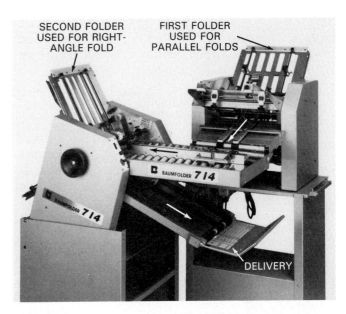

Fig. 29-6. Combination table-top folding unit that allows for parallel and right-angle folds. (Baumfolder Corp.)

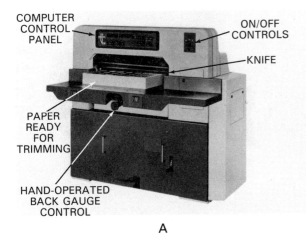

A

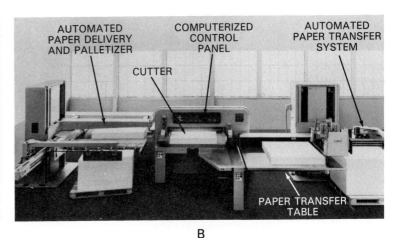

B

Fig. 29-4. A—A standard single-knife computerized paper cutter. B—This automated paper cutting and trimming system uses robotic components to deliver paper to the cutter. Paper travels to an automated palletizer when it has been cut. (Heidelberg U.S.A.)

Fig. 29-7. Commercial printers rely on heavy-duty folding equipment with parallel and right-angle folding capability. (Baumfolder Corp.)

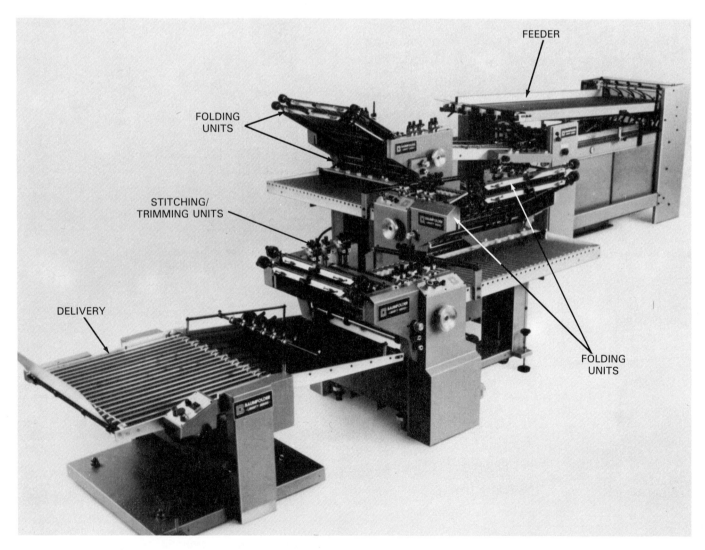

Fig. 29-8. A heavy-duty folder capabale of folding, stitching, and trimming in one operation. (Baumfolder Corp.)

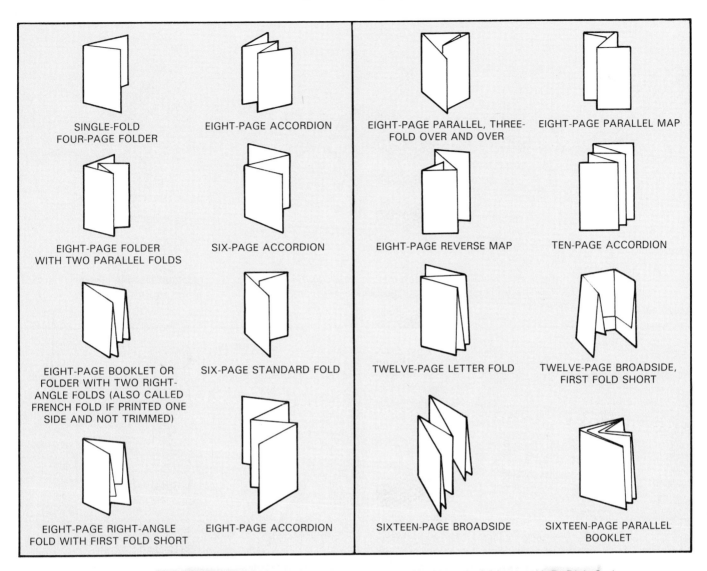

SINGLE-FOLD
FOUR-PAGE FOLDER

EIGHT-PAGE ACCORDION

EIGHT-PAGE FOLDER
WITH TWO PARALLEL FOLDS

SIX-PAGE ACCORDION

EIGHT-PAGE BOOKLET OR
FOLDER WITH TWO RIGHT-
ANGLE FOLDS (ALSO CALLED
FRENCH FOLD IF PRINTED ONE
SIDE AND NOT TRIMMED)

SIX-PAGE STANDARD FOLD

EIGHT-PAGE RIGHT-ANGLE
FOLD WITH FIRST FOLD SHORT

EIGHT-PAGE ACCORDION

EIGHT-PAGE PARALLEL, THREE-
FOLD OVER AND OVER

EIGHT-PAGE PARALLEL MAP

EIGHT-PAGE REVERSE MAP

TEN-PAGE ACCORDION

TWELVE-PAGE LETTER FOLD

TWELVE-PAGE BROADSIDE,
FIRST FOLD SHORT

SIXTEEN-PAGE BROADSIDE

SIXTEEN-PAGE PARALLEL
BOOKLET

Fig. 29-9. Study these typical applications of parallel and right-angle folders. (A.B. Dick Co.)

tually made, Fig. 29-10 shows how knife-type folders operate. Fig. 29-11 shows how buckle-type folders work.

With *knife-type folders* the folds are not actually made by the knife, but by the folding rollers. The *knife* pushes the paper into the rollers, which form the fold by rolling the paper between them. Although knife-type folders are slower operating, they are used almost exclusively for booklet work. This is because they are the most accurate for producing right-angle folds.

The *buckle-type folder* uses a set of rollers which allow the sheet to slide in or out without puckering the sheet. The folding rollers catch the sheet and buckle it in a pre-determined place. The location at which the buckle occurs can be adjusted for each job.

Types of folds

In designing a printed piece, the different types of folds and the limitations of mechanical folding should be carefully considered. The various types of common folds are illustrated in Fig. 29-12.

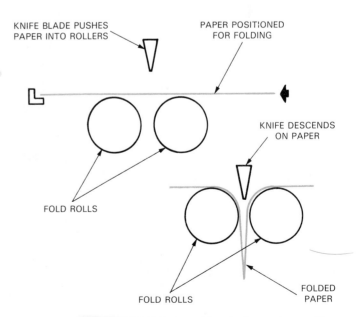

KNIFE BLADE PUSHES
PAPER INTO ROLLERS

PAPER POSITIONED
FOR FOLDING

FOLD ROLLS

KNIFE DESCENDS
ON PAPER

FOLD ROLLS

FOLDED
PAPER

Fig. 29-10. In the knife folding machine design, a descending knife blade creases the paper and forces it between the fold rolls. (John F. Cuneo Co.)

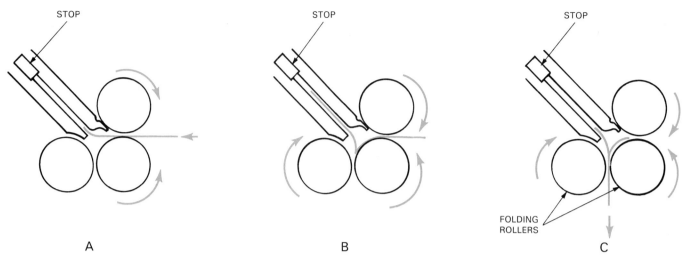

Fig. 29-11. Principle of the buckle folder. A—The sheet is propelled toward the stop. B—The sheet hits the stop and buckles downward. C—The folding rollers catch the buckle and complete the fold. (John F. Cuneo Co.)

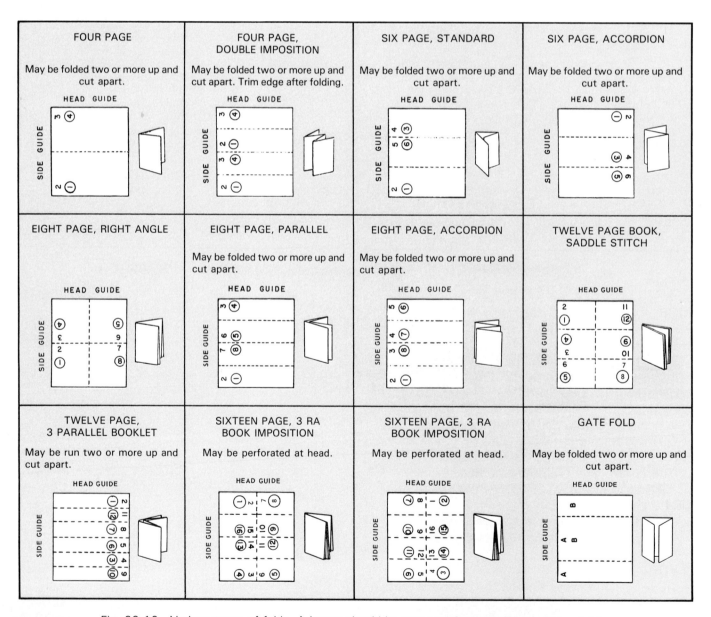

Fig. 29-12. Various types of folds. A layout should be prepared for every job requiring folding.
(U.S. Government Printing Office)

Four-page fold—The four-page fold is the simplest type. It can be used for price lists, programs, bill stuffers, menus, and instruction sheets.

Six-page fold—This type of fold is made with two parallel folds, either regular or accordion. It is used for envelope stuffers, letters, circulars, and brochures.

Eight-page fold—There are three types of eight-page folds. The first makes one parallel and one right-angle fold, also called a *French fold,* when printing is on one side of the paper. The second type makes two parallel folds. The third type makes a three-parallel accordion fold and is used for ease of opening.

Twelve-page fold—There are two ways to make a twelve-page fold. These include: regular and accordion. These are often used as a four-page letter, with the two right-angle folds folding lettersize to fit a mailing envelope.

Sixteen-page fold—A sixteen-page fold can be made with one parallel and two right-angle folds or with three parallel folds. The latter is used for easy-to-open transportation schedules. It can also be bound into a sixteen-page booklet.

Thirty-two-page and sixty-four-page folds—It is possible to make a thirty-two-page or sixty-four-page fold. This is done by simply following the same procedure as that for a sixteen-page fold. In all cases, make a signature *layout dummy* to check for accuracy of imposition.

WARNING! Keep hands and fingers away from all roller nip points on the folder. Serious accidents occur because of extreme folder speed.

PUNCHING

Punching involves cutting rectangular or specially shaped holes in paper to accommodate plastic and spiral bindings. Punching machines are designed for this purpose. Fig. 29-13 shows a punching-inserting machine used to cut rectangular holes in printed sheets and to insert plastic binding in the punched sheets. Automatic punching machines are used for long production runs, Fig. 29-14.

DRILLING, SLOTTING, AND CORNERING

Paper *drilling* makes round holes in paper. It is usually performed on a machine called a *paper drill.* Paper drilling machines vary in size and capacity from the single-drill models to multiple-drill models, Figs. 29-15 and 29-16. Some paper drilling machines can also slot and corner paper.

Slotting involves making holes that are not round. *Cornering* is any operation performed on the corners of paper. Fig. 29-17 shows common drilling, slotting, and cornering operations. Holes, slots, and cornering operations are also performed by punching machines, Fig. 29-18.

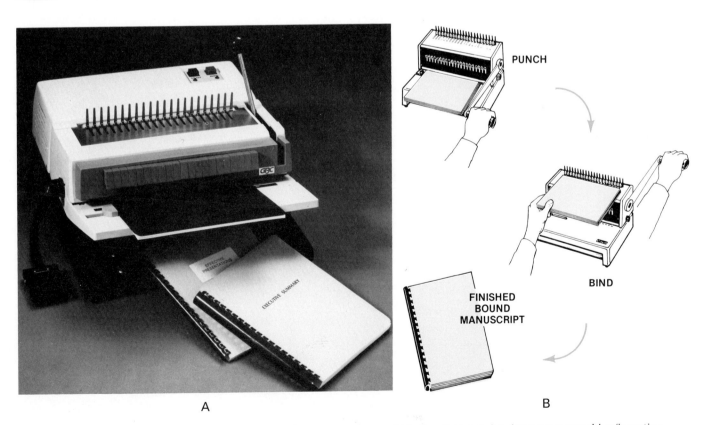

A   B

Fig. 29-13. A—Plastic strip binding is a semipermanent type of binding that can be done on a punching/inserting machine. B—After sheets and cover are gathered, they are punched. Special fingers on the binding machine open the backing strip and the punched sheets are inserted. Upon release, the backing strip closes, completing the binding operation. (General Binding Corp.)

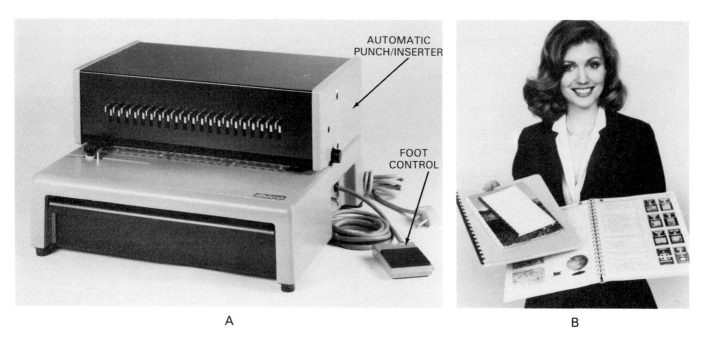

Fig. 29-14. A—Automatic plastic strip binding punching/inserting machines are used for long production runs. B—Plastic strip binding is used for many types of bound printed materials.   (Ibico Inc.)

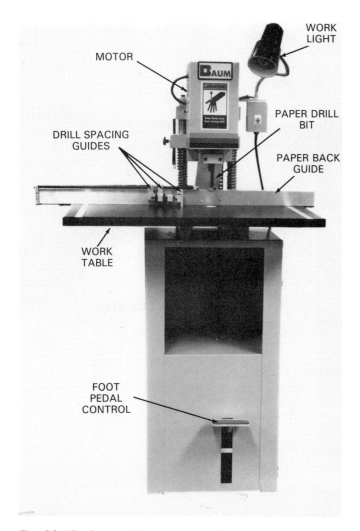

Fig. 29-15. Paper drilling machine. This is a single-spindle foot-operated model.   (Baumfolder Corp.)

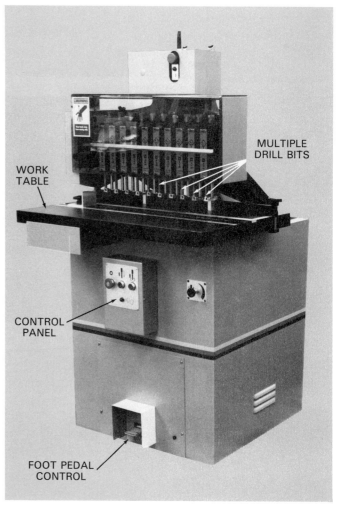

Fig. 29-16. A paper drilling machine equipped with multiple spindles.   (Baumfolder Corp.)

Paper drilling machines use a hollow drill, Fig. 29-19. *Drilling bits* can be fluted on the inside, beveled on the outside, or specially hardened on the tip. *Fluting* compresses the paper and permits it to travel up the barrel of the bit more freely.

Drilling paper

To drill paper on a single-spindled paper drill, follow the steps listed below:

1. Mark the proper location of the holes on the test sheet. Be sure to use an accurate *three-hole master* when attempting to drill three-hole paper.
2. Select the correct drill bit for the job.
3. Insert the drill bit into the machine following equipment manufacturer directions.
4. Align the bit with the hole location marks on the paper. Apply holding pressure with the bit and the paper holddown device to keep the test paper from moving. Place the edge of the paper guide against the test paper, and align the guide with the guide stop. Tighten the guide-stop device.

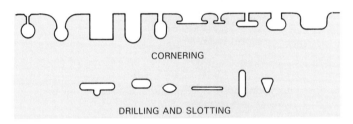

Fig. 29-17. Common examples of cornering, drilling, and slotting for cards and sheets.

Fig. 29-18. This machine performs drilling, slotting, and cornering operations. (General Binding Corp.)

5. Repeat step 4 for every hole location desired.
6. Set the back gauge for depth. The *back gauge* locates the distance the hole is drilled from the edge of the paper. The gauge markings are the distance from the center of the hole to the back gauge.
7. Adjust bit pressure. Use only enough pressure to drill cleanly through all paper while the foot pedal is at the lowest comfortable position.
8. Change the cutting block if it is worn.
9. Adjust the holddown device (clamp) to touch the top of the paper. The clamp holds the paper in place while the paper is being drilled.
10. To drill the paper, insert the pile of paper securely against both guides. Lower the rotating bit with the foot pedal. Use a small amount of stick lubricant on the bit at regular intervals.

Round cornering paper

Round corners can be made with a small hand-operated machine, Fig. 29-20. Round cornering can also be done on a paper drilling machine set up to corner paper. The paper is cornered on a paper drilling machine with a device that replaces the drill bit and paper holddown clamp. Two corner guide rails are added to the drill table to guide the paper accurately under the cornering device.

To corner paper, the operator sets up the machine, makes all final adjustments, and makes the corners.

Fig. 29-19. Paper drill bits are hollow, allowing waste pieces to be carried up through the drill, out a chute, and into a waste bin. (Challenge Machinery Co.)

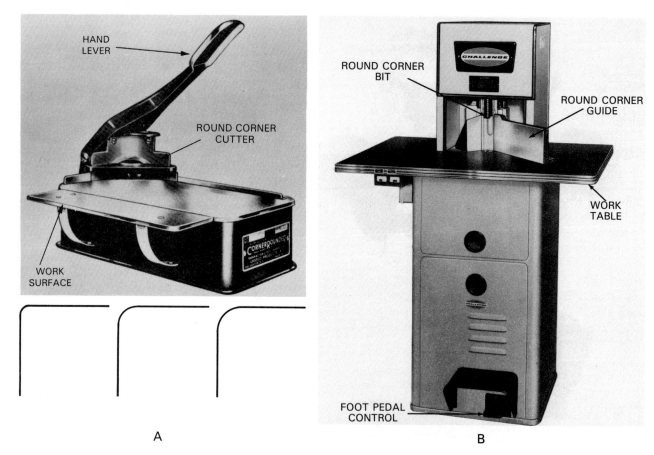

Fig. 29-20. A—Round corners can be made with a small, hand-operated, round cornering machine. B—Power drill can be converted to a round cornering machine. (Challenge Machinery Co.)

This is done by lowering the cornering blade through the pile of paper. Small piles should be cornered to avoid inaccuracy of the cut and premature dulling of the blade.

DANGER! Keep hands and fingers clear of the paper drill bit and paper holddown when operating this equipment.

GATHERING

Assembling individual sheets of paper or signatures in the correct sequence is called *gathering*. This operation is done by placing piles of paper or signatures in the correct order along the edge of a work surface. One sheet or signature at a time is picked up from each pile and assembled in proper sequence.

Automatic gathering machines are used for high-volume production, Fig. 29-21. Desk-top semi-automatic machines are used whenever a small volume of work is required.

COLLATING

The term collating is often confused with the term gathering, Fig. 29-22. *Collating* is an operation performed for the purpose of checking to see that the cor-

rect number of signatures has been gathered. *Collating codes* are printed in different positions on the binding fold of signatures, Fig. 29-23. After the magazine or book has been folded and gathered, a pattern is visible on the spine. Errors are quickly seen and easily corrected. Much of the work of collating is done on machines that gather flat sheets, stitch them, and then fold them into signatures. Refer to Fig. 29-24.

Hand-assembling booklets

The following steps can be used to *assemble* (gather and collate) a simple booklet consisting of a cover and a single signature. A more efficient procedure would involve a collating machine that would assemble and bind the cover and signature in a single run through the machine.

1. Lay the covers and signatures on a table with the signatures to the right.
2. Assemble the signatures into the covers by using the lip to flip the cover open. You can use a small amount of *glycerin* to help grasp the sheets. Insert the signatures into the cover from right to left.
3. For ease of operation, jog the booklets to the head in groups of approximately 10-15. The *head* is generally the guide edge on book work.
4. Stack the completed booklets in preparation for binding.

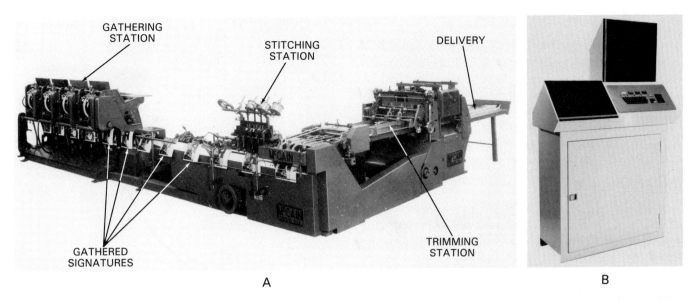

Fig. 29-21. A—High-speed binding machine equipped with stitcher and trimmer. B—This automated signature counting machine can be connected to binding unit. (McCain Manufacturing Corp.; Automation, Inc.)

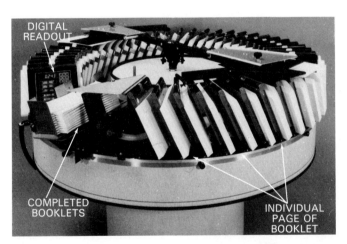

Fig. 29-22. A computerized, high-speed rotating collating machine with digital readout. (Challenge Machinery Co.)

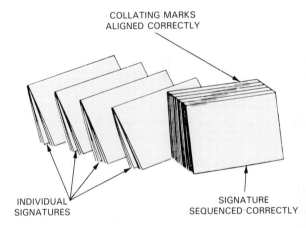

Fig. 29-23. Collating marks are printed on fold edges of signatures so a visible check of sequence can be made. Marks must align correctly to signify correct order of signatures.

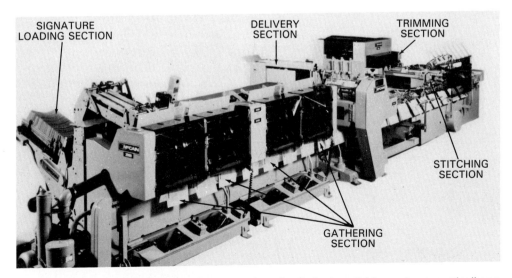

Fig. 29-24. These magazines are being gathered, stitched, and trimmed automatically on high-speed production equipment. (McCain Manufacturing Corp.)

JOGGING

An operation known as *jogging* involves the straightening of sheets of paper. Sheets must be jogged when loaded into a paper cutter, press, folder, or gathering machine. An automatic jogging machine is usually equipped with a two-sided slanting table, Fig. 29-25. The table vibrates and straightens the pile of sheets in a matter of seconds. Many different models and sizes of joggers are available, Fig. 29-26.

SCORING

Placing a crease in a sheet of thick paper or cardboard to aid in folding is referred to as *scoring*. The crease produces an embossed or raised ridge on the sheet. The fold is made with the ridge on the inside of the sheet to prevent stretching.

The width of the crease is varied according to the thickness of the paper being scored, Fig. 29-27. A thicker paper requires a wider crease, which gives a bigger groove to help make a cleaner fold. Scoring is usually done on automatic equipment. In some instances, scoring is done on letterpress equipment.

Several methods of scoring are used; these include:

1. Perforating wheels and scoring wheels on folding machines.
2. Rule and string.
3. Rule and wire.
4. Single and double-rule scoring on scoring and creasing presses or on cylinder or platen letterpress presses.

When using string, wire, or single and double-rule, the objective is to create a male-female die that will properly stretch the paper without cutting or otherwise damaging it, Fig. 29-28. The following points should be observed when scoring paper.

1. The distance between the two parts of the double die should be equal to the width of the scoring string, wire, or rule, plus twice the thickness of the paper.
2. The thicker the paper, the thicker the scoring element should be to prepare for a good fold.
3. Scoring wheels are usually NOT as effective as the above methods. To be as effective as possible, scoring wheels should ride over a notched roller into which they press the paper fibers.
4. Scoring wheels must be used with extreme care on coated papers or they will mar the surface and contribute to a poor-quality fold.

PERFORATING

Some printed products require that a part of it be removable, such as a ticket stub or receipt book. The operation of making a small series of holes for this purpose is referred to as *perforating,* Fig. 29-29. In this process, a series of very short slits or holes are cut in the paper leaving only a small bridge of paper remaining intact. The perforated sheet is then easily pulled or torn apart when required.

Perforating can be done on both offset and letterpress presses. Perforating is done on an offset press by means of a metal perforating strip attached to one of the cylinders.

WARNING! Care should be taken when using perforating material. The sharp cutting teeth can easily cut fingers and hands when being mounted on a press.

SLITTING

Slitting is the process of separating material by a rotary shearing action, Fig. 29-30. Printed materials can be slit to a specific size during folding or on special press setups. It is used to separate two or more products from a large sheet or web. It is also used in place

Fig. 29-25. Automatic jogging machines are used when sheets must be jogged before loading into a paper cutter, press, folder, or gathering machine.   (Autobond Ltd.)

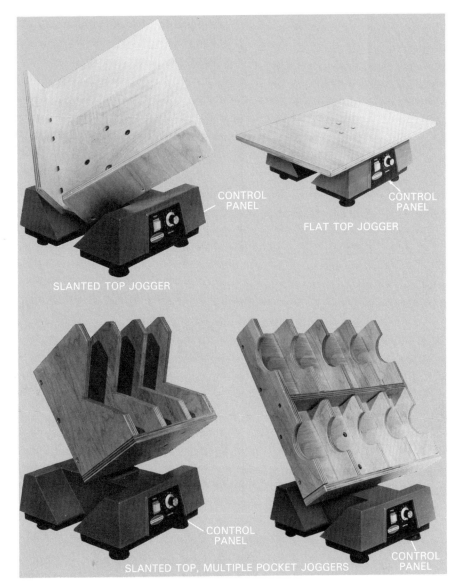

Fig. 29-26. Paper jogging machines are available in many models and sizes. (Challenge Machinery Co.)

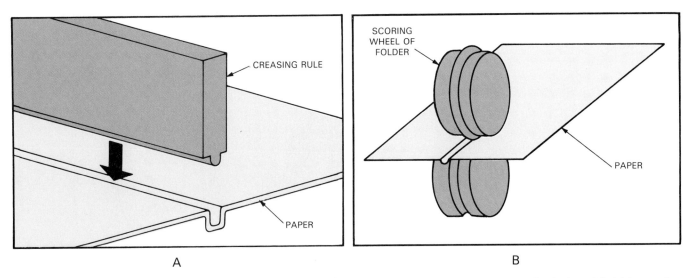

Fig. 29-27. A—Scoring or creasing can be done by steel rule using letterpress equipment or specially designed finishing equipment. B—Scoring can be done on a folder using scoring wheel. (A.B. Dick Co.)

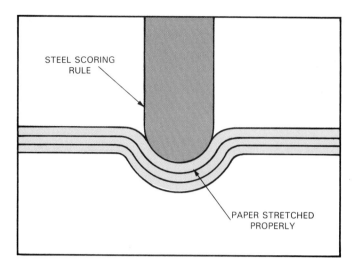

Fig. 29-28. The paper must be properly stretched without cutting or otherwise damaging it when scoring or creasing. (*Step-By-Step* magazine)

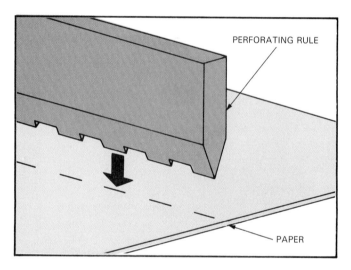

Fig. 29-29. Perforating operations place a series of small cuts or slits in the paper. A steel perforating rule is commonly used for this purpose. (A.B. Dick Co.)

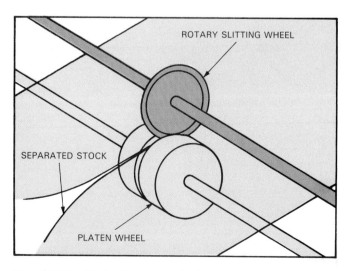

Fig. 29-30. Slitting is commonly done by a sharp wheel that cuts paper as it passes against wheel.

of cutting when it is more economical to integrate slitting as other operations are being performed.

DIE CUTTING

Any irregular shape or design, that cannot be cut with a straight cut, is done by a process known as *die cutting*. Since there is no standard design used in this operation, each metal die is custom-made to match the requirements of the job. See Fig. 29-31.

The *die* consists of a metal cutting edge that matches the outline shape of the design required. The die is mounted on a wooden dieboard with the cutting edge up, similar to a cookie cutter. Sponge rubber is glued near the cutting rule to assist in stripping the paper from the die on the press impression.

Die cutting presses are similar to letterpress printing presses. In some instances, traditional letterpress printing presses are used for die cutting. Examples of die cut products include boxes, cartons, and point-of-purchase containers. Fig. 29-32 shows a die cutting machine.

HOT STAMPING foil stamping

Gold, silver, or colored images of type or illustrations are commonly placed on covers of books, business letterheads, menus, business cards, and certificates. This process is known as *hot stamping*. It is a letterpress process that uses relief images and heat. Gold,

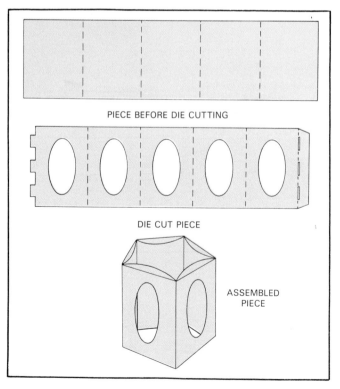

Fig. 29-31. Irregular shapes can be die cut. This is an example of a point-of-purchase display die. (Hammermill Paper Co.)

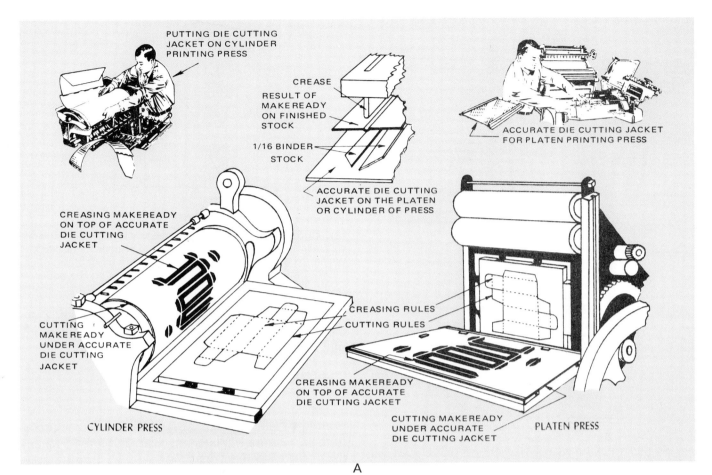

A

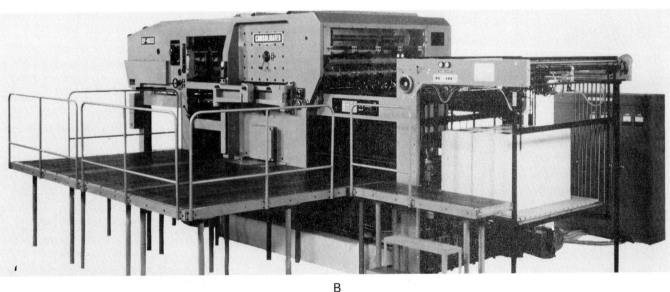

B

Fig. 29-32. A—Study these principles of letterpress die cutting. Each operation must be precise so that finished piece can be assembled to specifications. B—This combination die cutter, creaser and embosser machine will deliver 5,000 sheets per hour. (Accurate Steel Rule Die Manufacturing, Inc.; Consolidated Machinery Co.)

silver, or colored foil is positioned between the type and/or illustration and the surface to be stamped. Through the use of pressure and heat a permanent image is formed.

Hand-operated hot stamping machines for a small volume of work are available, Fig. 29-33. Letterpress printing presses are generally used for high volume operations, especially where large sheets of paper are required.

DANGER! Caution should be exercised when working around hot stamping equipment since pressure and extreme heat are used in the process.

657

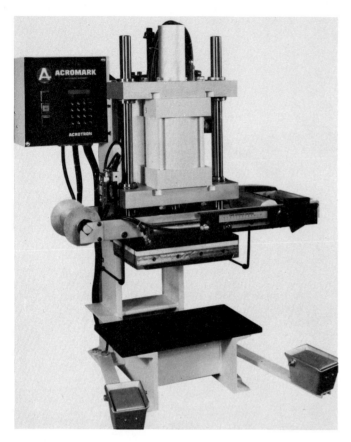

Fig. 29-33. This hand-operated hot stamping machine is used only for a small volume of work. (Acromark Co., Inc.)

EMBOSSING

Embossing creates a three-dimensional (raised) image on a printed product. The two types of embossings are regular and blind.

When the embossing is made over a printed image, it is known as a *regular embossing*. When the embossing is produced without any printing, it is known as *blind embossing*.

Embossings are made from a male and a female die, Fig. 29-34. The design is first cut or etched into the surface of a piece of brass, making the *female die*. The die is positioned and placed in a letterpress printing press. The *male die* is formed on the press using the female die to impress a paste-like substance.

VARNISHING

Varnishing is a process which places a coating or surface finish that imparts resistance to chemicals, heat, scuffing, water, or other elements that would damage paper. It also provides smoothness and luster, and thereby enhances color values. The surface becomes very smooth and shiny.

Overprint varnish can be placed on the paper using an offset press. If the whole area is to be coated, the dampener rollers can be removed and the varnish coating on the offset plate is transferred to the paper.

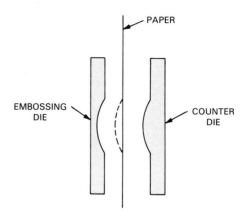

Fig. 29-34. Embossing requires the use of two dies; one die is in relief while the other is recessed. The paper is placed between the two dies. When the two dies are brought together, the clamping force creates a raised image on the paper.

Various kinds of coatings are available including print varnish, spirit varnish, Copal varnish, shellac solutions, and naphtha varnish.

THERMOGRAPHY

Thermography is a process that produces raised printing. The word "thermography" is a combination of the words *thermo* meaning heat, and *graphy* meaning to write. So the word thermography means literally, HEAT WRITING. It is a process which is widely used by printers to add an attractive quality to various kinds of printed materials. Some of these items include letterheads, business envelopes, and business cards.

The process known as thermography results from sprinkling fine resin powder over a freshly printed sheet. Any excess powder is removed from the non-inked areas and the powder that adheres to the wet ink is heated. The heat melts the powder, and, when cooled, a raised effect results.

Illustrations, halftones, and type are suitable for thermography. With care and the use of the correct ink and powder, finely detailed images are possible. Type less than six points in size should be avoided since it may fill in. Letters with delicate serifs should also be avoided since they tend to fill in with powder after heating.

Normally, *neutral* (clear) *powders* are used to permit the color of the base ink to show through. *Opaque colors* have complete hiding power. Opaque colors are available in green, yellow, orange, pink, red, and white. Fluorescent and metallic colors are also available. Powders are available from coarse to very fine. The choice is determined by the image. Images with fine lines use a *fine powder*. Images with heavy lines and elements use a *coarse powder*.

Almost any kind of ink can be used for thermography. This is because ink serves only as an adhesive for the powder. Special inks with little or no drier in them are used for quality work.

There are several models of commercial thermographic equipment. These are generally automatic machines for high-speed operations. For smaller applications, hand-fed thermography machines are available. See Fig. 29-35.

PACKAGING

The printer is responsible for packaging and delivery of the printed product. The main purpose of *packaging* is protection of the contents, Fig. 29-36. Special cartons are used for some printed materials. The product is usually banded or wrapped first and then placed in cartons. Bands are placed around the cartons. This makes them stronger and easier to handle.

Wrapping, banding, and boxing operations are usually done on automatic equipment in large firms. Small shops generally rely on hand-mechanical operations or semi-automatic types of equipment. Refer to Fig. 29-37.

Many printed products are tied with string. String is strong and not costly. Printed sheets and publications can be tied by hand. Tying equipment speeds the process. Fig. 29-38 shows a newspaper strapping machine.

Banding is a common method of holding printed materials together. Printed products are often placed in corrugated cartons. Bands are placed around the cartons to make them stronger and easier to handle. Automatic equipment is not always required to wrap, band, and box printed materials.

Shrink wrap

Shrink wrap is a transparent plastic film used to cover the printed parcels or products. It is one of the most efficient packaging methods used in the finishing

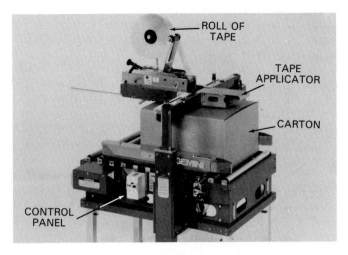

Fig. 29-36. After printed materials are packaged, an automatic taping machine seals the container flaps and readies the carton for shipment.   (Signode Corporation)

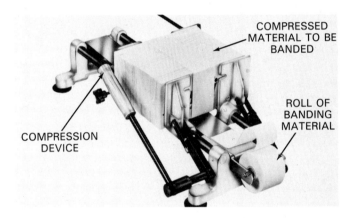

Fig. 29-37. Banding is a common method of holding printed materials together for delivery. This machine compresses the material into a packet while the operator bands it.
(Martin Yale Co.)

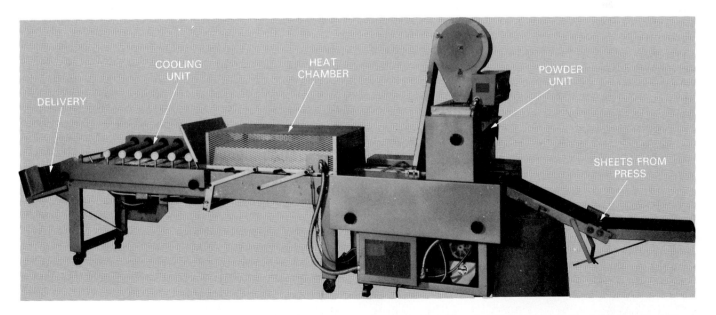

Fig. 29-35. Thermography results from sprinkling fine resin powder over a freshly printed sheet, such as a business card. Heat is applied to the inked image area. The heat melts the powder, and when cooled, a raised effect results.   (Virkotype Corp.)

department. The process makes the finished products easy to handle, easy to see, and attractive to the customer. See Fig. 29-39.

Special transparent plastic film is used in the shrink wrap packaging method. Heat is applied to the plastic material to cause it to shrink tightly around the printed material. Larger equipment can shrink-wrap packages approximately 12 inches wide, 18 inches long, and 8 inches high, Fig. 29-40. Individual packages are placed in cartons for shipment to the customer.

WARNING! Use caution around shrink wrap equipment since it utilizes extremely high temperatures to shrink and bond the plastic material to packages.

BINDING OPERATIONS

When a printed job is a book, magazine, or pamphlet, pages or signatures must be fastened or *bound* together. Binding serves two purposes. The binding holds everything together by fastening one element to the next. Secondly, the binding helps protect the printed material inside.

Common binding methods include:
1. Mechanical.
2. Loose-leaf.
3. Wire.
4. Sewn soft-cover.
5. Sewn case-bound.
6. Perfect.
7. Thermal.
8. Tipping.
9. Padding.

Mechanical binding

The mechanical binding method is very popular for fastening books, catalogs, manuals, and other similar volumes. *Mechanical binding* involves punching holes in the paper so that metal or plastic wire or strips can be threaded through the holes. Two common methods of mechanical binding include spiral and plastic. Mechanical binding allows pages to lie perfectly FLAT. However, adequate margin allowance must be provided for the holes. See Figs. 29-41 and 29-42.

Loose-leaf binding

Pages bound together with removable rings or posts are classified as a *loose-leaf binding*. Ring binding shown in Fig. 29-43 allows pages to be opened flat and removed or added as needed.

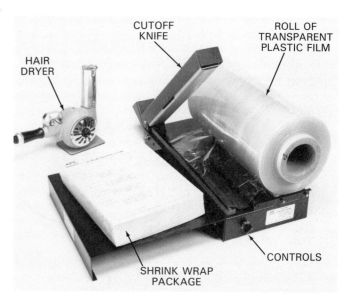

Fig. 29-39. This simple transparent plastic film shrink-wrapping unit uses a hair dryer to apply heat to the film material. (Autovend Packaging Systems)

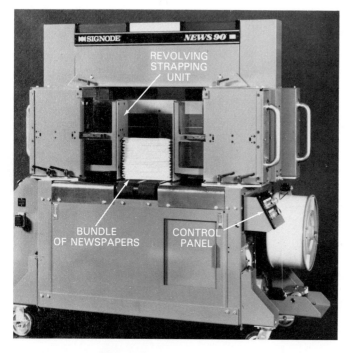

Fig. 29-38. Tying or strapping machines like this are used for newspaper operations. (Signode Corporation)

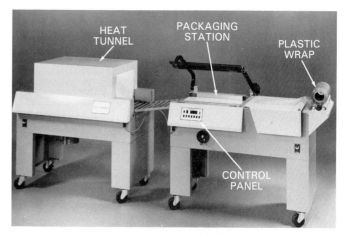

Fig. 29-40. This unit can shrink-wrap packages approximately 12 inches wide, 18 inches long, and 8 inches high. (X-Rite Co.)

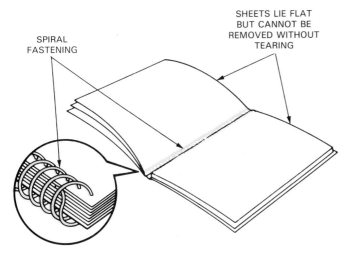

Fig. 29-41. Spiral fastening uses metal wire wound into a spring shape to form loose-leaf binding. Sheets lie flat but cannot be removed or installed from binding. (A.B. Dick Co.)

The binding post (screw post) type of loose-leaf binding does NOT lie flat when opened. Since holes must be punched for either the ring binding or binding post methods, correct clearance dimensions along the binding edges are essential.

Wire stitching binding

The *wire stitching binding* method of fastening sheets together is divided into saddle-wire and side-wire stitching.

The *saddle-wire method* consists of wires or staples that are inserted on the fold line, or saddle, of the pages. This method allows the booklet to be opened fully, but is not practical for binding thick volumes because it lacks strength. Normal margin allowances must be considered when using this method. Refer to Fig. 29-44.

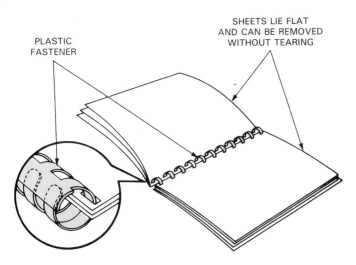

Fig. 29-42. Plastic bound book uses pronged-plastic device that inserts through holes in paper. This allows the book to lie flat and can be taken apart to add or remove pages. (A.B. Dick Co.)

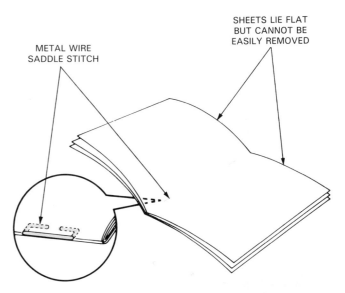

Fig. 29-44. Saddle-wire method of binding allows the booklet to be opened fully. This method is not practical for binding thick volumes because it lacks strength. (A.B. Dick Co.)

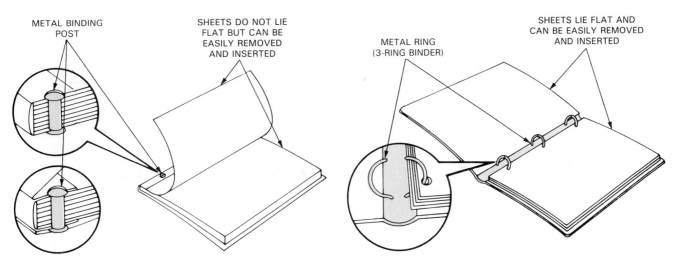

Fig. 29-43. Loose-leaf binding includes the use of binding posts (screw posts) or ring binding. Both methods allow pages to be inserted and removed. (A.B. Dick Co.)

WARNING! When operating stitching equipment, keep fingers clear of the stitching head(s). Staples are capable of being driven through material one inch thick on some types of equipment, and would easily penetrate your hands.

In the *side-wire method,* staples are inserted close to the fold and clinched at the back, Fig. 29-45. Pages of a side-wire stitched book cannot be opened flat. Extra margin allowance must be provided for the staples.

Fig. 29-46 shows a single-head stitcher. A multi-head stitcher is shown in Fig. 29-47.

Sewn soft-cover binding

With *sewn soft-cover binding,* strong thread is sewn through the binding edge of each sheet and securely holds all sheets in place between the covers of a book. Sewn soft-cover binding can withstand hard use for a limited time. Technical information is often bound in this manner because the book usually passes through many hands during its brief but valuable life. See Fig. 29-48.

Sewn case-bound binding

Sewn case-bound binding includes hard cover books assembled in signatures and then sewn together with strong thread, Fig. 29-49. Books with hard covers are sewn together when extremely hard use and long wear are required. The thickness of the signatures forms a wide spine to which the cover is attached. Sewn case-bound bindings are the most expensive method of binding, but provide the best durability. This book was manufactured with the sewn case-bound method.

Perfect binding

With *perfect binding,* the sheets of paper are held together with a flexible cement. Pocket-size books, telephone directories, magazines, and mail order catalogs are examples of perfect binding, Fig. 29-50.

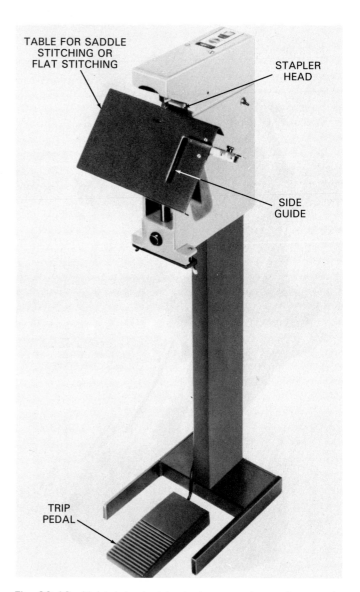

Fig. 29-46. Light-duty electric stapler uses wire staples to saddle or side stitch. (Michael Business Machines)

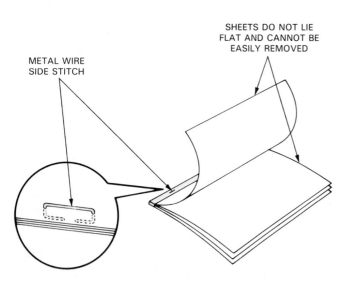

Fig. 29-45. Side-wire method of binding does not allow booklet to be opened fully. Extra margin allowance must be provided for the staples. (A.B. Dick Co.)

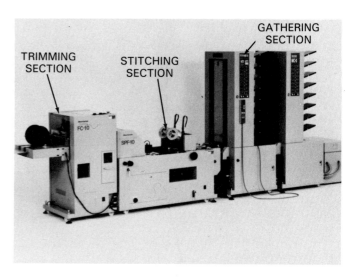

Fig. 29-47. This machine system gathers, stitches, and trims saddle-stitched pamphlets, magazines, and books. (Horizon Packaging Corp.)

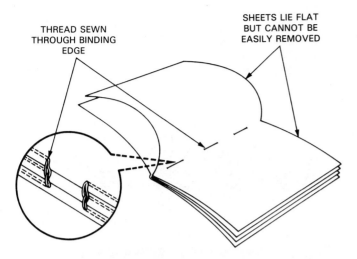

Fig. 28-48. Sewn soft-cover binding can withstand hard use for a limited time. This binding method allows pages to lie flat. (A.B. Dick Co.)

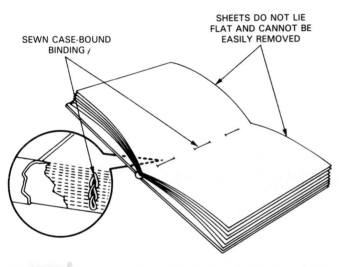

Fig. 29-49. Sewn case-bound binding is used for books with hard covers such as the textbook you are now reading. (A.B. Dick Co.)

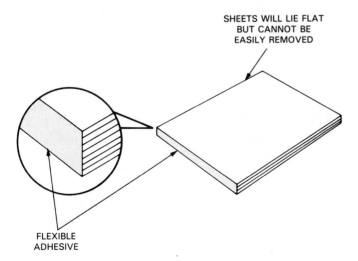

Fig. 29-50. Perfect binding holds sheets of paper together with a flexible adhesive. Pocket-size books are almost exclusively bound in this manner. (A.B. Dick Co.)

Books bound in this way lie flat when open. Perfect binding is fast, economical, and results in an attractive printed product. High-speed fully automatic equipment is used for perfect binding, Fig. 29-51.

Thermal binding

The *thermal binding* process is similar to perfect binding because it uses heat and a special adhesive applied to the edges of the sheets. Books or booklets made of individually-cut pages and covers are fed spine down through a thermal binding machine, Fig. 29-52. A strip of heated adhesive material is applied to the spine in one operation. A wrap-around cover can also be applied to the spine if desired.

In operation, pages are jogged and inserted into the cover. The book is placed in the thermal binding machine. The machine applies heat to activate the binding adhesive. Once the adhesive has been applied, a machine signal lets the operator know when to remove the finished book.

A

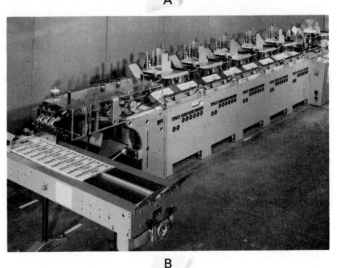

B

Fig. 29-51. A—This high-speed perfect binding machine can produce up to 12,000 books per hour. B—This perfect binding machine gathers single signature sheets, roughens the spine of the book body, applies flexible adhesive, adds the cover, and then clamps and trims three sides. (Sulby; Muller-Martini Corp.)

Fig. 29-52. The thermal binding process is similar to perfect binding. A strip of heated adhesive material is applied to the spine of the book in one operation. (General Binding Corp.)

Tipping

Tipping involves inserting a separate piece of printed material into the original pages of a printed product. An example of tipping-in would be pasting an advertising insert to a magazine signature with a thin strip of adhesive. Refer to Fig. 29-53.

In tipping, liquid adhesive is usually applied in a series of spots or a strip along one edge of the insert. The adhesive may be applied by machine or by hand. Once the adhesive has been applied, the insert is adhered to the magazine, book, or booklet.

Inserts are tipped either outside or inside a signature. Outside tipping can be done MECHANICALLY. Inside tipping, or tipping-in, must be done by HAND.

Tipping is generally performed before the gathering operation. Some tipping machines can handle tipping while gathering.

Padding

When tablets and notebooks are being assembled, padding is used to hold the sheets together. *Padding* is a form of perfect binding, Fig. 29-54. Usually, a piece of chipboard is included as a backing for each individual pad. A flexible cement coating is applied to one edge of the pile of sheets and backings.

The padding cement is a water soluble liquid with a thick, creamy appearance. It is available in clear or red colors to suit the requirements of the job. The cement is applied in two coats to the edges of the paper with an ordinary paint brush. When dry, the pads are cut apart, then trimmed to final size on a paper cutter. High-speed padding equipment is available for large bindery operations, Fig. 29-55.

The steps for padding are listed below:

1. Depending on what type of padding device is being used, prepare it for loading. If a device is not available, any flat surface, such as a counter top, can be used. In this method, weights must be placed on top of the paper to be padded.
2. Load the jogged paper into the guide edge of the padding device. Continue to load until the device is full. All sheets must be aligned perfectly whether using a padding device or a table top, Fig. 29-56.
3. Apply holding pressure to the paper pile. Remove or open the guide edge of the padding device.

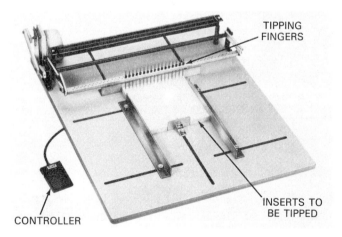

Fig. 29-53. Tipping involves inserting a separate piece of printed material into the original pages of a printed product. This is done by applying a thin strip of paste to hold the piece inside the original pages. (Martin Yale Co.)

Fig. 29-54. Paper to be padded is jogged and stacked in a padding frame. Pressure is applied to the pile and a flexible cement coating is applied to one edge of the pile of sheets. (Challenge Machinery Co.)

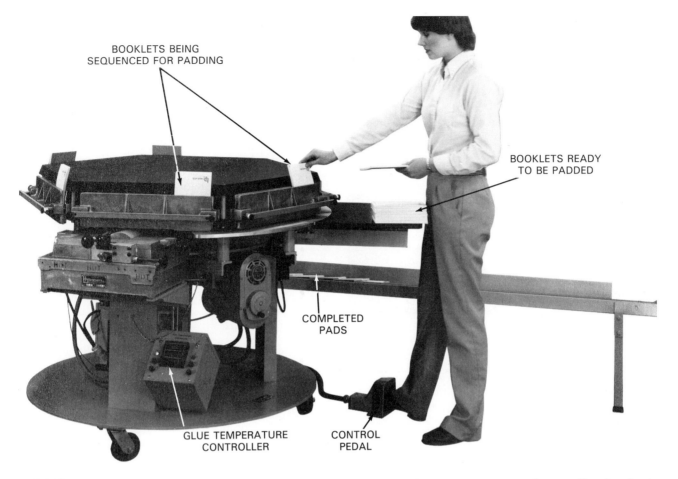

BOOKLETS BEING
SEQUENCED FOR PADDING

BOOKLETS READY
TO BE PADDED

COMPLETED
PADS

GLUE TEMPERATURE
CONTROLLER

CONTROL
PEDAL

Fig. 29-55. A multi-unit power padding press that can produce 1,400 single paper pads per hour.   (Brackett Inc.)

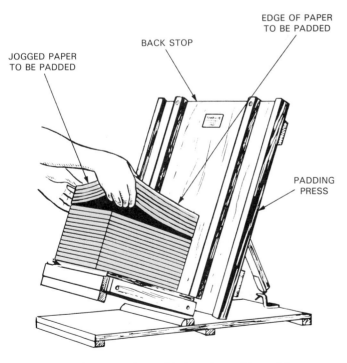

JOGGED PAPER
TO BE PADDED

BACK STOP

EDGE OF PAPER
TO BE PADDED

PADDING
PRESS

Fig. 29-56. Sheets to be padded are jogged in padding press against the back stop. Pressure is applied to leading edge of paper pile with clamp. The pile is then positioned with binding edge out. Adhesive is applied to the binding edge with a brush.   (Champion Manufacturing Company, Inc.)

4. With a brush, apply the padding compound evenly across the entire area to be padded, Fig. 29-57. Let the compound dry thoroughly for about 30 minutes. Apply a second coat of padding compound.
5. Remove the padded material from the padding device. Use a padding knife to separate each pad from the pile.
6. Use a paper cutter to trim the pads as necessary.

Padding carbonless papers

Carbonless papers have replaced many carbon interleaved forms. This is because they simplify paper-work, produce clean, smudge-free copies, and eliminate the mess associated with the use and disposal of carbon paper.

Carbonless papers are technically different from regular papers. They incorporate a chemical transfer system and use a reaction between two different chemical coatings to transfer images. The back side of the top sheet in the carbonless set is designated CB (coated back). The front and back sides of intermediate sheets have a receptor coating and an encapsulated coating respectively. These sheets are designated CFB (coated front and back). The last sheet of the set has a receptor coating only. This sheet is designated CF (coated front). A typical

3-part carbonless form is illustrated in Fig. 29-58.

When pressure is applied to the top sheet of the set by typing, writing, or crash printing, an image is formed. This occurs because of the reaction between the chemicals liberated from the collapsed capsules and the contacting receptor coating.

Special precautions must be taken not to damage the encapsulated chemicals by excessive pressure and friction during handling, printing, cutting, and trimming.

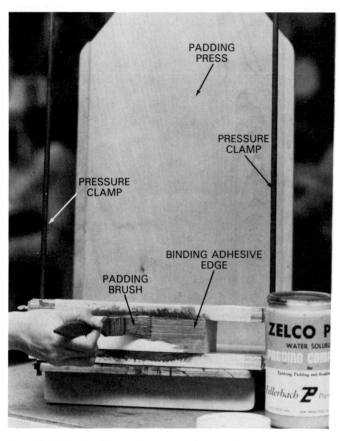

Fig. 29-57. Applying adhesive to the binding edge of several book piles.

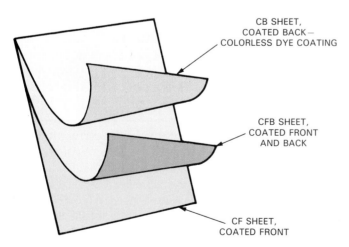

Fig. 29-58. Study the parts of a typical carbonless 3-part form. Sheets are designated as CB (coated back), CFB (coated front and back), and CF (coated front).

The proper side of each paper used in the set must be printed so that the sheets are collated in their required sequence. Manufacturers of carbonless papers provide specific instructions for the printing and processing of their carbonless papers.

The steps for carbonless padding are listed below:

1. Make sure the carbonless sheets are in the proper sequence (top ply CB, middle plies CFB, bottom ply CF). CF dry gum, blotter, or C2S ledger will NOT edge-pad satisfactorily. Do not pad edges with full ink coverage; ink inhibits (retards) proper penetration of fan-apart adhesive.
2. Jog the sheets first to obtain a smooth padding surface.
3. Stack the sheets no higher than about 18 inches in the padding device or on the edge of a level table surface.
4. The padding edge should be no wider than 17 inches.
5. Trim if necessary. The surface to be padded must be clean and free of paper lint and dust.
6. Use only one pound of weight per inch of edge. Use just enough so the stack will not shift. Weight should be even with the padding edge. If using CF tag or ledger in the sets, move the weight back one-quarter inch (6.4 mm) so sheets will separate.
7. Jog, if required, to make the padding surface perfectly smooth.
8. Apply the adhesive with a clean, dry bristle brush. Use plenty of adhesive and keep the brush saturated. Start at the center and brush toward the edges of the pile. Repeat until the entire pile is evenly coated. Three coats should be enough and do not wait too long between coats.
9. Rinse the brush in water. Do not use this brush with any other type of padding compound. Make sure you close the adhesive container properly so that it will not cure while in storage.
10. Allow the pile to dry thoroughly. Drying time is longer with tag or ledger paper or if forms are to be crash imprinted.
11. Fan the forms apart. Run your thumb upward from bottom of the pile and along the edges of the padded surface near the corners. Form sets will then separate from each other easily.

BINDERY PLANNING

Every job has specific requirements that must be performed. For example, with sheet-fed offset work, the first finishing operation is the placement and relationship of the press, folder, and cutter guides. Consideration is given to the placement of the press side guide and press gripper with respect to the edge of the press sheet. Proper planning will determine which sheet edge should be the press side guide edge and which sheet edge should be the press gripper edge. This will allow for the folding and cutting edges to be located.

Planning guide edges

When the press sheet must be folded or cut, the press guide edges are selected, and the bindery work is performed using these same guide edges. This ensures registration of the finished job. The printing, folding, or cutting distance from the guide edge will not vary if the same edge of the paper is used as the guide for both printing and bindery operations.

Often, when work requires folding, it is laid out with the press side guide edge corresponding to the folder lead edge. In this case, the press gripper edge corresponds to the folder guide side edge. This allows the guide corners of the press sheet and folder to correspond. If this procedure is followed, the fold and printing distances from the guide edges will be the same.

Note! The *press sheet's guide corner* is the intersection of the side guide and gripper edges. The *folder's guide corner* is the intersection of the edges of the guide side and lead edges.

If the guide edge is damaged and cannot be used, a new guide can be made. When a sheet is cut or fold-ed using the original guide edge, a new guide edge is created. The new guide edge is the edge on either side of the cut or fold. These new guides are used to make additional folds or cuts.

The relationship between the press and bindery edges is illustrated in Fig. 29-59. It should be noted that the guide corner of the press sheet corresponds to the guide corner for the folder.

Planning signature lip

It is necessary to consider the need for a signature assembling extender. The extender is known as a *lip* or *lap*. It is an extension of one side of a signature, Fig. 29-60. The lip is used to help open signatures in the center for assembling or saddle stitching. A lip is generally NOT required for side stitching.

A one-half inch (12.7 mm) lip is generally used for automatic assembly and binding machines. A one-quarter inch (6.4 mm) lip is used for hand assembling operations. Fig. 29-61 shows how automatic assembly and stitching machines use the lip to open signatures.

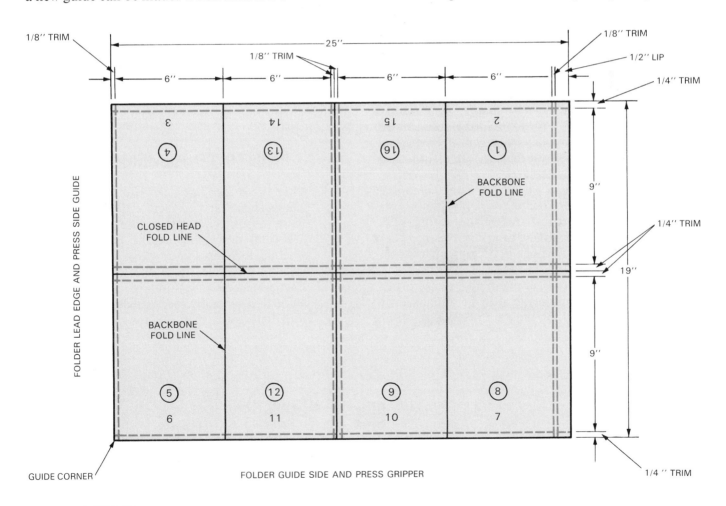

DASH LINE IS TRIM LINE.
SOLID LINE IS FOLD LINE.
THE NUMBER IN THE CIRCLES IS THE NUMBER OF
THE PAGE ON THE BACK OF THE SHEET.

Fig. 29-59. This layout illustrates the various elements of the press sheet as it relates to binding operations including trimming and folding.

667

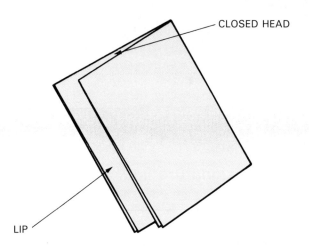

Fig. 29-60. A lip or lap is an extension of one side of a signature and is used to help open signatures in the center for assembling or saddle stitching.

Some binding machines, that use suction devices, may NOT require a lip.

When all elements of a booklet are assembled for hand-fed stitching, the lip on the center signature should extend beyond all other signature lips. This is done to ensure that the assembled booklet will open properly at the center.

Planning a bindery dummy

As you learned in a previous chapter, a *"dummy"* refers to a paper model of the job folded and assembled in the same way the finished product will be folded and assembled. The dummy shows the size, shape, form, and general plan of the printed piece.

Preparing a dummy is a very important planning step. The dummy is a production work plan that shows how a job should be processed. The dummy helps the planner visualize the complete job. It shows the order of assembly and the relationship of folds, trims, imposition, bleeds, and guides needed for the job.

A dummy requires careful planning. All the printing and finishing requirements must be visualized by the planner before making the dummy. Factors to be considered include: the number and size of pages, number of signatures, number and type of inserts, and requirements of the cover. Additionally, the basic imposition, folding requirements, assembly, and binding plan must be considered.

The preparation of a dummy was explained in detail in Chapter 7.

Planning is necessary at every stage in the printing process. Accurate and complete instructions on every job must be carried out and adhered to before going to the bindery. The bindery department should be given all essential information to handle a job efficiently.

A simple set of guidelines for the bindery is shown in Fig. 29-62.

Be sure that a plan is established for cutting and trimming paper, both before and after the job is printed. A simple form for this purpose is shown in Fig. 29-63.

Every job schedule for bindery operations should include full instructions and a dummy as to page numbers, trim size, crop marks, and side guide or gripper marks. If changes are made in original specifications or mistakes are found in positioning, the bindery should be alerted.

Most of the problems in the bindery are caused by poor planning, lack of quality control, lack of communication, and mistakes in other departments. When given a well-planned job, the bindery is capable of quality work. The bindery cannot, however, "perform miracles" and remedy mistakes made along the production path.

POINTS TO REMEMBER

1. After the image has been transferred, it is usually necessary to perform some other operation on the printed piece.
2. The final processing of the printed piece generally includes some type of finishing and binding.
3. Since finishing and binding equipment is varied and generally specialized, these operations tend to be highly labor intensive.

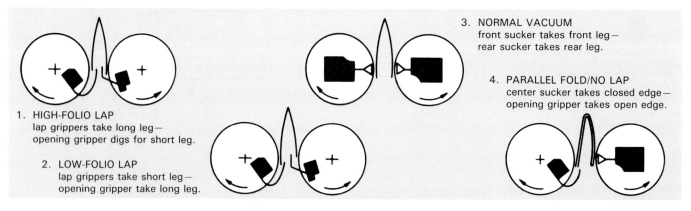

Fig. 29-61. This illustration shows how automatic assembly and stitching machines use the lap to open signatures. (McCain Manufacturing Co.)

BINDERY GUIDELINES
FOR PRINT SHOPS

1. Kind and weight of stock ....

2. Size of sheet to be run ....

3. Direction of the grain ....

4. Number of each page in signature ....

5. How job is to be run sheetwise
   One sided   Two sided   Work and Turn   Work and Tumble

   Other ....

6. Will there be bleeds .... split pages .... hairline line-up?
   (If so the pressroom should consult the bindery before work begins to determine the best way to lay out the job.)

7. What kind of packaging is desired? ....

8. Are deliveries to be bulk .... or split? What date ....

9. Is delivery to be in bulk .... or skids .... or packages ....

   Transported by elevator .... or stairs ....

10. Will addressed office supply labels be provided or be furnished by the bindery? ....

    Who supplies bill of lading or delivery ticket ....

    ....

11. Any other special instructions ....

    ....

    ....

    ....

    ....

Fig. 29-62. Guidelines used by the bindery department to plan and handle each job efficiently.   (Hammermill Paper Co.)

Date _____   **STOCK REQUISITION**   No. 000

The following stock is to be used on:

JOB ORDER NO._____   CUSTOMER _____   DESCRIPTION _____

| PAPER | | | BUY OUT | WAREHOUSE LOCATION | SHEET SIZE or ROLL WIDTH | WEIGHT (ROLLS) | NO. SHEETS | CUTTING INSTRUCTIONS | | | | AMOUNT RETURNED TO INVENTORY |
NAME	SUB.	COLOR						CUT TO	GRAIN	PCS. !OUT	TOTAL	

➡ Paper Bought From _____   P.O. # _____

DELIVER TO ☐ CUTTER ON _____   AND PRESS # _____   ON _____

Fig. 29-63. A stock requisition is used by the bindery department to cut and trim paper, both before and after the job is printed.   (National Association of Printers and Lithographers)

4. Finishing is the term used to describe all operations performed on printed materials from the delivery end of the printing press until the final shipment to the customer.
5. Finishing operations include cutting and trimming, folding, punching, drilling, gathering, collating, jogging, scoring, perforating, slitting, die cutting, hot stamping, embossing, and packaging of the product.
6. Binding operations serve two general purposes. First, the binding holds everything together by fastening one element to the next. Secondly, the binding helps protect the printed material inside.
7. When a printed piece is designed as a book, magazine, or pamphlet, various methods of fastening the printed sheets can be used.
8. Common binding methods include mechanical, loose-leaf, wire, sewn soft-cover, sewn case-bound, perfect, thermal, tipping, and padding.
9. Bindery planning is an important first phase of printing. It is here that specific requirements for the job must be determined.
10. Bindery planning requirements include the placement and relationship of the press, folder, and cutter guides. Consideration is also given to the placement of the press side guide and press gripper with respect to the edge of the press sheet.
11. A bindery dummy is prepared to show how the finished product will be folded and assembled.

KNOW THESE TERMS

Binding post, Collate, Die cut, Drill, French fold, Gather, Hot stamp, Loose-leaf binding, Mechanical binding, Padding, Perfect binding, Perforate, Punch, Ring binding, Saddle-wire stitch, Score, Sewn case-bound, Sewn soft-cover, Shrink wrap, Side-wire stitch, Thermal binding, Tipping, Signature lip.

REVIEW QUESTIONS

1. The equipment used for finishing and binding is usually automatic. True or false?
2. Printed sheets with several pages are prepared so that they can be folded into _____.
3. The two basic kinds of folding machines are the _____ folder and the _____ folder.
4. The cutting of rectangular or specially shaped holes for plastic binding and spiral binding is called _____.
5. Holes for a ring binder are made with a paper _____ _____ _____.

6. Placing individual sheets or signatures in the correct sequence is called _____.
7. The task of checking that the correct number of pages or signatures has been gathered is called:
   a. Jogging.
   b. Gathering.
   c. Collating.
   d. Finishing.
8. Placing a crease in sheets of thick paper or cardboard to aid in folding is called _____.
9. An irregular shape or design can be cut in paper or cardboard by:
   a. Scoring.
   b. Punching.
   c. Drilling.
   d. Die cutting.
10. A letterpress process that uses pressure and heat to make permanent images on objects with gold, silver, or colored foil is called _____ _____.
11. Transparent plastic film and heat are used in a packaging method called _____ _____.
12. Spiral binding and plastic binding are two common methods of:
    a. Mechanical binding.
    b. Wire binding.
    c. Loose-leaf binding.
    d. Perfect binding.
13. Pages that are bound together with removable rings or posts are called _____.
14. Explain saddle-wire stitching.
15. Because of its strength, saddle-wire stitching is recommended for binding _____.
16. In _____ stitching, staples are inserted close to the fold and clinched at the back.
17. Describe the process and applications of sewn bindings.
18. What is perfect binding?
19. _____ is a form of perfect binding used to hold the sheets of a tablet or notebook together.
20. Define the terms CB, CBF, and CF.
21. Summarize some of the reasons for problems in the bindery.

SUGGESTED ACTIVITIES

1. With your teacher's assistance, plan a class field trip to a local bindery.
2. Obtain samples of at least ten kinds of bindings described in this chapter. Be prepared to discuss the advantages and disadvantages of each binding method.
3. Prepare a paper cutting waste factor chart

that can be posted near the paper cutter in your shop. Use the standard sheet sizes used in your shop.

4. During this course in graphic communications, complete as many of the following finishing operations as possible: collating, gathering, folding, punching, drilling, round-cornering, jogging, perforating, die cutting, scoring, and slitting.

5. Prepare a chart or bulletin board display with samples of the finishing operations in Activity 4. Label each sample giving its name and application.

6. Familiarize yourself with carbonless papers. What is meant by precollated? For what purposes would a printer use several different reams and colors of uncollated carbonless paper?

7. Prepare a bindery dummy for a job scheduled to be produced in your shop.

Quick printer employees may be required to operate a variety of equipment.   (Sir Speedy, Inc.)

Section VIII

Issues for Class Discussion

1. Offset lithography presses are grouped by size and function. These include duplicators, offset presses, and web presses. Discuss the various production applications for each type of press along with major advantages and weaknesses of each piece of equipment.
2. Personal safety practices when operating offset press equipment should be of primary concern to all personnel in a printing operation. Discuss the various elements that should be included in a safety program for a school and/or commercial printing firm.
3. Offset press management devices, such as computerized consoles and automatic web splicers, have become common on medium and large size equipment. These components assist press personnel in maintaining quality control and product consistency. Discuss the various press management controls and systems in terms of their present application, advantages, future improvements, and use on small duplicator-size equipment.
4. Through individual research, investigation, and consultation with equipment sales representatives, discuss trends in offset press equipment which relate to operating features, sheet-size capabilities, speed, computerization, and overall design.
5. Through research and consultation with equipment sales representatives and people in the field, discuss trends in the finishing and bindery area. Determine how much automation, such as computers and robotics, can be expected to dominate this area of the printing and publishing industry.

Section IX

In-Plant Printers, Quick Printers, and Brokers

Chapter 30 covers three special areas of the printing and publishing industry. Although each is a separate enterprise, they make a vital contribution to graphic communications. These specialty printers and entrepreneurs are often overlooked by those studying graphic arts. They include in-plant printers, quick printers, and printing brokers.

The in-plant printer is generally housed within a company whose primary business is NOT printing. An example might be General Motors Corporation in which printing for the company's own requirements are met by an in-plant printing operation. The in-plant printing operation, therefore, supplies only graphic materials needed in the company's operation.

Quick printers are classified as being franchised or non-franchised. Their basic operating procedures are identical.

Printing brokers do NOT generally own equipment or facilities. Instead, they conduct business from an office facility. In some instances, brokers work from their home office. They arrange to have printing produced for their clients by various commercial printers.

Chapter 30

In-Plant Printers, Quick Printers, and Brokers

When you have completed the reading and assigned activities related to this chapter, you will be able to:
O Describe the function of in-plant printing departments.
O Describe the range of printed products produced by in-plant printing departments.
O Describe the primary differences between franchised quick printers, independent quick printers, and small commercial quick printers.
O Describe the operation of a printing broker.

You should be familiar with three specialty areas of the printing and publishing industry. These important enterprises make a vital contribution to graphic communications. They include: in-plant printing, quick printing, and printing brokers.

IN-PLANT PRINTING

In-plant printing consists of a printing facility that is owned and operated within its parent company or organization. It is one of the largest specialty groups in the printing and publishing industry. For example, firms such as IBM, General Motors, and Lockheed Aircraft maintain their own in-plant printing departments. These companies manufacture complex products that must be sold with some type of printed literature. Some of the printed products are business forms, brochures, service manuals, owner's manuals, and annual reports.

In-plant printing facilities differ from regular commercial printers in that they do not have an individual company name or advertise commercially. This means that they are housed within established firms and only service that organization. See Fig. 30-1.

The companies that maintain in-plant printing departments use this convenience to print needed office items such as invoices, letterheads, envelopes, and bulletins. Many companies print newsletters and promotional literature in black and white and also full color. Refer to Fig. 30-2.

Fig. 30-1. The in-plant printing facility consists of any printing operation that is owned by and serves the needs of a single company or corporation. (Rockwell International, Miehle Products, Graphic Systems Division)

675

Fig. 30-2. Companies that maintain in-plant printing departments may prepare needed office forms such as invoices, letterheads, envelopes, and bulletins for their own use. (AM International, Varityper Division)

In-plant industry size

The in-plant printing industry represents a $31 billion-a-year market segment. This is at least twice the value of sales generated by the commercial segment of the printing industry. A recent study indicates that there will be approximately 90,000 in-plant departments in the next few years.

The vast organization of in-plant printers recently formed the In-Plant Printing Management Association. Its logotype represents four main functions. These include: typesetting, art and copy preparation, copying, and printing.

Since its beginning, the in-plant industry has grown not only in size but in the scope of work and sophistication of equipment. During the industry's early years, most in-plant departments were similar to small commercial shops. In-plant operations generally had offset duplicators and light-duty bindery equipment. During those early years, the average in-plant department was an important part of the company it served. However, it was more of a money-saving convenience than a vital part of the organizational structure. Later, the need for printed information within companies increased. As a result, the responsibilities and importance of in-plant departments grew sharply.

In-plant departments now commonly handle all operations of the printing production cycle. These departments have expanded into the areas of word processing, computerized typesetting, electronic microfilming, computerized mailrooms, and four-color printing. See Fig. 30-3.

With the technology of most in-plant departments increasing, more of the printing needs of companies are being handled in-house. As a result, greater amounts of effective printed communications are being produced. This is largely due to the highly qualified people designing and printing the materials. This is because more technically qualified workers are being attracted to in-

A

B

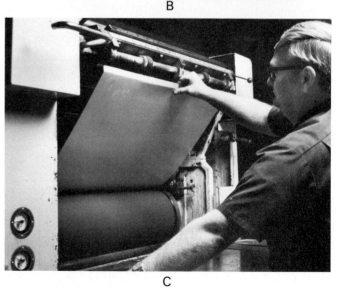

C

Fig. 30-3. A—In-plant printing departments are equipped to set type electronically. B—Computerized typesetting, electronic microfilming, and computerized mailing are a few of the expanded technology areas associated with in-plant departments. C—Some in-plant printers handle four-color process printing for brochures, annual reports, and advertising pieces. (Crossfield Corporation, Ltd.; Camex Corp.; 3M Company)

plant operations than in the past. Companies can now obtain the same professional results from their in-plant workers as they can by having the work done by an outside commercial firm. Look at Fig. 30-4.

In-plant departments are generally required to design and produce vital company documents. These include reports, office forms, labels, packaging, and numerous mailing pieces. Many in-plant departments have multicolor, sheet-fed offset presses. Companies, such as NCR, Burroughs, and IBM, operate web offset presses in the manufacture of forms used on the business machines they manufacture.

The increase in the vast quantity of products now produced by in-plant departments has affected the industry as a whole. This is due to the changing technology. Communications has become more of an electronic medium. Electronics control much of the printing and high-speed copying equipment. In many cases, electronics has become the actual carrier of information because of its speed and convenience.

The shift from printing-oriented to electronic-oriented in-plant departments is causing a great challenge to the industry. This change will affect the makeup, not only of the in-plant equipment, but the employees who operate the equipment. Unlike past in-plant departments, future employees will have data processing and/or computer backgrounds.

Most printing industry forecasters see a bright future for in-plant departments. They predict larger and faster production outputs, savings in paper, better communication of vital information, and overall cost reduction.

QUICK PRINTING

Quick printing consists primarily of small businesses with duplicator-size and small offset press equipment, Fig. 30-5. It is another specialty group within the printing and publishing industry. Some quick printers still provide while-you-wait service using camera-ready line copy originals. Press runs of the while-you-wait variety are generally 1,000 impressions or less. The copy is usually printed in black ink with maximum size of copy output being 11 x 17 inches.

Fig. 30-5. This duplicator-size offset press is typical of the equipment used by most quick printers. (A.B. Dick Co.)

Fig. 30-4. This mini-web offset press is used by an in-plant printer to produce vital company documents. (Harris Corp.)

Many quick printers are beginning to print jobs in color. They no longer restrict their work to black and white.

Quick printing is becoming increasingly "high-tech" as the industry moves toward multiple color reproduction and electronic printing. Speed, quality, and consistency are important to color reproduction. Sophisticated systems must be installed to monitor and control color. In addition to printing services, quick printers offer other services such as bindery and high-speed copying. See Fig. 30-6.

Most quick print shops are located in high-traffic areas of a city or community. However, greater numbers of quick printers are locating in manufacturing centers and industrial complexes. The quick printer may provide same-day printing services, but most of the work is completed in three to five days.

Quick printers often use photo-direct plates and offset duplicators as well as electrostatic plates and copying equipment, Fig. 30-7. Increasing numbers of quick printers are investing in small and medium size offset presses. This is because they have been able to diversify their client base to include larger and more sophisticated printed pieces. Quick printers accept every kind of job ranging from letterheads, reports, and brochures to business forms, cards, tickets, and invitations.

Many quick printers are currently in the process of adding all types of high-speed electronic reproduction equipment. These items include electronic and desktop publishing systems, word processors, long-run plates prepared on camera/platemakers, one- and four-color high-speed copiers, and second-color units for duplicators and small offset presses.

Fig. 30-7. This quick print operator is preparing the camera to make a photographic paper plate for the offset duplicator. (Sir Speedy, Inc.)

Recent improvements in high-speed copier equipment now make it possible to duplicate on both sides of a sheet of paper in one pass through the machine. This equipment is finding increasing use in quick print shops and office duplicating centers. These machines automatically feed originals, collate, and staple sheets in sets at a rate of up to 5,500 per hour. Two-sided duplicating is now a reality on these machines. Refer to Fig. 30-8.

New innovations in high-speed copier equipment give the quick printer added versatility and speed. Being able to produce solids and halftones makes the process even more attractive to quick printers.

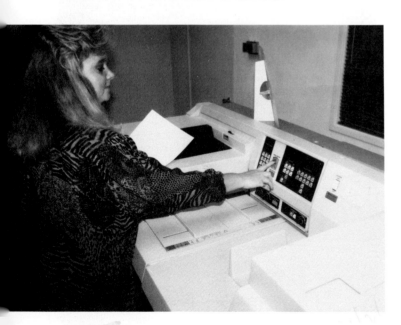

Fig. 30-6. Quick printers specialize in high-speed xerographic copying. This computerized copy machine can be programmmed to copy a pre-determined amount of documents including collating and stapling. (Sir Speedy, Inc.)

Fig. 30-8. This xerographic copying machine is capable of producing quality two-sided copies at over 5,500 per hour. (Xerox Corp.)

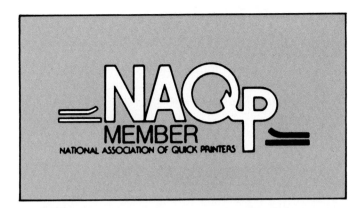

Fig. 30-9. The National Association of Quick Printers is an organization dedicated to serving quick printers in the United States. (National Association of Quick Printers)

Like the national In-Plant Printing Management Association, quick printers are served by two trade associations. One is called the National Association of Quick Printers, Fig. 30-9. The other is known as The American Quick Printing Association. Within these associations are three groups that together make up the quick printing industry. They include: franchised quick printers, independent quick printers, and small commercial quick printers.

Franchised quick printers

As defined by the Department of Labor, *franchising* is a licensing agreement. This is the principle of the franchised quick printer. The *franchisor* provides the franchisee with training not only in the technical areas of the business but in business management practices as well. The franchisor provides assistance in seeking adequate financing, proper site selection, business start-up promotions, and marketing. In most instances, the franchisor provides the equipment necessary for operation and offers on-going sales, promotional, and technical programs. For these services, the franchisee pays a lump sum amount in addition to a predetermined royalty on gross sales of the operation.

Prospective franchisees are recruited by various quick printing franchisors. Among these are the familiar companies such as Sir Speedy, Kwik-Kopy, Printmasters, and Postal Instant Press, Fig. 30-10.

Sales by franchised quick printers is over $1 billion annually. The quick printing industry as a whole represents a $5 billion market. The total number of franchised shops is steadily increasing. California, Texas, and Florida are the top three states in total number of franchised locations.

Following site selection and training, which generally covers all matters technical, managerial, financial and promotional interest, the franchisee begins operation, Fig. 30-11. Most franchised quick printing owners employ up to two or three people initially. The key employees are usually the press operator and counter person, Fig. 30-12.

Fig. 30-10. The franchisor helped this franchisee find a store location and also assisted in business start-up promotions and marketing. (Sir Speedy, Inc.)

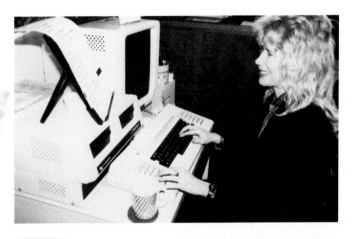

Fig. 30-11. Most franchised quick printing owners employ up to two or three people initially. (Sir Speedy, Inc.)

Fig. 30-12. The counter person along with the press operator in a quick printing operation are key employees. (Sir Speedy, Inc.)

679

The franchisors listed below were provided by the U.S. Department of Commerce in two reports entitled FRANCHISING IN THE ECONOMY and FRANCHISE OPPORTUNITIES HANDBOOK. This information may be of help to those who desire additional information related to franchising opportunities in the United States.

1. AlphaGraphics Printshops of the Future
   845 East Broadway
   Tucson, AZ 85719
2. American Speedy Printing Centers
   32100 Telegraph Road
   Birmingham, MI 48010
3. Big Red Q (Quickprint, Inc.)
   2545 W. 237 Street, Suite 2
   Torrance, CA 90505
4. Copy Mat
   48 Shattuck Square
   Berkeley, CA 94704
5. Franklin's Copy Service Inc.
   135 International Boulevard
   Atlanta, GA 30303
6. Ink Well, Inc.
   2323 Lake Club Drive
   Columbus, OH 43227
7. Insty-Prints Inc.
   1215 Marshall Road N.E.
   Minneapolis, MN 55413
8. Kwik-Kopy Corp.
   P.O. Box 777
   1 Kwik-Kopy Lane
   Cypress, TX 77429
9. Minuteman Press International
   1640 New Highway
   Farmingdale, NY 11735
10. Postal Instant Press
    8201 Beverly Boulevard
    Los Angeles, CA 90048
11. Printmasters, Inc.
    370 S. Crenshaw Boulevard, E 100
    Torrance, CA 90503
12. Print Shack
    500 N. Westshore Boulevard, Suite 610
    Tampa, FL 33609-1924
13. Quick Print
    3345 North Webb Road
    Wichita, KS 67226
14. Sir Speedy Inc.
    23131 Verdugo Drive
    Laguna Hills, CA 92654
15. Zippy Print
    260 W. Esplanade, Suite 300
    N. Vancouver, B.C. V7M 3G7

Independent quick printers

The *independent quick printer* differs from the franchised quick printer in that the facility is individually owned. Independent quick printers outnumber franchised and small commercial quick printers. The independents account for approximately 82 percent of all quick printing market sales.

Almost a third of independent quick printers have had prior printing experience. In contrast, first-time franchised quick printer owners have little printing background. However, the equipment and general operating procedure is similar to that of the franchisee.

The independents offer the same basic complement of quick printing services as the franchisees. These include services such as duplicating, high-speed copying, and simple binding. In addition, most independents offer services such as typesetting, photostat production, darkroom and camera operations, art and layout, facsimile transmission, and word processing. See Fig. 30-13.

The independent quick printer usually employs up to three people. The key employees are very much like the typical key people in a franchised shop. The majority of personnel are unskilled part-time employees. The owner or manager generally supervises a program of on-the-job training, Fig. 30-14.

Small commercial quick printers

The familiar small commercial printer who adds quick printing capabilities has much in common with the independent quick printer. These firms are referred to as *small commercial quick printers*. This generally occurs in response to the demands of customers for faster service. The commercial printer's quick printing work also improves cash-flow and provides fill-in work for otherwise idle employees.

Small commercial quick printers, like their independent counterparts, recognize the need to specialize. They feature such products as business forms, brochures, thermography, and four-color work. For

Fig. 30-13. Most independent quick printers offer services such as typesetting, photostat production, darkroom and camera operations, art and layout, facsimile transmission, and word processing. (Compugraphic Corp.)

Fig. 30-14. The majority of quick printing production workers are unskilled part-time employees. The owner or manager generally supervises a program of on-the-job training.

the small commercial quick printer, quality is of primary consideration.

Small commercial printers, with quick printing capabilities, must push marketing to remain competitive in the quick printing arena. This means that they must actively promote their services. As new technology gives quick printers a greater product range, they become more of a threat to the small commercial printer. As a result, the small commercial printer of the future may find it necessary to adopt the quick printer's marketing and fast service techniques. This will be essential if the commercial printer wants to continue offering quick printing services.

PRINTING BROKERS

Another specialty area of the printing and publishing industry is that of the printing broker. A *broker* is a person who acts as an agent in the transaction of purchases and the sale of printing. A *brokerage* is the business establishment or premise of a broker, typically an office.

Unlike a regular commercial printer who maintains visibility, the broker usually has no buildings, equipment, or personnel. Most brokers work directly out of their homes. The key to their business is personalized customer service. This means that personal contacts and telephone conversations are the principal means of communication and business success of a broker.

Brokers deal directly with commercial printers in arranging to have printing completed for their customers. A number of selected printing firms are used to match customer needs with available resources. A broker may communicate and work closely with as many as a dozen commercial printing firms, Fig. 30-15.

The printing broker must be an excellent business person to make a profit. Since the broker buys printing services for each customer's job, estimating and profit margins are critical factors. Brokers must learn to work with banks and other financial institutions for optimum cash flow.

Printing brokers are generally people who have had printing experience in the mechanical or managerial departments of printing firms. In addition, they have business and accounting experience along with a strong desire to work with people, Fig. 30-16. Previous sales experience is another asset in this field of work.

POINTS TO REMEMBER

1. Three separate, but closely related areas of the printing industry rely almost entirely on offset lithography. These important enterprises include in-plant printing, quick printing, and printing brokers.
2. In-plant printers form one of the largest specialty groups in the printing industry.

Fig. 30-15. This printing broker buys printing services for each customer's job. A broker may communicate and work closely with as many as a dozen commmercial printing firms.

Fig. 30-16. Most printing brokers are people who have had printing experience in the mechanical and/or managerial departments of printing firms.

3. In-plant printing operations are contained within large companies, such as: General Dynamics, Coca-Cola Bottling, and Ford Motor Company.
4. In-plant printers handle much of the work of the company, including manuals, business forms, brochures, and business cards.
5. Quick printing is another specialty group. These are generally small printing businesses with duplicator-size and small offset presses.
6. Quick printers are divided into three groups. These include: franchised, independent, and small commercial.
7. In a franchise operation, the franchisor provides most of the equipment, on-going sales and promotional aid programs.
8. A franchisee pays a lump sum amount along with a royalty fee on gross sales.
9. Independent quick printer facilities are individually owned. They outnumber franchised and small commercial quick printers.
10. Small commercial quick printers generally start by adding quick printing capabilities to an already well-established small commercial printing operation.
11. The printing broker is a person who acts as an agent in the transaction of purchases and the sale of printing.
12. Printing brokers most frequently work out of their homes since there is no equipment or additional personnel involved.
13. Printing brokers generally deal directly with commercial printers in arranging to have printing completed for their customers.
14. Printing brokers must be excellent sales and business people to make a profit.

KNOW THESE TERMS

Commercial quick printer, Franchised quick printer, Franchisee, Franchisor, Independent quick printer, In-plant printing, Printing broker, Quick printing.

REVIEW QUESTIONS

1. A printing operation that is owned by and serves the needs of a single company is called a:
   a. Quick printer.
   b. Printing broker.
   c. In-plant printer.
   d. Commercial printer.

2. _____ now controls much of the printing and duplicating equipment and has become the carrier of information because of its speed and convenience.
3. While-you-wait printing generally refers to _____ printers.
4. Two-sided copying is possible on some electrophotography copying machines. True or false?
5. A franchised quick printing operation involves the franchisor and the _____.
6. How does the independent quick printer differ from the franchised quick printer?
7. Compare commercial quick printers with independent quick printers.
8. The primary consideration of the commercial quick printer is:
   a. Competition.
   b. Speed.
   c. Quantity.
   d. Quality.
   e. All of the above.
9. A printing _____ is anyone who acts as an agent in the transaction of purchases and sale of printing.
10. Personalized customer service is the key to a printing broker's success in business. True or false?

SUGGESTED ACTIVITIES

1. With your teacher's assistance, arrange a class field trip to an in-plant printing facility. Prepare a short report covering the visit and use it to present a review to the class.
2. Visit a franchised quick printer, an independent quick printer, and a small commercial quick printer. Obtain a price sheet from each printer and compare the pricing. Prepare a short written report describing your findings.
3. With your teacher's assistance, locate a printing broker. Arrange for the broker to make a presentation to the class.
4. With your teacher's assistance, arrange to have an owner of a quick print shop talk to the class.
5. With your teacher's assistance, arrange for a quick print franchisor to discuss the various aspects of purchasing and operating a quick printing store. Topics such as cost of a franchise, location, possible profit centers, and growth potential are interesting areas for discussion.

Section IX

Issues for Class Discussion

1. An area of the printing and publishing industry showing tremendous growth is in-plant printing. Discuss some of the significant contributions that in-plant printers make to other areas of the commercial printing sector. One obvious example is subcontract or buy out work.

2. The quick printing sector of the printing and publishing industry is growing at a phenomenal pace. These shops are divided into franchised and non-franchised operations. Many economists, business leaders, and consultants have stated that franchising (not only in printing but in many other enterprises) is the "wave of the future." Discuss and prepare a list of advantages and disadvantages of owning a franchised print shop in contrast to that of a non-franchised shop. What is the general consensus of the class?

3. More and more people are becoming printing brokers. Many commercial printers are not happy with the idea of brokering because they feel that the broker undersells the commercial printer and thus takes business away. On the other hand, many commercial printers encourage brokers and realize that they present an added source of income. Discuss the pros and cons of this issue. How can both sides be satisfied? Is there a way to satisfy both sides?

APPENDIX

Appendix A

METRIC SYSTEM

The International System of Units (SI) provides a logical and interconnected framework for all measurement in science, industry, commerce, and education.

meter (m) — length
kilogram (kg) — mass
seconds (s) — time
ampere (A) — electric current
kelvin (K) — temperature
mole (mol) — amount of substance
candela (cd) — luminous intensity

CUSTOMARY AND METRIC CONVERSIONS

Known value	Symbol	Multiply by	Conversion value and symbol
inches	in	25.4	millimeters (mm)
feet	ft	0.3048	meter (m)
yards	yd	0.9114	meter (m)
quarts	qt	0.946 353	liter (l)
ounces	oz	28.349 5	gram (g)
pounds	lb	0.453 592	kilogram (kg)
Fahrenheit	°F	(°F − 32) x 5/9	Celsius (°C)
millimeter	mm	0.039 370 1	inches (in)
meter	m	3.280 84	feet (ft)
liter	l	1.056 69	quart (qt)
gram	g	0.035 274	ounce (oz)
kilogram	kg	2.204 62	pound (lb)
Celsius	°C	9/5 x °C + 32	Fahrenheit (°F)

METRIC UNITS

Measures of Length

10 millimeters	=	1 centimeter, cm
10 centimeters	=	1 decimeter, dm
10 decimeters	=	1 meter, m
10 meters	=	1 decameter, dam
10 decameters	=	1 hectometer, hm
10 hectometers	=	1 kilometer, km
10 kilometers	=	1 myriameter, mym

Measures of Weight

10 milligrams, mg	=	1 centigram, cg
10 centigrams	=	1 decigram, dg
10 decigrams	=	1 gram, g
10 grams	=	1 decagram, dag
10 decagrams	=	1 hectogram, hg
10 hectograms	=	1 kilogram, kg
10 kilograms	=	1 myriagram, myg
10 myriagrams	=	1 quintal, q
10 quintals	=	1 millier or ton, MT or t

Measures of Liquid Volume

10 milliliters, ml	=	1 centiliter, cl
10 centiliters	=	1 deciliter, dl
10 deciliters	=	1 liter, l
10 liters	=	1 decaliter, dal
10 decaliters	=	1 hectoliter, hl
10 hectoliters	=	1 kiloliter, kl

SI MULTIPLES AND PREFIXES

Prefix and symbol			Factor by which the unit is multiplied
exa, E	10^{18}	=	1 000 000 000 000 000 000.
peta, P	10^{15}	=	1 000 000 000 000 000.
tera, T	10^{12}	=	1 000 000 000 000.
giga, G	10^{9}	=	1 000 000 000.
mega, M	10^{6}	=	1 000 000.
kilo, k	10^{3}	=	1 000.
hecto, h	10^{2}	=	100.
deka, da	10^{1}	=	10.
	10^{0}	=	Base Unit
deci, d	10^{-1}	=	0.1
centi, c	10^{-2}	=	0.01
milli, m	10^{-3}	=	0.001
micro, μ	10^{-6}	=	0.000 001
nano, n	10^{-9}	=	0.000 000 001
pico, p	10^{-12}	=	0.000 000 000 001
femto, f	10^{-15}	=	0.000 000 000 000 001
atto, a	10^{-18}	=	0.000 000 000 000 000 001

Appendix B

DECIMAL CHART

inches fractional	decimal	inches fractional	decimal
1/64	.0156	33/64	.5156
1/32	.0312	17/32	.5312
3/64	.0469	35/64	.5469
1/16	.0625	9/16	.5625
5/64	.0781	37/64	.5781
3/32	.0938	19/32	.5938
7/64	.1094	39/64	.6094
1/8	.1250	5/8	.6250
9/64	.1406	41/64	.6406
5/32	.1562	21/32	.6562
11/64	.1719	43/64	.6719
3/16	.1875	11/16	.6875
13/64	.2031	45/64	.7031
7/32	.2188	23/32	.7188
15/64	.2344	47/64	.7344
1/4	.2500	3/4	.7500
17/64	.2656	49/64	.7656
9/32	.2812	25/32	.7812
19/64	.2969	51/64	.7969
5/16	.3125	13/16	.8125
21/64	.3281	53/64	.8281
11/32	.3438	27/32	.8438
23/64	.3594	55/64	.8594
3/8	.3750	7/8	.8750
25/64	.3906	57/64	.8906
13/32	.4062	29/32	.9062
27/64	.4219	59/64	.9219
7/16	.4375	15/16	.9375
29/64	.4531	61/64	.9531
15/32	.4688	31/32	.9688
31/64	.4844	63/64	.9844
1/2	.5000	1	1.0000

m/m	inches	m/m	inches	m/m	inches	m/m	inches	m/m	inches	m/m	inches
1	0.0394	51	2.0079	101	3.9764	151	5.9449	201	7.9134	251	9.8819
2	0.0787	52	2.0472	102	4.0157	152	5.9843	202	7.9527	252	9.9212
3	0.1181	53	2.0866	103	4.0551	153	6.0236	203	7.9921	253	9.9606
4	0.1575	54	2.1260	104	4.0945	154	6.0630	204	8.0315	254	10.0000
5	0.1969	55	2.1654	105	4.1339	155	6.1024	205	8.0709	255	10.0394
6	0.2362	56	2.2047	106	4.1732	156	6.1417	206	8.1102	256	10.0787
7	0.2756	57	2.2441	107	4.2126	157	6.1811	207	8.1496	257	10.1181
8	0.3150	58	2.2835	108	4.2520	158	6.2205	208	8.1890	258	10.1575
9	0.3543	59	2.3228	109	4.2913	159	6.2598	209	8.2283	259	10.1968
10	0.3937	60	2.3622	110	4.3307	160	6.2992	210	8.2677	260	10.2362
11	0.4331	61	2.4016	111	4.3701	161	6.3386	211	8.3071	261	10.2756
12	0.4724	62	2.4409	112	4.4094	162	6.3780	212	8.3464	262	10.3149
13	0.5118	63	2.4803	113	4.4488	163	6.4173	213	8.3858	263	10.3543
14	0.5512	64	2.5197	114	4.4882	164	6.4567	214	8.4252	264	10.3937
15	0.5906	65	2.5591	115	4.5276	165	6.4961	215	8.4646	265	10.4331
16	0.6299	66	2.5984	116	4.5669	166	6.5354	216	8.5039	266	10.4724
17	0.6693	67	2.6378	117	4.6063	167	6.5748	217	8.5433	267	10.5118
18	0.7087	68	2.6772	118	4.6457	168	6.6142	218	8.5827	268	10.5512
19	0.7480	69	2.7165	119	4.6850	169	6.6535	219	8.6220	269	10.5905
20	0.7874	70	2.7559	120	4.7244	170	6.6929	220	8.6614	270	10.6299
21	0.8268	71	2.7953	121	4.7638	171	6.7323	221	8.7008	271	10.6693
22	0.8661	72	2.8346	122	4.8031	172	6.7717	222	8.7401	272	10.7086
23	0.9055	73	2.8740	123	4.8425	173	6.8110	223	8.7795	273	10.7480
24	0.9449	74	2.9134	124	4.8819	174	6.8504	224	8.8189	274	10.7874
25	0.9843	75	2.9528	125	4.9213	175	6.8898	225	8.8583	275	10.8268
26	1.0236	76	2.9921	126	4.9606	176	6.9291	226	8.8976	276	10.8661
27	1.0630	77	3.0315	127	5.0000	177	6.9685	227	8.9370	277	10.9055
28	1.1024	78	3.0709	128	5.0394	178	7.0079	228	8.9764	278	10.9449
29	1.1417	79	3.1102	129	5.0787	179	7.0472	229	9.0157	279	10.9842
30	1.1811	80	3.1496	130	5.1181	180	7.0866	230	9.0551	280	11.0236
31	1.2205	81	3.1890	131	5.1575	181	7.1260	231	9.0945	281	11.0630
32	1.2598	82	3.2283	132	5.1969	182	7.1654	232	9.1338	282	11.1023
33	1.2992	83	3.2677	133	5.2362	183	7.2047	233	9.1732	283	11.1417
34	1.3386	84	3.3071	134	5.2756	184	7.2441	234	9.2126	284	11.1811
35	1.3780	85	3.3465	135	5.3150	185	7.2835	235	9.2520	285	11.2205
36	1.4173	86	3.3858	136	5.3543	186	7.3228	236	9.2913	286	11.2598
37	1.4567	87	3.4252	137	5.3937	187	7.3622	237	9.3307	287	11.2992
38	1.4961	88	3.4646	138	5.4331	188	7.4016	238	9.3701	288	11.3386
39	1.5354	89	3.5039	139	5.4724	189	7.4409	239	9.4094	289	11.3779
40	1.5748	90	3.5433	140	5.5118	190	7.4803	240	9.4488	290	11.4173
41	1.6142	91	3.5827	141	5.5512	191	7.5197	241	9.4882	291	11.4567
42	1.6535	92	3.6220	142	5.5906	192	7.5591	242	9.5275	292	11.4960
43	1.6929	93	3.6614	143	5.6299	193	7.5984	243	9.5669	293	11.5354
44	1.7323	94	3.7008	144	5.6693	194	7.6378	244	9.6063	294	11.5748
45	1.7717	95	3.7402	145	5.7087	195	7.6772	245	9.6457	295	11.6142
46	1.8110	96	3.7795	146	5.7480	196	7.7165	246	9.6850	296	11.6535
47	1.8504	97	3.8189	147	5.7874	197	7.7559	247	9.7244	297	11.6929
48	1.8898	98	3.8583	148	5.8268	198	7.7953	248	9.7638	298	11.7323
49	1.9291	99	3.8976	149	5.8661	199	7.8346	249	9.8031	299	11.7716
50	1.9685	100	3.9370	150	5.9055	200	7.8740	250	9.8425	300	11.8110

Appendix C

ISO—A-SIZE PAPER

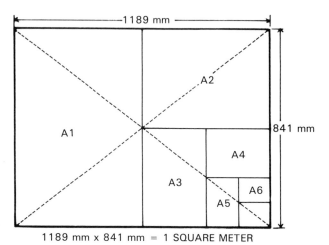

1189 mm x 841 mm = 1 SQUARE METER

RELATIONSHIP BETWEEN A-, B-,
and C-SIZE PAPER

COMPARISON CHART—A-SIZE PAPER

ISO—sizes	METRIC—millimeters	CUSTOMARY—inches
2A	1189 x 1682	46.81 x 66.22
A0	841 x 1189	33.11 x 46.81
A1	594 x 841	23.39 x 33.11
A2	420 x 594	16.54 x 23.39
A3	297 x 420	11.69 x 16.54
A4	210 x 297	8.27 x 11.69
A5	148 x 210	5.83 x 8.27
A6	105 x 148	4.13 x 5.83
A7	74 x 105	2.91 x 4.13
A8	52 x 74	2.05 x 2.91
A9	37 x 52	1.46 x 2.05
A10	26 x 37	1.02 x 1.46

COMPARISON CHART—B-SIZE PAPER

ISO—sizes	METRIC—millimeters	CUSTOMARY—inches
B0	1000 x 1414	39.37 x 55.67
B1	707 x 1000	27.83 x 39.37
B2	500 x 707	19.68 x 27.83
B3	353 x 500	13.90 x 19.68
B4	250 x 353	9.84 x 13.90
B5	175 x 250	6.93 x 9.84
B6	125 x 176	4.92 x 6.93
B7	88 x 125	3.46 x 4.92
B8	62 x 88	2.44 x 3.46
B9	44 x 62	1.73 x 2.44
B10	31 x 44	1.22 x 1.73

R20 SERIES OF PAPER WEIGHTS AND EQUIVALENT WEIGHTS

'R' series	Bond 17 x 22 432 x 559 mm	Cover 20 x 26 508 x 660 mm	Index 25½ x 30½ 648 x 775 mm	Newsprint 24 x 36 610 x 914 mm	Book 25 x 38 635 x 965 mm
g/m²	lb/ream	lb/ream	lb/ream	lb/ream	lb/ream
20.0	5.32	7.39	11.00	12.29	13.51
22.4	5.95	8.28	12.39	13.77	15.13
25.0	6.65	9.24	13.83	15.36	16.89
28.0	7.44	10.35	15.49	17.21	18.92
31.5	8.37	11.65	17.43	19.36	21.28
45.0	11.97	16.64	24.89	27.66	30.40
50.0	13.30	18.49	27.66	30.73	33.78
56.0	14.89	20.71	30.98	34.42	37.84
63.0	16.75	23.30	34.85	38.72	42.57
71.0	18.88	26.26	39.28	43.63	47.97
85.0	22.61	31.45	46.90	52.27	57.46
100.0	26.60	36.98	55.32	61.46	67.57
112.0	29.79	41.42	61.96	68.83	75.68
140.0	37.24	51.78	77.45	86.04	94.60
180.0	47.88	66.57	99.58	110.62	121.63
200.0	53.20	73.97	110.64	122.91	135.14
250.0	66.50	92.46	138.30	153.64	168.93
400.0	106.41	147.95	221.29	245.83	270.29

Appendix D

CUTTING CHARTS

A smart paper buyer always tries to use standard paper sizes that can be used without waste. Odd size pages can be wasteful and costly if the quantity is not large enough, or if there is not enough time to order a special size of paper.

This chart shows the number of pages to a standard paper size for several page sizes in use today. The paper size includes trim top, bottom, and side, but *not* bleed.

Trimmed Page Size	Number of Printed Pages	Number From Sheet	Standard Paper Size
4 x 9	4	12	25 x 38
	8	12	38 x 50
	12	4	25 x 38
	16	6	38 x 50
	24	2	25 x 38
4 1/4 x 5 3/8	4	32	35 x 45
	8	16	35 x 45
	16	8	35 x 45
	32	4	35 x 45
4 1/2 x 6	4	16	25 x 38
	8	8	25 x 35
	16	4	25 x 38
	32	2	25 x 38
5 1/2 x 8 1/2	4	16	35 x 45
	8	8	35 x 45
	16	4	35 x 45
	32	2	35 x 45
6 x 9	4	8	25 x 38
	8	4	25 x 38
	16	2	25 x 38
	32	2	38 x 50
8 1/2 x 11	4	4	23 x 35
	8	2	23 x 35
	16	2	35 x 45
9 x 12	4	4	25 x 38
	8	2	25 x 38
	16	2	38 x 50

Appendix E

EQUIVALENT WEIGHTS
In reams of 500 sheets, basis weights in bold type.

Grade of paper	BOOK 25 x 38	BOND 17 x 22	COVER 20 x 26	BRISTOL 22½ x 28½	INDEX 25½ x 30½	TAG 24 x 36	GRAMMAGE (g/m²)
BOOK	**30**	12	16	20	25	27	44
	40	16	22	27	33	36	59
	45	18	25	30	37	41	67
	50	20	27	34	41	45	74
	60	24	33	40	49	55	89
	70	28	38	47	57	64	104
	80	31	44	54	65	73	118
	90	35	49	60	74	82	133
	100	39	55	67	82	91	148
	120	47	66	80	98	109	178
BOND	33	**13**	18	22	27	30	49
	41	**16**	22	27	33	37	61
	51	**20**	28	34	42	46	75
	61	**24**	33	41	50	56	90
	71	**28**	39	48	58	64	105
	81	**32**	45	55	67	74	120
	91	**36**	50	62	75	83	135
	102	**40**	56	69	83	93	158
COVER	91	36	**50**	62	75	82	135
	110	43	**60**	74	90	100	163
	119	47	**65**	80	97	108	176
	146	58	**80**	99	120	134	216
	164	65	**90**	111	135	149	243
	183	72	**100**	124	150	166	271
BRISTOL	100	39	54	**67**	81	91	148
	120	47	65	**80**	98	109	178
	148	58	81	**100**	121	135	219
	176	70	97	**120**	146	162	261
	207	82	114	**140**	170	189	306
	237	93	130	**160**	194	216	351
INDEX	110	43	60	74	**90**	100	163
	135	53	74	91	**110**	122	203
	170	67	93	115	**140**	156	252
	208	82	114	140	**170**	189	328
TAG	110	43	60	74	90	**100**	163
	137	54	75	93	113	**125**	203
	165	65	90	111	135	**150**	244
	192	76	105	130	158	**175**	284
	220	87	120	148	180	**200**	326
	275	109	151	186	225	**250**	407

Appendix F

PRINTING PAPERS

All papers have certain properties and characteristics that affect printability and quality—terms that are not always related—for the finest quality of paper may not always print well. Weight, bulk and caliper, grain direction, color, opacity, surface texture, coatings, and strength are some of the factors to be considered before selecting paper.

Papers are generally classified and defined in terms of end use, as suggested by the grade names listed below. Basic sizes are shown in parentheses.

BOOK (25" x 38"). This grade encompasses the widest range of printing papers. As the name implies, they are widely used for books, magazines, folders, pamphlets, posters, and other commercial printing. The different grades of book paper are:

Coated (Enamel). Coatings of clay and other materials on base paper provide a smooth, glossy surface for quality print production. Types and grades include: No. 1 and No. 2 coated, machine coated, dull coated (matte), and cast coated.

Uncoated. Uncoated book papers are made in a wide range of weights and finishes: English Finish (E.F.), antique (wove and laid), supercalendered, vellum, eggshell, machine finish, and smooth.

Offset. Similar to coated and uncoated book papers except for sizing. Most offset papers can be printed letterpress, but nearly all book papers require a sizing to resist the moisture present in offset-lithography.

Text. Characterized by a wide range of attractive colors and surface textures. Used for announcements, brochures, and to create special effects. Most are sized to resist water penetration. Many are available in matching cover weights and matching envelopes.

Label. Intended for printing on only one side, the reverse side is either adhesive gummed at the mill, or treated to accept adhesives after printing. Available coated one side (C1S) or uncoated.

COVER (20" x 26"). Many grades of coated, text and book papers are made in matching cover weights. There are also many special cover papers with a variety of surface textures, coatings, and finishes.

BOND (17" x 22"). Primarily used for stationery and business forms, this category also includes ledger and writing grades. Available in a wide range of colors and weights. Surfaces accept typewriter and writing inks and erase easily. There are two types: sulphite and cotton fiber (rag content). Costlier bonds are made with 25% to 100% cotton fiber.

INDEX BRISTOL (25½" x 30½"). Characterized by stiffness and receptivity to printing inks, index is used wherever a stiff, inexpensive paper is required.

PRINTING (MILL) BRISTOL (22½" x 28½"). Generally stiffer than index, printing bristols are widely used for menus, greeting cards, covers, and tickets.

NEWSPRINT (24" x 36"). This inexpensive grade is limited primarily to cost-critical uses.

STANDARD PAPER SIZES AND WEIGHTS

It has been traditional practice to price papers on a ream (500 sheet) basis. This is being replaced by 1M (1,000 sheets) pricing. The following tables reflect this new trend, giving weight per 1,000 sheets for the most common sizes and weights of different paper grades.

BOOK, OFFSET, LABEL (Coated and Uncoated), TEXT
Weight per 1,000 sheets

Basis 25 x 38	30	35	40	45	50	60	70	80	90	100	120	150
17½ x 22½	25	29	33	37	41	50	58	66	75	83	99	124
19 x 25	30	35	40	45	50	60	70	80	90	100	120	150
20 x 26	33	38	44	49	55	66	77	88	99	109	131	164
22½ x 29	41	48	55	62	69	82	96	110	124	137	165	206
22½ x 35	50	58	66	75	83	99	116	133	149	166	199	249
23 x 29	42	49	56	63	70	84	98	112	126	140	169	211
23 x 35	51	59	68	76	85	102	119	136	153	169	203	254
24 x 36	55	64	73	82	91	109	127	146	164	182	218	273
25 x 38	60	70	80	90	100	120	140	160	180	200	240	300
26 x 40	66	77	88	99	109	131	153	175	197	219	263	328
28 x 42	74	87	99	111	124	149	173	198	223	248	297	371
28 x 44	78	91	104	117	130	156	182	207	233	259	311	389
30½ x 41	79	92	105	118	132	158	184	211	237	263	316	395
32 x 44	89	104	119	133	148	178	207	237	267	296	356	445
33 x 44	92	107	122	138	153	183	214	245	275	306	367	459
35 x 45	99	116	133	149	166	199	232	265	298	332	398	497
35 x 46	102	119	136	153	169	203	237	271	305	339	407	508
36 x 48	109	127	146	164	182	218	255	291	327	364	437	546
38 x 50	120	140	160	180	200	240	280	320	360	400	480	600
38 x 52	125	146	166	187	208	250	291	333	374	416	499	624
41 x 54	140	163	186	210	233	280	326	373	419	466	559	699
41 x 61	158	184	211	237	263	316	369	421	474	527	632	790
42 x 58	154	179	205	231	256	308	359	410	462	513	615	769
44 x 64	178	207	237	267	296	356	415	474	534	593	711	889
44 x 66	183	214	245	275	306	367	428	489	550	611	734	917
46 x 69	200	234	267	301	334	401	468	535	601	668	802	1002
46½ x 67½	198	231	264	297	330	396	463	529	593	661	793	991
52 x 76	250	291	333	374	416	499	582	666	749	832	998	1248

Appendix F (continued)

BOND, BUSINESS, WRITING, LEDGER
Weight per 1,000 sheets

Basis size 17 x 22	13	16	20	24	28	32	36	40
8½ x 11	6.5	8.0	10	12	14	16	18	20
8½ x 14	8.3	10.2	12.7	15.3	17.8	20.4	22.9	25.5
11 x 17	13	16	20	24	28	32	36	40
16 x 21	23	29	36	43	50	57	65	72
16 x 42	47	58	72	86	101	115	130	144
17 x 22	26	32	40	48	56	64	72	80
17 x 26	31	38	47	57	66	76	85	95
17½ x 22½	27	34	42	51	59	67	76	84
17 x 28	33	41	51	61	71	81	92	102
18 x 23	29	36	44	53	62	71	80	89
18 x 46	58	71	89	106	124	142	160	177
19 x 24	32	39	49	59	68	78	88	98
19 x 28	37	46	57	68	80	91	102	114
19 x 48	63	78	98	117	137	156	176	195
20 x 28	39	48	60	72	84	96	108	120
21 x 32	47	58	72	86	101	115	130	144
22 x 25½	39	48	60	72	84	96	108	120
22 x 34	52	64	80	96	112	128	144	160
22½ x 22½	35	43	54	65	76	87	97	108
22½ x 28½	45	55	69	82	96	110	123	137
22½ x 34½	54	66	83	100	116	133	149	166
22½ x 35	55	67	84	101	118	135	152	168
23 x 36	58	71	89	106	124	142	159	177
24 x 38	63	78	98	117	137	156	176	195
24½ x 24½	42	51	64	77	90	103	116	128
24½ x 28½	49	60	75	90	105	120	135	150
24½ x 29	50	61	76	91	106	122	137	152
24½ x 38½	66	81	101	121	141	161	182	202
24½ x 39	66	82	102	122	143	164	184	204
25½ x 44	78	96	120	144	168	192	216	240
26 x 34	61	76	94	113	132	151	170	189
28 x 34	66	82	102	122	143	163	184	204
28 x 38	74	91	114	136	159	182	205	228
34 x 44	104	128	160	192	224	256	288	320
35 x 45	109	135	168	202	236	270	303	337

COVER
Weight per 1,000 sheets

Basis 20 x 26	50	60	65	80	90	100	130
20 x 26	100	120	130	160	180	200	260
22½ x 28½	123	148	160	197	222	247	321
23 x 29	128	154	167	205	231	257	333
23 x 5	155	186	201	248	279	310	403
26 x 40	200	240	260	320	360	400	520
35 x 46	310	372	403	495	557	619	805

PRINTING BRISTOL
Weight per 1,000 sheets

Basis size 22½ x 28½	67	80	90	94	100	110	120	140	160
22½ x 28½	134	160	180	188	200	220	240	280	320
22½ x 35	165	196	221	231	246	270	295	344	393
23 x 35	168	201	226	236	251	276	301	352	402
26 x 40	217	259	292	305	324	357	389	454	519
28½ x 45	268	320	360	376	400	440	480	560	640

INDEX BRISTOL
Weight per 1,000 sheets

Basis size 25½ x 30½	90	120	140	170	220
20½ x 24¾	117	157	183	222	287
22½ x 28½	148	198	231	280	263
22½ x 35	182	243	284	344	446
25½ x 30½	180	240	280	340	440
28½ x 45	297	356	462	561	726

EQUIVALENT WEIGHTS
In reams of 500 sheets. Basis weights in bold face.

	BOND 17 x 22	BOOK 25 x 38	COVER 20 x 26	INDEX 25½ x 30½	BRISTOL 22½ x 28½
BOND	**13**	33	18	27	22
	16	41	22	34	28
	20	51	28	42	34
	24	61	33	50	42
	28	71	39	58	48
	32	81	45	66	55
	36	91	50	75	62
	40	102	56	83	68
BOOK	12	**30**	16	25	20
	16	**40**	22	33	27
	18	**45**	25	37	30
	20	**50**	27	41	33
	22	**55**	30	45	37
	24	**60**	33	49	41
	26	**65**	36	53	44
	28	**70**	38	57	47
	30	**75**	41	61	50
	31	**80**	44	65	54
	35	**90**	49	74	61
	39	**100**	55	82	68
	47	**120**	66	98	81
COVER	18	46	**25**	38	31
	25	64	**35**	52	43
	29	73	**40**	60	50
	36	91	**50**	75	62
	40	100	**55**	82	68
	43	110	**60**	90	74
	47	119	**65**	97	80
	58	146	**80**	120	99
	65	164	**90**	135	111
	72	183	**100**	150	123
INDEX	43	110	60	**90**	74
	53	135	74	**110**	91
	67	171	93	**140**	116
	82	208	114	**170**	140
BRISTOL	52	133	73	109	**90**
	58	148	81	121	**100**
	70	178	97	146	**120**
	82	207	114	170	**140**
	93	237	130	194	**160**
	105	267	146	218	**180**

Appendix F (continued)

HOW TO FIND THE WEIGHT OF A NUMBER OF SHEETS

FORMULA: Weight of 1,000 sheets times number of sheets divided by 1,000 results in total weight.

EXAMPLE: Determine the weight of 1,765 sheets of 25 x 38 x 80 (160M) lb. stock.

SOLUTION: 160 x 1,765 = 282,400 ÷ 1,000 = 282.4 lbs.

HOW TO FIND THE WEIGHT OF A ROLL OF PAPER

EXAMPLE: Determine weight of a roll of coated 2 sides book paper that is 30″ in diameter and 38″ wide.
1. Square the diameter. 30 x 30 = 900;
2. Multiply result by roll width. 900 x 38 = 34,200;
3. Multiply by the factor .034. 34,200 x .034 = 1,162.8 Answer—1,163 pounds.

FACTORS: These average factors apply for all weights:
Newsprint 016
Antique finish 018
Machine finish, English finish, offset, bond 027
Supercalendered, coated 1 side 030
Coated 2 sides 034

HOW TO FIND LINEAR YARDAGE IN PAPER ROLLS

EXAMPLE: Determine linear yardage of roll in previous example. Stock is 50 lb. basis weight.
1. Multiply weight of roll by 500 then by 24;
   1,163 x 500 = 581,500 x 24 = 13,956,000.
2. Divide that result by the sum of roll-width times basis weight.
   38 x 50 = 1,900. 13,956,000 ÷ 1,900 = 7345.3 yards.

APPROXIMATE WEIGHT OF ROLLS PER INCH OF WIDTH

Multiply roll width by figure below to get weight of roll.

PAPER TYPE	28″ DIAM.	30″ DIAM.	32″ DIAM.	34″ DIAM.	36″ DIAM.	38″ DIAM.	40″ DIAM.
Coated 2 sides	26.7	30.6	34.8	39.3	44.0	49.1	54.4
Coated 1 side	23.5	27.0	30.7	34.7	38.9	43.3	48.0
Supercalendered	23.5	27.0	30.7	34.7	38.9	43.3	48.0
English finish	21.2	24.3	27.6	31.2	35.0	39.0	43.2
Offset	17.1	19.7	22.5	25.4	28.4	31.6	35.1
Antique	14.1	16.2	18.4	20.8	23.3	26.0	28.8

PAPER STOCK ESTIMATOR

EXAMPLE: Job calls for 4,000 pieces which cut 16 out of one sheet. Follow first column to 16, then read across on that line to the 4,000 column, result, 250 sheets. No spoilage is included.

NO. OUT OF SHEET	QUANTITY OF PRESS RUN									
	500	1000	1500	2000	2500	3000	3500	4000	4500	5000
1	500	1000	1500	2000	2500	3000	3500	4000	4500	5000
2	250	500	750	1000	1250	1500	1750	2000	2250	2500
3	167	334	500	667	834	1000	1167	1334	1500	1667
4	125	250	375	500	625	750	875	1000	1125	1250
5	100	200	300	400	500	600	700	800	900	1000
6	84	167	250	334	417	500	584	667	750	834
7	72	143	215	286	358	429	500	572	643	715
8	63	125	188	250	313	375	438	500	563	625
9	56	112	167	223	278	334	389	445	500	556
10	50	100	150	200	250	300	350	400	450	500
11	46	91	137	182	228	273	319	364	410	455
12	42	84	126	168	209	250	292	334	375	417
13	39	77	116	154	193	231	270	308	347	385
14	36	72	108	144	179	215	250	286	322	358
15	34	67	100	134	167	200	234	267	300	334
16	32	63	94	125	157	188	219	250	282	313
17	30	59	89	118	148	177	206	236	265	295
18	28	56	84	112	139	167	195	223	250	279
19	27	53	79	106	132	158	185	211	237	264
20	25	50	75	100	125	150	175	200	225	250
21	24	48	72	96	120	143	167	191	215	239
22	23	46	69	91	114	137	160	182	205	228
23	22	44	66	87	109	131	153	174	196	218
24	21	42	63	84	105	125	146	167	188	209
25	20	40	60	80	100	120	140	160	180	200
26	20	39	58	77	97	116	135	154	174	193
27	19	38	56	75	93	112	130	149	167	186
28	18	36	54	72	90	108	125	143	161	179
29	18	36	54	72	87	103	121	138	156	173
30	17	34	51	67	84	100	117	134	150	167
31	17	33	49	65	81	97	113	130	146	162
32	16	32	47	63	79	94	110	125	141	157
33	16	31	46	61	76	91	107	122	137	152
34	15	30	45	59	74	89	103	118	133	148
35	15	29	43	58	72	86	100	115	129	143
36	14	28	42	56	70	84	98	112	125	139
37	14	28	41	55	68	82	95	109	122	136
38	14	27	40	53	66	79	93	106	119	132
39	13	26	39	52	65	77	90	103	116	131
40	13	25	38	50	63	75	88	100	113	125

CALIPER EQUIVALENTS

These are averages. Variations occur in mill runs.

	BASIS WEIGHT	PAPER FINISH			
		COATED	SMOOTH	VELLUM	ANTIQUE
BOOK	40		.0025	.0031	.0034
	45	.0021	.0028	.0035	.0037
	50	.0023	.0031	.0038	.0041
	60	.0028	.0038	.0046	.0050
	70	.0034	.0044	.0054	.0058
	80	.0040	.0050	.0059	.0065
	90	.0046	.0057	.0065	.0074
	100	.0052	.0063	.0071	.0082
	120	.0060	.0076	.0082	.0100
	150	.0072	.0095	.0106	.0123
COVER	50		.0058	.0070	.0075
	60	.0056			
	65		.0075	.0092	.0097
	80	.0072	.0093	.0113	.0120
	90		.0106	.0130	.0135
	100	.0092	.0116	.0140	.0150
	130		.0150	.0184	.0190
BOND	13		.0021	.0025	.0027
	16		.0026	.0031	.0033
	20		.0032	.0039	.0042
	24		.0038	.0047	.0050
INDEX	90		.0080	.0084	
	110		.0096	.0104	
	140		.0132	.0140	
	170		.0144	.0160	
BRISTOL	90	.0055	.0069	.0084	.0090
	100	.0061	.0076	.0093	.0100
	120	.0073	.0092	.0111	.0120
	140	.0085	.0107	.0130	.0140
	160	.0097	.0122	.0148	.0160
	180	.0110	.0137	.0167	.0180
	200	.0122	.0153	.0185	.0200
	220	.0134	.0167	.0204	.0220

GLOSSARY

ACCORDION FOLD: A term used in binding for two or more parallel folds that open like an accordion.

ACHROMATIC: Without color, a lens that reflects light of all colors equally is said to be achromatic.

ACTINIC LIGHT: Chemically active light obtained from carbon arc lamps and used to harden light-sensitive offset plate emulsions in photographic platemaking operations.

AGAINST THE GRAIN: Folding paper at right angles to grain of paper.

AGATE LINE: A standard of measurement for depth of columns of advertising space. Fourteen agate lines make one column inch.

ALTERATIONS: Changes made in copy after it has been set in type.

ANTIHALATION: Property of a film or plate, usually with an opaque backing, that prevents halation or reflection from back surface of film or plate base.

ANTIQUE FINISH: A term describing a surface, usually on book and cover papers, that has a natural rough finish.

APERTURE: In cameras, aperture is usually variable in form of an iris diaphragm and regulates amount of light that passes through lens.

ASCENDER: That part of letter that rises above main body, as in *b*.

AUTHOR'S ALTERATION (AA): Correction made by author on a proof sheet.

AUTOSCREEN FILM: A photographic film containing halftone screen. When exposed to a continuous tone image, it produces a dot pattern automatically—just as if a halftone screen had been used in camera.

BACKING UP: Printing reverse side of sheet already printed on one side.

BAD BREAK: In composition, setting of hyphenated line as first line of page. Also, incorrect word division.

BASIS WEIGHT: Weight in pounds of a ream (500 sheets) of paper cut to a given standard size for that grade: 25 x 38 inches for book papers, 20 x 26 inches for cover papers, etc. For example, 500 sheets of 25 x 38 inch, 80-pound coated paper will weigh eighty pounds.

BENDAY: Method of laying screen (dots, lines, and other textures) on artwork or plates to obtain various tones and shading.

BINDING: Attaching several pages of printed product together.

BLACK LETTER: A general term used to describe Gothic or Old English type design.

BLANKET: Rubber-surfaced material clamped around blanket cylinder of offset press that transfers image from plate to paper.

BLANKET CYLINDER: Cylinder on offset press on which blanket is mounted.

BLEED: Printed image extending to trim edge of sheet or page.

BLOCKING OUT: Operation of eliminating undesirable backgrounds and portions of film negative by opaquing image.

BLOWUP: An enlargement.

BLUELINE: Photoprint made from negatives or positives, used as proof.

BODY: Term referring to viscosity, consistency, or covering power of ink or vehicle.

BODY TYPE: Type used for main body of printed matter, as distinguished from larger headings.

BOLDFACE TYPE: Name given to type that is heavier than text type with which it is used.

BOND PAPER: Grade of writing or printing paper used where strength, durability, and permanence are essential.

BOOK PAPER: General term used to define a class or group of papers having common physical characteristics that, in general, are most suitable for graphic arts. Basic size of book paper is 25 x 38 inches.

BREAK FOR COLOR: In artwork and composition, to separate parts to be printed in different colors.

BROADSIDE: A large printed sheet, intended as a circular, folded into a size convenient for mailing.

BROCHURE: Pamphlet bound in form of a booklet.

BROWNLINE: Photoprint made from negatives or positives, used as a proof.

BULK: Degree of thickness of paper.

CALENDER: Set or stack of horizontal cast-iron rolls at end of a papermaking machine. Paper is passed between rolls to increase smoothness and gloss of its surface.

CALIPER: Thickness of paper, usually expressed in thousandths of an inch or hundredths of a millimeter.

CAMERA BACK: Back of process camera that holds photographic material during exposure.

CAMERA COPYBOARD: That part of a process camera on which copy to be photographed is placed.

CAMERA-READY COPY: Copy that is ready to be photographed without further alteration.

CAPS AND SMALL CAPS: Two sizes of capital letters made in one size of type, commonly used in most Roman typefaces.

CAPTION: Descriptive matter about an illustration.

CARBRO: A photograph in full color, frequently used for process color reproduction.

CASE: In bookbinding, covers of a hardbound book.

CASE-BOUND BOOK: A book with a stiff cover. Cover is made separately and sewn book is inserted into it.

CAST COATED: Coated paper dried under pressure against a polished cylinder to produce a highly glossed enamel finish.

CHALKING: Improper drying of offset ink. In chalking, pigment dusts off due to a lack of binding vehicle caused by too rapid absorption of vehicle into paper.

CHARACTERS PER PICA: Number of average characters that will fit in one pica.

COATED PAPER: Paper having a surface coating that produces a smooth finish.

COLD TYPE: A term denoting use of composition methods not involving hot-metal type.

COLLATE: In binding, assembling of sheets or signatures.

COLOR SEPARATION: Photographic negative exposed through one of tricolor filters and recording only one of primary colors.

COMBINATION: In lithography, joining of halftones and line negatives or positives in position to appear on plate as combinations.

COMPOSING MACHINE: Any of three categories of typesetting equipment: hot metal, typewriter, or photographic.

COMPOSITION: Setting of text copy into type either in hot metal or cold type or by photography.

COMPREHENSIVE: Detailed layout designed to show exactly how job will look when printed.

COMPUTERIZED TYPESETTING: Refers to perforated or magnetic tape or disks used to drive a computer which in turn produces photographic type on paper or film.

CONDENSED TYPE: A narrow or slender typeface.

CONTACT PRINT: A photographic print made from either negative or positive in contact with sensitized paper, film, or printing plate.

CONTACT SCREEN: Photographically made halftone screen having a dot structure of graded density and usually used in vacuum contact with film.

CONTINUOUS TONE: A photographic image that has not been screened and contains gradient tones from black to white.

CONTRAST: Tonal gradation between highlights and shadows in an original or reproduction.

COPY: Any furnished material (typewritten manuscript, pictures, artwork, etc.) to be used in production of printing.

COPYFITTING: Calculations made to determine size of type and width of line needed to fit copy into a given area of space.

COVER PAPER: Term applied to a great variety of papers used for outside covers of catalogs, brochures, and booklets to decorate and provide protection.

CROP: To eliminate portions of copy, usually on a photograph.

CRYSTALLIZATION: A condition in which a dried ink film repels a second ink that must be printed on top of it.

DANDY ROLL: Wire cylinder on papermaking machines that makes woven or laid effects as well as watermark itself.

DAMPENERS: Cloth-covered rollers on offset press that distribute dampening solution over plate. They receive solution from ductor rollers of dampening unit.

DARKROOM: Area that can be made devoid of light for working with photographic materials.

DECKLE EDGE: Untrimmed featheredge of sheet of paper formed where pulp flows against deckle.

DEEP-ETCH: Plate used for long runs where inked areas are slightly recessed below surface.

DENSITOMETER: Sensitive photo-electric instrument that measures density of photographic images or colors in color printing. Used in quality control to accurately determine whether color is consistent throughout run.

DENSITY: A photographic term often confused with "opacity" but correctly applied to quantity of metallic silver (or dyes) per unit area in negatives and positives.

DESCENDER: That part of letter that extends below main body, as in *p*.

DESKTOP PUBLISHING: Using a computer and special software to combine copy and artwork before making hardcopy.

DEVELOPER, DEVELOPMENT: Chemical agent and process used to render photographic images visible after exposure to light.

DIE CUTTING: Process of cutting openings or unusual shapes in paper and paperboard; examples of end results are egg cartons and counter displays.

DIE-STAMPING: An intaglio process for production of letterheads, cards, etc., printing from lettering or other designs engraved into copper or steel.

DIFFERENTIAL SPACING: Individual character widths required in quality typography.

DIRECT LITHOGRAPHY: Method of printing lithographically by direct transfer of ink from plate to paper.

DIRECT TO PLATE: Using electronic means of producing printing plate without using conventional layout sheets, flats, and film.

DIRTY PROOF: Proof that contains many errors.

DISK DRIVE: Device for reading information or program into computer memory, also provides method of outputting data to disk.

DISPLAY TYPE: In composition, type set larger than text, used to attract attention.

DOT: Individual element of a halftone.

DOT MATRIX PRINTER: Low quality printer that produces images with series of coarse dots.

DOUBLE-DOT HALFTONE: Two halftone negatives combined into one printing plate, having greater tonal range than a conventional halftone. One negative reproduces highlights and shadows; other reproduces middletones.

DPI: Dots per inch or measure of resolution. Three-hundred dpi is typical output of laser printer for example.

DRIER: In ink formulation, any substance added to hasten drying.

DRILLING: Piercing of paper with a rotating hollow drill to make round binder holes.

DROPDOWN MENU: Computer display that moves down and becomes visible when selected.

DROPOUT: A halftone with no screen dots in highlights.

DUCTOR ROLLER: On an offset press, roller in both inking and dampening systems that alternately contacts fountain roller and vibrating drum roller.

DUMMY: A preview of a proposed piece of printing containing several pages.

DUOTONE: A two-color halftone reproduction made from a one-color photograph.

DUPLEX PAPER: Paper having a different color or finish on each side.

ELECTRONIC PUBLISHING: Using a computer, special software, image setter, and other electronic devices to combine copy and artwork on a monitor before making hardcopy.

EM: In composition, square of a type body.

EMBOSSED FINISH: Paper with a raised or depressed surface resembling wood, cloth, leather, or another textured material.

EMULSION: Photographic term for a gelatin or collodion solution holding light-sensitive salts of silver in suspension.

EN: One-half width of an em.

ENAMEL: A term applied to a coated paper or to a coating material on a paper.

ENGLISH FINISH: A low-gloss paper adapted to fineline plates and diagrams. Widely used for letterpress-printed magazines.

ETCH: In offset lithography, an acidified gum solution used to desensitize nonprinted areas of plate.

EXPANDED TYPE: Type whose width is greater than normal, also called extended type.

FAMILY: A number of series of related type designs that have characteristics in common.

FEEDER: Section of an offset press that separates sheets and feeds them in position for printing.

FELT SIDE: Preferable side of paper for printing. Top side of sheet in paper manufacturing.

FILLING IN: Condition in offset lithography in which ink fills area between halftone dots or plugs up type.

FINISHING: Operations that enhance printed product — embossing, laminating, etc.

FIXING: Application of a chemical solution that removes unexposed silver salts in an emulsion without affecting metallic silver that has been deposited by developer. Fixing renders photographic image permanent.

FLASH EXPOSURE: Supplementary exposure given in halftone photography to strengthen dots in shadows of a negative.

FLAT: In offset lithography, assembled composite of negatives or positives, ready for platemaking. Also, a picture that is lacking contrast.

FLOW: Ability of an ink to spread over a surface or into a thin film.

FLUSH LEFT (OR RIGHT): In composition, type set to line up at left (or right) margin.

FLUSH PARAGRAPH: A paragraph with no indentation.

FLYING PASTER: In web printing, automatic pasting device that splices new roll of paper onto an expiring roll, without stopping press.

FOCAL LENGTH: Distance between optical center of a lens and point at which an object image is in sharp or critical focus. Focal length is usually engraved on front of lens barrel by manufacturer.

FOG: A photographic defect in which image is either locally or entirely veiled by a deposit of silver, defect due either to action of stray light or to improperly compounded chemical solutions.

FOLIO: Number of a page.

FONT: Complete assortment of type of one size and face.

FORM ROLLERS: Rollers, either ink or dampening, that contact plate.

FORMAT: Size, style, type page, margins, printing requirements, etc., of any magazine, catalog, book, or printed piece.

FOUNTAIN ROLLER: In water solution of an offset press, a nonferrous roller that revolves in fountain. This roller, in conjunction with ductor roller, meters water or fountain solution to press plate.

f-STOPS: Fixed sizes at which aperture of lens can be set, values of which are determined by ratio of aperture to focal length of lens.

GANGING: Process whereby various pieces of copy or art are combined so that the camera operator can shoot them all on one negative. Also, combining number of unrelated jobs for a single press run.

GATHERING: Assembling of folded signatures in proper sequence.

GOLDENROD FLAT: Method of assembling and positioning lithographic negatives or positives for exposure in contact with light-sensitized press plate.

GRAIN: In papermaking, direction in which most fibers lie.

GRAVURE: Process of printing in which sunken-image plate is generally run on a direct rotary press.

GRAY SCALE: Strip of standard gray tones, ranging from white to black, placed at side of original copy during photography to measure tonal range obtained.

GRIPPERS: Metal fingers that hold paper in place to delivery end of press.

GRIPPER EDGE: Leading edge of paper as it passes through printing press.

GRIPPER MARGIN: Unprintable blank edge of paper on which grippers bear, usually one-half inch or less.

GROUNDWOOD PULP: Mechanically-prepared coarse wood pulp used chiefly in manufacture of newsprint and low-cost grades of book paper.

GUM ARABIC: Gum obtained from either of two species of Acacia trees, used in all branches of graphic arts. Gum arabic is used to protect metal offset plates against oxidation.

GUMMING: In lithography, treating surfaces with a thin coating of gum arabic as a protection against oxidation.

GREEK: Words on computer screen are represented by symbols, box, etc.

HAIRLINE: Thin stroke on a type character.

HALATION: Photographic term for spreading of light action beyond proper boundaries in negatives, particularly in highlight areas of image.

HALFTONE: Reproduction of continuous-tone artwork, such as a photograph, with image formed by dots of various sizes.

HALFTONE TINT: Transformation of a solid area of a plate into a gray tone of any desired density by stripping in piece of film with uniform halftone density, i.e., 25 percent, 50 percent, etc.

HEAD MARGIN: White space above first line on page.

HICKEY: In lithography, imperfection in presswork due to many things, such as dirt on press, hardened specks of ink, etc.

HIGHLIGHT: Lightest or whitest parts in a printed picture, represented in halftone by smallest dots or absence of all dots.

ICON: Graphic symbol on computer monitor for tool, command, file, etc.

IMPOSITION: Laying out of type pages in a goldenrod flat so that they will be in correct order after printed sheet is folded.

IMPRESSION: Pressure of plate or blanket as it comes in contact with paper.

IMPRESSION CYLINDER: On an offset press, cylinder that carries paper sheet into contact with blanket.

INK FOUNTAIN: Container that supplies ink to inking rollers.

ITALIC: Style or form of letter that generally slants to right.

JOG: To align sheets of paper into a compact pile.

JUSTIFY: In composition, to space out lines uniformly to correct length.

KERN: That part of a letter of type that projects beyond metal body on which it is cast, such as head and tail of some italic letters; "*f*" is an example. Also, to selectively adjust unit spacing between characters.

KEY: To code copy to a dummy by means of symbols, usually letters. Insertions are sometimes "keyed" in like manner.

KEYLINING: Technique used in copy preparation by artist to handle copy for some simple types of color separations, or for indicating reverses or outline of backgrounds. Purpose is to provide copy for camera that avoids excessive opaquing of photographic negatives or positives.

KEYPAD: Typewriter type keyboard for inputting data in computer equipment.

KEY PLATE: In color printing, plate used as a guide for register of other colors. It normally contains most detail.

LAID PAPER: Paper having parallel lines watermarked at equal distances, giving a ribbed effect.

LAYOUT: Drawing or sketch of a proposed printed piece.

LASER PRINTER: High quality printer that uses narrow beam of light to deposit image on paper.

LEADERS: In composition, rows of dashes or dots used to guide eye across page. Used in tabular work, programs, tables of contents, etc.

LEADING: Additional space between lines of type.

LEDGER PAPER: Grade of business paper generally used for keeping records. Since it may be subjected to considerable wear, it requires a high degree of durability and permanence.

LEGEND: Descriptive matter below an illustration.

LENS: Photographic lens made up of several prisms or elements, both negative and positive, separated in lens barrel at definite distances by an air space.

LETTERSET (DRY OFFSET): Printing process that uses a blanket, like conventional offset, for transferring image from paper to plate. Unlike offset, it uses a relief plate and requires no dampening system.

LETTERSPACING: Spacing between each letter of a word.

LIGATURE: Two characters joined in one body or key, such as fi, fl, ffl, ffi, etc.

LIGHT FACE: Type used most commonly in composition of text or bookwork.

LINE COPY: Any copy suitable for reproduction without using a screen. Copy composed of lines or dots as distinguished from copy composed of continuous tones.

LINE GAUGE: Printer's rule used for measuring purposes.

LINE SPACING: Number of points between lines or number of lines per inch.

LOGOTYPE (OR LOGO): Name of a company or product in a special design used as a trademark in advertising.

LONG INK: An ink that has good flow in fountain.

LOWERCASE: Small letters in type, as distinguished from capital letters.

MACHINE COATED: Paper that is coated one or two sides on papermaking machine.

MAKEUP: In composition, arrangement of lines of type and illustrations into pages of proper length.

MARGINS: White space at sides, top, and bottom of copy.

MARKUP: To write up instructions, as on a dummy or manuscript copy.

MASK: In color separation photography, an intermediate photographic negative or positive used in color correction.

MASKING PAPER: Goldenrod paper to which negatives are fastened in stripping to produce a flat.

MEASURE: In composition, width of type, usually expressed in picas.

MECHANICAL (PASTE-UP): Method of assembling all copy elements into a unit for photographic platemaking (copy ready for camera).

MENU: Set of preprogrammed monitor commands, tools, symbols, etc., used to operate computer.

MIDDLETONES: Tonal range between highlights and shadows of a photograph or reproduction.

MOIRÉ: Undesirable screen pattern in color process printing caused by incorrect screen angles of halftones.

MOLLETON: A thick cotton fabric similar to flannel, having a long nap, and used on dampening form rollers.

MONITOR: Television type picture tube used with computerized equipment.

MORTISING: Cutting out areas in a plate or in cold-type artwork so that type can be inserted.

MOTTLE: Spotty or uneven appearance of printing. Most pronounced in solid areas.

MOUSE: Computer input device rolled over work surface to move cursor on monitor screen and to select objects quickly.

MULLEN TESTER: Machine for testing bursting strength of paper.

NEGATIVE: Photographic image on film in which black values in original subject are transparent; white values are opaque; light grays are dark; dark grays are light.

NEWSPRINT: A generic term used to describe kind of paper generally used in publication of newspapers.

NONPAREIL: Unit of measure equaling six points (one-half pica).

OFFSET: See SETOFF.

OFFSET LITHOGRAPHY: Printing process that utilizes a flat plate and oil-base ink. Water rolled on plate moistens nonprinting area and repels ink so that it adheres only to image area.

OLDSTYLE ROMAN: General classification of type derived from early Dutch, Venetian, and English designs. Caslon, Garamond, and Cloister are examples.

OPACITY: Property of sheet of paper that minimizes show-through of printing from backside or next sheet.

OPAQUE: To paint out unwanted areas on a negative so it will not be "burned" on plate. In paper, property that makes it less transparent.

OPAQUE INK: Special ink that conceals all color beneath it.

OPTICAL CENTER: Point where eye normally strikes page — approximately two-fifths down and slightly to left of center.

OVERPRINTING: Double printing; printing on an area that already has been printed.

OVERLAY: In artwork, a transparent or translucent covering over copy where color breaks, instructions, or corrections are marked. Also used over layout sheet for multicolor work.

ORTHOCHROMATIC: Side of photographic surfaces sensitive to ultraviolet, blue, yellow, green, and orange rays. Insensitive to red rays.

PACKING: In lithography, paper used to underlay a blanket, plate, or proof to bring surface to desired height.

PAGINATION: Number and arrangement of pages.

PANCHROMATIC: Photographic film sensitive to all colors.

PAPER MASTER: Paper printing plate used on offset duplicator. Image is made by hand drawing or typewriting.

PARALLEL FOLD: Second fold in a sheet parallel to first fold.

PASTE-UP: Preparation of copy for photographic reproduction by putting all elements in proper position.

PERFECT BINDING: Method of holding pages of book together without stitching or sewing. Backbones of gathered books are ground off leaving a rough surface of intermingled fibers to which adhesive is applied. Books are usually finished with a wraparound cover.

PERFECTING PRESS: Press that prints both sides of paper at one time.

PERFORATE: To punch a series or row of small holes or slots in a material to facilitate tearing.

pH: A scale used for expressing acidity or alkalinity of solutions. Also, degree of acidity or alkalinity of solutions. The pH scale ranges from 0 to 14. Determination is made by various electronic and colormetric devices designed for purpose.

PI: Type mixed and in an unusable condition.

PICA: Printer's unit of measurement used principally in measuring lines. One pica equals one-sixteenth of an inch.

PICKING: Lifting of paper surface during printing. It occurs when pulling force or tack of ink is greater than surface strength of paper.

PIGMENT: Fine solid particles used to give color, body, or opacity to printing inks.

PILING: Building up or caking of ink on rollers, plate, or blanket; ink will not transfer readily.

PLASTIC BINDING: Solid back comb rolled to make a binding cylinder of any thickness. Slots are punched along binding side of book, and plastic comb is inserted through slots.

PLATE FINISH: A term describing a smooth, hard finish of paper.

POINT: Printer's unit of measurement, used principally for designating type sizes. There are 12 points to a pica, 72 points to an inch.

POROSITY: Property of a paper sheet that allows permeation of air.

POSITIVE: Photographic image on film or glass that corresponds to original copy. Reverse side of negative.

PRIMARY COLORS: In printing inks, yellow, magenta (process red), and cyan (process blue). In light, primary colors are red, green, and blue.

PROCESS PRINTING: Printing from a series of two or more plates in halftone to produce other colors and shades. Usually in four-color process—magenta, cyan, yellow, and black.

PROGRESSIVE PROOFS: Proofs of each individual plate in a set of color process plates, pulled in proper colored inks, and also showing results of adding each color.

PROGRAM: Electronic instructions for computer to operate as word processor, proofing, graphics, desktop publishing, etc., system.

PROOF: Sample of copy and/or layout made at various stages of production of a printing job. Proofs are either checked internally or sent to customer for corrections or approval.

PROOFREAD: Act of reading and correcting copy internally at early stage of production to ensure accuracy before making plates and printing.

QUAD: In composition, blank spacing material, less than type high, used to fill out lines; also applies to unit of measure in photocomposition.

RAGGED COPY: Line of copy set with uneven left- or right-hand margins. Often used in advertising copy.

REAM: Five hundred sheets of paper.

REDUCERS: Varnishes, solvents, or oily, greasy compounds employed to bring ink or varnish to a softer consistency for use on press.

REGISTER: Fitting of two or more printing images upon same sheet of paper in exact alignment with each other.

REGISTER MARK: Crosses or other devices applied to original copy prior to photography. Used for positioning negatives in perfect register, or for color register of two or more colors in printing.

RELATIVE HUMIDITY: Amount of water vapor present in atmosphere expressed as a percentage of maximum that could be present at same temperature.

RESOLUTION: Refers to number of dots per inch or dpi and affects quality of image—high resolution would produce sharp, crisp image for instance.

RIGHT-ANGLE FOLD: In binding, a term used for two or more folds that are at 90° angles to each other.

RIPPLE FINISH: An irregular paper finish suggestive of ripples, produced by an embossing process.

ROLLER STRIPPING: A lithographic term denoting that ink does not stick to metal ink rollers on press.

ROMAN: Serified type in which upright strokes are vertical as distinguished from italic in which uprights are slanted.

ROTOGRAVURE: Intaglio or printing process with image area below surface of nonprinting area. (See GRAVURE.)

RUNAROUND: Copy arrangement that allows for positioning of pictures, diagrams, etc., within columns of type.

RUNNING HEAD: Title repeated at top of each page of a book.

SADDLE WIRE: To fasten a booklet by wiring it through middle fold of sheets.

SAFELIGHT: In photography, special darkroom lamp emitting illumination under which sensitized materials can be handled without danger of fogging through exposure to light.

SAN SERIF: A typeface lacking serifs.

SCANNER: An electronic device used in making of color-separation negatives and positives.

SCORE: To impress or indent a mark with a rule in paper to make folding easier.

SCRIBER: A small hand tool used for drawing lines on emulsion of an exposed photographic negative, tool scrapes off black emulsion.

SCUM: In offset lithography, greasy film that tends to sensitize nonimage areas of plate to accept ink.

SELF-COVER: Cover of same weight paper as inside text pages.

SENSITIVITY GUIDE: A narrow, calibrated, continuous-tone gray scale with each tone scale numbered. In platemaking operation, sensitivity guide is exposed on sensitized press plate with rest of work. Device measures tone values reproduced on plate.

SERIF: Short cross-lines at ends of main strokes in Roman typefaces.

SETOFF: In presswork, condition in which ink of a printed sheet rubs off or marks next sheet as it is being delivered. Formerly called offset.

SHADOW: Darkest parts in a photograph, represented in a printed halftone by largest dots.

SHADING SHEET: Art material, usually patterns or tones on an acetate sheet, that can be positioned on line artwork in selected areas to avoid tedious hand detail.

SHEETWISE: Term applied to method of printing each sheet first on one side with one form, then on other side with another form. Same edge of sheet is brought against guides for both printings.

SHORT INK: An ink that does not flow freely.

SHOW-THROUGH: Condition in which printing on one side of paper can be seen from other side when latter is viewed by reflected light.

SIDE WIRE: To wire sheets or signatures of a magazine or book on side near backbone.

SIGNATURE: In book, magazine, and catalog work, name given to a large printed sheet after it has been folded to required size.

SILHOUETTE HALFTONE: A halftone illustration from which screen surrounding any part of image has been eliminated.

SIZING: That property of paper that relates to its resistance to penetration of liquids (particularly to water) or vapors.

SKID: Platform support for a pile of cut sheets.

SLITTING: Cutting printed sheets or webs into two or more sections by means of cutting wheels on printing press, folding machine, or slitter-rewinder.

SMALL CAPITALS: An alphabet of small capital letters available in most Roman typefaces approximately size of lowercase letters. Used in combination with larger capital letters.

SOLID MATTER (SET SOLID): Lines of type leaving no extra space between them.

SPIRAL BINDING: A book bound with wires in spiral form inserted through holes punched along binding side.

STET: A proofreaders mark, normally written in nonimage area of margin, signifying that copy marked for corrections should stand as printed and should not be corrected.

STOCK: Paper to be cut and/or printed.

STOP: Photographic term for any type of lens aperture or diaphragm. Examples are iris, Waterhouse, and f-stops.

STREAM FEEDER: A type of press feeder that keeps several sheets of paper, overlapping each other, moving toward grippers.

STRIPPING: In offset lithography, arranging of negatives or positives in their proper position on a flat prior to platemaking.

SUBSTANCE: Weight in pounds of a ream (500 sheets) of paper cut to standard size (e.g., 17 x 22 inches) for business papers (bond, ledger, mimeograph, duplicator, and manifold). An example, 500 sheets of 20-pound bond cut to standard size will weigh 20 pounds.

SULPHATE WOOD PULP: Paper pulp made from wood chips cooked under pressure in a solution of caustic soda and sodium sulphide.

SULPHITE WOOD PULP: Paper pump made from wood chips cooked under pressure in a solution of bisulphite of lime.

SUPERCALENDER: Calender stack used to alter surface properties and appearance of paper. It has alternate metal and resilient rolls.

SURFACE PLATE: One of two basic types of lithographic press plates. A colloid image is formed on light-sensitized metal plate by action of actinic light through photographic negatives.

TACK: Property of cohesion between particles of an ink; pulling power of ink against another surface. A tacky ink does not break apart readily.

TEXT: Body matter of a page or book, as distinguished from heading.

TEXT PAPER: General term applied to antique, laid, or woven papers, sometimes watermarked and deckle-edged. Used for booklets, programs, announcements, and advertising printing.

THIRTY: Used in newspapers, the symbol "−30−" indicates end of story.

TINT: A light color, usually used for backgrounds.

TOOLBOX: Area on computer screen that contains icons, graphics tools, etc.

TOOTH: Quality of a paper that causes it to take ink readily.

TRANSPARENT: See SHOW-THROUGH.

TRANSPARENT INK: Ink that permits previous printing to show through, two colors blending to produce a third.

TRANSPOSE: To exchange position of a letter, word, or line with another letter, word, or line.

TRAPPING OF INKS: Property of a printing ink that makes it possible to superimpose one color on another, both in wet and dry printing.

TRIANGLE: Plastic triangle used to produce vertical or uniformly slanting lines, usually 45 and 30/60 degrees.

TRIM: Paper trimmed off a printed piece to square edges or remove bleed. Excess paper allowed around a printed piece for bleed. Actual operation of cutting off bleeds or folds with a three-knife trimmer.

TRIMMED SIZE: Final size of a printed piece after all bleeds and folds have been cut off.

T-SQUARE: T-shaped, precision straight edge used to square copy or edges of paper with edge of a drawing board or table.

TUSCHE: Liquid emulsion ink painted or drawn on a lithographic plate to form an image.

TYPOGRAPHY: Art of type selection that involves style, arrangement, and appearance of printed page.

VACUUM FRAME: In offset lithography, a printing frame in which negative or positive is held directly to plate by vacuum.

VARNISH: A thin, protective coating applied like ink on a printing press. Also, in inkmaking, it can be all or part of ink vehicle.

VEHICLE: Fluid component of printing ink that acts as a carrier for pigment.

VELLUM FINISH: In papermaking, a toothy finish that is relatively absorbent. Used when fast ink penetration is desired.

VELOX PRINT: Name for one of chloride printing papers made by Eastman Kodak Company, and sometimes erroneously used as name for similar developing papers.

VIBRATOR OR VIBRATING ROLLER: A metal or hand roller in an inking or dampening system.

VIGNETTE: An illustration in which background fades gradually away until it blends into unprinted paper.

VISCOSITY: In printing inks, a broad term encompassing properties of tack and flow.

WASHUP: Process of cleaning rollers, form or plate, and sometimes fountain of a press.

WATERMARK: In papermaking, name or design produced by raised pattern of dandy roll during manufacture.

WEB: Roll of paper used on web or rotary presses. Printed, and often folded, pasted, and counted in a continuous operation.

WEB PRESS: General term applied to high-speed presses that print from continuous rolls of paper.

WIDOW: In composition, single word in a line by itself, ending a paragraph; frowned upon in good typography.

WIRE SIDE: In papermaking, side of sheet next to wire in manufacturing; opposite from felt side.

WITH THE GRAIN: Term applied to folding paper parallel to grain of paper.

WORK AND TUMBLE: Printing second side of sheet by turning it over from gripper to back, using same side guide.

WORK AND TURN: Printing second side of sheet by turning it over left to right, using same edge of paper as gripper.

WRONG FONT: In proofreading, letters "wf" indicate a wrong typeface.

WYSIWYG: Computer term for "what you see is what you get," referring to monitor display.

XEROGRAPHY: Process used to make small lithographic plates, by transfer of image formed by a light-sensitive electrostatic charge on a special base.

INDEX

ACKNOWLEDGEMENTS

The authors and publisher wish to express their appreciation to the following individuals, companies, and organizations for their cooperation in furnishing photographs, line art, and technical information used in this book.

A.B. Dick Co.
Accurate Steel Rule Die Manufacturing, Inc.
Acromark Co., Inc.
Acti Manufacturing
Adobe Illustrator
AGFA Compugraphic Division
Agfa-Gavaert, Inc.
Aldus Corporation 1987. Aldus®, PageMaker®, and Aldus FreeHand® are registered trademarks of Aldus Corporation. All rights reserved.
A.M. International, Inc.
American Manufacturing Co.
ANSCO Graphic Arts Products
Apple Computer
Apple Macintosh
ATF Davidson Co.
Autobond Ltd.
AutoCount
Autologic, Inc.
Automation, Inc.
Autovend Packaging Systems
Baker Perkins Limited
Baldwin Technology Corp.
Baldwin-Gegenheimer Corp.
BARCO, Inc.
Baumfolder Corp.
Berthold Corp.
Bobst Group, Inc.
William Bockus
Boise Cascade Paper Group
Brackett, Inc.
Bob Bright
ByChrome Co.
Byers Corp.
Camex
Challenge Machinery Co.
Cham Pad Co.
Champion Manufacturing Company, Inc.
Chartpak
Chemco Photoproducts Co.
Chesley F. Carlson Co.
Color Service
Compugraphic Corp.
CompuScan
Consolidated International, Inc.
Consolidated Machinery Co.
Cosar Corp.
Crosfield Electronics
Crosfield Graphic Equipment
Daige Products, Inc.
Decco, Inc.
Didde Graphic Systems Corp.
Direct Image Corp.
Dirt Rider Magazine
DS America
DTK Computer, Inc.
Dunn Technology, Inc.
DuPont Imaging Systems
Dynamic Graphics, Inc.
Eastman Kodak Co.
E.I. DuPont de Nemours and Co.
Enco Products
Ethernet
Formatt Graphic Products Corp.
Foster Manufacturing Co.
Fotostar International, Inc.

General Binding Corp.
General Printing Ink Co.
Gerber Scientific Products, Inc.
Gilbert, Whitney, and Johns, Inc.
Graphic Arts Technical Foundation
Graphic Products Corp.
Grumbacher, Inc.
Gutenberg Museum, Mainz
Hamada of America, Inc.
Hammermill Paper Group
Harris Bindery Systems Division
Harris Graphics Corp.
Harris Intertype Corp.
Harris-Cottrell Company
Harry H. Rogers Company
Hart Bindery Services
HCM Corp.
Heidelberg Eastern, Inc.
Heidelberg U.S.A.
HELL
Hell-Color-Metal Corp.
Hewlett Packard
High Technology Solutions
Hunt Manufacturing Co.
Hunter Lab Corp.
Ibico, Inc.
Iconics Control, Inc.
Interchemical Corp.
Interlake, Inc.
International Business Machines
International Paper Co.
Itek Graphics Corp., Graphic Systems Division
John F. Cuneo Co.
John N. Schaedler, Inc.
Jomac, Inc.
Justrite Manufacturing Co.
Kansa Corp.
Kenro Photoproducts, Inc.
Keuffel and Esser Co.
Kimberly-Clark Corp.
Koh-I-Noor Rapidograph, Inc.
Krause
Kreonite, Inc.
Kroy, Inc.
Lab Safety Supply
Leedal, Inc.
The Lehigh Press, Inc.
Leslie Paper Co.
Letraset, Inc.
Linoterm
Linotype Co.
LogEtronics, Inc.
Martin Yale Co.
McCain Manufacturing Corp.
Allen McMakin
The Mead Corp.
M.E.G. Corp.
Mergenthaller Linotype Co.
Mergenthaller: Mycro-Tek Production Division
Metroliner Press
Michael Business Machines
Micro Essential Laboratory, Inc.
Microsoft Word
Miehle Product
Miller Printing Equipment Corp.
Minolta Corp.

M.P. Goodkin Co.
Muller-Martin Corp.
Multigraphics
NASA
National Association of Photo Lithographers
National Association of Printers and Lithographers
National Association of Printing Ink Manufacturers
National Association of Quick Printers
Nekoosa Papers, Inc.
nuArc Co.
Ohaus Corp.
Olec Corp.
Opti-Copy, Inc.
Pako Corp.
Pantone, Inc.
Paracomp Swivel 3D
Pinnacle Micro
Printer's Software, Inc.
Printing Developments, Inc.
Printing Industries Association, Southern California
Printing Industries of America, Inc.
Raden® C, Inc.
Regency Thermographers
Robertson Photo-Mechanix, Inc.
Rockwell International Corp., Graphic Systems Division
Rodel, Inc.
Rodenstock Precision Optics, Inc.
Royal Zenith Corp.
S.D. Warren Co.
Signode Corporation
Silicon Beach Software
Stephen Simms
Sir Speedy, Inc.
Smith System Manufacturing Co.
Smithsonian Institute
Southwestern Publishing Co.
Step-by-Step Graphics Magazine
Stewart Systems
Stouffer Graphic Arts Equipment Co.
Sulby
Sun Chemical Corp.
Superpaint
TEC Systems
3G Graphics
3M Company
Tobias Associates, Inc.
Townsend Industries
U.S. Government Printing Office
Van Son Holland Ink Corp.
Vandercook, Division of Illinois Tool Works
Varn Product Company, Inc.
Virkotype Corp.
Visual Graphics Corp.
Volkwagen of America, Inc.
W.A. Krueger Co.
Walden-Mott Corp.
Westvaco Corp.
Xerox Corp.
X-Rite Co.
Xyvision, Inc.
Zipatone, Inc.